Seventh Edition

Understanding Social Problems

Linda A. Mooney

David Knox

Caroline Schacht

East Carolina University

WADSWORTH
CENGAGE Learning™

Australia • Brazil • Japan • Korea • Mexico • Singapore • Spain • United Kingdom • United States

![Wadsworth Cengage Learning logo]

Understanding Social Problems,
Seventh Edition
Linda A. Mooney, David Knox, and
Caroline Schacht

Senior Publisher: Linda Schreiber-Ganster

Sociology Editor: Erin Mitchell

Developmental Editor: Tangelique Williams

Assistant Editor: Erin Parkins

Editorial Assistant: Pamela Simon

Media Editor: Melanie Cregger

Marketing Manager: Andrew Keay

Marketing Assistant: Jillian Meyers

Marketing Communications Manager: Laura
    Localio

Content Project Manager: Cheri Palmer

Creative Director: Rob Hugel

Art Director: Caryl Gorska

Print Buyer: Becky Cross

Rights Acquisitions Account Manager, Text:
    Roberta Broyer

Rights Acquisitions Account Manager,
    Image: Robyn Young

Production Service: MPS Limited,
    A Macmillan Company

Text Designer: Lisa Buckley

Photo Researcher: Jaime Jankowski,
    PrePress

Illustrator: MPS Limited, A Macmillan
    Company

Cover Designer: Larry Didona

Cover Image (clockwise from top to
    bottom): © Chip East/Reuters/Corbis;
    © Thomas S. England/Photo Researchers,
    Inc.; © Tom Pennington/MCT/Landov;
    © Stephen Shaver/UPI/Landov; © Amy
    Smith/iStockphoto; © John Kuntz/The
    Plain Dealer/Landov.

Compositor: MPS Limited, A Macmillan
    Company

For product information and technology assistance, contact us at
Cengage Learning Customer & Sales Support, 1-800-354-9706.
For permission to use material from this text or product,
submit all requests online at www.**cengage.com/permissions**
Further permissions questions can be emailed to
permissionrequest@cengage.com

Library of Congress Control Number: 2009939349

ISBN-13: 978-0-495-81296-8

ISBN-10: 0-495-81296-X

Wadsworth
20 Davis Drive
Belmont, CA 94002-3098
USA

Cengage Learning is a leading provider of customized learning solutions with office locations around the globe, including Singapore, the United Kingdom, Australia, Mexico, Brazil, and Japan. Locate your local office at **www.cengage.com/global.**

Cengage Learning products are represented in Canada by Nelson Education, Ltd.

To learn more about Wadsworth, visit **www.cengage.com/wadsworth**

Purchase any of our products at your local college store or at our preferred online store **www.CengageBrain.com.**

Printed in the United States of America
1 2 3 4 5 6 7 13 12 11 10

Brief Contents

Contents

PART 4   Problems of Globalization

Features

The Human Side

Preface

Understanding Social Problems is intended for use in a college-level sociology course. We recognize that many students enrolled in undergraduate sociology classes are not sociology majors. Thus, we have designed our text with the aim of inspiring students—no matter what their academic major or future life path may be—to care about the social problems affecting people throughout the world. In addition to providing a sound theoretical and research basis for sociology majors, Understanding Social Problems also speaks to students who are headed for careers in business, psychology, health care, social work, criminal justice, and the nonprofit sector, as well as to those pursuing degrees in education, fine arts, and the humanities or to those who are "undecided." Social problems, after all, affect each and every one of us, directly or indirectly. And everyone—whether a leader in business or politics, a stay-at-home parent, or a student—can become more mindful of how his or her actions (or inactions) perpetuate or alleviate social problems. We hope that Understanding Social Problems not only informs but also inspires, planting seeds of social awareness that will grow no matter what academic, occupational, and life path students choose.

New to this Edition

The seventh edition of Understanding Social Problems has been streamlined by shortening some of the longer chapters and reducing the number of chapters from 16 to 15. We eliminated the chapter on problems of youth and aging, and incorporated material from this chapter into other chapters. We also eliminated the margin quotes, replacing them with pullout quotes that are offset for emphasis. This edition includes two new photo essays: Chapter 1 now has a photo essay on "Students Making a Difference" and Chapter 4 features a new photo essay titled "Prison Programs That Work." Most of the opening vignettes are new, as are many of the What Do You Think? sections, which are designed to engage students in critical thinking. Many of the chapter features (The Human Side, Social Problems Research Up Close, and Self and Society) have been updated or replaced with new content. The seventh edition has retained pedagogical features that students and professors find useful, including a running glossary, list of key terms, chapter review, and Test Yourself sections. Finally, each chapter has new photos and new and updated figures and tables, as well as new and revised material, detailed as follows.

    Chapter 1 (Thinking about Social Problems) features a new photo essay on student activism that includes discussions of the U.S. civil rights movement, antiwar protests, animal rights, the fight for marriage equality, and Tiananmen Square and Chinese students' battle for democratic reform. An updated Self and Society feature presents the newest statistics available on U.S. freshman attitudes toward select social problems, and the Social Problems Research Up Close feature includes new data on the sexual behavior of high school students. This revised chapter has a new section on "Seven Good Reasons to Read This Book."

Chapter 2 (Problems of Illness and Health Care) includes a new *Social Problems Research Up Close* feature: "The National College Health Assessment," and a new *The Human Side* feature*:* "A Former Insurance Industry Insider Speaks Out for Health Care Reform." This chapter includes new and updated material on the "revolving door" in the health industry and government, health care reform, the uninsured, and HIV/AIDS. The section on stigma and mental health has also been expanded. New *What Do You Think?* sections pose the questions, "Should veterans with PTSD be eligible to receive the Purple Heart medal?" and "Do you favor or oppose amending the Constitution to guarantee health care as a right for every American?"

Chapter 3 (Alcohol and Other Drugs) features all new data from the most recent reports available on alcohol and drug use including the World Drug Report, the Monitoring the Future survey, the National Survey on Drug Use and Health (NSDUH), the European Monitoring Centre for Drugs and Drug Addiction (EMCDDA), the European School Survey Project on Alcohol and Other Drugs (ESPAD), and the World Health Organization's report on the global tobacco epidemic.

This revised chapter includes an analysis of college environmental conditions that lead to binge drinking, updated information on the economic costs of alcohol and other drugs, and a new discussion on Mexican cartels and violence at the Mexican-U.S. border. The Strategies for Action section has been reorganized into two parts—strategies dealing with legal drugs and strategies dealing with illegal drugs, and a discussion of the Merida Initiative, the FDA's control of the tobacco industry, and the Obama administration's approach to the drug problem is also included. There are several new *What Do You Think?* sections on such topics as "Should drug cartels be able to post videos on YouTube?"; "Does beer pong encourage binge drinking?"; and "Should Red Bull Cola (which according to German officials has trace levels of cocaine) be banned?"

Chapter 4 (Crime and Social Control) presents updated statistics from the Uniform Crime Report, the National Crime Victimization Survey, the Department of Justice, the Federal Trade Commission, the Internet Crime Complaint Center, the National Gang Intelligence Center, and the National Gang Threat Assessment report. There is also a new photo essay on prison programs that help reduce recidivism rates. The *Self and Society* feature—"Criminal Activities Survey"—has been updated and there is a new *The Human Side* feature entitled, "The Hidden Consequences of Rape." New material has been added from Innocence Lost, a domestic child sex trafficking prevention initiative, and the Innocent Images National Initiative, which investigates online child sexual exploitation and pornography. Topics that have been expanded include white-collar crime, the economic marginalization of women hypothesis, and the use of lethal injection. New material has also been added on the impact of the economic downturn on criminal justice policy. Three new *What Do You Think?* sections have been added which ask readers, "Should maritime pirates be treated as punitively as airline hijackers?"; "Does exposure to violent toys, games, and videos lead to aggressive behavior?"; and "Will criminalizing sex offenders' use of social networking sites reduce the number of sexual predators?"

Chapter 5 (Family Problems) begins with a new opening vignette about Michelle and Barack Obama and their families. This revised chapter includes updated information on changing structures and patterns in U.S. families and households. The section on intimate partner violence includes new data on abusive relationships reported by college students. This chapter also features updated information on the rise in teen pregnancies and unmarried childbearing and includes a new *The Human Side* feature: "A Teen Mom Tells Her Story." New topics in this chapter include foster care, how weak social ties contribute to the high divorce rate, and the role of forgiveness in post-divorce parenting relationships. A new *What Do You Think?* section asks readers if the criminalization of polygamy violates freedom of religion.

Chapter 6 (Poverty and Economic Inequality) includes new data on U.S. and global income, wealth inequality, and CEO compensation. This updated chapter includes coverage of the housing crisis: housing bubbles, subprime mortgages, upside-down mortgages, and foreclosures. The section on homelessness has been expanded and includes recent data on hate crimes against the homeless and "bum bashing" videos. A section on poverty rates for same-sex couples has been added to the section on Family Structure and Poverty. The section on myths about welfare has been expanded. In a new *The Human Side* feature titled "Poverty Never Takes a Holiday," a poor African woman describes poverty from her point of view. This chapter now includes a section on the effects of economic development on indigenous people. New *What Do You Think?* sections ask, "Why do you think Americans perceive the level of inequality in the United States to be much less than what it actually is?" and "Just as there is a federal minimum wage, do you think that there should be a federal maximum wage?"

Since the last edition of *Understanding Social Problems* was published, the United States and other parts of the world have experienced a prolonged economic downturn. In this revised edition, **Chapter 7** (Work and Unemployment) provides coverage of the global economic crisis, TARP, and the American Recovery and Reinvestment Act. This chapter features a new *Social Problems Research Up Close* feature: Job Loss at "Mid-Life"; a new *The Human Side* feature: "Down but Not Out: From Hedge Funds to Pizza Delivery"; and a new *Self and Society* feature: "How Do Your Spending Habits Change in Hard Economic Times?" This chapter now includes a section on employment and retirement concerns of older Americans, and a section on child labor. There is new coverage of outsourcing and the 2008 amendment to the Family and Medical Leave Act. A new table compares U.S. work policies with those of other countries, and another new table presents data on U.S. employer-based work-family benefits and policies. New *What Do You Think?* sections include questions about whether the world will ever achieve a post-petroleum economy, and whether cities should cancel the Fourth of July fireworks show and instead use the money on other needs, such as feeding the hungry.

Chapter 8 (Problems in Education) includes new statistics from the Organization for Economic Cooperation and Development, the Condition of Education, the National Center for Educational Statistics, the Educational Research Center, the Schott Report, and *Quality Counts 2009*. A new section on "The Challenges of Higher Education" has been added. Other new material includes the effects of the economic stimulus package on such programs as Head Start, Title I, and the federal student loan program, and the impact of state budget cuts on local school districts. The section formally called "Character Education" now also includes service learning, civic engagement, public service, and the Edward M. Kennedy Serve America Act of 2009. New concepts such as the stereotype threat hypothesis, green schools, e-learning, and the four assurances have also been introduced. New educational policies discussed in this chapter include the Safe Schools Improvement Act, the Improving Head Start for School Readiness Act of 2007, the DREAM Act, and the Higher Education Act. A new *The Human Side* feature, "You Want Heroes?" has been added as well as three new *What Do You Think?* sections ("Does English-only instruction hurt or help non-native English speaking students?"; "Should high schools have exit examinations as a requirement for graduation?"; "Do you think the DREAM Act should be passed?").

Chapter 9 (Race, Ethnicity, and Immigration) includes a discussion of the racial hate backlash against President Obama, and provides updated information on hate crimes and hate groups. Material on immigration has been expanded and includes a new section on myths about immigration and immigrants and an updated *Self and Society* feature on "Attitudes Toward U.S. Immigrants and Immigration." This revised chapter also discusses a new form of racism known as "Racism 2.0." New *What Do You Think?* sections include those that ask, "If blacks and whites honestly expressed their true feelings about race relations, do you think this would do more to bring races together or cause greater racial division?" and "If you honestly assess yourself, would you say that you have at least some feelings of racial prejudice?"

New to **Chapter 10** (Gender Inequality) are added sections on "Health and Gender" and "Gendered Violence," as well as a new *The Human Side* feature "On Gender and Marriage" and a new *Self and Society* feature: "An Inventory of Attitudes Toward Sexist/Nonsexist Language (IASNL)." This revised chapter now has greater emphasis on global issues and on the men's movement including a discussion of The Mankind Project. There are added discussions on Michael Kimmel's *Guyland* (2008) and Sadker and Zittleman's, *Still Failing at Fairness: How Gender Bias Cheats Boys and Girls in Schools* (2009). The section on media and gender has been expanded including a gender analysis of the 2008 presidential election. New information on the global gender gap, the impact of the economic crisis on men and women's employment, and on the "motherhood penalty" in relationship to politics, work, and wages has also been added. New *What Do You Think?* sections include "What do you think of a men's advocacy group on campus?"; "Have gender roles really changed that much over the last 40 years?"; and "Do you think dowry killings can be prevented through law?"

In this seventh edition, the name of **Chapter 11** has been changed from *Issues in Sexual Orientation* to *Sexual Orientation and the Struggle for Equality* to more accurately reflect the chapter's content and focus. This revised chapter includes updated information on the legal status of homosexuality and same-sex relationships globally and in the United States, a new section on "The Role of 'Coming Out' in the Struggle for Equality," and new data on gay-friendly workplace policies and employees who are "out" at their workplace. We have added new research findings on how family rejection affects the health and well-being of LGBT individuals, and a new *The Human Side* feature: "A Letter to My Gay Son." We have also added material on Harvey Milk and the movie made about him, as well as new Gallup Poll data on attitudes toward homosexuality. The section on "Effects of Antigay Bias and Discrimination on Heterosexuals" has been expanded to include a discussion of how antigay discrimination results in the loss of talented and dedicated professionals. New *What Do You Think?* sections include the following questions: (1) What are the implications of using the term *homosexual lifestyle?*; (2) "Don't ask, don't tell" policy is the only law in the country that requires people to be dishonest about themselves, or else lose their job. What do you think about a law that requires people to be dishonest?;

and (3) Should public opinion polls on same-sex marriage determine government's legislative decisions regarding same-sex marriage?

Chapter 12 (Population Growth and Urbanization) presents new statistics on current and projected population growth and fertility rates worldwide. This chapter includes a new section on the growing elderly population and a new *The Human Side* feature titled "One Man's Decision to Not Have Children." New data on urban poverty in the United States is presented along with coverage of a pioneering "shrink to survive" strategy of de-urbanization. A new *What Do You Think?* section asks whether meeting the "growing demands of a growing population" is solving the problem or perpetuating the problem.

Chapter 13 (Environmental Problems) includes a new chapter opening vignette about Greenpeace activists "Kingsnorth Six." This revised chapter has a new section on light pollution; a new *Self and Society* feature titled "Outcomes Expected from National Action to Reduce Global Warming," and a new *Social Problems Research Up Close* feature called "The Seven Sins of Greenwashing." There is new data on a wide range of topics, including global warming and climate change, chemicals in the environment, disappearing species, environmental injustice, and new fuel economy standards.

The revised **Chapter 14** (Science and Technology) highlights global and U.S. Internet trends and new information on computer security and computer threats including a discussion of China's continued censorship of the Internet. The reorganized and expanded Internet section now includes subheadings on e-commerce and finances, politics and e-government, social networking and communication, the search for knowledge, games and entertainment, and the malicious use of the Internet (e.g., cyber-bullying). This revised chapter has updated information on the debate over stem cells, cloning, genetically modified organisms, and abortion. A new *The Human Side* feature has been added ("For the Love of Andy") as well as a new *Self and Society* ("What Is Your Science IQ?").

Chapter 15 (Conflict, War, and Terrorism) presents updated statistics on global military spending, U.S. military spending since 9/11, the cost of the Iraq War, and attitudes among Americans about defense spending. This revised chapter also features updated information on terrorist attacks worldwide, and U.S. attitudes toward terrorism, security, and the Iraq war. We have expanded the sections on feminism and war; the Guantanamo detention center, torture, and waterboarding; and have updated the discussion of suicides rates in the armed forces, including new material on the backlog of PTSD and other medical claims in the Veterans Administration. This chapter presents updated information on North Korea's, Iran's, and South Asia's nuclear programs with particular emphasis on Pakistan. The section on arms control and disarmament now emphasizes nuclear nonproliferation and major treaties. Finally, there is a new *Social Problems Research Up Close* feature ("The Survey of War-Affected Youth"), a new *The Human Side* feature ("Taking Chance"), and a new *What Do You Think?* section that asks "Should the ban on media coverage of coffins being returned from war be lifted by Defense Secretary Gates?"

Features and Pedagogical Aids

We have integrated a number of features and pedagogical aids into the text to help students learn to think about social problems from a sociological perspective. Our mission is to help students not only apply sociological concepts to observed situations in their everyday lives and to think critically about social problems and their implications, but also learn to assess how social problems relate to their lives on a personal level.

Exercises and Boxed Features

Self and Society. Each chapter includes a social survey designed to help students assess their own attitudes, beliefs, knowledge, or behaviors regarding some aspect of the social problem under discussion. In Chapter 5 (Family Problems), for example, the "Abusive Behavior Inventory" invites students to assess the frequency of various abusive behaviors in their own relationships. The *Self and Society* feature in Chapter 3 (Alcohol and Other Drugs) allows students to measure the consequences of their own drinking behavior and compare it to respondents in a national sample.

The Human Side. In addition to the *Self and Society* boxed features, each chapter includes a boxed feature that further personalizes the social problems under discussion by describing personal experiences of individuals who have been affected by them. *The Human Side* feature in Chapter 4 (Crime and Social Control), for example, describes the horrific consequences of being a victim of rape, and *The Human Side* feature in Chapter 9 (Race, Ethnicity, and Immigration) describes the experiences of an immigrant day laborer who was victimized by a violent hate crime.

Social Problems Research Up Close. This feature, found in every chapter, presents examples of social science research and illustrates the sociological enterprise, from theory and data collection to findings and conclusions, thus exposing students to various studies and research methods. The *Social Problems Research Up Close* feature in Chapter 1 discusses social science research, frequently found sections in a research article, and how to read a contingency table. Other *Social Problems Research Up Close* topics include bullying, job loss in midlife, computer hacking, and young children's perceptions of cigarette smoking.

Photo Essay. Chapter 2 (Problems of Illness and Health Care) includes a photo essay titled "Modern Animal Food Production: Health and Safety Issues." In Chapter 6 (Poverty and Economic Inequality), a photo essay covers the topic "Lack of Clean Water and Sanitation among the Poor." In Chapter 13 (Environmental Problems), a photo essay depicts "Effects of Global Warming and Climate Change." This edition features two new photo essays. In Chapter 1, a new photo essay looks at "Students Making a Difference," and Chapter 4 features a new photo essay chronicling "Prison Programs that Work."

In-Text Learning Aids

Vignettes. Each chapter begins with a vignette designed to engage students and draw them into the chapter by illustrating the current relevance of the topic under discussion. Chapter 2 (Problems of Illness and Health Care), for instance, begins with a description of a family with no health insurance that must resort to receiving medical care from an annual free health clinic. Chapter 9 (Race, Ethnicity, and Immigration) begins with the story of an Obama campaign volunteer and supporter who was attacked by three white men who shouted, "Nigger president!"

Key Terms. Important terms and concepts are highlighted in the text where they first appear. To reemphasize the importance of these words, they are listed at the end of every chapter and are included in the glossary at the end of the text.

Running Glossary. This seventh edition continues the running glossary that highlights the key terms in every chapter by putting the key terms and their definitions in the text margins.

What Do You Think? Sections. Each chapter contains several sections called *What Do You Think?* These sections invite students to use critical thinking skills to answer questions about issues related to the chapter content. For example, one *What Do You Think?* feature in Chapter 3 (Alcohol and other Drugs) asks students "Should marijuana be legalized to raise revenues for ailing state economies?" A *What Do You Think?* feature in Chapter 11 (Sexual Orientation and the Struggle for Equality) asks why male homosexuality is illegal in many countries, but female homosexuality is not. In Chapter 12 (Population Growth and Urbanization), a *What Do You Think?* feature asks readers if the U.S. birth rate would increase if the U.S. government instituted paid parenting leave and government-supported child care.

Glossary. All key terms are defined in the end-of-text glossary.

Understanding [Specific Social Problem] Sections. All too often, students, faced with contradictory theories and study results, walk away from social problems courses without any real understanding of their causes and consequences. To address this problem, chapter sections titled "Understanding . . . [specific social problem]" cap the body of each chapter just before the chapter summaries. Unlike the chapter summaries, these sections synthesize the material presented in the chapter, summing up the present state of knowledge and theory on the chapter topic.

Supplements

The seventh edition of *Understanding Social Problems* comes with a full complement of supplements designed with both faculty and students in mind.

Supplements for the Instructor

Instructor's Resource Manual with Test Bank. This supplement, written by Shannon Carter of University of Central Florida, offers instructors learning objectives, key terms, lecture outlines, student projects, classroom activities, Internet and InfoTrac® College Edition exercises, and video suggestions. Test items include multiple-choice and true-false questions with answers and page references, as well as short-answer and essay questions for each chapter. Each multiple-choice item has the question type (factual, applied, or conceptual) indicated. All questions are labeled as new, modified, or pickup, so instructors know if the question is new to this edition of the test bank, modified but picked up from the previous edition of the test bank, or picked up straight from the previous edition of the test bank. Concise user guides for InfoTrac College Edition and InfoMarks® are included as appendices.

PowerLecture with JoinIn™ and ExamView®. This easy-to-use, one-stop digital library and presentation tool includes the following:

- Preassembled **Microsoft® PowerPoint® lecture slides**, prepared by Gary Titchener of Des Moines Area Community College, with graphics from the text, making it easy for you to assemble, edit, publish, and present custom lectures for your course.
- The PowerLectures CD-ROM, which includes video-based polling and quiz questions that can be used with the **JoinIn™ on TurningPoint®** personal response system.
- PowerLectures that also feature **ExamView testing software,** which includes all the test items from the printed test bank in electronic format, enabling you to create customized tests of up to 250 items that can be delivered in print or online.

Videos. Adopters of *Understanding Social Problems* have several different video options available with the text.

ABC® Videos/DVD: Social Problems. This series of videos, comprised of footage from ABC broadcasts, is specially selected and arranged to accompany your Social Problems course. The segments may be used in conjunction with Wadsworth, Cengage Learning's Social Problems texts to help provide real-world examples to illustrate course concepts, or to instigate discussion. ABC Videos feature short, high-interest clips from current news events as well as historic raw footage going back 40 years. Clips are drawn from such programs as *World News Tonight, Good Morning America, This Week, PrimeTime Live, 20/20,* and *Nightline,* as well as numerous ABC News specials and material from the Associated Press Television News and British Movietone News collections. Contact your Cengage Learning representative for a complete listing of videos and policies.

AIDS in Africa DVD. Southern Africa has been overcome by a pandemic of unparalleled proportions. This documentary series focuses on Namibia, a new democracy, and the many actions that are being taken to control HIV/AIDS there. Included in this series are four documentary films created by the Project Pericles scholars at Elon University.

Wadsworth Sociology Video Library. This large selection of thought-provoking films is available to adopters based on adoption size.

Supplements for the Student

Study Guide. Each chapter of this critically updated study guide, written by Gary Titchener of Des Moines Area Community College, includes a brief chapter outline, learning objectives, key terms, matching exercise, a chapter review fill-in-the-blank exercise, worksheets that students can complete directly in the study guide to help them prepare for exams, Internet activities, InfoTrac College Edition exercises, and a practice test, consisting of multiple-choice and true-false questions with answers and page references, as well as short-answer questions and essay questions with page references to enhance and test student understanding of chapter concepts.

Online Resources

CengageNOW™ Personalized Study, a diagnostic tool (including a chapter-specific *Pre-Test, Individualized Study Plan,* and *Post-Test* written by Lois Sabol of Yakima Valley Community College) helps students master concepts and prepare for exams by creating a study plan based on the students' performance on the Pre-Test. Easily assign Personalized Study for the entire term, and, if you want, results will automatically post to your grade book. Order new student texts packaged with the access code to ensure that your students have four months of free access from the moment they purchase the text. Contact your local Wadsworth, Cengage Learning representative for ordering details. CengageNOW also features the most intuitive, easy-to-use online course management and study system on the market. It saves you time through its automatic grading and easy-to-use grade book and provides your students with an efficient way to study.

Extension: Wadsworth's Sociology Reader Collection. Create your own customized reader for your sociology class, drawing from dozens of classic and contemporary articles found on the exclusive Wadsworth, Cengage Learning TextChoice database. Using the TextChoice website (www.TextChoice.com), you can preview articles, select your content, and add your own original material. TextChoice will then produce your materials as a printed supplementary reader for your class.

Wadsworth's Sociology Home Page (www.cengage.com/sociology). Here you will find a wealth of sociology resources such as Census 2000: A Student Guide for Sociology, Breaking News in Sociology, Guide to Researching Sociology on the Internet, Sociology in Action, and much more. Contained on the home page is the companion website for *Understanding Social Problems,* seventh edition.

Mooney/Knox/Schacht's *Understanding Social Problems* Companion Website (www. cengage.com/sociology/mooney). This site provides access to useful learning resources for each chapter of the book. Instructors can also access password-protected instructor's manuals, PowerPoint lectures, and important sociology links. Click on the companion website to find useful learning resources for each chapter of the book. Some of these resources include

- Tutorial practice quizzes that can be scored and e-mailed to the instructor
- Web links
- Internet exercises
- Flash cards of the text's glossary
- Crossword puzzles
- Essay questions
- Learning objectives
- Virtual explorations

And much more!

WebTutor™ for WebCT® or Blackboard®. Preloaded with content and available via access code when packaged with this text, WebTutor pairs all the content of this text's rich book companion website with all the sophisticated course management functionality of a WebCT or Blackboard product. You can assign materials (including online quizzes) and have the results flow automatically to your grade book.

InfoTrac College Edition. Four months' access to this online database—featuring reliable, full-length articles from thousands of academic journals and periodicals—is available with this text at no additional charge! This fully searchable database now features stable, topically bookmarked InfoMarks URLs to assist in research, plus InfoWrite critical thinking and writing tools. The database also offers 20 years' worth of full-text articles from almost 5,000 diverse sources, such as academic journals, newsletters, and up-to-the-minute periodicals, including *Time, Newsweek, Science, Forbes,* and *USA Today.* This incredible depth and breadth of material—available 24 hours a day from any computer with Internet access—makes conducting research so easy that your students will want to use it to enhance their work in every course!

Acknowledgments

This text reflects the work of many people. We would like to thank the following for their contributions to the development of this text: Chris Caldeira and Erin Mitchell, Acquisitions Editors; Tangelique Williams, Development Editor; Rachael Krapf, Editorial Assistant; Erin Parkins, Assistant Editor; Cheri Palmer, Content Project Manager; Jill Traut, Project Manager at MPS Limited; and Jaime Jankowski, Senior Photo Researcher of PrePress. We would also like to acknowledge the support and assistance of Carol L. Jenkins, John T. Crist, Marieke Van Willigen, Leon Wilson, Kelly Bristol, and Ronnie Miller. To each we send our heartfelt thanks. Special thanks also to George Glann, whose valuable contributions have assisted in achieving the book's high standard of quality from edition to edition.

Additionally, we are indebted to those who read the manuscript in its various drafts and provided valuable insights and suggestions, many of which have been incorporated into the final manuscript:

Linda Kaye Larrabee
Texas Tech University

Vickie Holland Taylor
Danville Community College

J. Meredith Martin
University of New Mexico

Jay Watterworth
University of Colorado at Boulder

Jason Wenzel
Valencia Community College

We are also grateful to the reviewers of the first, second, third, fourth, fifth, and sixth editions: David Allen, *University of New Orleans;* Patricia Atchison, *Colorado State University;* Wendy Beck, *Eastern Washington University;* Walter Carroll, *Bridgewater State College;* Deanna Chang, *Indiana University of Pennsylvania;* Roland Chilton, *University of Massachusetts;* Verghese Chirayath, *John Carroll University;* Margaret Chok, *Pellissippi State Technical Community College;* Kimberly Clark, *DeKalb College–Central Campus;* Anna M. Cognetto, *Dutchess Community College;* Robert R. Cordell, *West Virginia University at Parkersburg;* Barbara Costello, *Mississippi State University;* William Cross, *Illinois College;* Kim Davies, *Augusta State University;* Doug Degher, *Northern Arizona University;* Katherine Dietrich, *Blinn College;* Jane Ely, *State University of New York–Stony Brook;* William Feigelman, *Nassau Community College;* Joan Ferrante, *Northern Kentucky University;* Robert Gliner, *San Jose State University;*

Roberta Goldberg, *Trinity College;* Roger Guy, *Texas Lutheran University;* Julia Hall, *Drexel University;* Millie Harmon, *Chemeketa Community College;* Madonna Harrington-Meyer, *University of Illinois;* Sylvia Jones, *Jefferson Community College;* Nancy Kleniewski, *University of Massachusetts, Lowell;* Daniel Klenow, *North Dakota State University;* Sandra Krell-Andre, *Southeastern Community College;* Pui-Yan Lam, *Eastern Washington University;* Mary Ann Lamanna, *University of Nebraska;* Phyllis Langton, *George Washington University;* Cooper Lansing, *Erie Community College;* Tunga Lergo, *Santa Fe Community College, Main Campus;* Dale Lund, *University of Utah;* Lionel Maldonado, *California State University, San Marcos;* Judith Mayo, *Arizona State University;* Peter Meiksins, *Cleveland State University;* JoAnn Miller, *Purdue University;* Clifford Mottaz, *University of Wisconsin–River Falls;* Lynda D. Nyce, *Bluffton College;* Frank J. Page, *University of Utah;* James Peacock, *University of North Carolina;* Barbara Perry, *Northern Arizona University;* Ed Ponczek, *William Rainey Harper College;* Donna Provenza, *California State University at Sacramento;* Cynthia Reynaud, *Louisiana State University;* Carl Marie Rider, *Longwood University;* Jeffrey W. Riemer, *Tennessee Technological University;* Cherylon Robinson, *University of Texas at San Antonio;* Rita Sakitt, *Suffolk County Community College;* Mareleyn Schneider, *Yeshiva University;* Paula Snyder, *Columbus State Community College;* Lawrence Stern, *Collin County Community College;* John Stratton, *University of Iowa;* D. Paul Sullins, *The Catholic University of America;* Joseph Trumino, *St. Vincent's College of St. John's University;* Robert Turley, *Crafton Hills College;* Alice Van Ommeren, *San Joaquin Delta College;* Joseph Vielbig, *Arizona Western University;* Harry L. Vogel, *Kansas State University;* Robert Weaver, *Youngstown State University;* Rose Weitz, *Arizona State University;* Bob Weyer, *County College of Morris;* Oscar Williams, *Diablo Valley College;* Mark Winton, *University of Central Florida;* Diane Zablotsky, *University of North Carolina;* Joan Brehm, *Illinois State University;* Doug Degher, *Northern Arizona University;* Heather Griffiths, *Fayetteville State University;* Amy Holzgang, *Cerritos College;* Janét Hund, *Long Beach City College;* Kathrin Parks, *University of New Mexico;* Craig Robertson, *University of North Alabama;* Matthew Sanderson, *University of Utah;* Jacqueline Steingold, *Wayne State University;* William J. Tinney, Jr., *Black Hills State University.*

Finally, we are interested in ways to improve the text, and invite your feedback and suggestions for new ideas and material to be included in subsequent editions.

You can contact us at mooneyl@ecu.edu, knoxd@ecu.edu, or schachtc@ecu.edu.

About the Authors

Linda A. Mooney, PhD, is an associate professor of sociology at East Carolina University in Greenville, North Carolina. In addition to social problems, her specialties include law, criminology, gender, and issues in sexuality. She has published more than 30 professional articles in such journals as *Social Forces, Sociological Inquiry, Sex Roles, Sociological Quarterly,* and *Teaching Sociology.* She has won numerous teaching awards, including the University of North Carolina Board of Governor's Distinguished Professor for Teaching Award.

David Knox, PhD, is professor of sociology at East Carolina University. He has taught Social Problems, Introduction to Sociology, and Sociology of Marriage Problems. He is the author or co-author of 10 books and more than 80 professional articles. His research interests include marriage, family, intimate relationships, and sexual values and behavior.

Caroline Schacht, MA, is a teaching instructor of sociology at East Carolina University. She has taught Introduction to Sociology, Deviant Behavior, Sociology of Food, Sociology of Education, Individuals in Society, and Courtship and Marriage. She has co-authored several textbooks in the areas of social problems, introductory sociology, courtship and marriage, and human sexuality.

Thinking about Social Problems

"Unless someone like you cares a whole awful lot, nothing is going to get better. It's not."

Dr. Seuss, *The Lorax*

In a March 2008 Gallup Poll, a random sample of Americans were asked, "What do you think is the most important problem facing this country today?" Leading problems included the economy, the war in Iraq, health care, the energy crisis, immigration, unemployment, government corruption, inflation, poverty, terrorism, and crime and violence (Jacobs 2008). Moreover, survey results indicate that, overall, just 17 percent of Americans were satisfied "with the way things are going in the United States at this time" (Saad 2008). Compared with previous years, this number is quite low, tying the lowest-ever recorded satisfaction rate in 1992. Dissatisfaction with "the way things are going" no doubt played a role in the 1992 election of Bill Clinton, a Democrat, after eight years of a Republican administration. Similarly, President Obama's election was fueled by a general dissatisfaction with the direction of the country and a sense that a change was needed. In his inaugural address in 2009, President Obama acknowledged many of the social problems our country has to face:

After the economic turndown of 2008, the U.S. Congress passed and President Obama signed into law the $787 billion American Recovery and Reinvestment Act of 2009. The stimulus package was designed to help failing industries, create jobs, promote consumer spending, rescue the failed housing market, and encourage energy-related investments.

> That we are in the midst of crisis is now well understood. Our nation is at war against a far-reaching network of violence and hatred. Our economy is badly weakened, a consequence of greed and irresponsibility on the part of some, but also our collective failure to make hard choices and prepare the nation for a new age. Homes have been lost, jobs shed, businesses shuttered. Our health care is too costly, our schools fail too many—and each day brings further evidence that the ways we use energy strengthen our adversaries and threaten our planet. (Obama 2009)

A global perspective on social problems is also troubling. In 1990, the United Nations Development Programme published its first annual *Human Development Report*, which measured the well-being of populations around the world according to a "human development index." This index measures three basic dimensions of human development: longevity, as measured by life expectancy at birth; knowledge (i.e., literacy, educational attainment); and a decent standard of living. The most recent report, focusing on climate change and its impact on poverty, the destruction of the environment, economic disaster, crime, health, and the like, states that the "...the world has less than a decade to change its course. Actions taken—or not taken—in the years ahead will have a profound bearing on the future course of human development. The world lacks neither the financial resources nor the technological capabilities to act. What is missing is a sense of urgency, human solidarity, and collective interest" (United Nations Development Programme 2008, p. 1).

Problems related to poverty and malnutrition, inadequate education, acquired immunodeficiency syndrome (AIDS), inadequate health care, crime, conflict, oppression of minorities, environmental destruction, and other social issues are both national and international concerns. Such problems present both a threat and a challenge to our national and global society. The primary goal of this textbook is to facilitate increased awareness and understanding of problematic social conditions in U.S. society and throughout the world.

Although the topics covered in this book vary widely, all chapters share common objectives: to explain how social problems are created and maintained; to indicate how they affect individuals, social groups, and societies as a whole; and to examine programs and policies for change. We begin by looking at the nature of social problems.

What Is a Social Problem?

There is no universal, constant, or absolute definition of what constitutes a social problem. Rather, social problems are defined by a combination of objective and subjective criteria that vary across societies, among individuals and groups within a society, and across historical time periods.

Objective and Subjective Elements of Social Problems

Although social problems take many forms, they all share two important elements: an objective social condition and a subjective interpretation of that social condition. The **objective element of a social problem** refers to the existence of a social condition. We become aware of social conditions through our own life experience, through the media, and through education. We see the homeless, hear gunfire in the streets, and see battered women in hospital emergency rooms. We read about employees losing their jobs as businesses downsize and factories close. In television news reports, we see the anguished faces of parents whose children have been killed by violent youths.

The **subjective element of a social problem** refers to the belief that a particular social condition is harmful to society or to a segment of society and that it should and can be changed. We know that crime, drug addiction, poverty, racism, violence, and pollution exist. These social conditions are not considered social problems, however, unless at least a segment of society believes that these conditions diminish the quality of human life.

By combining these objective and subjective elements, we arrive at the following definition: A **social problem** is a social condition that a segment of society views as harmful to members of society and in need of remedy.

Variability in Definitions of Social Problems

Individuals and groups frequently disagree about what constitutes a social problem. For example, some Americans view the availability of abortion as a social problem, whereas others view restrictions on abortion as a social problem. Similarly, some Americans view homosexuality as a social problem, whereas others view prejudice and discrimination against homosexuals as a social problem. Such variations in what is considered a social problem are due to differences in values, beliefs, and life experiences.

Definitions of social problems vary not only within societies but also across societies and historical time periods. For example, before the 19th century, a husband's legal right and marital obligation was to discipline and control his wife through the use of physical force. Today, the use of physical force is regarded as a social problem rather than a marital right.

Tea drinking is another example of how what is considered a social problem can change over time. In 17th- and 18th-century England, tea drinking was regarded as a "base Indian practice" that was "pernicious to health, obscuring industry, and impoverishing the nation" (Ukers 1935, cited by Troyer & Markle 1984). Today, the English are known for their tradition of drinking tea in the afternoon.

Because social problems can be highly complex, it is helpful to have a framework within which to view them. Sociology provides such a framework. Using a sociological perspective to examine social problems requires knowledge of the basic concepts and tools of sociology. In the remainder of this chapter, we discuss

objective element of a social problem Awareness of social conditions through one's own life experiences and through reports in the media.

subjective element of a social problem The belief that a particular social condition is harmful to society, or to a segment of society, and that it should and can be changed.

social problem A social condition that a segment of society views as harmful to members of society and in need of remedy.

some of these concepts and tools: social structure, culture, the "sociological imagination," major theoretical perspectives, and types of research methods.

What Do You Think? People increasingly are using information technologies (e.g., blogs, web portals, online news feeds) to get their daily news with "traditional media outlets…struggling to hold their market share" (Saad 2007, p. 1). For example, in 2009, a number of major dailies closed their doors, including *Denver's Rocky Mountain News*, the *Seattle Post Intelligencer*, and the *San Francisco Chronicle* (Shaw 2009). If your local print and/or online newspaper folded, where would you go for news? What role do the various media play in our awareness of social problems? Will definitions of social problems change as sources of information change and, if so, in what way?

Saul Porto/ AP Photo

Whereas some individuals view homosexual behavior as a social problem, others view homophobia as a social problem. Here, participants carry a giant rainbow flag during a gay pride parade in Toronto, Canada. The 2006 Canadian census was revamped to include "same-sex married spouse" as a response option (Beeby 2005).

structure The way society is organized including institutions, social groups, statuses, and roles.

institution An established and enduring pattern of social relationships.

Elements of Social Structure and Culture

Although society surrounds us and permeates our lives, it is difficult to "see" society. By thinking of society in terms of a picture or image, however, we can visualize society and therefore better understand it. Imagine that society is a coin with two sides: On one side is the structure of society and on the other is the culture of society. Although each side is distinct, both are inseparable from the whole. By looking at the various elements of social structure and culture, we can better understand the root causes of social problems.

Elements of Social Structure

The **structure** of a society refers to the way society is organized. Society is organized into different parts: institutions, social groups, statuses, and roles.

Institutions. An **institution** is an established and enduring pattern of social relationships. The five traditional institutions are family, religion, politics, economics, and education, but some sociologists argue that other social institutions, such as science and technology, mass media, medicine, sports, and the military, also play important roles in modern society. Many social problems are generated by inadequacies in various institutions. For example, unemployment may be influenced by the educational institution's failure to prepare individuals for the job market and by alterations in the structure of the economic institution.

Social Groups. Institutions are made up of social groups. A **social group** is defined as two or more people who have a common identity, interact, and form a social relationship. For example, the family in which you were reared is a social group that is part of the family institution. The religious association to which you may belong is a social group that is part of the religious institution.

Social groups can be categorized as primary or secondary. **Primary groups,** which tend to involve small numbers of individuals, are characterized by intimate and informal interaction. Families and friends are examples of primary groups. **Secondary groups,** which may involve small or large numbers of individuals, are task-oriented and are characterized by impersonal and formal interaction. Examples of secondary groups include employers and their employees and clerks and their customers.

Statuses. Just as institutions consist of social groups, social groups consist of statuses. A **status** is a position that a person occupies within a social group. The statuses we occupy largely define our social identity. The statuses in a family may consist of mother, father, stepmother, stepfather, wife, husband, child, and so on. Statuses can be either ascribed or achieved. An **ascribed status** is one that society assigns to an individual on the basis of factors over which the individual has no control. For example, we have no control over the sex, race, ethnic background, and socioeconomic status into which we are born. Similarly, we are assigned the status of child, teenager, adult, or senior citizen on the basis of our age—something we do not choose or control.

An **achieved status** is assigned on the basis of some characteristic or behavior over which the individual has some control. Whether you achieve the status of college graduate, spouse, parent, bank president, or prison inmate depends largely on your own efforts, behavior, and choices. One's ascribed statuses may affect the likelihood of achieving other statuses, however. For example, if you are born into a poor socioeconomic status, you may find it more difficult to achieve the status of college graduate because of the high cost of a college education.

Every individual has numerous statuses simultaneously. You may be a student, parent, tutor, volunteer fund-raiser, female, and Hispanic. A person's *master status* is the status that is considered the most significant in a person's social identity. In the United States, a person's occupational status is typically regarded as a master status. If you are a full-time student, your master status is likely to be student.

Roles. Every status is associated with many **roles,** or the set of rights, obligations, and expectations associated with a status. Roles guide our behavior and allow us to predict the behavior of others. As a student, you are expected to attend class, listen and take notes, study for tests, and complete assignments. Because you know what the role of teacher involves, you can predict that your teacher will lecture, give exams, and assign grades based on your performance on tests.

A single status involves more than one role. For example, the status of prison inmate includes one role for interacting with prison guards and another role for interacting with other prison inmates. Similarly, the status of nurse involves different roles for interacting with physicians and with patients.

Elements of Culture

Whereas social structure refers to the organization of society, **culture** refers to the meanings and ways of life that characterize a society. The elements of culture include beliefs, values, norms, sanctions, and symbols.

social group Two or more people who have a common identity, interact, and form a social relationship.

primary groups Usually small numbers of individuals characterized by intimate and informal interaction.

secondary groups Involving small or large numbers of individuals, groups that are task-oriented and are characterized by impersonal and formal interaction.

status A position that a person occupies within a social group.

ascribed status A status that society assigns to an individual on the basis of factors over which the individual has no control.

achieved status A status that society assigns to an individual on the basis of factors over which the individual has some control.

role The set of rights, obligations, and expectations associated with a status.

culture The meanings and ways of life that characterize a society, including beliefs, values, norms, sanctions, and symbols.

Beliefs. **Beliefs** refer to definitions and explanations about what is assumed to be true. The beliefs of an individual or group influence whether that individual or group views a particular social condition as a social problem. Does second-hand smoke harm nonsmokers? Are nuclear power plants safe? Does violence in movies and on television lead to increased aggression in children? Our beliefs regarding these issues influence whether we view the issues as social problems. Beliefs influence not only how a social condition is interpreted but also the existence of the condition itself. For example, police officers' beliefs about their supervisors' priorities affected officers' problem-solving behavior and the time devoted to it (Engel & Worden 2003). The Self and Society feature in this chapter allows you to assess your own beliefs about various social issues and to compare your beliefs with a national sample of first-year college students.

Values. **Values** are social agreements about what is considered good and bad, right and wrong, desirable and undesirable. Frequently, social conditions are viewed as social problems when the conditions are incompatible with or contradict closely held values. For example, poverty and homelessness violate the value of human welfare; crime contradicts the values of honesty, private property, and nonviolence; racism, sexism, and heterosexism violate the values of equality and fairness.

Values play an important role not only in the interpretation of a condition as a social problem but also in the development of the social condition itself. Sylvia Ann Hewlett (1992) explains how the American values of freedom and individualism are at the root of many of our social problems:

> There are two sides to the coin of freedom. On the one hand, there is enormous potential for prosperity and personal fulfillment; on the other are all the hazards of untrammeled opportunity and unfettered choice. Free markets can produce grinding poverty as well as spectacular wealth; unregulated industry can create dangerous levels of pollution as well as rapid rates of growth; and an unfettered drive for personal fulfillment can have disastrous effects on families and children. Rampant individualism does not bring with it sweet freedom; rather, it explodes in our faces and limits life's potential. (pp. 350–51)

Absent or weak values may contribute to some social problems. For example, many industries do not value protection of the environment and thus contribute to environmental pollution.

Norms and Sanctions. **Norms** are socially defined rules of behavior. Norms serve as guidelines for our behavior and for our expectations of the behavior of others.

There are three types of norms: folkways, laws, and mores. *Folkways* refer to the customs and manners of society. In many segments of our society, it is customary to shake hands when being introduced to a new acquaintance, to say "excuse me" after sneezing, and to give presents to family and friends on their birthdays. Although no laws require us to do these things, we are expected to do them because they are part of the cultural tradition, or folkways, of the society in which we live.

Laws are norms that are formalized and backed by political authority. It is normative for a Muslim woman to wear a veil. However, in the United States, failure to remove the veil for a driver's license photo is grounds for revoking the permit. Such is the case of a Florida woman who brought suit against the state, claiming that her religious rights were being violated because she was required to remove her veil for the driver's license photo (Candey 2002). She appealed the decision

beliefs Definitions and explanations about what is assumed to be true.

values Social agreements about what is considered good and bad, right and wrong, desirable and undesirable.

norms Socially defined rules of behavior including folkways, mores, and laws.

Indicate whether you agree or disagree with each of the following statements:

Statement	Agree	Disagree
1. Federal military spending should be increased.	_____	_____
2. The federal government is not doing enough to control environmental pollution.	_____	_____
3. There is too much concern in the courts for the rights of criminals.	_____	_____
4. Abortion should be legal.	_____	_____
5. The death penalty should be abolished.	_____	_____
6. Undocumented immigrants should be denied access to public education.	_____	_____
7. Marijuana should be legalized.	_____	_____
8. It is important to have laws prohibiting homosexual relationships.	_____	_____
9. Colleges have the right to ban extreme speakers.	_____	_____
10. The federal government should do more to control the sale of handguns.	_____	_____
11. Racial discrimination is no longer a major problem in America.	_____	_____
12. Realistically, an individual can do little to bring about changes in our society.	_____	_____
13. Wealthy people should pay a larger share of taxes than they do now.	_____	_____
14. Affirmative action in college admissions should be abolished.	_____	_____
15. Same-sex couples should have the right to legal marital status.	_____	_____

Percentage of First-Year College Students Agreeing with Belief Statements*

STATEMENT NUMBER	PERCENTAGE AGREEING IN 2006		
	TOTAL	WOMEN	MEN
1. Military spending should be increased.	28	24	32
2. Federal government is not doing enough to stop pollution.	79	82	76
3. There is too much concern for criminals' rights.	57	55	60
4. Abortion should be legal.	58	58	59
5. The death penalty should be abolished.	35	38	31
6. Immigrants should be denied access to public schools.	47	43	53
7. Marijuana should be legalized.	41	37	47
8. It is important to have laws prohibiting gay relationships.	23	18	30
9. Colleges should be able to ban speakers on campus.	41	38	44
10. Federal government should do more to control the sale of handguns.	72	79	64
11. Racial discrimination is no longer a problem.	20	16	25
12. Individuals can't influence social change.	27	24	31
13. The wealthy should pay higher taxes.	60	61	60
14. Affirmative action in college admissions should be abolished.	48	43	53
15. Same-sex couples should have a legal right to marry.	66	72	59

*Percentages are rounded.

Source: Pryor et al. 2008.

to Florida's District Court of Appeal and lost. The Court recognized, however, "the tension created as a result of choosing between following the dictates of one's religion and the mandates of secular law" (Associated Press 2006).

Mores are norms with a moral basis. Violations of mores may produce shock, horror, and moral indignation. Both littering and child sexual abuse are violations

TABLE 1.1 Types and Examples of Sanctions		
	POSITIVE	NEGATIVE
Informal	Being praised by one's neighbors for organizing a neighborhood recycling program	Being criticized by one's neighbors for refusing to participate in the neighborhood recycling program
Formal	Being granted a citizen's award for organizing a neighborhood recycling program	Being fined by the city for failing to dispose of trash properly

of law, but child sexual abuse is also a violation of our mores because we view such behavior as immoral.

All norms are associated with **sanctions,** or social consequences for conforming to or violating norms. When we conform to a social norm, we may be rewarded by a positive sanction. These may range from an approving smile to a public ceremony in our honor. When we violate a social norm, we may be punished by a negative sanction, which may range from a disapproving look to the death penalty or life in prison. Most sanctions are spontaneous expressions of approval or disapproval by groups or individuals—these are referred to as informal sanctions. Sanctions that are carried out according to some recognized or formal procedure are referred to as formal sanctions. Types of sanctions, then, include positive informal sanctions, positive formal sanctions, negative informal sanctions, and negative formal sanctions (see Table 1.1).

Symbols. A **symbol** is something that represents something else. Without symbols, we could not communicate with each other or live as social beings.

The symbols of a culture include language, gestures, and objects whose meaning the members of a society commonly understand. In our society, a red ribbon tied around a car antenna symbolizes Mothers Against Drunk Driving; a peace sign symbolizes the value of nonviolence; and a white-hooded robe symbolizes the Ku Klux Klan. Sometimes people attach different meanings to the same symbol. The Confederate flag is a symbol of southern pride to some and a symbol of racial bigotry to others.

The elements of the social structure and culture just discussed play a central role in the creation, maintenance, and social response to various social problems. One of the goals of taking a course in social problems is to develop an awareness of how the elements of social structure and culture contribute to social problems. Sociologists refer to this awareness as the "sociological imagination."

The Sociological Imagination

The **sociological imagination,** a term C. Wright Mills (1959) developed, refers to the ability to see the connections between our personal lives and the social world in which we live. When we use our sociological imagination, we are able to distinguish between "private troubles" and "public issues" and to see connections between the events and conditions of our lives and the social and historical context in which we live.

For example, that one person is unemployed constitutes a private trouble. That millions of people are unemployed in the United States constitutes a public issue. Once we understand that other segments of society share personal troubles

sanctions Social consequences for conforming to or violating norms.

symbol Something that represents something else.

sociological imagination The ability to see the connections between our personal lives and the social world in which we live.

such as human immunodeficiency virus (HIV) infection, criminal victimization, and poverty, we can look for the elements of social structure and culture that contribute to these public issues and private troubles. If the various elements of social structure and culture contribute to private troubles and public issues, then society's social structure and culture must be changed if these concerns are to be resolved.

Rather than viewing the private trouble of being unemployed as a result of an individual's faulty character or lack of job skills, we may understand unemployment as a public issue that results from the failure of the economic and political institutions of society to provide job opportunities to all citizens, as exemplified by the 2009 U.S. recession. Technological innovations emerging from the Industrial Revolution led to machines replacing individual workers. During the economic recession of the 1980s, employers fired employees so the firms could stay in business. Thus, in both these cases, social forces rather than individual skills largely determined whether a person was employed.

> When we use our sociological imagination, we are able to distinguish between "private troubles" and "public issues" and to see connections between the events and conditions of our lives and the social and historical context in which we live.

Theoretical Perspectives

Theories in sociology provide us with different perspectives with which to view our social world. A perspective is simply a way of looking at the world. A theory is a set of interrelated propositions or principles designed to answer a question or explain a particular phenomenon; it provides us with a perspective. Sociological theories help us to explain and predict the social world in which we live.

Sociology includes three major theoretical perspectives: the structural-functionalist perspective, the conflict perspective, and the symbolic interactionist perspective. Each perspective offers a variety of explanations about the causes of and possible solutions to social problems.

Structural-Functionalist Perspective

The structural-functionalist perspective is based largely on the works of Herbert Spencer, Emile Durkheim, Talcott Parsons, and Robert Merton. According to structural functionalism, society is a system of interconnected parts that work together in harmony to maintain a state of balance and social equilibrium for the whole. For example, each of the social institutions contributes important functions for society: Family provides a context for reproducing, nurturing, and socializing children; education offers a way to transmit a society's skills, knowledge, and culture to its youth; politics provides a means of governing members of society; economics provides for the production, distribution, and consumption of goods and services; and religion provides moral guidance and an outlet for worship of a higher power.

The structural-functionalist perspective emphasizes the interconnectedness of society by focusing on how each part influences and is influenced by other parts. For example, the increase in single-parent and dual-earner families has contributed to the number of children who are failing in school because parents have become less available to supervise their children's homework. As a result of changes in technology, colleges are offering more technical programs, and many adults are returning to school to learn new skills that are required in the workplace. The increasing number of women in the workforce has contributed to the formulation of policies against sexual harassment and job discrimination.

Structural functionalists use the terms *functional* and *dysfunctional* to describe the effects of social elements on society. Elements of society are functional if they contribute to social stability and dysfunctional if they disrupt social stability. Some aspects of society can be both functional and dysfunctional. For example, crime is dysfunctional in that it is associated with physical violence, loss of property, and fear. But according to Durkheim and other functionalists, crime is also functional for society because it leads to heightened awareness of shared moral bonds and increased social cohesion.

Sociologists have identified two types of functions: manifest and latent (Merton 1968). **Manifest functions** are consequences that are intended and commonly recognized. **Latent functions** are consequences that are unintended and often hidden. For example, the manifest function of education is to transmit knowledge and skills to society's youth. But public elementary schools also serve as babysitters for employed parents, and colleges offer a place for young adults to meet potential mates. The babysitting and mate-selection functions are not the intended or commonly recognized functions of education; hence, they are latent functions.

What Do You Think? In viewing society as a set of interrelated parts, structural functionalists argue that proposed solutions to social problems may lead to other social problems. For example, urban renewal projects displace residents and break up community cohesion. Racial imbalance in schools led to forced integration, which in turn generated violence and increased hostility between the races. What are some other "solutions" that lead to social problems? Do all solutions come with a price to pay? Can you think of a solution to a social problem that has no negative consequences?

Structural-Functionalist Theories of Social Problems

Two dominant theories of social problems grew out of the structural-functionalist perspective: social pathology and social disorganization.

Social Pathology. According to the social pathology model, social problems result from some "sickness" in society. Just as the human body becomes ill when our systems, organs, and cells do not function normally, society becomes "ill" when its parts (i.e., elements of the structure and culture) no longer perform properly. For example, problems such as crime, violence, poverty, and juvenile delinquency are often attributed to the breakdown of the family institution; the decline of the religious institution; and inadequacies in our economic, educational, and political institutions.

Social "illness" also results when members of a society are not adequately socialized to adopt its norms and values. People who do not value honesty, for example, are prone to dishonesties of all sorts. Early theorists attributed the failure in socialization to "sick" people who could not be socialized. Later theorists recognized that failure in the socialization process stemmed from "sick" social conditions, not "sick" people. To prevent or solve social problems, members of society must receive proper socialization and moral education, which may be accomplished in the family, schools, churches, or workplace and/or through the media.

manifest functions
Consequences that are intended and commonly recognized.

latent functions
Consequences that are unintended and often hidden.

Social Disorganization. According to the social disorganization view of social problems, rapid social change (e.g., the cultural revolution of the 1960s) disrupts the norms in a society. When norms become weak or are in conflict with each other, society is in a state of **anomie,** or normlessness. Hence, people may steal, physically abuse their spouses or children, abuse drugs, commit rape, or engage in other deviant behavior because the norms regarding these behaviors are weak or conflicting. According to this view, the solution to social problems lies in slowing the pace of social change and strengthening social norms. For example, although the use of alcohol by teenagers is considered a violation of a social norm in our society, this norm is weak. The media portray young people drinking alcohol, teenagers teach each other to drink alcohol and buy fake identification cards (IDs) to purchase alcohol, and parents model drinking behavior by having a few drinks after work or at a social event. Solutions to teenage drinking may involve strengthening norms against it through public education, restricting media depictions of youth and alcohol, imposing stronger sanctions against the use of fake IDs to purchase alcohol, and educating parents to model moderate and responsible drinking behavior.

Conflict Perspective

Contrary to the structural-functionalism perspective, the conflict perspective views society as composed of different groups and interests competing for power and resources. The conflict perspective explains various aspects of our social world by looking at which groups have power and benefit from a particular social arrangement. For example, feminist theory argues that we live in a patriarchal society—a hierarchical system of organization controlled by men. Although there are many varieties of feminist theory, most would hold that feminism "demands that existing economic, political, and social structures be changed" (Weir and Faulkner 2004, p. xii).

The origins of the conflict perspective can be traced to the classic works of Karl Marx. Marx suggested that all societies go through stages of economic development. As societies evolve from agricultural to industrial, concern over meeting survival needs is replaced by concern over making a profit, the hallmark of a capitalist system. Industrialization leads to the development of two classes of people: the bourgeoisie, or the owners of the means of production (e.g., factories, farms, businesses); and the proletariat, or the workers who earn wages.

The division of society into two broad classes of people—the "haves" and the "have-nots"—is beneficial to the owners of the means of production. The workers, who may earn only subsistence wages, are denied access to the many resources available to the wealthy owners. According to Marx, the bourgeoisie use their power to control the institutions of society to their advantage. For example, Marx suggested that religion serves as an "opiate of the masses" in that it soothes the distress and suffering associated with the working-class lifestyle and focuses the workers' attention on spirituality, God, and the afterlife rather than on worldly concerns such as living conditions. In essence, religion diverts the workers so that they concentrate on being rewarded in heaven for living a moral life rather than on questioning their exploitation.

> Industrialization leads to the development of two classes of people: the bourgeoisie, or the owners of the means of production (e.g., factories, farms, businesses); and the proletariat, or the workers who earn wages.

anomie A state of normlessness in which norms and values are weak or unclear.

Conflict Theories of Social Problems

There are two general types of conflict theories of social problems: Marxist and non-Marxist. Marxist theories focus on social conflict that results from economic inequalities; non-Marxist theories focus on social conflict that results from competing values and interests among social groups.

Marxist Conflict Theories. According to contemporary Marxist theorists, social problems result from class inequality inherent in a capitalistic system. A system of haves and have-nots may be beneficial to the haves but often translates into poverty for the have-nots. As we will explore later in this textbook, many social problems, including physical and mental illness, low educational achievement, and crime, are linked to poverty.

In addition to creating an impoverished class of people, capitalism also encourages "corporate violence." Corporate violence can be defined as actual harm and/or risk of harm inflicted on consumers, workers, and the general public as a result of decisions by corporate executives or managers. Corporate violence can also result from corporate negligence; the quest for profits at any cost; and willful violations of health, safety, and environmental laws (Reiman 2007). Our profit-motivated economy encourages individuals who are otherwise good, kind, and law-abiding to knowingly participate in the manufacturing and marketing of defective brakes on American jets, fuel tanks on automobiles, and contraceptive devices (e.g., intrauterine devices [IUDs]). The profit motive has also caused individuals to sell defective medical devices, toxic pesticides, and contaminated foods in the United States and abroad. In 2009, according to the U.S. Food and Drug Administration, the Peanut Corporation of America "knowingly shipped peanut products that could have been tainted" with salmonella poisoning (Stark et al. 2009).

Preschooler Jacob Hurley, who became seriously ill after eating peanut butter manufactured by the Peanut Corporation of America, is shown sitting with his father Peter Hurley, who is testifying before a House Energy and Commerce Committee hearing on Capitol Hill in Washington, DC, January 2009.

Mark Wilson/Getty Images

Marxist conflict theories also focus on the problem of **alienation,** or powerlessness and meaninglessness in people's lives. In industrialized societies, workers often have little power or control over their jobs, a condition that fosters in them a sense of powerlessness in their lives. The specialized nature of work requires workers to perform limited and repetitive tasks; as a result, the workers may come to feel that their lives are meaningless.

Alienation is bred not only in the workplace but also in the classroom. Students have little power over their education and often find that the curriculum is not meaningful to their lives. Like poverty, alienation is linked to other social problems, such as low educational achievement, violence, and suicide.

alienation A sense of powerlessness and meaninglessness in people's lives.

Marxist explanations of social problems imply that the solution lies in eliminating inequality among classes of people by creating a classless society. The nature of work must also change to avoid alienation. Finally, stronger controls must be applied to corporations to ensure that corporate decisions and practices are based on safety rather than on profit considerations.

Non-Marxist Conflict Theories. Non-Marxist conflict theorists, such as Ralf Dahrendorf, are concerned with conflict that arises when groups have opposing values and interests. For example, antiabortion activists value the life of unborn embryos and fetuses; pro-choice activists value the right of women to control their own bodies and reproductive decisions. These different value positions reflect different subjective interpretations of what constitutes a social problem. For antiabortionists, the availability of abortion is the social problem; for pro-choice advocates, the restrictions on abortion are the social problem. Sometimes the social problem is not the conflict itself but rather the way that conflict is expressed. Even most pro-life advocates agree that shooting doctors who perform abortions and blowing up abortion clinics constitute unnecessary violence and lack of respect for life. Value conflicts may occur between diverse categories of people, including nonwhites versus whites, heterosexuals versus homosexuals, young versus old, Democrats versus Republicans, and environmentalists versus industrialists.

Solving the problems that are generated by competing values may involve ensuring that conflicting groups understand each other's views, resolving differences through negotiation or mediation, or agreeing to disagree. Ideally, solutions should be win-win, with both conflicting groups satisfied with the solution. However, outcomes of value conflicts are often influenced by power; the group with the most power may use its position to influence the outcome of value conflicts. For example, when Congress could not get all states to voluntarily increase the legal drinking age to 21, it threatened to withdraw federal highway funds from those that would not comply.

Symbolic Interactionist Perspective

Both the structural-functionalist and the conflict perspectives are concerned with how broad aspects of society, such as institutions and large social groups, influence the social world. This level of sociological analysis is called *macrosociology*: It looks at the big picture of society and suggests how social problems are affected at the institutional level.

Microsociology, another level of sociological analysis, is concerned with the social-psychological dynamics of individuals interacting in small groups. Symbolic interactionism reflects the microsociological perspective and was largely influenced by the work of early sociologists and philosophers such as Max Weber, Georg Simmel, Charles Horton Cooley, G. H. Mead, W. I. Thomas, Erving Goffman, and Howard Becker. Symbolic interactionism emphasizes that human behavior is influenced by definitions and meanings that are created and maintained through symbolic interaction with others.

Sociologist W. I. Thomas (1931/1966) emphasized the importance of definitions and meanings in social behavior and its consequences. He suggested that humans respond to their definition of a situation rather than to the objective situation itself. Hence, Thomas noted that situations that we define as real become real in their consequences.

Symbolic interactionism also suggests that social interaction shapes our identity or sense of self. We develop our self-concept by observing how others interact with us and label us. By observing how others view us, we see a reflection of ourselves that Cooley calls the "looking-glass self."

Last, the symbolic interactionist perspective has important implications for how social scientists conduct research. German sociologist Max Weber argued that, to understand individual and group behavior, social scientists must see the world through the eyes of that individual or group. Weber called this approach *verstehen,* which in German means "to understand." *Verstehen* implies that, in conducting research, social scientists must try to understand others' views of reality and the subjective aspects of their experiences, including their symbols, values, attitudes, and beliefs.

Symbolic Interactionist Theories of Social Problems

A basic premise of symbolic interactionist theories of social problems is that a condition must be *defined or recognized* as a social problem for it to *be* a social problem. Three symbolic interactionist theories of social problems are based on this general premise.

Blumer's Stages of a Social Problem. Herbert Blumer (1971) suggested that social problems develop in stages. First, social problems pass through the stage of *societal recognition*—the process by which a social problem, for example, drunk driving, is "born." Second, *social legitimation* takes place when the social problem achieves recognition by the larger community, including the media, schools, and churches. As the visibility of traffic fatalities associated with alcohol increased, so did the legitimation of drunk driving as a social problem. The next stage in the development of a social problem involves *mobilization for action,* which occurs when individuals and groups, such as Mothers Against Drunk Driving, become concerned about how to respond to the social condition. This mobilization leads to the *development and implementation of an official plan* for dealing with the problem, involving, for example, highway checkpoints, lower legal blood-alcohol levels, and tougher regulations for driving drunk.

Blumer's stage-development view of social problems is helpful in tracing the development of social problems. For example, although sexual harassment and date rape occurred throughout the 20th century, these issues did not begin to receive recognition as social problems until the 1970s. Social legitimation of these problems was achieved when high schools, colleges, churches, employers, and the media recognized their existence. Organized social groups mobilized to develop and implement plans to deal with these problems. Groups successfully lobbied for the enactment of laws against sexual harassment and the enforcement of sanctions against violators of these laws. Groups also mobilized to provide educational seminars on date rape for high school and college students and to offer support services to victims of date rape.

Some disagree with the symbolic interactionist view that social problems exist only if they are recognized. According to this view, individuals who were victims of date rape in the 1960s may be considered victims of a problem, even though date rape was not recognized at that time as a social problem.

Labeling Theory. Labeling theory, a major symbolic interactionist theory of social problems, suggests that a social condition or group is viewed as problematic if it is labeled as such. According to labeling theory, resolving social problems sometimes involves changing the meanings and definitions that are attributed to people and situations. For example, so long as teenagers define drinking alcohol as "cool" and "fun," they will continue to abuse alcohol. So long as our society defines providing sex education and contraceptives to teenagers as inappropriate or immoral, the teenage pregnancy rate in the United States will continue to be higher than that in other industrialized nations.

Social Constructionism. Social constructionism is another symbolic interactionist theory of social problems. Similar to labeling theorists and symbolic interactionism in general, social constructionists argue that individuals who interpret the social world around them socially construct reality. Society, therefore, is a social creation rather than an objective given. As such, social constructionists often question the origin and evolution of social problems. For example, most Americans define "drug abuse" as a social problem in the United States but rarely include alcohol or cigarettes in their discussion. A social constructionist would point to the historical roots of alcohol and tobacco use as a means of understanding their legal status. Central to this idea of the social construction of social problems are the media, universities, research institutes, and government agencies, which are often responsible for the public's initial "take" on the problem under discussion.

Table 1.2 summarizes and compares the major theoretical perspectives, their criticisms, and social policy recommendations as they relate to social problems. The study of social problems is based on research as well as on theory, however. Indeed, research and theory are intricately related. As Wilson (1983) stated:

> Most of us think of theorizing as quite divorced from the business of gathering facts. It seems to require an abstractness of thought remote from the practical activity of empirical research. But theory building is not a separate activity within sociology. Without theory, the empirical researcher would find it impossible to decide what to observe, how to observe it, or what to make of the observations. (p. 1)

Social Problems Research

Most students taking a course in social problems will not become researchers or conduct research on social problems. Nevertheless, we are all consumers of research that is reported in the media. Politicians, social activist groups, and organizations attempt to justify their decisions, actions, and positions by citing research results. As consumers of research, we need to understand that our personal experiences and casual observations are less reliable than generalizations based on systematic research. One strength of scientific research is that it is subjected to critical examination by other researchers (see this chapter's Social Problems Research Up Close feature). The more you understand how research is done, the better able you will be to critically examine and question research rather than to passively consume research findings. In the remainder of this section, we discuss the stages of conducting a research study and the various methods of research that sociologists use.

The more you understand how research is done, the better able you will be to critically examine and question research rather than to passively consume research findings.

Stages of Conducting a Research Study

Sociologists progress through various stages in conducting research on a social problem. In this section, we describe the first four stages: (1) formulating a research question, (2) reviewing the literature, (3) defining variables, and (4) formulating a hypothesis.

TABLE 1.2 Comparison of Theoretical Perspectives

	STRUCTURAL FUNCTIONALISM	CONFLICT THEORY	SYMBOLIC INTERACTIONISM
Representative theorists	Emile Durkheim Talcott Parsons Robert Merton	Karl Marx Ralf Dahrendorf	George H. Mead Charles Cooley Erving Goffman
Society	Society is a set of interrelated parts; cultural consensus exists and leads to social order; natural state of society—balance and harmony.	Society is marked by power struggles over scarce resources; inequities result in conflict; social change is inevitable; natural state of society—imbalance.	Society is a network of interlocking roles; social order is constructed through interaction as individuals, through shared meaning, make sense out of their social world.
Individuals	Individuals are socialized by society's institutions; socialization is the process by which social control is exerted; people need society and its institutions.	People are inherently good but are corrupted by society and its economic structure; institutions are controlled by groups with power; "order" is part of the illusion.	Humans are interpretive and interactive; they are constantly changing as their "social beings" emerge and are molded by changing circumstances.
Cause of social problems?	Rapid social change; social disorganization that disrupts the harmony and balance; inadequate socialization and/or weak institutions.	Inequality; the dominance of groups of people over other groups of people; oppression and exploitation; competition between groups.	Different interpretations of roles; labeling of individuals, groups, or behaviors as deviant; definition of an objective condition as a social problem.
Social policy/solutions	Repair weak institutions; assure proper socialization; cultivate a strong collective sense of right and wrong.	Minimize competition; create an equitable system for the distribution of resources.	Reduce impact of labeling and associated stigmatization; alter definitions of what is defined as a social problem.
Criticisms	Called "sunshine sociology"; supports the maintenance of the status quo; needs to ask "functional for whom?"; does not deal with issues of power and conflict; incorrectly assumes a consensus.	Utopian model; Marxist states have failed; denies existence of cooperation and equitable exchange; cannot explain cohesion and harmony.	Concentrates on micro issues only; fails to link micro issues to macro-level concerns; too psychological in its approach; assumes label amplified problem.

Formulating a Research Question. A research study usually begins with a research question. Where do research questions originate? How does a particular researcher come to ask a particular research question? In some cases, researchers have a personal interest in a specific topic because of their own life experiences. For example, a researcher who has experienced spouse abuse may wish to do research on such questions as "What factors are associated with domestic violence?" and "How helpful are battered women's shelters in helping abused women break the cycle of abuse in their lives?" Other researchers may ask a particular research question because of their personal values—their concern for humanity and the desire to improve human life. Researchers who are concerned about the spread of HIV infection and AIDS may conduct research on questions such as "How does the use of alcohol influence condom use?" and "What educational strategies are effective for increasing safer sex behavior?" Researchers may also want to test a particular sociological theory, or some aspect of it, to establish its validity or conduct studies to evaluate the effect of a social policy or program. Research questions may also be formulated by the concerns of community groups and social activist organizations in collaboration with academic researchers. Government and industry also hire researchers to answer questions such as "How many children are victimized by episodes of violence at school?" and "What types of computer technologies can protect children against being exposed to pornography on the Internet?"

What Do You Think? In a free society, there must be freedom of information. That is why the U.S. Constitution and, more specifically, the First Amendment protect journalists' sources. If journalists are compelled to reveal their sources, their sources may be unwilling to share information, and this would jeopardize the public's right to know. A journalist cannot reveal information given in confidence without permission from the source or a court order. Do you think sociologists should be granted the same protections as journalists? If a reporter at your school newspaper uncovered a scandal at your university, should he or she be protected by the First Amendment?

Reviewing the Literature. After a research question is formulated, the researcher reviews the published material on the topic to find out what is already known about it. Reviewing the literature also provides researchers with ideas about how to conduct their research and helps them formulate new research questions. A literature review serves as an evaluation tool, allowing a comparison of research findings and other sources of information, such as expert opinions, political claims, and journalistic reports.

Defining Variables. A **variable** is any measurable event, characteristic, or property that varies or is subject to change. Researchers must operationally define the variables they study. An *operational definition* specifies how a variable is to be measured. For example, an operational definition of the variable "religiosity" might be the number of times the respondent reports going to church or synagogue. Another operational definition of "religiosity" might be the respondent's answer to the question "How important is religion in your life?" (for example, 1 is not important; 2 is somewhat important; 3 is very important).

Operational definitions are particularly important for defining variables that cannot be directly observed. For example, researchers cannot directly observe concepts such as "mental illness," "sexual harassment," "child neglect," "job satisfaction," and "drug abuse." Nor can researchers directly observe perceptions, values, and attitudes.

variable Any measurable event, characteristic, or property that varies or is subject to change.

hypothesis A prediction or educated guess about how one variable is related to another variable.

dependent variable The variable that the researcher wants to explain; the variable of interest.

independent variable The variable that is expected to explain change in the dependent variable.

Formulating a Hypothesis. After defining the research variables, researchers may formulate a **hypothesis,** which is a prediction or educated guess about how one variable is related to another variable. The **dependent variable** is the variable that the researcher wants to explain; that is, it is the variable of interest. The **independent variable** is the variable that is expected to explain change in the dependent variable. In formulating a hypothesis, the researcher predicts how the independent variable affects the dependent variable. For example, Kmec (2003) investigated the impact of segregated work environments on minority wages, concluding that "minority concentration in different jobs, occupations, and establishments is a considerable social problem because it perpetuates racial wage inequality" (p. 55). In this example, the independent variable is workplace segregation, and the dependent variable is wages.

In studying social problems, researchers often assess the effects of several independent variables on one or more dependent variables. Jekielek (1998) examined the impact of parental conflict and marital disruption (two independent

Each chapter in this book contains a *Social Problems Research Up Close* box that describes a research report or journal article that examines some sociologically significant topic. Some examples of the more prestigious journals in sociology include the *American Sociological Review*, the *American Journal of Sociology*, and *Social Forces*. Journal articles are the primary means by which sociologists, as well as other scientists, exchange ideas and information. Most journal articles begin with *an introduction and review of the literature*. Here, the investigator examines previous research on the topic, identifies specific research areas, and otherwise "sets the stage" for the reader. Often in this section, research hypotheses are set forth, if applicable. A researcher, for example, might hypothesize that the sexual behavior of adolescents has changed over the years as a consequence of increased fear of sexually transmitted diseases and that such changes vary on the basis of sex.

The next major section of a journal article is *sample and methods*. In this section, an investigator describes the characteristics of the sample, if any, and the details of the type of research conducted. The type of data analysis used is also presented in this section (see Appendix). Using the sample research question, a sociologist might obtain data from the Youth Risk Behavior Surveillance Survey collected by the Centers for Disease Control and Prevention.

This self-administered questionnaire is distributed biennially to more than 10,000 high school students across the United States.

The final section of a journal article includes the *findings and conclusions*. The findings of a study describe the results, that is, what the researcher found as a result of the investigation. Findings are then discussed within the context of the hypotheses and the conclusions that can be drawn. Often, research results are presented in tabular form. Reading tables carefully is an important part of drawing accurate conclusions about the research hypotheses. In reading a table, you should follow the steps listed here (see table within this box):

1. *Read the title of the table and make sure that you understand what the table contains*. The title of the table indicates the unit of analysis (high school students), the dependent variable (sexual risk behaviors), the independent variables (sex and year), and what the numbers represent (percentages).
2. *Read the information contained at the bottom of the table, including the source and any other explanatory information*. For example, the information at the bottom of this table indicates that the data are from the Centers for Disease Control and Prevention, that "sexually active" was defined as having

intercourse in the last three months, and that data on condom use were only from those students who were defined as being currently sexually active.
3. *Examine the row and column headings*. This table looks at the percentage of males and females, over four years, who reported ever having sexual intercourse, having four or more sex partners in a lifetime, being currently sexually active, and using condoms during the last sexual intercourse.
4. *Thoroughly examine the data in the table carefully, looking for patterns between variables*. As indicated in the table, in general, "risky" sexual behavior of males has gone down between 2001 and 2005. However, in 2007, a higher percentage of males: (1) ever had sexual intercourse, (2) had four or more sex partners, and/or (3) were currently sexually active when compared to 2005. Further, of males who were sexually active, fewer reported using a condom during last intercourse in 2007 than in 2005. Similarly, the percentage of females who were ever or currently sexually active has increased over the time period studied, and the percentage of females using a condom decreased between 2005 and 2007.
5. *Use the information you have gathered in Step 4 to address the hypotheses*. Clearly, sexual practices, as hypothesized, have changed over time. For

variables) on the emotional well-being of children (the dependent variable). Her research found that both parental conflict and marital disruption (separation or divorce) negatively affect children's emotional well-being. However, children in high-conflict, intact families exhibit lower levels of well-being than children who have experienced high levels of parental conflict, but whose parents divorce or separate.

Methods of Data Collection

After identifying a research topic, reviewing the literature, defining the variables, and developing hypotheses, researchers decide which method of data collection to use. Alternatives include experiments, surveys, field research, and secondary data.

example, both males and females, when comparing data from 2001 to 2007, have a general increase in condom use during sexual intercourse. On the other hand, the percentage of males and females reporting four or more sex partners has also increased in the same time period. Look at the table and see what patterns you detect, and how these patterns address the hypothesis.

6. *Draw conclusions consistent with the information presented*. From the table, can we conclude that sexual practices have changed over time? The answer is probably yes, although the limitations of the survey, the sample, and the measurement techniques used always should be considered. Can we conclude that the observed changes are a consequence of the fear of sexually

transmitted diseases? The answer is *no*. Having no measure of fear of sexually transmitted diseases over the time period studied, we are unable to come to such a conclusion. More information, from a variety of sources, is needed. The use of multiple methods and approaches to study a social phenomenon is called *triangulation*.

		FOUR OR MORE SEX	CURRENTLY	CONDOM USED
Percentage of High School Students Reporting Sexual Risk Behaviors, by Sex and Survey Year				
SURVEY YEAR	EVER HAD SEXUAL INTERCOURSE	FOUR OR MORE SEX PARTNERS DURING LIFETIME	CURRENTLY SEXUALLY ACTIVE*	CONDOM USED DURING LAST INTERCOURSE†
MALE				
2001	48.5	17.2	33.4	65.1
2003	48.0	17.5	33.8	68.8
2005	47.9	16.5	33.3	70.0
2007	49.8	17.9	34.3	68.5
FEMALE				
2001	42.9	11.4	33.4	51.8
2003	45.3	11.2	34.6	57.4
2005	45.7	12.0	34.6	55.9
2007	45.9	11.8	35.6	54.9

*Sexual intercourse during the three months preceding the survey
†Among currently sexually active students
Source: Centers for Disease Control and Prevention 2008.

Experiments. **Experiments** involve manipulating the independent variable to determine how it affects the dependent variable. Experiments require one or more experimental groups that are exposed to the experimental treatment(s) and a control group that is not exposed. After the researcher randomly assigns participants to either an experimental group or a control group, the researcher measures the dependent variable. After the experimental groups are exposed to the treatment, the researcher measures the dependent variable again. If participants have been randomly assigned to the different groups, the researcher may conclude that any difference in the dependent variable among the groups is due to the effect of the independent variable.

An example of a "social problems" experiment on poverty would be to provide welfare payments to one group of unemployed single mothers (experimental

experiment A research method that involves manipulating the independent variable to determine how it affects the dependent variable.

© renewed 1993 by Alexandra Milgram, and distributed by Penn State Media Sales.

In one of the most famous experiments in the social sciences, Stanley Milgram found that 65 percent of a sample of ordinary citizens were willing to use harmful electric shocks—up to 450 volts—on an elderly man with a heart condition simply because the experimenter instructed them to do so. It was later revealed that the man was not really receiving the shocks and that he had been part of the experimental manipulation. The experiment, although providing valuable information, raised many questions on the ethics of scientific research.

group) and no such payments to another group of unemployed single mothers (control group). The independent variable would be welfare payments; the dependent variable would be employment. The researcher's hypothesis would be that mothers in the experimental group would be less likely to have a job after 12 months than mothers in the control group.

The major strength of the experimental method is that it provides evidence for causal relationships, that is, how one variable affects another. A primary weakness is that experiments are often conducted on small samples, usually in artificial laboratory settings; thus the findings may not be generalized to other people in natural settings.

Surveys. **Survey research** involves eliciting information from respondents through questions. An important part of survey research is selecting a sample of those to be questioned. A **sample** is a portion of the population, selected to be representative so that the information from the sample can be generalized to a larger population. For example, instead of asking all abused spouses about their experience, the researcher could ask a representative sample of them and assume that those who were not questioned would give similar responses. After selecting a representative sample, survey researchers either interview people, ask them to complete written questionnaires, or elicit responses to research questions through computers.

What Do You Think? Imagine that you are doing research on the prevalence of cheating on examinations at your university or college. How would you get a random sample of the population? What variables do you think predict cheating; that is, what are some of the independent variables you would examine? How would you operationalize these variables? What are some of the problems associated with doing research on such a topic?

Interviews. In interview survey research, trained interviewers ask respondents a series of questions and make written notes about or tape-record the respondents' answers. Interviews may be conducted over the telephone or face-to-face.

One advantage of interview research is that researchers are able to clarify questions for the respondent and follow up on answers to particular questions. Researchers often conduct face-to-face interviews with groups of individuals who might otherwise be inaccessible. For example, some AIDS-related research attempts to assess the degree to which individuals engage in behavior that places them at high risk for transmitting or contracting HIV. Street youth and intravenous drug users, both high-risk groups for HIV infection, may not have a telephone or address because of their transient lifestyle. These groups may be accessible, however, if the researcher locates their hangouts and conducts face-to-face interviews. Research on drug addicts may also require a face-to-face interview survey design (Jacobs 2003).

The most serious disadvantages of interview research are cost and the lack of privacy and anonymity. Respondents may feel embarrassed or threatened when asked questions that relate to personal issues such as drug use, domestic violence, and sexual behavior. As a result, some respondents may choose not to

survey research A research method that involves eliciting information from respondents through questions.

sample A portion of the population, selected to be representative so that the information from the sample can be generalized to a larger population.

participate in interview research on sensitive topics. Those who do participate may conceal or alter information or give socially desirable answers to the interviewer's questions (e.g., "No, I do not use drugs").

Questionnaire. Instead of conducting personal or phone interviews, researchers may develop questionnaires that they either mail or give to a sample of respondents. Questionnaire research offers the advantages of being less expensive and less time-consuming than face-to-face or telephone surveys. In addition, questionnaire research provides privacy and anonymity to the research participants. This reduces the likelihood that they will feel threatened or embarrassed when asked personal questions and increases the likelihood that they will provide answers that are not intentionally inaccurate or distorted. A study on the relationship between minority composition of the workplace and the likelihood of workplace drug testing is a case in point. Questionnaires were sent to union leaders of the Communication Workers of America (CWA), asking them about drug-testing policies at their local job sites. Analysis indicated that, as the minority composition of the workplace goes up, the likelihood of *preemployment testing* and *testing with cause* increases, whereas the likelihood of *random* drug testing decreases (Gee et al. 2006).

The major disadvantage of mail questionnaires is that it is difficult to obtain an adequate response rate. Many people do not want to take the time or make the effort to complete and mail a questionnaire. Others may be unable to read and understand the questionnaire.

"Talking" Computers. A new method of conducting survey research is asking respondents to provide answers to a computer that "talks." Newman et al. (2002) found that syringe exchange program participants were more likely to report "stigmatized behavior" using computer-assisted self-interviewing, but less likely to report "psychological distress" when compared to face-to-face interview respondents. Thus, as in research in general, the reliability of data collected may depend on the interaction between the information sought and the method used.

Field Research. **Field research** involves observing and studying social behavior in settings in which it occurs naturally. Two types of field research are participant observation and nonparticipant observation.

In participant observation research, the researcher participates in the phenomenon being studied so as to obtain an insider's perspective on the people and/or behavior being observed. Palacios and Fenwick (2003), two criminologists, attended dozens of raves over a 15-month period to investigate the South Florida drug culture. In nonparticipant observation research, the researcher observes the phenomenon being studied without actively participating in the group or the activity. For example, Simi and Futrell (2009) studied white power activists by observing and talking to organizational members but did not participate in any of their unconventional activities.

Sometimes sociologists conduct in-depth detailed analyses or case studies of an individual, group, or event. For example, Fleming (2003) conducted a case study of young auto thieves in British Columbia. He found that, unlike professional thieves, the teenagers' behavior was primarily motivated by thrill-seeking—driving fast, the rush of a possible police pursuit, and the prospect of getting caught.

The main advantage of field research on social problems is that it provides detailed information about the values, rituals, norms, behaviors, symbols, beliefs, and emotions of those being studied. A potential problem with field research is

field research Research that involves observing and studying social behavior in settings in which it occurs naturally.

that the researcher's observations may be biased (e.g., the researcher becomes too involved in the group to be objective). In addition, because field research is usually based on small samples, the findings may not be generalizable.

Secondary Data Research. Sometimes researchers analyze secondary data, which are data that other researchers or government agencies have already collected or that exist in forms such as historical documents; police reports; school records; and official records of marriages, births, and deaths. Caldas and Bankston (1999) used information from Louisiana's 1990 Graduation Exit Examination to assess the relationship between school achievement and television-viewing habits of more than 40,000 tenth graders. The researchers found that, in general, television viewing is inversely related to academic achievement for whites but has little or no effect on school achievement for Black Americans. A major advantage of using secondary data in studying social problems is that the data are readily accessible, so researchers avoid the time and expense of collecting their own data. Secondary data are also often based on large representative samples. The disadvantage of secondary data is that the researcher is limited to the data already collected.

Seven Good Reasons to Read This Book

This textbook approaches the study of social problems with several student benefits in mind:

1. *Understanding that the social world is too complex to be explained by just one theory will expand your thinking about how the world operates.* For example, juvenile delinquency doesn't have just one cause—it is linked to (1) an increased number of youths living in inner-city neighborhoods with little or no parental supervision (social disorganization theory); (2) young people having no legitimate means of acquiring material wealth (anomie theory); (3) youths being angry and frustrated at the inequality and racism in our society (conflict theory); and (4) teachers regarding youths as "no good" and treating them accordingly (labeling theory).

2. *Developing a sociological imagination will help you see the link between private troubles and public issues.* In a society that values personal responsibility, there is a tendency to define failure as a consequence of individual free will. The *sociological imagination* enables us to understand how social forces underlie personal misfortunes and failures, and contribute to personal successes and achievements.

3. *Understanding globalization can help you become a safe, successful, and productive world citizen.* Whether the spread of HIV, war, environmental destruction, human trafficking, or overpopulation, social problems in one part of the world affect other parts of the world. Today's problems call for collective action involving world citizens. There is some indication that students are already responding. In 2009, the fastest growing minor at the University of California at Berkeley was "Global Poverty and Practice" (Anwar 2009).

4. *The Self and Society exercises increase self-awareness and allow you to position yourself within the social landscape.* For example, earlier in this chapter, you had the opportunity to assess your beliefs about a number of social problems, and to compare your responses to a national sample of first-year college students.

Both structural functionalism and conflict theory address the nature of social change, although in different ways. Durkheim, a structural functionalist, argued that social change, if rapid, was disruptive to society and that the needs of society should take precedence over the desires of individuals; that is, social change should be slow and methodical regardless of popular opinion.

To conflict theorists, social change is a result of the struggle for power by different groups. Specifically, Marx argued that social change was a consequence of the struggle between different economic classes as each strove for supremacy. Marx envisioned social change as primarily a revolutionary process ultimately leading to a utopian society.

Social movements are one means by which social change is realized. A **social movement** is an organized group of individuals with a common purpose to either promote or resist social change through collective action. Some people believe that, to promote social change, one must be in a position of political power and/or have large financial resources. However, the most important prerequisite for becoming actively involved in improving levels of social well-being may be genuine concern and dedication to a social "cause." The following vignettes provide a sampler of college student's making a difference.

- Neha Patel of the University of Miami was shocked to realize that recycling was not part of the culture of South Beach, Florida. With a $1,000 grant from Starbucks, she initiated a pilot recycling program called "Raise the Bar" by providing local taverns with recycling bins and an incentive to use them—10 cents per bottle or can up to $100 (Liebowitz 2009).

- In 2006, 33 student activists "hit the road for a seven-week tour of 19 religious and military colleges that discriminate against gay and lesbian students" (Ferrara et al. 2006, p. 1). Although the bus was "tagged" with homophobic slogans and activists were arrested on six campuses, leader Jacob Reitan said the trip was "hugely positive." Activists met and talked with over 10,000 people, including ten school presidents.

- Sean Sellers, a recent graduate of the University of Texas at Austin successfully led a "Boot the Bell" campaign in which "22 colleges and high schools either managed to remove a Taco Bell franchise from the campus or prevent one from being built" as part of a general protest against sweatshops in the field (Berkowitz 2005, p. 1). Yum! Brands Inc., which owns Taco Bell as well as Kentucky Fried Chicken, Long John Silver's, and Pizza Hut, has agreed to increase the pay of tomato pickers and to improve working conditions of farm workers in general.

- At Middlebury College in Vermont, students successfully convinced the administration that global warming is a real problem that needs to be addressed—immediately. Student activists convinced university officials that the college should invest $11 million in a biomass plant—a plant "fueled by wood chips, grass pellets, and a self-sustaining willow forest" (James 2007, p. 1). Further, after five days of protesting by 1,000 vocal activists, one Middlebury group convinced Vermont Senator Bernie Sanders to reintroduce legislation in Congress that would reduce carbon emissions by 80 percent by the year 2050.

- Students at several colleges are petitioning their university administrations to buy "fair-trade coffee"—coffee that is certified by monitors to have come from farmers who were paid a fair price for their beans. Many of these students are members of Students for Fair Trade. As one student said, "This is easy activism." Students make their voices heard by buying coffee with a fair-trade certified label or by not buying coffee at all (Batsell 2002).

- Students at Grinnell College in Iowa, in response to accusations of human rights violations of union workers in Coca-Cola bottling plants in Colombia (South America), formed an anti-Coke campaign. Using the official boycotting policy of the college, the student initiative passed a boycott on all Coca-Cola products in November 2004. Because Coca-Cola had an exclusive contract with Grinnell, Coke and Coke products continued to be sold on campus. However, wherever they were sold, there were signs reading, "The Grinnell College student body has voted to boycott Coca-Cola products. This is a Coca-Cola product" (KillerCoke.org 2005).

- Students at hundreds of campuses are members of anti-sweatshop groups such as Worker Rights Consortium (WRC). The WRC is a student-run watchdog organization that inspects factories worldwide, monitoring the monitors, as part of the anti-sweatshop movement. In 2009, as a result of the WRC and other anti-sweatshop groups, over a dozen schools (e.g., Harvard, Cornell, Georgetown) ended their contracts with collegiate apparel manufacturer Russell Athletics for violations of labor standards (Burns 2009).

5. *The Human Side features make you a more empathetic and compassionate human being.* The study of social problems is always about the quality of life of individuals. By conveying the private pain and personal triumphs associated with social problems, we hope to elicit a level of understanding that may not be attained through the academic study of social problems alone.

social movement An organized group of individuals with a common purpose to either promote or resist social change through collective action.

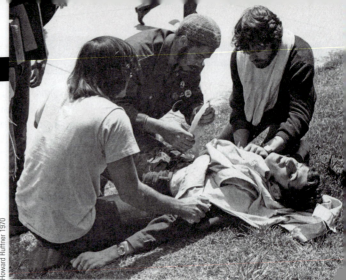

Howard Ruffner 1970

Student activism is not new nor is it unique to the United States. In the 1930s, the American Youth Congress (AYC) protested racial injustice, educational inequality, and the looming involvement of the United States in WWII. Called the "student brain of the New Deal" by some, the political power of the AYC would not be felt again until the student demonstrations of the 1960s (The Eleanor Roosevelt Papers 2008). Today, however, there is a new activism as students all over the world protest perceived injustices (Rifkind 2009). Aided by new technologies, social networking sites such as Facebook and MySpace allow for "virtual activism" as hundreds of thousands of students join online causes such as "Stop Global Warming" and "Save Darfur."

▲ The events of May 4, 1970, led singer/songwriter Neil Young to compose "Ohio" ("Tin soldiers and Nixon coming, we're finally on our own. This summer I hear the drumming, four dead in Ohio. ...."). Here, an injured student lies on the ground as onlookers stare in disbelief.

This chapter's photo essay highlights some of the most prominent examples of student activism, past and present. Although the faces have changed over time, the passion and dedication with which students voice their concerns has not.

Antiwar Demonstrations

During the Vietnam War era, students across the United States were vocal about their opposition to America's involvement in the war. In 1970, at Kent State University, the Ohio National Guard opened fire on unarmed student demonstrators, resulting in four deaths and nine injuries. The "Kent State Massacre" sparked campus protests across the nation, leading to the only nationwide student strike in U.S. history. A government report on antiwar demonstrations concluded that the shootings of students by the Ohio National Guard were unjustified (The Scranton Report 1971). No criminal charges were ever filed.

Animal Rights

Across the nation, student advocacy groups are speaking out against animal cruelty. Whether protesting the treatment of circus animals, the use of animals for research, dissection in school classrooms, the auction and shipment of horses to be slaughtered for meat, the confined and cramped spaces in which chickens, pigs, cattle, sheep, and other animals are kept before slaughter, or the abandonment and inhumane treatment

Courtesy of Ioana Samartinean

◄ The Animal Welfare Association of Arizona State University hosts "Meatout Day" to promote veganism and awareness that, as one placard reads, "Flesh is Flesh and Meat is Murder."

24

of companion animals, student groups such as Students for the Ethical Treatment of Animals (SETA) are organizing to be the voice of rights for animals. Through such venues as weblogs, Facebook, rallies, and community outreach, students are effectively speaking up for those who cannot.

Political and Economic Oppression

In 1989, thousands of students from universities across China sat peacefully in Tiananmen Square protesting for democratic reforms and social justice. On June 3rd, tanks entered the square and opened fired on the unarmed students, killing or injuring hundreds, perhaps thousands. There is no official tally of the casualties due to the Chinese government's subsequent clampdown on media and the reporting of any dissident activities, a policy that continues today. The protests at Tiananmen Square have been described "as the greatest challenge to the

communist state in China since the 1949 revolution" (BBC 2008, p. 1).

Marriage Equality

The term *marriage equality*, is fairly new but has become the rallying call of many college and high school students alike. Founded in 1998, the Gay-Straight Alliance Network (GSAN) "is a youth leadership organization that connects school-based Gay-Straight Alliances to each other and to community resources" (GSAN 2009, p. 1). Among other initiatives, GSAN, a member of the grassroots consortium Marriage Equality USA, actively promotes the marriage rights of same-sex couples. Despite some successes (e.g., it is now legal for same-sex partners to marry in several states), the passage of Proposition 8 led to a ban on same-sex marriages in California, leading to protests throughout the country (Garrison 2009; Dolan 2009).

American Civil Rights Movement

On February 1, 1960, four African American students entered a Greensboro Woolworth's store to buy school supplies (Sykes 1960; Schlosser 2000). If their money was good enough to buy school supplies, why not a cup of coffee,

Darron R. Silva/Aurora Photos

▲ Students from local high schools and Western Michigan University gather to protest discrimination and abuse against gays.

they reasoned? At 4:30 p.m., they sat at the "whites only" lunch counter, intending to place an order. The four young men sat at the counter until closing but were never served. The next day more students sat at the counter—they too were never served. As news of the "sit-in" spread, students returned to the Greensboro Woolworth's and to other lunch counters across the South. White and Black American students alike from New York to San Francisco began picketing Woolworth's in support of the "Greensboro Four." This one act by four students was the pivotal step in propelling forward what became known as the American civil rights movement (Schlosser 2000).

Jeff Widener/AP Photo

▲ In 1989, an unknown man brings to a halt the People's Liberation Army as they advanced to disburse peaceful student demonstrations near Tiananmen Square in Peking (Beijing), China. The government's response to the student protests sparked global outrage.

Four Black American students from what was then called ▶ North Carolina Agricultural and Technical College returned to sit at a "whites only" counter at Woolworth's in Greensboro, North Carolina, on February 2, 1960, setting off one of the most significant protests of the civil rights movement. The counter now sits on display in the Smithsonian Institute in Washington, DC.

Jack Moebes/CORBIS

6. *The Social Problems Research Up Close features teach you the basics of scientific inquiry, making you a smarter consumer of "pop" sociology, psychology, anthropology, and the like.* These boxes demonstrate the scientific enterprise from theory and data collection to findings and conclusions. Examples of research topics covered include college students' health, young children's attitudes toward smoking, bullying and victimization among minority youth, and computer hackers.

7. *Learning about social problems and their structural and cultural origins helps you, individually or collectively, make a difference in the world.* Individuals can make a difference in society through the choices they make. You may choose to vote for one candidate over another, demand the right to reproductive choice or protest government policies that permit it, drive drunk or stop a friend from driving drunk, repeat a homophobic or racist joke or chastise the person who tells it, and practice safe sex or risk the transmission of sexually transmitted diseases. Collective social action is another, often more powerful way to make a difference. This chapter's photo essay visually portrays students acting collectively to change the world.

What Do You Think? The 2009 Serve America Act "dramatically increases the size of the AmeriCorps service program…, expands ways for students to earn money for college, and creates opportunities for all Americans to serve in their communities" (Hass 2009, p. 1). One way college students can serve their communities is through service learning. In its simplest form, service learning entails students volunteering in the community and receiving academic credit for their efforts. Universities and colleges are increasingly requiring service learning credits as a criterion for graduation. Do you think that all students should be required to engage in service learning? Why or why not?

Understanding Social Problems

At the end of each chapter, we offer a section titled "Understanding" in which we reemphasize the social origin of the problem being discussed, the consequences, and the alternative social solutions. Our hope is that readers will end each chapter with a "sociological imagination" view of the problem and with an idea of how, as a society, we might approach a solution.

Sociologists have been studying social problems since the Industrial Revolution. Industrialization brought about massive social changes: The influence of religion declined, and families became smaller and moved from traditional, rural communities to urban settings. These and other changes have been associated with increases in crime, pollution, divorce, and juvenile delinquency. As these social problems became more widespread, the need to understand their origins and possible solutions became more urgent. The field of sociology developed in response to this urgency. Social problems provided the initial impetus for the development of the field of sociology and continue to be a major focus of sociology.

There is no single agreed-on definition of what constitutes a social problem. Most sociologists agree, however, that all social problems share two important elements: an objective social condition and a subjective interpretation of that condition. Each of the three major theoretical perspectives in sociology—structural-functionalist, conflict, and symbolic interactionist—has its own notion of the causes, consequences, and solutions of social problems.

CHAPTER REVIEW

■ **What is a social problem?**

Social problems are defined by a combination of objective and subjective criteria. The objective element of a social problem refers to the existence of a social condition; the subjective element of a social problem refers to the belief that a particular social condition is harmful to society or to a segment of society and that it should and can be changed. By combining these objective and subjective elements, we arrive at the following definition: A social problem is a social condition that a segment of society views as harmful to members of society and in need of remedy.

■ **What is meant by the structure of society?**

The structure of a society refers to the way society is organized.

■ **What are the components of the structure of society?**

The components are institutions, social groups, statuses, and roles. Institutions are an established and enduring pattern of social relationships and include family, religion, politics, economics, and education. Social groups are defined as two or more people who have a common identity, interact, and form a social relationship. A status is a position that a person occupies within a social group and that can be achieved or ascribed. Every status is associated with many roles, or the set of rights, obligations, and expectations associated with a status.

■ **What is meant by the culture of society?**

Whereas social structure refers to the organization of society, culture refers to the meanings and ways of life that characterize a society.

■ **What are the components of the culture of society?**

The components are beliefs, values, norms, and symbols. Beliefs refer to definitions and explanations about what is assumed to be true. Values are social agreements about what is considered good and bad, right and wrong, desirable and undesirable. Norms are socially defined rules of behavior. Norms serve as guidelines for our behavior and for our expectations of the behavior of others. Finally, a symbol is something that represents something else.

■ **What is the sociological imagination, and why is it important?**

The sociological imagination, a term that C. Wright Mills (1959) developed, refers to the ability to see the connections between our personal lives and the social world in which we live. It is important because, when we use our sociological imagination, we are able to distinguish between "private troubles" and "public issues" and to see connections between the events and conditions of our lives and the social and historical context in which we live.

■ **What are the differences between the three sociological perspectives?**

According to structural functionalism, society is a system of interconnected parts that work together in harmony to maintain a state of balance and social equilibrium for the whole. The conflict perspective views society as composed of different groups and interests competing for power and resources. Symbolic interactionism reflects the microsociological perspective and emphasizes that human behavior is influenced by definitions and meanings that are created and maintained through symbolic interaction with others.

■ **What are the first four stages of a research study?**

The first four stages of a research study are formulating a research question, reviewing the literature, defining variables, and formulating a hypothesis.

■ **How do the various research methods differ from one another?**

Experiments involve manipulating the independent variable to determine how it affects the dependent variable. Survey research involves eliciting information from respondents through questions. Field research involves observing and studying social behavior in settings in which it occurs naturally. Secondary data are data that other researchers or government agencies have already collected or that exist in forms such as historical documents, police reports, school records, and official records of marriages, births, and deaths.

■ **What is a social movement?**

Social movements are one means by which social change is realized. A social movement is an organized group of individuals with a common purpose to either promote or resist social change through collective action.

TEST YOURSELF

1. Definitions of social problems are clear and unambiguous.
   a. True
   b. False
2. The social structure of society contains
   a. statuses and roles
   b. institutions and norms
   c. sanctions and social groups
   d. values and beliefs
3. The culture of society refers to its meaning and the ways of life of its members.
   a. True
   b. False
4. Alienation
   a. refers to a sense of normlessness
   b. is focused on by symbolic interactionist
   c. can be defined as the powerlessness and meaninglessness in people's lives
   d. is a manifest function of society
5. Blumer's stages of a social problems begins with
   a. mobilization for action
   b. societal recognition
   c. social legitimation
   d. development and implementation of a plan
6. The independent variable comes first in time; i.e., it precedes the dependent variable.
   a. True
   b. False
7. The third stage in defining a research study is
   a. Formulating a hypothesis
   b. Reviewing the literature
   c. Defining the variables
   d. Formulating a research question
8. A sample is a subgroup of the population—the group to whom you actually give the questionnaire.
   a. True
   b. False
9. Studying police behavior by riding along with patrol officers would be an example of
   a. Participant observation
   b. Nonparticipant observation
   c. Field research
   d. Both a and c
10. Student benefits of the book include
    a. Providing global coverage of social problems
    b. Highlighting social problems research
    c. Encouraging students to take pro-social action
    d. All of the above

Answers: 1. b; 2. a; 3. a; 4. c; 5. b; 6. a; 7. c; 8. a; 9. d; 10. d.

KEY TERMS

achieved status 5
alienation 12
anomie 11
ascribed status 5
belief 6
culture 5
dependent variable 17
experiment 19
field research 21
hypothesis 17
independent variable 17
institution 4

latent function 10
manifest function 10
norm 6
objective element of a social
    problem 3
primary group 5
role 5
sample 20
sanction 8
secondary group 5
social group 5
social movement 23

social problem 3
sociological imagination 8
status 5
structure 4
subjective element of a social
    problem 3
survey research 20
symbol 8
value 6
variable 17

MEDIA RESOURCES

Understanding Social Problems,
Seventh Edition Companion Website
www.cengage.com/sociology/mooney
Visit your book companion website, where you will find flash cards, practice quizzes, Internet links, and more to help you study.

CENGAGENOW™

Just what you need to know NOW! Spend time on what you need to master rather than on information you already have learned. Take a pre-test for this chapter, and CengageNOW will generate a personalized study plan based on your results. The study plan will identify the topics you need to review and direct you to online resources to help you master those topics. You can then take a post-test to help you determine the concepts you have mastered and what you will need to work on. Try it out! Go to www.cengage.com/login to sign in with an access code or to purchase access to this product.

Gideon Mendel/Documentary/CORBIS

2

Problems of Illness and Health Care

"The defense this nation seeks involves a great deal more than building airplanes, ships, guns, and bombs. We cannot be a strong nation unless we are a healthy nation."

U.S. President Franklin Roosevelt, 1940

29

At this annual three-day free medical clinic in Virginia, rural families, most with little or no health insurance, line up for hours to receive free health care. All services and medical supplies are donated.

Suzy Allman/Getty Images News /Getty Images

At 2:00 A.M., a car drives into a field in rural southwestern Virginia. At the crack of dawn, the driver and two passengers get out of the car, walk a half mile, and join a line of people that stretches a quarter of a mile across the field. Betty, a 29-year-old mother of six who works at a restaurant, is seriously overweight. Her 14-year-old daughter, Molly, has had such terrible tooth pain that she is unable to eat and has lost 15 pounds. Betty's boyfriend, Jake, who works at a dry cleaner, has had pain in his side off and on for nearly a year. Betty, Molly, and Jake wait in line for the gates to open for the Rural Area Medical Clinic—a weekend event that occurs once a year for anyone who has no health insurance. During this event, about 1,500 volunteer doctors, nurses, dentists, and staff provide services to more than 6,000 women, men, and children without insurance. When Betty is examined, she finds out she has diabetes. Molly had such a severe dental infection that she had to have eight teeth pulled. Jake learned that he has abdominal cancer that could have been detected much earlier with a physical examination (Garson 2007).

In the United States, lack of health insurance and the high cost of health care is a pressing concern for millions of Americans, and literally a matter of life or death for some. In this chapter, we address problems of illness and health care in the United States and throughout the world. Taking a sociological look at health issues, we examine why some social groups experience more health problems than others and how social forces affect and are affected by health and illness. We begin by looking at patterns of health and illness around the world.

developed countries
Countries that have relatively high gross national income per capita and have diverse economies made up of many different industries.

The Global Context: Patterns of Health and Illness Around the World

In making international comparisons, countries are often classified into one of three broad categories according to their economic status: (1) **developed countries** (also known as *high-income countries*) have relatively high gross national income per capita and have diverse economies made up of many different industries; (2) **developing countries** (also known as *middle-income countries*) have relatively low gross national income per capita, and their economies are much simpler, often relying on a few agricultural products; and (3) **least developed countries** (known as *low-income countries*) are the poorest countries of the world.

In this section that focuses on health and illness from a global perspective, we reveal the striking disparities in patterns of health and illness among developed, developing, and least developed nations. We then discuss three worldwide health problems: human immunodeficiency virus (HIV)/acquired immunodeficiency syndrome (AIDS), obesity, and mental illness.

Morbidity, Life Expectancy, and Mortality

Three measures of the health of populations are morbidity, life expectancy, and mortality. **Morbidity** refers to illnesses, symptoms, and the impairments they produce. In less developed countries, where poverty and chronic malnutrition are widespread, infectious and parasitic diseases, such as HIV disease, tuberculosis, diarrheal diseases (caused by bacteria, viruses, or parasites), measles, and malaria are much more prevalent than in developed countries, where chronic health problems such as cardiovascular disease and cancer are the major health threats (Weitz 2010).

Wide disparities in **life expectancy**—the average number of years that individuals born in a given year can expect to live—exist between regions of the world (see Figure 2.1). Japan has the longest life expectancy (83 years), Swaziland has the lowest life expectancy (40 years), and 18 countries (primarily in Africa) have life expectancies of less than 50 years (UNICEF 2008).

Today, the leading cause of **mortality,** or death, worldwide is cardiovascular disease (including heart disease and stroke), accounting for 30 percent of all deaths (World Health Organization 2006). In the United States, the leading cause of death for both women and men is heart disease, followed by cancer and stroke (National Center for Health Statistics 2008). As shown in Table 2.1, U.S. mortality patterns vary by age. Later, we discuss how patterns of mortality are related to social factors, such as social class, sex, race or ethnicity, and education.

Mortality Rates Among Infants and Children. The **infant mortality rate,** the number of deaths of live-born infants under 1 year of age per 1,000 live births (in any given year), provides an important measure of the health of a population. In 2007, infant mortality rates ranged from an average of 84 in least developed nations to an average of 5 in industrialized nations. The U.S. infant mortality rate was 7; 37 countries had infant mortality rates that were lower than that of the United States (UNICEF 2008). The **under-5 mortality rate**—deaths of children under age 5—range from an average of 153 in least developed nations to an average of 6 in industrialized countries.

One of the major causes of infant and child death worldwide is diarrhea, resulting from poor water quality and sanitation. Only two-thirds (62 percent) of the global population has access to adequate sanitation (UNICEF 2008). Another major contributing factor to deaths of infants and children is undernutrition. In the developing world, one in four children under age 5 is underweight (UNICEF 2006).

Maternal Mortality Rates. The **maternal mortality rate is** a measure of deaths that result from complications associated with pregnancy, childbirth, and unsafe abortion. Women in the United States and other developed countries

developing countries Countries that have relatively low gross national income per capita, with simpler economies that often rely on a few agricultural products.

least developed countries The poorest countries of the world.

morbidity Illnesses, symptoms, and the impairments they produce.

life expectancy The average number of years that individuals born in a given year can expect to live.

mortality Death.

infant mortality rate The number of deaths of live-born infants under 1 year of age per 1,000 live births (in any given year).

under-5 mortality rate The rate of deaths of children under age 5.

maternal mortality rate A measure of deaths that result from complications associated with pregnancy, childbirth, and unsafe abortion.

Figure 2.1 Life Expectancy and Under-5 Mortality Rate by Region, 2007
Source: UNICEF 2008.

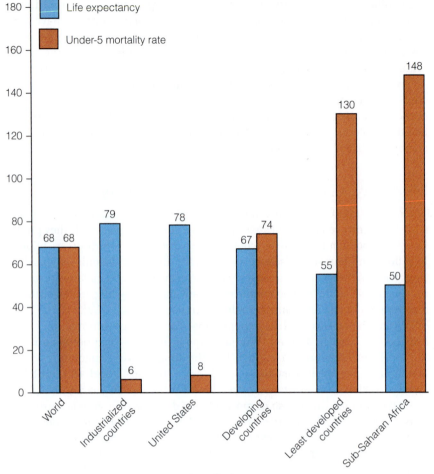

TABLE 2.1 **Top Three Causes of Death by Selected Age Groups, United States**

| AGE (YEARS) | FIRST | LEADING CAUSES OF DEATH | |
		SECOND	THIRD
1–4	Unintentional injuries	Congenital/chromosomal abnormalities	Cancer
5–14	Unintentional injuries	Cancer	Congenital/Chromosomal abnormalities
15–24	Unintentional injuries	Homicide	Suicide
25–44	Unintentional injuries	Cancer	Heart disease
45–64	Cancer	Heart disease	Stroke
65 and older	Heart disease	Cancer	Stroke

Source: National Center for Health Statistics (2008, Table 312).

generally do not experience pregnancy and childbirth as life-threatening. But for women ages 15 to 49 in developing countries, maternal mortality is the leading cause of death and disability. When Tanzanian mothers are in labor, they often say to their older children, "I'm going to go and fetch the new baby; it is a dangerous journey and I may not return" (Grossman 2009). The most common causes of maternal death are hemorrhage (severe loss of blood), infection, and complications related to unsafe abortion.

Rates of maternal mortality show a greater disparity between rich and poor countries than any of the other societal health measures. Of the 529,000 annual maternal deaths worldwide, including 68,000 deaths from unsafe abortion, only 1 percent occurs in high-income countries (World Health Organization 2005). Women's lifetime risk of dying from pregnancy or childbirth is highest in sub-Saharan Africa, where one in sixteen women dies of pregnancy-related causes, compared with one in 4,000 women in developed countries (UNICEF 2006). High maternal mortality rates in less developed countries are related to poor-quality and inaccessible health care; most women give birth without the assistance of trained personnel (see Table 2.2). High maternal mortality rates are also linked to malnutrition and poor sanitation and to higher rates of pregnancy and childbearing at early ages. Women in many countries also lack access to family planning services and/or do not have the support of their male partners to use contraceptive methods such as condoms. Consequently, many women resort to abortion to limit their childbearing, even in countries where abortion is illegal and unsafe.

> When Tanzanian mothers are in labor, they often say to their older children, "I'm going to go and fetch the new baby; it is a dangerous journey and I may not return."

TABLE 2.2 Trained Childbirth Assistance and Lifetime Chance of Maternal Mortality Region

	PERCENTAGE OF BIRTHS ATTENDED BY SKILLED PERSONNEL	LIFETIME CHANCE OF MATERNAL MORTALITY
Developed countries	99	1 in 4,000
Developing countries	57	1 in 61
Sub-Saharan Africa	41	1 in 16

Source: UNICEF (2006).

Patterns of Burden of Disease

Another approach to measuring the health status of a population provides an indicator of the overall burden of disease on a population through a single unit of measurement that combines not only the number of deaths but also the impact of premature death and disability on a population (Murray & Lopez 1996). This comprehensive unit of measurement, called the disability-adjusted life year (DALY), reflects years of life lost to premature death and years lived with a disability. More simply, one DALY is equal to one lost year of healthy life.

Worldwide, tobacco is the leading cause of burden of disease (World Health Organization 2002). Hence, tobacco has been called "the world's most lethal weapon of mass destruction" (SmokeFree Educational Services 2003, p. 1).

What Do You Think? Data on deaths from international terrorism and tobacco-related deaths in 37 developed and eastern European countries revealed that tobacco-related deaths outnumbered terrorist deaths by about a whopping 5,700 times (Thomson & Wilson 2005). The number of tobacco deaths was equivalent to the impact of a September 11, 2001, type terrorist attack every 14 hours! Given that tobacco-related deaths grossly outnumber terrorism-related deaths, why hasn't the U.S. government waged a "war on tobacco" on a scale similar to its "war on terrorism"?

Sociological Theories of Illness and Health Care

Next, we discuss how the three major sociological theories—structural functionalism, conflict theory, and symbolic interactionism—contribute to our understanding of illness and health care.

Structural-Functionalist Perspective

The structural-functionalist perspective examines how changes in society affect health. As societies develop and increase the standard of living for their members, life expectancy increases and birthrates decrease (Weitz 2010). At the same time, the main causes of death and disability shift from infectious disease and high death rates among infants and women of childbearing age (owing to complications of pregnancy, unsafe abortion, or childbearing) to chronic, noninfectious illness and disease. This shift is referred to as the **epidemiological transition,** whereby low life expectancy and predominance of parasitic and infectious diseases shift to high life expectancy and predominance of chronic and degenerative diseases. As societies make the epidemiological transition, birthrates decline and life expectancy increases, so diseases that need time to develop, such as cancer, heart disease, Alzheimer's disease, arthritis, and osteoporosis, become more common, and childhood- and pregnancy-related health problems and deaths become less common.

Just as social change affects health, health concerns may lead to social change. The emergence of HIV and AIDS in the U.S. gay male population was a force that helped unite and mobilize gay rights activists. Concern over the effects of exposure to tobacco smoke—the greatest cause of disease and death in the United States and other developed countries—has led to legislation banning smoking in public places.

According to the structural-functionalist perspective, health care is a social institution that functions to maintain the well-being of societal members and, consequently, of the social system as a whole. Illness is dysfunctional in that it interferes with people performing needed social roles. To cope with nonfunctioning members and to control the negative effects of illness, society assigns a temporary and unique role to those who are ill—the sick role (Parsons 1951). This role carries with it an expectation that the person who is ill will seek and receive competent medical care, adhere to the prescribed regimen, and return as soon as possible to normal role obligations.

Finally, the structural-functionalist perspective draws attention to latent dysfunctions, or unintended and often unrecognized negative consequences of social patterns or behavior. For example, a latent dysfunction of widespread use of some prescription drugs is the emergence of drug resistance, which occurs when drugs kill the weaker disease-causing germs while allowing variants resistant to the drugs to flourish. For generations, the drug chloroquine was added to table salt to prevent malaria. But overuse led to drug-resistant strains of malaria, and now chloroquine is useless in preventing malaria (McGinn 2003).

Conflict Perspective

The conflict perspective focuses on how wealth, status, power, and the profit motive influence illness and health care. Worldwide, populations living in poverty experience more health problems and have less access to quality medical

epidemiological transition
A societal shift from low life expectancy and predominance of parasitic and infectious diseases to high life expectancy and predominance of chronic and degenerative diseases.

care (Feachum 2000). The conflict perspective points to ways in which powerful groups and wealthy corporations influence health-related policies and laws through lobbying and financial contributions to politicians and political candidates. Private health insurance companies have much to lose if the United States adopts a national public health insurance program or even a public insurance option, and have spent millions of dollars opposing such proposals. Nine Republicans on the Senate Finance Committee who opposed Obama's proposal of a public-option health insurance plan have received $2.6 million from health industries (Mayer 2009). Another industry with a vested interest in health-related legislation is the pharmaceutical and health products industry, which spent $1.2 million a day on lobbying during the first three months of 2009, when health care reform was a top priority in Congress (Beckel 2009).

The powerful relationship that health-related industries have with government is reflected in the **revolving door**, the practice of employees cycling between roles in an industry, and roles in government that influence that industry. In 2009, three dozen former members of Congress were employed by pharmaceutical and health product industries (Beckel 2009).

The conflict perspective criticizes the pharmaceutical and health care industry for placing profits above people. In her book *Money-Driven Medicine,* Maggie Mahar (2006) explains that power in our health care system has shifted from physicians, who are committed to putting their patients' interests ahead of their own financial interests, to corporations that are legally bound to put their shareholders' interests first. "Thus, many decisions about how to allocate health care dollars have become marketing decisions. Drugmakers, device makers, and insurers decide which products to develop based not on what patients need, but on what their marketers tell them will sell—and produce the highest profit" (Mahar 2006, p. xviii). For example, pharmaceutical companies' research and development budgets are spent not according to public health needs but rather according to calculations about maximizing profits. Because the masses of people in developing countries lack the resources to pay high prices for medication, pharmaceutical companies do not consider the development of drugs for diseases of poor countries as a profitable investment.

Profits also compromise drug safety. Most pharmaceutical companies outsource their clinical drug trials (which assess drug effectiveness and safety) to contract research organizations (CROs) in developing countries where operating costs are low and regulations are lax (Allen 2007). The validity of the clinical trial results from CROs is questionable, however, because CROs can earn more money (in royalties and in future contracts) when the clinical trial results are favorable.

The profit motive also affects health via the food industry. See this chapter's Photo Essay for a look at health problems associated with modern food animal production—problems that stem largely from the concern of the food animal industry for economic efficiency and profit.

Finally, conflict theorists also point to the ways in which male domination and bias influence health care and research. When the male erectile dysfunction drug Viagra made its debut in 1998, women across the United States were outraged by the fact that some insurance policies covered Viagra (or were considering covering it), even though female contraceptives were not covered. The male-dominated medical research community has also been criticized for neglecting women's health issues and excluding women from major health research studies (Johnson & Fee 1997).

revolving door The practice of employees cycling between roles in an industry, and roles in government that influence that industry.

Symbolic Interactionist Perspective

Symbolic interactionists focus on (1) how meanings, definitions, and labels influence health, illness, and health care; and (2) how such meanings are learned through interaction with others and through media messages and portrayals. According to the symbolic interactionist perspective of illness, "there are no illnesses or diseases in nature. There are only conditions that society, or groups within it, has come to define as illness or disease" (Goldstein 1999, p. 31). Psychiatrist Thomas Szasz (1961/1970) argued that what we call "mental illness" is no more than a label conferred on those individuals who are "different," that is, those who do not conform to society's definitions of appropriate behavior.

Defining or labeling behaviors and conditions as medical problems is part of a trend known as **medicalization.** Initially, medicalization was viewed as occurring when a particular behavior or condition deemed immoral (e.g., alcoholism, masturbation, or homosexuality) was transformed from a legal problem into a medical problem that required medical treatment. The concept of medicalization has expanded to include (1) any new phenomena defined as medical problems in need of medical intervention, such as post-traumatic stress disorder, premenstrual syndrome, and attention-deficit/hyperactivity disorder; and (2) "normal" biological events or conditions that have come to be defined as medical problems in need of medical intervention, including childbirth, menopause, and death.

Conflict theorists view medicalization as resulting from the medical profession's domination and pursuit of profits. A symbolic interactionist perspective suggests that medicalization results from the efforts of sufferers to "translate their individual experiences of distress into shared experiences of illness" (Barker 2002, p. 295). In her study of women with fibromyalgia (a pain disorder that has no identifiable biological cause), Barker suggested that the medicalization of symptoms and distress through a diagnosis of fibromyalgia gives sufferers a framework for understanding and validating their experience of distress.

According to symbolic interactionism, conceptions of health and illness are socially constructed. It follows, then, that definitions of health and illness vary over time and from society to society. In some countries, being fat is a sign of health and wellness; in others, it is an indication of mental illness or a lack of self-control. Among some cultural groups, perceiving visions or voices of religious figures is considered a normal religious experience, whereas such "hallucinations" would be indicative of mental illness in other cultures. In 18th- and 19th-century America, masturbation was considered an unhealthy act that caused a range of physical and mental health problems. Individuals caught masturbating were often locked up in asylums, treated with drugs (such as sedatives and poisons), or subjected to a range of interventions designed to prevent masturbation by stimulating the genitals in painful ways, preventing genital sensation, or deadening it. These physician-prescribed interventions included putting ice on the genitals; blistering and scalding the penis, vulva, inner thighs, or perineum; inserting electrodes into the rectum and urethra; cauterizing the clitoris by applying pure carbolic acid; circumcising the penis; and surgically removing the clitoris, ovaries, and testicles (Allen 2000). Today, most health professionals agree that masturbation is a normal, healthy aspect of sexual expression.

Symbolic interactionism draws attention to the effects that meanings and labels have on health behaviors and health-related policies. For example, as tobacco sales have declined in developed countries, transnational tobacco companies have looked for markets in developing countries, using advertising strategies that depict smoking as "an inexpensive way to buy into glamorous lifestyles of

medicalization Defining or labeling behaviors and conditions as medical problems.

the upper or successful social class" (Egwu 2002, p. 44). In 2004, the Centers for Medicare and Medicaid Services decided to remove language in Medicare's coverage manual that states that obesity is not an illness (Stein & Connolly 2004). Labeling obesity as an illness means that Medicare can cover treatment for obesity, ranging from joining weight-loss or fitness clubs to surgery and counseling.

Symbolic interactionists also focus on the stigmatization of individuals who are in poor health or who lack health insurance. A **stigma** refers to a discrediting label that affects an individual's self-concept and disqualifies that person from full social acceptance. (Originally, the word *stigma* referred to a mark burned into the skin of a criminal or slave.) The stigma associated with poor health often results in prejudice and discrimination against individuals with mental illnesses, drug addictions, physical deformities and impairments, missing or decayed teeth, obesity, HIV infection and AIDS, and other health conditions. Further, a study of U.S. adults without insurance found that "uninsured Americans…noted the stigma of lacking health insurance, citing medical providers who treat them like 'losers' because they are uninsured" (Sered & Fernandopulle 2005, p. 16).

The stigma associated with health problems and/or lack of health insurance implies that individuals—rather than society—are responsible for their health. In U.S. culture, "sickness increasingly seems to be construed as a personal failure—a failure of ethical virtue, a failure to take care of oneself 'properly' by eating the 'right' foods or getting 'enough' exercise, a failure to get a Pap smear, a failure to control sexual promiscuity, genetic failure, a failure of will, or a failure of commitment—rather than society's failure to provide basic services to all of its citizens" (Sered & Fernandopulle 2005, p. 16). One stigmatized population is those with HIV/AIDS, a topic we discuss in the next section.

HIV/AIDS: A Global Health Concern

One of the most urgent worldwide public health concerns is the spread of HIV, which causes AIDS. Since the first reported cases of HIV/AIDS in 1981, more than 25 million people have died of AIDS, and 33 million are living with HIV/AIDS (Kaiser Family Foundation 2009a). An estimated 8 in 10 people infected with HIV do not know it.

HIV is transmitted through sexual intercourse, through sharing unclean intravenous needles, through perinatal transmission (from infected mother to fetus or newborn), through blood transfusions or blood products, and, rarely, through breast milk. Worldwide, the predominant mode of HIV transmission is through heterosexual contact (World Health Organization 2004).

HIV/AIDS in Africa and Other Regions

HIV/AIDS is most prevalent in Africa, particularly sub-Saharan Africa, where one in twenty adults is infected with HIV. Two-thirds of people with HIV live in sub-Saharan Africa (Kaiser Family Foundation 2009a). But HIV/AIDS also affects millions of people living in India and hundreds of thousands of people in China, the Mediterranean region, Western Europe, and Latin America. Eastern European countries and central Asia are experiencing increasing rates of HIV infection, mainly from drug-injecting behavior and to a lesser extent from unsafe sex.

The high rates of HIV in developing countries, particularly sub-Saharan Africa, are having alarming and devastating effects. HIV/AIDS has reversed the

stigma A discrediting label that affects an individual's self-concept and disqualifies that person from full social acceptance.

Modern Animal Food Production: Health and Safety Issues

▲ The hogs in this concentrated animal feeding operation (CAFO) will live their entire lives, from birth to slaughter, inside a crowded, controlled indoor environment.

Many health problems have been associated with modern methods of raising and processing food animals. Increasingly, food animals are not raised in expansive meadows or pastures; rather, they are raised in concentrated animal feeding operations (CAFOs), also known as "factory farms," giant corporate-controlled livestock farms where large numbers (sometimes tens or hundreds of thousands) of animals—typically cows, hogs, turkeys, or chickens—are "produced" in factory-like settings, often indoors, to maximize production and profits.

The diet of factory-farmed animals consists largely of corn, which is cheap and efficient in fattening the animals. Corn-fed beef is less healthy than grass-fed beef, as it contains more saturated fat, which contributes to heart disease (Pollan 2006). The digestive system of cows is designed for grass; corn makes cows sick and susceptible to disease. Factory-farmed chickens, turkeys, and hogs, and farm-raised fish are also susceptible to disease because of the crowded and unsanitary living conditions. To prevent the spread of disease in CAFOs or fish farms, food animals are fed antibiotics, which contributes to the emergence of super-resistant bacterial infections that will not respond to treatment.

Another animal food health threat is bovine spongiform encephalopathy (BSE), commonly known as "mad cow disease." The disease spreads when infected cows are used in livestock feed. People who eat BSE-infected beef may develop a human form of the disease—Creutzfeldt-Jakob disease (CJD), which is fatal.

About a third of U.S. dairy cows are given recombinant bovine growth hormone (rBGH), manufactured by Monsanto and sold under the trade name Posilac, to increase milk production. Although

▲ Carrie Mahan, 29, died from Creutzfeldt-Jakob disease.

▲ To prevent disease from spreading among animals living in crowded conditions, factory-farmed animals are fed diets laced with antibiotics.

Cows are given hormones (through ▶ shots or ear implants) to increase milk production.

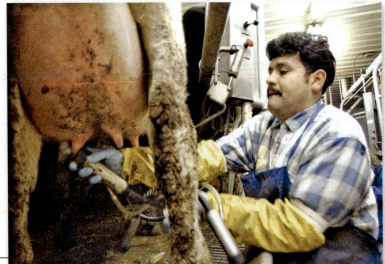

the Food and Drug Administration (FDA) approved Posilac in 1993, and has supported Monsanto's claims that milk containing rBGH is safe for consumers, some experts warn that rBGH raises the risk of breast, colon, and prostate cancer (Epstein 2006). Other countries, including all of Europe, Canada, Australia, New Zealand, and Japan, have banned milk containing rBGH.

A major health problem related to modern slaughterhouse and meatpacking production techniques is the contamination of meat with fecal matter. In the slaughterhouse, if the animal's hide has not been adequately cleaned, chunks of manure may fall from it onto the meat.

◀ In modern meat-processing plants, the meat from hundreds of different cows is mixed up to be ground, so a single animal infected with *E. coli* can contaminate 32,000 pounds of ground beef that may be shipped throughout the United States.

One of the worst outbreaks of *E. coli* 0157:H7 food poisoning occurred in 1996, when nearly 6,000 cases were recorded in Japan. ▼

When the cow's stomach and intestines are removed, the fecal matter in the digestive system may spill out and contaminate the meat as well. Fecal matter contains the microbe *Escherichia coli* 0157:H7, which can cause serious illness and death in humans. Because of the mass production techniques of modern meat processing, a single cow infected with *E. coli* 0157:H7 can contaminate 32,000 pounds of ground beef (Schlosser 2002).

Finally, a number of health problems result from the massive quantities of animal waste that are produced and stored around factory farms. A single dairy cow produces more than 20 tons of manure annually, and a hog can produce more than two tons (Weeks 2007). Manure is often stored in lagoons, which can leak, break, or be washed away by big storms, contaminating

▲ As workers remove the stomach and intestines of beef cattle, fecal matter can spill out and contaminate the meat.

groundwater. Manure sprayed on fields as fertilizer can also pollute groundwater.

Livestock factories also pose a threat to air quality. People who live near large livestock farms complain about headaches, runny noses, sore throats, nausea, stomach cramps, diarrhea, burning eyes, coughing, bronchitis, and shortness of breath (Singer & Mason 2006; Weeks 2007).

This hog manure lagoon near Milford, ▶ Utah, holds 3 million gallons of hog waste. A tarp has been spread over the 30-foot-deep lagoon in an attempt to protect nearby residents from foul smell of the manure.

Millions of children whose parents died of AIDS grow up in orphanages.

gains in life expectancy made in sub-Saharan Africa, which peaked at 49 years in the late 1980s and fell to 46 years in 2005 (World Health Organization 2004). The HIV/AIDS epidemic creates an enormous burden for the limited health care resources of poor countries. Economic development is threatened by the HIV epidemic, which diverts national funds to health-related needs and reduces the size of a nation's workforce. The epidemic has orphaned 15 million children (one or both parents has died of AIDS); most live in sub-Saharan Africa (Kaiser Family Foundation 2009a). Some scholars fear that AIDS-affected countries could become vulnerable to political instability as the growing number of orphans exacerbates poverty and produces masses of poor young adults who are vulnerable to involvement in criminal activity and recruitment for insurgencies (Mastny & Cincotta 2005).

HIV/AIDS in the United States

According to the Centers for Disease Control and Prevention (2009), more than 700,000 people in the United States are living with HIV/AIDS. Among U.S. adults and adolescents, three-quarters (74 percent) of new HIV/AIDS diagnoses in 2007 were among men, and half (51 percent) were among Black Americans. Among men with HIV/AIDS, the primary mode of transmission is through male-to-male sexual contact, followed by heterosexual contact and injection drug use. Among women with HIV/AIDS, the primary mode of transmission is through heterosexual contact, followed by injection drug use.

Despite the widespread concern about HIV, many Americans—especially adolescents and young adults—engage in high-risk behavior. A national survey of college students found that only half reported having used a condom the last time they had vaginal intercourse, and only one-quarter reported having used a condom the last time they had anal intercourse (American College Health Association 2008).

The Growing Problem of Obesity

Obesity is a major health problem throughout the industrialized world and is increasing in developing countries, especially in urban areas. Two-thirds of U.S. adults are either overweight or obese. Since 1980, U.S. adult obesity rates have doubled from 15 percent to 30 percent today, and childhood obesity rates have nearly tripled from 6.5 percent to 16.3 percent (Trust for America's Health 2008). Obesity, which can lead to heart disease, hypertension, diabetes, and other health problems, is the second biggest cause of preventable deaths in the United States (second only to tobacco use) (Stein & Connolly 2004). A report published in the *New England Journal of Medicine* suggested that obesity will shorten the average

U.S. life expectancy by at least two to five years over the next fifty years, reversing the mostly steady increase in life expectancy that has occurred over the past two centuries (Olshansky et al. 2005).

Although genetics and certain medical conditions contribute to many cases of overweight and obesity, two social and lifestyle factors that play a major role in the obesity epidemic are patterns of food consumption and physical activity level. Less than one-third (30 percent) of U.S. adults (age 18 or older) engage in regular leisure-time physical activity (National Center for Health Statistics 2008). More than one-third of youths in grades 9 through 12 do not engage in regular vigorous physical activity (National Center for Chronic Disease Prevention and Health Promotion 2004).

Americans are increasingly eating out at fast-food and other restaurants where foods tend to contain more sugars and fats than foods consumed at home. Fast food consumption is strongly associated with weight gain and insulin resistance, suggesting that fast food increases the risk of obesity and type 2 diabetes (Niemeier et al. 2006; Pereira et al. 2005). Consumption of snack foods and sugary soft drinks has also increased. Among children aged 6 to 11 years, consumption of chips, crackers, popcorn, and/or pretzels tripled from the mid-1970s to the mid-1990s. Consumption of soft drinks doubled during the same period (Sturm 2005). As processed foods are increasingly marketed throughout the world, the consumption of foods high in fats and sweeteners is also increasing in developing nations. This changing pattern of food consumption, known as the *nutrition transition,* is contributing to a rapid rise in obesity and diet-related chronic diseases worldwide (Hawkes 2006).

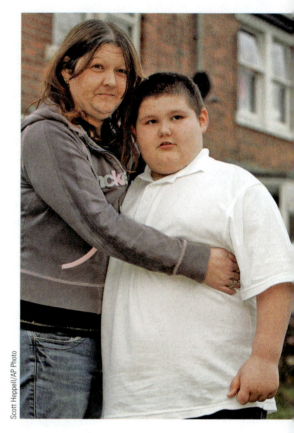

Scott Heppell/AP Photo

Childhood obesity is becoming more common throughout the developed world. At 8 years of age, Connor McCreaddie, shown here with his mother, weighed 218 pounds.

What Do You Think? In 2007, 8-year-old Connor McCreaddie of the United Kingdom weighed 218 pounds. A child protection conference was held to determine whether Connor should be removed from his home and placed into foster care, where his diet would be carefully controlled. This decision involved determining whether Connor's mother was abusing him by providing Connor with excessive high-calorie food. In this case, Connor's mother was allowed to keep custody of her son (*Guardian* 2007). In a similar case in North Carolina, a mother whose 7-year-old son weighed more than 250 pounds reported that the local Division of Social Services threatened to take her child away if he did not lose weight (Associated Press 2007). Do you think that severely obese children should be considered as victims of child abuse and taken from their parents and placed in foster care?

Obesity is also related to socioeconomic status. In less developed countries, poverty is associated with undernutrition and starvation. In the United States, however, being poor is associated with an increased risk of being overweight or obese. High-calorie processed foods tend to be more affordable than fresh vegetables, fruits, and lean meats or fish. Residents of low-income areas often lack access to large grocery stores that sell a variety of foods, and instead rely on neighborhood fast-food chains and convenience stores that sell mostly high-calorie processed food.

Mental Illness: The Hidden Epidemic

What it means to be mentally healthy varies across cultures. In the United States, **mental health** is defined as the successful performance of mental function, resulting in productive activities, fulfilling relationships with other people, and the ability to adapt to change and to cope with adversity (U.S. Department of Health and Human Services 2001). **Mental illness** refers collectively to all mental disorders, which are health conditions that are characterized by alterations in thinking, mood, and/or behavior associated with distress or impaired functioning and that meet specific criteria (such as level of intensity and duration) specified in the classification manual used to diagnose mental disorders, *The Diagnostic and Statistical Manual of Mental Disorders* (American Psychiatric Association 2000) (see Table 2.3).

Mental illness is a "hidden epidemic" because the shame and embarrassment associated with mental problems discourage people from acknowledging and talking about them. Negative stereotypes of people with mental illness contribute to its stigma. People are twice as likely today than they were in 1950s to believe that mentally ill people are violent. The reality is that the vast majority of people with mental illness are not violent, though they are 2.5 times more likely to be victims of violence than members of the general population (Dingfelder 2009a). Due to negative

mental health The successful performance of mental function, resulting in productive activities, fulfilling relationships with other people, and the ability to adapt to change and to cope with adversity.

mental illness All mental disorders, which are health conditions that are characterized by alterations in thinking, mood, and/or behavior associated with distress or impaired functioning and that meet specific criteria (such as level of intensity and duration) specified in *The Diagnostic and Statistical Manual of Mental Disorders*.

TABLE 2.3 Disorders Classified by the American Psychiatric Association

CLASSIFICATION	DESCRIPTION
Anxiety disorders	Disorders characterized by anxiety that is manifest in phobias, panic attacks, or obsessive-compulsive disorder
Dissociative disorders	Problems involving a splitting or dissociation of normal consciousness, such as amnesia and multiple personality
Disorders first evident in infancy, childhood, or adolescence	Disorders including mental retardation, attention-deficit/hyperactivity, and stuttering
Eating or sleeping disorders	Disorders including anorexia, bulimia, and insomnia
Impulse control disorders	Problems involving the inability to control undesirable impulses, such as kleptomania, pyromania, and pathological gambling
Mood disorders	Emotional disorders such as major depression and bipolar (manic-depressive) disorder
Organic mental disorders	Psychological or behavioral disorders associated with dysfunctions of the brain caused by aging, disease, or brain damage (such as Alzheimer's disease)
Personality disorders	Maladaptive personality traits that are generally resistant to treatment, such as paranoid and antisocial personality types
Schizophrenia and other psychotic disorders	Disorders with symptoms such as delusions or hallucinations
Somatoform disorders	Psychological problems that present themselves as symptoms of physical disease, such as hypochondria
Substance-related disorders	Disorders resulting from abuse of alcohol and/or drugs, such as barbiturates, cocaine, or amphetamines

attitudes toward mental illness, the majority of U.S. adults don't want someone with a mental illness in their workplace or marrying into their family (see Figure 2.2).

Extent and Impact of Mental Illness

One-quarter (26 percent) of U.S. adults have a diagnosable mental disorder in any given year. Mental disorders are the leading cause of disability for individuals between ages 15 and 44 (National Institute of Mental Health 2008).

Untreated mental illness can lead to poor educational achievement, lost productivity, unsuccessful relationships, significant distress, violence and abuse, incarceration, unemployment, homelessness, and poverty. Half of students identified as having emotional disturbances drop out of high school (Gruttadaro 2005). As many as one in five adults in U.S. prisons and as many as 70 percent of youth incarcerated in juvenile justice facilities are mentally ill (Honberg 2005; Human Rights Watch 2003). Most suicides in the United States (more than 90 percent) are committed by individuals with a mental disorder, most commonly a depressive or substance abuse disorder (National Institute of Mental Health 2008).

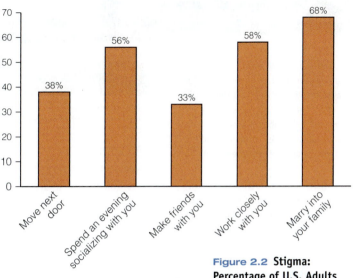

Figure 2.2 Stigma: Percentage of U.S. Adults Reporting They Are Unwilling to Have Various Levels of Contact with a Person with Mental Illness
Source: Martin et al. 2000.

What Do You Think? The age group with the highest rate of "serious psychological distress" is between ages 18 and 25. Why do you think this is so?

Causes of Mental Disorders

Some mental illnesses are caused by genetic or neurological pathological conditions. However, social and environmental influences, such as poverty, relationship abuse, job loss, divorce, the death of a loved one, the onset of illness or disabling injury, and war, also can trigger mental health problems. For example, war contributes to post-traumatic stress disorder that many military personnel serving in war zones experience. In conclusion, most mental disorders are caused by a combination of genetic, biological, and environmental factors (U.S. Department of Health and Human Services 2001).

Social Factors and Lifestyle Behaviors Associated with Health and Illness

Health problems are linked to lifestyle behaviors such as excessive alcohol consumption, cigarette smoking, unprotected sexual intercourse, and inadequate consumption of fruits and vegetables (see this chapter's Social Problems Research Up Close feature). However, health and illness are also affected by social

The National College Health Assessment

The National College Health Assessment is a survey developed by the American College Health Association to assess the health status of college students across the country. After briefly describing the sample and methods, we present selected findings of the 2008 National College Health Assessment.

Sample and Methods

A total of 48 postsecondary institutions self-selected to participate in the Fall 2007 National College Health Assessment. Data from institutions that did not use random sampling techniques were not used, yielding a final sample of 20,507 students at 39 campuses (American College Health Association 2008). The average response rate was 31 percent: 63 percent for schools using paper surveys and 21 percent for schools conducting web-based surveys. The survey contains questions that assess student health status and health problems, risk and protective behaviors, and health impediments to academic performance.

Selected Findings and Conclusions

- *Most commonly reported health problems.* The most commonly reported

Table 1 Top Six Self-Reported Health Problems Students Experienced in the Past School Year

HEALTH PROBLEM	RANK	PERCENTAGE
Back pain	1	46.6
Allergy problems	2	45.5
Sinus infection	3	28.8
Depression	4	17.8
Strep throat	5	13.2
Anxiety disorder	6	12.4

Adapted from: American College Health Association. 2008. American College Health Association National College Health Assessment Fall 2007 reference group data report. Baltimore: American College Health Association.

health problems of college students are allergy problems and back pain (see Table 1 within this box).

- *Alcohol, tobacco, and marijuana use.* The majority of college students (62 percent) reported having consumed alcohol in the past 30 days. Among students who drink alcohol, nearly one in three (31 percent) said that, in the past year, they did something they later regretted as a result of drinking alcohol. Nearly one in five (19 percent) of college students said that they used

cigarettes in the past year, and 13 percent had used marijuana.

- *Sexual health and condom use.* Nearly half (45 percent) of college students reported that they had at least one sexual partner in the past year. Among sexually active students, only 50 percent said they used a condom the last time they had vaginal intercourse; 25 percent used a condom during anal intercourse. Twelve percent of students reported using (or their partner used) emergency

factors such as globalization, social class and poverty, education, race, and gender.

Globalization

Broadly defined as the growing economic, political, and social interconnectedness among societies throughout the world, **globalization** has had both positive and negative effects on health. On the positive side, globalized communications technology enhances the capacity to monitor and report on outbreaks of disease, disseminate guidelines for controlling and treating disease, and share medical knowledge and research findings (Lee 2003). On the negative side, aspects of globalization such as increased travel and the expansion of trade and transnational corporations have been linked to a number of health problems.

globalization The growing economic, political, and social interconnectedness among societies throughout the world.

Effects of Increased Travel on Health. Increased business travel and tourism facilitates the spread of infectious disease. In just the first two months

MENTAL HEALTH DIFFICULTY	NEVER	1–10	11+
	PERCENTAGE	PERCENTAGE	PERCENTAGE
Felt things were hopeless	39	52	10
Felt so depressed it was difficult to function	57	36	7
Seriously considered attempting suicide	90	9	1
Attempted suicide	98	2	<1

Adapted from: American College Health Association. 2008. American College Health Association National College Health Assessment Fall 2007 reference group data report. Baltimore: American College Health Association.

Note: Percentages are rounded.

contraception (the "morning-after pill") within the past year.

- *Nutrition, exercise, and weight.* Only 6 percent of college students reported that they ate the recommended five or more servings of fruits and vegetables daily; 40 percent reported that they exercised vigorously for at least 20 minutes or moderately for at least 30 minutes at least three days a week. Based on estimated body mass index (BMI) calculated from students' reported height and weight, 23 percent of College students are overweight, and 14 percent are obese.

- *Mental health.* Sixteen percent of college students report having been diagnosed with depression sometime in their lifetimes. Of these, 24 percent are currently receiving therapy for depression, and 36 percent are currently taking medication for depression. Table 2 within this box presents the number of times students reported experiencing various mental health difficulties in the past school year.

Discussion

The results of the National College Health Assessment reveal the extent and types

of health problems and risk behaviors of college students. These data can be used to help colleges and universities design and implement health services that meet the needs of college students.

of the swine flu pandemic of 2009, the disease spread to infect nearly 600,000 people in more than 70 countries around the world.

Effects of Increased Trade and Transnational Corporations on Health. Increased international trade has expanded the range of goods available to consumers, but at a cost to global health. The increased transportation of goods by air, sea, and land contributes to pollution caused by the burning of fossil fuels. In addition, the expansion of international trade of harmful products such as tobacco, alcohol, and processed or "fast" foods is associated with a worldwide rise in cancer, heart disease, stroke, and diabetes (World Health Organization 2002).

Expanding trade has also facilitated the growth of transnational corporations that set up shop in developing countries to take advantage of lower labor costs and lax environmental and labor regulations (see also Chapter 7). Because of lax labor and human rights regulations, factory workers in transnational corporations are often exposed to harmful working conditions that increase the risk of illness, injury, and mental anguish (Hippert 2002). Weak environmental laws in developing countries enable transnational corporations to produce high levels of pollution and environmental degradation, which negatively affect the health of entire populations.

Social Class and Poverty

Poverty is associated with malnutrition, indoor air pollution, hazardous working conditions, lack of access to medical care, and unsafe water and sanitation (see also Chapter 6). In the United States, low socioeconomic status is associated with higher incidence and prevalence of health problems, disease, and death. Poverty is associated with higher rates of health-risk behaviors such as smoking, alcohol drinking, being overweight, and being physically inactive. The poor are also exposed to more environmental health hazards, and have unequal access to and use of medical care (Lantz et al. 1998). In addition, members of the lower class are subject to the most stress and have the fewest resources to cope with it (Cockerham 2007). Stress has been linked to a variety of physical and mental health problems, including high blood pressure, cancer, chronic fatigue, and substance abuse.

Just as poverty contributes to health problems, health problems contribute to poverty. Health problems can limit one's ability to pursue education or vocational training and to find or keep employment. The high cost of health care not only deepens the poverty of people who are already barely getting by but also can financially devastate middle-class families. Later in this chapter we look more closely at the high cost of health care and its consequences for individuals and families.

Poverty and Mental Health. Adults living below the poverty level were more than four times as likely to report serious psychological distress as adults in families with an income at least twice the poverty level (National Center for Health Statistics 2008). Two explanations for the link between social class and mental illness are the *causation* explanation and the *selection* explanation (Cockerham 2007). The selection explanation suggests that mentally ill individuals have difficulty achieving educational and occupational success and thus tend to drift to the lower class, whereas the mentally healthy are upwardly mobile. The *causation* explanation suggests that lower-class individuals experience greater adversity and stress as a result of their deprived and difficult living conditions, and this stress can reach the point at which individuals can no longer cope with daily living. Research that tested these two explanations found more support for the causation explanation (Hudson 2005).

Education

Although economic resources are important influences on health, education seems to be the strongest single predictor of good health. A *New York Times* report on the link between education and lifespan concluded the following: "The one social factor that researchers agree is consistently linked to longer lives in every country where it has been studied is education" (Kolata 2007, p. A1). Individuals with low levels of education are more likely to engage in health-risk behaviors such as smoking and heavy drinking. Women with less education are less likely to seek prenatal care and are more likely to smoke during pregnancy, which helps explain why low birth weight and infant mortality are more common among children of less educated mothers (Children's Defense Fund 2000).

In some cases, lack of education means that individuals do not know about health risks or how to avoid them. A national survey in India found that only 18 percent of illiterate women had heard of AIDS, compared with 92 percent of women who had completed high school (Ninan 2003).

> Education seems to be the strongest single predictor of good health.

Gender

Another social factor that affects health is gender. In many developing countries, the low status of females results in their being nutritionally deprived and having less access to medical care than men have (UNICEF 2006).

Sexual violence and gender inequality contribute to growing rates of HIV among girls and women. Although women are more susceptible to HIV infection for physical and biological reasons, increasing rates of HIV among women are also due to the fact that many women, especially in African countries, do not have the social power to refuse sexual intercourse and/or to demand that their male partners use condoms (Lalasz 2004).

In the United States, at least one in three women has been beaten, coerced into sex, or abused in some way—most often by someone the woman knows (Alan Guttmacher Institute 2004). "Although neither health care workers nor the general public typically thinks of battering as a health problem, it is a major cause of injury, disability, and death among American women, as among women worldwide" (Weitz 2010, p. 55).

In the United States before the 20th century, the life expectancy of U.S. women was shorter than that of men because of the high rate of maternal mortality that resulted from complications of pregnancy and childbirth. In the United States today, life expectancy of women (80.4 years) is greater than that of men (75.2 years) (National Center for Health Statistics 2008). Lower life expectancy for U.S. men is due to a number of factors. Men tend to work in more dangerous jobs than women, such as agriculture, construction, and the military. In addition, "beliefs about masculinity and manhood that are deeply rooted in culture…play a role in shaping the behavioral patterns of men in ways that have consequences for their health" (Williams 2003, p. 726). Men are socialized to be strong, independent, competitive, and aggressive and to avoid expressions of emotion or vulnerability that could be construed as weakness. These male gender expectations can lead men to take actions that harm themselves or to refrain from engaging in health-protective behaviors. For example, socialization to be aggressive and competitive leads to risky behaviors (such as dangerous sports, fast driving, and violence) that contribute to men's higher risk of injuries and accidents. Men are more likely than women to smoke cigarettes and to abuse alcohol and drugs but are less likely than women to visit a doctor and to adhere to medical regimens (Williams 2003).

Gender and Mental Health. One national survey found that U.S females were more likely than males to experience severe psychological distress (Substance Abuse and Mental Health Services Administration 2008). However, rates of mental illness are similar for women and men, although they differ in the types of mental illness they experience; women have higher rates of mood and anxiety disorders, and men have higher rates of personality and substance-related disorders. Although women are more likely to attempt suicide, men are more likely to succeed at it because they use deadlier methods.

Biological factors may account for some of the gender differences in mental health. Hormonal changes during menstruation and menopause, for example, may predispose women to depression and anxiety. High testosterone and androgen levels in males may be linked to the greater prevalence of personality disorders in men, but research is not conclusive. Other explanations for gender differences in mental health focus on ways in which gender roles contribute to different types of mental disorders. For example, the unequal status of women

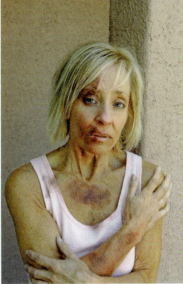

Robin Nelson / PhotoEdit

Physical abuse is a major cause of injury, disability, and death among women.

and the strain of doing the majority of housework and child care may predispose women to experience greater psychological distress.

Racial and Ethnic Minority Status

In the United States, racial and ethnic minorities are more likely than non-Hispanic whites to rate their health as fair or poor (see Figure 2.3). Black Americans, especially men, have a lower life expectancy than white men (see Table 2.4). Black Americans are more likely than white Americans to die from stroke, heart disease, cancer, HIV infection, unintentional injuries, diabetes, cirrhosis, and homicide. The highest rates of obesity are among Black Americans, followed by Hispanics (Trust for America's Health 2008). Black Americans also have the highest rate of infant mortality of all racial and ethnic groups, largely because of higher rates of prematurity and low birth weight.

Compared with white Americans, Native Americans have more health problems related to heavy alcohol use and diabetes. Due to their higher rates of poverty, Native Americans are more likely to lack adequate sanitation and clean water, which causes higher rates of infectious diseases. Like Black Americans and Native Americans, Hispanics also have lower life expectancies and more illness related to their lower socioeconomic status, including HIV, diabetes, and liver disease (often caused by alcohol abuse) (Weitz 2010). Asian Americans typically have better health than do other U.S. minority groups, in large part due to the fact that they have the highest levels of income and education of any racial or ethnic U.S. minority group. Traditional Asian diets, which include lots of fish and vegetables, may also account for their higher levels of health.

Figure 2.3 Fair or Poor Health Status by Race/Ethnicity
Source: James et al. 2007.

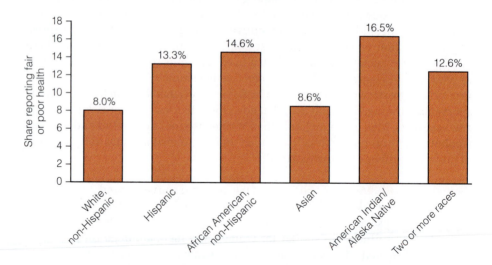

TABLE 2.4 Life Expectancy in the United State by Sex and Black/White Race*

ALL RACES		BLACK		WHITE	
FEMALE	MALE	FEMALE	MALE	FEMALE	MALE
80	75	77	70	81	76

Source: National Center for Health Statistics (2008).

*For individuals born in 2005.

Racial and ethnic differences in health status are largely due to differences in socioeconomic status. As we discuss later, U.S. minorities are less likely to have health insurance, and are therefore less likely to get timely and routine care and are more likely to be hospitalized for preventable conditions. The poorer health of minorities is also related to the fact that minorities are more likely than whites to live in environments where they are exposed to hazards such as toxic chemicals and other environmental hazards (see also Chapter 13). In addition, discrimination contributes to poorer health among oppressed racial and ethnic populations by restricting access to the quantity and quality of public education, housing, and health care. For example, a study of patients who had experienced heart attacks at 658 U.S. hospitals found that Black American patients were much less likely than white patients to get basic diagnostic tests, clot-busting drugs, or angioplasties (Vaccarino et al. 2005).

Race, Ethnicity, and Mental Health. Asian Americans have relatively low rates of mental illness. Research finds no significant difference among races in their overall rates of mental illness (Cockerham 2007). Differences that do exist are often associated more with social class than with race or ethnicity. However, some studies suggest that minorities have a higher risk for mental disorders, such as anxiety and depression, in part because of racism and discrimination, which adversely affect physical and mental health (U.S. Department of Health and Human Services 2001). Minorities also have less access to mental health services, are less likely to receive needed mental health services, often receive lower-quality mental health care, and are underrepresented in mental health research (U.S. Department of Health and Human Services 2001).

Family and Household Factors

Family and household factors are related to both physical and mental health. Married adults are healthier and have lower levels of depression and anxiety compared with adults who are single, divorced, cohabiting, or widowed (Mirowsky & Ross 2003; Schoenborn 2004). Two explanations for the association between marital status and health are the *selection* and the *causation* theories (also discussed earlier in explaining the link between poverty and health). In this context, the selection theory suggests that healthy individuals are more likely to marry and to stay married. The causation theory says that better health among married individuals results from the economic advantages of marriage and from the emotional support provided by most marriages—the sense of being cared about, loved, and valued (Mirowsky & Ross 2003).

For children, living in a two-parent household is associated with better health outcomes. A Swedish study found that children living with only one parent have a higher risk of death, mental illness, and injury than those in two-parent families, even when their socioeconomic disadvantage is taken into account (Hollander 2003).

Problems in U.S. Health Care

The World Health Organization's analysis of the world's health systems found that, although the United States spends a higher portion of its gross domestic product on health care than any other country, it ranks 37 out of 191 countries

> **Although the United States spends a higher portion of its gross domestic product on health care than any other country, it ranks 37 out of 191 countries according to its performance.**

according to its performance (World Health Organization 2000). The report concluded that France provides the best overall health care among major countries, followed by Italy, Spain, Oman, Austria, and Japan. In a comparison of health care in six countries—Australia, Canada, Germany, New Zealand, the United Kingdom, and the United States—the United States ranked last on dimensions of access, patient safety, efficiency, and equity (Davis et al. 2007).

After presenting a brief overview of U.S. health care, we address some of the major health care problems in the United States—inadequate health insurance coverage, the high cost of medical care and insurance, the managed care crisis, and inadequate mental health care.

U.S. Health Care: An Overview

In the United States, there is no one health care system; rather, health care is offered through various private and public means (see Figure 2.4).

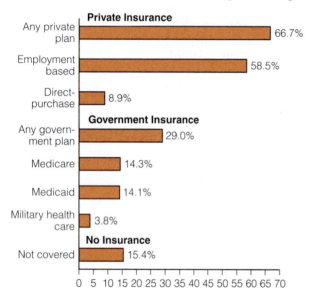

Figure 2.4 **Coverage by Type of Health Insurance, 2008**
Source: DeNavas-Walt et al. 2009.

In traditional health insurance plans, the insured choose their health care provider, who is reimbursed by the insurance company on a fee-for-service basis. Individuals with insurance typically must pay an out-of-pocket "deductible" (usually ranging from a few hundred to a thousand dollars or more per year per person) and then are often required to pay a percentage of medical expenses (e.g., 20 percent) until a maximum out-of-pocket expense amount is reached (after which insurance will cover 100 percent of medical costs up to a limit).

Health maintenance organizations (HMOs) are prepaid group plans in which people pay a monthly premium for comprehensive health care services. HMOs attempt to minimize hospitalization costs by emphasizing preventive health care. *Preferred provider organizations (PPOs)* are health care organizations in which employers who purchase group health insurance agree to send their employees to certain health care providers or hospitals in return for cost discounts. In this arrangement, health care providers obtain more patients but charge lower fees to buyers of group insurance.

Managed care refers to any medical insurance plan that controls costs through monitoring and controlling the decisions of health care providers. In many plans, doctors must call a utilization review office to receive approval before they can hospitalize a patient, perform surgery, or order an expensive diagnostic test. Although the terms *HMO* and *managed care* are often used interchangeably, HMOs are only one form of managed care. Most Americans who have private insurance belong to some form of managed care plan. Recipients of Medicaid and Medicare may also belong to a managed care plan.

Medicare. **Medicare** is funded by the federal government and reimburses the elderly and the disabled for their health care. Individuals contribute payroll taxes to Medicare throughout their working lives and generally become eligible for Medicare when they reach 65, regardless of their income or health status.

managed care Any medical insurance plan that controls costs through monitoring and controlling the decisions of health care providers.

Medicare A federally funded program that provides health insurance benefits to the elderly, disabled, and those with advanced kidney disease.

Medicare consists of four separate programs: Part A is hospital insurance for inpatient care, which is free, but enrollees may pay a deductible and a co-payment. Part B is a supplementary medical insurance program, which helps pay for physician, outpatient, and other services. Part B is voluntary and is not free; enrollees must pay a monthly premium as well as a co-payment for services. Medicare does not cover long-term nursing home care, dental care, eyeglasses, and other types of services, which is why many individuals who receive Medicare also enroll in Part C, which allows beneficiaries to purchase private supplementary insurance that receives payments from Medicare. Part D is an outpatient drug benefit that is voluntary and requires enrollees to pay a monthly premium, meet an annual deductible, and pay coinsurance for their prescriptions.

Medicaid and SCHIP. **Medicaid,** which provides health care coverage for the poor, is jointly funded by the federal and state governments (see also Chapter 6). Contrary to the belief that Medicaid covers all poor people, it does not. Eligibility rules and benefits vary from state to state, and, in many states, Medicaid provides health care only for those who are well below the federal poverty level.

In 1997, the **State Children's Health Insurance Program (SCHIP)** was created to expand health coverage to children without insurance, many of whom come from families with incomes too high to qualify for Medicaid but too low to afford private health insurance. Under this initiative, states receive matching federal funds to provide medical insurance to children without insurance.

Workers' Compensation. **Workers' compensation** (also known as *workers' comp*) is an insurance program that provides medical and living expenses for people with work-related injuries or illnesses. Employers pay a certain amount into their state's workers' compensation insurance pool, and workers injured on the job can apply to that pool for medical expenses and for compensation for work days lost. In exchange for that benefit, workers cannot sue their employers for damages. However, not all employers acquire workers' compensation insurance, even in states where it is legally required. Further, many employees with work-related illness or injuries do not apply for workers' compensation benefits because (1) they fear getting fired for making a claim, (2) they are not aware that they are covered by workers' comp, and/or (3) the employer offers incentives (i.e., bonuses) to employees when no workers' comp claims are filed in a given period of time (Sered & Fernandopulle 2005). Even when employees file a workers' comp claim, the coverage the employee receives rarely covers the cost of the employee's injury or illness. "The typical scenario . . . is one in which the workers' comp insurance company delays accepting and paying the worker's claim. In the meantime, the injured employee racks up medical and other bills and then, in desperation, accepts a lump sum monetary settlement that does not come close to covering medical expenses or replacing lost wages" (Sered & Fernandopulle 2005, p. 94).

Military Health Care. Military health care includes Civilian Health and Medical Program for Uniformed Services (CHAMPUS), Civilian Health and Medical Program of the Department of Veterans Affairs (CHAMPVA), and care provided by the Department of Defense and the Department of Veterans Affairs. A series of 2007 *Washington Post* reports brought attention to the abysmal conditions at military medical facilities and Veterans Administration (VA) hospitals around the country. Walter Reed's Army Medical Center Building 18 was found to have mice, mold, and rot. Soldiers and veterans around the country reported

Medicaid A public health insurance program, jointly funded by the federal and state governments, that provides health insurance coverage for the poor who meet eligibility requirements.

State Children's Health Insurance Program (SCHIP) A public health insurance program, jointly funded by the federal and state governments, that provides health insurance coverage for children whose families meet income eligibility standards.

workers' compensation Also known as workers' comp, an insurance program that provides medical workers' compensation and living expenses for people with work-related injuries or illnesses.

that their home post medical treatment facility was characterized by "indifferent, untrained staff; lost paperwork; medical appointments that drop from computers; and long waits for consultations" (Hull & Priest 2007). Other reports of military medical facilities described peeling paint, asbestos, overflowing trash, fruit fly infestations, no nurses, and lack of blankets and linens.

Many veterans have complained that they have not received the benefits they deserve or that they have had long waits to get benefits. One veteran, who was wounded twice on two tours of duty, and awarded two Purple Hearts and three Bronze Stars, fought for six years to get disability compensation because his service records had been lost (CNN Heroes 2009).

Mental health care for military personnel and military veterans is also inadequate. Only about half of veterans with post-traumatic stress disorder seek treatment, either because of the stigma of having mental health problems or because help was not available due to a shortage of military mental health professions (Dingfelder 2009b).

Inadequate Health Insurance Coverage

According to U.S. census data, 15.4 percent of Americans (46.3 million people) did not have health insurance coverage in 2008 (DeNavas-Walt et al. 2009). Those who have health insurance have no guarantee that their coverage will continue when they it most need. To increase their profits, health insurance companies frequently deny or limit treatment, cover only less expensive drugs, and even terminate policies to avoid paying claims. For example, when a retired nurse from Texas discovered she had breast cancer, her insurer scrutinized her medical records for any technicality that would allow refusal of payment. They discovered that she earlier had treatment for acne, which the dermatologist mistakenly noted as precancerous, so they canceled her policy three days before a scheduled double mastectomy, claiming that she had misinformed them about her medical history (Harris 2009). An investigation of the three largest insurers found they canceled coverage of more than 20,000 people over a five-year period, saving more than $300 million in medical claims (Orr 2009).

Disparities in Health Insurance Coverage.

Whites are more likely than racial and ethnic minorities to have health insurance. Hispanics have the largest percentage of uninsured, followed by American Indians and Alaska Natives, blacks Native Hawaiians and other Pacific Islanders, and Asians (DeNavas-Walt et al. 2009).

Of all age groups, young adults ages 18 to 24 are the least likely to have health insurance. In 2008, more than a quarter (28.6 percent) of young adults was uninsured (DeNavas-Walt et al. 2009).

> Of all age groups, young adults ages 18 to 24 are the least likely to have health insurance.

What Do You Think? Many college students do not have health insurance, and each year hundreds of college students withdraw from school because of their inability to pay medical bills from accidents or unexpected illnesses. Some universities require students to have health insurance. Although mandatory health insurance may keep students from dropping out, it also adds to college bills and may prevent some individuals who cannot afford health insurance from enrolling in college. Do you think that universities should require students to have health insurance? Why or why not?

Health insurance status also varies by income and employment. The higher an individual's income, the more likely the individual will have health insurance. Employed individuals are also more likely than unemployed individuals to be insured. However, employment is no guarantee of health care coverage; in 2008, 17.2 percent of full-time workers were uninsured (DeNavas-Walt et al. 2009). More than two-thirds of those without insurance live in a household with one full-time worker, and more than one-third of those without insurance in the United States have annual family incomes of more than $40,000 (Zeldin & Rukavina 2007). Not all businesses offer health benefits to their employees; in 2008, 63 percent of businesses offered health benefits to at least some of their employees (Kaiser Family Foundation 2008). Even when employers offer health insurance, some employees are not eligible for health benefits because of waiting periods or part-time status. Some employees who are eligible may not enroll in employer-provided health insurance because they cannot afford to pay their share of the premiums.

Inadequate Insurance for the Poor. Many Americans believe that Medicaid and SCHIP—public health insurance programs for the poor—cover all low-income children, adults, and families. But Medicaid eligibility levels are set so low that many low-income adults are not eligible. Some states have waiting lists for Medicaid. Another problem is that, because of the low reimbursement payments from Medicaid, many health care providers do not accept Medicaid patients.

The 2009 reauthorization of SCHIP extended the program through 2013, and provided $32.8 billion additional money (paid for by an increase in tobacco tax) to expand eligibility and mental health and dental coverage (Sack 2009). The 2009 reauthorization also expands coverage by permitting states to (1) drop the requirement that legal immigrants wait five years before joining the program, (2) provide coverage to pregnant women, and (3) raise the income cutoff for eligibility to up to three times the poverty level. Despite state budget deficits hit hard by the recent recession, several states have spent millions of dollars to extend SCHIP to 250,000 additional children. But budget shortfalls have prevented other states from extending SCHIP coverage. California, with the country's largest budget deficit, imposed a freeze on new SCHIP enrollments (Sack 2009).

Consequences of Inadequate Health Insurance. An estimated 45,000 deaths per year in the United States are attributable to lack of health insurance (Yang 2009). Individuals who lack health insurance are less likely to receive preventive care, are more likely to be hospitalized for avoidable health problems, and are more likely to have disease diagnosed in the late stages (Kaiser Commission on Medicaid and the Uninsured 2004). In a study of individuals who experienced an unintentional injury or a new chronic health problem, individuals without insurance reported receiving less medical care and poorer short-term changes in health than those with insurance (Hadley 2007).

> An estimated 45,000 deaths per year in the United States are attributable to lack of health insurance.

Because most health care providers do not accept patients who do not have insurance, many individuals without insurance resort to using the local hospital emergency room. In a study of young adults with chronic health conditions, loss of health insurance resulted in decreased use of office-based physician services and a dramatic increase in visits to hospital emergency rooms (Scal & Town 2007). A federal law called the Emergency Medical Treatment and Active Labor Act requires hospitals to assess all patients who come to their emergency rooms to determine whether an emergency medical condition exists and, if it does, to stabilize patients before transferring them to another facility. Hospital patients without insurance are almost always billed at a much higher cost than the prices negotiated by insurance companies.

Individuals who lack dental insurance commonly have untreated dental problems, which can lead to or exacerbate other health problems.

> Because they affect the ability to chew, untreated dental problems tend to exacerbate conditions such as diabetes or heart disease....Missing and rotten teeth make it painful if not impossible to chew fruits, whole-grain foods, salads, or many of the fiber-rich foods recommended by doctors and nutrition experts. (Sered & Fernandopulle 2005, pp. 166–67)

In their book *Uninsured in America*, Sered and Fernandopulle (2005) described one interviewee who "covered her mouth with her hand during our entire interview because she was embarrassed about her rotting teeth" and another interviewee "used his pliers to yank out decayed and aching teeth" (p. 166). The authors note that "almost every time we asked interviewees what their first priority would be if the president established universal health coverage tomorrow, the immediate answer was 'my teeth'" (p. 166).

The High Cost of Health Care

The United States spends over $2.2 trillion a year on health care, which translates into $7,421 per person and more than 16 percent of gross domestic product (GDP)—far more than any other industrialized nation (Centers for Medicare and Medicaid Services 2009). Yet virtually every other wealthy nation has better health outcomes, as measured by life expectancy and infant mortality. Why does the United States spend so much on health care? How do the high costs of health care affect individuals and families?

Several factors have contributed to escalating medical costs. These include increased longevity; excessive and inappropriate medical care; and the high costs of health care administration, drugs, doctors' fees, hospital services, medical technology, and health insurance.

Increased Longevity. Because of improved sanitation and medical advances, people are living longer today than in previous generations. People older than age 65 use medical services more than younger individuals and are also more likely to take prescription medicine on a daily basis. The average health care expense for the elderly U.S. population was $14,797 in 2007, compared with $4,511 per year for nonelderly adults (ages 19 to 64) (Centers for Medicare and Medicaid Services 2009).

People are not only living longer, they are also spending more of their lives with chronic diseases. A century ago, the average adult in Western nations spent only 1 percent of life in illness, but today the average adult spends more than 10 percent of life sick (Robbins 2006). Today people survive illnesses, conditions, and injuries that would have killed them a generation ago. Infants born prematurely who would not have survived a generation ago are kept alive today in hospital incubators. Individuals with HIV/AIDS are living longer today, owing to the availability of new (and expensive) drugs. Individuals with kidney disease are receiving dialysis treatment and kidney transplants. People with heart disease are undergoing bypass surgery and other treatments that are extending their lives and their medical expenses.

Cost of Hospital Services, Doctors' Fees, and Medical Technology.

High hospital costs and doctors' fees contribute to rising costs of health care. The use of expensive medical technology, unavailable just decades ago, also contributes to high medical bills. Consider the advancements in the treatment of preterm babies, for which very little could be done in 1950. By 1990, special ventilators, artificial pulmonary surfactant to help infant lungs develop, neonatal intensive care, and steroids for mother and/or baby became standard treatment for preterm babies in the United States (Kaiser Family Foundation 2007). Such technology greatly improved the life expectancy of preterm babies but also adds to health care costs.

Cost of Drugs.

The high cost of drugs also contributes to health care costs. The United States pays 81 percent more for patented brand-name prescription drugs than Canada and six western European nations (Sager & Socolar 2004). The high prices that Americans pay for prescription drugs partly explain why the pharmaceutical industry is among the most profitable industries in the United States.

Manufacturers argue that, in countries where governments regulate prices, consumers pay too little for drugs. U.S. drug prices are high, claim drugmakers, because of the high cost of researching and developing new drugs. But most large drug companies pay substantially more for marketing, advertising, and administration than for research and development (Families USA 2007).

Cost of Health Insurance.

In 2008, the average annual premiums for employer-sponsored coverage were $4,704 for an individual and $12,680 for a family, with workers contributing, on average, $721 annually toward the cost of individual coverage and $3,354 toward family coverage. Since 1999, average insurance premiums have increased 119 percent (Kaiser Family Foundation 2008).

With the rising cost of medical insurance, companies are increasing the employees' share of the cost, decreasing the benefits, or not providing insurance at all. The cost of health care to businesses also affects the prices that consumers pay for goods and services. For example, in 2007, the cost of health benefits for employees at General Motors accounted for $1,783 of the sticker price of each new vehicle (Specter & Stoll 2007). U.S. businesses struggle to compete in a global economy where businesses in many other countries are not saddled with health care costs of their employees.

Cost of Health Care Administration.

Health care administrative expenses in the United States per capita are six times higher than in western European nations (The National Coalition on Health Care 2009). Harrison (2008) explains the reason:

> The United States has the most bureaucratic health care system in the world, including over 1,500 different companies, each offering multiple plans, each with its own marketing program and enrollment procedures, its own paperwork and policies, its CEO salaries, sales commissions, and other nonclinical costs—and, of course, if it is a for-profit company, its profits.

Consequences of the High Cost of Health Care for Individuals and Families.

When an uninsured driver hit 12-year-old Candice Jackson while she was getting off a school bus, she spent four months in the hospital and incurred about $90,000 in uncovered medical bills. Her mother needed two knee replacements, adding another $20,000 to the family's medical bills. A few years later, at age 16, Candice swerved off the road to avoid hitting a deer and sustained a head injury that required brain surgery. In anticipation of the medical bills from Candice's latest mishap, Candice's dad, Lanny Jackson, who works in the service center of a car dealership, feels forced to file for bankruptcy (Springen 2006).

Lanny Jackson is not alone. One study found that medical bills, as well as income lost due to illness, contributed to two-thirds of all bankruptcies in 2007 (Himmelstein et al. 2009). Most medical debtors were well-educated, homeowners with middle-class jobs, and three-fourths had health insurance. Having insurance does not guarantee that one is protected against financial devastation resulting from illness or injury, because even the insured typically must pay co-payments, deductibles, and exclusions. In addition, the link between coverage and employment means that insurance is often lost when it is needed the most—when workers lose their jobs because of medical problems. Although the Consolidated Omnibus Budget Reconciliation Act (COBRA) allows people to continue their insurance coverage when they lose a job, the premiums for a family typically exceed $1,000, whereas the average unemployment insurance payment is $1,425. The American Recovery and Reinvestment Act includes a subsidy of 65 percent of individuals' COBRA costs for up to nine months, but this is a temporary measure, and millions of workers will be unemployed for longer than nine months. Even with the 65 percent subsidy, COBRA costs for an average family are $370 a month, or a quarter of the average monthly unemployment benefit, not including deductibles or co-pays (Orr 2009).

> Having insurance does not guarantee that one is protected against financial devastation resulting from illness or injury, because even the insured typically must pay co-payments, deductibles, and exclusions.

Many individuals forgo needed medicine and/or medical care when they cannot afford to pay for it (see Table 2.5). Forgoing medicine or medical care often exacerbates the medical condition, leading to even higher medical costs, or tragically, leading to death.

TABLE 2.5 Cutting Back on Medical Care Due to Cost

In the past 12 months, have you or another family member living in your household . . . because of cost, or not?

	PERCENT SAYING "YES"
Relied on home remedies or over-the-counter (OTC) drugs instead of seeing a doctor	37%
Skipped dental care or checkups	35%
Put off or postponed getting health care you needed	31%
Skipped recommended medical test or treatment	27%
Not filled a prescription for medicine	26%
Cut pills in half or skipped doses of medicine	19%
Had problems getting mental health care	8%
Did ANY of the above	55%

*Source: Kaiser Family Foundation 2009b.

A more drastic measure involves breaking the law to receive free medical care in prison. This was the case for Larry Causey, age 57, who called the FBI and told them he was going to rob the post office in West Monroe, Louisiana. Then he went to the post office and handed a note to a teller demanding money. He left empty-handed and sat in his car until officers arrested him. Larry had no intention of committing robbery. Larry had cancer and could not afford cancer treatment, so he staged a robbery to get arrested and be put in jail, where he would receive medical treatment for his cancer (*Grand Island Independent* 2001). Larry Causey's story is not as uncommon as we may think. "Sheriffs nationwide say they're also arresting people willing to trade their freedom for a free visit to the doctor" (*Grand Island Independent* 2001).

The Managed Care Crisis

In an attempt to control medical care costs in the last few decades, the U.S. health care system has seen a dramatic rise in managed care. But Americans are concerned about the reduced quality of health care resulting from the emphasis on cost containment in managed care. Surveys found that more people have said that managed care plans do a "bad job" than a "good job" in serving customers (Kaiser Health Poll Report 2004). In a survey of physicians' views on the effects of managed care, the majority responded that managed care has negative effects on the quality of patient care because of limitations on diagnostic tests, length of hospital stay, and choice of specialists (Feldman et al. 1998). One former director of an HMO described the situation:

> I've seen from the inside how managed care works. I've been pressured to deny care, even when it was necessary. I have seen the bonus checks given to nurses and doctors for their denials. I have seen the medical policies that keep patients from getting care they need...and the inadequate appeal procedures. (Peeno 2000, p. 20)

Inadequate Mental Health Care

Since the 1960s, U.S. mental health policy has focused on reducing costly and often neglectful institutional care and on providing more humane services in the community. This movement, known as **deinstitutionalization,** had good intentions but has largely failed to live up to its promises. Only one in five U.S. children with mental illness is identified and receives treatment, and fewer than half of adults with "serious psychological distress" received mental health treatment or counseling during the past year (Gruttadaro 2005; Substance Abuse and Mental Health Services Administration 2008). The most common reason for not seeking treatment is "could not afford the cost" (see Figure 2.5).

Mental health services are often inaccessible, especially in rural areas. In most states, services are available from "9 to 5"; the system is "closed" in the evenings and on weekends when many people with mental illness experience the greatest need. Across the nation, people with severe mental illness end up in jails and prisons, homeless shelters, and hospital emergency rooms. Many children with untreated mental disorders drop out of school or end up in foster care or the juvenile justice system. As many as 70 percent of youths incarcerated in juvenile justice facilities have mental disorders (Honberg 2005). In a survey of 367 colleges and universities in the United States and Canada, most (92 percent) counseling center

deinstitutionalization The removal of individuals with psychiatric disorders from mental hospitals and large residential institutions to outpatient community mental health centers.

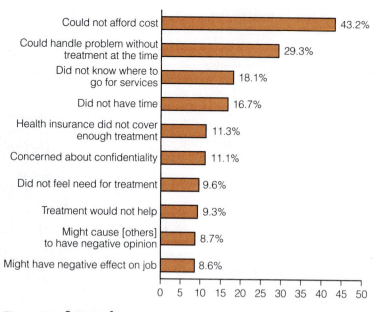

Could not afford cost — 43.2%
Could handle problem without treatment at the time — 29.3%
Did not know where to go for services — 18.1%
Did not have time — 16.7%
Health insurance did not cover enough treatment — 11.3%
Concerned about confidentiality — 11.1%
Did not feel need for treatment — 9.6%
Treatment would not help — 9.3%
Might cause [others] to have negative opinion — 8.7%
Might have negative effect on job — 8.6%

0 5 10 15 20 25 30 35 40 45 50

Figure 2.5 Reasons for Not Receiving Mental Health Services in the Past Year Among Adults with an Unmet Need for Mental Health Care, 2007
Source: Substance Abuse and Mental Health Services Administration 2008.

directors believe that the number of college students with severe psychological problems has increased in recent years. Yet only 58 percent of colleges and universities offer psychiatric services on campus (Gallagher 2006).

Given the increasing growth of minority populations, another deficit in the mental health system is the inadequate number of mental health clinicians who speak the client's language and who are aware of cultural norms and values of minority populations (U.S. Department of Health and Human Services 2001).

The mental health system is also plagued by inadequate federal and state funding of public mental health centers, which results in rationing care to those most in need. Thus, people must "hit bottom" before they can receive services.

What Do You Think? In 2004, Jordan Nott, a former student at George Washington University, was suspended and barred from campus after hospitalizing himself for depression and suicidal thoughts. The university charged Nott with violating its code of conduct by engaging in "endangering behavior" (Capriccioso 2006). In 2007, Virginia passed a bill to prevent public colleges and universities from dismissing students for attempting suicide or seeking mental health treatment for suicidal thoughts or behaviors (Sampson 2007). Do you think that suicidal students might avoid seeking mental health treatment if that puts them at risk for being dismissed from school?

Strategies for Action: Improving Health and Health Care

selective primary health care An approach to health care that focuses on using specific interventions to target specific health problems.

comprehensive primary health care An approach to health care that focuses on the broader social determinants of health, such as poverty and economic inequality, gender inequality, environment, and community development.

Two broad approaches to improving the health of populations are selective primary health care and comprehensive primary health care (Sanders & Chopra 2003). **Selective primary health care** focuses on interventions that target specific health problems, such as promoting condom use to prevent HIV infections and providing immunizations against childhood diseases to promote child survival. **Comprehensive primary health care** focuses on the broader social determinants of health, such as poverty and economic inequality, gender inequality, environment, and community development.

As you read the following sections on improving maternal and infant health, preventing and alleviating HIV/AIDS, and fighting obesity, see whether you can identify which strategies represent selective primary health care approaches and which strategies are comprehensive. Strategies to alleviate social problems discussed in subsequent chapters of this textbook are also important elements to a comprehensive primary health care approach.

Improving Maternal and Infant Health

Access to family planning services, affordable methods of contraception, medical care, and safe abortion services are important determinants of the well-being of mothers and their children (Save the Children 2002). Family planning reduces maternal mortality by reducing the number of unintended pregnancies and by enabling women to space births two to three years apart, which decreases infant mortality significantly (Murphy 2003). Providing women and children access to medical care is also critical for improving their health. For example, providing pregnant women with an inexpensive drug called misoprostol can stop postpartum hemorrhage or excessive bleeding—the leading cause of maternal mortality (Grossman 2009).

Improving the status and power of women is an important strategy in improving their health. Promoting women's education increases the status and power of women to control their reproductive lives, exposes women to information about health issues, and also delays marriage and childbearing. In many developing countries, women's lack of power and status means that they have little control over health-related decisions (UNICEF 2006). Men make the decisions about whether or when their wives (or partners) will have sexual relations, use contraception, or use health services.

Improving maternal and infant health globally requires funding, but the cost is not prohibitive. The price of providing basic health services for mothers and infants in low-income countries is only $3 per person (Oxfam GB 2004). The question is, do the rich countries of the world have the political will to support efforts to protect the health and lives of women and infants in the developing world?

HIV/AIDS Prevention and Alleviation Strategies

As of this writing, there is no vaccine to prevent HIV infection. As researchers continue to work on developing such a vaccine, a number of other strategies are available to help prevent and treat HIV/AIDS.

HIV/AIDS Education and Access to Condoms. HIV/AIDS prevention efforts include educating populations about how HIV is transmitted and how to protect against HIV transmission, and providing access to condoms as a means of preventing HIV transmission. A survey of U.S. adults found that significant numbers say that they do not know how HIV is transmitted or mistakenly believe that it is possible to transmit HIV through kissing, sharing a drinking glass, and touching a toilet seat (Kaiser Family Foundation 2004b). More than half did not know that a pregnant woman with HIV infection can take medication to reduce the risk of her baby being born with HIV infection.

HIV/AIDS education occurs through media and public service announcements, faith-based groups, health care providers, and schools. With the HIV infection rate growing among the older-than-50 population, HIV/AIDS education is also taking place in some senior centers (Goldberg 2005). Some HIV/AIDS education is based on the ABC approach—a prevention strategy that involves three elements: A = Abstain; B = Be faithful; C = Use condoms (Halperin et al. 2004).

Providing education that advocates condom use and providing youth with access to condoms are controversial topics. Many conservatives believe that promoting use of condoms sends the "wrong message" that sex outside marriage is OK.

Public health messages encourage people to get tested for HIV.

Another controversy involves the question of whether to provide condoms to prison inmates. Vermont and Mississippi allow condom distribution in prisons, as do Canada, most of western Europe, and parts of Latin America. One deputy at the Los Angeles Sheriff's Department, which allows only homosexual inmates to receive condoms provided by a local nonprofit organization, said, "We're not promoting sex; we're promoting health" (Sanders 2005).

HIV Testing. Another strategy to curb the spread of HIV involves encouraging individuals to get tested for HIV infection so that they can modify their behavior (to avoid transmitting the virus to others) and so that they can receive early medical intervention, which can slow or prevent the onset of AIDS. An estimated one-fourth to one-third of HIV-infected Americans do not know that they are infected (Kaiser Family Foundation 2004a). More than a third of U.S. adults (37 percent) reported having ever been tested for HIV (MMWR 2009). Unfortunately, many individuals who have HIV infection continue to engage in risky behaviors, such as unprotected anal, genital, or oral sex and needle sharing (Diamond & Buskin 2000; Hollander 2005).

The Fight Against HIV/AIDS Stigma and Discrimination. The HIV/AIDS-related stigma stems from societal views that people with HIV/AIDS are immoral and shameful, and results in discrimination in employment, housing, social relationships, and medical care. A survey of 1,000 physicians and nurses in Nigeria found that 1 in 10 admitted to refusing care for an HIV/AIDS patient or had denied HIV/AIDS patients admission to a hospital, and 20 percent believed that people living with HIV/AIDS have behaved immorally and deserved their fate (AVERT 2004). The stigma surrounding HIV/AIDS has also led to acts of violence against people perceived to be infected with HIV.

HIV/AIDS stigma and discrimination can deter people from getting tested for the disease, can make them less likely to acknowledge their risk of infection, and can discourage those who are HIV-positive from discussing their HIV status with their sexual and needle-sharing partners. One U.S. study found that HIV-infected teens rarely disclose their HIV status to even their close friends because they believe that having HIV is stigmatizing and that disclosure would cause their friends to fear, judge, and/or reject them (Suris et al. 2007). Combating the stigma and discrimination against people who are affected by HIV/AIDS is crucial to improving care, quality of life, and emotional health for people living with HIV and AIDS and to reducing the number of new HIV infections.

Fighting antigay prejudice and discrimination is also important in efforts to support the well-being of individuals diagnosed with HIV/AIDS. In Africa, where about one-half of the nations have laws that criminalize same-sex sexual behavior, "fear of arrest prevents people from attending meetings or socializing in locations where their sexual identities become suspect. These are precisely the locations, however, where HIV prevention training, counseling, and materials (informational brochures, condoms...etc.) are available" (Johnson 2007, pp. 46–47).

Needle Exchange Programs. To reduce transmission of HIV among injection drug users, their sex partners, and their children, some countries and U.S. communities have established **needle exchange programs** (also known as syringe exchange programs), which provide new, sterile syringes in exchange for used, contaminated syringes. Many needle exchange programs also provide other social and health services, such as referrals to drug counseling and treatment, HIV testing and screening for other sexually transmissible diseases, hepatitis vaccinations, and condoms. The American Medical Association, the American Public Health Association, and the World Health Organization have endorsed needle exchange as an effective means of HIV prevention. Needle exchange programs also protect public health by providing safe disposal of potentially infectious syringes.

In Canada, sterile injection equipment is available to drug users in pharmacies and through numerous needle exchange programs. In contrast, most U.S. states prohibit the sale or possession of sterile needles or syringes without a medical prescription.

In 2007, 185 needle exchange programs were operating in 36 states (Centers for Disease Control and Prevention 2007). Although state and local public funding for needle exchange programs has increased in recent years, the United States is the only country in the world to explicitly ban the use of federal funds for needle exchange (Human Rights Watch 2005).

Cynthia Leshomo is the 2005 winner of the Miss HIV Stigma Free Pageant for HIV-positive women. First held in 2002, the Miss HIV pageant is a way of showing that HIV-positive individuals need not be ashamed and that with treatment; they can look good and lead productive lives.

Financial and Medical Aid to Developing Countries. Life-extending treatment for individuals infected with HIV is not affordable for many people in the developing world. Developing countries—those hardest hit by HIV/AIDS—depend on aid from wealthier countries to help provide medications, HIV/AIDS education programs, and condoms. In 2002, the United Nations helped to create the Global Fund to Fight AIDS, Tuberculosis, and Malaria to help poor countries fight these diseases. Total global spending on HIV rose from 300 million in 1996, to 13.7 billion in 2008, and the number of people receiving HIV treatment in poor countries has increased tenfold since 2002 (Kaiser Family Foundation 2009a).

Fighting the Growing Problem of Obesity

In general, reducing and preventing obesity requires encouraging people to (1) eat a diet with sensible portions, with lots of high-fiber fruits and vegetables and with minimal sugar and fat, and (2) engage in regular physical activity. Some of the strategies to achieve these goals include the following:

- *Restrictions on advertisements.* The food industry spends an enormous amount of money advertising to children. In response to concerns about childhood obesity, Ireland has banned advertising of candy and fast food on television, Great Britain has banned advertising for junk food on

needle exchange programs
Programs designed to reduce transmission of HIV by providing intravenous drug users with new, sterile syringes in exchange for used, contaminated syringes.

children's television programming, and Sweden and Norway prohibit advertising that targets children. In the 1970s and 1980s, the U.S. Federal Trade Commission considered restrictions on advertising for junk food aimed at children, but those efforts were opposed by the food and advertising industries.

- *Public education.* A variety of public education strategies are being used to inform the public about the importance of exercise and diet and their effects on health. France requires that advertisements promoting processed, sweetened, or salted food and drinks on television, radio, billboards, and the Internet must include one of four health messages: "For your health, eat at least five fruits and vegetables a day," "For your health, undertake regular physical activity," "For your health, avoid eating too much fat, too much sugar, too much salt," or "For your health, avoid snacking between meals" (Combes 2007).

  Because many Americans do not know how many calories, fat grams, sodium, and so on, are in the foods they eat, consumers are encouraged to read the nutritional labels on packaged foods. Menu labeling—the posting of nutrition information on menus and menu boards—is required in Seattle, New York City, and San Francisco, and several other localities and states have considered similar legislation (Trust for America's Health 2008). Proposed federal legislation called the Menu Education and Labeling (MEAL) Act, if passed, would require all chain restaurants to list nutritional information for all meals on the menu.

- *School nutrition and physical activity programs.* A number of states and school districts have implemented or considered bans or restrictions on vending machine sales of sugary drinks and "junk food" snacks. In an effort to increase consumption of produce, some schools have "Farm to School" programs that consist of school gardens and the opportunity to purchase fresh, locally grown produce to use in school breakfast and lunch programs. And many states have recess and/or physical activity requirements, although these requirements are often inadequate or not enforced.

  Proposed federal legislation to combat obesity includes the Child Nutrition Promotion and School Lunch Protection Act of 2006, which would redefine what are considered "foods of minimal nutritional value" and restrict the sale of such foods in schools. The Childhood Obesity Reduction Act would create a Congressional Council to Combat Childhood Obesity. This council would provide grants to schools to develop and implement programs to increase exercise and improve nutritional choices.

- *Interventions to treat obesity.* Interventions to treat obesity include weight-loss or fitness clubs, nutrition and weight-loss counseling, weight-loss medications, and surgical procedures. In 2004, the U.S. government classified obesity as an illness, so Medicare could cover obesity treatments. The proposed Medicaid Obesity Treatment Act would require Medicaid coverage of prescription drugs to treat obesity. Some private insurers also cover treatment for obesity; at least seven states require it (Cotton et al. 2007).

Strategies to Improve Mental Health Care

Two areas for improving mental health care in the United States are eliminating the stigma associated with mental illness and improving health insurance coverage for treating mental disorders.

Eliminating the Stigma of Mental Illness. The first White House conference on mental health called for a national campaign to eliminate the stigma associated with mental illness. Fearing the negative label of "mental illness" and the social rejection and stigmatization associated with mental illness, individuals are reluctant to seek psychological services (Komiya et al. 2000). In a study of eighth graders, boys were more likely than girls to agree with the statement, "Seeing a counselor for emotional problems makes people think you are weird or different" (Chandra & Minkovitz 2006). In the same study, more boys than girls (38 percent versus 23 percent) reported that they were not willing to use mental health services. The most frequently cited reason was "Too embarrassed by what other kids would say." To assess your attitudes toward seeking professional psychological help, see this chapter's Self and Society feature.

Reducing the stigma associated with mental illness might be achieved through encouraging individuals to seek treatment and making treatment accessible and affordable. The surgeon general's report on mental health explained the following:

> Effective interventions help people to understand that mental disorders are not character flaws but are legitimate illnesses that respond to specific treatments, just as other health conditions respond to medical interventions. (U.S. Department of Health and Human Services 1999, p. viii)

The National Alliance on Mental Illness (NAMI) has a Stigma-Busters campaign, whereby the public submits instances of media content that stigmatize individuals with mental illness to StigmaBusters, which then conveys their concerns to media organizations and corporations, urging them to avoid stigmatizing portrayals of mental illness.

"Real Men, Real Depression" is an anti-stigma public education campaign that includes print, television, and radio public service announcements. "Breaking the Silence," a curriculum for elementary, middle, and high schools available through NAMI, uses true stories, activities, a board game, and posters to debunk myths about mental illness and sensitize students to the pain that words such as *psycho* and *schizo* and frightening or comic media images of mentally ill people can cause (Harrison 2002). The Department of Defense has launched an anti-stigma campaign called "Real Warriors, Real Battles, Real Strength" designed to assure military personnel that seeking mental health treatment will not harm their career and to publicize stories of military personnel who have been successfully treated for mental health problems (Dingfelder 2009b). Effective anti-stigma campaigns not only focus on eradicating negative stereotypes of people with mental illness, but also emphasize the positive accomplishments and contributions of people with mental illness.

Fearing the negative label of "mental illness" and the social rejection and stigmatization associated with mental illness, individuals are reluctant to seek psychological services.

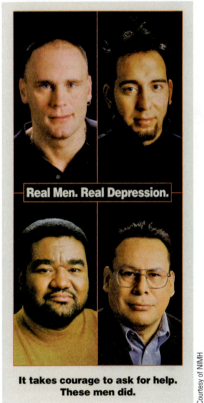

Real Men. Real Depression.

It takes courage to ask for help. These men did.

Courtesy of NIMH

This public education brochure on men and depression is available from the National Institute of Mental Health (NIMH), www.nimh.nih.gov.

What Do You Think? The Pentagon has ruled that the Purple Heart medal given to soldiers wounded or killed in combat may not be awarded to war veterans with post-traumatic stress disorder (PTSD) because it is not a physical wound (Alvarez & Eckholm 2009). Do you think veterans with PTSD should be eligible to receive the Purple Heart medal? Why or why not?

Eliminating Inequalities in Health Care Coverage for Mental Disorders. In 2008, President G. W. Bush signed landmark legislation that requires health insurance plans to treat mental illness and physical illness equally—a concept known as **parity**. The new law prohibits higher co-pays or deductibles for mental health and substance abuse treatment. The law also prohibits insurance plans from lowering mental illness benefit levels or limiting the number of therapy sessions or hospital treatment days. The parity law has significant limitations; it does not require insurance companies to offer mental health treatment coverage, or to cover every condition listed in the American Psychiatric Association's *Diagnostic and Statistical Manual of Mental Disorders* (DSM-IV) (Jenkins 2008). The parity law also does not apply to employers with fewer than 50 employees. Some states have enacted mental health parity laws, which vary in their scope and application, but many of these laws do not address substance abuse, are limited to the more serious mental illnesses, or apply only to government employees.

parity In health care, a concept requiring equality between mental health care insurance coverage and other health care coverage.

U.S. State and Federal Health Care Reform

The United States is the only country in the industrialized world that does not have any mechanism for guaranteeing health care to its citizens. Other countries, such as Canada, Great Britain, France, Sweden, Germany, and Italy, have national health insurance systems that provide **universal health care** or socialized medicine—health care to all citizens. National health insurance is typically administered and paid for by government. Despite differences in how national health insurance works in various countries, typically, the government (1) directly controls the financing and organization of health services, (2) directly pays providers, (3) owns most of the medical facilities (Canada is an exception), (4) guarantees universal access to health care, and (5) allows private care for individuals who are willing to pay for their medical expenses (Cockerham 2007). Most countries with national health insurance allow or encourage private insurance as an upgrade to a higher class of service and a fuller range of services (Quadagno 2004). To the extent that health care is rationed in countries with national health insurance, rationing is done on the basis of medical need, not ability to pay.

The goals of health care reform efforts in the United States generally fall into one of three categories: (1) creation of a universal health program; (2) expansion of existing government health insurance programs; and (3) making private insurance more affordable through tax credits or deductions, and other means. Various health care reform efforts are being made at the state and federal levels.

State-Level Health Care Reform. In 2006, Massachusetts passed landmark legislation requiring residents to be insured, similar to the requirement that all automobiles must be insured. The plan expanded Medicaid coverage for the poor, provided subsidized coverage for those who are near-poor, and mandated that middle-income people without insurance either purchase private insurance (using pretax dollars) through their employers or through a new state agency—the Commonwealth Care Health Insurance Connector. Individuals who remain uninsured face monetary penalties, as do employers who fail to provide health insurance for their employees. Although at least half of the previously uninsured in Massachusetts now have health insurance, the reform has failed to achieve universal coverage, and it has been more expensive than expected (Nadin et al. 2009). Nevertheless, other states have also taken steps to increase health insurance coverage of their residents. Some states have increased benefits to Medicaid clients who demonstrate healthy behavior, such as showing up for doctor's appointments, getting their children immunized, or following disease management programs. Some states offer tax credits to small businesses that offer insurance to their employees, and others offer small businesses and people without insurance discounted coverage through the state.

Federal Health Care Reform. Since 1912, when Theodore Roosevelt first proposed a national health insurance plan, the Truman, Nixon, Carter, Clinton, and Obama administrations have advocated the idea of health care for all Americans. The majority of U.S. adults (61 percent) favors the government guaranteeing health insurance for all citizens, even if it means raising taxes (Pew Research Center 2009).

A number of federal measures to reform health care have been proposed. The National Health Insurance Act, introduced by Representative John Conyers, would expand Medicare to every U.S. resident, creating a **single-payer health care** system in which a single tax-financed public insurance program replaces private insurance companies. Under this plan, all U.S. residents would be issued a national health insurance card, would receive all medically

universal health care
A system of health care, typically financed by the government, that ensures health care coverage for all citizens.

single-payer health care
A health care system in which a single tax-financed public insurance program replaces private insurance companies.

necessary services (including dentistry, eye care, mental health services, substance abuse treatment, prescription drugs, and long-term care), would have no co-payments or deductibles, and would see the doctor of their choice. If a single-payer plan is adopted, it is estimated to save enough on administrative costs to provide coverage for all those without insurance and to substantially help the underinsured.

There is considerable support for single-payer health care among doctors, nurses, and the general public. The insurance industry opposes the adoption of such a system because the private health insurance industry would be virtually eliminated, and so spends a great deal of money on lobbying, political contributions, and public relations to influence the health reform debate (see this chapter's The Human Side feature). During Senate Finance Committee hearings on health care reform in May 2009, 41 witnesses were called to testify, but not one was a single-payer advocate. Thirteen doctors, lawyers, nurses, and other advocates stood up before Senator Max Baucus, chairman of the committee, and demanded that a single-payer advocate be allowed to testify. Baucus refused to hear their testimony, and the 13 single-payer advocates, known as "the Baucus 13," were handcuffed, arrested, and charged with "disruption of Congress" (Single Payer Action 2009). Not surprisingly, Senator Baucus, who has played a leading role in health reform efforts in Congress, has received more campaign contributions from health insurance and pharmaceutical industries than any other current Democrat in Congress (Single Payer Action 2009).

What Do You Think? In 2005, Representative Pete Stark (D-California) proposed an amendment to the U.S. Constitution to guarantee health care as a right for every American. Stark argued that "the health of every American is vital to their unalienable rights of 'life, liberty, and the pursuit of happiness.' ... To ensure these rights are fully enjoyed, we must be certain that every American can access quality health care—regardless of their income, race, education or job status" ("Stark Introduces Constitutional Amendment" 2005). Do you favor or oppose amending the Constitution to guarantee health care as a right for every American?

At the time of this writing, President Obama is asking the U.S. Congress to pass health care reform legislation that would create a public health insurance option. Many are threatened by the prospect of government intervention in health care, particularly the insurance industry, which has consistently opposed health insurance reform. Yet many of the problems discussed in this chapter stem from for-profit corporations who, ironically, warn that a government-run insurance program will lead to the very conditions corporatized medicine has created: higher prices, less choice, rationing, and excessive bureaucracy (Nader 2009).

Some have argued that partisan politics have played a role in blocking health care reform, citing as evidence remarks made by Republican Senator Jim DeMint, who, in referring to Obama's health reform proposal said, "If we're able to stop Obama on this, it will be his Waterloo. It will break him" (quoted in Moyers & Winship 2009). Skepticism over health care reform also stems from questions about how much reform will cost and who will pay for it. Although the answers to the questions of cost are still being worked out, what is certain is that we are already paying a high price for the system we have and, in the words of Obama, "If we don't change, we can't expect a different result."

Understanding Problems of Illness and Health Care

Although human health has probably improved more over the past half-century than over the previous three millennia, the gap in health between rich and poor remains wide, and the very poor suffer appallingly (Feachum 2000). Poor countries need economic and material assistance to alleviate problems such as HIV/AIDS, high maternal and infant mortality rates, and malaria. The wealthy countries of the world do have resources to make a difference in the health of the world. When a tsunami took the lives of thousands of people in 2004, media attention to the tragedy elicited an outpouring of aid throughout the world to help affected regions. Meanwhile, malaria, which has been referred to as "a silent tsunami," takes the lives of more than 150,000 African children each month, which is about the same as the death toll of the southern Asia tsunami disaster (Sachs 2005). Yet malaria continues to receive comparatively scant public attention even though it is largely preventable with $5 mosquito bed nets, and treatable with medicines at roughly $1 per dose.

Although poverty may be the most powerful social factor affecting health, other social factors that affect health include globalization, increased longevity, family structure, gender, education, and race or ethnicity. Although individuals make choices—choices such as whether to smoke, exercise, eat a healthy diet, engage in risky sexual activity, use condoms, wear a seat belt, and so on—that affect their health those choices are also influenced by social, economic, and political forces that must be addressed if the goal is to improve the health not only of individuals but also of entire populations. By focusing on individual behaviors that affect health and illness, we often overlook social causes of health problems (Link & Phelan 2001).

A sociological view of illness and health care looks not only at the social causes, but also of the social *consequences* of health problems—consequences that potentially affect us all. In *Uninsured in America* (2005), Sered and Fernandopulle explain:

> If millions of American children do not have reliable, basic health care, all children who attend American schools are at risk through daily exposure to untreated disease. If millions of restaurant and food industry workers do not have health insurance, people preparing food and waiting tables are sharing their health problems with everyone they serve. . . . If tens of millions of Americans go without basic and preventive care, we all pay the bill when their health problems turn into complex medical emergencies necessitating expensive . . . treatment. (p. 20)

Although certain changes in medical practices and policies may help to improve world health, "the health sector should be seen as an important, but not the sole, force in the movement toward global health" (Lerer et al. 1998, p. 18). Just as the social causes of problems of illness and health care are diverse, so must be the social solutions. Thus, a comprehensive approach to improving the health of a society requires addressing diverse issues such as poverty and economic inequality,

Courtesy of PR Watch, Center for Media and Democracy

Wendell Potter is the former vice president of corporate communications at CIGNA, one of the largest health insurance companies in the United States. In the following account, Mr. Potter describes how the insurance industry influences the health reform debate and explains why he decided to speak out against the insurance industry.

I'm the former insurance industry insider now speaking out about how big for-profit insurers have hijacked our health care system and turned it into a giant ATM for Wall Street investors, and how the industry is using its massive wealth and influence to determine what is (and is not) included in the health care reform legislation members of Congress are now writing.

Although . . . I had a great career in the insurance industry (four years at Humana and nearly 15 at CIGNA), in recent years I had grown increasingly uncomfortable serving as one of the industry's top PR executives. My responsibilities at CIGNA. . . . included serving as the company's chief spokesman to the media on all corporate and financial matters. I also served on a lot of trade association committees and industry-financed coalitions, many of which were essentially front groups for insurers. So I was in a unique position to see not only how Wall Street analysts and investors influence decisions insurance company executives make but also how the industry has carried out behind-the-scenes PR and lobbying campaigns to kill or weaken any health care reform efforts that threatened insurers' profitability.

I also have seen how the industry's practices . . . have contributed to the tragedy of nearly 50 million people being uninsured as well as to the growing number of Americans who, because insurers now require them to pay thousands of dollars out of their own pockets before their coverage kicks in—are underinsured. An estimated 25 million of us now fall into that category.

What I saw happening over the past few years was a steady movement away from the concept of insurance and toward "individual responsibility," a term used a lot by insurers. . . . This is playing out as a continuous shifting of the financial burden of health care costs away from insurers and employers and onto the backs of individuals. As a result, more and more sick people are not going to the doctor or picking up their prescriptions because of costs. If they are unfortunate enough to become seriously ill or injured, many people enrolled in these plans find themselves on the hook for such high medical bills that they are losing their homes to foreclosure or being forced into bankruptcy.

As an industry spokesman, I was expected to put a positive spin on this trend that the industry created and euphemistically refers to as "consumerism" and to promote so-called "consumer-driven"

gender inequality, population growth, environmental issues, education, housing, energy, water and sanitation, agriculture, and workplace safety.

Improving the health of the world also means seeking nonmilitary solutions to international conflicts. In addition to the deaths, injuries, and illnesses that result from combat, war diverts economic resources from health programs, leads to hunger and disease caused by the destruction of infrastructure, causes psychological trauma, and contributes to environmental pollution (Sidel & Levy 2002). Thus "the prevention of war . . . is surely one of the most critical steps mankind can make to protect public health" (White 2003, p. 228).

The World Health Organization (1946) defined **health** as "a state of complete physical, mental, and social well-being" (p. 3). Based on this definition, we conclude this chapter with the suggestion that the study of social problems is, essentially, the study of health problems, as each social problem is concerned with the physical, mental, and social well-being of humans and the social groups of which they are a part. As you read other chapters in this book, consider how the problems in each chapter affect the health of individuals, families, populations, and nations.

health According to the World Health Organization, "a state of complete physical, mental, and social well-being."

health plans....I thought I could live with being a well-paid huckster and hang in there a few more years until I could retire. I probably would have if I hadn't made a . . . decision a couple of years ago that changed the direction of my life. While visiting my folks in northeast Tennessee . . . , I read in the local paper about a health "expedition" being held that weekend a few miles up U.S. 23 in Wise, Virginia. Doctors, nurses, and other medical professionals were volunteering their time to provide free medical care to people who lived in the area. . . . Remote Area Medical, a nonprofit group whose original mission was to provide free care to people in remote villages in South America, was organizing the expedition. I decided to check it out. . . .

Nothing could have prepared me for what I saw when I reached the Wise County Fairgrounds, where the expedition was being held. Hundreds of people had camped out all night in the parking lot to be assured of seeing a doctor or dentist when the gates opened. By the time I got there, long lines of people stretched from every animal stall and tent where the volunteers were treating patients.

That scene was so visually and emotionally stunning, it was all I could do to hold back tears. How could it be that citizens of the richest nation in the world were being treated this way?

A couple of weeks later I was boarding a corporate jet. . . . When the flight attendant served my lunch on gold-rimmed china and gave me a gold-plated knife and fork to eat it with, I realized for the first time that someone's insurance premiums were paying for me to travel in such luxury. I also realized that one of the reasons those people in Wise County had to wait in long lines to be treated in animal stalls was because our Wall Street-driven health care system has created one of the most inequitable health care systems on the planet.

Although I quit my job last year, I did not make a final decision to speak out as a former insider until recently when it became clear to me that the insurance industry and its allies (often including drug and medical device makers, business groups, and even the American Medical Association) were succeeding in shaping the current debate on health care reform. . . . Remember this: Whenever you hear a politician or pundit use the term *government-run health care* and warn that the creation of a public health insurance option that would compete with private insurers (or heaven forbid, a single-payer system like the one Canada has) will "lead us down the path to socialism," know that the original source of the sound bite most likely was some flack like I used to be.

Bottom line: I ultimately decided the stakes are too high for me to just sit on the sidelines and let the special interests win again. So I have joined forces with thousands of other Americans who are trying to persuade our lawmakers to listen to us for a change, not just to the insurance and drug company executives who are spending millions to shape reform to benefit them and the Wall Street hedge fund managers they are beholden to. . . .

The people of Wise County and every county deserve much better than to be left behind to suffer or die ahead of their time due to Wall Street's efforts to keep our government from ensuring that all Americans have real access to first-class health care.

Source: Potter 2009.

CHAPTER REVIEW

■ **What are three measures that serve as indicators of the health of populations? Which health measure reveals the greatest disparity between developed and developing countries?**

Measures of health that serve as indicators of the health of populations include morbidity, life expectancy, and mortality rates (including infant and under-5 childhood mortality rates and maternal mortality rates). Maternal mortality rates reveal the greatest disparity between developed and developing countries.

■ **Which theoretical perspective criticizes the pharmaceutical and health care industry for placing profits above people?**

The conflict perspective criticizes the pharmaceutical and health care industry for placing profits above people. For example, pharmaceutical companies' research and development budgets are spent not according to public health needs but rather according to calculations about maximizing profits. Because the masses of people in developing countries lack the resources to pay high prices for medication, pharmaceutical companies do not see the development of drugs for diseases of poor countries as a profitable investment.

■ **Where is HIV/AIDS most prevalent in the world?**
HIV/AIDS is most prevalent in Africa, particularly sub-Saharan Africa, where one in twenty adults is infected with HIV. Two-thirds of people with HIV live in sub-Saharan Africa.

■ **What is the second biggest cause of preventable deaths in the United States (second only to tobacco)?**
Obesity, which can lead to heart disease, hypertension, diabetes, and other health problems, is the second biggest cause of preventable deaths in the United States.

■ **Why is mental illness referred to as a "hidden epidemic"?**
Mental illness is a "hidden epidemic" because the shame and embarrassment associated with mental problems discourage people from acknowledging and talking about mental illness. Because male gender expectations associate masculinity with emotional strength, men are particularly prone to deny or ignore mental problems.

■ **What features of globalization have contributed to health problems?**
Features of globalization that have been linked to problems in health are increased travel and the expansion of trade and transnational corporations.

■ **What are some health concerns related to modern food animal production?**
Health concerns related to modern food animal production include (1) higher levels of saturated fat in corn-fed beef; (2) use of antibiotics in animal feed, which contributes to super-resistant bacterial infections; (3) Creutzfeldt-Jakob disease, the human form of mad cow disease; (4) possible health risks associated with growth hormones in milk; (5) *E. coli* contamination of meat; and (6) contamination of water and air by animal manure.

■ **According to a World Health Organization analysis of the world's health systems, which country provides the best overall health care?**
The World Health Organization found that France provides the best overall health care among major countries, followed by Italy, Spain, Oman, Austria, and Japan. The United States ranked 37 out of 191 countries, despite the fact that the United States spends a higher portion of its gross domestic product on health care than any other country.

■ **What is the difference between selective primary health care and comprehensive primary health care?**
Selective primary health care focuses on using specific interventions to target specific health problems, such as promoting condom use to prevent HIV infections and providing immunizations against childhood diseases to promote child survival. In contrast, comprehensive primary health care focuses on the broader social determinants of health, such as poverty and economic inequality, gender inequality, environment, and community development.

■ **What are the categories of health care reform efforts in the United States?**
The goals of health care reform efforts in the United States generally fall into one of three categories: (1) creation of a universal health program; (2) expansion of existing government health insurance programs; and (3) implementing tax incentives and other strategies to make private insurance more affordable.

■ **How does the World Health Organization define health?**
Health, according to the World Health Organization, is "a state of complete physical, mental, and social well-being." Based on this definition, we suggest that the study of social problems is, essentially, the study of health problems, because each social problem is concerned with the physical, mental, and social well-being of humans and the social groups of which they are a part.

TEST YOURSELF

1. Worldwide, the leading cause of death is
   a. HIV/AIDS
   b. traffic accidents
   c. heart disease
   d. cancer
2. The United States has the lowest infant mortality rate in the world.
   a. True
   b. False

3. Worldwide, the predominant mode of HIV transmission is through
   a. heterosexual contact
   b. needle-sharing
   c. same-sex sexual contact
   d. blood transfusions
4. In the United States, _____ of adults are either overweight or obese.
   a. 10 percent
   b. One-third
   c. Two-thirds
   d. 20 percent

5. What age group has the highest rate of "serious psychological stress?"
   a. 12 to 17
   b. 18 to 25
   c. 26 to 64
   d. 65 and over
6. The United States spends far more per person on health care than does any other industrialized nation.
   a. True
   b. False
7. All U.S. children living below the poverty line are covered by publicly funded health insurance.
   a. True
   b. False
8. Of all age groups in the United States, young adults ages 18 to 24 are the least likely to have health insurance.
   a. True
   b. False
9. The United States is the only country in the industrialized world that does not have any mechanism for guaranteeing health care to its citizens.
   a. True
   b. False
10. Why did the "Baucus 13" get arrested?
    a. They illegally imported prescription drugs from Canada and sold them for a profit.
    b. They sold bogus health insurance to the elderly.
    c. They knew they were HIV-infected and they did not inform their sexual partners, who later became infected with HIV.
    d. They disrupted Congress by demanding that an advocate of single-payer health insurance be allowed to testify at senate hearings on health care reform.

Answers: 1, c; 2, b; 3, a; 4, c; 5, b; 6, a; 7, b; 8, a; 9, a; 10, d.

KEY TERMS

MEDIA RESOURCES

Understanding Social Problems,
Seventh Edition Companion Website
www.cengage.com/sociology/mooney

Visit your book companion website, where you will find flash cards, practice quizzes, Internet links, and more to help you study.

Just what you need to know NOW! Spend time on what you need to master rather than on information you already have learned. Take a pre-test for this chapter, and CengageNOW will generate a personalized study plan based on your results. The study plan will identify the topics you need to review and direct you to online resources to help you master those topics. You can then take a post-test to help you determine the concepts you have mastered and what you will need to work on. Try it out! Go to www.cengage.com/login to sign in with an access code or to purchase access to this product.

Kwame Zikomo/SuperStock

Alcohol and Other Drugs

Substance abuse, the nation's number one preventable health problem, places an enormous burden on American society, harming health, family life, the economy, and public safety, and threatening many other aspects of life.

Robert Wood Johnson Foundation, Institute for Health Policy, Brandeis University

Carson loved the outdoors. He ran his first 10K race at six, finishing second in his age group (Osborn 2009). He lettered in tennis each of his four years in high school, ran on the cross-country team, played lacrosse, and was an avid cycle enthusiast (Osborn 2009). But Carson Starkey wasn't just an athlete; he graduated in the top 10 percent of his high school class and attended California Polytechnic State University, one of the top architectural engineering schools in the country. Like thousands of other students, Carson came home over Thanksgiving vacation his freshman year. It was the last time his family would ever see him.

On December 2, just hours after attending a fraternity party, Carson Starkey died of alcohol poisoning. According to one observer who had gone through a similar initiation process, a typical pledge drinks "a 44-ounce can of beer, 20 to 24 ounces of hard alcohol, and likely an additional 20 to 24 ounces of a 20-proof chaser within a couple of hours" (Wilson 2009, p. 1). An autopsy report concluded that Carson Starkey's blood alcohol level was between 0.39 and 0.45. Police have officially declared Carson's death a result of criminal hazing.

Drug-induced death is just one of the many negative consequences that can result from alcohol and drug abuse. The abuse of alcohol and other drugs is a social problem when it interferes with the well-being of individuals and/or the societies in which they live—when it jeopardizes health, safety, work and academic success, family, and friends. But managing the drug problem is a difficult undertaking. In dealing with drugs, a society must balance individual rights and civil liberties against the personal and social harm that drugs promote—crack babies, suicide, drunk driving, industrial accidents, mental illness, unemployment, and teenage addiction. When to regulate, what to regulate, and who should regulate are complex social issues. Our discussion begins by looking at how drugs are used and regulated in other societies.

The Global Context: Drug Use and Abuse

Pharmacologically, a **drug** is any substance other than food that alters the structure or functioning of a living organism when it enters the bloodstream. Using this definition, everything from vitamins to aspirin is a drug. Sociologically, the term *drug* refers to any chemical substance that (1) has a direct effect on the user's physical, psychological, and/or intellectual functioning; (2) has the potential to be abused; and (3) has adverse consequences for the individual and/ or society. Societies vary in how they define and respond to drug use. Thus, drug use is influenced by the social context of the particular society in which it occurs.

drug Any substance other than food that alters the structure or functioning of a living organism when it enters the bloodstream.

Drug Use and Abuse around the World

Globally, 5 percent of the world's population between the ages of 15 and 64—208 million people—reported using at least one illicit drug in the previous year (WDR 2008). About half that number reported *regular* illicit drug use, that is, use at least once a month. According to the most recent report, cannabis (i.e., marijuana and hashish) remains by far the most widely used illegal drug, with 3.9 percent of the world's population between 15 and 64 (166 million people) using the drug at least once in the last 12 months. By comparison, only 0.8 percent of the world's 15 to 64 population (34 million) is estimated to have used amphetamine-type stimulants (e.g., Ecstasy) annually.

> Over two-thirds of all smokers in the world live in 10 countries, with China alone having nearly 30 percent of the global smoking population.

The prevalence of drug use also varies by country. According to a survey conducted by the World Health Organization, alcohol use is much higher in the Americas, Europe, New Zealand, and Japan than in the Middle East, Africa, and China (Degenhardt et al. 2008). Tobacco use also varies by region. Over two-thirds of all smokers in the world live in 10 countries, with China alone having nearly 30 percent of the global smoking population (WHO 2008).

Illicit drug use varies by region as well. For example, the annual prevalence of cannabis ranges from less than 1 percent of the 15- to 64-year-old population in Malta to 17 percent in Canada (WDR 2008; EMCDDA 2008). Annual prevalence of cocaine use varies from 0.1 percent (e.g., in Greece) to 3 percent (e.g., in Spain), and annual Ecstasy use ranges from 0.2 percent in Greece to 1.8 percent of the 15- to 64-year-old population in the United Kingdom (WDR 2008; EMCDDA 2008). Moreover, lifetime use of *any* illicit drug by 15- and 16-year-olds ranges from 46 percent in the Czech Republic, to 5 percent in Armenia, with U.S. rates at 34.1 percent (ESPAD 2009; MTF 2009). Finally, when a sample of U.S. tenth graders were asked about alcohol and cigarette use, 58.3 percent and 31.7 percent respectively reported using the drug at least once in their lifetime compared to 89.0 percent and 58.0 percent respectively of European 15- and 16-year-olds (ESPAD 2009; MTF 2009).

Some have argued that differences in drug use can be attributed to variations in drug policies. The Netherlands, for example, has had an official government policy of treating the use of drugs such as marijuana and hashish as a public health issue rather than a criminal justice issue since the mid-1970s. Treatment of the drug user and prevention of future drug use are prioritized over the more punitive response of imprisonment found in many other countries.

In the first decade of the policy, drug use did not appear to increase. However, increases in marijuana use were reported in the early 1990s, with the advent of "cannabis cafés." These coffee shops sell small amounts of marijuana

There are over 1,000 cannabis cafes in the Netherlands. As pictured above, each cafe or coffee shop as they are sometimes called has a menu identifying the price per quantity for various types of marijuana and hashish. Other legal cannabis products in the Netherlands include cannabis beer, candy, biscuits, chocolate and tea.

Paul Rogers/ Getty Images

for personal use and, presumably, prevent casual marijuana users from coming into contact with "hard drug" dealers (MacCoun & Reuter 2001; Drug Policy Alliance 2003). Some evidence suggests that marijuana use among Dutch youth is decreasing, rebutting those who would argue that liberal drug policies result in increased drug abuse (Sheldon 2000; Burke 2006). Nonetheless, concerns over the increasing strength of marijuana and the millions of drug tourists who visit the Netherlands annually have led to two recent proposals: (1) close some or all of the coffee shops, and (2) establish government run cannabis plantations to ensure the quality of the drug, regulate prices, and crack down on illegal suppliers (Mock 2008).

Great Britain has also adopted a "medical model," particularly in regard to heroin and cocaine. As early as the 1960s, English doctors prescribed opiates and cocaine for their drug-addicted patients who were unlikely to quit using drugs on their own and for the treatment of withdrawal symptoms. By the 1970s, however, British laws had become more restrictive, making it difficult for either physicians or users to obtain drugs legally. Today, British government policy provides for limited distribution of drugs by licensed drug treatment specialists to addicts who might otherwise resort to crime to support their habits (Abadinsky 2008). In recent years, Canada has been moving toward a medical model of drug abuse. For example, Vancouver has the "first supervised heroin-injection clinic in North America" and there are plans to expand such sites to Montreal and Quebec City (*The Economist* 2008).

What Do You Think? On February 23, 2009, Tom Ammiano, a California State Assembly member, introduced legislation that would "regulate the cultivation and sale of marijuana, and then tax it" (Mieszkowski 2009, p. 1). According to U.S. Department of Agriculture (USDA) estimates, marijuana is California's largest cash crop, bringing in $14 billion a year (Stateman 2009). If passed, the Marijuana Control, Regulation, and Education Act would allow California to control marijuana in the same way it controls alcohol and, in the same way, would generate much needed state revenues. Do you think marijuana should be legalized? If so, what age limits, if any, would you impose?

In stark contrast to such health-based policies, many other countries execute drug users and/or dealers or subject them to corporal punishment, which may include whipping, stoning, beating, and torture. Such policies are found primarily in less developed nations, where religious and cultural prohibitions condemn the use of any type of drug, including alcohol and tobacco. For example, convicted drug offenders in Bangladesh, China, India, Laos, and Pakistan are subject to severe punishments including the threat of death (CNN 2009).

Drug Use and Abuse in the United States

In the United States, cultural definitions of drug use are contradictory—condemning it on the one hand (e.g., heroin), yet encouraging and tolerating it on the other (e.g., alcohol). At various times in U.S. history, many drugs that are illegal today were legal and readily available. In the 1800s and the early 1900s, opium was routinely used in medicines as a pain reliever, and morphine was taken as a treatment for dysentery and fatigue. Amphetamine-based inhalers were legally available until 1949, and cocaine was an active ingredient in Coca-Cola until 1906, when it was replaced with another drug—caffeine (Witters et al. 1992; Abadinsky 2008). Not surprisingly, Americans' concerns with drugs have varied over the years. In the 1970s, when drug use was at its highest, concern over drugs was relatively low. However, in 2007,

DRUG	TIME PERIOD		
	LIFETIME	PAST YEAR	PAST MONTH
Any illicit drug*	46.1	14.4	8.0
Marijuana and hashish	40.6	10.1	5.8
Cocaine	14.5	2.3	0.8
Crack	3.5	0.6	0.2
Heroin	1.5	0.1	0.06
Hallucinogens	13.8	1.5	0.4
LSD	9.1	0.3	0.1
PCP	2.5	0.1	0.0
Ecstasy	5.0	0.9	0.2
Inhalants	9.1	0.8	0.2
Nonmedical use of any psychotherapeutic drug†	20.3	6.6	2.8
Pain relievers	13.3	5.0	2.1
Tranquilizers	8.2	2.1	0.7
Stimulants	8.7	1.2	0.4
Methamphetamine	5.3	0.5	0.3
Sedatives	3.4	0.3	0.1
Cigarettes	65.3	28.5	24.2
Alcohol	N/A	N/A	51.1

*"Any illicit drug" includes marijuana/hashish, cocaine (including crack), heroin, hallucinogens, inhalants, or any prescription-type psychotherapeutic used nonmedically.

†"Nonmedical use" of any prescription-type pain reliever, tranquilizer, stimulant, or sedative; does not include over-the-counter drugs.

N/A = not available

Source: NSDUH 2008.

when a sample of Americans was asked to assess the drug problem in the United States, 73 percent described the problem as either "extremely" or "very" serious (Carroll 2007).

As Table 3.1 indicates, use of illicit drugs in a person's lifetime is fairly common. Of people 12 years old and older, 46.1 percent reported using an illicit drug sometime in their lives. In 2007, marijuana and hashish had the highest occurrence in lifetime use (40.6 percent), with heroin having the lowest use (1.5 percent). Despite these relatively high numbers, particularly for marijuana and hashish, use of alcohol and tobacco is much more widespread than use of illicit drugs. Half of Americans age 12 and older reported being *current* alcohol drinkers, and an estimated 71 million Americans 12 and older reported *current* use of a tobacco product (NSDUH 2008).

Sociological Theories of Drug Use and Abuse

drug abuse The violation of social standards of acceptable drug use, resulting in adverse physiological, psychological, and/or social consequences.

chemical dependency A condition in which drug use is compulsive and users are unable to stop because of physical and/or psychological dependency.

Drug abuse occurs when acceptable social standards of drug use are violated, resulting in adverse physiological, psychological, and/or social consequences. For example, when an individual's drug use leads to hospitalization, arrest, or divorce, such use is usually considered abusive. Drug abuse, however, does not always entail drug addiction. Drug addiction, or **chemical dependency**, refers to a condition in which drug use is compulsive—users are unable to stop because of their dependency. The dependency may be psychological (the individual needs the drug to achieve a feeling of well-being) and/or physical (withdrawal symptoms occur when the individual stops taking the drug). For example, withdrawal from marijuana includes depression, anger, decreased appetite, and restlessness (Zickler 2003). In 2007, more than 22 million Americans, 9 percent of the population 12 or older, were defined as being dependent on or abusers of alcohol and/or other drugs. Of that number, 15.5 million (70.4 percent) were dependent on or abused alcohol only, 3.7 million (16.8 percent) were dependent on or abused illicit drugs but not alcohol, and 3.2 million (14.5 percent) were dependent on or abused both illicit drugs and alcohol (NSDUH 2008). Individuals who are dependent on or abuse illicit drugs and/or alcohol are disproportionately male, American Indians, or Alaska natives between the ages of 18 and 25 (NSDUH 2008).

Various theories provide explanations for why some people use and abuse drugs. Drug use is not simply a matter of individual choice. Theories of drug use explain how structural and cultural forces as well as biological and psychological factors influence drug use and society's responses to it.

Structural-Functionalist Perspective

Structural functionalists argue that drug abuse is a response to weakening societal norms. As society becomes more complex and as rapid social change occurs, norms and values become unclear and ambiguous, resulting in anomie—a state of normlessness. Anomie may exist at the societal level, resulting in social strains and inconsistencies that lead to drug use. For example, research indicates that increased alcohol consumption in the 1830s and the 1960s was a response to rapid social change and the resulting stress (Rorabaugh 1979). Anomie produces inconsistencies in cultural norms regarding drug use. For example, although public health officials and health care professionals warn of the dangers of alcohol and tobacco use, advertisers glorify the use of alcohol and tobacco, and the U.S. government subsidizes the alcohol and tobacco industries. Furthermore, cultural traditions, such as giving away cigars to celebrate the birth of a child and toasting a bride and groom with champagne, persist.

Anomie may also exist at the individual level, as when a person suffers feelings of estrangement, isolation, and turmoil over appropriate and inappropriate behavior. An adolescent whose parents are experiencing a divorce, who is separated from friends and family as a consequence of moving, or who lacks parental supervision and discipline may be more vulnerable to drug use because of such conditions. Thus, from

> Anomie produces inconsistencies in cultural norms regarding drug use.... [P]ublic health officials and health care professionals warn of the dangers of alcohol and tobacco use, advertisers glorify the use of alcohol and tobacco, and the U.S. government subsidizes the alcohol and tobacco industries.

a structural-functionalist perspective, drug use is a response to the absence of a perceived bond between the individual and society and to the weakening of a consensus regarding what is considered acceptable.

Consistent with this perspective, in a national poll of Americans 18 years or older, peer pressure and lack of parental supervision were the two most common responses given for why teenagers take drugs (Pew Research Center 2002). Similarly, the importance of the family in deterring drug use is highlighted in the national youth media campaign—"Parents. The Anti-Drug" (ONDCP 2009).

Conflict Perspective

The conflict perspective emphasizes the importance of power differentials in influencing drug use behavior and societal values concerning drug use. From a conflict perspective, drug use occurs as a response to the inequality perpetuated by a capitalist system. Societal members, alienated from work, friends, and family as well as from society and its institutions, turn to drugs as a means of escaping the oppression and frustration caused by the inequality they experience. Furthermore, conflict theorists emphasize that the most powerful members of society influence the definitions of which drugs are illegal and the penalties associated with illegal drug production, sales, and use.

This poster from the Office of National Drug Control Policy's National Youth Anti-Drug Media Campaign emphasizes the importance of a close relationship between parent and child in the fight against drug use by youths.

For example, alcohol is legal because it is often consumed by those who have the power and influence to define its acceptability—white males (NSDUH 2008). This group also disproportionately profits from the sale and distribution of alcohol and can afford powerful lobbying groups in Washington D.C. to guard the alcohol industry's interests. Because this group also commonly uses tobacco and caffeine, societal definitions of these substances are also relatively accepting. Conversely, minority group members disproportionately use crack cocaine rather than powder cocaine (Mauer 2009). Although the pharmacological properties of the two drugs are the same, possession of 5 grams of crack cocaine carries the same penalty under federal law as possession of 500 grams of powdered cocaine (Taifia 2006). In 2007, recognizing the disparity, the U.S. Supreme Court held that federal judges have the power to give "reasonably" shorter sentences than federal guidelines would indicate (Mears 2007).

The use of opium by Chinese immigrants in the 1800s provides a historical example. The Chinese, who had been brought to the United States to work on the railroads, regularly smoked opium as part of their cultural tradition. As unemployment among white workers increased, however, so did resentment of Chinese laborers. Attacking the use of opium became a convenient means of attacking the Chinese, and in 1877, Nevada became the first of many states to prohibit opium use. As Morgan (1978) observed:

The first opium laws in California were not the result of a moral crusade against the drug itself. Instead, it represented a coercive action directed against a vice that was merely an appendage of the real menace—the Chinese—and not the Chinese per se, but the laboring "Chinamen" who threatened the economic security of the white working class. (p. 59)

The criminalization of other drugs, including cocaine, heroin, and marijuana, follows similar patterns of social control of the powerless, political opponents, and/or minorities. In the 1940s, marijuana was used primarily by minority group members, and users faced severe criminal penalties. However, after white middle-class college students began to use marijuana in the 1970s, the government reduced the penalties associated with its use. Although the nature and pharmacological properties of the drug had not changed, the population of users was now connected to power and influence. Thus, conflict theorists regard the regulation of certain drugs, as well as drug use itself, as a reflection of differences in the political, economic, and social power of various interest groups.

Symbolic Interactionist Perspective

Symbolic interactionism, which emphasizes the importance of definitions and labeling, concentrates on the social meanings associated with drug use. If the initial drug use experience is defined as pleasurable, it is likely to recur, and the individual may earn the label of "drug user" over time. If this definition is internalized so that the individual assumes an identity of a drug user, the behavior will probably continue and may even escalate. Conversely, Copes et al. (2008) observed that respondents who self-identified as "hustlers" rather than "crackheads" were less likely to fall prey to the debilitating effects of the drug for ". . . [S]lipping into uncontrollable addiction is antithetical to the hustler identity. . . " (p. 256).

Drug use is also learned through symbolic interaction in small groups. In a study of binge drinking, researchers found that students who believed that their friends were binge drinking were more likely to also engage in binge drinking themselves (Weitzman et al. 2003). First-time users learn not only the motivations for drug use and its techniques but also what to experience. Becker (1966) explained how marijuana users learn to ingest the drug. A novice being coached by a regular user reported the experience:

> I was smoking like I did an ordinary cigarette. He said, "No, don't do it like that." He said, "Suck it, you know, draw in and hold it in your lungs . . . for a period of time." I said, "Is there any limit of time to hold it?" He said, "No, just till you feel that you want to let it out, let it out." So I did that three or four times. (p. 47)

Marijuana users not only learn the way to ingest the smoke but also to label the experience positively. When peers define certain drugs, behaviors, and experiences as not only acceptable but also pleasurable, drug use is likely to continue.

Interactionists also emphasize that symbols can be manipulated and used for political and economic agendas. The popular DARE (Drug Abuse Resistance Education) program, with its antidrug emphasis fostered by local schools and police, carries a powerful symbolic value with which politicians want the public to identify. "Thus, ameliorative programs which are imbued with these potent symbolic qualities (like DARE's links to schools and police) are virtually assured widespread public acceptance (regardless of actual effectiveness) which in turn advances the interests of political leaders who benefit from being associated with highly visible, popular symbolic programs" (Wysong et al. 1994, p. 461). Ironically, a **meta-analysis** of the program led West and O'Neal (2004) to conclude that DARE does not significantly prevent drug use among school-aged children.

meta-analysis Meta-analysis combines the results of several studies addressing a research question, i.e., it is the analysis of analyses.

Biological and Psychological Theories

Drug use and addiction are probably the result of a complex interplay of social, psychological, and biological forces. Biological research has primarily concentrated on the role of genetics in predisposing an individual to drug use. Research indicates that severe, early-onset alcoholism may be genetically predisposed, with some men having 10 times the risk of addiction as those without a genetic predisposition. Interestingly, other problems such as depression, chronic anxiety, and attention deficit disorder are also linked to the likelihood of addiction. Nonetheless, researchers warn, "Nobody is predestined to be an alcoholic" (Firshein 2003).

Biological theories of drug use also hypothesize that some individuals are physiologically predisposed to experience more pleasure from drugs than others and, consequently, are more likely to be drug users. According to these theories, the central nervous system, which is composed primarily of the brain and spinal cord, processes drugs through neurotransmitters in a way that produces an unusually euphoric experience. Individuals not so physiologically inclined reported less pleasant experiences and are less likely to continue use (Jarvik 1990; National Institute on Alcohol Abuse and Alcoholism 2000).

Psychological explanations focus on the tendency of certain personality types to be more susceptible to drug use. Individuals who are particularly prone to anxiety may be more likely to use drugs as a way to relax, gain self-confidence, or ease tension. For example, research indicates that child maltreatment, particularly among females, contributes to alcohol and drug abuse that extends into adulthood (Gilbert et al. 2009).

Psychological theories of drug abuse also emphasize that drug use may be maintained by positive or negative reinforcement. Thus, for example, cocaine use may be maintained as a result of the rewarding "high" it produces—a positive reinforcement. Alternatively, heroin use, often associated with severe withdrawal symptoms, may continue as a result of a negative reinforcement, that is, the distress the user feels when faced with withdrawal (Abadinsky 2008). Reinforcement may come from a variety of sources including the media. In a study of the portrayal of alcohol use in 601 contemporary movies, researchers found that exposure to alcohol use in the movies was positively associated with early-onset drinking (Sargant et al. 2006).

Frequently Used Legal Drugs

Social definitions regarding which drugs are legal or illegal vary over time, circumstance, and societal forces. In the United States, two of the most dangerous and widely abused drugs, alcohol and tobacco, are legal.

Alcohol: The Drug of Choice

Americans' attitudes toward alcohol have a long and varied history. Although alcohol was a common beverage in early America, by 1920, the federal government had prohibited its manufacture, sale, and distribution through the passage of the Eighteenth Amendment to the U.S. Constitution. Many have argued that Prohibition, like the opium regulations of the late 1800s, was in fact a "moral crusade" (Gusfield 1963) against immigrant groups who were more likely to use alcohol. The amendment had little popular support and was repealed in

Indicate whether you have or have not experienced any of the following in the last 12 months as a consequence of your own drinking. When finished, compare your responses to those of a national sample of college students.

Consequence	Yes	No
1. Did something you later regretted	——	——
2. Forgot where you were or what you did	——	——
3. Got in trouble with the police	——	——
4. Had sex with someone without giving your consent	——	——
5. Had sex with someone without getting their consent	——	——
6. Had unprotected sex	——	——
7. Physically injured yourself	——	——
8. Physically injured another person	——	——
9. Seriously considered suicide	——	——
10. Reported one or more of the above	——	——

These survey items are from the American College Health Association's National College Health Assessment (2009). The following data are for 2008, and represent "only those institutions that surveyed all students, or used a random sampling technique" (American College Health Association 2009, p. 18).

The final number of students for this reference group is 26,685 students from 40 public and private, 2-year and 4-year colleges and universities.

Consequence	Percentage Reporting Consequence		
	Females	Males	Total
1. Did something you later regretted	33.4	34.0	33.7
2. Forgot where you were or what you did	28.1	31.7	29.2
3. Got in trouble with the police	3.5	6.6	4.5
4. Had sex with someone without giving your consent	2.1	1.7	2.0
5. Had sex with someone without getting their consent	0.3	0.5	0.4
6. Had unprotected sex	14.0	17.2	15.1
7. Physically injured yourself	14.5	17.5	15.4
8. Physically injured another person	1.5	4.2	2.4
9. Seriously considered suicide	1.4	2.0	1.6
10. Reported one or more of the above	48.1	52.0	49.4

Source: American College Health Association. 2009. *National College Health Assessment II*. Reference Group Executive Summary, Fall 2008. Baltimore, Maryland: American College Health Association.

1933. Today, the U.S. population is experiencing a resurgence of concern about alcohol. What has been called a "new temperance" has manifested itself in federally mandated 21-year-old drinking age laws, warning labels on alcohol bottles, increased concern over fetal alcohol syndrome and underage drinking, stricter enforcement of drinking and driving regulations (e.g., checkpoint traffic stops), and zero-tolerance policies. Such practices may have had an effect on drinking norms. Between 2006 and 2007, the rate of alcohol use in the past month among 12- to 20-year-olds significantly decreased (NSDUH 2008).

Despite such restrictive policies, alcohol remains the most widely used and abused drug in America. According to a recent poll, 29 percent of U.S. adults drink alcohol at least once a week and 6 percent report drinking daily (Corso 2009). Although most people who drink alcohol do so moderately and experience few negative effects (see this chapter's Self and Society feature), alcoholics are psychologically and physically addicted to alcohol and suffer various degrees of physical, economic, psychological, and personal harm.

The National Survey on Drug Use and Health, conducted by the U.S. Department of Health and Human Services, reported that 127 million Americans age 12 and older consumed alcohol at least once in the month preceding the survey; that is, they were *current users* (NSDUH 2008). Of this number, 6.9 percent reported **heavy drinking** and 23.3 percent reported **binge drinking** 58 million people.

heavy drinking As defined by the U.S. Department of Health and Human Services, five or more drinks on the same occasion on each of five or more days in the past thirty days prior to the National Survey on Drug Use and Health.

binge drinking As defined by the U.S. Department of Health and Human Services, drinking five or more drinks on the same occasion on at least one day in the past 30 days prior to the National Survey on Drug Use and Health.

Even more troubling were the 10.7 million current users of alcohol who were 12 to 20 years old—underage drinkers—many of whom got their alcohol for free from adults—often parents. More than half of underage drinkers reported binge or heavy drinking (NSDUH 2008). Although teen drinking has decreased in recent years, binge drinking in college continues to attract the public's attention. The likelihood of a college student binge drinking is impacted by environmental variables including place of residence (e.g., on campus versus off campus); cost and availability of alcohol; campus, local, and state alcohol policies; age, gender, and ethnic and racial makeup of the student population, prevention strategies, and the college drinking culture (Wechsler and Nelson 2008).

What Do You Think? There are many different drinking games, although beer pong is one of the most popular. Throwing a ping pong ball across a table into a cup of beer (and, if it's yours, guess what? Drink up!) has become such a hit that it's now a quasi-sport with a national tournament that carries a $50,000 first prize (Keegan and Haire 2008). There's even a virtual version on Nintendo's Wii system. However, fears of it leading to binge drinking have led some communities and college campuses to ban the "sport." Still other venues have instituted precautions such as filling the cups only half full of beer or having a separate bartender for players to keep track of who is consuming how much. Do you think that beer pong encourages binge drinking and, if so, what recommendations would you make to limit the severity of the consequences?

Research indicates that the younger the age of onset, the higher the probability that an individual will develop a drinking disorder at some time in his or her life.

Many binge drinkers began drinking in high school, with almost one-third having their first drink before age 13. Research indicates that the younger the age of onset, the higher the probability that an individual will develop a drinking disorder at some time in his or her life (Hingson et al. 2006; Behrendt et al. 2009). For example, an individual's chance of becoming dependent on alcohol is 40 percent if the person's drinking began before the age of 13. Additional results from the National Survey on Drug Use and Health (2008) include the following:

- The highest levels of both heavy and binge drinking are among 21- to 25-year-olds; people 65 or older had the lowest rates of binge drinking.
- Rates of alcohol use are higher among the employed than among the unemployed; however, patterns of heavy or binge drinking are highest among the unemployed.
- Among adults 26 years of age or older, college graduates are less likely to be binge and heavy alcohol users than those who did not graduate from college.
- Asians are least likely to report binge drinking, and American Indians and Alaska natives are most likely to report it.
- Full-time college students between the ages of 18 and 22 years old are more likely to have used alcohol in the past month, binge drink, and drink more heavily than their peers not enrolled in college full-time.
- More males than females age 12 to 20 years reported binge drinking and heavy drinking, although rates of current alcohol use by males and females were similar.

Despite the fact that males are more likely than females to abuse alcohol, some evidence suggests that female drinking is on the rise (Armstrong and McCarroll 2004; Zailckas 2005).

The Tobacco Epidemic

Native Americans first cultivated tobacco and introduced it to the European settlers in the 1500s. The Europeans believed that tobacco had medicinal properties, and its use spread throughout Europe, ensuring the economic success of the colonies in the New World. Tobacco was initially used primarily through chewing and snuffing, but smoking became more popular in time, even though scientific evidence that linked tobacco smoking to lung cancer existed as early as 1859 (Feagin & Feagin 1994). However, the U.S. Surgeon General did not conclude that tobacco products are addictive and that nicotine causes dependency until 1989.

Tobacco is one of the most widely used drugs in the United States. According to a U.S. Department of Health and Human Services survey, 60.1 million American—24.2 percent of those 12 and older—are current cigarette smokers (NSDUH 2008). Use of all tobacco products, including smokeless tobacco (8.1 million users), cigars (13.3 million users), pipe tobacco (2.0 million users), and cigarettes, is higher for high school graduates than for college graduates, males, and Americans Indians and Alaska natives (NSDUH 2008).

In 2007, 3.1 million youths between the ages of 12 and 17 reported use of a tobacco product in the past month (NSDUH 2008). Research evidence suggests that youth develop attitudes and beliefs about tobacco products at an early age (see this chapter's Social Problems Research Up Close feature). Advertising of tobacco products continues to have an influence on youth despite a 1998 legal settlement between the tobacco companies and the states that prohibited any tobacco company from taking "action, directly or indirectly, to target youth . . . in the advertising, promotion or marketing of tobacco products" (Tobacco Free Kids 2005, p. 1). For example, although R. J. Reynolds, maker of Kool, Camel, and Winston cigarettes, in a 2006 agreement with state attorneys general, agreed not to market or advertise candy flavored cigarettes (e.g., a pineapple-and-coconut-flavored cigarette called Kauai Kolada and a rum-dipped cigarette called Al Capone Slims), legal loopholes allow for their continued production and distribution (American Heart Association 2008a; Lombardi 2009).

The Campaign for Tobacco-Free Kids calls the introduction of candy-flavored cigarettes and smokeless tobacco an "outrageous" tactic to lure youth into using tobacco products. Note the appeal to African American youth and women in some of the packaging.

Campaign for Tobacco-Free Kids®/www.tobaccofreekids.org

Most research on smoking onset, i.e., the age at which an individual begins to smoke, is focused on young adults or adolescents. Research on those younger than age 12 is rare but important—the younger people are when they begin to smoke, the higher the probability of developing a nicotine addiction (American Cancer Society 2009). The present research examines "children's attitudes toward, belief about and lifestyle associations with cigarette smoking" (Freeman et al. 2005, p. 1537).

Sample and Methods

Children were recruited from three racially, ethnically, and economically diverse elementary schools in the southwestern United States. Sample demographics were representative of the surrounding community with half of the respondents being female and 72.1 percent being non-Hispanic whites. All children were in the second (n = 100) or fifth grade (n = 141); i.e., ages 7 and 8, or 11 and 12.

Using a child-friendly computer program, children were randomly assigned to view a set of six photographs and then were asked to pick the picture of the person who would like to smoke cigarettes the *most* and the person who would like to smoke cigarettes the *least*. Within each set of photographs, the images portrayed a range of physical attractiveness, sociability, and independence. For example, one of the sets included pictures of (1) a heavy-set woman standing in a field, (2) a thin, attractive female standing by a car, (3) a middle-aged man reading by a fireplace, (4) a young man playing lacrosse with some friends, (5) a young girl riding her bicycle, and (6) a young man sitting in a boat fishing by himself (p. 1539).

After each photo selection, trained interviewers asked the children to respond to probes concerning lifestyle associations such as "Tell me about him/her," and "What does he/she like best about smoking cigarettes?" The interviewer followed up the initial responses until perceiving "that the participant had thoroughly articulated their perceptions of the person's motivations for liking (not liking) to smoke cigarettes" (p. 1540).

Finally, respondents were asked a series of structured survey items about the presence or absence of cigarette smoking in the home as well as a set of items designed to tap attitudes about the age-appropriateness of smoking, the negative consequences of smoking, and the instrumental benefits of smoking. Given the age of the respondents, possible agree-disagree responses included "yes," "maybe," "don't think so," and "no way."

Findings and Conclusions

The results indicated that both second and fifth graders' selections of those who would like to smoke the *most* were influenced by lifestyle associations. For example, the picture of a slender, professionally dressed woman leaning against a car elicited such responses as, "She thinks she's all that," and "She's too cool not to smoke." There were no significant patterns of picture selections (only one picture was selected) for those who would like to smoke the *least*. Further, the presence of a smoker in the respondent's home was unrelated to picture selection.

Responses to the open-ended probes indicated that pictures of people who liked to smoke cigarettes the *most* were associated with certain motivations—they smoked to "feel better, alleviate stress, or overcome a bad mood; [to] feel cool or look cool; and to have something to do" (p. 1542). Conversely, the only picture of an individual that was consistently selected as someone who liked to smoke cigarettes the *least* (i.e., a young girl on a bicycle) was described as being too young to smoke, being already happy, and knowing that smoking has negative health consequences.

Finally, analysis of the survey items indicated that both second and fifth graders strongly agree with the negative consequences of smoking—that it looks "dumb," is "gross," and "makes people feel sick." However, fifth graders, when compared with second graders, were significantly more likely to agree that smoking helps relax people and "cheers people up when they are in a bad mood" (p. 1542).

In summarizing the results, the authors concluded that, at an early age, children have already developed lifestyle associations with smokers although not with non-smokers (p. 1543). The lifestyle association attributed to the one picture consistently selected as someone who would like to smoke the *least* was in part the fact that the woman was "too young" to smoke cigarettes. As the authors noted, this is not a desirable outcome in terms of smoking prevention, for it implies that there is an age at which smoking is appropriate and may "predispose young children to initiate tobacco use when they reach adolescence" (p. 1544). Thus, developing age-specific prevention programs may facilitate reducing tobacco use.

Source: Freeman, Dan, Merrie Brucks, and Melanie Wallendorf, 2005 Young children's understandings of cigarette smoking. *Addiction* 100(6):1537–45.

There is also considerable evidence that cigarette advertisers target minorities. For example, Primack et al. (2007) found that tobacco advertisements in African American communities were 2.6 times higher per person than in white communities. Further, the likelihood of tobacco-related billboards was 70 percent higher in African American rather than white communities. Advertising works.

Overall, rates of smoking among African American youth have increased since the early 1990s (American Heart Association 2008b).

Finally, the tobacco industry has developed advertising campaigns targeting women, including Philip Morris's "purse packs" modeled after cosmetic cases and half the size of a regular package of cigarettes (Tobacco Free Kids 2009). Such initiatives are also taking place in developing countries where

> Because most women currently do not use tobacco, the tobacco industry aggressively markets to them to tap this potential new market. Advertising, promotion, and sponsorship, including charitable donations to women's causes, weaken cultural opposition to women using tobacco. Product design and marketing, including the use of attractive models in advertising and brands marketed specifically to women, are explicitly crafted to encourage women to smoke (WHO 2008, p. 16–17).

Frequently Used Illegal Drugs

More than 19.9 million people in the United States are current illicit drug users, representing 8 percent of the population age 12 and older. Users of illegal drugs, although varying by type of drug used, are more likely to be male, to be young, and to be a member of a minority group (NSDUH 2008).

Marijuana Madness

Marijuana is the most commonly used and most heavily trafficked illicit drug in the world (see Table 3.2 for a list of commonly abused drugs, their commercial and street names, and their intoxication and health effects). Globally, there are 166 million marijuana users representing 3.9 percent of the world's 15- to 64-year-old population. Regionally, marijuana is also the most dominant illicit drug, and its cultivation and consumption are particularly high in the Americas. For example, the largest producers of marijuana are Mexico, followed by the United States and Canada (WDR 2008).

Marijuana's active ingredient is THC (Δ^9-tetrahydrocannabinol), which in varying amounts can act as a sedative or a hallucinogen. When just the top of the marijuana plant is sold, it is called hashish. Hashish is much more potent than marijuana, which comes from the entire plant. Marijuana use dates back to 2737 B.C. in China, and marijuana has a long tradition of use in India, the Middle East, and Europe. In North America, hemp, as it was then called, was used to make rope and as a treatment for various ailments. Nevertheless, in 1937, Congress passed the Marijuana Tax Act, which restricted the use of marijuana; the law was passed as a result of a media campaign that portrayed marijuana users as "dope fiends" and, as conflict theorists note, was enacted at a time of growing sentiment against Mexican immigrants (Witters et al. 1992).

There are more than 14.4 million current marijuana users, representing 5.8 percent of the U.S. population age 12 and older (NSDUH 2008). According to the most recent Monitoring the Future (MTF) survey, 12.5 percent of eighth, tenth, and twelfth graders in the United States reported current marijuana use. The study further found that 14.6 percent reported use of an illicit drug in the past month, with marijuana being the most commonly reported (MTF 2009). Further, of first-time users of an illicit drug, over half initiate with marijuana (NSDUH 2008) (see Figure 3.1).

TABLE 3.2 Commonly Abused Drugs

SUBSTANCES: CATEGORY AND NAME	EXAMPLES OF COMMERCIAL AND STREET NAMES	DEA SCHEDULE*/ HOW ADMINISTERED**	INTOXICATION EFFECTS/POTENTIAL HEALTH CONSEQUENCES
Cannabinoids			
hashish	boom, chronic, gangster, hash, hash oil, hemp	I/swallowed, smoked	*euphoria, slowed thinking and reaction time, confusion, impaired balance and coordination*/cough, frequent respiratory infections; impaired memory and learning; increased heart rate, anxiety, panic attacks; tolerance, addiction
marijuana	blunt, dope, ganja, grass, herb, joints, Mary Jane, pot, reefer, sinsemilla, skunk, weed	I/swallowed, smoked	
Depressants			
barbiturates	*Amytal, Nembutal, Seconal, Phenobarbital:* barbs, reds, red birds, phennies, tooies, yellows, yellow jackets	II, III, V/injected, swallowed	*reduced anxiety; feeling of well-being; lowered inhibitions; slowed pulse and breathing; lowered blood pressure; poor concentration*/fatigue; confusion; impaired coordination, memory, judgment; addiction; respiratory depression and arrest; death
benzodiazepines (other than flunitrazepam)	*Ativan, Halcion, Librium, Valium, Xanax:* candy, downers, sleeping pills, tranks	IV/swallowed, injected	*sedation, drowsiness*/depression, unusual excitement, fever, irritability, poor judgment, slurred speech, dizziness, life-threatening withdrawal
flunitrazepam***	*Rohypnol:* forget-me pill, Mexican Valium, R2, Roche, roofies, roofinol, rope, rophies	IV/swallowed, snorted	*sedation, drowsiness*/dizziness
GHB***	*gamma-hydroxybutyrate:* G, Georgia home boy, grievous bodily harm, liquid Ecstasy	I/swallowed	visual and gastrointestinal disturbances, urinary retention, memory loss for the time under the drug's effects
methaqualone	*Quaalude, Sopor, Parest:* ludes, mandrex, quad, quay	I/injected, swallowed	*drowsiness, nausea*/vomiting, headache, loss of consciousness, loss of reflexes, seizures, coma, death
Dissociative Anesthetics			*euphoria*/depression, poor reflexes, slurred speech, coma
ketamine	*Ketalar SV:* cat Valiums, K, Special K, vitamin K	III/injected, snorted, smoked	*increased heart rate and blood pressure, impaired motor function*/memory loss; numbness; nausea/vomiting
PCP and analogs	*phencyclidine:* angel dust, boat, hog, love boat, peace pill	I, II/injected, swallowed, smoked	at high doses, delirium, depression, respiratory depression and arrest
Hallucinogens			*possible decrease in blood pressure and heart rate, panic, aggression, violence*/loss of appetite, depression
LSD	*lysergic acid diethylamide:* acid, blotter, boomers, cubes, microdot, yellow sunshines	I/swallowed, absorbed through mouth tissues	*altered states of perception and feeling; nausea*/persisting perception disorder (flashbacks)
			Also, for LSD and mescaline—increased body temperature, heart rate, blood pressure; loss of appetite, sleeplessness, numbness, weakness, tremors for LSD—persistent mental disorders

Substances	Examples of Commercial and Street Names	DEA Schedule/How Administered	Acute Effects/Health Risks
mescaline	buttons, cactus, mesc, peyote	I/swallowed, smoked	
psilocybin	magic mushroom, purple passion, shrooms	I/swallowed	*nervousness, paranoia*
Opioids and Morphine Derivatives			*pain relief, euphoria, drowsiness*/nausea, constipation, confusion, sedation, respiratory depression and arrest, tolerance, addiction, unconsciousness, coma, death
codeine	*Empirin with Codeine, Fiorinal with Codeine, Robitussin A-C, Tylenol with Codeine:* Captain Cody, Cody, schoolboy; (with glutethimide) doors & fours, loads, pancakes and syrup	II, III, IV, V/injected, swallowed	
fentanyl and fentanyl analogs	*Actiq, Duragesic, Sublimaze:* Apache, China girl, China white, dance fever, friend, goodfella, jackpot, murder 8, TNT, Tango and Cash	I, II/injected, smoked, snorted	*less analgesia, sedation, and respiratory depression thanmorphine*
heroin	*diacetylmorphine:* brown sugar, dope, H, horse, junk, skag, skunk, smack, white horse	I/injected, smoked, snorted	*staggering gait*
morphine	*Roxanol, Duramorph:* M, Miss Emma, monkey, white stuff	II, III/injected, swallowed, smoked	
opium	*laudanum, paregoric:* big O, black stuff, block, gum, hop	II, III, V/swallowed, smoked	
oxycodone HCL	*OxyContin:* Oxy, O.C., killer	II/swallowed, snorted, injected	
hydrocodone bitartrate, acetaminophen	*Vicodin:* vike, Watson-387	II/swallowed	
Stimulants			*increased heart rate, blood pressure, metabolism; feelings of exhilaration, energy, increased mental alertness*/rapid or irregular heart beat; reduced appetite, weight loss, heart failure, nervousness, insomnia
amphetamine	*Biphetamine, Dexedrine:* bennies, black beauties, crosses, hearts, LA turn-around, speed, truck drivers, uppers	II/injected, swallowed, smoked, snorted	*rapid breathing*/tremor, loss of coordination; irritability, anxiousness, restlessness, delirium, panic, paranoia, impulsive behavior, aggressiveness, tolerance, addiction, psychosis
cocaine	*Cocaine hydrochloride:* blow, bump, C, candy, Charlie, coke, crack, flake, rock, snow, toot	II/injected, smoked, snorted	*increased temperature*/chest pain, respiratory failure, nausea, abdominal pain, strokes, seizures, headaches, malnutrition, panic attacks

TABLE 3.1 Continued

SUBSTANCES: CATEGORY AND NAME	EXAMPLES OF COMMERCIAL AND STREET NAMES	DEA SCHEDULE*/ HOW ADMINISTERED*	INTOXICATION EFFECTS/POTENTIAL HEALTH CONSEQUENCES
MDMA (methyl-enedioxymeth-amphetamine)	Adam, clarity, Ecstasy, Eve, lover's speed, peace, STP, X, XTC	I/swallowed	*mild hallucinogenic effects, increased tactile sensitivity, empathic feelings/* impaired memory and learning, hyperthermia, cardiac toxicity, renal failure, liver toxicity
methamphetamine	*Desoxyn:* chalk, crank, crystal, fire, glass, go fast, ice, meth, speed	II/injected, swallowed, smoked, snorted	*aggression, violence, psychotic behavior/* memory loss, cardiac and neurological damage; impaired memory and learning, tolerance, addiction
methylphenidate (safe and effective for treatment of ADHD)	*Ritalin:* JIF, MPH, R-ball, Skippy, the smart drug, vitamin R	II/injected, swallowed, snorted	
nicotine	cigarettes, cigars, smokeless tobacco, snuff, spit tobacco, bidis, chew	not scheduled/smoked, snorted, taken in snuff and spit tobacco	additional effects attributable to tobacco exposure: adverse pregnancy outcomes; chronic lung disease, cardiovascular disease, stroke, cancer; tolerance, addiction
Other Compounds			
anabolic steroids	*Anadrol, Oxandrin, Durabolin, Depo-Testosterone, Equipoise:* roids, juice	III/injected, swallowed, applied to skin	*no intoxication effects/* hypertension, blood clotting and cholesterol changes, liver cysts and cancer, kidney cancer, hostility and aggression, acne; in adolescents, premature stoppage of growth; in males, prostate cancer, reduced sperm production, shrunken testicles, breast enlargement; in females, menstrual irregularities, development of beard and other masculine characteristics
Dextromethorphan (DXM)	Found in some cough and cold medications; Robotripping, Robo, Triple C	not scheduled/swallowed	*dissociative effects, distorted visual perceptions to complete dissociative effects/* for effects at higher doses see "dissociative anesthetics"
inhalants	*Solvents (paint thinners, gasoline, glues), gases (butane, propane, aerosol propellants, nitrous oxide), nitrites (isoamyl, isobutyl, cyclohexyl):* laughing gas, poppers, snappers, whippets	not scheduled/inhaled through nose or mouth	*stimulation, loss of inhibition; headache; nausea or vomiting; slurred speech, loss of motor coordination; wheezing/* unconsciousness, cramps, weight loss, muscle weakness, depression, memory impairment, damage to cardiovascular and nervous systems, sudden death

*Schedule I and II drugs have a high potential for abuse. They require greater storage security and have a quota on manufacturing, among other restrictions. Schedule I drugs are available for research only and have no approved medical use; Schedule II drugs are available only by prescription (unrefillable) and require a form for ordering. Schedule III and IV drugs are available by prescription, may have five refills in six months, and may be ordered orally. Some Schedule V drugs are available over the counter.

**Taking drugs by injection can increase the risk of infection through needle contamination with staphylococci, HIV, hepatitis, and other organisms.

***Associated with sexual assaults.

Source: Adapted from the NIDA (2009).

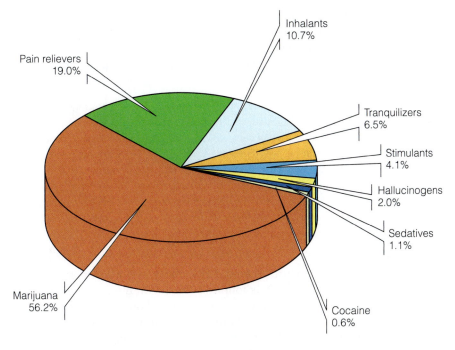

Figure 3.1 **Specific Drug Used when Initiating Illicit Drug Use Aged 12 and Older, 2007***
Source: NSDUH 2008.
*Among past-year first-time users of illicit drugs

Inhalants 10.7%

Pain relievers 19.0%

Tranquilizers 6.5%

Stimulants 4.1%

Hallucinogens 2.0%

Sedatives 1.1%

Cocaine 0.6%

Marijuana 56.2%

NOTE: The percentages add to greater than 100 percent because of a small number of respondents initiating multiple drugs on the same day.

In the present debate over the legalization of marijuana, many express fears that it is a **gateway drug**, the use of which causes progression to other drugs. "The gateway hypothesis holds that consumption of abusable drugs progresses in orderly fashion through several discrete stages. The entire sequence, which is exhibited by only a small minority of drug users, begins with beer or wine and moves progressively through hard liquor or tobacco, marijuana, and finally hard drugs" (Tarter et al. 2006, p. 2134). Most research suggests, however, that people who experiment with one drug are more likely to experiment with another. Indeed, most drug users use several drugs at the same time. As Lee and Abdel-Ghany (2004) note, there is a strong "contemporaneous relationship between smoking cigarettes, drinking alcohol, smoking marijuana, and using cocaine" (p. 454).

Cocaine: From Coca-Cola to Crack

Cocaine is classified as a stimulant and, as such, produces feelings of excitation, alertness, and euphoria. Although prescription stimulants such as methamphetamine and dextroamphetamine are commonly abused, over the last 20 years societal concern over drug abuse has focused on cocaine. Its increased use, addictive qualities, physiological effects, and worldwide distribution have fueled such concerns. More than any other single substance, cocaine led to the early phases of the war on drugs.

Cocaine, which is made from the coca plant, has been used for thousands of years. Coca leaves were used in the original formula for Coca-Cola but, in the early 1900s, anti-cocaine sentiment emerged as a response to the heavy use of cocaine among urban blacks, poor whites, and criminals (Witters et al. 1992; Thio 2007; Friedman-Rudovsky 2009). Cocaine was outlawed in 1914 by the Harrison Narcotics Act, but its use and effects continued to be misunderstood.

gateway drug A drug (e.g., marijuana) that is believed to lead to the use of other drugs (e.g., cocaine).

For example, a 1982 *Scientific American* article suggested that cocaine was no more habit-forming than potato chips (Van Dyck & Byck 1982). As demand and then supply increased, prices fell from $100 a dose to $10 a dose, and "from 1978 to 1987, the U.S. experienced the largest cocaine epidemic in history" (Witters et al. 1992, p. 256).

What Do You Think? Energy drinks such as Red Bull, Monster, and Full Throttle have been popular for years. Health concerns have been raised (some energy drinks have as much caffeine as 14 cans of Coca-Cola!), but German officials have now raised a legal concern as well (Friedman-Rudovsky 2009). After chemically analyzing Red Bull Cola, the soft drink version of Red Bull, it was discovered that the drink contains cocaine. Although not posing a serious health threat, international law prohibits trade of the coca plant outside of the Andean region of Bolivia unless the cocaine alkaloid has been removed—a point that is in dispute. Would you, as German officials are likely to, recommend a nationwide ban of the drink? Why or why not?

Globally, cocaine use has decreased, particularly in North America where the largest number of cocaine users is found (WDR 2008). Europe is the exception, where cocaine use has increased with average annual prevalence rates of 3.6 percent for the 15- to 64-year-old population. European users are diversified, from the most marginalized to the most privileged reflecting the different forms of the drug—powder cocaine versus crack. Cocaine, after marijuana, is the second most widely used illegal drug in the Americas (EMCDDA 2008).

According to the National Survey on Drug Use and Health, 2.1 million Americans 12 years and older are current cocaine users—a slight although not statistically significant decrease from 2007 (NSDUH 2008). In 2008, the current prevalence of cocaine use by eighth, tenth, and twelfth graders was 1.3 percent. The percentage of twelfth graders indicating that getting cocaine is "fairly easy" or "very easy" decreased between 2007 (47.1 percent) and 2008 (42.4 percent), with half of twelfth graders responding that trying cocaine once or twice is a "great risk" (MTF 2009).

Crack is a crystallized product made by boiling a mixture of baking soda, water, and cocaine. The result, also called rock, base, and gravel, is relatively inexpensive and was not popular until the mid-1980s. Crack is one of the most dangerous drugs to surface in decades. Crack dealers often give drug users their first few "hits" free, knowing the drug's intense high and addictive qualities are likely to produce returning customers. An addiction to crack can take six to ten weeks; an addiction to pure cocaine can take three to four years (Thio 2007).

According to the National Survey on Drug Use and Health, there are an estimated 610,000 current crack users between the ages of 15 and 64 in the United States, down from 702,000 in 2006 (NSDUH 2008). Results from the Monitoring the Future survey indicate that 0.6 percent of eighth, tenth, and twelfth graders are current crack users. When high school seniors were asked, "*How much do you think people risk harming themselves (physically or in other ways) if they try crack once or twice?*— 47.5 percent responded at "great risk." Finally, 35.2 percent of twelfth graders report that crack is "fairly easy" or "very easy" to get.

crack A crystallized illegal drug product produced by boiling a mixture of baking soda, water, and cocaine.

Methamphetamine

Methamphetamine is a central nervous system stimulant that is highly addictive. Although the drug has only recently become popular, it is not new.

During the Second World War, soldiers on both sides used it to reduce fatigue and enhance performance. Hitler was widely believed to be a meth addict. Later, in the 1960s, President John Kennedy also used the drug and soon it caught on among so-called "speed freaks." But, because it was extremely expensive as well as difficult to obtain, meth was never close to being as widely used as cocaine. (Thio 2007, p. 276)

Today, methamphetamine is relatively inexpensive and easily obtained, with more than 23.3 percent of high school seniors reporting that "crystal meth," i.e., methamphetamine in its crystalline form, is "fairly easy" or "very easy" to get (MTF 2009). Ease of obtaining the drug is, in part, a result of the number of clandestine laboratories in the United States, the large amounts of methamphetamine smuggled into the United States from Mexico, and the ease of production. Because methamphetamine can be made from cold medications such as Sudafed, the U.S. Congress passed the Comprehensive Methamphetamine Control Act of 1996 that made obtaining the chemicals needed to make methamphetamine more difficult (ONDCP 2006; Thio 2007). In 2006, the Combat Methamphetamine Epidemic Act, which further articulated standards for selling over-the-counter medications used in methamphetamine production, went into effect.

> Ease of obtaining the drug is … a result of the number of clandestine laboratories in the United States, the large amounts of methamphetamine smuggled into the United States from Mexico, and the ease of production.

Globally, in terms of consumption, there are an estimated 15 million to 16 million methamphetamine users—a rate similar to the number of cocaine or heroin users worldwide (WDR 2008). In the United States, 5.3 percent of the population 12 and older—13.1 million people—reported using methamphetamine at least once in their lifetime, a slight decline from 2006. During the same time period, the current rate of methamphetamine use by full-time college students (18- to 22-year-olds) also decreased (NSDUH 2008). In 2007, 44.3 percent of past-year methamphetamine users aged 12 and older reported that they obtained the methamphetamine they most recently used from a friend or relative for free. Current methamphetamine use by eighth, tenth, and twelfth graders is 0.7 percent (MTF 2009).

Other Illegal Drugs

Other drugs abused throughout the world include **club drugs** (e.g., LSD and Ecstasy), heroin, prescription drugs (e.g., tranquilizers and amphetamines), and inhalants (e.g., glue).

Worldwide, over nine million people between the ages of 15 and 64 use Ecstasy, the most common name for MDMA, at least once a year. Of those 9 million users, the largest numbers are from Europe (33 percent), followed by North America (26 percent), Asia (23 percent), South America and Oceania (8 percent), and Africa (2 percent) (WDR 2008). In 2007, 5 percent of the American population age 12 and older had used Ecstasy at least once in their lifetime.

Ketamine and LSD both produce visual effects when ingested. Use of ketamine, an animal tranquilizer, is growing globally with, for example, rates for the under-21 population in Hong King reaching 73 percent in 2006 (Brownell 2009). In the United States, ketamine use by eighth, tenth, and twelfth graders has decreased in recent years (MTF 2009). LSD is a synthetic hallucinogen, although many other hallucinogens are produced naturally (e.g., salvia). In 2008, 9.1 percent of the American population age 12 and older reported lifetime use of LSD. Although twelfth graders have a higher past-month use of LSD (1.1 percent) than

club drugs A general term for illicit, often synthetic, drugs commonly used at nightclubs or all-night dances called "raves."

the general population age 12 and older (0.1 percent), LSD use among teenagers has decreased in recent years (NSDUH 2008).

GHB and Rohypnol are often called **date-rape drugs** because of their use in rendering victims incapable of resisting sexual assaults. In 1990, the Food and Drug Administration banned GHB, a central nervous system depressant, although kits containing all the necessary ingredients to manufacture the drug continued to be available on the Internet. On February 18, 2000, President Clinton signed a bill that made GHB a controlled substance and thus illegal to manufacture, possess, or sell. Nonetheless, 1.2 percent of twelfth graders, 0.5 percent of tenth graders, and 1.1 percent of eight graders reported past-year use of GHB (MTF 2009).

Rohypnol, presently illegal in the United States, is lawfully sold by prescription in more than 70 countries for the short-term treatment of insomnia (Abadinsky 2008). It belongs to a class of drugs known as benzodiazepines, which also includes common prescription drugs such as Valium, Halcion, and Xanax. Rohypnol is a tasteless and odorless depressant. The effects of Rohypnol begin within 20 minutes and can incapacitate a victim for up to 12 hours (Abadinsky 2008). In 2008, 0.5 percent, 0.4 percent, and 1.3 percent of eighth, tenth, and twelfth graders, respectively, reported past-year use of Rohypnol (MTF 2009).

Heroin. Heroin is an analgesic—that is, a painkiller—and is the most commonly abused of a class of drugs called opiates. Most heroin comes from the poppy fields of Afghanistan, which controls 90 percent of the global opium market and provides the Taliban with an estimated $300 million a year (Filkins 2009). Highly addictive, heroin can be injected, snorted, or smoked. If intravenous injection is used, the onset of the euphoric effects is felt within 7 to 8 seconds; if heroin is snorted or smoked, peak effects are felt within 10 to 15 minutes (NIDA 2005).

Overall, use of heroin in Europe (0.6 percent) is higher than in North America (0.4 percent), although both annual prevalence rates are above the global average of 0.3 percent of the 15- to 64-year-old population (WDR 2008). According to the National Survey on Drug Use and Health, current U.S. heroin users decreased by nearly 200,000 between 2006 and 2007 (NSDUH 2008). In 2008, the rate of heroin use among eighth, tenth, and twelfth graders remained the same from the previous year, although substantially reduced from the record highs of the 1990s.

Psychotherapeutic Drugs Estimates of lifetime use of psychotherapeutic drugs—that is, nonmedical use of any prescription pain reliever, stimulant, sedative, or tranquilizer—remained essentially the same between 2006 and 2007 (NSDUH 2008). More than 6 million people, 2.8 percent of the U.S. population 12 and older, reported current use of a psychotherapeutic drug for nonmedical reasons in 2007. Of these users, 5.2 million used pain relievers, 1.8 million used tranquilizers, 1.1 million used stimulants, and 346,000 used sedatives. Over half of those who reported prescription drug use for nonmedical reasons reported that they received the drugs from friends and relatives for free (NSDUH 2008).

Recently, there has been an increase in the use of pain relieving synthetic opioids such as codeine and OxyContin (WDR 2008). In 2007, 1.8 percent of Americans 12 and older reported using OxyContin for nonmedical reasons at least once in their lifetime (NSDUH 2008). Lifetime OxyContin use was even higher for eighth, tenth, and twelfth graders—3.4 percent of the students surveyed (MTF 2009). Several studies indicate that psychotherapeutic drug abuse among college students is particularly problematic. As Figure 3.2 indicates, of the eleven most commonly abused drugs by high school seniors, seven of them are prescriptions or over-the-counter (MTF 2009).

date-rape drugs Drugs that are used to render victims incapable of resisting sexual assaults.

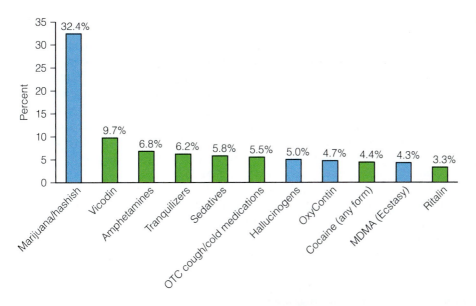

Figure 3.2 **Prevalence of Past-Year Drug Abuse among 12th Graders, 2008**
In 2008, seven of the top 11 drugs that high school seniors most commonly abused were either prescribed or over-the-counter medications (green columns). *Source*: Data from MTF 2009; adapted from U.S. Department of Health and Human Services.

Inhalants Inhalants act on the central nervous system with users reporting a psychoactive, mind-altering effect (Abadinsky 2008). Common inhalants include adhesives (e.g., rubber cement), food products (e.g., vegetable cooking spray), aerosols (e.g., hair spray and air fresheners), anesthetics (ether), and cleaning agents (e.g., spot remover). In total, more than 1,000 household products are currently abused. More than 22.5 million people 12 and older (9.1 percent) have reported trying inhalants at least once in their lifetime (NSDUH 2008). Youth, however, are particularly prone to inhalant use, erroneously believing it is harmless or that prolonged use is necessary for any harm to result. In 2008, 13.1 percent of eighth, tenth, and twelfth grade students reported lifetime use of an inhalant, nearly twice the rate of European 15- and 16-year-olds (MTF 2009; ESPAD 2009).

Sakchai Lalit/ AP Photo

Throughout much of the developing world, homeless children "huff" glue to escape hunger pains and the horrors of living on the street. Globally, UNICEF estimates that there are more than 200 million homeless children—the equivalent of two-thirds the population of the United States.

Societal Consequences of Drug Use and Abuse

Drugs are a social problem not only because of their adverse effects on individuals but also because of the negative consequences their use has for society as a whole. Everyone is a victim of drug abuse. Drugs contribute to problems within the family and to crime rates, and the economic costs of drug abuse are enormous. Drug abuse also has serious consequences for health at both the individual and the societal level.

The Cost to the Family

The cost of drug abuse to families is incalculable. It is estimated that in the United States, one in 10 children under the age of 18 lives with at least one parent in need of treatment for drug or alcohol dependency (SAMHSA 2009). Children raised in such homes are more likely to (1) live in an environment riddled with conflict, (2) have a higher probability of physical illness including injuries or death from an automobile accident, and (3) are more likely to be victims of child abuse and neglect (SAMHSA 2007; 2009). Children of alcoholics, if using alcohol or other drugs, are also four times more likely to have alcohol or drug problems than children of nonalcoholics (SAMHSA 2007).

Parents who reported abusing alcohol in the past year are also more likely to report cigarette and illicit drug use than parents who did not report alcohol abuse in the previous year. They were also more likely to report "household turbulence," including yelling, serious arguments, and violence (NSDUH 2004). Moreover, alcohol abuse is the single most common trait associated with wife abuse (Flanzer 2005). The more violent the interaction, the more likely that the husband has been drinking excessively. According to official statistics, on the average, alcohol or other drugs are present in an estimated 42 percent of all incidences of nonfatal intimate partner violence (U.S. Department of Justice 2007).

Crime and Violence

At the fifth Summit of the Americas, President Obama addressed the escalating violence surrounding the drug trade and, more specifically, the need to "to stop the flow of guns and bulk cash south across our borders" (Obama 2009, p. 1). In 2008, over 6,000 people were killed in drug violence in Mexico; drug cartels, fueled by American guns and money, have extended their operations to over 200 U.S. cities (Archibold 2009). Not surprisingly, the result is that 79 percent of Americans say they are either "very concerned" or "somewhat concerned" about the drug violence in Mexico (Jones 2009).

The drug behavior of individuals arrested, incarcerated, and in drug treatment programs also provides evidence of a link between drugs and crime. For example, surveys indicate that about 27 percent of victims of violent crime report that the offender was involved with alcohol or drugs. Further, in 2004, 32 percent of state prisoners and 26 percent of federal prisoners said they had committed their current offense while under the influence of drugs (Bureau of Justice Statistics 2008). Similarly, juvenile delinquency is associated with drug use. In 2007, 12- to 17-year-olds who used an illicit drug in the past year were more than twice as likely to get in a serious fight at work or at school than their counterparts who did not report past-year use (NSDUH 2008).

The relationship between crime and drug use, however, is complex. Sociologists disagree as to whether drugs actually "cause" crime or whether, instead, criminal activity leads to drug involvement. Alternatively, as Siegel (2006) noted, criminal involvement and drug use can occur at the same time; that is, someone can take drugs and commit crimes out of the desire to engage in risk-taking behaviors. Furthermore, because both crime and drug use are associated with low socioeconomic status, poverty may actually be the more powerful explanatory variable.

In addition to the hypothesized crime–drug use link, some criminal offenses are defined by use of drugs: possession, cultivation, production, and sale of controlled substances; public intoxication; drunk and disorderly conduct; and driving while intoxicated. Driving while intoxicated is one of the most common drug-related crimes. According to the National Highway Traffic Safety Administration (NHTSA 2008a), 13,000 people were killed in alcohol-impaired driving crashes in 2007, accounting for 32 percent of all motor vehicle traffic fatalities in the United States. Alcohol is not the only drug that impairs driving. Ten million people 12 years of age and older reported driving while under the influence of an illicit drug. As with alcohol-impaired drivers, the rate is highest for 18- to 25-year-olds (NSDUH 2008).

The High Price of Alcohol and Other Drugs

Using data from 2005, a recent report by the National Center on Addiction and Substance Abuse at Columbia University (CASA 2009) set the total cost of substance abuse and addiction in the United States at $467.7 billion. More importantly, the report contends that, in 2005, for every dollar spent on drug abuse by federal and state governments

> . . . 95.6 cents went to shoveling up the wreckage and only 1.9 cents on prevention and treatment, 0.4 cents on research, 1.4 cents on taxation or regulation and 0.7 cents on interdiction. Under any circumstances spending more than 95 percent of taxpayer dollars on the consequences of tobacco, alcohol, and other drug abuse and addiction and less than two percent to relieve individuals and taxpayers of this burden would be considered a reckless misallocation of public funds. In these economic times, such upside-down-cake public policy is unconscionable. (CASA 2009, p. i)

At the federal level, the cost of "shoveling up the wreckage" includes the cost of (1) health care due to substance abuse and addiction (the highest proportion of wreckage spending), (2) adult and juvenile crime (e.g., corrections), (3) child and family assistance programs (e.g., welfare), (4) education (e.g., Safe School Initiatives), (5) public safety (e.g., drug enforcement), (6) mental health and developmental disabilities (e.g., treatment of addiction), and (7) the federal workforce (e.g., loss of productivity). The report concludes that prevention programs must become a priority to reduce the economic costs of drug abuse.

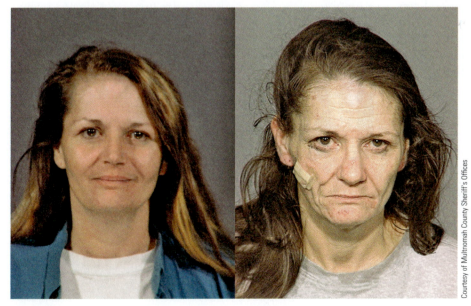

For those who use methamphetamine, the physical transformation is remarkable. The time lapse between the before (left) and after (right) pictures of this methamphetamine user is only three years five months.

Courtesy of Multnomah County Sheriff's Offices

Physical and Mental Health Costs

Cigarette smoking is the leading preventable cause of disease and deaths in the world (see Figure 3.3). According to the World Health Organization, the global

> According to the World Health Organization, the global tobacco epidemic kills 5.4 million people a year and, unless significant policy changes occur, by 2030, there will be more than 8 million deaths a year, 80 percent in developing countries.

tobacco epidemic kills 5.4 million people a year and, unless significant policy changes occur, by 2030, there will be more than 8 million deaths a year, 80 percent in developing countries (WHO 2008). In the United States, an estimated 443,000 people die from smoking or exposure to secondhand smoke each year, and another 8.6 million people suffer from serious smoke-related illnesses (CDC 2009).

Annually, alcohol abuse is responsible for over 2 million deaths worldwide, 80,000 in the United States (CDC 2008). Early and continued abuse of alcohol is related to cancer, heart disease, cirrhosis of the liver, alcoholic

Figure 3.3 Tobacco Use Is a Risk Factor for Six of the Eight Leading Causes of Death in the World
Source: World Health Organization 2008.

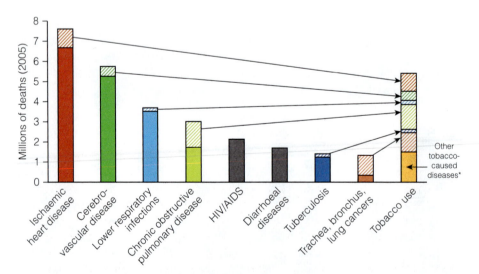

*Includes mouth and oropharyngeal cancers, oesophageal cancer, stomach cancer, liver cancers, as well as cardiovascular diseases other than ischaemic heart disease and cerebrovascular disease.

NOTE: Striped areas indicate proportions of deaths that are related to tobacco use and are colored according to the column of the respective cause of death.

hepatitis, high blood pressure, and pancreatitis (HHS 2009). Maternal prenatal alcohol use is associated with one of the leading preventable causes of birth defects and developmental disabilities in children—**fetal alcohol syndrome**—a syndrome characterized by serious physical and mental handicaps, including low birth weight, facial deformities, mental retardation, and hearing and vision problems (CDC 2006).

Heavy alcohol and drug use are also associated with negative consequences for an individual's mental health. For example, nonsmokers exposed to second-hand smoke are twice as likely to suffer from depression as nonsmokers not exposed to secondhand smoke (Elias 2009). In 2007, adults who reported a past-year major depressive episode were more likely to have reported past-year use of an illegal drug (27.4 percent) when compared to those without a major depressive episode (12.8 percent) (NSDUH 2008). Similar patterns for users of specific drugs (cocaine, marijuana, heroin, cocaine, hallucinogens, inhalants, and nonmedical use of prescription-type psychotherapeutics) were also detected.

Treatment Alternatives

In 2007, 3.9 million people aged 12 or older were treated for some kind of problem associated with the use of alcohol or illicit drugs (NSDUH 2008) (see Figure 3.4). Individuals who are interested in overcoming problem drug use have a number of treatment alternatives from which to choose. Some options include family therapy, counseling, private and state treatment facilities, community care programs, pharmacotherapy (i.e., use of treatment medications), behavior modification, drug maintenance programs, and employee assistance programs. Two commonly used techniques are inpatient or outpatient treatment and peer support groups.

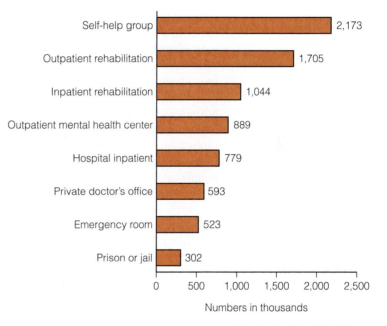

Figure 3.4 Location Where Last-Year Substance Use Treatment Was Received among Persons 12 and Older, 2007
Source: NSDUH 2008.

Inpatient and Outpatient Treatment

Inpatient treatment refers to treatment of drug dependence in a hospital and, most importantly, includes medical supervision of detoxification. Most inpatient programs last between 30 and 90 days and target individuals whose withdrawal symptoms require close monitoring (e.g., alcoholics, cocaine addicts). Some drug-dependent patients, however, can be safely treated as outpatients. Outpatient treatment allows individuals to remain in their home and work environments and is often less expensive. In outpatient treatment, patients are under the care of a physician who evaluates their progress regularly, prescribes needed medication, and watches for signs of a relapse.

The longer patients stay in treatment, the greater the likelihood of a successful recovery (NIDA 2006). Other variables that predict success include

fetal alcohol syndrome A syndrome characterized by serious physical and mental handicaps as a result of maternal drinking during pregnancy.

support of family and friends, employer intervention, a positive relationship with therapeutic staff, and a program of recovery that addresses many of the needs of the patients. Although often assumed to be so, internal motivation may not be a prerequisite for change. Researchers supported by the National Institute on Drug Abuse and the Department of Veterans Affairs (VA) studied 2,095 men who were treated for alcohol or drug problems in 15 VA hospitals and then followed for a period of five years. Some of the men were voluntarily being treated while the treatment of others was court-mandated. Investigators concluded that, although the internal motivation of those who received court-ordered treatment was initially lower, there were few differences between the two groups in terms of abstinence, recidivism, and employment five years later (Kelly et al. 2005).

Peer Support Groups

Twelve-Step Programs. Both Alcoholics Anonymous (AA) and Narcotics Anonymous (NA) are voluntary associations whose only membership requirement is the desire to stop drinking or taking drugs. AA and NA are self-help groups in that nonprofessionals operate them, offer "sponsors" to each new member, and proceed along a continuum of 12 steps to recovery. Members are immediately immersed in a fellowship of caring individuals with whom they meet daily or weekly to affirm their commitment. Some have argued that AA and NA members trade their addiction to drugs for feelings of interpersonal connectedness by bonding with other group members. In a survey of recovering addicts, more than 50 percent reported using a self-help program such as AA in their recovery (Willing 2002). AA boasts over 100,000 groups where over 2 million members meet in 150 countries (Alcoholics Anonymous 2007).

Symbolic interactionists emphasize that AA and NA provide social contexts in which people develop new meanings. Others who offer positive labels, encouragement, and social support for sobriety surround abusers. Sponsors tell the new members that they can be successful in controlling alcohol and/or drugs "one day at a time" and provide regular interpersonal reinforcement for doing so. Some research indicates that mutual support programs work. For example, in a study assessing the effectiveness of such groups, Kelly et al. (2006) concluded that involvement in such groups may be very successful—for both males and females—even when participation is limited.

Therapeutic Communities. In **therapeutic communities,** which house between 35 and 500 people for up to 15 months, participants abstain from drugs, develop marketable skills, and receive counseling. Synanon, which was established in 1958, was the first therapeutic community for alcoholics and was later expanded to include other drug users. More than 400 residential treatment centers are now in existence, including Daytop Village and Phoenix House, the largest therapeutic communities in the country. Phoenix Houses serve more than 7,000 men, women, and teens a day in over 120 locations in nine states (Phoenix House 2008). The longer a person stays at such a facility, the greater the chance of overcoming dependency. Living with a partner before entering the program and having a strong self-concept are also predictive of success (Dekel et al. 2004). Symbolic interactionists argue that behavioral changes appear to be a consequence of revised self-definition and the positive expectations of others.

therapeutic communities
Organizations in which approximately 35 to 500 individuals reside for up to 15 months to abstain from drugs, develop marketable skills, and receive counseling.

Strategies for Action: America Responds

Drug use is a complex social issue that is exacerbated by the structural and cultural forces of society that contribute to its existence. Although the structure of society perpetuates a system of inequality, creating in some the need to escape, the culture of society, through the media and normative contradictions, sends mixed messages about the acceptability of drug use. Thus, trying to end drug use by developing programs, laws, or initiatives may be unrealistic. Nevertheless, numerous social policies have been implemented or proposed to help control drug use and its negative consequences with various levels of success.

Alcohol and Tobacco

Although there may be some overlap (e.g., education), strategies to deal with alcohol and tobacco abuse are often different from those initiated to deal with illegal drugs. Prohibition, the largest social policy attempt to control a drug in the United States, was a failure, and criminalizing tobacco is likely to be just as successful. However, research has identified several promising strategies in reducing alcohol and tobacco use including economic incentives, government regulations, legal sanctions, and education and treatment.

Economic Incentives. One method of reducing alcohol and tobacco use is to increase the cost of the product. Dinno and Glantz (2009), after examining the cost of cigarettes by state, conclude that increased cigarette prices are associated with reduced prevalence and consumption rates. The benefit of increasing the cost of cigarettes, however, varies by a state's current smoking and tax rate— "A $1.00 increase in South Carolina's $0.07 cigarette tax, which is the lowest in the nation and has not increased since 1977, could increase the state's annual revenue by $180 million . . . and in 5 years would result in 78,200 fewer smokers and prevent more than 15,700 smoking-related deaths" (CASA 2009, p. 55). Similarly, Williams et al. (2005) note that "increasing the price of alcohol which can be achieved by eliminating price specials and promotions, or by raising price excise taxes, would lead to a reduction in both moderate and heavy drinking by college students" (p. 88). Other examples of economic incentives include reimbursement for smoking or alcohol cessation programs, reduced health insurance premiums for nonsmokers, and reduced car insurance premiums for nondrinkers.

Government Regulation. Whether federal, state, or local, governmental regulations have had some success in reducing tobacco and alcohol use and the problems associated with them. In 1984, states raised the legal drink age to 21 under threat of losing federal highway funds and, according to National Highway Traffic Safety Administration, over 4,000 drunken driving deaths have been prevented in last five years (NHTSA 2008b). Clean air laws restrict smoking in the workplace, bars, restaurants, and the like, and reduce consumption rates as well as secondhand smoke exposure.

One of the most important pieces of legislation in recent years is the Family Smoking Prevention and Tobacco Control Act of 2009. The law gives authority to the Food and Drug Administration to regulate the manufacturing (e.g., tobacco companies must now disclose ingredients in their products), marketing (e.g., tobacco names or logos may no longer be used to sponsor sporting events), and sale (e.g., terms like *light, mild*, and *low tar* may no longer be used) of tobacco products. Additionally, the law requires that warning labels must be strengthened and cover at least the top half of the front and back panels of the packaging by 2011 (Myers 2009a; Fact Sheet 2009).

Legal Action. Federal and state governments, as well as smokers, ex-smokers, and the families of victims of smoking have taken legal action against tobacco companies. The 1990s brought billion-dollar judgments against the tobacco industry on behalf of states suing for reimbursement of smoking-related health care costs and family's seeking punitive damages for the deaths of loved ones (Timeline 2001). In 1998, tobacco manufacturers reached a settlement with 46 states, agreeing to pay billions of dollars for reimbursement of state smoking-related health costs. The settlement also restricted the marketing, promotion, and advertising of tobacco products directed toward minors (Wilson 1999). Most recently, after years of litigation, the District of Columbia U.S. Court of Appeals held that "the tobacco industry had and continues to engage in a massive, decades-long campaign to defraud the American public . . . including falsely denying that nicotine is addictive, falsely representing that 'light,' and 'low tar' cigarettes present fewer health risks, falsely denying that they market to kids, and falsely denying that secondhand smoke causes disease" (Myers 2009b, p. 1).

Suits against retailers, distributors, and manufacturers of alcohol are more recent and often modeled after tobacco litigation. These suits primarily concern accusations of unlawful marketing, sales to underage drinkers, and failure to adequately warn of the risks of alcohol. To date, legal action against alcohol manufacturers have been much less successful than suits against the tobacco industry (Willing 2004; Marin Institute 2006)

Prevention. Although impossible to discuss all the programs and policies that are successful it is in reducing alcohol and tobacco consumption, one is worth noting. There is a large body of evidence that exposure to cigarette or alcohol advertisements increase drug use and, conversely, exposure to anti-smoking or anti-drinking advertisements reduce drug use (Terry-McElrath et al. 2007; Aloise-Young et al. 2006; Wakefield et al. 2008; Snyder et al. 2006). Despite the effectiveness of anti-smoking and anti-drinking campaigns, worldwide, for example, less than 5 percent of countries have total bans on the marketing, promotion, and sponsorship of tobacco products (WHO 2008). Anti-smoking and anti-drinking campaigns should be directed toward children. According to the National Center on Addiction and Substance Abuse at Columbia University, "a child who reached age 21 without smoking, using illicit drugs, or abusing alcohol is virtually certain never to do so" (CASA 2009, p. 6).

WEAPONS OF MASS DESTRUCTION
Tobacco killed over four million people last year.

© 2003 buttout.com

Recognizing the power of the media, in 1998, the U.S. Congress created the National Youth Anti-Drug Media Campaign under the control of the Office of National Drug Control Policy. Anti-smoking and anti-drinking campaigns are particularly successful when it comes to the prevention of tobacco and alcohol use.

harm reduction A recent public health position that advocates reducing the harmful consequences of drug use for the user as well as for society as a whole.

Illegal Drugs

The War on Drugs. In the 1980s, the federal government declared a "war on drugs," which was based on the belief that controlling drug availability would limit drug use and, in turn, drug-related problems. In contrast to a **harm reduction** position, which focuses on minimizing the costs of drug use for both user and society (e.g., distributing clean syringes to decrease the risk of HIV infection), this "zero-tolerance" approach advocates get-tough law enforcement policies, and is responsible for the dramatic increase in the jail and prison population. In 1980, there were an estimated 40,000 drug offenders in jail or prison; today, there are about 500,000 (Mauer 2009).

The harsher penalties enacted as part of the "war on drugs" required prison sentences for almost all drug offenders—first-time or repeat—and limited judicial discretion in deciding what best served the public's interest. The "Rockefeller drug laws" as they were called resulted in a disproportionate number of Hispanics and African Americans receiving "excessively long and unnecessary prison sentences" for even the most minor drug offenders (Human Rights Watch 2007, p. 1). In response to public outcries and accusations of institutional racism (see Chapter 9), reform of such laws began (Peters 2009).

What Do You Think? To recruit members and intimidate enemies, Mexican drug cartel leaders have posted videos on YouTube (Jervis 2009). Many are so gruesome that they are preceded by a warning that the content may be inappropriate for viewers under the age of 18. Although YouTube officials have alerted law enforcement agencies about the videos, it could be argued that allowing them to be broadcast is "aiding and abetting" the enemy in the war against drugs. Do you think drug cartel leaders should be able to post such videos on the YouTube website? Why or why not?

There is also the question as to whether or not the war on drugs is working—is it stopping the flow of illegal drugs into the United States and lowering drug-related problems? Several countries, including Canada, have declared the war on drugs a failure, arguing that it places too much emphasis on law enforcement, which has siphoned off money that could have been used for a more "balanced approach" (Mickleburgh & Galloway 2007). Similarly, a recent report by the National Center on Addiction and Substance Abuse at Columbia University concluded that policies that allocate billions of dollars to "disrupt and deter" the flow of illegal drugs into the United States have little impact on drug use and addiction (CASA 2009, p. 58).

There are also concerns that present policies are not only ineffective but create collateral damage. *In America's Longest War* (1994), Yale law professor Steven Duke and coauthor Albert C. Gross argued that the war on drugs, much like Prohibition, has intensified other social problems: drug-related gang violence and turf wars, the creation of syndicate-controlled black markets, unemployment, the spread of AIDS, overcrowded prisons, corrupt law enforcement officials, and the diversion of police from other serious crimes. Further, the war on drugs has had a tremendous impact on families: conviction of a drug offense can lead to a parent being in prison, eviction from public housing, deportation, and permanent exclusion from public assistance (Caulkins et al. 2005).

Finally, the war on drugs continues at an astronomical societal cost—a projected $14.1 billion in 2009 (ONDCP 2008). The U.S. policy on fighting drugs is two-pronged. First is **demand reduction,** which entails reducing the demand for drugs through treatment and prevention (see Figure 3.5). The second strategy is **supply reduction.** A much more punitive

> **demand reduction** One of two strategies in the U.S. war on drugs (the other is supply reduction), demand reduction focuses on reducing the demand for drugs through treatment, prevention, and research.

> **supply reduction** One of two strategies in the U.S. war on drugs (the other is demand reduction), supply reduction concentrates on reducing the supply of drugs available on the streets through international efforts, interdiction, and domestic law enforcement.

... [T]he war on drugs, much like Prohibition, has intensified other social problems: drug-related gang violence and turf wars, the creation of syndicate-controlled black markets, unemployment, the spread of AIDS, overcrowded prisons, corrupt law enforcement officials, and the diversion of police from other serious crimes.

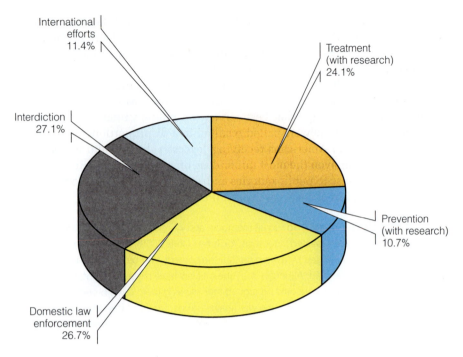

International efforts 11.4%

Treatment (with research) 24.1%

Interdiction 27.1%

Prevention (with research) 10.7%

Domestic law enforcement 26.7%

strategy, supply reduction relies on international efforts, interdiction, and domestic law enforcement to reduce the supply of illegal drugs. In 2009, 35 percent of drug control spending was focused on demand reduction and 65.0 percent on supply reduction.

One of the problems with a supply reduction strategy is that it is often dependent upon the cooperation of other governments to intervene in drug-related matters. In 2008, the U.S. Congress passed the Merida Initiative—an agreement between the U.S. and Mexico, Haiti, the Dominican Republic, and Central America—pledging $700 million in U.S. aid to help fight drug trafficking organizations. Nearly half of the initiative is focused on drug interdiction efforts and border security (Johnson 2009).

Interestingly, when American adults were asked in a national poll about financial assistance to foreign countries to fight drug trafficking, 42 percent responded that the United States was providing too much assistance (Pew Research Center 2002). Rather than foreign aid and military assistance, many argue that trade sanctions should be imposed in addition to crop eradication programs and interdiction efforts. Others, however, noting the relative failure of such programs in reducing the supply of illegal drugs entering the United States, argue that the war on drugs should be abandoned and that deregulation is preferable to the side effects of regulation.

Deregulation or Legalization. Given the questionable successes of the "war on drugs," it is not surprising that many advocate alternative measures to the rather punitive emphasis of the last several decades.

Deregulation is the reduction of government control over certain drugs. For example, although individuals must be 21 years old to purchase alcohol and 18 to purchase cigarettes, both substances are legal and can be purchased freely. In some states, possession of marijuana in small amounts is a misdemeanor rather

deregulation The reduction of government control over, for example, certain drugs.

than a felony, and in 13 states marijuana is lawfully used for medical purposes. Deregulation is popular in other countries as well. For example, personal possession of any drug, even those considered the most dangerous, is legal in Spain, Italy, the Baltic States, or the Czech Republic (*The Economist* 2009).

Proponents for the **legalization** of drugs affirm the right of adults to make informed choices. They also argue that the tremendous revenues realized from drug taxes could be used to benefit all citizens, that purity and safety controls could be implemented, and that legalization would expand the number of distributors, thereby increasing competition and reducing prices. Drugs would thus be safer, drug-related crimes would be reduced, and production and distribution of previously controlled substances would be taken out of the hands of the underworld.

Those in favor of legalization also suggest that the greater availability of drugs would not increase demand, pointing to countries where some drugs have already been decriminalized. **Decriminalization,** or the removing of penalties for certain drugs, would promote a medical rather than criminal approach to drug use that would encourage users to seek treatment and adopt preventive practices. For example, making it a criminal offense to sell or possess hypodermic needles without a prescription encourages the use of non-sterile needles that spread infections such as HIV and hepatitis.

Opponents of legalization argue that it would be construed as government approval of drug use and, as a consequence, drug experimentation and abuse would increase. Furthermore, although the legalization of drugs would result in substantial revenues for the government, drug trafficking and black markets would still flourish because all drugs would not be decriminalized (e.g., crack). Legalization would also require an extensive and costly bureaucracy to regulate the manufacture, sale, and distribution of drugs. Finally, the position that drug use is an individual's right cannot guarantee that others will not be harmed. It is illogical to assume that a greater availability of drugs will translate into a safer society.

State Initiatives. Several initiatives have resulted in statewide referendums concerning the cost-effectiveness of government policies. For example, as a result of the Substance Abuse and Crime Prevention Act of 2000 (Proposition 36), California (as well as many other states) now requires that nonviolent first- and second-time minor drug offenders receive treatment, including job training, therapy, literacy education, and family counseling rather than jail time. The act, passed by 61 percent of California voters, permanently changed state law. Every year, over 30,000 drug offenders enter treatment, about half of them for the first times (Wheeler 2008).

Lastly, over the past decade, voters and state governments have enacted significant drug policy reforms. For example, Connecticut passed significant overdose prevention legislation, following in the footsteps of New Mexico. Texas and Kansas passed legislation providing for treatment instead of incarceration for first-time drug offenders, and Illinois passed legislation allowing for the sale of sterile syringes without a prescription (Drug Policy Alliance 2007). Further, many states have active drug policy reform organizations. A Better Way Foundation in Connecticut, for example, is a nonprofit organization "dedicated to shifting current drug policy from a paradigm that prioritizes incarceration to one that prioritizes public health, treatment, and public safety" (Better Way Foundation 2008, p. 1).

legalization Making prohibited behaviors legal; for example, legalizing drug use or prostitution.

decriminalization The removal of criminal penalties for a behavior, as in the decriminalization of drug use.

According to journalist Scott Kraft, new "fencing and high-tech devices make it difficult for drug traffickers to cross the [Mexican-U.S.] border" (Kraft 2009, p. 1). As a result, smugglers have resorted to carrying their illegal imports on their backs and crossing the desert on foot. This chapter's The Human Side excerpt vividly describes the efforts of border patrol agents who have had to use "tracking skills borrowed a century ago from Native Americans: 'cutting for sign' to detect where someone has crossed the Earth's surface, and 'pushing sign,' tracking that person down."

N.M. — Bill Fraley knelt to examine the brown, pebbled soil, like an art professor studying a familiar drawing.

"See those two fine-lines?" he said, passing a finger over two shoe prints, each with washboard rows of ridges. His hand moved to another heel print a few inches away. "And there's a doper lug," the heel imprint of a boot sometimes worn by drug smugglers. . . .

A few steps away, a 5-foot barbed-wire fence cut through the cactus and greasewood, separating the United States from Mexico. The Border Patrol agent stood and tipped back the brim of his Stratton cowboy hat, eyes hidden behind aviator sunglasses. A satisfied expression hung on his chiseled face.

There were at least three of them, he figured. "It rained all day yesterday and these signs are on top of the rain," he said. "So I'd say they crossed yesterday, between 6 and 7. And it looks like they've got heavy loads of dope on them."

Tuesday, 10 a.m.

To Fraley, 50, sign-cutting is both art and science. He looks for footprints, though he usually finds just fragments. He looks for disturbances: turned-over rocks, broken twigs, bent barbed wire. He looks for chewed gum, a cigarette butt, the residue of a line of cocaine snorted on a rock. He looks for clues to fix the time. . . .

"What I really love," Fraley said," is when you come across an ant pile that's been stepped in. Ants will rebuild an anthill in an hour, so if you see a footprint in an anthill you'd better look up—because you're likely to be looking right at your adversaries.". . .

This group, Parra and Fraley agreed, would probably spend tonight moving through the Animas Mountains, which appeared in the distance, bathed in the blue shadows of puffy clouds. Tracking them through that terrain at night would be impossible. But eventually, the smugglers would have to drop down into the valley and cross westward to the Peloncillo Mountains.

That's when the Border Patrol would have its best chance to catch them.

Wednesday, 1 p.m.

Parra's radio crackled with good news. Seismic sensors in the Animas Mountains had recorded movement overnight, and that morning, agents on horseback had picked up the sign. It was the same group. . . .

About 3 p.m., Parra, 43, pulled to a stop and met up with the horse patrol unit leader, Lawrence "Junior" Helbig. Helbig and his partner had spent much of the day tracking the footprints.

"It's the same group," Helbig said. "The odds of having several fine-lines and a doper lug together are just too high."

Fraley had confirmed three sets of footprints; now Helbig had spotted two more. A late-morning shower had mucked up the trail, but the color of the soil in the prints suggested they were only about four hours old. . . .

Two weeks earlier, Helbig had tracked five smugglers into a nearby thicket. The smugglers jumped out and "quailed," running in all directions. Agents caught them and found several hundred pounds of marijuana as well as an AK-47 rifle. One of the smugglers said the weapon was for protection—from other drug smugglers.

"Thank goodness they're not real violent toward us yet," Helbig said. "But there's a reason we carry a sidearm."

Wednesday, 8:15 p.m.

Parra, tired after a 14-hour day, headed home. . . . Jose Portillo, 36, the night supervisor, set a trap.

He assigned two agents to hide on one side of County Road 1. Using thermal

Understanding Alcohol and Other Drug Use

In summary, substance abuse—that is, drugs and their use—is socially defined. As the structure of society changes, the acceptability of one drug or another changes as well. As conflict theorists assert, the status of a drug as legal or illegal is intricately linked to those who have the power to define acceptable and unacceptable drug use. There is also little doubt that rapid social change, anomie, alienation, and inequality further drug use and abuse. Symbolic interactionism

imaging binoculars, they would try to pick up the smugglers as they descended into the valley. Then they would radio another two-man unit, this one armed with M-4 rifles. If all went according to plan, the smugglers would never make it across County Road 1. . . .

At 9:15 p.m., Portillo checked in with his two teams.

Nothing.

"Tonight is the night to catch them," Portillo said, gazing out his windshield, Orion shimmering in the sky. "It's harder after this."

At 1:46 a.m., Portillo reached for the radio handset.

"Let's pull out," he told his surveillance units.

He didn't try to hide his disappointment. "They've crossed by now," he said. "They must have taken a different route."

Pause.

"Tomorrow, it's do or die."

Thursday, 10 a.m.
Parra read the overnight report: The smugglers had crossed County Road 1 several miles from the stakeout, and made it to the Peloncillo Mountains.

Rogelio Villa and his partner, on foot in the Peloncillo range, picked up the sign. "We've got our guys over here," Villa radioed Parra. "They were definitely here late last night or early this morning. Looks like there are four or five of them."

The footprints were different, but Parra wasn't worried. Smugglers often swap out their boots. "It's likely these are the same guys," he said.

Traffickers know that footprints can give them away. So they walk on rocks, where they don't leave prints. They walk backward. They wear boots like those worn by Border Patrol agents. They tie strips of carpet to their soles to avoid leaving clear prints on dirt roads. ("I've even seen them take the hoofs from cattle and glue them to their shoes," one agent said.)

Two agents jumped ahead to see how far the smugglers had gotten. At a cattle watering tank, they came upon a rancher's motion-activated game camera. An agent took the memory card out of the camera and put it in his own.

The photo that popped up was clear: a muscular, dark-haired man with a short beard, wearing black jeans and a sweat shirt under his striped shirt. A water bottle in his hand had been shrouded in black cloth, to avoid a reflection that might give his position away. On his back was a parcel, about 3 feet square. Marijuana. . . .

Friday, noon
Two infrared scopes aimed at Weatherby Canyon had detected no movement the night before. Parra returned to look for signs that the smugglers had come down the mountain. Two younger agents searched for an hour and found nothing.

Within 10 minutes of arriving, though, Parra picked up the footprints. . . .

"I'm leaning toward thinking they're still up on top of those hills," he said, lighting a Marlboro. "We were keeping this area very hot last night, and they could have stayed up."

Later, Parra reconsidered when a nearby rancher reported that his dogs had barked loudly at 3 a.m. "I don't see how we would have missed them," he said. "But the dopers must have been moving."

Saturday morning
Tim Lowe, the day supervisor, dispatched two agents on ATVs to Weatherby.

A scope had been deployed there briefly the night before, "but they 10-3ed it," Lowe said, using the code for terminating an operation. No one had been able to pick up the sign again.

Sunday afternoon
The exit for Steins Ghost Town on Interstate 10 leads to a cemetery of weathered crosses. Next to the cemetery, pieces of cloth and shoulder straps made of old blankets lay on the ground.

The drugs were gone, likely bound for Tucson and points west.

So were the smugglers, headed back to Mexico to collect their paychecks and pick up another load.

Back on the border, the day shift was out—cutting for new sign.

Source: Kraft, Scott. 2009. Pursuing smugglers, border agents become trackers. *Los Angeles Times*, May 12. http://www.latimes.com/news/nationworld/nation.

also plays a significant role in the process: If people are labeled "drug users" and are expected to behave accordingly, then drug use is likely to continue. If people experience positive reinforcement of such behaviors and/or have a biological predisposition to use drugs, the probability of their drug involvement is even higher. Thus, the theories of drug use complement rather than contradict one another.

There are two issues that need to be addressed in understanding drug use. The first is at the micro level—why does a given individual use alcohol or other drugs? Many individuals at high risk for drug use have been "failed by society"—they are living in poverty, unemployed, victims of abuse, dependents of addicted and neglectful parents, and the like. Despite the social origins of drug use, many treatment alternatives, emanating from a clinical model of drug use, assume that the origin of the problem lies within the individual rather than in the structure and culture of society. Although admittedly the problem may lie within the individual when treatment occurs, policies that address the social causes of drug abuse must be a priority in dealing with the drug problem in the United States.

The second question, related to the first, asks why drug use varies so dramatically across societies, often independent of a country's drug policies. The United States metes out some of the most severe penalties for drug violations in the world, but has one of the highest rates of marijuana and cocaine use. On the other hand, five years after Portugal decriminalized all drugs, youth drug use is down and the number of people seeking drug treatment has more than doubled (Szalavitz 2009). Most compellingly, a 2008 survey of 17 countries conducted by the World Health Organization concluded that there is no link between the harshness of drug policies and the consumption rates of its citizenry (Degenhardt et al. 2008).

That said, what is needed is a more balanced approach—one that acknowledges that not all drugs have the same impact on society or on the individuals that use them. The present administration appears to be leaning this way. For example, in a break from the previous administration, President Obama supported federally funded needle exchange programs—a staple of harm reduction advocates—and has stated that it is "entirely appropriate" to use marijuana for the same purposes and under the same controls as other drugs prescribed by a physician (Heinrich 2009; Dinan and Conery 2009). On the other hand, there is little doubt that the present administration is serious about reducing the supply of illegal drugs entering the United States, as evidenced by the President's support of the Merida Initiative and the imposition of financial sanctions against Mexican drug cartels (Hsu 2009). Only time will tell if this new approach to drug control, one that reflects both a public health and criminal justice position, will be more successful than the policies of previous administrations.

> . . . [A] 2008 survey of 17 countries conducted by the World Health Organization concluded that there is no link between the harshness of drug policies and the consumption rates of its citizenry.

CHAPTER REVIEW

- **What is a drug, and what is meant by drug abuse?**
  Sociologically, the term *drug* refers to any chemical substance that (1) has a direct effect on the user's physical, psychological, and/or intellectual functioning; (2) has the potential to be abused; and (3) has adverse consequences for the individual and/or society. Drug abuse occurs when acceptable social standards of drug use are violated, resulting in adverse physiological, psychological, and/or social consequences.

- **How do the three sociological theories of society explain drug use?**
  Structural functionalists argue that drug abuse is a response to the weakening of norms in society, leading to a condition known as anomie or normlessness. From a conflict perspective, drug use occurs as a response to the inequality perpetuated by a capitalist system as societal members respond to alienation from their work, family, and friends. Symbolic interactionism concentrates on the social meanings associated with

drug use. If the initial drug use experience is defined as pleasurable, it is likely to recur, and over time the individual may earn the label of "drug user."

■ What are the most frequently used legal and illegal drugs?

Alcohol is the most commonly used and abused legal drug in America. The use of tobacco products is also very high, with 24 percent of Americans reporting that they currently smoke cigarettes. Marijuana is the most commonly used illicit drug, with 166 million marijuana users, representing 3.9 percent of the world's population between the ages of 15 and 64.

■ What are the consequences of drug use?

The consequences of drug use are fourfold. First is the cost to the family, often manifesting itself in higher rates of divorce, spouse abuse, child abuse, and child neglect. Second is the relationship between drugs and crime. Those arrested have disproportionately higher rates of drug use. Although drug users commit more crimes, sociologists disagree as to whether drugs actually "cause" crime or whether, instead, criminal activity leads to drug involvement. Third are the economic costs (e.g., loss of productivity), which are in the billions. Last are the health costs of abusing drugs, including shortened life expectancy; higher morbidity (e.g., cirrhosis of the liver and lung cancer); exposure to HIV infection, hepatitis, and other diseases through shared needles; a weakened immune system; birth defects such as fetal alcohol syndrome; drug addiction in children; and higher death rates.

■ What treatment alternatives are available for drug users?

Although there are many ways to treat drug abuse, two methods stand out. The inpatient-outpatient model entails medical supervision of detoxification and may or may not include hospitalization. Twelve-step programs such as Alcoholics Anonymous (AA) and Narcotics Anonymous (NA) are particularly popular, as are therapeutic communities. Therapeutic communities are residential facilities where drug users learn to redefine themselves and their behavior as a response to the expectations of others and self-definition.

■ What can be done about the drug problem?

First, there are government regulations limiting the use (e.g., the law establishing the 21-year-old drinking age) and distribution (e.g., prohibitions about importing drugs) of legal and illegal drugs. The government also imposes sanctions on those who violate drug regulations and provides treatment facilities for other offenders. Economic incentives (e.g., cost) and prevention programs have also been found to impact consumption rates. Finally, legal action holding companies responsible for the consequences for their product—for example, class action suits against tobacco producers—have been fairly successful.

TEST YOURSELF

1. "Cannabis cafés" are commonplace throughout England.
   a. True
   b. False
2. The most used illicit drug in the world is
   a. heroin
   b. marijuana
   c. cocaine
   d. methamphetamine
3. What theory would argue that the continued legality of alcohol is a consequence of corporate greed?
   a. Structural functionalism
   b. Symbolic interactionism
   c. Reinforcement theory
   d. Conflict theory
4. Cigarettes smoking is
   a. the third leading cause of preventable death in the United States
   b. not addictive
   c. the most common use of tobacco products
   d. increasing in the United States
5. In the United States drinking is highest among young, nonwhite males.
   a. True
   b. False
6. According to the National Survey on Drug Use and Health, binge drinking is defined as five or more drinks per occasion on _____ or more days in a one-month period.
   a. 1
   b. 2
   c. 5
   d. 10
7. The active ingredient in marijuana, THC, can act as a sedative or a hallucinogen.
   a. True
   b. False
8. In 2009, most federal drug control dollars were allocated to
   a. international efforts
   b. domestic law enforcement
   c. prevention and research
   d. treatment and research
9. Decriminalization refers to the removal of penalties for certain drugs.
   a. True
   b. False
10. The two-pronged drug control strategy of the U.S. government entails supply reduction and harm reduction.
    a. True
    b. False

Answers: 1: b; 2: b; 3: d; 4: c; 5: b; 6: a; 7: a; 8: b; 9: a; 10: b.

KEY TERMS

binge drinking   81
chemical dependency   77
club drugs   91
crack   90
date-rape drugs   92
decriminalization   103

demand reduction   101
deregulation   102
drug   73
drug abuse   77
fetal alcohol syndrome   97
gateway drug   89

harm reduction   100
heavy drinking   81
legalization   103
meta-analysis   79
supply reduction   101
therapeutic communities   98

MEDIA RESOURCES

Understanding Social Problems,
Seventh Edition Companion Website
www.cengage.com/sociology/mooney
Visit your book companion website, where you will find flash cards, practice quizzes, Internet links, and more to help you study.

 Just what you need to know NOW! Spend time on what you need to master rather than on information you already have learned. Take a pretest for this chapter, and CengageNOW will generate a personalized study plan based on your results. The study plan will identify the topics you need to review and direct you to online resources to help you master those topics. You can then take a posttest to help you determine the concepts you have mastered and what you will need to work on. Try it out! Go to www.cengage.com/login to sign in with an access code or to purchase access to this product.

Burke/Triolo/Brand X Pictures/Jupiter Images

4

Crime and Social Control

"Unjust social
arrangements are
themselves a kind of
extortion, even
violence."

John Rawls, *A Theory of Justice*

109

Jessica Lunsford was kidnapped, raped, and murdered by John Evander Couey. In 2009, he died of natural causes while awaiting execution.

Steve Cannon/AP Photo

Jessica Lunsford was a vivacious, happy third grader, until her body was found buried in a dark hole stuffed in two black trash bags with her favorite purple toy dolphin. Her hands had been tied behind her back with speaker wire. Her small fingers were poking out of the garbage bags that served as her coffin—signs of her efforts to get air. Kidnapped, raped, and buried alive by a convicted sex offender, 9-year-old Jessica Lunsford died in the way those who loved her most tried to protect her from—in the dark, abused, and afraid (Jessica Lunsford Foundation 2006; MSNBC 2007; CNN 2007).

No one will ever know what it was like for Jessica to be abruptly awakened at 3:00 on that February 24 morning in 2005. The previous night, she had done the things she normally did on Wednesday nights and then prepared for bed. She kissed her father good night and was tucked into bed by her grandmother. The door was left open a crack so the light from the family room could come in as she fell asleep (Kruse 2007).

High on cocaine, convicted sex offender John Evander Couey entered Jessie's dark bedroom in Homosassa, Florida, put his hand over her mouth, and told her she was going with him. Three weeks later, after Couey was captured in Georgia, Jessie's body was found not far from where she lived. In 2007, Couey was convicted of murder, rape, and kidnapping. Throughout the proceedings, Couey sat at the defense table and colored in a coloring book—something Jessica Lunsford will never again do.

Adam Walsh, Amber Hagerman, Megan Kanka, and Jessica Lunsford: tragically, these are the names of children whose abductions have led to changes in the criminal justice system. "The impacts of Jessica Lunsford's story have been huge. . . . Enormous good has come and thousands of children's lives are going to be saved because Jessica Lunsford lived," said Ernie Allen, president and chief executive officer of the National Center for Missing and Exploited Children (CNN 2007, p. 1). On April 19, 2005, the state of Florida passed the Jessica Lunsford Act, a tough sex offender law that requires a "25-year minimum prison term for people convicted of certain sex crimes against children, and lifetime tracking by global positioning satellite once they're outside of prison" (Associated Press 2005, p. 1). Despite some evidence that enforcing the law is resource intensive and may have unwanted side effects, at least 43 states have passed similar laws (Rothfeld 2009).

Federal laws such as the Jessica Lunsford Act are just one part of the massive bureaucracy called the criminal justice system, a system that often comes under public scrutiny, particularly in high-profile cases, such as the kidnapping, rape, and murder of Jessica Lunsford, the unsolved murders of five women at a clothing store, and the stalking death of college student Johanna Justin-Jinich (Stafford & Johnson 2009; Nelson 2009).

In this chapter, we examine the criminal justice system as well as theories, types, and demographic patterns of criminal behavior. The economic, social, and psychological costs of crime are also examined. The chapter concludes with a discussion of social control, including policies and prevention programs designed to reduce crime in the United States.

The Global Context: International Crime and Social Control

Several facts about crime are true throughout the world. First, crime is ubiquitous—there is no country where crime does not exist. Second, most countries have the same components in their criminal justice systems: police, courts, and prisons. Third adult males make up the largest category of crime suspects worldwide; and fourth, in all countries, theft is the most common crime committed, whereas violent crime is a relatively rare event.

Even so, dramatic differences do exist in international crime rates, although comparisons are made difficult by variations in measurement and crime definitions (Siegel 2006). Because of these difficulties, Winslow and Zhang (2008, pp. 30–32) created a global crime database from data from the United Nations and Interpol—the international police agency. In their discussion of the "United States versus the World," the authors reveal some interesting and often counterintuitive findings. First, the United States does not have the highest crime rate in the world. Using their database, the United States ranks 12th among 165 nations, with Sweden, Denmark, Australia, and Great Britain, in rank order, each having a higher crime rate than the United States.

Winslow and Zhang (2008) also examined crime rates by dividing them into types of crime—violent crime or property crime. Violent crime, as discussed later in the chapter, includes murder, rape, robbery, and aggravated assault. When one compares the United States to other countries, the United States once again is not in the top 10. Several developing countries (e.g., Namibia and Swaziland) as well as developed countries (e.g., Australia and Sweden) have higher violent crime rates than the United States. Property crimes show a similar pattern. Based on the global crime database created by Winslow and Zhang (2008), the United States ranks 13th in property crimes (car theft, burglary, and larceny) with Sweden, Denmark, Australia, and Great Britain topping the list.

Violent crime and property crimes represent just two types of crime that take place worldwide. Although we are concerned about these types of crimes and the possibility of victimization, Interpol has identified six global priority areas (Interpol 2009): (1) drugs and criminal organizations (e.g., drug trafficking), (2) financial and high-tech crimes (e.g., counterfeiting, fraud, and cybercrime), (3) tracing of fugitives, (4) countering terrorism (discussed in Chapter 15), (5) trafficking in human beings, and 6) fighting corruption (e.g. enforcing the rule of law). Each of these priority areas contains a relatively new category of crimes—transnational crime. As defined by the U.S. Department of Justice, **transnational crime** is "organized criminal activity across one or more national borders" (U.S. Department of Justice 2003). The significance of transnational crime should not be minimized. As Shelley states (2007):

> Transnational crime will be a defining issue of the 21st century for policy makers—as defining as the Cold War was for the 20th century and colonialism was for the 19th. Terrorists and transnational crime groups will proliferate because these crime groups are major beneficiaries of globalization. They take advantage of increased travel, trade, rapid money movements, telecommunications and computer links, and are well positioned for growth. (p. 1)

For example, the Internet has led to an explosive growth in child pornography. In 2005, a United Nations expert on the subject told the 53-nation U.N.

transnational crime Criminal activity that occurs across one or more national borders.

According to the U.S. Department of State, between 600,000 and 800,000 people are trafficked across international borders annually—80 percent are women and girls and 50 percent are minors—and millions more are trafficked within their own country.

Commission on Human Rights that governments must act now to curb the proliferation of child pornography (Klapper 2005). Of late, several successes have been recorded. In 2007, the Austrian authorities reported uncovering an international pornography ring in which suspects viewed online videos of children being sexually abused. The pornography ring was estimated to involve more than 2,300 suspects from 77 countries including the United States, Germany, France, South Africa, and Russia (Associated Press 2007).

Human trafficking is another example of transnational crime. According to the U.S. Department of State, between 600,000 and 800,000 people are trafficked across international borders annually—80 percent are women and girls and 50 percent are minors—and millions more are trafficked within their own country. Although the majority of people are trafficked into commercial sexual exploitation, others are trafficked into forced labor, sexual servitude, and for use as child soldiers (U.S. State Department 2008).

What Do You Think? According to the International Maritime Bureau (IMB), hijacking of ships and other vessels reached an all time high in 2008—293 incidents reported, 49 ships hijacked, and 889 crew taken hostage, of which 32 were injured, 11 killed, and 21 missing and presumed dead (IMB 2009). Somali pirates alone hijacked more than 40 ships in 2008, resulting in tens of millions of dollars in ransom (Gettleman 2009). In federal law, death resulting from aircraft piracy, or attempted hijacking of an aircraft, carries the death penalty. Do you think that deaths that result from maritime hijacking should carry the death penalty? What if the death is not of a hostage but of a fellow hijacker?

Sources of Crime Statistics

The U.S. government spends millions of dollars annually to compile and analyze crime statistics. A **crime** is a violation of a federal, state, or local criminal law. For a violation to be a crime, however, the offender must have acted voluntarily and with intent and have no legally acceptable excuse (e.g., insanity) or justification (e.g., self-defense) for their behavior. The three major types of statistics used to measure crime are official statistics, victimization surveys, and self-report offender surveys.

crime An act, or the omission of an act, that is a violation of a federal, state, or local criminal law for which the state can apply sanctions.

crime rate The number of crimes committed per 100,000 population.

clearance rate The percentage of crimes in which an arrest and official charge have been made and the case has been turned over to the courts.

Official Statistics

Local sheriffs' departments and police departments throughout the United States collect information on the number of reported crimes and arrests and voluntarily report them to the Federal Bureau of Investigation (FBI). The FBI then compiles these statistics annually and publishes them, in summary form, in the Uniform Crime Reports (UCR). The UCR lists **crime rates** or the number of crimes committed per 100,000 population, the actual number of crimes and the percentage of change over time, and clearance rates. **Clearance rates** measure the percentage of cases in which an arrest and official charge have been made and the case has been turned over to the courts.

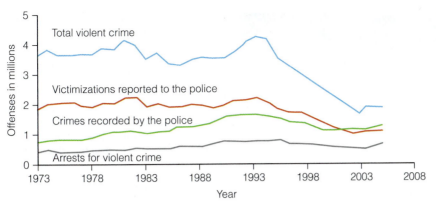

Figure 4.1 Four Measures of Serious Violent Crime
Source: BJS 2006a.
The serious violent crimes included are rape, robbery, aggravated assault, and homicide.

These statistics have several shortcomings. For example, many incidents of crime go unreported. It is estimated that, in 2005, only 38 percent of rapes and sexual assaults, 52 percent of robberies, 47 percent of assaults, and 56 percent of household burglaries were actually reported to the police (U.S. Census Bureau 2007). Even if a crime is reported, the police may not record it (see Figure 4.1). Alternatively, some rates may be exaggerated. Motivation for such distortions may come from the public (e.g., demanding that something be done), from political officials (e.g., election of a sheriff), and/or from organizational pressures (e.g., budget requests). For example, a police department may "crack down" on drug-related crimes in an election year. The result is an increase in the recorded number of these offenses for that year. Such an increase reflects a change in the behavior of law enforcement personnel, not a change in the number of drug violations. Thus, official crime statistics may be a better indicator of what police are doing rather than of what criminals are doing.

In the 1970s, the "law enforcement community called for a thorough evaluation of the UCR Program to recommend an expanded and enhanced data collection system to meet the needs of law enforcement in the 21st century" (FBI 2009a, p. 1). The result, the National Incident-Based Reporting System (NIBRS), requires that law enforcement agencies provide extensive information on each criminal incident and arrest for 46 offenses (Group A) and arrestee information on eleven lesser offenses (Group B). Thus far, the FBI has certified 31 states for NIBRS participation. The hope is that, once implemented nationally, the NIBRS will provide more reliable and comprehensive crime data.

Victimization Surveys

Acknowledging "the dark figure of crime," that is, the tendency for so many crimes to go unreported and thus undetected by the UCR, the U.S. Department of Justice conducts the National Crime Victimization Survey (NCVS). Begun in 1973 and conducted annually, the NCVS interviews over 130,000 people about their experiences as victims of crime. Interviewers collect a variety of information, including the victim's background (e.g., age, race and ethnicity, sex, marital status, education, and area of residence), relationship to the offender (stranger or non-stranger), and the extent to which the victim was harmed. In 2007, the latest year for which victimization data are available, teens and young adults were the most likely to be the victims of violent crime, and those older than age 65 were most likely to be the victims of property crime (BJS 2008a). Although victimization surveys provide detailed information about crime victims, they provide less reliable data on offenders.

Read each of the following questions. If, since the age of 16, you have ever engaged in the behavior described, place a "1" in the space provided. If you have not engaged in the behavior, put a "0" in the space provided. After completing the survey, read the section on interpretation to see what your answers mean.

Questions

1. Have you ever been in possession of drug paraphernalia? _____

2. Have you ever lied about your age or about anything else when making application to rent an automobile? _____

3. Have you ever obtained a false ID to gain entry to a bar or event? _____

4. Have you ever tampered with a coin-operated vending machine or parking meter? _____

5. Have you ever shared, given, or shown pornographic material to someone under 18? _____

6. Have you ever begun and/or participated in a basketball, baseball, or football pool? _____

7. Have you ever used "filthy, obscene, annoying, or offensive" language while on the telephone? _____

8. Have you ever given or sold a beer to someone under the age of 21? _____

9. Have you ever been on someone else's property (land, house, boat, structure, and so on) without that person's permission? _____

10. Have you ever forwarded a chain letter with the intent to profit from it? _____

11. Have you ever improperly gained access to someone else's e-mail or other computer account? _____

12. Have you ever written a check for over $150 when you knew it was bad? _____

Interpretation

Each of the activities described in these questions represents criminal behavior that was subject to fines, imprisonment, or both under the laws of Florida in 2008. For each activity, the following table lists the maximum prison sentence and/or fine for a first-time offender. To calculate your "prison time" and/or fines, sum the numbers corresponding to each activity you have engaged in.

Offense	Maximum Prison Sentence	Maximum Fine
1. Possession of drug paraphernalia	1 year	$1,000
2. Fraud	5 years	$5,000
3. Possession of false ID or driver's license	5 years	$5,000
4. Larceny	2 months	$500
5. Protection of minors from obscenity	5 years	$5,000
6. Illegal gambling	2 months	$500
7. Harassing/obscene telecommunications	2 months	$500
8. Illegal distribution of alcohol	2 months	$500
9. Trespassing	1 year	$1,000
10. Illegal gambling	1 year	$1,000
11. Illegal misappropriation of cyber communication	5 years	$5,000
12. Worthless check	5 year	$5,000

Source: Florida Criminal Code (2008).

Self-Report Offender Surveys

Self-report surveys ask offenders about their criminal behavior. The sample may consist of a population with known police records, such as a prison population, or it may include respondents from the general population, such as college students.

Self-report data compensate for many of the problems associated with official statistics but are still subject to exaggerations and concealment. The Criminal Activities Survey in this chapter's Self and Society feature asks you to indicate whether you have engaged in a variety of illegal activities.

Self-report surveys reveal that virtually every adult has engaged in some type of criminal activity. Why then is only a fraction of the population labeled criminal? Like a funnel, which is large at one end and small at the other, only a small proportion of the total population of law violators are ever convicted of a crime. For individuals to be officially labeled criminals, their behavior (1) must become known to have occurred; (2) must come to the attention of the police who then file a report, conduct an investigation, and make an arrest; and finally, (3) the arrestee must go through a preliminary hearing, an arraignment, and a trial and may or may not be convicted. At every stage of the process, offenders may be "funneled" out. As Figure 4.1 indicates, the measures of crime used at various points in time lead to different results.

Sociological Theories of Crime

Some explanations of crime focus on psychological aspects of the offenders, such as psychopathic personalities, unhealthy relationships with parents, and mental illness. Other crime theories focus on the role of biological variables, such as central nervous system malfunctioning, stress hormones, vitamin or mineral deficiencies, chromosomal abnormalities, and a genetic predisposition toward aggression. Sociological theories of crime and violence emphasize the role of social factors in criminal behavior and societal responses to it.

Structural-Functionalist Perspective

According to Durkheim and other structural functionalists, crime is functional for society. One of the functions of crime and other deviant behavior is that it strengthens group cohesion: "The deviant individual violates rules of conduct that the rest of the community holds in high respect; and when these people come together to express their outrage over the offense . . . they develop a tighter bond of solidarity than existed earlier" (Erikson 1966, p. 4).

Crime can also lead to social change. For example, an episode of local violence may "achieve broad improvements in city services . . . be a catalyst for making public agencies more effective and responsive, for strengthening families and social institutions, and for creating public-private partnerships" (National Research Council 1994, pp. 9–10).

Although structural functionalism as a theoretical perspective deals directly with some aspects of crime, it is not a theory of crime per se. Three major theories of crime have developed from structural functionalism, however. The first, called strain theory, was developed by Robert Merton (1957) and uses Durkheim's concept of *anomie,* or normlessness. Merton argued that, when the structure of society limits legitimate means (e.g., a job) of acquiring culturally defined goals (e.g., money), the resulting strain may lead to crime.

Individuals, then, must adapt to the inconsistency between means and goals in a society that socializes everyone into wanting the same thing but provides opportunities for only some (see Table 4.1). *Conformity* occurs when individuals accept the culturally defined goals and the socially legitimate means of achieving them. Merton suggested that most individuals, even those who do not have easy access to the means and goals, remain conformists. *Innovation* occurs when an individual accepts the goals of society but rejects or lacks the socially legitimate means of achieving them. Innovation, the mode of adaptation most associated with criminal

TABLE 4.1 Merton's Strain Theory

MODE OF ADAPTATION	CULTURALLY DEFINED GOALS	STRUCTURALLY DEFINED MEANS
1. Conformity	+	+
2. Innovation	+	–
3. Ritualism	–	+
4. Retreatism	–	–
5. Rebellion	±	±

+ = acceptance of/access to; – = rejection of/lack of access to; ± = rejection of culturally defined goals and structurally defined means and replacement with new goals and means

Source: Adapted with permission of The Free Press, a Division of Simon & Schuster Adult Publishing Group, from Robert K. Merton's *Social theory and social structure* (1957). Copyright © 1957 by The Free Press; copyright renewed 1985 by Robert K. Merton. All rights reserved.

behavior, explains the high rate of crime committed by uneducated and poor individuals who do not have access to legitimate means of achieving the social goals of wealth and power.

Another adaptation is *ritualism*, in which, for example, individuals accept a lifestyle of hard work but reject the cultural goal of monetary rewards. Ritualists go through the motions of getting an education and working hard, yet they are not committed to the goal of accumulating wealth or power. *Retreatism* involves rejecting both the cultural goal of success and the socially legitimate means of achieving it. Retreatists withdraw or retreat from society and may become alcoholics, drug addicts, or vagrants. Finally, *rebellion* occurs when individuals reject both culturally defined goals and means and substitute new goals and means. For example, rebels may use social or political activism to replace the goal of personal wealth with the goal of social justice and equality.

Whereas strain theory explains criminal behavior as a result of blocked opportunities, subcultural theories argue that certain groups or subcultures in society have values and attitudes that are conducive to crime and violence. Members of these groups and subcultures, as well as other individuals who interact with them, may adopt the crime-promoting attitudes and values of the group. For example, Kubrin and Weitzer (2003) found that retaliatory homicide is a response to subcultural norms of violence that exist in some neighborhoods.

However, if blocked opportunities and subcultural values are responsible for crime, why don't all members of the affected groups become criminals? Control theory may answer that question. Hirschi (1969), consistent with Durkheim's emphasis on social solidarity, suggests that a strong social bond between individuals and the social order constrains some individuals from violating social norms. Hirschi identified four elements of the social bond: attachment to significant others, commitment to conventional goals, involvement in conventional activities, and belief in the moral standards of society. Several empirical tests of Hirschi's theory support the notion that the higher the attachment, commitment, involvement, and belief, the higher the social bond and the lower the probability of criminal behavior. For example, Ford (2005), using data from the National Youth Survey, concludes that a strong family bond lowers the probability of adolescent substance use and delinquency, and Bell (2009) reports that weaker attachment to parents is associated with a greater likelihood of gang membership for both males and females.

Conflict Perspective

Conflict theories of crime suggest that deviance is inevitable whenever two groups have differing degrees of power; in addition, the more inequality there is in a society, the greater the crime rate in that society. Social inequality leads individuals to commit crimes such as larceny and burglary as a means of economic survival. Other individuals, who are angry and frustrated by their low position in the socioeconomic hierarchy, express their rage and frustration through crimes such as drug use, assault, and homicide. In Argentina, for example, the soaring violent crime rate is hypothesized to be "a product of the enormous imbalance in income distribution . . . between the rich and the poor" (Pertossi 2000).

According to the conflict perspective, those in power define what is criminal and what is not, and these definitions reflect the interests of the ruling class. Laws against vagrancy, for example, penalize individuals who do not contribute to the capitalist system of work and consumerism. Furthermore, D'Alessio and

Stolzenberg (2002, p. 178) found that "in cities with high unemployment, unemployed defendants have a substantially higher probability of pretrial detention" than employed defendants. Rather than viewing law as a mechanism that protects all members of society, conflict theorists focus on how laws are created by those in power to protect the ruling class. For example, wealthy corporations contribute money to campaigns to influence politicians to enact tax laws that serve corporate interests (Reiman & Leighton 2010).

In addition, conflict theorists argue that law enforcement is applied differentially, penalizing those without power and benefiting those with power. For example, a 2009 report by the National Council on Crime and Delinquency found that ". . . in arrests, court processing and sentencing, new admissions and ongoing populations in prison and jails, probation and parole, capital punishment, and recidivism . . . persons of color, particularly African Americans, are more likely to receive less favorable results than their white counterparts" (Hartney & Vuong 2009, p. 2). Female prostitutes are more likely to be arrested than are the men who seek their services. Rape laws originated to serve the interests of husbands and fathers who wanted to protect their property—wives and unmarried daughters. Finally, unlike street criminals, corporate criminals are most often punished by fines rather than lengthy prison terms.

Societal beliefs also reflect power differentials. For example, "rape myths" are perpetuated by the male-dominated culture to foster the belief that women are to blame for their own victimization, thereby, in the minds of many, exonerating the offenders. Such myths include the notion that when a woman says *no* she means *yes,* that "good girls" don't get raped, that appearance indicates willingness, and that women secretly want to be raped. Not surprisingly, there is less rape in societies where women and men have greater equality.

Yadid Levy/Alamy

To Marxists, the cultural definition of women as property contributes to the high rates of female criminality and, specifically, involvement in prostitution, drug abuse, and petty theft. In the Netherlands, prostitution has been legal since 2000. Amsterdam's "red-light district" is famous for its displays of "window prostitutes."

Symbolic Interactionist Perspective

Two important theories of crime emanate from the symbolic interactionist perspective. The first, labeling theory, focuses on two questions: How do crime and deviance come to be defined as such, and what are the effects of being labeled criminal or deviant? According to Howard Becker (1963):

> Social groups create deviance by making rules whose infractions constitute deviance, and by applying those rules to particular people and labeling them as outsiders. From this point of view, deviance is not a quality of the act a person commits, but rather a consequence of the application by others of rules and sanctions to an "offender." The deviant is one to whom the label has successfully been applied; deviant behavior is behavior that people so label. (p. 238)

Labeling theorists make a distinction between **primary deviance,** which is deviant behavior committed before a person is caught and labeled an offender, and **secondary deviance,** which is deviance that results from being caught and labeled. After a person violates the law and is apprehended, that person is

primary deviance Deviant behavior committed before a person is caught and labeled an offender.

secondary deviance Deviant behavior that results from being caught and labeled as an offender.

stigmatized as a criminal. This deviant label often dominates the social identity of the person to whom it is applied and becomes the person's "master status," that is, the primary basis on which the person is defined by others.

Being labeled as deviant often leads to further deviant behavior because (1) the person who is labeled as deviant is often denied opportunities for engaging in nondeviant behavior, and (2) the labeled person internalizes the deviant label, adopts a deviant self-concept, and acts accordingly. For example, a teenager who is caught selling drugs at school may be expelled and thus denied opportunities to participate in nondeviant school activities (e.g., sports and clubs) and to associate with nondeviant peer groups. The labeled and stigmatized teenager may also adopt the self-concept of a "druggie" or "pusher" and continue to pursue drug-related activities and membership in the drug culture.

The assignment of meaning and definitions learned from others is also central to the second symbolic interactionist theory of crime, differential association. Edwin Sutherland (1939) proposed that, through interaction with others, individuals learn the values and attitudes associated with crime as well as the techniques and motivations for criminal behavior. Individuals who are exposed to more definitions favorable to law violation (e.g., "crime pays") than to unfavorable ones (e.g., "do the crime, you'll do the time") are more likely to engage in criminal behavior. Thus, children who see their parents benefit from crime or who live in high-crime neighborhoods where success is associated with illegal behavior are more likely to engage in criminal behavior.

> Individuals who are exposed to more definitions favorable to law violation (e.g., "crime pays") than to unfavorable ones (e.g., "do the crime, you'll do the time") are more likely to engage in criminal behavior.

Unfavorable definitions come from a variety of sources. Of particular concern of late is the role of video games in promoting criminal or violent behavior. *Grand Theft Auto* has players "head bashing, looting, drug-dealing, drive-by shooting, and running over innocent bystanders with a taxi" (Richtel 2003). In response to this and other violent video games, many states now require a video rating system that differentiates between cartoon violence, fantasy violence, intense violence, and sexual violence. In 2009, a federal court of appeals struck down a California law that prohibited the sale or rental of violent video games to anyone under 18 as unconstitutional (*USA Today* 2009).

What Do You Think? Othón Cuevas Córdova, a congressman from Oaxaca in Mexico, was shocked over the Christmas holidays when his nephew pointed a toy gun at him and said, "Tío, I'm going to kill you" (Lacey 2009, p. 1). In a country overrun with violent crime, ". . . from infancy, children are learning the culture of violence and we need to do something about it" said the congressman (p. 1). In 2009, he introduced a proposal in Mexico's National Assembly to ban the "fabrication, importation, and sale of toy guns and other warlike toys" (p. 1). Do you think that exposure to violent toys, games, movies, and the like leads to aggressiveness? If so, could that explain gender differences in violent crime rates?

index offenses Crimes identified by the FBI as the most serious, including personal or violent crimes (homicide, assault, rape, and robbery) and property crimes (larceny, motor vehicle theft, burglary, and arson).

Types of Crime

The FBI identifies eight index offenses as the most serious crimes in the United States. The **index offenses,** or street crimes as they are often called, can be against a person (called violent or personal crimes) or against property (see Table 4.2).

Other types of crime include vice crime (such as drug use, gambling, and prostitution), organized crime, white-collar crime, computer crime, and juvenile delinquency. Hate crimes are discussed in Chapter 9.

Street Crime: Violent Offenses

The most recent data available from the FBI's Uniform Crime Reports indicate that the 2007 violent crime rate decreased from the previous year by 1.4 percent. Remember, however, that crime statistics represent only those crimes *reported* to the police: 1.4 million violent crimes in 2007. Victim surveys indicate that a little over half of all violent crimes are actually reported to the police (U.S. Census Bureau 2007).

Violent crime includes homicide, assault, rape, and robbery. *Homicide* refers to the willful or nonnegligent killing of one human being by another individual or group of individuals. Although homicide is the most serious of the violent crimes, it is also the least common, accounting for 1.2 percent of all violent crimes (FBI 2008a). A typical homicide scenario includes a male killing a male with a handgun after a heated argument. The victim and offender are disproportionately young and of minority status. When a woman is murdered and the victim-offender relationship is known, she is most likely to have been killed by her husband or boyfriend (FBI 2008a).

Mass murders have more than one victim in a killing event. In 2007, Cho Seung-Hui, dubbed the "Virginia Tech killer," was responsible for one of the largest mass murders in U.S. history, killing more than 30 students and faculty on the college campus. Unlike mass murder, serial murder is the "unlawful killing of two or more victims by the same offender(s), in separate events" (U.S. Department of Justice 2008, p. 1). The most well-known serial killers, who were responsible for some of the most horrific episodes of homicide, are Ted Bundy, Kenneth Bianchi, and Jeffrey Dahmer. More recently, Dennis Rader, the self-proclaimed "BTK" (bind, torture, kill) killer was captured. Accused of killing 10 people (2 men and 8 women) between 1974 and 1991, Rader was convicted of murder and received 10 consecutive life sentences with no chance of parole for 175 years (Coates 2005; Romano 2005).

Another form of violent crime, *aggravated assault*, involves attacking a person with the intent to cause serious bodily injury. Like homicide, aggravated assault occurs most often between members of the same race and, as with violent crime in general, is more likely to occur in warm weather months. In 2007, the assault rate was over 50 times the murder rate, with assaults making up an estimated 60.8 percent of all violent crimes (FBI 2008a).

Rape is also classified as a violent crime and is also intraracial; that is, the victim and offender tend to be from the same racial group. The FBI definition of *rape* contains three elements: sexual penetration, force or the threat of force,

TABLE 4.2 Index Crime Rates, Percentage Change, and Clearance Rates, 2007

	RATE PER 100,000, 2007	PERCENTAGE CHANGE IN RATE (2006–2007)	PERCENTAGE CLEARED, 2007
Violent crime			
Murder	5.6	−1.3	61.2
Forcible rape	30.0	−3.2	40.0
Robbery	147.6	−1.2	25.9
Aggravated assault	283.8	−1.3	54.1
Total	466.9	−1.4	44.5
Property crime			
Burglary	722.5	−0.9	12.4
Larceny/theft	2,177.8	−1.3	18.6
Motor vehicle theft	363.3	−8.8	12.6
Arson	24.7*	−6.7	18.3
Total[†]	3,263.5	−2.1	16.5

Source: FBI 2008b.

*Arson rates per 100,000 are calculated independently because population coverage for arson is lower than for the other index offenses. Percent change is for volume, not rate.

[†] Property crime totals do not include arson.

In many jurisdictions, victims are allowed to make statements to the court describing the impact of their victimization on their life and the lives of friends and family. Here, 23-year-old Jessica poignantly describes the often unseen consequences of violent crime, in this case, the impact of a brutal rape.

My name is Jessica. I once knew what that meant. Now all I can tell you is that I am still Jessica, however, this no longer holds any meaning to me. I have lost my identity in the cruelest of ways.

I was a 23-year-old single mother, a sister, a daughter, a girlfriend, partner, friend, and confidant to many people. I was strong, independent, and willing to make an effort. I was attempting to hold down a second job to make a better life for my child and a better person out of me. I was destroyed.

I was raped. Not once, not twice, but so many times and in so many ways that it all becomes a blur. These images haunt my days, my nights, my dreams, and my realities. I am no longer the person I was before. I was once the person that people could rely on. Now I am a shell of my former self, a speck of the brave person that was Jessica. I had my way of life, my self-esteem, my respect, and my dignity stripped from me in the most terrifying of situations.

My trust in people is all but destroyed. I even have trouble enjoying a quiet drink with my partner, family, or friends without feeling anxious and wary. I am constantly looking over my shoulder, fearing there is someone there who wants to hurt me. This is only the beginning. I fluctuate from extreme insomnia to extreme fatigue. My motivation is gone. My joy of motherhood is waning. My ability to love and care for others is disappearing. My trust in the justice system is all but gone. I feel like I tried so hard and was beaten down. I feel weak and vulnerable.

I suffered physical trauma to my shoulder, knees, and feet from being dragged along carpet. I suffered cuts and bruising to my genitals from continuous rapes. I suffered marks to my ankles and wrists from being bound, and I lost chunks of hair from being gagged with tape around my head. I thought I was going to die. I take medication to sleep, to relax, and to stay relatively sane. I live as a corpse with no visible future to aim for. I have cut myself several times to try and take away the pain in my heart. I don't have the strength to end my own life.

What I experienced was like looking into the eyes of Satan himself. I am not a religious person but I know that I have seen the very depths of hell.

I wish that my loved ones didn't know what happened to me. To see the horror in their eyes and feel the pain in their hearts is unbearable. My beautiful daughter doesn't know the beginning of what I suffered but she suffered too. She feels like she was abandoned by her mother and required counseling.

The trauma happened to my partner. I can see it in the way he looks at me. Our communication, love, sex, and friendship is going to take a long time to repair. He has been so deeply affected by my experience but I can't help him and he can't help me. Nobody can.

I was burgled at work on my third day of my second job. I was bound and gagged. I was abducted. I was lied to. I was confused. I was cold and alone. I WAS RAPED. This man, who does not deserve a name, hurt me in the most unimaginable of ways. Yes I survived. Yes I am alive, I just don't live. I am existing. I am empty. . . .

I am a real person who went through torture. I am not a statistic or a nameless face on the street. I am your mother, your sister, your daughter, and your friend. What happened to me was real. At least give me the satisfaction of seeing this man put in jail for the maximum term of 25 years. After all, if I don't qualify as a victim of the most violent, serious, and heinous of crimes then who the hell does?

I am Jessica.

Source: The Herald Sun 2008.

---

and nonconsent of the victim. In 2007, 90,427 forcible rapes were reported in the United States (FBI 2008a). Rapes are more likely to occur in warm months, in part because of the greater ease of victimization. People are outside more and later, doors are open, windows are unlocked, and so forth.

Perhaps as much as 80 percent of all rapes are **acquaintance rapes**—rapes committed by someone the victim knows. Although acquaintance rapes are the most likely to occur, they are the least likely to be reported and the most difficult to prosecute. Unless the rape is what Williams (1984) calls a **classic rape**—that is, the rapist was a stranger who used a weapon and the attack resulted in serious bodily injury—women hesitate to report the crime out of fear of not being believed. The increased use of "rape drugs," such as Rohypnol, may lower reporting levels even further (see Chapter 3). This chapter's The Human Side poignantly describes the impact of rape on one woman's life.

acquaintance rape Rape committed by someone known to the victim.

classic rape Rape committed by a stranger, with the use of a weapon, resulting in serious bodily injury to the victim.

Robbery, unlike simple theft, also involves force or the threat of force or putting a victim in fear and is thus considered a violent crime. Officially, in 2007, more than 445,125 robberies took place in the United States. Robberies are most often (42.8 percent) committed with the use of a firearm and occur disproportionately in southern states (FBI 2008a). Robbers and thus robberies vary dramatically in type, from opportunistic robberies whose victims are easily accessible and that yield only a small amount of money, to professional robberies of commercial establishments, such as banks, jewelry stores, and convenience stores. According to the FBI, in 2007, the average dollar value lost per robbery was $1,321. Of different victim categories, the highest average dollar loss was $4,201 per bank robbery (FBI 2008a).

Street Crime: Property Offenses

Property crimes are those in which someone's property is damaged, destroyed, or stolen; they include larceny, motor vehicle theft, burglary, and arson. The number of property crimes has gone down since 1998, with a 10.1 percent decrease in the last decade. *Larceny*, or simple theft, accounts for more than two-thirds of all property arrests (FBI 2008a). In 2007, the average dollar value lost per larceny incident was $886. Examples of larcenies include purse snatching, theft of a bicycle, pickpocketing, theft from a coin-operated machine, and shoplifting. In 2007, an estimated 6.6 million larcenies were reported in the United States (FBI 2008a), the most common index offense.

Larcenies involving automobiles and auto accessories are the largest category of thefts. However, because of the cost involved, *motor vehicle theft* is considered a separate index offense. Numbering more than 1.1 million in 2007, the motor vehicle theft rate has decreased 11.8 percent since 1998 (FBI 2008a). Because of insurance requirements, vehicle theft is one of the most highly reported index crimes, and, consequently, estimates between the FBI's Uniform Crime Reports and the National Crime Victimization Survey are fairly compatible. Less than 13 percent of motor vehicle thefts are cleared.

Burglary, which is the second most common index offense after larceny, entails entering a structure, usually a house, with the intent to commit a crime while inside. Official statistics indicate that, in 2007, more than 2.1 million burglaries occurred, a rate of 722 per 100,000 population (FBI 2008a). Most burglaries are residential rather than commercial and take place during the day when houses are unoccupied. The most common type of burglary is forcible entry, followed by unlawful entry.

Arson involves the malicious burning of the property of another. Estimating the frequency and nature of arson is difficult given the legal requirement of "maliciousness." Of the reported cases of arson, 42.9 percent involved structures (most of which were residential), and 27.9 percent involved movable property (e.g., boat or car), with the remainder being miscellaneous property (e.g., crops or timber). In 2007, the average dollar amount of damage as a result of arson was $17,289 (FBI 2008a).

Looking across index offense categories, it should be noted that all violent and property crime rates have decreased between 2006 and 2007. Interestingly, despite the economic downturn and the oft-cited relationship between the economy and crime, preliminary analysis of 2008 crime data indicates a decrease in crime rates for each of the eight index offenses (FBI 2009b).

Vice Crime

Vice crimes, often thought of as crimes against morality, are illegal activities that have no complaining participant(s) and are often called **victimless crimes.** Examples of vice crimes include using illegal drugs, engaging in or soliciting prostitution, illegal gambling, and pornography.

Most Americans view drug use as socially disruptive (see Chapter 3). There is less consensus, however, nationally or internationally, that gambling and prostitution are problematic. For example, the Netherlands legalized prostitution in 2000, hoping to cut the ties between the sex trade and organized crime—a link that remains. Alternatively, Sweden penalizes clients of prostitutes and treats sex workers as victims, an approach several European countries including England and Wales are now adopting (*The Economist* 2008).

In the United States, prostitution is illegal with the exception of several counties in Nevada. Despite its illegal status, it is a multimillion-dollar industry with 77,607 arrests for prostitution and commercial vice in 2007 (FBI 2008a). Motivated by profit, smugglers traffic thousands of women and children into the United States for purposes of prostitution. Trafficking *within* the United States also occurs. In 2003, the Innocence Lost National Initiative was established to address domestic sex trafficking of children. Federal, state, and local law enforcement efforts have resulted in the recovery of 577 child victims. It is estimated that there are "tens of thousands" of child sex trafficking victims in the United States (FBI 2008b).

What Do You Think? In May 2005, Brazil, where prostitution is legal, turned down $40 million in U.S. AIDS prevention grants to protest what is being called a "loyalty oath against prostitution" (Kaplan 2005). The loyalty oath requires that U.S. and foreign nongovernmental organizations (NGOs) take an explicit stance opposed to prostitution in return for U.S. funding. Despite several court findings that the policy violates the First Amendment rights of the plaintiff organizations (Brennan Center for Justice 2009), the loyalty oath remains in effect for agencies in foreign countries wishing to receive AIDS prevention grants from the United States government. Do you think that U.S. funding for AIDS prevention to other countries should be linked to and contingent on a loyalty oath against prostitution?

Gambling is legal in many U.S. states including casinos in Nevada, New Jersey, Connecticut, North Carolina, and other states, as well as state lotteries, bingo parlors, horse and dog racing, and jai alai. In addition, although illegal in the United States, online gambling flourishes on offshore gambling sites. Some have argued that there is little difference, other than societal definitions of acceptable and unacceptable behavior, between gambling and other risky ventures such as investing in the stock market. Conflict theorists are quick to note that the difference is who is making the wager.

Pornography, particularly Internet pornography, is a growing international problem. Regulation is made difficult by fears of government censorship and legal wrangling as to what constitutes "obscenity." For many, the concern with pornography is not its consumption per se but the possible effects of viewing or reading pornography—increased sexual aggression. Although the literature on this topic is mixed, Conklin (2007, p. 221) concluded that there is no "consistent evidence that nonviolent pornography causes sex crimes."

Organized crime refers to criminal activity conducted by members of a hierarchically arranged structure devoted primarily to making money through illegal

victimless crimes Illegal activities that have no complaining participant(s) and are often thought of as crimes against morality, such as prostitution.

organized crime Criminal activity conducted by members of a hierarchically arranged structure devoted primarily to making money through illegal means.

means. Although discussed under victimless crimes because of its association with prostitution, drugs, and gambling, organized crime groups often use coercive tactics. For example, organized crime groups may force legitimate businesses to pay "protection money" by threatening vandalism or violence.

The traditional notion of organized crime is the Mafia, a national band of interlocked Italian families. However, members of many ethnic groups engage in organized crime in the United States:

> Chinese, Vietnamese, Korean, and Japanese gangs have been found on the East and West coasts, active in smuggling drugs and extorting money from businesses in their communities. Scores of other groups can be found in various cities: Israelis dominating insurance fraud in Los Angeles, Cubans running illegal gambling operations in Miami, Canadians engaging in gun smuggling and money laundering in Miami, Russians carrying out extortion and contract murders in New York. (Thio 2007, p. 374)

Organized crime also occurs in other countries. For example, with more than 90,000 members and associates in 3,000 crime groups, the Japanese Yakuza are one of the largest crime organizations in the world. The young men who join the Yakuza tend to be from the lower class and must undergo a training period of five years. During this apprenticeship, members learn absolute loyalty to their superiors as well as the other norms and values of the group. The Yakuza are involved in drugs, illegal gambling, and prostitution, as well as several legitimate businesses. Interestingly, the Yakuza proudly display their name at their "corporate" headquarters, and recruits wear lapel pins identifying themselves as members (Thio 2004; Winslow & Zhang 2008).

Unlike traditional crime organizations that are hierarchically arranged, transnational crime organizations tend to be decentralized and less likely to operate through legitimate businesses. Transnational crime organizations, like transnational crime in general, directly or indirectly involve more than one country. Transnational crime organizations are a growing threat to the United States and to global security. As Wagley (2006) explained:

> The end of the Cold War—along with increasing globalization beginning in the 1990s—has helped criminal organizations expand their activities and gain global reach. Criminal networks are believed to have benefited from the weakening of certain government institutions, more open borders, and the resurgence of ethnic and regional conflicts across the former Soviet Union and many other regions. Transnational criminal organizations have also exploited expanding financial markets and rapid technological developments. (p. 1)

Transactional crime organizations are also less likely than the traditional crime "families" to develop around a family or ethnic structure. Transnational crime organizations are involved in many types of transnational crime including money laundering, narcotics, arms smuggling, and trafficking in people. Further, terrorists are increasingly supporting themselves through transnational organized crime groups. For example, the 2003 bombing of a Madrid commuter train was financed through drug dealing (Wagley 2006).

Transnational crime organizations are involved in many types of transnational crime including money laundering, narcotics, arms smuggling, and trafficking in people.

White-Collar Crime

White-collar crime includes both *occupational crime,* in which individuals commit crimes in the course of their employment, and *corporate crime,* in which corporations violate the law in the interest of maximizing profit. Occupational crime is motivated by individual gain. Employee theft of merchandise, or pilferage, is one of the most common types of occupational crime. Other examples include embezzlement, forgery and counterfeiting, and insurance fraud. Price fixing, antitrust violations, and security fraud are all examples of corporate crime, that is, crime that benefits the organization.

In recent years, several officers of major corporations, including Enron, Worldcom, Adelphia, and ImClone, have been charged with securities fraud, tax evasion, and insider trading. Joseph Cassano, a former executive at AIG (American International Group, Inc.), is presently being investigated by the FBI, the Securities and Exchange Commission, and the British Serious Fraud Office for misrepresenting the extent of AIG's losses after the company insured a trillion dollars worth of risky loans held by banks. AIG has received over $180 billion in bailout money from the U.S. government to meet its financial obligations, and is often "credited" with playing a major role in the U.S. economic crisis (Schecter et al. 2009; Salow 2009).

In 2002, President Bush signed the Sarbanes-Oxley Act, which significantly increased penalties for white-collar offenders. Nonetheless, many white-collar criminals go unpunished. First, many companies, not wishing the bad publicity surrounding a scandal, simply dismiss the parties involved rather than press charges. Second, many white-collar crimes, as traditional crimes, go undetected. In a survey of a representative sample of 1,600 U.S. households, the National White Collar Crime Center (NWCCC) found that nearly one in two households (46.5 percent) had been the victim of some type of white-collar crime in the last year. However, only 14 percent of the crimes were brought to the attention of a law enforcement agency (NWCCC 2006).

Third, federal prosecutions of white-collar criminals have decreased recently. Few believe the decrease is a result of a lower prevalence of white-collar crime offenses. Two forces appear to be in operation. First, white-collar crimes are becoming increasingly complex, making prosecution a time- and resource-intensive endeavor. Second, the decrease in white-collar crime prosecutions represents a shift in priorities (Marks 2006). After the events of 9/11, nearly one-third of FBI agents were moved from criminal programs to terrorism and intelligence duties leaving "the bureau seriously exposed in investigating areas such as white-collar crime. . ." (Lichtblau et al. 2008, p. 1). Today, no doubt triggered by investigations of mortgage fraud on the heels of the subprime mortgage crisis, the Obama

On March 12, 2009, Bernard Madoff pled guilty to 11 felony counts related to a massive Ponzi scheme run through his investment firm. The scheme to defraud investors of $65 billion dollars took place over a 20-year period and involved thousands of victims. In June of 2009, the 71-year-old Madoff was sentenced to 150 years, a term usually reserved for "terrorists, traitors, and the most violent criminals" (Hayes & Neumeister 2009, p. 1). Madoff was also ordered to forfeit $171 billion in personal property.

Kathy Willens/AP Photo

white-collar crime Includes both *occupational crime*, in which individuals commit crimes in the course of their employment, and *corporate crime*, in which corporations violate the law in the interest of maximizing profit.

administration is focusing its attention on financial crimes. The 2010 federal budget provides resources to expand the number of FBI agents and government attorneys fighting white-collar crime (U.S. Department of Justice 2009).

Corporate violence, a form of corporate crime, refers to the production of unsafe products and the failure of corporations to provide a safe working environment for their employees. Corporate violence is the result of negligence, the pursuit of profit at any cost, and intentional violations of health, safety, and environmental regulations. For example, in 2007, contaminated pet food was responsible for the deaths of hundreds of dogs and cats. An investigation led to the discovery that the combination of two ingredients, melamine (used in the manufacturing of plastics) and cyanuric acid (used to chlorinate pools), was responsible for the deaths (McCormick 2007; FDA 2008). In 2008, two Chinese executives and a U.S. company president and CEO were indicted for scheming to import tainted products used in the manufacturing of pet food (FDA 2008). Similarly, in 2009, a Chinese court sentenced two men to death and one to life in prison for endangering public safety by producing and selling contaminated dairy products. Three hundred thousand children became ill as a result of melamine-laced milk, and at least six died (Barboza 2009). Finally, Peanut Corporation of America (PCA) is being investigated for its part in the largest outbreak of salmonella poisoning in U.S. history. Allegedly, PCA continued to ship peanut products after they tested positive for salmonella. The outbreak resulted in 3,000 products being recalled, 700 reported illnesses, and nine deaths (Schmit 2009). Table 4.3 summarizes some of the major categories of white-collar crime.

TABLE 4.3 Types of White-Collar Crime

CRIMES AGAINST CONSUMERS	CRIMES AGAINST EMPLOYEES
Deceptive advertising	Health and safety violations
Antitrust violations	Wage and hour violations
Dangerous products	Discriminatory hiring practices
Manufacturer kickbacks	Illegal labor practices
Physician insurance fraud	Unlawful surveillance practices
CRIMES AGAINST THE PUBLIC	**CRIMES AGAINST EMPLOYERS**
Toxic waste disposal	Embezzlement
Pollution violations	Pilferage
Tax fraud	Misappropriation of government funds
Security violations	Counterfeit production of goods
Police brutality	Business credit fraud

Computer Crime

Computer crime refers to any violation of the law in which a computer is the target or means of criminal activity. Sometimes called cybercrime, computer crime is one of the fastest-growing types of crime in the United States. Hacking, or unauthorized computer intrusion, is one type of computer crime. In 2008, eleven people in an "international ring of thieves" were charged with stealing over 41 million credit and debit card numbers from national retail outlets such as Barnes and Noble, T.J. Maxx, and BJ's Wholesale Club (Stone 2008).

In 2008, the Consumer Sentinel Network (CSN) logged nearly one million consumer complaints, with identity theft being the most common complaint category (FTC 2009). **Identity theft** is the use of someone else's identification (e.g., social security number or birth date) to obtain credit or other economic rewards. Although mail theft is one of the most common modes of obtaining the needed information, new technologies have contributed to the increased rate of identity theft. For example, in 2006, a Department of Veterans Affairs official downloaded the personnel records of more than 26 million veterans to his laptop computer, which was then stolen, "exposing all the information necessary to swipe the identity of virtually every person released from military service since 1975" (Levy 2006).

Identity theft is just one category of computer crime. Another category is Internet fraud. According to a 2008 report by the Internet Crime Complaint Center (ICCC

corporate violence The production of unsafe products and the failure of corporations to provide a safe working environment for their employees.

computer crime Any violation of the law in which a computer is the target or means of criminal activity.

identity theft The use of someone else's identification (e.g., social security number, birth date) to obtain credit or other economic rewards.

TABLE 4.4 Amount Lost by Selected Fraud Type for Individuals Reporting Monetary Loss 2008

COMPLAINT TYPE	% OF REPORTED TOTAL LOSS	OF THOSE WHO REPORTED A LOSS THE AVERAGE (MEDIAN) $ LOSS PER COMPLAINT
Check Fraud	7.8%	$3,000.00
Confidence Fraud	14.4%	$2,000.00
Nigerian Letter Fraud	5.2%	$1,650.00
Computer Fraud	3.8%	$1,000.00
Nondelivery (merchandise and payment)	28.6%	$800.00
Auction Fraud	16.3%	$610.00
Credit/Debit Card Fraud	4.7%	$223.00

Source: ICCC 2009.

2008), the most common type of Internet fraud was nondelivery of merchandise or payment (32.9 percent), followed by Internet auction fraud (e.g., eBay), which comprised 25.5 percent of all complaints. Other Internet fraud categories include check fraud, confidence fraud, Nigerian letter fraud (i.e., a letter offering the recipient the "opportunity" to share in millions of dollars being illegally transferred to the United States—just send us your bank account numbers!), computer fraud, and credit or debit card fraud. Monetary losses and the percent of total losses attributed to each fraud category are displayed in Table 4.4.

Another type of computer crime is online child sexual exploitation. According to the National Center for Missing and Exploited Children (NCMEC), more than 30 million children younger than age 18 are on the "Net," and one in seven "receives a sexual solicitation online which includes a request to engage in sexual activity, a request to engage in sexual talk, or a request to give out personal sexual information" (NCMEC 2007, p. 1). The FBI's Innocent Images National Initiative investigates online child sexual exploitation as well as Internet child pornography (FBI 2008c). Online child pornography/child sexual exploitation investigations comprised 39 percent of all investigations in the FBI's Cyber Division in 2007. Between 1996 and 2007, there was a 2,062 percent increase in the number of online child sexual exploitation/child pornography investigations opened.

What Do You Think? With increased concern over child sexual predators, many states are legislating new and tougher laws. In 2009, Minnesota legislators approved a bill that limits Internet use by sex offenders who are released from prison under intensive supervision. The new law, which takes effect in 2010, prohibits offenders to "log on to, create or maintain a personal Web page or social networking account if it permits contact with anyone under 18" (Brunswick 2009, p. 1). Probation officers will now have the authority to make unannounced inspections of an offender's computer. Do you think such a law will reduce the number of child predators? Do you think the new law violates the civil rights of the offender?

Juvenile Delinquency

In general, children younger than age 18 are handled by the juvenile courts, either as status offenders or as delinquent offenders. A *status offense* is a violation that can be committed only by a juvenile, such as running away from home, truancy, and underage drinking. A *delinquent offense* is an offense that would be a crime if committed by an adult, such as the eight index offenses. The most common status offenses handled in juvenile court are underage drinking, truancy, and running away. In 2007, 15.4 percent of all arrests (excluding traffic violations) were of offenders younger than age 18 (FBI 2008a). As is the case with adults, juveniles commit more property crimes than violent crimes.

Although the number of juveniles arrested for violent crimes decreased by 14.1 percent between 1998 and 2007, Americans remain concerned about juvenile violence and, specifically, the high rate of gang-related violence. The growth of gangs is, in part, a function of two interrelated social forces: the increased availability of guns in the 1980s, and the lucrative and expanding drug trade. In 2008, there were an estimated one million gang members in more than 20,000 local, regional, and national gangs throughout the United States (National Gang Intelligence Center 2009). According to law enforcement agencies, gangs are becoming more organized and are expanding their criminal involvement from urban communities to suburban and rural areas. Gangs are estimated to be responsible for as much as 80 percent of all crime in some communities, and are becoming more violent. Between 2003 and 2007, 94.3 percent of all gang-related homicides were committed with the use of a firearm. Some gangs, such as Mara Salvatrucha (MS-13), "deal" in weapons, selling them to other gang members for a profit as well as keeping them for their own personal use.

Demographic Patterns of Crime

Although virtually everyone violates a law at some time, individuals with certain demographic characteristics are disproportionately represented in the crime statistics. Victims, for example, are disproportionately young, lower-class, minority males from urban areas. Similarly, the probability of being an offender varies by gender, age, race, social class, and region (see Figure 4.2).

Gender and Crime

It is a universal truth that women everywhere are less likely to commit crime than men. In the United States, both official statistics and self-report data indicate that females commit fewer violent crimes than males. In 2007, males accounted for 75.8 percent of all arrests, 81.8 percent of all arrests for violent crime, and 66.6 percent of all arrests for property crimes (FBI 2008a). Not only are females less likely than males to commit serious crimes, but also the monetary value of female involvement in theft, property damage, and illegal drugs is typically less than that for similar offenses committed by males.

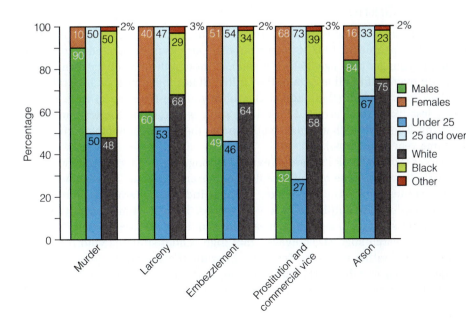

Figure 4.2 **Percentage of Arrests by Sex, Age, and Race, 2007**
Source: FBI 2008a.

Nonetheless, rates of female criminality have increased dramatically over the last decade. For example, between 1998 and 2007, arrest rates for women increased for robbery (20.3 percent), burglary (10.9 percent), and drug abuse violations (28.6 percent) (FBI 2008a). Heimer et al. (2005) argue that such increases are a function of the *economic marginalization* of women relative to men, i.e., female criminality goes up when women's economic circumstances in relation to men's decline.

The recent increase in crimes committed by females has led to growth of feminist criminology. **Feminist criminology** focuses on how the subordinate position of women in society affects their criminal behavior and victimization. For example, Chesney-Lind and Shelden (2004) reported that arrest rates for runaway juvenile females are higher than those for males not only because girls are more likely to run away as a consequence of sexual abuse in the home but also because police with paternalistic attitudes are more likely to arrest female runaways than male runaways. Feminist criminology thus adds insights into understanding crime and violence that are often neglected by traditional theories concentrating on gender inequality in society. Feminist criminology has also had an impact on public policy. Mandatory arrest for domestic violence offenders, the development of rape shield laws, public support for battered women's shelters, laws against sexual harassment, and the repeal of the spousal exception in rape cases are all, according to Winslow and Zhang (2008), outcomes of feminist criminology.

The subordinate position of women in the United States also affects their victimization rates. A report from the Harvard School of Public Health revealed that 70 percent of all female homicide victims in industrial countries are American (Harvard School of Public Health 2002). A female in the United States is five times more likely to be murdered than a female in Germany, eight times more likely to be murdered than a female in England, and three times more likely to be murdered than a female in Canada. Although males are almost four times more likely to be murdered than females, when a woman is murdered, it is most likely by an ex-boyfriend, husband, or other intimate partner (BJS 2006b).

Age and Crime

In general, criminal activity is more prevalent among younger people than among older people. In 2007, 44.4 percent of all arrests in the United States were of people younger than age 25 (FBI 2008a). Although those younger than age 25 made up over half of all arrests in the United States for crimes such as robbery, burglary, motor vehicle theft, and arson, those younger than age 25 were significantly less likely to be arrested for crimes such as fraud, and forgery and counterfeiting. Those older than age 65 made up less than 1 percent of total arrests for the same year (FBI 2008a).

Why is criminal activity more prevalent among individuals in their late teens and early twenties? One reason is that juveniles are insulated from many of the legal penalties for criminal behavior. Younger individuals are also more likely to be unemployed or employed in low-wage jobs. Thus, as strain theorists argue, they have less access to legitimate means for acquiring material goods.

Some research suggests, however, that high school students who have jobs become more, rather than less, involved in crime (Felson 2002). In earlier generations, teenagers who worked did so to support themselves and/or their families. Today, teenagers who work typically spend their earnings on recreation and

feminist criminology An approach that focuses on how the subordinate position of women in society affects their criminal behavior and victimization.

"extras," including car payments and gasoline. The increased mobility associated with having a vehicle also increases the opportunity for criminal behavior and reduces parental control.

Other hypothesized reasons for the age-crime relationship are also linked to specific theories of criminal behaviors. For example, conflict theorists would argue that teenagers and young adults have less power in society than their middle-aged and elderly counterparts. One manifestation of this lack of power is that the police, using a mental map of who is a "typical offender," are more likely to have teenagers and young adults in their suspect pool. With increased surveillance of teenagers and young adults comes increased detection of criminal involvement—a self-fulfilling prophecy.

What Do You Think? Crime statistics are sensitive to demographic changes. For example, crime rates in the United States began to rise in the 1960s as baby boom teenagers entered high school. That said, recent FBI arrest data indicate that offenders are disproportionately young males (FBI 2008a; FBI 2009b). Do you think that crime rates are a function of the number of young males in the general population? If so, do you think their elevated numbers reflect their higher rates of criminal involvement or institutional bias in the criminal justice system?

Race, Social Class, and Crime

Race is a factor in who gets arrested. Minorities are disproportionately represented in official statistics. For example, African Americans represent about 13 percent of the population but account for 39 percent of all arrests for violent index offenses, and 29.8 percent of all arrests for property index offenses (FBI 2008a). They have 3.5 times the arrest rate of whites for drug offenses, are six times more likely to be admitted to prison, and, if admitted to prison for a violent crime, receive longer sentences than their white counterparts (Hartney & Vuong 2009).

Nevertheless, it is inaccurate to conclude that race and crime are causally related. First, official statistics reflect the behaviors and policies of criminal justice actors. Thus, the high rate of arrests, conviction, and incarceration of minorities may be a consequence of individual and institutional bias against minorities. For example, **racial profiling**—the practice of targeting suspects on the basis of race—may be responsible for their higher arrest rates. Proponents of the practice argue that because race, like gender, is a significant predictor of who commits crime, the practice should be allowed. Opponents hold that racial profiling is little more than discrimination and should, therefore, be abolished.

Mark Richards/Photo Edit

Most street gangs are local or neighborhood gangs that operate from a single location, ranging in size from just a few to several hundred members (National Gang Intelligence Center 2009). Here members of the Death Squad gang flash signs while at a city park in Los Angeles.

racial profiling The law enforcement practice of targeting suspects on the basis of race.

[African Americans] ... have 3.5 times the arrest rate of whites for drug offenses, are six times more likely to be admitted to prison, and, if admitted to prison for a violent crime, receive longer sentences than their white counterparts.

Second, race and social class are closely related in that nonwhites are overrepresented in the lower classes. Because lower-class members lack legitimate means to acquire material goods, they may turn to instrumental, or economically motivated, crimes. In addition, although the "haves" typically earn social respect through their socioeconomic status, educational achievement, and occupational role, the "have-nots" more often live in communities where respect is based on physical strength and violence, as subcultural theorists argue. For example, Kubrin (2005) examined the "street code" of inner-city black neighborhoods by analyzing rap music lyrics. Her results indicate that song "lyrics instruct listeners that toughness and the willingness to use violence are central to establishing viable masculine identity, gaining respect, and building a reputation" (p. 375). This chapter's Social Problems Research Up Close feature examines violence and "being tough" in a sample of minority youth.

A third hypothesis is that criminal justice system contact, which is higher for nonwhites, may actually act as the independent variable; that is, it may lead to a lower position in the stratification system. Kerley and colleagues (2004) found that "contact with the criminal justice system, especially when it occurs early in life, is a major life event that has a deleterious effect on individuals' subsequent income level" (p. 549).

Some research indicates, however, that even when social class backgrounds of blacks and whites are comparable, blacks have higher rates of criminality. Sampson et al. (2005), using data from nearly 3,000 respondents aged 18 to 25, found that the likelihood of self-reported violence by blacks was 85 percent higher than for whites. Interestingly, the likelihood of Latino self-reported violence was 10 percent less than that reported by whites.

Region and Crime

In general, crime rates and, in particular, violent crime rates, are higher in metropolitan areas than in nonmetropolitan areas. In 2007, the violent crime rate in metropolitan statistical areas was 504 per 100,000 population; in cities in non-metropolitan statistical areas, it was 395 per 100,000 population (FBI 2008a).

Higher crime rates in urban areas result from several factors. First, social control is a function of small intimate groups that socialize their members to engage in law-abiding behavior, expressing approval for their doing so and disapproval for their noncompliance. In large urban areas, people are less likely to know each other and thus are not influenced by the approval or disapproval of strangers. Demographic factors also explain why crime rates are higher in urban areas: large cities have large concentrations of poor, unemployed, and minority individuals. Finally, some of the nation's most violent cities, including the ten most dangerous cities in the United States, have been identified by the U.S. Department of Justice as transit points for Mexican drug cartels (Greenburg 2009).

Although large urban areas, in general, have higher crime rates, they have been decreasing over the last 15 years (Rosin 2008). For example, crime in New York City is the lowest it has been in 40 years, with murder rates continuing to decline in 2009 (*Associated Press* 2009). Recent violent crime trends may indicate a shift from large urban areas to small towns with populations of less than 10,000. In 2008, small towns were the only community type that had increases in the numbers of murders (5.5 percent), rapes (1.4 percent), and robberies (3.9 percent) (FBI 2009a).

A large body of research documents that fighting in adolescence is a fairly common event although frequencies vary significantly by race, age, and sex. For example, black youth are more likely to report being involved in at least one physical fight in the previous year when compared to their white counterparts (CDCP 2008). Thus the present study is an important one for it examines the relationship between select explanatory variables (both risks and assets) and the likelihood of fighting among a sample of at-risk minority youth (Wright & Fitzpatrick 2006).

Sample and Methods

All respondents were in grades 5 through 12 and were enrolled in a central Alabama school system (Wright & Fitzpatrick 2006). The school district sent home a letter detailing the purpose of the study, and parents were asked to give permission for their child's inclusion in the study and referred to a sample questionnaire that was on file and available for review. The final sample consisted of 1,642 African American youth (51 percent female) with a median age of 14 years. Participation in the survey was voluntary, and the response rate was 65 percent.

The dependent variable is fighting and was measured by asking respondents the frequency of their fighting in the last 30 days. Socio-demographic variables include sex, age, mother's and father's education, and mother's and father's occupational status. Risk factors, that is, factors associated with an increase in the likelihood of fighting, include academic performance (poor grades), family intactness (zero or one parent in the home), parental violence (physical assault from parent or other adult guardian), and a composite measure of gang affiliation (being a member, being asked to be a member, or having friends who are members).

Asset variables are variables that are predicted to decrease incidents of fighting. The eight asset variables were divided into three categories. The first category is self-esteem measured by a respondent's sense of satisfaction, pride, worth, and respect. Parental involvement was measured by a respondent's (1) parents monitoring of where their child goes with their friends, (2) frequency of talking to parents about problems, and (3) frequency of eating dinner with the family. School involvement includes teacher attention, respondent's involvement in school activities and clubs, and self-reported happiness with school.

Results and Conclusions

Frequency of fighting was higher for elementary and middle school students than for high school students, with the highest fighting frequency occurring in middle schools. Across school type, 15 percent or more of the students reported the highest response category of fighting—six or more times over the last 30 days.

Data analysis also indicates that fighting is negatively associated with family intactness and self-esteem, with two-parent homes and high self-esteem leading to lower probabilities of fighting. Alternatively, parental violence and gang affiliation is associated with increased probabilities of fighting. Talking to parents about problems and having parents who monitor activities with friends are significantly associated with decreased rates of fighting. Additional asset variables associated with decreased levels of fighting include being happy with school, attention from teachers, and involvement with school clubs. As the authors note, interestingly, higher involvement in sports was associated with higher rates of fighting.

When multiple variables were analyzed at the same time, three of the four risk factors were significantly associated with higher rates of fighting—lower grades, exposure to violence in the family, and gang affiliation. Of the asset variables, a lack of parental monitoring and being unhappy at school were predictive of increased fighting behavior. Note that low self-esteem, when in the presence of other variables, is not associated with youthful fighting.

The authors concluded that the "risk and asset" model they present has practical implications in terms of organizing intervention techniques. Risk factors need to be "suppressed or eliminated" and asset factors need to be "encouraged or facilitated" (Wright & Fitzpatrick 2006, p. 260). For example, results from the present study show that "parental monitoring and being happy at school were associated with lower frequency of fighting, suggesting the importance of continued support for outreach to parents and further efforts to reduce or eliminate the community factors that promote proliferation of gangs" (Wright & Fitzpatrick 2006, p. 251).

Source: Wright and Fitzpatrick 2006.

Crime rates also vary by region of the country. In 2007, both violent and property crimes were highest in southern states, followed by western, midwestern, and northeastern states. Violent crime is particularly high in the South, with 45.8 percent of all murders and 44.9 percent of all aggravated assaults recorded in southern states (FBI 2008a). The high rate of southern lethal violence has been linked to high rates of poverty and minority populations in the South, a southern "subculture of violence," higher rates of gun ownership, and a warmer climate that facilitates victimization by increasing the frequency of social interaction.

The Costs of Crime and Social Control

The costs of crime and violence are difficult to quantify but minimally include physical injury and loss of life, economic losses, and social and psychological costs.

Physical Injury and Loss of Life

Crime often results in physical injury and loss of life. For example, homicide is the second most common cause of death among 15- to 25-year-olds, exceeded only by accidental death (U.S. Census Bureau 2009). In 2007, there were nearly 17,000 victims of homicide (FBI 2008a). That number is dwarfed, however, by the deaths that take place as a consequence of white-collar crime. Criminologist Steven Barkan (2006), who collected data from a variety of sources, reported that annually there are (1) 56,425 workplace-related deaths from illness or injury; (2) 9,600 deaths from unsafe products; (3) 35,000 deaths from environmental pollution; and (4) 12,000 deaths from unnecessary surgery. Adding these figures together, 113,025 people a year die from corporate and professional crime and misconduct (p. 388).

Moreover, the U.S. Public Health Service now defines violence as one of the top health concerns facing Americans. Health initiatives related to crime include reducing drug and alcohol use and the deaths and diseases associated with them, lowering rates of domestic violence, preventing child abuse and neglect, and reducing violence through public health interventions. Finally, it must be noted that crime has mental as well as physical health consequences. For example, violent crime, and particularly rape and sexual assault, are related to post-traumatic stress disorder.

Economic Costs

Conklin (2007, p. 50) suggested that the financial costs of crime can be classified into at least six categories. First are *direct losses* from crime, such as the destruction of buildings through arson, of private property through vandalism, and of the environment by polluters. In 2007, the average dollar loss of destroyed or damaged property as a result of arson was $17,289 (FBI 2008a). Second are costs associated with the *transferring of property*. Bank robbers, car thieves, and embezzlers have all taken property from its rightful owner at tremendous expense to the victim and society. For example, it is estimated that, in 2007, $7.4 billion was lost as a result of motor vehicle theft; the average value per vehicle at the time of the theft was $6,775 (FBI 2008a).

A third major cost of crime is that associated with *criminal violence,* including the medical cost of treating crime victims. The National Crime Prevention Council (NCPC 2005) estimates that the average cost for *each* criminal incident of rape or sexual assault is $7,700, including expenses related to medical and mental health care, law enforcement, and victim and social services.

Fourth are the costs associated with the production and sale of illegal goods and services, that is, *illegal expenditures*. The expenditure of money on drugs, gambling, and prostitution diverts funds away from the legitimate economy and enterprises, and lowers property values in high-crime neighborhoods. Fifth is the cost of *prevention and protection*—the billions of dollars spent on locks and safes, surveillance cameras, guard dogs, and the like. It is estimated that Americans spend $65 billion annually on self-protection items (Surgeon General 2002).

Finally, there is the cost of *controlling crime.* In 2005, federal, state, and local governments spent over $200 billion for law enforcement, judicial services, and corrections (NCPC 2008). Correction costs alone have soared 127 percent in the last 20 years, with one in every $18 state general fund spent on corrections in 2008 (Lawrence 2009).

Although the costs from "street crimes" are staggering, the costs from "crimes in the suites," such as tax evasion, fraud, false advertising, and antitrust violations, are greater than the cost of the FBI index crimes combined (Reiman & Leighton 2010). Further, Barkan (2006), using an FBI estimate, reported that the total cost of property crime and robbery is $17.1 billion annually. This is less than the $44 billion price tag for employee theft alone.

Social and Psychological Costs

Crime entails social and psychological costs as well as economic costs. One such cost—fear—is dependent upon individual perceptions of crime as a problem. When a random sample of Americans were asked, "Is there more crime in the United States than there was a year ago, or less?," 67 percent responded more, 15 percent responded less, and 9 percent responded crime rates were the same (Gallup Poll 2009a). Thus, despite continuing declines in crime rates, the public's perception is that crime has increased.

Such misconceptions are fueled by media presentations that may not accurately reflect the crime picture. For example, Krisberg et al. (2009), when comparing youth crime rates to newspaper coverage of juvenile crime, concluded that the media doesn't "provide a balanced perspective on crime or youth issues," instead focusing on "crime increases and 'crime emergencies'" (p. 7).

Not only do Americans worry about crime at the aggregate level, but they also worry about crime at the individual level. When a random sample of Americans were asked the extent to which they worry about crime, 46 percent reported that they frequently or occasionally worry about having their house burglarized, 43 percent frequently or occasionally worry about having their car stolen or broken into, and 31 percent frequently or occasionally worry about a school-aged child being physically harmed at school (Gallup Poll 2009a).

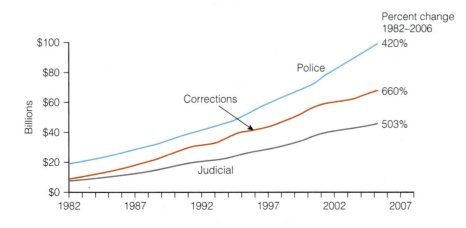

Figure 4.3 Direct Expenditure by Criminal Justice Function, 1982–2006
Source: BJS 2008b.

How do most Americans deal with their fear of street crime? According to a Gallup Poll, the most common method of dealing with fear of victimization is to "avoid going to certain places/neighborhoods you might otherwise want to go to" (Carlson 2005, p. 2). Such behavioral changes are just one category of social and psychological costs that Shapland and Hall (2007) identified. Others include a sense of shock, a loss of trust, feelings of guilt for being victimized, anger, and a sense of vulnerability. Although these responses vary by type of offense, it should not be concluded that white-collar crimes do not carry a social and psychological toll. Rosoff et al. (2002) state that white-collar crime can produce "feelings of cynicism among the public, remove an essential element of trust from everyday interaction, de-legitimatize political institutions, and weaken respect for the law" (p. 346). In addition, the authors argue that white-collar crime "encourages and facilitates" other types of crime; that is, "there is a connection, both direct and indirect, between 'crime in the suites' and 'crime in the streets'" (p. 346).

Strategies for Action: Crime and Social Control

Clearly, one way to combat crime is to attack the social problems that contribute to its existence. Moreover, when a random sample of Americans were asked which of two views came closer to their own in dealing with the crime problem, increasing law enforcement or resolving social problems, the majority of respondents (65 percent) selected resolving social problems (Gallup Poll 2007).

In addition to policies that address social problems, numerous social programs have been initiated to alleviate the crime problem. These policies and programs include local initiatives, criminal justice policies, legislative action, and international efforts in the fight against crime.

Local Initiatives

Youth programs such as the Boys and Girls Club, and community programs that involve families and schools are an effective "first line of defense" against crime and juvenile delinquency.

Youth Programs. Early intervention programs acknowledge that preventing crime is better than "curing" it once it has occurred. Fight Crime: Invest in Kids is a nonpartisan, nonprofit anticrime organization made up of more than 3,000 law enforcement leaders and violence survivors (Fight Crime 2009). The organization

> takes a hard-nosed look at crime prevention strategies, informs the public and policy makers about those findings, and urges investment in programs proven effective by research. . . . [The] organization focuses on high-quality early education programs, prevention of child abuse and neglect, after-school programs for children and teens, and interventions to get troubled kids back on track. (p. 1)

One such program is the Perry Preschool Project. After a sample of 123 African American children were randomly assigned to either a control group or an experimental group, the experimental group members received academically oriented interventions for one to two years, frequent home visits, and weekly parent-teacher conferences. The control and experimental groups were compared on 14 occasions from age 3 to 40. As adults, experimental group members had higher employment and homeownership rates, and significantly lower violent and property crime rates (Schweinhart 2007).

In recognition of the link between juvenile delinquency and adult criminality, many anti-crime programs are directed toward at-risk youths. The Helping Families Initiative in Mobile County, Alabama, targets children who are "at risk" and intervenes with the assistance of the district attorney's office, social workers, police officers, teachers, and parents. The program focuses on ". . . the roughly 60 percent of children who have been suspended for serious violations such as fighting and bringing drugs to school, but haven't been arrested or adjudicated . . ." (Maxwell 2006, p. 28). Intervention entails everything from family counseling to transportation services. Although assessment of the program is difficult, school officials note that suspension rates have decreased since the program began.

Finally, many youth programs are designed to engage juveniles in noncriminal activities and integrate them into the community. In Weed and Seed, a program under the Department of Justice, "law enforcement agencies and prosecutors cooperate in 'weeding out' violent criminals and drug abusers . . . and 'seed' much-needed human services including prevention, intervention, treatment, and neighborhood restoration programs" (Weed and Seed 2009, p. 1). As part of the program, "safe havens" are established in, for example, schools, where multi-agency services are provided for youth.

Community Programs. Neighborhood watch programs involve local residents in crime prevention strategies. For example, MAD DADS (Men Against Destruction—Defending Against Drugs and Social Disorder) patrol the streets in high-crime areas of the city on weekend nights, providing positive adult role models and fun community activities for troubled children. Members also report crime and drug sales to police, paint over gang graffiti, organize gun buyback programs, and counsel incarcerated fathers. At present, 100,000 men and women are in MAD DADS in 67 chapters in 17 states (MAD DADS 2009). In 2008, 15,000 communities in 50 states—more than 37 million people—participated in "National Night Out," a crime prevention event in which citizens, businesses, neighborhood organizations, and local officials joined together in outdoor activities to heighten awareness of neighborhood problems, promote anticrime messages, and strengthen community ties (National Night Out 2009).

Mediation and victim-offender dispute-resolution programs are also increasing, with thousands of such programs worldwide. The growth of these programs is a reflection of their success rate: two-thirds of cases referred result in face-to-face meetings, 95 percent of these cases result in a written restitution agreement, and 90 percent of the written restitution agreements are completed within one year (VORP 2009).

Criminal Justice Policy

The criminal justice system is based on the principle of **deterrence**—the use of harm or the threat of harm to prevent unwanted behaviors. The criminal justice system assumes that people rationally choose to commit crime, weighing the rewards and consequences of their actions. Thus "get-tough" measures hold that maximizing punishment will increase deterrence and cause crime rates to go down. Thirty years of "get-tough" policies, however, have created other criminal justice problems (e.g., prison crowding) and have not significantly reduced **recidivism** rates, leaving many experts to ask, "what works?"

deterrence The use of harm or the threat of harm to prevent unwanted behaviors.

recidivism A return to criminal behavior by a former inmate, most often measured by re-arrest, re-conviction, or re-incarceration.

Law Enforcement Agencies. In 2007, the United States had 699,850 full-time law enforcement officers and 318,104 full-time civilian employees (e.g., clerks, meter attendants, correctional guards), yielding an estimated three law enforcement personnel per 1,000 inhabitants (FBI 2008a). There are over 18,000 law enforcement agencies in the United States, including municipal (e.g., city police), county (e.g., sheriff's department), state (e.g., highway patrol), and federal agencies (e.g., FBI), often with overlapping jurisdictions (Siegel 2009). In 2008, 60 percent of a random sample of adults responded that they had a "great deal" or "quite a lot" of confidence in the police (Gallup Poll 2009b).

In 2005, the latest year for which national data is available, 19 percent of the U.S. population 16 and older experienced face-to face contact with a police officer. The most common reason given, over 50 percent of the total, was traffic-related. Although white, black, and Hispanic drivers were equally likely to be stopped by the police, blacks and Hispanics were more likely to searched. Further, 1.6 percent of people having face-to-face police contact reported force or the threat of force being used against them. Blacks and Hispanics reported higher rates of police force than whites (Durose et al. 2007).

Accusations of racial profiling, police brutality, and discriminatory arrest practices have made police–citizen cooperation in the fight against crime difficult. In response to such trends, the Crime Control Act of 1994 established the Office of Community Oriented Policing Services (COPS). By addressing "the root causes of criminal and disorderly behavior, rather than simply responding to crimes once they have been committed, community policing concentrates on preventing both crime and the atmosphere of fear it creates" (COPS 2009 p. 1). Such an approach often includes "problem-oriented policing," a model that includes analyzing the underlying causes of crime, looking for solutions, and actively seeking out alternatives to standard law enforcement practices.

Rehabilitation versus Incapacitation. An important debate concerns the primary purpose of the criminal justice system: Is it to rehabilitate offenders or to incapacitate them through incarceration? Both rehabilitation and incapacitation are concerned with *recidivism rates,* or the extent to which criminals commit another crime. Advocates of **rehabilitation** believe that recidivism can be reduced by changing the criminal, whereas proponents of **incapacitation** think that recidivism can best be reduced by placing offenders in prison so that they are unable to commit further crimes against the general public.

Fear of crime has led to a public emphasis on incapacitation and a demand for tougher mandatory sentences, a reduction in the use of probation and parole, support of a "three strikes and you're out" policy, and truth-in-sentencing laws. However, these tough measures have recently come under attack for three reasons. First, research indicates that incarceration may not deter crime. Nationally, over 700,000 people are released from confinement every year and over half of them will be back in prison within three years (Pew 2008).

Second is the accusation that get-tough measures, such as California's "three strikes and you're out" policy, are not equally applied. Chen's (2008) analysis of over 170,000 California inmates indicates that African Americans compared to whites and Hispanics are more likely to receive "third-strike sentences," with the greatest racial disparities being for property and drug offenses. Similarly, males are more likely to receive third-strike sentencing than females.

Finally, in an environment of budget deficits and legislative cuts, states simply can no longer afford the policies of decades ago. At a total cost of $50 billion in 2008,

rehabilitation A criminal justice philosophy that argues that recidivism can be reduced by changing the criminal through such programs as substance abuse counseling, job training, education, and so on.

incapacitation A criminal justice philosophy that argues that recidivism can be reduced by placing offenders in prison so that they are unable to commit further crimes against the general public.

state corrections spending outpaced budget increases in education, transportation, and public assistance (Pew 2009). As a response to the economic downturn and concerns over the effectiveness of get-tough policies, many states are rethinking correctional policies—closing prisons, eliminating mandatory sentencing, replacing jail time with community-based programs, and providing treatment rather than punishment for nonserious drug offenders (Steinhauer 2009).

Clearly, sentencing more offenders for longer periods of time to confinement enhances incapacitation. However, faced with budget cuts, states—as well as the Obama administration—are revisiting the ideals of rehabilitation. Rehabilitation assumes that criminal behavior is caused by sociological, psychological, and/or biological forces rather than being solely a product of free will. If such forces can be identified, the necessary change can be instituted. Many rehabilitation programs focus on helping the inmate reenter society. In 2008, President Bush signed the Second Chance Act, which supports reentry programs in the hopes of reducing recidivism (Greenblatt 2008). This chapter's Photo Essay highlights successful rehabilitation programs including reentry initiatives.

Ted S. Warren/A.P. Photo

According to the U.S. Bureau of Justice Statistics, 2,310,984 prisoners were in federal or state prisons and local jails in June 2008. Typically, individuals convicted of felonies are confined in prisons, whereas people who have committed misdemeanors and have sentences of one year or less are confined in jails.

Corrections. The United States incarcerates more people than any other country in the world—over 2.3 million in 2009. An examination of global rates is even more revealing. The U.S. incarceration rate of 750 per 100,000 population exceeds many times over those of other countries; for example, the rate in England is 148, Germany's rate is 93, and France's is 85 (Pew 2008). Incarceration rates also vary by demographics. In 2006, 1 in 106 white men, 1 in 36 Hispanic men, and 1 in 15 African American men over the age of 18 were incarcerated (Pew 2008).

The U.S. incarceration rate has grown at an alarming rate—700 percent between 1950 and 2005, and, despite a general decrease in crime, it is expected to continue to grow, with the greatest increases being in the West, the South, and the Midwest (Pew 2007). Growth means more needed funds and, at an average prisoner cost of $29,000 a year, states are relying more and more on community alternatives, including probation and parole (Moore 2009).

> The U.S. incarceration rate of 750 per 100,000 population exceeds many times over those of other countries; for example, the rate in England is 148, Germany's rate is 93, and France's is 85.

Probation entails the conditional release of an offender who, for a specific time period and subject to certain conditions, remains under court supervision in the community. **Parole** entails release from prison, for a specific time period and subject to certain conditions, before the inmate's sentence is finished. Although varying by race, age, and gender, over five million people were on probation or parole in the United States in 2008 (Pew 2009) (see Figure 4.4). Some argue that new technologies (e.g., global positioning systems) coupled with research-based treatment and reentry programs "can produce double-digit reductions in recidivism and save states money along the way" (Pew 2009, p. 2).

probation The conditional release of an offender who, for a specific time period and subject to certain conditions, remains under court supervision in the community.

parole Parole entails release from prison, for a specific time period and subject to certain conditions, before the inmate's sentence is finished.

By 2011, an estimated 1.7 million people will be behind bars, with increases in all but four states (Pew 2007). The number of female inmates will increase by 16 percent, and there will be more elderly, both groups adding to the already skyrocketing cost of corrections. There are alternatives, however, and many of them have proven to be successful. Drug courts are a case in point. Begun in 1989, today there are over 2,000 drug courts in the United States (NCJRS 2009). Drug courts provide treatment and judical supervision (e.g., random drug testing) to nonviolent offenders as an alternative to incarceration. A review of the literature on drug courts reveals that such "programs decrease recidivism, increase treatment retention, and are a cost-effective alternative to incarceration" (Nored & Carlan 2008, p. 338).

▲ The goals of the Prison University Program are to "educate and challenge students intellectually; to prepare them to lead thoughtful and productive lives inside and outside of prison; to provide them with skills needed to obtain meaningful employment and economic stability post-release; and to prepare them to become providers, leaders, and examples for their families and communities" (PUP 2009, p. 1). Pictured above, a San Quentin inmate writing a paper for a class.

Each of the following correctional practices described has been empirically evaluated and found to be associated with positive changes in the inmate participants. From reducing recidivism rates and enhancing self-esteem, to lowering aggression and increasing the likelihood of post-release employment, these programs not only are cost-effective, but they are also humane. As we face one of the greatest economic crises in history, few could argue that such opportunities to alter the direction of failed criminal justice policies should be ignored.

Prison University Project

Thousands of inmates across the United States participate in postsecondary education programs. Research documents their benefits: reduced recidivism, enhanced problem-solving skills, safer prison conditions, a more marketable post-release inmate, and taxpayer savings (Correctional Association 2009). San Quentin's Prison University Project (PUP) is just such a program. Taught by college professors, graduate students, and other volunteers, PUP provides 12 college courses each semester—classes in the humanities, liberal arts, social sciences, math, and science—leading to an associate of arts degree (PUP 2009).

Indiana Canine Assistant and Adolescent Network

Programs such as Puppies Behind Bars (see photo), Puppies in Prison, Pen Pals, Project Pooch, and Prison Pet Partnership have been instrumental in changing the lives of inmates, breeder and shelter dogs, and the beneficiaries of the inmate-trainer's months of hard work and discipline. The men and women in the Indiana

◄ Puppies Behind Bars (PBB) was recently showcased on the Oprah Winfrey television show. For 16 months, puppies live in a prison cell with their inmate-trainer who teaches them basic obedience skills. At the end of that time, the dogs are assessed for suitability as service dogs for the disabled or as explosive detection dogs for law enforcement agencies. If approved, they are returned to schools where they continue their formal training. In 2006, PBB began a program called Dog Tags, whereby service dogs are donated to injured soldiers coming home from Iraq and Afghanistan (PBB 2009).

Canine Assistant and Adolescent Network (ICAAN) program train service and therapy dogs. An empirical evaluation of ICAAN documents the positive impact on the rehabilitation of participating offenders—higher self-esteem and better communication skills, and a marked improvement in patience and trust leading to better inmate-prison employee relations (Turner 2007).

Habitat for Humanity Prison Partnership

Started in Texas in 1999, prisons throughout the United States today partner with Habitat for Humanity to build homes for needy and low-income families. Although some inmates work within the prison to build prefabricated house frames, low security–risk prisoners work side by side with other Habitat for Humanity volunteers at the construction site. The Partnership allows prisoners to build confidence, provides them with marketable skills, and

fosters pride in their accomplishments. Sometimes called "factories behind fences," such programs are not new. Whether building furniture, making garments, or manufacturing signs, getting work experience helps "inmates acquire the skills they need to secure gainful employment upon release and avoid recidivism" (Moses & Smith 2007, p. 32).

Nebraska's Prison Nursery Program

Between 1977 and 2007, the number of women incarcerated in the United States increased by 832 percent (WPA 2009). With about 4 percent of female inmates pregnant at the time of incarceration, there has been a growing trend to provide prison nursery programs (see photo). Nebraska's Prison Nursery Program provides inmates with prenatal care education, parenting skills, information on child development, "hands-on training" for new and expectant mothers, and community resources upon release. After 10 years of operation, an evaluation of the program revealed lower misconduct and recidivism rates when compared to inmates who were required to give up their infants (Carlson 2009).

Prison Entrepreneurship Program

When Catherine Rohr, a Wall Street investor, toured a Texas prison, she had an epiphany—criminals and people in business are a lot alike. They both assess risks, live

John Gaines/The Hawk Eye/AP Photo

▲ "The prisoners who participate with Habitat for Humanity will gain a renewed belief in themselves," said Millard Fuller, founder and president of Habitat for Humanity International. "Habitat for Humanity is building more than houses. We are building lives." Pictured, two Iowa State Penitentiary inmates and correctional trade leader Mike Peters (right) put up a wall for a Habitat for Humanity home in Madison, Iowa.

by their instincts, share profits, network, and compete with one another. It was then she founded the Prison Entrepreneurship Program (PEP). Today, corporate leaders and faculty volunteers teach business skills to PEP participants—former drugs dealers, gang leaders, hustlers, and felons—by equipping them with the tools for success. Over 90 percent of the 440 graduates have found jobs within four weeks of being released, 57 have started their own businesses, and recidivism rates are as low as 5 percent (PEP 2009; Beiser 2009).

Shaul Schwarz/Getty Images

▲ Here, inmates wait for a parenting class at Indiana Women's Prison. Only nine states presently provide nursery facilities for female prisoners. Indiana's program also allows nonviolent offenders who have experience with child-rearing to live and work as "nannies" in the nursery unit, earning $1.30 a day (Smalley 2009).

Trevor Kobrin/Prison Entrepreneurship program

Since its creation in 2004, the Prison Entrepreneurship Program ▶ (PEP) has been instrumental in helping inmates transition to successful businessmen. Here, PEP students talk with community business leaders as part of their five-month series of classes (PEP 2009).

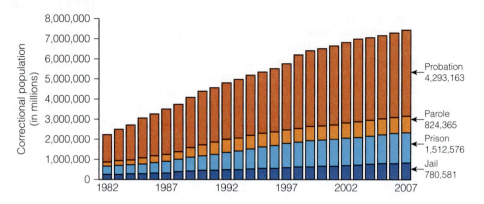

NOTE: Due to offenders with dual status, the sum of these four correctional categories slightly overstates the total correctional population.

What Do You Think? An estimated 2,000 inmates who committed their crimes while under the age of 18 are presently serving time in U.S. prisons. The majority of these inmates, either as offenders or accomplices, were involved in a homicide (Liptak 2009). Nonetheless, in 2009 the U.S. Supreme Court agreed to hear a case concerning life in prison without parole sentencing for *non-homicide offenders* who committed their crimes while under the age of 18. Do you think offenders who were not involved in a homicide event should be sentenced to life in prison without parole? Why or why not?

Capital Punishment. With **capital punishment,** the state (the federal government or a state) takes the life of a person as punishment for a crime. In 2008, China, Iran, Saudi Arabia, Pakistan, and the United States were responsible for 93 percent of all executions worldwide. In 2008, the United States recorded the lowest number of executions since 1995, with 37 executions (Amnesty International 2009). In 2009, 3,300 men and women were on death row across the United States.

Of the 35 states that have the death penalty, 34 and the federal government almost exclusively use lethal injection as the method of execution. Two concerns, however, have been raised with the use of lethal injection, leading some states to halt executions. First is the question of whether or not death by lethal injection violates the Eighth Amendment's prohibition against cruel and unusual punishment. In 2007, a district court judge held that Tennessee's lethal injection procedures "present a substantial risk of unnecessary pain" that could "result in a terrifying, excruciating death" (Schelzig 2007, p. 1). However, in 2008, the U.S. Supreme Court held that Kentucky's use of lethal injection was not a violation of the Eighth Amendment (*Baze v. Rees* 2008).

The second issue concerns the role of physicians in state executions. According to Vu (2007, p. 1), the "American Medical Association is adamant that it is a violation of medical ethics for doctors to participate in, or even be present at, executions." Nonetheless, in 2009, the North Carolina Supreme Court ruled against the state's medical board for punishing physicians who took part in executions (Ovaska & Bonner 2009).

Proponents of capital punishment argue that executions of convicted murderers are necessary to convey public disapproval and intolerance for such heinous

capital punishment The state (the federal government or a state) takes the life of a person as punishment for a crime.

crimes. Those against capital punishment believe that no one, including the state, has the right to take another person's life and that putting convicted murderers behind bars for life is a "social death" that conveys the necessary societal disapproval.

Proponents of capital punishment also argue that it deters individuals from committing murder. Critics of capital punishment hold, however, that because most homicides are situational and are not planned, offenders do not consider the consequences of their actions before they commit the offense. Critics also point out that the United States has a higher murder rate than many western European nations that do not practice capital punishment and that death sentences are racially discriminatory. A study of capital punishment in the United States, between 1973 and 2002, found that ". . . minority death row inmates convicted of killing whites face higher execution probabilities than other capital offenders" (Jacobs et al. 2007, p. 610).

Capital punishment advocates suggest that executing a convicted murderer relieves taxpayers of the costs involved in housing, feeding, guarding, and providing medical care for inmates. Opponents of capital punishment argue that financial considerations should not determine the principles that decide life and death issues. In addition, taking care of convicted murderers for life may actually be less costly than sentencing them to death because of the lengthy and costly appeals process for capital punishment cases. In 2009, Maryland, Colorado, Kansas, Nebraska, and New Hampshire each considered abolishing the death penalty due to cost (Urbina 2009). California is also considering abolishing the death penalty. Former attorney general of California, John Van de Kamp (2009) argues:

> Confinement on death row (with all the attendant security requirements) adds $90,000 per inmate per year to the normal cost of incarceration. Appeals and habeas corpus proceedings add tens of thousands more. In all, it costs [California] $125 million a year more to prosecute and defend death penalty cases and to keep inmates on death row than it would simply to put all those people in prison for life without parole.

Nevertheless, those in favor of capital punishment argue that it protects society by preventing convicted individuals from committing another crime, including the murder of another inmate or prison official. One study of the deterrent effect of capital punishment concluded that each execution is associated with at least eight fewer homicides (Rubin 2002). Opponents contend that capital punishment may result in innocent people being sentenced to death. According to the Innocence Project, there have been 239 post-conviction exonerations using DNA evidence since 1989 (Innocence Project 2009).

Legislative Action

Legislative action is one of the most powerful methods of fighting crime. Federal and state legislatures establish criminal justice policy by the laws they pass, the funds they allocate, and the programs they embrace.

Gun Control. In 2009, when a national sample of U.S. adults were asked, "In general, do you feel that the laws covering the sale of firearms should be made more strict, less strict, or kept as they are now?" 49 percent responded more strict, 41 percent responded less strict, and 10 percent responded that the laws should be kept as they are (Gallup Poll 2009c).

Those against gun control argue that not only do citizens have a constitutional right to own guns but also that more guns may actually lead to less crime as would-be offenders retreat in self-defense when confronted (Lott 2003). Advocates of gun control, however, insist that the 250 million privately owned firearms in the United States, one-third of them handguns (Conklin 2007), significantly contribute to the violent crime rate in the United States and distinguish the country from other industrialized nations.

After a seven-year battle with the National Rifle Association (NRA), gun control advocates achieved a small victory in 1993, when Congress passed the Brady Bill. The law initially required a five-day waiting period on handgun purchases so that sellers can screen buyers for criminal records or mental instability. The law was amended in 1998, to include an instant check of buyers and their suitability for gun ownership. Today, the law requires background checks of not just handgun users but also those who purchase rifles and shotguns.

In addition to federal regulations, cities and states can create other restrictions. In Washington, DC, gun ownership has virtually been banned for the last 30 years (Williams 2007). Only residents with permits such as police, security guards, and the like, can possess handguns and rifles. In what has been called the most important gun control case decades, in 2008, the U.S. Supreme Court in *District of Columbia v. Heller* affirmed the right of an individual to own a firearm for private use (*District of Columbia v. Heller 2008*). The recent killing of a security guard at the U.S. Holocaust Memorial Museum, concern over the illegal flow of weapons to Mexico, and a change in administrations may, however, signal a renewed interest in gun legislation.

Other Legislation. Major legislative initiatives have been passed in recent years, including the 1994 Violent Crime Control and Law Enforcement Act, which created community policing, "three strikes and you're out," and truth-in-sentencing laws. A sample of significant crime-related legislation presently before Congress includes the following bills (U.S. Congress 2009):

- *Drug Sentencing Reform and Cocaine Kingpin Trafficking Act.* This act, among other things, eliminates the five-year mandatory minimum prison term for first-time offenders convicted of possession of crack cocaine.
- *Save Our Children: Stop the Violent Predators Against Children DNA Act.* If passed, the act would create a database solely for collecting DNA information on violent child predators.
- *Child Gun Safety and Gun Access Prevention Act.* This bill, as proposed, would increase the age of handgun eligibility from 18 to 21, and would prohibit those under 21 from possessing semiautomatic assault weapons.
- *No Parole for Sex Offenders Act.* This legislation would require that states, under certain conditions, deny parole for all offenders convicted of a crime against a minor and all sexually violent predators.

International Efforts in the Fight Against Crime

Europol is the European law enforcement organization that handles criminal intelligence. Unlike the FBI, Europol officers do not have the power to arrest; they predominantly provide support services for law enforcement agencies of counties that are members of the European Union. For example, Europol coordinates the dissemination of information, provides operational analysis and technical support, and generates strategic reports (Europol 2009). Europol, in conjunction with law enforcement agencies in member states, fights against transnational crimes such as illicit drug trafficking, child pornography, human trafficking, money laundering, and counterfeiting of the euro.

Interpol, the International Criminal Police Organization, was established in 1923, and is the world's largest international police organization, with 187 member countries (Interpol 2009). Similar to Europol, Interpol provides support services for law enforcement agencies of member nations. It has four core functions (Interpol 2009). First, Interpol operates a worldwide police communications network that operates 24 hours a day, 7 days a week. Second, Interpol's extensive databases (e.g., DNA profiles, fingerprints, suspected terrorists) ensure that police get the information they need to investigate existing crime and prevent new crime from occurring. Third, Interpol provides emergency support services and operational activities to law enforcement personnel in the field. Finally, Interpol provides police training and development to help member states better fight the increasingly complex and globalized nature of crime.

Finally, the International Centre for the Prevention of Crime (ICPC) is a consortium of policy makers, academicians, police, governmental officials, and non-governmental agencies from all over the world (ICPC 2009). Located in Montreal, Canada, members of ICPC "exchange experience, consider emerging knowledge, and improve policies and programmes in crime prevention and community safety" (p. 1). In fulfilling such tasks, the ICPC seeks to (1) raise awareness of and access to crime prevention knowledge; (2) enhance community safety; (3) facilitate the sharing of crime prevention information between countries, cities, and justice systems; and (4) respond to calls for technical assistance.

Understanding Crime and Social Control

What can we conclude from the information presented in this chapter? Research on crime and violence supports the contentions of both structural functionalists and conflict theorists. Inequality in society, along with the emphasis on material well-being and corporate profit, produces societal strains and individual frustrations. Poverty, unemployment, urban decay, and substandard schools—the symptoms of social inequality—in turn lead to the development of criminal subcultures and conditions favorable to law violation. Furthermore, criminal behavior is encouraged by the continued weakening of social bonds among members of society and between individuals and society as a whole, the labeling of some acts and actors as "deviant," and the differential treatment of minority groups by the criminal justice system.

Recently, there has been a general decline in crime, making it tempting to conclude that get-tough criminal justice policies are responsible for the reductions. Other valid explanations exist and are likely to have contributed to the falling rates: changing demographics, community policing, stricter gun control, and a reduction in the use of crack cocaine.

Concerns over the cost of "nail'em and jail'em" policies, overcrowded prisons, and high recidivism rates have some policy makers looking elsewhere. Several states are already expanding the use of community-based initiatives and developing evidence-based reentry programs. Further, in 2009, The National Criminal Justice Commission Act of 2009 was introduced into the U.S. Senate. If passed, the act would "create a blue-ribbon commission charged with undertaking an 18-month, top-to-bottom review of our entire criminal justice system" (Fact Sheet 2009).

Interpol The largest international police organization in the world.

Rather than getting tough on crime after the fact, some advocate getting serious about prevention. Prevention programs are not only preferable to dealing with the wreckage crime leaves behind, they are also cost-effective. For example, the Perry Preschool Project, as discussed earlier, cost $15,166 per participant but produced savings of $258,888 per participant. Of that savings, 88 percent was associated with a reduction in costs related to criminal justice (Schweinhart 2007).

Lastly, the movement toward **restorative justice,** a philosophy primarily concerned with repairing the victim-offender community relation, is in direct response to the concerns of an adversarial criminal justice system that encourages offenders to deny, justify, or otherwise avoid taking responsibility for their actions.

Restorative justice holds that the justice system, rather than relying on "punishment, stigma, and disgrace" (Siegel 2006, p. 275), should "repair the harm" (Sherman 2003, p. 10). Key components of restorative justice include restitution to the victim, remedying the harm to the community, and mediation. In a meta-analysis of 35 restorative justice programs, offenders in restorative justice programs were less likely to become recidivists and more likely to meet restitution obligations than offenders not in restorative justice programs (Latimer et al. 2005).

restorative justice A philosophy primarily concerned with reconciling conflict between the victim, the offender, and the community.

CHAPTER REVIEW

■ **Are there any similarities between crime in the United States and crime in other countries?**
All societies have crime and have a process by which they deal with crime and criminals; that is, they have police, courts, and correctional facilities. Worldwide, most offenders are young males, and the most common offense is theft; the least common offense is murder.

■ **How can we measure crime?**
There are three primary sources of crime statistics. First are official statistics, for example, the FBI's Uniform Crime Reports, which are published annually. Second are victimization surveys designed to get at the "dark figure" of crime; crime that official statistics miss. Finally, self-report studies have all the problems of any survey research. Investigators must be cautious about whom they survey and how they ask the questions.

■ **What sociological theory of criminal behavior blames the schism between the culture and structure of society for crime?**
Strain theory was developed by Robert Merton (1957) and uses Durkheim's concept of *anomie,* or

normlessness. Merton argued that, when the structure of society limits legitimate means (e.g., a job) of acquiring culturally defined goals (e.g., money), the resulting strain may lead to crime. Individuals, then, must adapt to the inconsistency between means and goals in a society that socializes everyone into wanting the same thing but provides opportunities for only some.

■ **What are index offenses?**
Index offenses, as defined by the FBI, include two categories of crime: violent crime and property crime. Violent crimes include murder, robbery, assault, and rape; property crimes include larceny, car theft, burglary, and arson. Property crimes, although less serious than violent crimes, are the most numerous.

■ **What is meant by white-collar crime?**
White-collar crime includes two categories: occupational crime, that is, crime committed in the course of one's occupation; and corporate crime, in which corporations violate the law in the interest of maximizing profits. In occupational crime, the motivation is individual gain.

How do social class and race affect the likelihood of criminal behavior?

Official statistics indicate that minorities are disproportionately represented in the offender population. Nevertheless, it is inaccurate to conclude that race and crime are causally related. First, official statistics reflect the behaviors and policies of criminal justice actors. Thus, the high rate of arrests, conviction, and incarceration of minorities may be a consequence of individual and institutional bias against minorities. Second, race and social class are closely related in that nonwhites are overrepresented in the lower classes. Because lower-class members lack legitimate means to acquire material goods, they may turn to instrumental, or economically motivated, crimes. Thus, the apparent relationship between race and crime may, in part, be a consequence of the relationship between these variables and social class.

What are some of the economic costs of crime?

First are direct losses from crime, such as the destruction of buildings through arson or of the environment by polluters. Second are costs associated with the transferring of property (e.g., embezzlement). A third major cost of crime is that associated with criminal violence (e.g., the medical cost of treating crime victims). Fourth are the costs associated with the production and sale of illegal goods and services. Fifth is the cost of prevention and protection. Finally, there is the cost of the criminal justice system, law enforcement, litigation and judicial activities, corrections, and victims' assistance.

What is the present legal status of capital punishment in this country?

Of the 35 states that have the death penalty, 34 and the federal government almost exclusively use lethal injection as the method of execution. Questions concerning the constitutionality of lethal injection and the role of physicians in state executions have led to several court cases.

TEST YOURSELF

1. The United States has the highest violent crime rate in the world.
   a. True
   b. False
2. The Uniform Crime Reports is a compilation of data from
   a. U.S. Census Bureau
   b. law enforcement agencies
   c. victimization surveys
   d. the Department of Justice
3. According to _____, crime results from the absence of legitimate opportunities as limited by the social structure of society.
   a. Hirschi
   b. Marx
   c. Merton
   d. Becker
4. Which of the following is not an index offense?
   a. Drug possession
   b. Homicide
   c. Rape
   d. Burglary
5. The economic costs of white-collar crime outweigh the costs of traditional street crime.
   a. True
   b. False
6. Women everywhere commit less crime than men.
   a. True
   b. False
7. Probation entails
   a. early release from prison
   b. a suspended sentence
   c. court supervision in the community in lieu of incarceration
   d. incapacitation of the offender
8. Europol is an advisory and support law enforcement agency for European Union members.
   a. True
   b. False
9. Lethal injection
   a. violates the Eighth Amendment of the U.S. Constitution
   b. has been determined to be painless
   c. by physicians is approved by the American Medical Association
   d. is the most common method of execution in the United States
10. The United States has the highest incarceration rate in the world.
    a. True
    b. False

Answers: 1: b; 2: b; 3: c; 4: a; 5: a; 6: a; 7: c; 8: a; 9: d; 10: a.

KEY TERMS

MEDIA RESOURCES

Understanding Social Problems,
Seventh Edition Companion Website
www.cengage.com/sociology/mooney
Visit your book companion website, where you will find flash cards, practice quizzes, Internet links, and more to help you study.

Just what you need to know NOW! Spend time on what you need to master rather than on information you already have learned. Take a pretest for this chapter, and CengageNOW will generate a personalized study plan based on your results. The study plan will identify the topics you need to review and direct you to online resources to help you master those topics. You can then take a posttest to help you determine the concepts you have mastered and what you will need to work on. Try it out! Go to www.cengage.com/login to sign in with an access code or to purchase access to this product.

Roy Morsch/Surf/CORBIS

5

Family Problems

"We must recognize that there are healthy as well as unhealthy ways to be single or to be divorced, just as there are healthy and unhealthy ways to be married."

Stephanie Coontz, Family historian

Throughout his presidential campaign, Barack Obama and his wife Michelle emphasized the priority of family relationships in their lives. In a speech he gave after winning the presidential election, Obama proclaimed, "And I would not be standing here tonight without the unyielding support of my best friend for the last 16 years . . . the rock of our family, the love of my life. . . ." Referring to his wife, Barack Obama said, "When I go home, she wants me to be a good father and a good husband. And everything else is secondary to that." Michelle Obama, having earned degrees from Princeton and Harvard Law School, is not apologetic about prioritizing her parental role. In a speech she gave at the 2008 Democratic National Convention, Michelle Obama said, "I come here as a Mom whose girls are the heart of my heart and the center of my world— they're the first thing I think about when I wake up in the morning, and the last thing I think about when I go to bed at night." Michelle described her first role in the White House as "mom-in-chief," whose job it is to help her daughters through the transition. The family Barack Obama has created with Michelle is quite different from the one he came from. His mother was 18 when he was born; she married 6 months before his birth. When Barack was a young boy, his parents divorced and his father moved to Kenya and became an absent parent, and his mother remarried, moved overseas, and let her son be raised by his grandparents.

Referring to his wife, Barack Obama said, "When I go home, she wants me to be a good father and a good husband. Everything else is secondary to that."

Mark Wilson/Getty Images News/Getty Images

However, not all families are as stable and loving as Barack and Michelle Obama's appears to be. As well, not all individuals are resilient to the challenges of teen parenting and divorce, as in the case of Barack Obama and his mother. In this chapter, we turn our attention to family problems, including teen pregnancy, divorce and its aftermath, and violence and abuse in intimate and family relationships. We begin by examining the diversity in family life across the globe and the changing patterns in U.S. families. Note that many of the problems families face, such as health problems, poverty, job-related problems, drug and alcohol abuse, discrimination, and military deployment of a spouse are dealt with in other chapters in this text.

The Global Context: Families of the World

family A kinship system of all relatives living together or recognized as a social unit, including adopted people.

The U.S. Census Bureau defines *family* as a group of two or more people related by blood, marriage, or adoption. Sociology offers a broader definition of family: A **family** is a kinship system of all relatives living together or recognized as a social unit. This broader definition recognizes foster families, unmarried same-sex and opposite-sex couples with or without children, and any relationships that function and feel like a family. As we describe in the following section, family forms and patterns vary worldwide.

Monogamy and Polygamy. In many countries, including the United States, the only legal form of marriage is **monogamy**—a marriage between two partners. A common variation of monogamy is **serial monogamy**—a succession of marriages in which a person has more than one spouse over a lifetime but is legally married to only one person at a time.

Polygamy—a form of marriage in which one person may have two or more spouses—is practiced on all continents throughout the world (Zeitzen 2008). The most common form of polygamy, known as **polygyny**, involves one husband having more than one wife. A less common form of polygamy is **polyandry**—the concurrent marriage of one woman to two or more men.

In the United States, Congress outlawed polygamy in 1892; thus, being married to more than one spouse is a crime referred to as **bigamy.** Although the Mormon Church has officially banned polygamy, it is still practiced among members of the Fundamentalist Church of Jesus Christ Latter-Day Saints (FLDS)—a Mormon splinter group with about 10,000 members who believe that a man must marry at least three wives to go to heaven (Zeitzen 2008).

Polygamy in the United States also occurs among some immigrants who come from countries where polygamy is accepted, such as Mali and Ghana and other West African countries. It is estimated, for example, that thousands of New Yorkers are involved in polygamous marriages (Bernstein 2007). Immigrants who practice polygamy generally keep their lifestyle a secret because polygamy is grounds for deportation under U.S. immigration law.

The issue of greatest concern regarding polygamy in the United States is the forcing of underage girls into polygamous marriages. In 2008, Texas state officials raided a FLDS ranch in Eldorado, Texas, and removed more than 400 children, placing them in temporary state custody to protect them from allegedly abusive conditions. Although 31 of these children were girls aged 14 to 17 who had children or were pregnant, a Court of Appeals ruled that the state did not have sufficient evidence of imminent danger to remove the children, and the court ordered the return of the children.

HBO/The Kobal Collection/Picture Desk

The HBO series *Big Love* gave visibility to the illegal practice of polygamy among some religious fundamentalist groups.

monogamy Marriage between two partners; the only legal form of marriage in the United States.

serial monogamy A succession of marriages in which a person has more than one spouse over a lifetime but is legally married to only one person at a time.

polygamy A form of marriage in which one person may have two or more spouses.

polygyny A form of marriage in which one husband has more than one wife.

polyandry The concurrent marriage of one woman to two or more men.

bigamy The criminal offense of marrying one person while still legally married to another.

What Do You Think? Fundamentalist Mormons, who practice polygamy as an extension of their religious beliefs, have attempted to justify polygamy by referring to the Constitution's First Amendment right to freedom of religion. In 1879, the Supreme Court in *Reynolds v. United States* ruled that, although people are free to believe in their religious principles, religious belief is not justification for acting against the law. Do you think that the criminalization of polygamy violates the right to freedom of religion? Should the government be tolerant of polygamy?

Role of Women and Men in the Family. The roles of women and men in families also vary across societies. In some societies, wives are expected to be subservient to their husbands. In sub-Saharan Africa, the cultural view of wives as inferior to their husbands is linked to widespread wife abuse. According to Nigeria's minister for women's affairs, "It is like it is a normal thing for women to be treated by their husbands as punching bags. . . . The Nigerian man thinks that a woman is his inferior" (quoted by LaFraniere 2005, p. A1). In most societies in East Africa, the roles of men and women tend to be separate and prescribed, with men having the power and authority in the family. In many societies, traditional roles prescribe that men are the head of the household, economic provider, and disciplinarian of children; and women take care of the housework and child care. In developed Western countries, marriages tend to be egalitarian, which means women and men view each other as equal partners who share decision making and assign family roles based on choice rather than on traditional beliefs about gender (Lindsey 2005).

Deseret Morning News/Getty Images News/Getty Images/

These women and children of the Fundamentalist Church of Jesus Christ of Latter-Day Saints were removed from a compound in Eldorado, Texas, where polygamy is practiced, after allegations of abuse were reported.

Social Norms Related to Childbearing In every society, women learn that their role includes having children. However, compared with less developed societies where social expectations for women to have children are strong, social norms concerning childbearing are more flexible in developed countries with high levels of gender equality; women may view having children as optional—as a personal choice.

Norms about childbirth out of wedlock also vary across the globe. Although more than one-third of U.S. births are to unmarried women, compared with some countries, the United States has a low proportion of births outside of marriage. Two-thirds of births in Iceland and at least half of births in Norway and Sweden are out of wedlock. Denmark, France, the United Kingdom, and Finland are other countries with higher proportions of births outside of marriage than the United States. In other countries, such as Germany, Italy, Greece, and Japan, less than 15 percent of births occur out of wedlock (Ventura & Bachrach 2000). In India, it is almost unheard of for a Hindu woman to have a child outside marriage; unwed childbearing would bring great shame to a woman and her family (Laungani 2005).

Same-Sex Couples. Norms and policies concerning same-sex intimate relationships also vary around the world. In some countries, homosexuality is punishable by imprisonment or even death. In other countries, and in some U.S.

states, same-sex couples are granted legal rights to marry (see Chapter 10). Several countries, and some U.S. states, grant same-sex couples legal rights and protections that are more limited than marriage. For example, in France, registered same-sex (and opposite-sex) couples can enter a type of marital arrangement called *Pacte civil de solidarite* (civil solidarity pact), or PAC, which grants them the right to file joint tax returns, extend their Social Security coverage to each other, and receive the same health, employment, and welfare benefits as legal spouses.

It is clear from the previous discussion that families are shaped by the social and cultural context in which they exist. As we discuss the family issues addressed in this chapter—violence and abuse, divorce, and teenage childbearing—we refer to social and cultural forces that shape these events and the attitudes surrounding them. Next, we look at changing patterns and structures of U.S. families and households.

Changing Patterns in U.S. Families

Some of the significant changes in U.S. families and households that have occurred over the past several decades include the following:

- *Increased singlehood and older age at first marriage.* U.S. women and men are staying single longer. From 1960 to 2008, the median age at first marriage increased from about 20 to 26 for women and from about 23 to 27 for men. Today, 13.5 percent of women and 19.7 percent of men ages 40 to 44 have never been married—the highest figures in this nation's history (U.S. Census Bureau 2009).

- *Increased heterosexual and same-sex cohabitation.* Many U.S. adults who are technically "single" are in long-term committed cohabiting relationships with a partner. From 1960 to 2008, the number of cohabiting unmarried couples skyrocketed (see Figure 5.1). The percentage of people who cohabited with their spouses before marriage more than doubled between 1980 and 2000, rising from 16 percent to 41 percent (Amato et al. 2007).

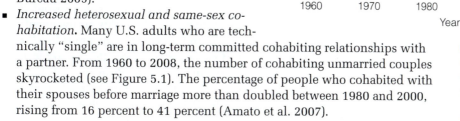

Figure 5.1 Number of Cohabiting, Unmarried U.S. Couples of the Opposite Sex, by Year
Source: U.S. Census Bureau 2009 and earlier reports in the same series.

What Do You Think? Adults with divorced parents are more likely to cohabit before marriage than are adults with continuously married parents (Amato et al. 2007). Why do you think this is so?

Most Americans view cohabitation as a normal, acceptable stage in the courtship process (Thornton & Young-DeMarco 2001). Couples live together to assess their relationship, to reduce or share expenses, or to avoid losing pensions from previously deceased partners. For some, including many same-sex couples who live in states where they cannot legally marry, cohabitation is an alternative to marriage. Most of the 5.5 million cohabiting

Brad Pitt implied solidarity with gays and lesbians when he said that he and Angelina Jolie will "consider tying the knot when everyone . . . who wants to be married is legally able."

couples in 2000 were heterosexual couples, but about one in nine had partners of the same sex (Simmons & O'Connell 2003). Some heterosexual couples do not marry out of solidarity with gays and lesbians, who can't legally wed in most states. Brad Pitt said that he and Angelina Jolie will "consider tying the knot when everyone . . . who wants to be married is legally able" (quoted in Davis 2009, p. 58).

Increased cohabitation among adults means that children are increasingly living in families that may function as two-parent families but do not have the social or legal recognition that married-couple families have. About 4 in 10 (43 percent) opposite-sex unmarried partner households, one-fifth (22 percent) of gay male couples, and one-third (34 percent) of lesbian couples have children present in the home (Simmons & O'Connell 2003). When children are denied a legal relationship to both parents because of the parents' unmarried status, they may be denied Social Security survivor benefits, health care insurance, or the ability to have either parent authorize medical treatment in an emergency, among other protections. Some states, cities, counties, and employers allow unmarried partners (same-sex and/or heterosexual partners) to apply for a **domestic partnership** designation, which grants them some legal entitlements, such as health insurance benefits and inheritance rights that have traditionally been reserved for married couples.

- *A new family form: Living apart together.* Some couples live apart in different cities or states because of their employment situation. Known as "commuter marriages," these couples generally would prefer to live together, but their jobs require them to live apart. However, other couples (married or unmarried) live apart in separate residences out of choice. Family scholars have identified this arrangement as an emerging family form known as **living apart together (LAT) relationships**. Couples may choose this family form for a number of reasons, including the desire to maintain a measure of independence and to avoid problems that may arise from living together. This new social phenomenon has been observed in several western European countries as well as in the United States (Lara 2005; Levin 2004).

domestic partnership A status that some states, counties, cities, and workplaces grant to unmarried couples, including gay and lesbian couples, which conveys various rights and responsibilities.

living apart together (LAT) relationships An emerging family form in which couples—married or unmarried—live apart in separate residences.

What Do You Think? What do you think some of the advantages of living apart together could be? How about the disadvantages? Would you consider living apart together with your partner or spouse?

- *Increased births to unmarried women.* The percentage of births to unmarried women rose to historic levels in 2007: nearly four in ten (39.7 percent) births in 2007 were to unmarried women (Hamilton et al. 2009). The highest rates of nonmarital births are among blacks, Native Americans/Alaskan natives, and Hispanics (see Figure 5.2).

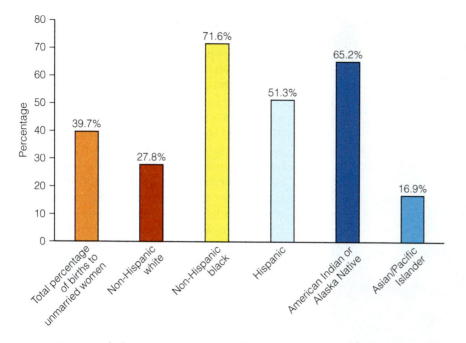

Figure 5.2 **Percentage of All Births to Unmarried U.S. Women by Race and Hispanic Origin, 2007**
Source: Hamilton et al. 2009.

Having a baby outside marriage has become more socially acceptable and does not carry the stigma it once did. In a 2008 national survey of U.S. adults, more than half (55 percent) said that having a baby outside of marriage is "morally acceptable" (Gallup Organization 2009).

Family scholar Stephanie Coontz (1997) emphasized the fact that we must be careful not to overdramatize the increase because "much illegitimacy was covered up in the past" (p. 29). In addition, not all unwed mothers are single parents. Half of new unwed mothers are cohabiting with the fathers when their children are born (Sigle-Rushton & McLanahan 2002).

■ *Increased divorce and blended families.*
The **refined divorce rate**—the number of divorces per 1,000 married women—increased dramatically from 1960, to its peak around 1980, then decreased until 2005, before increasing again (see Figure 5.3) (The National Marriage Project 2009). More than one-quarter (28 percent) of U.S. adults have been divorced; among 50- to 64-year-olds, 45 percent have been divorced (Saad 2006).

Most divorced individuals remarry and create blended families, traditionally referred to as stepfamilies. An estimated one-quarter of all children born in the United States will live with a stepparent before they reach adulthood (Mason 2003). In most states, stepparents have no obligation to support their stepchildren during the marriage, nor do they have any right of custody or control. In the event of divorce, stepparents usually have no rights to custody or even visitation and no obligation to pay child support. Stepchildren have no right of inheritance in the event of a stepparent's death (unless the stepparent has adopted or specified such inheritance in a will) (Mason 2003). However, stepparents who adopt their stepchildren have the same parental rights and obligations as biological parents.

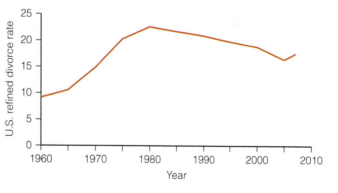

Figure 5.3 **U.S. Refined Divorce Rate,* by Year**
Source: National Marriage Project 2009.
*Number of divorces per 1,000 married women

refined divorce rate The number of divorces per 1,000 married women.

Getty Images Entertainment/Getty Images

Not all nonmarital births imply the absence of a father. Actress Goldie Hawn and actor Kurt Russell have been in a committed cohabiting relationship for more than 20 years. Although their child Wyatt was born "out of wedlock," he has been raised in a stable, loving family with his mother and his father. Kurt Russell also helped Goldie raise her two children (Kate and Oliver) from a previous marriage.

■ *Increased employment of mothers.* Employment of married women with children under age 18 rose from 24 percent in 1950, to 67 percent in 2008 (Gilbert 2003; Bureau of Labor Statistics 2009). In 2008, in nearly two-thirds (62 percent) of U.S. married-couple families with children under age 18, both parents were employed (Bureau of Labor Statistics 2009). The idea of the traditional two-parent family in which the husband is a breadwinner and the wife is a homemaker does not reflect the reality of most U.S. families. Only one in five U.S. married-couple families has an employed husband and an unemployed wife (Bureau of Labor Statistics 2009). Yet work, school, and medical care in the United States tend to be organized around the expectation that every household has a full-time mother at home who is available to transport children to medical appointments, pick up children from school on early dismissal days, and stay home when a child is sick (Coontz 1992).

The Marital Decline and Marital Resiliency Perspectives on the American Family

Do the recent transformations in American families signify a collapse of marriage and family in the United States? Does the trend toward diversification of family forms mean that marriage and family are disintegrating, falling apart, or even disappearing? Or has family simply undergone transformations in response to changes in socioeconomic conditions, gender roles, and cultural values? The answers to these questions depend on whether we adopt the marital decline perspective or the marital resilience perspective.

According to the **marital decline perspective,** (1) personal happiness has become more important than marital commitment and family obligations, and (2) the decline in lifelong marriage and the increase in single-parent families have contributed to a variety of social problems, such as poverty, delinquency, substance abuse, violence, and the erosion of neighborhoods and communities (Amato 2004). According to the **marital resiliency perspective,** "poverty, unemployment, poorly funded schools, discrimination, and the lack of basic services (such as health insurance and child care) represent more serious threats to the well-being of children and adults than does the decline in married two-parent families" (Amato 2004, p. 960). According to this perspective, many marriages in the past were troubled, but because divorce was not socially acceptable, these problematic marriages remained intact. Rather than the view of

> The idea of the traditional two-parent family in which the husband is a breadwinner and the wife is a homemaker does not reflect the reality of most U.S. families.

divorce as a sign of the decline of marriage, divorce provides adults and children an escape from dysfunctional home environments.

Although the high rate of marital dissolution seems to suggest a weakening of marriage, divorce may also be viewed as resulting from placing a high value on marriage, such that a-less-than-satisfactory marriage is unacceptable. In effect, people who divorce may be viewed not as incapable of commitment but as those who would not settle for a bad marriage. Indeed, the expectations that young women and men have of marriage have changed. Whereas once the main purpose of marriage was to have and raise children, today women and men want marriage to provide adult intimacy and companionship (Coontz 2000).

The high rate of childbirths out of wedlock and single parenting is also not necessarily indicative of a decline in the value of marriage. In interviews with a sample of low-income single women with children, most women said they would like to be married but just have not found "Mr. Right" (Edin 2000). Low-income single mothers in Edin's study were reluctant to marry the father of their children because these men had low economic status, traditional notions of male domination in household and parental decisions, and patterns of untrustworthy and even violent behavior. Given the low level of trust these mothers have of men and given their view that husbands want more control than the women are willing to give them, women realize that a marriage that is also economically strained is likely to be conflictual and short-lived. "Interestingly, mothers say they reject entering into economically risky marital unions out of respect for the institution of marriage, rather than because of a rejection of the marriage norm" (Edin 2000, p. 130).

Is the well-being of a family measured by the degree to which that family conforms to the idealized married, two-parent, stay-at-home mom model of the 1950s? Or is family well-being measured by function rather than form? As suggested by family scholars Mason et al. (2003), "the important question to ask about American families . . . is not how much they conform to a particular image of the family, but rather how well do they function—what kind of love, care, and nurturance do they provide?" (p. 2).

Finally, it is important to have a perspective that takes into account the historical realities of families. Family historian Stephanie Coontz (2004) explained:

> . . . many things that seem new in family life are actually quite traditional. Two-provider families, for example, were the norm through most of history. Stepfamilies were more numerous in much of history than they are today. There have been several times and places when cohabitation, out-of-wedlock births, or nonmarital sex were more widespread than they are today. (p. 974)

In sum, it is clear that the institution of marriage and family has undergone significant changes in the last few generations. What is not as clear is whether these changes are for the better or for the worse. As this chapter's Social Problems Research Up Close feature discusses, changes in marriage and family may be viewed as neither all good nor all bad, but are perhaps a more complex mix. Coontz (2005a) noted, "Marriage has become more joyful, more loving, and more satisfying for many couples than ever before in history. At the same time it has become optional and more brittle" (p. 306).

Whereas once the main purpose of marriage was to have and raise children, today women and men want marriage to provide adult intimacy and companionship.

marital decline perspective A pessimistic view of the current state of marriage that includes the beliefs that (1) personal happiness has become more important than marital commitment and family obligations, and (2) the decline in lifelong marriage and the increase in single-parent families have contributed to a variety of social problems.

marital resiliency perspective A view of the current state of marriage that includes the beliefs that (1) poverty, unemployment, poorly funded schools, discrimination, and the lack of basic services (such as health insurance and child care) represent more serious threats to the well-being of children and adults than does the decline in married two-parent families, and (2) divorce provides adults and children an escape from dysfunctional home environments.

In *Alone Together: How Marriage in America Is Changing*, Paul Amato and colleagues (2007) examined survey data to see how marital quality changed between 1980 and 2000. One of the key questions the authors attempted to answer is: Has marriage become less satisfying and stable? Or has marriage become stronger and more satisfying for couples today compared to in the past?

Sample and Methods

Data for this study came from two national random samples of married adults 55 years and younger. Response rates for both the 1980 and the 2000 sample were over 60 percent. Data were collected through telephone interviews designed to measure five dimensions of marital quality: (1) marital happiness; (2) marital interaction; (3) marital conflict; (4) marital problems; and (5) divorce proneness. To assess whether marital quality changed between 1980 and 2000, the researchers compared the responses from 1980 with those from 2000. They also looked at gender differences to see whether marital quality for wives and husbands differed.

Selected Findings and Discussion

Analysis of the 1980 and 2000 data suggests that marriages have become stronger and more satisfying in some respects and weaker and less satisfying in other respects. In the following selected findings, note that two dimensions of marital quality improved (conflict and problems), one dimension deteriorated (interaction), and two dimensions were unchanged (happiness and divorce proneness).

1. *Marital happiness:* The percentage of spouses who rated their marriage as better than average was nearly equivalent in 1980 (74 percent) as in 2000 (68 percent).

2. *Marital interaction:* All five measures of marital interaction showed a decline from 1980 to 2000. Compared with spouses in 1980, spouses in 2000 were less likely to report that their husband or wife almost always accompanied them while visiting friends, shopping, eating their main meal of the day, going out for leisure, and working around the home.

3. *Marital conflict:* Of the five items that measure marital conflict, three declined significantly between 1980 and 2000: reports of disagreeing with one's spouse "often" or "very often," reports of violence ever occurring in the marriage, and reports of marital violence within the previous three years. No change was observed in the percentage of spouses who reported arguments over the division of household labor or in the number of serious quarrels in the previous two months.

4. *Marital problems:* Compared with spouses in 1980, spouses in 2000 were less likely to report marital problems in the following four areas: getting angry easily, having feelings that are hurt easily, experiencing jealousy, and being domineering. There was little change in the percentage of spouses who reported that getting in trouble with police, drinking or drugs, or extramarital sex were problems in their marriage.

5. *Divorce proneness:* The average level of divorce proneness was stable between 1980 and 2000. However, the percentage of spouses who reported either a low or a high degree of divorce proneness increased from 1980 to 2000, whereas the percentage reporting moderate divorce proneness declined.

6. *Gender differences:* Responses to items measuring marital interaction, marital problems, and divorce proneness were similar for husbands and wives. Two items measuring marital happiness revealed a gender difference: Between 1980 and 2000, wives (but not husbands) reported greater happiness with the amount of understanding received from their spouses. Wives also reported more happiness with their husbands' work around the house, whereas husbands reported less happiness with th wives' work around the house. Regard the measure of marital conflict, both husbands and wives reported declines in violence.

In sum, this research suggests that betwe 1980 and 2000, "marriages became more peaceful, with fewer disagreements, less aggression, and fewer interpersonal sour of tension between spouses" (Amato et a 2007, p. 68). Although overall levels of marital satisfaction did not seem to chan from 1980 to 2000, the results for divorc proneness were mixed, with an increase i the proportion of both stable and unstab marriages. The finding that most concern the researchers was that, between 1980 and 2000, the lives of husbands and wive became more separate, as spouses shared fewer activities. Citing previous research that has found that spouses who spend less time together are less happy in their marriage and more likely to divorce, Ama et al. suggested that "it is possible that the gradual decline in marital interaction between 1980 and 2000 will erode future marital happiness and increase subseque levels of marital instability" (p. 69).

Sociological Theories of Family Problems

Three major sociological theories—structural functionalism, conflict theory, and symbolic interactionism—help to explain different aspects of the family institution and the problems in families today.

Structural-Functionalist Perspective

The structural-functionalist perspective views the family as a social institution that performs important functions for society, including producing new members, regulating sexual activity and procreation, socializing the young, and providing physical and emotional care for family members. According to the structural-functionalist perspective, traditional gender roles contribute to family functioning: Women perform the "expressive" role of managing household tasks and providing emotional care and nurturing to family members, and men perform the "instrumental" role of earning income and making major family decisions.

According to the structural-functionalist perspective, the high rate of divorce and the rising number of single-parent households constitute a "breakdown" of the family institution that has resulted from rapid social change. The structural-functionalist perspective views the breakdown of the family as a primary social problem that leads to secondary social problems such as crime, poverty, and substance abuse.

Structural-functionalist explanations of family problems examine how changes in other social institutions contribute to family problems. For example, a structural-functionalist view of divorce examines how changes in the economy (such as more dual-earner marriages) and in the legal system (such as the adoption of "no-fault" divorce) contribute to high rates of divorce. Changes in the economic institution, specifically falling wages among unskilled and semiskilled men, also contribute to both intimate partner abuse and the rise in female-headed single-parent households (Edin 2000).

Conflict and Feminist Perspectives

Conflict theory focuses on how capitalism, social class, and power influence marriages and families. Feminist theory is concerned with how gender inequalities influence and are influenced by marriages and families. Feminists are critical of the traditional male domination of families—a system known as **patriarchy**—that is reflected in the tradition of wives taking their husband's last name and children taking their father's name. Patriarchy implies that wives and children are the property of husbands and fathers.

The overlap between conflict and feminist perspectives is evident in views on how industrialism and capitalism have contributed to gender inequality. With the onset of factory production during industrialization, workers—mainly men—left the home to earn incomes and women stayed home to do unpaid child care and domestic work. This arrangement resulted in families founded on what Engels calls "domestic slavery of the wife" (quoted by Carrington 2002, p. 32). Modern society, according to Engels, rests on gender-based slavery, with women doing household labor for which they receive neither income nor status, whereas men leave the home to earn an income. Times have certainly changed since Engels made his observations, with most wives today leaving the home to earn incomes. However, wives employed full-time still do the bulk of unpaid domestic labor, and women are more likely than men to compromise their occupational achievement to take on child care and other domestic responsibilities. The continuing unequal distribution of wealth that favors men contributes to inequities in power and fosters economic dependence of wives on husbands. When wives do earn more money than their husbands (which is the case in 30 percent of marriages), the divorce rate is higher—the women can afford to leave abusive or inequitable relationships (Jalovaara 2003).

Economic factors have also influenced norms concerning monogamy. In societies in which women and men are expected to be monogamous within marriage, there is a double standard that grants men considerably more tolerance for being

patriarchy A male-dominated family system that is reflected in the tradition of wives taking their husband's last name and children taking their father's name.

nonmonogamous. Engels explained that monogamy arose from the concentration of wealth in the hands of a single individual—a man—and from the need to bequeath this wealth to children of his own, which requires that his wife be monogamous. The "sole exclusive aims of monogamous marriage were to make the man supreme in the family and to propagate, as the future heirs to his wealth, children indisputably his own" (quoted by Carrington 2002, p. 32).

Feminist and conflict perspectives on domestic violence suggest that the unequal distribution of power among women and men and the historical view of women as the property of men contribute to wife battering. When wives violate or challenge the male head-of-household's authority, the male may react by "disciplining" his wife or using anger and violence to reassert his position of power in the family.

Although modern gender relations within families and within society at large are more egalitarian than in the past, male domination persists, even if less obvious. Lloyd and Emery (2000) noted that "one of the primary ways that power disguises itself in courtship and marriage is through the 'myth of equality between the sexes.' . . . The widespread discourse on 'marriage between equals' serves as a cover for the presence of male domination in intimate relationships . . . and allows couples to create an illusion of equality that masks the inequities in their relationships" (pp. 25–26).

Conflict theorists emphasize that powerful and wealthy segments of society largely shape social programs and policies that affect families. The interests of corporations and businesses are often in conflict with the needs of families. Corporations and businesses strenuously fought the passage of the 1993 Family and Medical Leave Act, which gives people employed full-time for at least 12 months in companies with at least 50 employees up to 12 weeks of unpaid time off for parenting leave, illness or death of a family member, and elder care. Government, which corporate interests largely influence through lobbying and political financial contributions, enacts policies and laws that serve the interests of for-profit corporations rather than families.

Symbolic Interactionist Perspective

The symbolic interactionist perspective is concerned with how labels affect meaning and behavior. For example, when a noncustodial divorced parent (usually a father) is awarded "visitation" rights, he may view himself as a visitor in his children's lives. The meaning attached to the visitor status can be an obstacle to the father's involvement because the label *visitor* minimizes the importance of the noncustodial parent's role (Pasley & Minton 2001). Fathers' rights advocates suggest replacing the term *visitation* with terms such as *parenting plan* or *time-sharing arrangement,* because these terms do not minimize either parent's role.

Symbolic interactionists also point to the effects of interaction on one's self-concept, especially the self-concept of children. In a process called the "looking-glass self," individuals form a self-concept based on how others interact with them. Family members, such as parents, grandparents, siblings, and spouses, have a powerful effect on our self-concepts. For example, negative self-concepts may result from verbal abuse in the family, whereas positive self-concepts may develop in families in which interactions are supportive and loving. The importance of social interaction in children's developing self-concept suggests a compelling reason for society to accept rather than stigmatize nontraditional family forms. Imagine the effect on children who are called "illegitimate" or who are teased for having two moms or dads.

The symbolic interactionist perspective is useful in understanding the dynamics of domestic violence and abuse. For example, some abusers and their victims learn to define intimate partner violence as an expression of love (Lloyd 2000). Emotional abuse often involves using negative labels (e.g., *stupid, whore,* or *bad*) to define a partner or family member. Such labels negatively affect the self-concept of abuse victims, often convincing them that they deserve the abuse. In the next section, we examine violence and abuse in intimate and family relationships.

Violence and Abuse in Intimate and Family Relationships

Although intimate and family relationships provide many individuals with a sense of intimacy and well-being, these relationships involve physical violence, verbal and emotional abuse, sexual abuse, and/or neglect for others. In U.S. society, people are more likely to be physically assaulted, abused and neglected, sexually assaulted and molested, or killed in their own homes rather than anywhere else, and by other family members rather than by anyone else (Gelles 2000). Before reading further, you may want to take the Abusive Behavior Inventory in this chapter's Self and Society feature.

> In U.S. society, people are more likely to be physically assaulted, abused and neglected, sexually assaulted and molested, or killed in their own homes rather than anywhere else, and by other family members rather than by anyone else.

Intimate Partner Violence and Abuse

Abuse in relationships can take many forms, including emotional and psychological abuse, physical violence, and sexual abuse. **Intimate partner violence (IPV)** refers to actual or threatened violent crimes committed against individuals by their current or former spouses, cohabiting partners, boyfriends, or girlfriends.

Prevalence and Patterns of Intimate Partner Violence.

Globally, one woman in every three has been subjected to violence in an intimate relationship (United Nations Development Programme 2000). In the United States, most acts of intimate partner violence (85 percent) are committed against women; however, 22 percent of men have experienced physical, sexual, or psychological IPV (compared with 29 percent of women) (National Center for Injury Prevention and Control 2006). When women assault their male partners, these assaults tend to be acts of retaliation or self-defense (Johnson 2001; Swan et al. 2008). Most research on intimate partner abuse has been conducted on heterosexuals, but more than one-third of gay women and gay men in one study reported physical violence in their relationships in the past year (McKenry et al. 2004).

Johnson and Ferraro (2003) identified the following four patterns of partner violence:

1. *Common couple violence* refers to occasional acts of violence arising from arguments that get "out of hand." Common couple violence usually does not escalate into serious or life-threatening violence.
2. *Intimate terrorism* is motivated by a wish to control one's partner and involves not only violence but also economic subordination, threats, isolation, verbal and emotional abuse, and other control tactics. Intimate terrorism

intimate partner violence (IPV) Actual or threatened violent crimes committed against individuals by their current or former spouses, cohabiting partners, boyfriends, or girlfriends.

Circle the number that best represents your closest estimate of how often each of the behaviors happened in your relationship with your partner or former partner during the previous six months (1, never; 2, rarely; 3, occasionally; 4, frequently; 5, very frequently).

1. Called you a name and/or criticized you	1	2	3	4	5
2. Tried to keep you from doing something you wanted to do (e.g., going out with friends, going to meetings)	1	2	3	4	5
3. Gave you angry stares or looks	1	2	3	4	5
4. Prevented you from having money for your own use	1	2	3	4	5
5. Ended a discussion with you and made the decision himself/herself	1	2	3	4	5
6. Threatened to hit or throw something at you	1	2	3	4	5
7. Pushed, grabbed, or shoved you	1	2	3	4	5
8. Put down your family and friends	1	2	3	4	5
9. Accused you of paying too much attention to someone or something else	1	2	3	4	5
10. Put you on an allowance	1	2	3	4	5
11. Used your children to threaten you (e.g., told you that you would lose custody, said he/she would leave town with the children)	1	2	3	4	5
12. Became very upset with you because dinner, housework, or laundry was not done when he/she wanted it done or done the way he/she thought it should be	1	2	3	4	5
13. Said things to scare you (e.g., told you something "bad" would happen, threatened to commit suicide)	1	2	3	4	5
14. Slapped, hit, or punched you	1	2	3	4	5
15. Made you do something humiliating or degrading (e.g., begging for forgiveness, having to ask permission to use the car or to do something)	1	2	3	4	5
16. Checked up on you (e.g., listened to your phone calls, checked the mileage on your car, called you repeatedly at work)	1	2	3	4	5
17. Drove recklessly when you were in the car	1	2	3	4	5
18. Pressured you to have sex in a way you didn't like or want	1	2	3	4	5
19. Refused to do housework or child care	1	2	3	4	5
20. Threatened you with a knife, gun, or other weapon	1	2	3	4	5
21. Spanked you	1	2	3	4	5
22. Told you that you were a bad parent	1	2	3	4	5
23. Stopped you or tried to stop you from going to work or school	1	2	3	4	5
24. Threw, hit, kicked, or smashed something	1	2	3	4	5
25. Kicked you	1	2	3	4	5
26. Physically forced you to have sex	1	2	3	4	5
27. Threw you around	1	2	3	4	5
28. Physically attacked the sexual parts of your body	1	2	3	4	5
29. Choked or strangled you	1	2	3	4	5
30. Used a knife, gun, or other weapon against you	1	2	3	4	5

Scoring

Add the numbers you circled and divide the total by 30 to find your score. The higher your score, the more abusive your relationship.

Source: Shepard, Melanie F., and James A. Campbell. 1992 (September). The Abusive Behavior Inventory: A Measure of Psychological and Physical Abuse. *Journal of Interpersonal Violence* 7(3):291–305. Used by permission of Sage Publications, 2455 Teller Road, Newbury Park, CA 91320.

is almost entirely perpetrated by men and is more likely to escalate over time and to involve serious injury.

3. *Violent resistance* refers to acts of violence that are committed in self-defense, usually by women against a male partner.
4. *Mutual violent control* is a rare pattern of abuse "that could be viewed as two intimate terrorists battling for control" (Johnson & Ferraro 2003, p. 169).

Intimate partner abuse also takes the form of sexual aggression, which refers to sexual interaction that occurs against one's will through use of physical force, threat of force, pressure, use of alcohol or drugs, or use of position of authority. In 2007, 61 percent of female rape or sexual assault victims (12 and older) reported that the offender was an intimate partner, friend, or acquaintance (Rand 2008).

A survey of more than 20,000 college students at 39 colleges and universities found that, in the past 12 months, students were much more likely to have experienced an emotionally abusive relationship than a physically or sexually abusive relationship (see Table 5.1) (American College Health Association 2008).

TABLE 5.1 Abusive Relationships Reported by College Students

PERCENT OF COLLEGE STUDENTS REPORTING ABUSIVE RELATIONSHIP IN PAST 12 MONTHS

TYPE OF ABUSE	MALE	FEMALE	TOTAL
Emotional	10.4	16.0	14.1
Physical	1.9	2.6	2.4
Sexual	1.5	2.1	2.0

Source: American College Health Association 2008.

Effects of Intimate Partner Violence and Abuse. Each year, intimate partner violence results in about 2 million injuries and 1,300 deaths nationwide (National Center for Injury Prevention and Control 2006). More than three-fourths of murder victims killed by an intimate partner are women (Federal Bureau of Investigation 2008). Many battered women are abused during pregnancy, resulting in a high rate of miscarriage and birth defects. Psychological consequences for victims of intimate partner violence can include depression, anxiety, suicidal thoughts and attempts, lowered self-esteem, inability to trust men, fear of intimacy, and substance abuse (National Center for Injury Prevention and Control 2006).

Battering also interferes with women's employment. Some abusers prohibit their partners from working. Other abusers "deliberately undermine women's employment by depriving them of transportation, harassing them at work, turning off alarm clocks, beating them before job interviews, and disappearing when they promise to provide child care" (Johnson & Ferraro 2003, p. 508). Battering also undermines employment by causing repeated absences, impairing women's ability to concentrate, and lowering their self-esteem and aspirations.

Abuse is also a factor in many divorces, which often results in a loss of economic resources. Women who flee an abusive home and who have no economic resources may find themselves homeless.

Children who witness domestic violence are at risk for emotional, behavioral, and academic problems as well as future violence in their own adult relationships (Parker et al. 2000; Kitzmann et al. 2003). Children may also commit violent acts against a parent's abusing partner.

Why Do Some Adults Stay in Abusive Relationships? Adult victims of abuse are commonly blamed for tolerating abusive relationships and for not leaving the relationship as soon as the abuse begins. However, from the point of view of the victims, there are compelling reasons to stay. These reasons include love, emotional dependency, commitment to the relationship, hope that things will get better, the view that violence is legitimate because they "deserve"

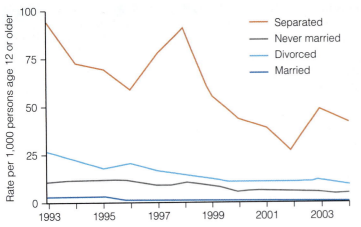

Figure 5.4 Nonfatal Intimate Partner Victimization Rate for Females by Marital Status, 1993–2004
Source: Catalano 2006.

it, guilt, fear, economic dependency, and feeling stuck.

Many victims of intimate partner abuse stay because they fear retribution from their abusive partner if they leave. Indeed, the rate of intimate partner violence is highest among separated couples (see Figure 5.4). Some victims also delay leaving a violent home because they fear the abuser will hurt or neglect a family pet (Fogle 2003).

Victims also stay because abuse in relationships is usually not ongoing and constant but rather occurs in cycles. The **cycle of abuse** involves a violent or abusive episode followed by a makeup period when the abuser expresses sorrow and asks for forgiveness and "one more chance." The makeup period may last for days, weeks, or even months before the next violent outburst occurs.

Child Abuse

Child abuse refers to the physical or mental injury, sexual abuse, negligent treatment, or maltreatment of a child under the age of 18 by a person who is responsible for the child's welfare. The most common form of child maltreatment is **neglect**—the caregiver's failure to provide adequate attention and supervision, food and nutrition, hygiene, medical care, and a safe and clean living environment (see Figure 5.5).

The highest rates of victimization are for the youngest children (birth to 1 year), and for children with disabilities. Although half of child abuse and neglect

cycle of abuse A pattern of abuse in which a violent or abusive episode is followed by a makeup period when the abuser expresses sorrow and asks for forgiveness and "one more chance," before another instance of abuse occurs.

child abuse The physical or mental injury, sexual abuse, negligent treatment, or maltreatment of a child younger than age 18 by a person who is responsible for the child's welfare.

neglect A form of abuse involving the failure to provide adequate attention, supervision, nutrition, hygiene, health care, and a safe and clean living environment for a minor child or a dependent elderly individual.

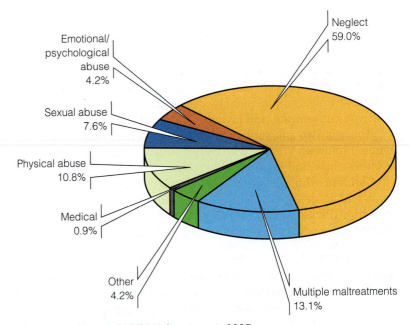

Figure 5.5 Types of Child Maltreatment, 2007
Source: U.S. Department of Health and Human Services, Administration on Children, Youth, and Families 2009.

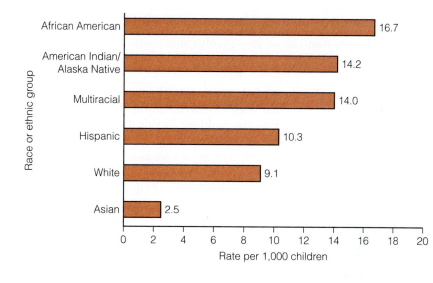

Figure 5.6 Rates of Child Abuse and Neglect by Race and Ethnicity: United States, 2007
Source: U.S. Department of Health and Human Services, Administration on Children, Youth, and Families 2009.

Chart data (Rate per 1,000 children):
- African American: 16.7
- American Indian/Alaska Native: 14.2
- Multiracial: 14.0
- Hispanic: 10.3
- White: 9.1
- Asian: 2.5

victims are white, *rates* of victimization are higher among some minority children (see Figure 5.6).

Perpetrators of child abuse are most often the parents of the victim. Parents who are at the greatest risk of child maltreatment are those who are socially isolated, poor or unemployed, or young and single, have a mental illness, lack an understanding of children's needs and child development, and have a history of domestic abuse (National Center for Injury Prevention and Control 2007).

Effects of Child Abuse. Physical injuries sustained by child abuse cause pain, disfigurement, scarring, physical disability, and death. In 2007, an estimated 1,760 U.S. children died of abuse or neglect (U.S. Department of Health and Human Services, Administration on Children, Youth, and Families 2009). Most of these children were younger than age 4. Most murders of children younger than age 5 are committed by a parent or other family member. Head injury is the leading cause of death in abused children (Rubin et al. 2003). **Shaken baby syndrome**, whereby a caregiver shakes a baby to the point of causing the child to experience brain or retinal hemorrhage, most often occurs in response to a baby, typically younger than 6 months, who will not stop crying (Ricci et al. 2003; Smith 2003). Battered or shaken babies often suffer permanent disabilities.

Abuse during childhood is also associated with depression, low academic achievement, smoking, alcohol and drug abuse, eating disorders, obesity, teen pregnancy, sexually transmitted diseases, sexual promiscuity, low self-esteem, aggressive behavior, juvenile delinquency, adult criminality, suicide, and experiencing abuse victimization as an adult (Administration for Children and Families 2003; National Center for Injury Prevention and Control 2007). Sexual abuse of young girls is associated with decreased self-esteem, increased levels of depression, running away from home, alcohol and drug use, and multiple sexual partners (Jasinski et al. 2000; Whiffen et al. 2000). A review of the research suggests that sexual abuse of boys produces many of the same reactions that sexually abused girls experience, including depression, sexual dysfunction, anger, self-blame, suicidal feelings, guilt, and flashbacks (Daniel 2005). Married adults who were physically and sexually abused as children report lower marital satisfaction, higher stress, and lower family cohesion than married adults with no abuse history (Nelson & Wampler 2000).

Matthew Herrnkind, left, and Julie Herrnkind, right, were charged with second-degree murder in the death of their 3-year-old girl, Sylena. The couple had a long history of abusing their children.

shaken baby syndrome A form of child abuse whereby the caretaker shakes a baby to the point of causing the child to experience brain or retinal hemorrhage.

Elder Abuse, Parent Abuse, and Sibling Abuse

Domestic violence and abuse may involve adults abusing their elderly parents or grandparents, children abusing their parents, and siblings abusing each other.

Elder Abuse. A survey of abuse of U.S. adults age 60 and older found that there are 8.3 reports of elder abuse annually for every 1,000 older Americans (Teaster et al. 2006). **Elder abuse** includes physical abuse, sexual abuse, psychological abuse, financial abuse (such as improper use of the elder person's financial resources), and neglect. The most common form of elder abuse is neglect—failure to provide basic health and hygiene needs, such as clean clothes, doctor visits, medication, and adequate nutrition. Neglect also involves unreasonable confinement, isolation of elderly family members, lack of supervision, and abandonment.

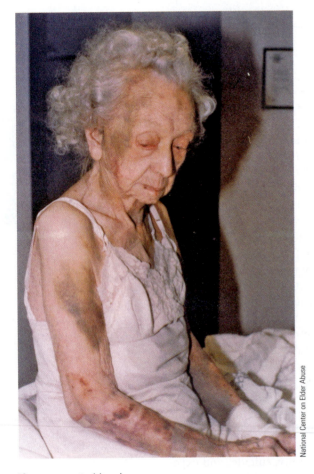

Older women are far more likely than older men to suffer from abuse or neglect. Two of three cases of elder abuse reported to state adult protective services involved women. Although elder abuse also occurs in nursing homes, most cases of elder abuse occur in a domestic setting. The most likely perpetrators are adult children, followed by other family members and spouses or intimate partners (Teaster et al. 2006).

Parent Abuse. Some parents are victimized by their children's violence, ranging from hitting, kicking, and biting to pushing a parent down the stairs and using a weapon to inflict serious injury to or even kill a parent. More violence is directed against mothers than against fathers, and sons tend to be more violent toward parents than are daughters (Ulman 2003). In most cases of children being violent toward their parents, the parents had been violent toward the children.

Sibling Abuse. The most prevalent form of abuse in families is sibling abuse. Ninety-eight percent of the females and 89 percent of the males in one study reported having been emotionally abused by a sibling, and 88 percent of the females and 71 percent of the males reported having been physically abused by a sibling (Simonelli et al. 2002). Sexual abuse also occurs in sibling relationships.

If you suspect elder abuse or are concerned about the well-being of an older person, call your state abuse hotline immediately.

elder abuse The physical or psychological abuse, financial exploitation, or medical abuse or neglect of the elderly.

Factors Contributing to Intimate Partner and Family Violence and Abuse

Various individual, family, and social factors contribute to domestic violence and abuse.

Individual and Family Factors. Individual and family factors associated with intimate partner and family violence and abuse include a history of family violence, drug and alcohol abuse, and poverty. Men who witnessed their fathers abusing their mothers and women who witnessed their mothers abusing their fathers are more likely to become abusive partners themselves (Heyman & Slep 2002; Babcock et al. 2003). Individuals who were abused as children are more likely to report being abused in an adult domestic relationship (Heyman & Slep 2002). Mothers who have been sexually abused as children are more likely to physically abuse their own children (DiLillo et al. 2000). Although a history of abuse is associated with an increased likelihood of being abusive as an adult, most adults who experienced or witnessed violence in their families of origin are not likely to be aggressive with their partners (Busby et al. 2008).

Alcohol use is reported as a factor in 50 percent to 75 percent of incidents of physical and sexual aggression in intimate relationships (Lloyd & Emery 2000). Alcohol and other drugs increase aggression in some individuals and enable the offender to avoid responsibility by blaming the violent behavior on drugs or alcohol.

Although abuse in adult relationships occurs among all socioeconomic groups, it is more prevalent among the poor. Studies show that at least 50 percent to 60 percent of women receiving welfare have experienced physical abuse by an intimate partner, compared with 22 percent of the general population (Family Violence Prevention Fund 2005). However, Kaufman and Zigler (1992) noted that, "although most poor people do not maltreat their children, and poverty, per se, does not cause abuse and neglect, the correlates of poverty, including stress, drug abuse, and inadequate resources for food and medical care, increase the likelihood of maltreatment" (p. 284).

Gender Inequality and Gender Socialization.
In the United States before the late 19th century, a married woman was considered the property of her husband. A husband had a legal right and marital obligation to discipline and control his wife through the use of physical force. This traditional view of women as property may contribute to men's doing with their "property" as they wish. In a study of men in battering intervention programs, about half of the men viewed battering as acceptable in certain situations (Jackson et al. 2003).

The view of women and children as property also explains marital rape and father-daughter incest. Historically, the penalties for rape were based on property rights laws designed to protect a man's property—his wife or daughter—from rape by other men (Russell 1990). Although a husband or father "taking" his own property in the past was not considered rape, today, marital rape is considered a crime in all 50 states.

Traditional male gender roles have taught men to be aggressive and to be dominant in male-female relationships. "Because it is so clearly associated with masculinity in American culture, violence is a social practice that enables men to express a masculine identity" (Anderson 1997, p. 667). Traditional male gender socialization also discourages men from verbally expressing their feelings, which increases the potential for violence and abusive behavior (Umberson et al. 2003). Traditional female gender roles have also taught women to be submissive to their male partner's control.

Acceptance of Corporal Punishment.
Corporal punishment is the intentional infliction of pain for a perceived misbehavior (Block 2003). Many mental health professionals and child development specialists argue that it is ineffective and damaging to children. Children who experience corporal punishment display

corporal punishment The intentional infliction of pain for a perceived misbehavior.

more antisocial behavior, are more violent, and have an increased incidence of depression as adults (Straus 2000). Yet many parents accept the cultural tradition of spanking as an appropriate form of child discipline.

More than 90 percent of parents of toddlers reported using corporal punishment (Straus 2000). Parents of black children are more likely to use corporal punishment than parents of white or Latino children (Grogan-Kaylor & Otis 2007). Although not everyone agrees that all instances of corporal punishment constitute abuse, some episodes of parental "discipline" are undoubtedly abusive.

Inaccessible or Unaffordable Community Services. Many cases of child and elder abuse and neglect are linked to inaccessible or unaffordable health care, day care, elder care, and respite care facilities. Failure to provide medical care to children and elderly family members (a form of neglect) is sometimes a result of the lack of accessible or affordable health care services in the community. Failure to provide supervision for children and adults may result from inaccessible day care and elder care services. Without elder care and respite care facilities, socially isolated families may not have any help with the stresses of caring for elderly family members and children with special needs.

Strategies for Action: Preventing and Responding to Domestic Violence and Abuse

Next, we look at strategies for preventing and responding to violence and abuse in intimate and family relationships.

Prevention Strategies

Abuse-prevention strategies include public education and media campaigns, which may help to reduce domestic violence by conveying the criminal nature of domestic assault and offering ways to prevent abuse. Other abuse-prevention efforts focus on parent education to teach parents realistic expectations about child behavior and methods of child discipline that do not involve corporal punishment. For example, Mental Health America (2003) distributes a fact sheet on alternatives to spanking (see Table 5.2).

What Do You Think? In 1979, Sweden became the first country in the world to ban corporal punishment in all settings, including the home. By 2009, 24 countries banned corporal punishment in all settings (Global Initiative to End All Corporal Punishment of Children 2009). In the United States, it is legal in all 50 states for a parent to spank, hit, belt, paddle, whip, or otherwise inflict punitive pain on a child, so long as the corporal punishment does not meet the individual state's definition of child abuse. Corporal punishment is also permitted in public schools in 22 states and in private schools in every state except Iowa and New Jersey. Do you think that the United States should ban corporal punishment? Why or why not?

Another abuse-prevention strategy involves reducing violence-provoking stress by reducing poverty and unemployment and providing adequate housing, child care programs and facilities, nutrition, medical care, and educational opportunities. Strengthening the supports for poor families with children reduces violence-provoking stress and minimizes neglect that results from inaccessible or unaffordable community services.

Responding to Domestic Violence and Abuse

After domestic violence and abuse has occurred, victims may seek safety in a shelter or safe house and take legal action against the abuser. Children may be placed in foster care, and the abuser may participate in mandatory or voluntary treatment.

Shelters and Safe Houses. Between 1993 and 2004, about 21 percent of female victims and 10 percent of male victims of nonfatal IPV contacted a private or government agency for assistance (Catalano 2006). The National Domestic Violence Hotline (1-800-799-SAFE) is a 24-hour, toll-free service that provides crisis assistance and local domestic violence shelter and safe house referrals for callers across the country.

Shelters provide abused women and their children with housing, food, and counseling services. Safe houses are private homes of individuals who volunteer to provide temporary housing to abused people who decide to leave their violent homes. Some communities have abuse shelters for victims of elder abuse. Because one in four victims reports a delay in leaving dangerous domestic situations because of concerns over the safety of a pet, some programs offer a safe shelter for pets of victims of domestic violence (Fogle 2003).

Arrest and Restraining Order. Domestic violence and abuse are crimes for which individuals can be arrested, jailed, and/or ordered to leave the home or enter a treatment program. About half of the states and Washington, DC, now have mandatory arrest policies that require police to arrest abusers, even if the victim does not want to press charges. Abuse victims may obtain a restraining order that prohibits the perpetrator from going near the abused partner. However, many victims of intimate partner violence do not report the violence to law enforcement authorities because victims (1) believe that such violence is a private or personal matter, (2) fear retaliation, (3) view the violence as a "minor" crime, (4) want to protect the offender, and (5) believe that the police will not help or will be ineffective (Catalano 2006). Fewer than half (42 percent) of rape or sexual assault incidents were reported to police (Rand 2008).

Foster Care Placement. Children who are abused in the family may be removed from their homes and placed in government-supervised foster care. Foster care placements include other family members, certified foster parents, group homes, and other institutional facilities. More than 700,000 children are in foster care, waiting to be reunited with their families or adopted.

Due to the economic recession, more prospective adoptive parents are considering adopting foster children because they cannot afford private adoptions. However, there continues to be a shortage of people willing to adopt foster children because foster children tend to be older and are more likely to have emotional or physical problems (Koch 2009). Every year, more than 20,000 children who have not been adopted or reunited with their families must leave foster care because they turn 18. In some states, youth who age out of the foster care system can receive government aid such as housing assistance and Medicaid. However, many youth who age out of foster care struggle to fend for themselves, and many become homeless.

Another problem that plagues the foster care system is that, although it is intended to protect children from abuse, foster parents or caregivers sometimes abuse the children. Excluding extreme cases of abuse and neglect, some evidence suggests that children whose families are investigated for abuse or neglect are better off staying with their families than entering foster care, as foster care children are more likely to drop out of school, commit crimes, abuse drugs, and become teen parents (Doyle 2007).

Treatment for Abusers. Treatment for abusers—which may be voluntary or mandated by the court—typically involves group and/or individual counseling, substance abuse counseling, and/or training in communication, conflict resolution, and anger management. Treatment for men who sexually abuse children typically involves cognitive behavior therapy (changing the thoughts that lead to sex abuse) and medication to reduce the sex drive (Stone 2004). Men who stop abusing their partners learn to take responsibility for their abusive behavior, develop empathy for their partner's victimization, reduce their dependency on their partners, and improve their communication skills (Scott & Wolfe 2000).

Problems Associated with Divorce

The United States has the highest divorce rate among Western nations. Despite the decline in divorce rates in recent years, 40 percent of first marriages end in divorce and 60 percent of those marriages involve children (Kimmel 2004). Divorce is considered problematic not only because of the negative effects it has on children but also because of the difficulties it causes for adults. However, in some societies, legal and social barriers to divorce are considered problematic because such barriers limit the options of spouses in unhappy and abusive marriages. Ireland did not allow divorce under any condition until 1995, and Chile did not allow divorce until 2004.

Even when divorce is a legal option, social barriers often prevent spouses from divorcing. Hindu women, for example, experience great difficulty leaving a marriage, even when the husband is abusive, because divorce leads to loss of status, possible loss of custody of the children, homelessness, poverty, and being labeled a "loose" woman (Laungani 2005, p. 88).

Social Causes of Divorce

When we think of why a particular couple gets divorced, we typically think of a number of individual and relationship factors that might have contributed to the marital breakup: incompatibility in values or goals, poor communication, lack of conflict resolution skills, sexual incompatibility, extramarital relationships, substance abuse, emotional or physical abuse or neglect, boredom, jealousy, and difficulty coping with change or stress related to parenting, employment, finances, in-laws, and illness. However, understanding the high rate of divorce in U.S. society requires awareness of how the following social and cultural factors contribute to marital breakup:

1. *Changing function of marriage.* Before the Industrial Revolution, marriage functioned as a unit of economic production and consumption that was largely organized around producing, socializing, and educating children. However, the institution of marriage has changed over the last few generations:

   > Marriage changed from a formal institution that meets the needs of the larger society to a companionate relationship that meets the needs of the couple and their children and then to a private pact that meets the psychological needs of individual spouses. (Amato et al. 2007, p. 70)

   When spouses do not feel that their psychological needs—for emotional support, intimacy, affection, love, or personal growth—are being met in the marriage, they may consider divorce with the hope of finding a new partner to fulfill these needs.

2. *Increased economic autonomy of women.* Before 1940, most wives were not employed outside the home and depended on their husband's income. Today, the majority of married women are in the labor force. A wife who is unhappy in her marriage is more likely to leave the marriage if she has the economic means to support herself (Jalovaara 2003). An unhappy husband may also be more likely to leave a marriage if his wife is self-sufficient and can contribute to the support of the children.

TABLE 5.3 Factors That Decrease Women's Risk of Separation or Divorce During the First 10 Years of Marriage	
FACTOR	PERCENT DECREASE IN RISK OF DIVORCE OR SEPARATION
Annual income over $50,000 (versus under $25,000)	30
Having a baby 7 months or more after marriage (versus before marriage)	24
Marrying over 25 years of age (versus under 18)	24
Having an intact family of origin (versus having divorced parents)	14
Religious affiliation (versus none)	14
Some college (versus high school dropout)	13

Source: National Marriage Project 2007.

3. *Increased work demands and economic stress.* Another factor influencing divorce is increased work demands and the stresses of balancing work and family roles. Some workers are putting in longer hours, often working overtime or taking second jobs, while others face job loss and unemployment. As discussed in Chapters 6 and 7, many families struggle to earn enough money to pay for rising housing, health care, and costs for child care. Financial stress can cause marital problems. Couples with an annual income under $25,000 are 30 percent more likely to divorce than couples with incomes over $50,000 (see Table 5.3).

4. *Inequality in marital division of labor.* Many employed parents, particularly mothers, come home to work a **second shift**—the work involved in caring for children and household chores (Hochschild 1989). Wives are more likely than husbands to perceive the marital division of labor—household chores and child care—as unfair (Nock 1995). This perception of unfairness can lead to marital tension and resentment, as reflected in the following excerpt:

My husband's a great help watching our baby. But as far as doing housework or even taking the baby when I'm at home, no. He figures he works five days a week; he's not going to come home and clean. But he doesn't stop to think that I work seven days a week. Why should I have to come home and do the housework without help from anybody else? My husband and I have been through this over and over again. Even if he would just pick up from the kitchen table and stack the dishes for me, that

Wives tend to be less happy in marriage than husbands when they perceive the division of household labor to be unfair.

second shift The household work and child care that employed parents (usually women) do when they return home from their jobs.

Radius Images/Jupiter Images

would make a big difference. He does nothing. . . . He'll help out if I'm not here, but the minute I am, all the work at home is mine. (quoted by Hochschild 1997, pp. 37–38)

Women want to be equal partners in their marriages, not just in earning income but also in sharing the work of household chores, child rearing, marital communication, and in making decisions for the family. Frustrated by men's lack of participation in marital work, women who desire relationship egalitarianism may see divorce as the lesser of two evils (Hackstaff 2003).

5. *Liberalized divorce laws.* Before 1970, the law required a couple who wanted a divorce to prove that one of the spouses was at fault and had committed an act defined by the state as grounds for divorce—adultery, cruelty, or desertion. In 1969, California became the first state to initiate **no-fault divorce,** which permitted a divorce based on the claim that there were "irreconcilable differences" in the marriage. Today, all 50 states recognize some form of no-fault divorce. No-fault divorce law has contributed to the U.S. divorce rate by making divorce easier to obtain.

6. *Increased individualism.* U.S. society is characterized by **individualism**—the tendency to focus on one's individual self-interests and personal happiness rather than on the interests of one's family and community. "Marital commitment lasts only as long as people are happy and feel that their own needs are being met" (Amato 2004, p. 960). Belief in the right to be happy, even if it means getting divorced, is reflected in social attitudes toward divorce: more than two-thirds (70 percent) of U.S. adults report that divorce is morally acceptable (Gallup Organization 2009). **Familism**—the view that the family unit is more important than individual interests—is still prevalent among Asian Americans and Mexican Americans, which helps to explain why the divorce rate is lower among these groups than among whites and Black Americans (Mindel et al. 1998).

7. *Weak social ties.* Couples with strong social ties have a network of family and friends who provide support during difficult times, and who express disapproval for behavior that threatens the stability of the marriage. Couples who live in the same community for a long time have the opportunity to develop and maintain strong social ties. But many Americans move from place to place during their adult years, more so than people in other countries, which may help explain why the U.S. divorce rate is higher than in other countries. Cherlin (2009) explains that "moving from one community to another could affect marriages because it disrupts social ties. Migration can separate people from friends and relatives who could help them through family crises" (pp. 148–149).

8. *Increased life expectancy.* Finally, more marriages today end in divorce, in part, because people live longer than they did in previous generations and "till death do us part" involves a longer commitment than it once did. Indeed, one can argue that "marriage once was as unstable as it is today, but it was cut short by death not divorce" (Emery 1999, p. 7).

Consequences of Divorce

When parents have bitter and unresolved conflict and/or if one parent is abusing a child or the other parent, divorce may offer a solution to family problems. However, divorce often has negative effects for ex-spouses and their children and also contributes to problems that affect society as a whole.

no-fault divorce A divorce that is granted based on the claim that there are irreconcilable differences within a marriage (as opposed to one spouse being legally at fault for the marital breakup).

individualism The tendency to focus on one's individual self-interests and personal happiness rather than on the interests of one's family and community.

familism The view that the family unit is more important than individual interests.

Physical and Mental Health Consequences. Numerous studies show that divorced individuals have more health problems and a higher risk of mortality than married individuals; divorced individuals also experience lower levels of psychological well-being, including more unhappiness, depression, anxiety, and poorer self-concepts (Amato 2003). Both divorced and never-married individuals are, on average, more distressed than married people because unmarried people are more likely than married people to have low social attachment, low emotional support, and increased economic hardship (Walker 2001). Some research suggests that divorce leads to higher levels of depressive symptoms for women, but not for men (Kalmijn & Monden 2006), especially when young children are in the family (Williams & Dunne-Bryant 2006). This finding is probably due to the increased financial and parenting strains experienced by divorced mothers who have custody of young children.

However, some studies have found that divorced individuals report higher levels of autonomy and personal growth than married individuals do (Amato 2003). For example, many divorced mothers report improvements in career opportunities, social lives, and happiness after divorce; some divorced women report more self-confidence, and some men report more interpersonal skills and a greater willingness to self-disclose. For people in a poor-quality marriage, divorce has a less negative or even a positive effect on well-being (Amato 2003; Kalmijn & Monden 2006). However, leaving a bad marriage does not always result in increased well-being because "divorce is a trigger for even more problems after the divorce" (Kalmijn & Monden 2006, p. 1210). In sum, some men and women experience a decline in well-being after divorce; others experience an improvement.

> . . .Some men and women experience a decline in well-being after divorce; others experience an improvement.

Economic Consequences. Following divorce, there tends to be a dramatic drop in women's income and a slight drop in men's income (Gadalla 2009). Compared with married individuals, divorced individuals have a lower standard of living, have less wealth, and experience greater economic hardship, although this difference is considerably greater for women than for men (Amato 2003). The economic costs of divorce are often greater for women and children because women tend to earn less than men (see Chapter 10) and because mothers devote substantially more time to household and child care tasks than fathers do. The time women invest in this unpaid labor restricts their educational and job opportunities as well as their income. Men are less likely than women to be economically disadvantaged after divorce because they continue to profit from earlier investments in education and career.

After divorce, both parents are responsible for providing economic resources to their children. However, some nonresident parents fail to provide child support. In some cases, failure to pay child support is not due to fathers being "deadbeats" but rather to the fact that many fathers are "dead broke." Fathers who are unemployed or who have low-wage jobs may be unable to make child support payments.

Effects on Children and Young Adults. Parental divorce is a stressful event for children and is often accompanied by a variety of stressors, such as continuing conflict between parents, a decline in the standard of living, moving and perhaps changing schools, separation from the noncustodial parent (usually the father), and parental remarriage. These stressors place children of divorce at higher risk for a variety of emotional and behavioral problems. Reviews of research on the consequences of divorce for children have found that children with divorced parents score lower on measures of academic success, psychological adjustment, self-concept, social competence, and long-term health; they also have higher levels of aggressive behavior and depression (Amato 2003; Wallerstein 2003).

Many of the negative effects of divorce on children are related to the economic hardship associated with divorce. Economic hardship is associated with less effective and less supportive parenting, inconsistent and harsh discipline, and emotional distress in children (Demo et al. 2000). Despite the adverse effects of divorce on children, research findings suggest that "most children from divorced families are resilient, that is, they do not suffer from serious psychological problems" (Emery et al. 2005, p. 24). Other researchers conclude that "most offspring with divorced parents develop into well-adjusted adults," despite the pain they feel associated with the divorce (Amato & Cheadle 2005, p. 191).

Divorce can also have positive consequences for children and young adults. In highly conflictual marriages, divorce may actually improve the emotional well-being of children relative to staying in a conflicted home environment (Jekielek 1998). In interviews with 173 grown children whose parents divorced years earlier, Ahrons (2004) found that most of the young adults reported positive outcomes for their parents and for themselves. Although many young adults who have divorced parents fear that they too will have an unhappy marriage (Dennison & Koerner 2008), such a fear can also lead young adults to think carefully about their choices regarding marriage.

Effects on Father-Child Relationships. Children who live with their mothers may have a damaged relationship with their nonresidential father, especially if he becomes disengaged from their lives. Some research has found that young adults whose parents divorced are less likely to report having a close relationship with their father compared with children whose parents are together (DeCuzzi et al. 2004). However, in another study of 173 adult children of divorce, more than half felt that their relationships with their fathers improved after the divorce (Ahrons 2004). Children may benefit from having more quality time with their fathers after parental divorce. Some fathers report that they became more active in the role of father after divorce.

One study has found that the older the child is at the time of parental separation, the greater the amount of time children spend with their fathers (Swiss & Bourdais 2009). This study also found that fathers earning $50,000 or more per year are likely to see their children more often than fathers who earn less than $30,000 per year.

Women who have primary custody of the children serve as gatekeepers for the relationship their children have with their fathers (Trinder 2008). As gatekeepers, custodial mothers can either keep the gate open and encourage contact between the children and their father, or make every effort to keep the gate closed, cutting off the children's contact with their father.

Some divorced mothers not only fail to encourage their children's relationships with their fathers but also actively attempt to alienate the children from their father. (We note that some divorced fathers do likewise.) Thus, some children of divorce suffer from **parental alienation syndrome (PAS)**, defined as an emotional and psychological disturbance in which children engage in exaggerated and unjustified denigration and criticism of a parent (Family Court Reform Council of America 2000). Parental alienation syndrome has been described as a form of "psychological kidnapping," whereby one parent manipulates children's psyches to make them hate and reject the other parent. Children who suffer from

parental alienation syndrome (PAS) An emotional and psychological disturbance in which children engage in exaggerated and unjustified denigration and criticism of a parent.

PAS are victims of a form of child abuse in which one parent essentially brainwashes the child to hate the other parent. Parents may alienate their child from the other parent by engaging in the following behaviors:

- Minimizing the importance of contact and relationship with the other parent
- Being rude to the other parent; refusing to speak to or tolerate the presence of the other parent, even at events important to the child; refusing to allow the other parent near the home for drop-off or pick-up visitations
- Failing to display any positive interest in the child's activities or experiences during visits
- Expressing disapproval or dislike of the child's spending time with the other parent and refusing to discuss anything about the other parent ("I don't want to hear about . . .") or selective willingness to discuss only negative matters
- Making innuendos and accusations against the other parent, including statements that are false
- Demanding that the child keep secrets from the other parent
- Destruction of gifts or memorabilia from the other parent
- Promoting loyalty conflicts (e.g., offering an opportunity for a desired activity that conflicts with scheduled visitation) (Schacht 2000)

Long-term effects of PAS on children can include depression, guilt, hostility, alcoholism and other drug abuse, and other symptoms of internal distress (Family Court Reform Council of America 2000). The effects on the rejected parent are equally devastating.

Some noncustodial divorced fathers discontinue contact with their children as a coping strategy for managing emotional pain (Pasley & Minton 2001). Many divorced fathers are overwhelmed with feelings of failure, guilt, anger, and sadness over the separation from their children (Knox 1998). Hewlett and West (1998) explained that "visiting their children only serves to remind these men of their painful loss, and they respond to this feeling by withdrawing completely" (p. 69). Divorced fathers commonly experience the legal system as favoring the mother in child-related matters. One divorced father commented:

> I believe that the system [judges, attorneys, etc.] have [sic] little or no consideration for the father. At some point the system creates an environment where the father loses any natural desire to see his children because it becomes so difficult, both financially and emotionally. At that point, he convinces himself that the best thing to do is wait until they are older. (quoted by Pasley & Minton 2001, p. 242)

As we have seen, the effects of divorce on adults and children are mixed and variable. In a review of research on the consequences of divorce for children and adults, Amato (2003) concluded that "divorce benefits some individuals, leads others to experience temporary decrements in well-being that improve over time, and forces others on a downward cycle from which they might never fully recover" (p. 206).

Strategies for Action: Strengthening Marriage and Alleviating Problems of Divorce

Two general strategies for responding to the problems of divorce are those that prevent divorce by strengthening marriages and those that strengthen postdivorce families.

"Our pre-nup guarantees that we'll be together forever.
In the event of divorce, you and I get custody of each other!"

Strategies to Strengthen Marriage and Prevent Divorce

A growing "marriage movement" involves efforts to strengthen marriage and prevent divorce through a number of strategies, including premarital and marriage education, covenant marriage and divorce law reform, and provision of workplace and economic supports. Amato et al. (2007) explained,

> Policies to strengthen marital quality and stability are based on consistent evidence that happy and stable marriages promote the health, psychological well-being, financial security of adults . . . as well as children. . . . Moreover, recent research suggests that a large proportion of marriages that end in divorce are not deeply troubled, and that many of these marriages might be salvaged if spouses sought assistance for relationship problems . . . and stayed the course through difficult times. (pp. 245–46)

Marriage Education. Marriage education, also known as family life education, includes various types of workshops and classes that (1) teach relationship skills, communication, and problem solving; (2) convey the idea that sustaining healthy marriages requires effort; and (3) convey the importance of having realistic expectations of marriage, commitment, and a willingness to make personal sacrifices (Hawkins et al. 2004). An alternative or supplement to face-to-face family life education is web-based education, such as the Forever Families website, a faith-based family education website (Steimle & Duncan 2004).

The Healthy Marriage Initiative provides federal funds to support research and programs to encourage healthy marriages and promote involved and responsible fatherhood. Funds may be used for a variety of activities, including marriage and premarital education, public advertising campaigns that promote healthy marriage, high school programs on the value of marriage, marriage mentoring programs, and parenting skills programs. Some states have passed or considered legislation that requires marriage education in high schools or that provides incentives (such as marriage license fee reductions) to couples who complete a marriage education program.

Covenant Marriage and Divorce Law Reform. With the passing of the 1996 Covenant Marriage Act, Louisiana became the first state to offer two types of marriage contracts: (1) the standard marriage contract that allows a no-fault divorce (after a six-month separation), or (2) a **covenant marriage**, which permits divorce only under condition of fault (e.g., abuse, adultery, or felony conviction) or after a two-year separation. Couples who choose a covenant marriage must also get premarital counseling. Variations of the covenant marriage have also been adopted in Arizona and Arkansas. Only 3 percent of couples in states with covenant marriage laws have chosen the covenant marriage option (Coontz 2005b).

The covenant marriage option was designed to strengthen marriages and decrease divorce. However, critics argue that covenant marriage may increase family problems by making it more emotionally and financially difficult to terminate a problematic marriage and by prolonging the exposure of children to parental conflict (Applewhite 2003).

What Do You Think? Several states have considered reforming divorce laws to make divorce harder to obtain by extending the waiting period required before a divorce is granted or requiring proof of fault (e.g., adultery or abuse). Do you think that making divorce more difficult to obtain would help keep families together? Or would it hurt families by increasing the conflict between divorcing spouses (which harms the children as well as the adults involved) and the legal costs of getting a divorce (which leaves less money to support any children)?

Workplace and Economic Supports. The most important pro-marriage and divorce-prevention measures may be those that maximize employment and earnings. Given that research finds a link between financial hardship and marital quality, policies to strengthen marriage should include a focus on the economic well-being of poor and near-poor couples and families (Amato et al. 2007). "Policy makers should recognize that any initiative that improves the financial security and well-being of married couples is a pro-marriage policy" (Amato et al. 2007, p. 256). Supports such as job training, employment assistance, flexible workplace policies that decrease work-family conflict, affordable child care, and economic support, such as the earned income tax credit are discussed in Chapters 6 and 7. In addition, policy makers should take a hard look at policies that penalize poor couples for marrying. Poor couples who marry are often penalized by losing Medicaid benefits, food stamps, and other forms of assistance.

> The most important pro-marriage and divorce-prevention measures may be those that maximize employment and earnings.

covenant marriage A type of marriage (offered in a few states) that requires premarital counseling and that permits divorce only under condition of fault or after a marital separation of more than two years.

Strategies to Strengthen Families During and After Divorce

When one or both marriage partners decide to divorce, what can the couple do to minimize the negative consequences of divorce for themselves and their children? According to Ahrons (2004), the post-divorce conflict between parents and not the divorce itself is most traumatic for children. A review of the literature on the effects of parental conflict on children suggests that children who are exposed to high levels of parental conflict are at risk for anxiety, depression, and disruptive behavior; they are more likely to be abusive toward romantic partners in adolescence and adulthood and are likely to have higher rates of divorce and maladjustment in adulthood (Grych 2005). Next, we discuss ways in which divorced (or divorcing) parents can minimize the conflict with their ex-spouse and develop a cooperative parenting relationship.

Forgiveness. Research suggests that divorced parents who forgive each other are more likely to have positive and cooperative co-parenting after divorce (Bonach 2009). Forgiveness does not mean accepting, condoning, or excusing the offender. Rather, forgiveness involves making a choice to think, feel, and behave less negatively toward someone who has hurt or offended you and to act with good will toward the offender (Fincham et al. 2006).

Divorce Mediation. In **divorce mediation**, divorcing couples meet with a neutral third party, a mediator, who helps them resolve issues of property division, child support, child custody, and spousal support (i.e., alimony) in a way that minimizes conflict and encourages cooperation. In a longitudinal study, researchers compared two groups of divorcing parents who were petitioning for a court custody hearing: parents who were randomly assigned to try mediation and those who were randomly assigned to continue the adversarial court process (Emery et al. 2005). If mediation did not work, the parents in the mediation group could still go to court to resolve their case. The parents who participated in mediation were much more likely to settle their custody dispute outside court than the parents who did not. The researchers found that mediation can not only speed settlement, save money on attorney and court fees, and increase compliance, it can also result in improved relationships between nonresidential parents and children as well as between divorced parents 12 years after the dispute settlement. An increasing number of jurisdictions and states have mandatory child custody mediation programs, whereby parents in a custody or visitation dispute must attempt to resolve their dispute through mediation before a court will hear the case.

Divorce Education Programs. Divorce education programs are designed to help parents who are divorced or planning to divorce reduce parental conflict and educate them about the factors that affect their children's adjustment. Parents are taught how to respond to their children's reactions to divorce, and how to cooperate in co-parenting. An experimental research study on the effectiveness of a court-ordered education program for divorcing parents found that participation in the program had a significant positive effect on co-parenting skills and parent relationship for both mothers and fathers (Whitehurst et al. 2008).

What Do You Think? Many counties and some states (e.g., Arizona and Hawaii) require divorcing spouses to attend a divorce education program, whereas it is optional in other jurisdictions. Do you think that parents of minor children should be required to complete a divorce education program before they can get a divorce? Why or why not?

divorce mediation A process in which divorcing couples meet with a neutral third party (mediator) who assists the individuals in resolving issues such as property division, child custody, child support, and spousal support in a way that minimizes conflict and encourages cooperation.

Teenage Childbearing

In the 1950s, when most teen mothers were married and were expected to be stay-at-home wives, teenage childbearing was not a public concern. Teenage births today are considered problematic because most teenage births occur outside wedlock and because early parenthood is associated with a higher risk for negative outcomes for teen parents and their children (Mauldon 2003).

The U.S. teenage birthrate (per 1,000 teens ages 15 to 19) peaked in 1991 at 61.8, and then dropped steadily until 2005, but then began to increase again. In 2007, the teen birthrate was 42.5—much lower than in 1991, but still a 5 percent increase from 2005, which leaves us to wonder if the upward trend will continue. Teenage birth rates are highest for Hispanic teens and lowest for Asian or Pacific Islander teens (see Figure 5.7).

After years of steadily decreasing, the U.S. teen birthrate has recently begun to increase.

Causes and Consequences of Teen Childbearing

Most teens do not want to get pregnant or to cause a pregnancy, and most teen pregnancies are unplanned. Although some factors, such as low educational attainment and minority status, are associated with a higher risk for teenage childbearing, teen pregnancy and childbearing happen to all segments of the population (see this chapter's Human Side feature).

Figure 5.7 Birth Rates (per 1,000) of U.S. Teenage Females, Ages 15 to 19, by Race and Hispanic Origin, 2007
Source: Hamilton et al. 2009.

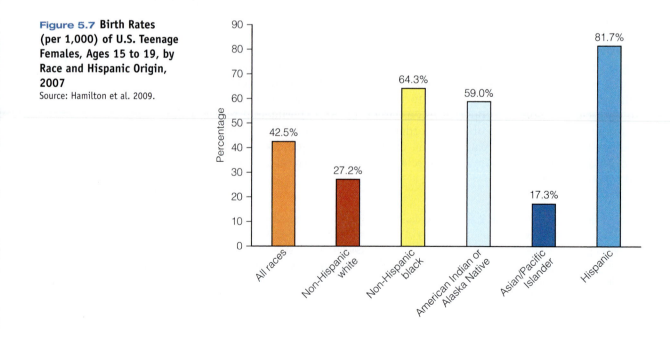

I had always been one of the "good" kids. One of the kids who wouldn't smoke, do drugs, drink alcohol, or have sex. At least that's what everyone, including me, thought. Then I fell in love with another "good" kid—the type every mother wishes her daughter would fall in love with.

We were going to graduate from high school, go to college, get married, and then have a child or two. Eventually our physical relationship began to reflect our determination to wed someday. When we began to have sex, I was 16 and he was 17. Three months after we began having sex, we decided we could not handle a sexual relationship. We felt guilty, because premarital sex is against everything we believed in. We agreed to stop having sex. Then four months later, my mother . . . realized I had not had a period lately. She asked if I could possibly be pregnant. I told her no. I couldn't be pregnant. Not me. Something like that wouldn't happen to me.

We bought a pregnancy test and left the store. We carefully followed the instructions. She told my father that afternoon. That night, my boyfriend told his parents. The next day, my parents, his parents, he, and I met and talked for a few hours. Our parents were hurt, shocked, disappointed. Three days later my mother, my boyfriend, his parents, and I went to my family doctor. I was indeed pregnant. The following day I went to a crisis pregnancy counselor and to my mother's obstetrician. I was approximately 21 weeks pregnant. The next day, my mother, boyfriend, and I went for the ultrasound and we were given a picture of the baby. . . .

The Decisions

On learning I was pregnant, there were several decisions to make. The first was what to do with the baby. Abortion was not an option. My first instinct was to give the baby up for adoption. My parents told me I could never live with that and they are right. So, we were keeping the baby.

The next decision was how to support the baby. We learned that in New Jersey, because I was under 19 and my father made too much money, to get health insurance for the baby I would have to move out and go on welfare. Moving out was out of the question, so my parents would have to become the legal guardians of the baby.

The next decision was what my boyfriend and I should do. . . . We decided the best decision was for us to continue our education. This way, we will be able to fully support ourselves and our baby in five or six years.

Then my boyfriend and I had to decide if we were going to get married now, wait until I graduated from high school, or wait for five or six years. We decided to get married now. We're going to live with my parents until we graduate from college and can support ourselves.

The Costs

Being pregnant and having a baby is expensive. A normal pregnancy costs $3,000 to $5,000. If there are any complications, the cost may rise to $8,000 or more. Then there's the maternity clothes, baby furniture, baby clothes, baby toys, baby bottles, formula, diapers, and a mountain of other costs. . . . Pregnancy is more than just financial problems. There's lost trust and respect between my parents and me, his parents and me, my parents and him, and with adults in our church, our friends, and our peers. Because of us, our close friends will be watched more carefully. . . .

Reflections

In some ways, I'm lucky my parents are here to help so much. They did all of the night feedings. However, in some ways, it reinforces my feelings of incompetence. When I can't take care of the baby's needs, they can. They know better than I do what he needs and wants. There are days I feel more like a sister than a mother. . . .

Teenagers my age think, "Oh, it's so neat that they have a kid," or "They're married. That's so cool." It's really not. They don't know the emotions involved or the problems.

People ask me if we used contraceptives. They seem relieved when I answer, "No." It's almost as if it's OK then. It will never happen to them. I can understand that. I felt that way once, too. It's easy for people to preach abstinence or "safe sex." Safe sex is a myth. I know a 7-month-old baby conceived while the mother was on the Pill and the father used a condom. Abstinence is an unrealistic ideal. I don't believe there is a cure for the epidemic of teenage pregnancy. Then again, I don't believe in a lot of things anymore. Losing my childhood too early has left me too bitter.

Source: Anonymous, 16, Contributor. 2007 (April 30). "Teen Mom Tells Her Story." Available at http://www.sexetc.org.

Early Sexual Activity. A national study of U.S. youth in grades 9 through 12 found that nearly half (48 percent) had had sexual intercourse, and more than a third (35 percent) were currently sexually active. Among sexually active youth, 39 percent reported that they or their partner had not used a condom during the last sexual intercourse (Centers for Disease Control and Prevention 2008).

Low Educational Achievement.
Low academic achievement is both a contributing factor and a potential outcome of teenage parenthood. Teenage females who do poorly in school may have little hope of success and achievement in pursuing educational and occupational goals, and they may think that their only remaining option for a meaningful role in life is to become a parent.

Becoming a teen parent also tends to curtail future academic achievement. In one study, eighth grade students who later became teenage parents had significantly lower test scores, were more likely to have had behavior problems in school, had lower educational aspirations for the future, and were more likely to have been held back at least one grade in school versus their peers who did not become teenage parents (Mollborn 2007). When these teens reached age 26, those who had become teen parents had, on average, a high school degree, whereas others had attained two years of postsecondary education.

The children of teen parents are also at risk for low academic achievement. One study found that children born to mothers ages 17 and younger began kindergarten with lower levels of school readiness—including lower math and reading scores, language and communication skills, social skills, and physical and emotional well-being—compared with children born to older mothers (Terry-Humen et al. 2005). When the researchers controlled for the mother's marital status and socioeconomic status, these effects were diminished but still important. However, other researchers who studied the effects of teen parenting on children's well-being found that "teen parenting had little or no effect on children's performance on standardized academic tests; correlations between early fertility and children's test scores reflect individual and family background factors rather than the causal influence of early childbearing itself" (Levine et al. 2007, p. 116).

> Teenage females who do poorly in school may have little hope of success and achievement in pursuing educational and occupational goals, and they may think that their only remaining option for a meaningful role in life is to become a parent.

Poverty.
Teens from low socioeconomic backgrounds are more likely than those from higher socioeconomic backgrounds to become teenage parents. Although many teen mothers experience poverty before becoming pregnant, teenage childbearing tends to exacerbate the problems that these disadvantaged young women already face. Most teenagers who become parents are not married and have no means of economic support or have limited earning capacity. Teen mothers often rely on the support of their own parents or rely on public assistance. Having a child at a young age makes it difficult for the teen parent to attain academic credentials, compete in the job market, and break out of what for many is a cycle of poverty that includes their children.

Poor Health Outcomes.
Compared with older pregnant women, pregnant teenagers are less likely to receive timely prenatal care and to gain adequate weight. Teens are also more likely to smoke and use alcohol and drugs during pregnancy (Ventura et al. 2000). As a consequence, infants born to teenagers are at higher risk of low birth weight, of premature birth, and of dying in the first year of life.

Strategies for Action: Interventions in Teenage Childbearing

Although some teen pregnancies end in abortion or miscarriage, more than half of pregnant teens give birth (Guttmacher Institute 2006). Interventions in teenage childbearing include efforts to prevent teenage pregnancies through sex education and access to contraceptive services and to provide various types of support to teenage parents and their children.

Sexuality Education and Access to Contraceptive Services

The United States has one of the highest teen pregnancy rates in the developed world—almost twice as high as those of England, Wales, and Canada, and eight times as high as those of the Netherlands and Japan (Guttmacher Institute 2006). In northern and western Europe, low teen pregnancy and birth rates are attributed to the widespread availability and use of effective contraception among sexually active teens (Singh & Darroch 2000).

The G. W. Bush administration emphasized abstinence-only sex education. To receive federal funding for abstinence sex education, states could not use the funds to promote condom or contraceptive use, and teachers were required to teach ideas such as bearing children outside of wedlock is harmful to society and "likely to have harmful psychological and physical effects" (Huffstutter 2007).

Supporters of abstinence-only programs believe that promoting condoms sends the "wrong message" that sex outside of marriage is OK. Critics of abstinence-only sex education argue that such programs do not protect youth against pregnancy or sexually transmissible disease. Critics of abstinence-only programs often cite a report that found that more than 80 percent of the abstinence-only curricula contain false, misleading, or distorted information about reproductive health (Waxman 2004). Critics also point to a scientific evaluation of abstinence education programs commissioned by the Congress that found that youth who received abstinence education were no more likely than control group youth to have abstained from sex and, among those who reported having had sex, they had similar numbers of sexual partners and had initiated sex at the same mean age (Trenholm et al. 2007).

Most U.S. adults want schools to provide **comprehensive sexuality education** that includes topics such as contraception, sexually transmitted diseases, HIV/AIDS, and disease-prevention methods, as well as the benefits of abstinence (Boonstra 2009). Yet comprehensive sex education is not universally taught. As of Fall 2009, only 15 states and the District of Columbia require that sex education programs cover contraception (Guttmacher Institute 2009a).

Pregnancy: a lifetime commitment Are you ready?

El Embarzo: un compromiso para toda la vida ¿Ya estás listo tú?

LATINO MEN'S HEALTH SERIES
SERIE SOBRE LA SALUD DEL HOMBRE LATINO

The East Los Angeles Men's Health Center, a project of Bienvenidos Children's Center, Inc. http://www.bienvenidos.org

This brochure, distributed by the National Latino Fatherhood and Family Initiative, is a teen pregnancy prevention effort targeting young Latino men.

comprehensive sexuality education Sex education that includes topics such as contraception, sexually transmitted diseases, HIV/AIDS, and disease-prevention methods as well as the benefits of abstinence.

The Responsible Education About Life (REAL) Act proposes providing funds for states that provide a comprehensive approach to sex education.

Most sexually active teens in the United States cannot obtain contraceptive services in their schools and find significant barriers to obtaining contraception elsewhere. In 2009, 21 states and the District of Columbia allowed all minors to consent to contraceptive services; 25 states permitted minors to consent to contraceptive services only under certain conditions (e.g., if the minor is married, has had a previous pregnancy, or is referred by a physician or clergy), and four states have no explicit policy on minors' authority to consent (Guttmacher Institute 2009b).

Teens younger than age 17 are also restricted from over-the-counter access to Plan B (also called the "morning-after pill"), the only Food and Drug Administration (FDA)–approved product for emergency contraception. This type of contraception is used to prevent pregnancy after unprotected intercourse occurs and must be taken within 72 hours of unprotected intercourse. In 2006, the FDA approved Plan B as an over-the-counter medication for those 18 and older, and in 2009, the FDA ruled that 17-year-olds may purchase Plan B without a prescription. Teens younger than age 17 must have a prescription to purchase Plan B.

What Do You Think? Another barrier to access to contraception for all U.S. women is the refusal of some pharmacists to fill prescriptions for birth control. Four states—Arkansas, Georgia, Mississippi, and South Dakota—permit pharmacists to refuse to dispense contraceptives (Guttmacher Institute 2009c). Do you think pharmacists should have the legal right to refuse to fill a prescription for products such as birth control pills or Plan B emergency contraception?

Computerized Infant Simulators

Some teen pregnancy prevention programs use computerized infant simulators to give adolescents a realistic view of parenting. Computerized infant simulators are realistic, life-sized computerized "dolls" that are programmed to cry at random

Computerized infant simulators, such as the one pictured here, are used in parenting education as well as teenage pregnancy prevention programs.

Mark Goldstein/ Independent Record/ AP Photo

intervals (typically between 8 and 12 times in 24 hours), with crying periods lasting typically between 10 and 15 minutes. The "baby" stops crying only when the caregiver "attends" to the doll by inserting a key into a slot in the infant simulator's back until it stops crying. The infant simulator records data, including the amount of time the caregiver took to attend to the infant (insert the key) and any instances of "rough handling," such as dropping, hitting, or shaking the doll. Participants who are found to neglect or handle the doll roughly may receive a private counseling session and may be required to take a parenting class.

An evaluation of a computerized infant simulator program with adolescents found that the program was effective in changing perceptions of the time and effort involved in caring for an infant and in recognizing the significant effect having a baby has on all aspects of one's life (de Anda 2006). Nearly two-thirds of the adolescent participants reported that the program helped change their minds about using birth control.

Resources and Assistance to Teenage Parents

Teenage parents can benefit from assistance with health care, housing, child care, and financial support. Although teenage parents tend to fall behind in educational attainment, "If they are provided with enough material resources, contemporary teenage parents may be able to go quite far in school, despite their initial socioeconomic and educational disadvantage" (Mollborn 2007, p. 102).

Increase Men's Involvement with Children

Strategies to increase and support fathers' involvement with their children are relevant to both children of teen mothers and children of divorce. At the federal and state levels, fatherhood initiative programs encourage fathers' involvement with children through a variety of means (U.S. Department of Health and Human Services 2000). These include promoting responsible fatherhood by improving work opportunities for low-income fathers, increasing child support collections, providing parent education training for men, supporting access and visitation by noncustodial parents, and involving boys and young men in teenage pregnancy prevention and early parenting programs. Because teenage parents are less likely than older parents to use positive and effective child-rearing techniques, parent education programs for teen mothers and fathers are an important component of improving the lives of young parents and their children.

Understanding Family Problems

Family problems can best be understood within the context of the society and culture in which they occur. Although domestic violence, divorce, and teenage pregnancy and parenthood may appear to result from individual decisions, myriad social and cultural forces influence these decisions.

The impact of family problems, including divorce, abuse, and teenage childbearing, is felt not only by family members but also by society at large. Family members experience life difficulties such as poverty, school failure, low self-esteem, and mental and physical health problems. Each of these difficulties contributes to a cycle of family problems in the next generation. The impact on society includes public expenditures to assist single-parent families and victims of domestic violence and neglect, increased rates of juvenile delinquency, and lower worker productivity.

For some, the solution to family problems implies encouraging marriage and discouraging other family forms, such as single parenting, cohabitation, and same-sex unions. However, many family scholars argue that the fundamental issue is making sure that children are well cared for, regardless of their parents' marital status or sexual orientation. Some even suggest that marriage is part of the problem, not part of the solution. Martha Fineman of Cornell Law School said, "This obsession with marriage prevents us from looking at our social problems and addressing them. . . . Marriage is nothing more than a piece of paper, and yet we rely on marriage to do a lot of work in this society: It becomes our family policy, our police in regard to welfare and children, the cure for poverty" (quoted by Lewin 2000, p. 2).

Strengthening marriage is a worthy goal because strong marriages offer many benefits to individuals and their children. However, "strengthening marriage does not have to mean a return to the patriarchal family of an earlier era. . . . Indeed, greater marital stability will only come about when men are willing to share power, as well as housework and child care, equally with women" (Amato 1999, p. 184). Strengthening marriage does not mean that other family forms should not also be supported. In their book *Joined at the Heart,* Al and Tipper Gore (2002) suggested that the first and most important step to helping families is to change our way of thinking about families so that our view of family encompasses those who are connected emotionally and committed to one another as family—those who are "joined at the heart" (p. 327). The reality is that the post-modern family comes in many forms, each with its strengths, needs, and challenges. Given the diversity of families today, social historian Stephanie Coontz (2004) suggested that "the appropriate question . . . is not what single family form or marriage arrangement we would prefer in the abstract, but how we can help people in a wide array of different committed relationships minimize their shortcomings and maximize their solidarities" (p. 979). She further argued that

> If we withdrew our social acceptance of alternatives to marriage, marriage itself might suffer. . . . The same personal freedoms that allow people to expect more from their married lives also allow them to get more out of staying single and give them more choice than ever before in history about whether or not to remain together. (Coontz 2005a, p. 310)

Efforts to prevent teenage childbearing are aimed to protect both teen parents and their children from negative outcomes such as poverty and poor health. However, it is important to keep in mind that negative outcomes of teen childbearing are largely the result of preexisting social disadvantages of the teens who become parents. If we mistakenly assume that negative outcomes are caused by teen parenting by itself, we may neglect other teens with disadvantaged backgrounds who do not become teen parents, but who nevertheless suffer and whose children later suffer similar negative outcomes (Levine et al. 2007). The underlying causes of teen childbearing—poverty, economic inequality, and problems in education—need to be addressed for the well-being of *all* youth, not just teen parents. Finally, an important aspect of preventing teen childbearing is to prevent teen pregnancy. Research, as well as the experience of many European countries, suggests that providing teens with contraceptives is effective in reducing teen pregnancy.

The three family problems emphasized in this chapter—domestic violence and abuse, problems of divorce, and teenage parenthood—have something in common: Economic hardship and poverty can be a contributing factor and a consequence of each of these problems. In the next chapter, we turn our attention to poverty and economic inequality—problems that are at the heart of many other social ills.

CHAPTER REVIEW

■ **What are some examples of diversity in families around the world?**
Some societies recognize monogamy as the only legal form of marriage, whereas other societies permit polygamy. Societies also vary in their policies regarding same-sex couples and their norms regarding childbearing and the roles of women and men in the family.

■ **What are some of the major changes in U.S. families that have occurred in the past several decades?**
Some of the major changes in U.S. families that have occurred in recent decades include increased singlehood and older age at first marriage, increased heterosexual and same-sex cohabitation, the emergence of living apart together (LAT) relationships, increased births to unmarried women, increased divorce and blended families, and increased employment of married mothers. According to the marital decline perspective, the recent transformations in American families signify a collapse of marriage and family in the United States. According to the marital resiliency perspective, poverty, unemployment, poorly funded schools, discrimination, and the lack of basic services (such as health insurance and child care) are more harmful to the well-being of children and adults than is the decline in married two-parent families.

■ **Feminist theories of family are most similar to which of the three main sociological theories: structural functionalism, conflict theory, or symbolic interactionism?**
Feminist theories of family are most aligned with conflict theory. Both feminist and conflict theories are concerned with how gender inequality influences and results from family patterns.

■ **What are the four patterns of partner violence that Johnson and Ferraro (2003) identified?**
The four patterns of partner violence are (1) common couple violence (occasional acts of violence arising from arguments that get "out of hand"); (2) intimate terrorism (violence that is motivated by a wish to control one's partner); (3) violent resistance (acts of violence that are committed in self-defense); and (4) mutual violent control (both partners battling for control).

■ **Why do many abused adults stay in abusive relationships?**
Adult victims of abuse are commonly blamed for choosing to stay in their abusive relationships. From the point of view of the victim, reasons to stay in the relationship include love, emotional dependency, commitment to the relationship, hope that things will get better, the view that violence is legitimate because they "deserve" it, guilt, fear, economic dependency, feeling stuck, and fear of loneliness. Some victims stay because they fear the abuser will abuse or neglect a pet.

■ **What are some of the effects of divorce on children?**
Reviews of recent research on the consequences of divorce for children find that children with divorced parents score lower on measures of academic success, psychological adjustment, self-concept, social competence, and long-term health and that they have higher levels of aggressive behavior and depression. Such effects are related to the economic hardship associated with divorce, the reduced parental supervision resulting from divorce, and parental conflict during and after divorce. In highly conflictual marriages, divorce may actually improve the emotional well-being of children relative to staying in a conflicted home environment.

■ **What is divorce mediation?**
In divorce mediation, divorcing couples meet with a neutral third party, a mediator, who helps them resolve issues of property division, child custody, child support, and spousal support in a way that minimizes conflict and encourages cooperation. In some states, counties, and jurisdictions, divorcing couples who are disputing child custody issues are required to participate in divorce mediation before their case can be heard in court.

■ **Why is teenage childbearing considered a social problem?**
Teenage childbearing is considered a social problem because of the adverse consequences for teenage mothers and their children, including (1) increased risk of poverty for single mothers and their children, (2) risk of poor health outcomes for babies born to teenage

women, and (3) risk of dropping out of school for teenage mothers and for low academic achievement of their children. Poverty and low educational achievement are also factors that contribute to teenage childbearing.

■ **How does the European approach to teenage sexuality compare with the U.S. approach?**
The European approach to teenage sexual activity involves providing widespread confidential and accessible contraceptive services to adolescents. Although sex education is provided in schools throughout the United States, most programs emphasize abstinence and do not provide students with access to contraception. Research suggests that comprehensive sexuality education that includes topics such as abstinence, sexually transmitted diseases, HIV/AIDS, contraception, and disease-prevention methods is more effective for preventing pregnancy, as well as disease.

TEST YOURSELF

1. The United States has the highest nonmarital birthrate of any country in the world.
   a. True
   b. False
2. Two perspectives on the state of marriage in the United States are the marital decline perspective and the marital _____ perspective.
   a. health
   b. resiliency
   c. incline
   d. stability
3. A study on how marriage has changed found that, between 1980 to 2000, marriages
   a. have become more prone to divorce
   b. have become more conflictual
   c. involve less interaction between husband and wife
   d. all of the above
4. In the United States, people are more likely to be physically assaulted, sexually assaulted and molested, or killed by _____ than by anyone else.
   a. A family member
   b. An employee
   c. A stranger
   d. A friend
5. A survey of more than 20,000 college students at 39 colleges and universities found that, in the past 12 months, students were much more likely to have experienced an emotionally abusive relationship than a physically or sexually abusive relationship.
   a. True
   b. False
6. Which of the following is the most prevalent form of abuse in families?
   a. Sexual abuse by a father
   b. Sexual abuse by an uncle or cousin
   c. Verbal abuse by a mother
   d. Sibling abuse
7. How many states recognize no-fault divorce?
   a. None
   b. 5
   c. 25 and the District of Columbia
   d. All 50
8. In states with covenant marriage laws, most couples getting married choose the covenant marriage option.
   a. True
   b. False
9. A national study of U.S. youth in grades 9 through 12 found that nearly half had had sexual intercourse.
   a. True
   b. False
10. Research suggests that adolescents who are exposed to abstinence-only sex education are less likely to become sexually active in their teens than are adolescents who do not receive abstinence-only sex education.
    a. True
    b. False

Answers: 1: b; 2: b; 3: c; 4: a; 5: a; 6: d; 7: d; 8: b; 9: a; 10: b.

KEY TERMS

MEDIA RESOURCES

Understanding Social Problems,
Seventh Edition Companion Website
www.cengage.com/sociology/mooney
Visit your book companion website, where you will find
flash cards, practice quizzes, Internet links, and more to
help you study.

 Just what you need to know NOW!
Spend time on what you need to
master rather than on information
you already have learned. Take a pretest for this chapter,
and CengageNOW will generate a personalized study
plan based on your results. The study plan will identify
the topics you need to review and direct you to online
resources to help you master those topics. You can then
take a posttest to help you determine the concepts you
have mastered and what you will need to work on. Try
it out! Go to www.cengage.com/login to sign in with an
access code or to purchase access to this product.

© Irving Olson, 2007

6

Poverty and Economic Inequality

"We are the first generation that can look extreme poverty in the eye, and say this and mean it—we have the cash, we have the drugs, we have the science. Do we have the will to make poverty history?"

Bono, U2 (rock music group)

The Global Context: Poverty and Economic Inequality Around the World | The Human Side: Poverty Never Takes a Holiday | Sociological Theories of Poverty and Economic Inequality | Patterns of Poverty in the United States | Consequences of Poverty and Economic Inequality | Photo Essay: Lack of Clean Water and Sanitation among the Poor | Self and Society: Food Security Scale | Strategies for Action: Alleviating Poverty | Social Problems Research Up Close: Making Ends Meet: Survival Strategies among Low-Income and Welfare Single Mothers | Understanding Poverty and Economic Inequality | Chapter Review

Not far from the Capitol Building, 60-year-old John Treece pondered his life in deep poverty as he left a local food pantry with two bags of free groceries. Plagued by arthritis and back problems from years of manual labor, Treece has been unable to find a full-time job for 15 years. He's tried to get Social Security disability benefits, but the Social Security Administration disputes his injuries and work history. Treece earns a little more than $5,000 a year doing odd jobs. His clothes are tattered and he lives in a $450-a-month room in a boarding house in a high-crime neighborhood. Treece does not go hungry, thanks to food stamps, the food pantry, and help from relatives. But items that require cash, such as toothpaste, soap, and toilet paper, are harder to come by. "Sometimes it makes you want to do the wrong thing, you know," said Treece, referring to crime. "But I ain't a kid no more. I can't do no time. At this point, I ain't got a lotta years left." Despite his poor circumstances, Treece is positive and grateful for what he has. "I don't ask for nothing. . . . I just thank the Lord for this day and ask that tomorrow be just as blessed" (quoted in Pugh 2007).

Bettmann/CORBIS

Washington, DC, the capital of one of the wealthiest nations in the world, has one of the highest rates of poverty in the United States.

In this chapter, we examine the extent of poverty globally and in the United States, focusing on the consequences of poverty for individuals, families, and societies. We present theories of poverty and economic inequality and consider strategies for rectifying economic inequality and poverty.

The Global Context: Poverty and Economic Inequality Around the World

Who are the poor? The answer depends on how we define and measure poverty.

Defining and Measuring Poverty

Absolute poverty refers to the lack of resources necessary for well-being—most importantly food and water, but also housing, sanitation, education, and health care. In contrast, **relative poverty** refers to the lack of material and economic resources compared with some other population. If you are a struggling college student living on a limited budget, you may feel as though you are "poor" compared with the middle- or upper-middle-class lifestyle to which you may aspire.

absolute poverty The lack of resources necessary for material well-being—most importantly food and water, but also housing, sanitation, education, and health care.

relative poverty The lack of material and economic resources compared with some other population.

However, if you have a roof over your head; access to clean water, toilets, and medical care; and enough to eat, you are not absolutely poor; indeed, you have a level of well-being that millions of people living in absolute poverty may never achieve.

Members of this African village live in extreme poverty.

Measures of Poverty. The World Bank sets a "poverty threshold" of $1.25 per day for the least developed countries, and $2 per day for middle-income countries and regions, including Latin America and eastern Europe. Those living on less than $1.25 a day live in **extreme poverty**.

Another measure of poverty is the proportion of people who consume less than a minimum amount of calories to maintain health—who are chronically hungry or undernourished. Calorie requirements differ by gender and age, and for different levels of physical activity, so minimum calorie requirements vary by country, and from year to year, depending on the age and gender structure of the population.

In industrial countries, national poverty lines are sometimes based on the median household income of a country's population. According to this relative poverty measure, members of a household are considered poor if their household income is less than 50 percent of the median household income in that country.

The concept of poverty is multidimensional, and includes food insecurity, poor housing, unemployment, psychological distress, powerlessness, hopelessness, vulnerability, and lack of access to health care, education, and transportation (Narayan 2000). To capture the multidimensional nature of poverty, the United Nations Development Programme (UNDP) developed a composite measure of poverty: the **human poverty index (HPI)** (UNDP 1997). Rather than measure poverty by income, three measures of deprivation are combined to yield the HPI: (1) deprivation of a long, healthy life; (2) deprivation of knowledge; and (3) deprivation in decent living standards. As shown in Table 6.1, the HPI for developing countries (HPI-1) is measured differently from the HPI for industrialized countries (HPI-2). Among the 19 industrialized countries for which the HPI-2 was calculated, Sweden has the lowest level of human poverty (6.0 percent), followed by Norway, the Netherlands, Finland, and Denmark (UNDP 2009). The industrialized countries with the highest rates of human poverty are Italy (29.8 percent), Ireland (15.9 percent), and the United States (15.2 percent).

Dictionary definitions and objective measures of poverty are useful, but a true understanding of poverty requires seeing poverty through the eyes of someone who has lived in poverty. In this chapter's Human side feature, a poor African woman describes what poverty means to her.

extreme poverty Living on less than $1.25 a day.

human poverty index (HPI) A measure of poverty based on measures of deprivation of a long, healthy life; deprivation of knowledge; and deprivation in decent living standards.

TABLE 6.1 Measures of Human Poverty in Developing and Industrialized Countries

	LONGEVITY	KNOWLEDGE	DECENT STANDARD OF LIVING
For developing countries	Probability at birth of not surviving to age 40	Adult illiteracy	A composite measure based on 1. Percentage of people without access to safe water 2. Percentage of people without access to health services 3. Percentage of children younger than 5 who are underweight
For industrialized countries	Probability at birth of not surviving to age 60	Adult functional illiteracy rate	Percentage of people living below the income poverty line, which is set at 50 percent of median disposable income

Source: Adapted from UNDP 2000.

U.S. Measures of Poverty. In 1964, the Social Security Administration devised a poverty index based on data that indicated that families spent about one-third of their income on food. The official poverty level was set by multiplying food costs by three. Since then, the poverty level has been updated annually for inflation but has otherwise remained unchanged. Poverty thresholds differ by the number of adults and children in a family and by the age of the family head of household, but is the same across the continental United States (see Table 6.2). Anyone living in a household with pretax income below the official poverty line is considered "poor." Individuals living in households with incomes that are above the poverty line, but not very much above it, are

The Human Side | Poverty Never Takes a Holiday

I know poverty because poverty was there before I was born and it has become part of life like the blood through my veins. . . . Poverty is going empty with no hope for the future.

Poverty is getting nobody to feel your pain and poverty is when your dreams go in vain because nobody is there to help you. Poverty is watching your mothers, fathers, brothers, and sisters die in pain and in sorrow just because they couldn't get something to eat. Poverty is hearing your grandmothers and grandfathers cry out to death to come take them because they are tired of this world. Poverty is watching your own children and grandchildren die in your arms but there is nothing you can do. Poverty is watching your children and grandchildren share tears in their deepest sleep. Poverty is suffering from HIV/AIDS and dying a shameful death but nobody seems to care. Poverty is when you hide your face and wish nobody could see you just because you feel less than a human being. Poverty is when you dream of bread and fish you never see in the day light. . . . Poverty is when the hopes of your fathers and grandfathers just vanish within a blink of an eye. I know poverty and I know poverty just like I know my father's name. Poverty never sleeps. Poverty works all day and night. Poverty never takes a holiday.

Source: "Poverty Never Takes a Holiday." n.d. Available at http://cozay.com

TABLE 6.2 Poverty Thresholds: 2008 (Householder Younger Than 65 Years Old)	
HOUSEHOLD MAKEUP	POVERTY THRESHOLD
One adult	$11,201
Two adults	$14,417
One adult, one child	$14,840
Two adults, one child	$17,330
Two adults, two children	$21,834

Source: U.S. Census Bureau 2009a.

classified as "near-poor," and those living in households with income below 50 percent of the poverty line live in "deep poverty," also referred to as "severe poverty." A common working definition of "low-income" households is households with incomes that are between 100 percent and 200 percent of the federal poverty line or up to twice the poverty level.

The U.S. poverty line has been criticized on several grounds. First, the official poverty line is based on pretax income so tax burdens that affect the amount of disposable income available to meet basic needs are disregarded. On the other hand, it underestimates income for some families because it does not count the federal earned income tax credit (EITC) many families receive. Family assets, such as savings and property, are also excluded in official poverty calculations, and noncash government benefits that assist low-income families—food stamps, Medicaid, and housing and child care assistance—are not taken into account.

Overall, the poverty line underestimates the extent of material hardship in the United States because it is based on the assumption that low-income families spend one-third of their household income on food. That was true in the 1950s, but because housing, medical care, child care, and transportation costs have risen more rapidly than food costs, low-income families today spend far less than one-third of their income on food. In addition, the current poverty measure is a national standard that does not reflect the significant variation in the cost of living from state to state and between urban and rural areas.

Researchers have determined that, across the country, families on average need a minimum income that is about twice the poverty line (roughly $40,000 for a family of four) (Cauthen & Fass 2007). In cities, the figure is higher ($50,000); in rural areas, the figure is lower ($30,000). When a 2007 Gallup Poll asked the American public to estimate the minimum amount of yearly income a family of four would need "to get along in your local community," the average answer was $52,000 (rounded to the nearest thousand) (Jones 2007). This figure varied by region: $61,000 in the East, $46,000 in the Midwest, and $49,000 in the South. There was also variation between urban areas ($48,000), suburban areas ($58,000), and rural areas ($42,000).

TABLE 6.3 Percent of Population Living in Poverty, by Region, 2005		
	<$1.25	<$2.00
East Asia and Pacific	17.9	39.7
China	15.9	36.3
Eastern Europe and Central Asia	5.0	10.6
Latin America and Caribbean	8.2	17.9
Middle East and North Africa	4.6	19.0
South Asia	40.4	74.0
India	41.6	75.6
Sub-Saharan Africa	50.4	72.2
TOTAL:	25.7	47.6

Source: Chen and Ravallion 2008.

The Extent of Global Poverty and Economic Inequality

Nearly half of the population in the developing world—2.6 billion people—live in poverty on less than $2.00 a day, and about a quarter—1.4 billion people—live in extreme poverty on less than $1.25 a day (see Table 6.3) (Chen & Ravallion 2008). The highest rate of poverty is in sub-Saharan Africa, where half of the population lives on less than $1.25 a day. Due to lags in data availability, these data do not reflect the recent global financial crisis, which means that the rate of poverty is most likely higher than the data reported here.

"Money can't buy happiness, but if I had a big house, fancy car and a giant plasma TV, I wouldn't mind being unhappy!"

With so much wealth in the world, it is hard to fathom that so many people live in poverty. Economic inequality is a well-established feature of our social world. In 2000, the average income of the richest 20 countries was 37 times that of the poorest 20 countries—a gap that doubled since 1960 (World Bank 2001). From 1960 to 2002, income per person in the world's poorest countries rose only slightly, from $212 to $267, whereas income in the richest 20 nations tripled from $11,417 to $32,339 (Schifferes 2004). Although inequality between nations accounts for most of the inequality in global distribution of income, differences are growing for incomes within nations (Goesling 2001).

Although the United States is a wealthy nation, second only to Norway in per capita income, it has the greatest degree of income inequality and the highest rate of poverty of any industrialized nation. In the United States, workers in the top 10 percent earn nearly five times more than those at the bottom 10 percent—a higher ratio than in any other industrialized country (Mishel et al. 2009). In 2006, average after-tax income of the top 1 percent of U.S. households was 23 times higher than that of the middle fifth—nearly triple the income gap since 1970. The income gap between the top 1 percent and the poorest fifth of Americans has widened even more dramatically, from 22.6 times higher in 1979, to 72.7 times higher in 2006. As shown in Figure 6.1, between 1979 and 2006, average after-tax income of the top 1 percent more than tripled, from $337,000 to more than $1.2 million, whereas income of the middle fifth rose only 21 percent, from $42,900 to $52,100 (Sherman 2009).

Another example of economic inequality in the United States is the gap between the compensation

> Although the United States is a wealthy nation, second only to Norway in per capita income, it has the greatest degree of income inequality and the highest rate of poverty of any industrialized nation.

Figure 6.1 Change in Average Real After-Tax Income (1979–2006)

Source: Sherman 2009.

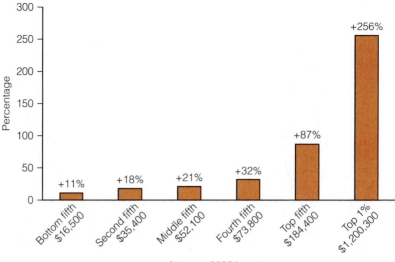

Change in Average Real After-Tax Income (1979–2006)

TABLE 6.4	Ratio of CEO Pay to Average Worker Pay, U.S., 1965–2007
1965	24:1
1978	35:1
1989	71:1
2000	298:1
2002	143:1
2007	275:1

Source: Mishel et al. 2009.

(salaries, bonuses, stock options, and so on) of chief executive officers (CEOs) and the average employee. In 2007, CEOs at 350 large U.S. corporations received an average of $12.3 million, 275 times what the typical worker earned, and more than twice what CEOs in other advanced countries earn (Mishel et al. 2009). That means that a CEO earned more in one workday (there are 260 in a year) than the typical worker earned all year. Although average worker pay increased 10 percent from 1989 to 2007, CEO pay rose 167 percent. Table 6.4 shows the dramatic increase in the ratio of CEO pay to average worker pay since 1965.

What Do You Think? Researchers asked citizens of various nations about their perceptions of income inequality in their country, and then compared those responses to the actual measurements of inequality (Forster & d'Ercole 2005). Among 17 advanced countries, U.S. citizens had the largest gap—by a wide margin—between their perception of inequality and its reality. Why do you think Americans perceive the level of inequality in the United States to be much less than what it actually is?

Global wealth is also distributed unequally. **Wealth** refers to the total assets of an individual or household minus liabilities (mortgages, loans, and debts). Wealth includes the value of a home, investment real estate, cars, unincorporated business, life insurance (cash value), stocks, bonds, mutual funds, trusts, checking and savings accounts, individual retirement accounts (IRAs), and valuable collectibles.

wealth The total assets of an individual or household minus liabilities.

In the United States, the wealthiest 1 percent of U.S. households owns a larger share of national wealth than the entire bottom 90 percent (Mishel et al.

2009). A comprehensive study on the world distribution of household wealth revealed that:

- The richest 1 percent of adults in the world own 40 percent of global household wealth; the richest 2 percent of adults own more than half of global wealth; and the richest 10 percent of adults own 85 percent of total global wealth.
- The poorest half of the world adult population owns barely 1 percent of global wealth.
- Households with per adult assets of $2,200 are in the top half of the world wealth distribution; assets of $61,000 per adult places a household in the top 10 percent, and assets of more than $500,000 per adult places a household in the richest 1 percent worldwide.
- Although North America has only 6 percent of the world adult population, it accounts for one-third (34 percent) of all household wealth worldwide. More than one-third (37 percent) of the richest 1 percent of individuals in the world reside in the United States (Davies et al. 2006).

Sociological Theories of Poverty and Economic Inequality

The three main theoretical perspectives in sociology—structural functionalism, conflict theory, and symbolic interactionism—offer insights into the nature, causes, and consequences of poverty and economic inequality.

Structural-Functionalist Perspective

According to the structural-functionalist perspective, poverty results from institutional breakdown: economic institutions that fail to provide sufficient jobs and pay, educational institutions that fail to provide adequate education in low-income school districts, family institutions that do not provide two parents, and government institutions that do not provide sufficient public support. These institutional breakdowns create a "culture of poverty," whereby people living in persistent poverty, known as the **underclass**, develop norms, values, beliefs, and self-concepts that contribute to their own plight. The **culture of poverty** that develops among the underclass is characterized by single-parent families, an emphasis on gratification in the present rather than in the future, and a relative lack of participation in society's major institutions (Lewis 1966). Those who grow up in a culture of poverty learn to view early sexual activity, unmarried parenthood, joblessness, reliance on public assistance, illegitimate income-producing activities (e.g., selling drugs), and substance abuse as common and "normal." Caught in this culture of poverty, members of the underclass develop feelings of marginality, helplessness, dependency, powerlessness, inferiority, and personal unworthiness (Lewis 1998). The culture of poverty is transmitted from one generation to the next, perpetuating the cycle of poverty. The structural-functionalist perspective reminds us that the culture of poverty persists because of the constraints and blocked opportunities that have resulted largely from the failures of the educational institution and of the economic institution to provide employment (Wilson 1996; Jargowsky 1997; Van Kempen 1997). Sociologist William Julius Wilson explains:

underclass A persistently poor and socially disadvantaged group.

culture of poverty The set of norms, values, beliefs, and self-concepts that contribute to the persistence of poverty among the underclass.

Where jobs are scarce . . . and where there is a disruptive or degraded school life purporting to prepare youngsters for eventual participation in the workforce, many people eventually lose their feeling of connectedness to work in the formal economy; they no longer expect work to be a regular, and regulating, force in their lives. . . . These circumstances also increase the likelihood that the residents will rely on illegitimate sources of income, thereby further weakening their attachment to the legitimate labor market. (Wilson 1996, pp. 52–53)

From a structural-functionalist perspective, economic inequality within a society can be beneficial for society, as a system of unequal pay motivates people to achieve higher levels of training and education and to take on jobs that are more important and difficult by offering higher rewards for higher achievements (Davis & Moore 1945). However, this argument is criticized on the grounds that many important occupational roles, such as child care workers and nurse assistants, have low salaries, whereas many individuals in nonessential roles (e.g., professional sports stars and entertainers) earn outrageous sums of money. The structural-functionalist argument that CEO pay is high because of the risks and responsibilities of the job falls apart when one considers that the average CEO pay is 56 times the pay of a U.S. Army general with 20 years of experience (Anderson et al. 2004). If pay is based on risk and responsibility, does it make sense that the annual pay of the first 919 U.S. soldiers who were killed in Iraq was about equal to the combined pay of just five average U.S. CEOs? The argument that CEO pay is high to motivate and reward high performance is shattered by the

News that AIG was paying top executives a total of $165 million in bonuses, after receiving $182.5 billion in federal aid, sparked protests around the country.

fact that CEOs are paid huge salaries and bonuses even when they contribute to the economic failure of their corporation. The public outrage over executive bonuses at companies receiving bailout money from the Troubled Asset Relief Program (TARP) led President Obama to impose a $500,000 limit on the CEO pay at companies receiving TARP money.

Conflict Perspective

Karl Marx (1818–1883) proposed that economic inequality results from the domination of the *bourgeoisie* (owners of the means of production) over the *proletariat* (workers). The bourgeoisie accumulate wealth as they profit from the labor of the proletariat, who earn wages far below the earnings of the bourgeoisie. Modern conflict theorists recognize that the power to influence economic outcomes comes not only from ownership of the means of production but also from management position, interlocking board memberships, control of media, and financial contributions to politicians.

For example, wealthy corporations use financial political contributions to influence politicians to enact policies that benefit the wealthy. Laws and policies that favor the rich, such as tax breaks that benefit the wealthy, are sometimes referred to as **wealthfare**. For example, tax and accounting loopholes save CEOs a total of $20 billion per year (Anderson et al. 2008).

Laws and policies that benefit corporations, such as low-interest government loans to failing businesses and special subsidies and tax breaks to corporations, are known as **corporate welfare.** Tax loopholes enable corporations to avoid paying their full tax rate. Between 2001 and 2003, 252 of America's largest and most profitable corporations avoided paying state income taxes on nearly two-thirds of their U.S. profits—costing state governments $42 billion (McIntyre 2005a). In 2002, 275 large U.S. corporations paid, on average, only 17.3 percent of their U.S. profits in federal income taxes—less than half the 35 percent rate that the tax code requires (McIntyre 2005b).

The Obama administration has given billions of bailout and economic stimulus dollars to U.S. businesses, primarily banks, automobile manufacturers, and insurance companies, with the intention of freeing up credit, reversing the economic recession of 2008/2009, and creating jobs. Critics view these handouts, especially the bailout monies to corporations that failed due to their own greed and poor judgment, as corporate welfare, and have called President Obama "King of Corporate Welfare" (Edsall 2009).

Conflict theorists also note that, throughout the world "free market," economic reform policies have been hailed as a solution to poverty. Yet, although such economic reform has benefited many wealthy corporations and investors, it has also resulted in increasing levels of global poverty. As companies relocate to countries with abundant supplies of cheap labor, wages decline. Lower wages lead to decreased consumer spending, which leads to more industries closing plants, going bankrupt, and/or laying off workers (downsizing). These actions result in higher unemployment rates and a surplus of workers, enabling employers to lower wages even more.

Symbolic Interactionist Perspective

Symbolic interactionism focuses on how meanings, labels, and definitions affect and are affected by social life. This view calls attention to ways in which wealth and poverty are defined and the consequences of being labeled "poor." Individuals who are viewed as poor—especially those receiving public assistance (i.e., welfare)—are often stigmatized as lazy, irresponsible, and lacking in abilities, motivation, and moral values. Wealthy individuals, on the other hand, tend to be viewed as capable, motivated, hardworking, and deserving of their wealth.

The symbolic interactionist perspective also focuses on the meanings of being poor. A qualitative study of more than 40,000 poor women and men in 50 countries around the world explored the meanings of poverty from the perspective of those who live in poverty (Narayan 2000). One of the study's findings is that the experience of poverty involves psychological dimensions such as powerlessness, voicelessness, dependency, shame, and humiliation.

Meanings and definitions of wealth and poverty vary across societies and across time. Although many Americans think of poverty in terms of income level, for millions of people, poverty is not primarily a function of income but of their alienation from sustainable patterns of consumption and production. For indigenous women living in the least developed areas of the world, poverty

wealthfare Laws and policies that benefit the rich.

corporate welfare Laws and policies that benefit corporations.

Among the Dinka, wealth is based on how many cattle a family owns.

and wealth are determined primarily by access to and control of their natural resources (such as land and water) and traditional knowledge, which are the sources of their livelihoods (Susskind 2005).

By global standards, the Dinka, the largest ethnic group in the sub-Saharan African country of Sudan, are among the poorest of the poor, being among the least modernized peoples of the world. In Dinka culture, wealth is measured in large part by how many cattle a family owns. Although modernized populations might label the Dinka as poor, the Dinka view themselves as wealthy. As one Dinka elder explained, "It is for cattle that we are admired, we, the Dinka. . . . All over the world, people look to us because of cattle . . . because of our great wealth; and our wealth is cattle" (Deng 1998, p. 107).

TABLE 6.5 U.S. Poverty Rates by Age, 2008

AGE (YEARS)	POVERTY RATE
Younger than 18	19.0
18–24	18.4
25–34	13.2
35–44	10.4
45–54	9.1
55–59	8.8
60–64	9.7
65 and older	9.7
All ages	13.2

Source: DeNavas-Walt et al. 2009.

Patterns of Poverty in the United States

Although poverty is not as widespread or severe in the United States as it is in many other parts of the world, poverty is nevertheless a significant social problem in the United States. In 2008, 39.8 million Americans—13.2 percent of the U.S. population—lived below the poverty line (DeNavas-Walt et al. 2009). According to research by Mark Rank at the University of Wisconsin, more than half (58 percent) of Americans between the ages of 20 and 75 will spend at least one year in poverty, and one in three Americans will experience a full year of extreme poverty at some point in adult life (Pugh 2007).

Age and Poverty

Children are more likely than adults to live in poverty (see Table 6.5). More than one-third (35.3 percent) of the U.S. poor population are children (DeNavas-Walt et al. 2009). Compared with other industrialized countries, the United States has the highest child poverty rate.

Children are more likely than adults to live in poverty.

Photo courtesy of the author

What Do You Think? In our sociology classes, we introduce the topic of U.S. poverty by asking students to think of an image of a person who represents poverty in America and to draw that imaginary person. Students are asked to give the person a name (to indicate their sex) and to write down the age of the person. Most students draw a picture of a middle-aged man. Yet U.S. poverty statistics reveal that the higher poverty rates are among women, not men, and among youth, not middle-aged adults. Why do you think the most common image of a U.S. poor person is a middle-aged man?

Sex and Poverty

Women are more likely than men to live below the poverty line—a phenomenon referred to as the **feminization of poverty.** The 2008 poverty rates for U.S. women and men were 14.4 and 12.0, respectively (U.S. Census Bureau 2009b). As discussed in Chapter 10, women are less likely than men to pursue advanced educational degrees and tend to have low-paying jobs, such as service and clerical jobs. However, even with the same level of education and the same occupational role, women still earn significantly less than men. Women who are minorities and/or who are single mothers are at increased risk of being poor.

> Education is one of the best insurance policies for protecting an individual against living in poverty.

Education and Poverty

Education is one of the best insurance policies for protecting an individual against living in poverty. In general, the higher a person's level of educational attainment, the less likely that person is to be poor (see also Chapter 8) (see Figure 6.2).

feminization of poverty The disproportionate distribution of poverty among women.

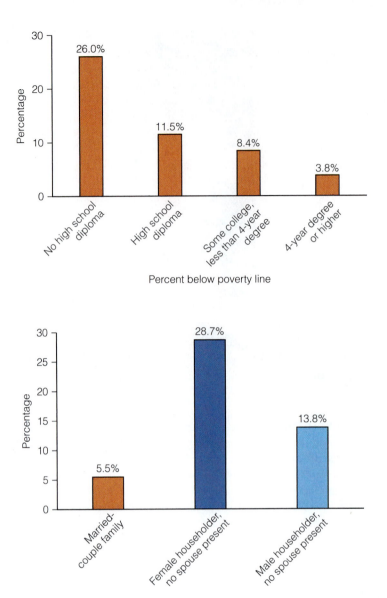

Figure 6.2 Poverty Status, by Years of School Completed
Source: U.S. Census Bureau 2009b.

Figure 6.3 U.S. Poverty Rates by Family Structure, 2008
Source: DeNavas-Walt et al. 2009.

Family Structure and Poverty

Poverty is much more prevalent among female-headed single-parent households than among other types of family structures (see Figure 6.3). In other industrialized countries, poverty rates of female-headed families are lower than those in the United States. Unlike the United States, other developed countries offer a variety of supports for single mothers, such as income supplements, tax breaks, universal child care, national health care, and higher wages for female-dominated occupations.

Lesbian couples and their families are more likely to be poor than heterosexual couples and their families; gay male couple families have the lowest poverty rate of all family structures (see Table 6.6). Children in same-sex couple families are twice as likely to be poor as children of married heterosexual couples (Albelda et al. 2009).

Race or Ethnicity and Poverty

As displayed in Figure 6.4, poverty rates are higher among blacks, Hispanics, and Asians than among non-Hispanic whites. As discussed in Chapter 9, past and present discrimination has contributed to the persistence of poverty among minorities. Other contributing factors include the loss of manufacturing jobs from the inner city, the movement of whites and middle-class blacks out of the inner city, and the resulting concentration of poverty in predominantly minority inner-city neighborhoods (Massey 1991; Wilson 1987, 1996). Finally, blacks and Hispanics are more likely to live in female-headed households with no spouse present—a family structure that is associated with high rates of poverty.

TABLE 6.6 Percent of Poor Adults and Children in Coupled Families, by Type of Household*

	ADULTS	CHILDREN
Married heterosexual couples	5.4	9.4
Male couples	4.0	20.9
Female couples	6.9	19.7

Source: Albelda et al. 2009.
*Based on Census 2000 data.

Labor Force Participation and Poverty

A common image of the poor is that they are jobless and unable or unwilling to work. Although the poor in the United States are primarily children and adults who are not in the labor force, many U.S. poor are classified as **working poor**—individuals who spend at least 27 weeks per year in the labor force (working or looking for work), but whose income falls below the official poverty level.

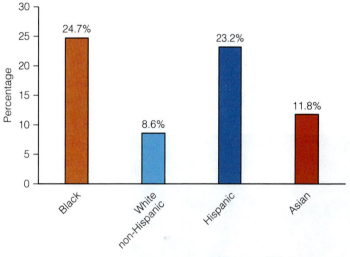

Figure 6.4 **U.S. Poverty Rates by Race and Hispanic Origin, 2008**
Source: DeNavas-Walt et al. 2009.

Consequences of Poverty and Economic Inequality

Poverty is associated with health problems and hunger, increased vulnerability from natural disasters, problems in education, problems in families and parenting, and housing problems. These various problems are interrelated and contribute to the perpetuation of poverty across generations, feeding a cycle of intergenerational poverty. In addition, poverty and economic inequality breed social conflict and war.

Health Problems, Hunger, and Poverty

In developing countries, absolute poverty is associated with hunger and malnutrition, high rates of maternal and infant deaths, indoor air pollution from heating and cooking fumes, and unsafe water and sanitation (see this chapter's Photo Essay) (World Health Organization 2002). In 2009, more than a billion people—more than the combined population of the United States, Canada, and the European Union—were undernourished (FAO 2009). Although chronic hunger decreased in the 1980s and 1990s, it has been increasing in recent years as a result of high food prices and the current global economic downturn, and now has reached a record high. The highest rate of hunger is in sub-Saharan Africa where one in three people are undernourished.

working poor Individuals who spend at least 27 weeks per year in the labor force (working or looking for work) but whose income falls below the official poverty level.

More than 1 billion people lack access to clean water, and 2.6 billion lack access to adequate sanitation (United Nations Development Programme 2006). Worldwide, 1.2 billion people—nearly one in five people—practice open defecation in public places (United Nations 2009). Water-related diseases cause 3 million deaths per year (Oxfam 2006). Each year, 1.5 million children under age 5 die from diarrhea resulting from unsafe water and lack of sanitation (UNICEF 2006). Other diseases associated with lack of clean water and sanitation include trachoma (which causes blindness), typhoid fever, intestinal worms, and guinea worm.

▲ The *Musca sorbens fly*, which breeds in human feces, causes 2 million new cases of blindness-causing trachoma each year in the developing world. The flies burrow into human eyes, which causes decades of repeat infections. Victims describe the infections as feeling as if they have thorns in their eyes.

Charles O. Cecil / Alamy

In many poor areas of the world, people drink whatever water is available, and they urinate and defecate in the street, along the roadside, in buckets, or in plastic bags that are tied up and thrown in ditches or along the road. Used in this way, plastic bags are known as "flying toilets."

Three-quarters of the world's rural population obtain water from a communal source, which means family members (usually women and girls) must walk to the water source and carry water back to the family. To collect enough water for drinking, food preparation, personal hygiene, house cleaning, and laundry, a household of five needs at least 32 gallons of water per day (a little more than 6 gallons per person) (Satterthwaite & McGranahan 2007). This is equivalent to carrying six heavy suitcases of water every day.

Cultural norms in many countries require that women not be seen urinating or defecating, which forces them to limit their food and water intake so they can relieve themselves in the dark of night in fields or roadsides. One woman in Bangladesh explained, "Men can answer the call of nature anytime they want . . . but women have to wait

◄ A person gets guinea worm, a parasite, by drinking water contaminated with the larvae. The worm grows up to three feet long in the body and eventually erupts through the skin, causing extremely painful blisters.

In many poor areas of the world, it ▶ is common for residents to defecate in plastic bags that they dump in ditches or throw on the roadside.

until darkness" (United Nations Development Programme 2006, p. 48). Delaying bodily functions can cause liver infection and acute constipation, and going out in darkness to eliminate places women at risk for physical attack.

Improving access to clean water and sanitation for poor populations is one of the most important priorities in the fight against poverty. For every $1 invested in water and sanitation, $3 to $4 are saved in health spending or through increased productivity (Oxfam 2006). More importantly, access to clean water and sanitation is a basic human right that should be enjoyed by every woman, man, and child.

▲ In poor regions of the world, many people drink whatever water is available.

▲ Whereas Americans and other wealthy populations take running water for granted, poor populations must leave their homes to get water, often from a water kiosk or standpipe where they may have to wait in a long line.

▲ Because of the time demands involved in getting water, many women and girls are not able to go to school or get a job.

LifeStraws® are portable ▶ water purifiers that can be carried around for easy access to clean, safe drinking water.

In the United States, low socioeconomic status is associated with higher incidence and prevalence of health problems, disease, and death (Malatu & Schooler 2002). Hunger in the United States is measured by the percentage of households that are "food insecure," which means that the household had difficulty providing enough food for all its members due to a lack of resources. In 2007, 11.1 percent of U.S. households (13 million households) were food insecure at some time during the year, and 4.1 percent of households had very low food security (some household members had reduced food intake and disrupted normal eating patterns due to lack of resources) (Nord et al. 2008). Assess your own degree of food security in this chapter's Self and Society feature.

Poor U.S. children and adults tend to receive inadequate and inferior health care, which exacerbates their health problems. Finally, poverty is linked to higher levels of mental health problems, including stress, depression, and anxiety (Leventhal & Brooks-Gunn 2003).

Economic inequality also affects psychological and physical health. Poor and middle-income adults who live in states with the greatest gap between the rich and the poor are much more likely to rate their own health as poor or fair than people who live in states where income is more equitably distributed (Kennedy et al. 1998). Subjective perceptions of happiness depend more on how an individual's income compares with other people's income than on the actual amount of their income (International Labour Organization 2008a).

Liba Taylor/Robert Harding Picture Library Ltd/Alamy

Every year, 6 million children die from malnutrition before they reach their first birthday.

Natural Disasters and Poverty

Although natural disasters such as hurricanes, tsunamis, floods, and earthquakes strike indiscriminately—rich and poor alike—the poor are more vulnerable to devastation from such disasters. After a tsunami devastated a large part of South and Southeast Asia in December 2004—killing thousands and tearing apart families, villages, and communities—Oxfam director Barbara Stocking noted that "it is not mere chance that most of those who died or have been left homeless and destitute were already among the world's poorest. Poor families are always much more severely affected by natural disasters. . . . They live in flimsier homes" (Stocking 2005).

The poor have few to no resources to help them avoid or cope with natural disasters. Many poor people in the path of Hurricane Katrina lacked a means of transportation to evacuate. When the poor lose their homes and their livelihoods in a natural disaster, they do not have the resources to rebuild. For example, when poor fishing communities lost their boats and nets—their very means of survival—in the Asian tsunami, they had no bank accounts or insurance policies to replace their losses.

Poverty also affects natural disaster relief efforts. Because the poorest of the tsunami victims lived in areas with weak or nonexistent infrastructure, hundreds of thousands of tsunami victims were cut off from aid because they did not live near a road or airport. In addition, just as upper-class passengers

The U.S. Department of Agriculture conducts national surveys to assess the degree to which U.S. households experience food security, food insecurity, and food insecurity with hunger. To assess your own level of food security, respond to the following items and use the scoring key to interpret your results:

1. In the last 12 months, the food that (I/we) bought just didn't last, and (I/we) didn't have money to get more.
   (a) Often true
   (b) Sometimes true
   (c) Never true

2. In the last 12 months, (I/we) couldn't afford to eat balanced meals.
   (a) Often true
   (b) Sometimes true
   (c) Never true

3. In the last 12 months, did you ever cut the size of your meals or skip meals because there wasn't enough money for food?
   (a) Yes
   (b) No (skip Question 4)

4. If you answered yes to Question 3, how often did this happen in the last 12 months?
   (a) Almost every month
   (b) Some months but not every month
   (c) Only 1 or 2 months

5. In the last 12 months, did you ever eat less than you felt you should because there wasn't enough money to buy food?
   (a) Yes
   (b) No

6. In the last 12 months, were you ever hungry but didn't eat because you couldn't afford enough food?
   (a) Yes
   (b) No

Scoring and Interpretation

The answer responses in boldface type indicate affirmative responses. Count the number of affirmative responses you gave to the items, and use the following scoring key to interpret your results.

Number of Affirmative Responses and Interpretation

0 or 1 item: *Food secure* (In the last year, you have had access to enough food for an active, healthy life.)

2, 3, or 4 items: *Food insecure* (In the last year, you have had limited or uncertain availability of food and have been worried or unsure you would get enough to eat.)

5 or 6 items: *Food insecure with hunger evident* (In the last year, you have experienced more than isolated occasions of involuntary hunger as a result of not being able to afford enough food.)

If you scored as food insecure (with or without hunger), you might consider exploring whether you are eligible for public food assistance (e.g., food stamps) or whether there is a local food assistance program (e.g., food pantry or soup kitchen) that you could use.

Source: Based on the short form of the 12-month Food Security Scale found in Bickel et al. 2000.

Amit Dave/Reuters/Landov

Natural disasters, such as the December 2004 tsunami, are more devastating to the poor, who live in flimsier housing, have little to no infrastructure, and lack resources to cope with and recover from devastation.

Many of the more than 1,300 people who died in the wake of Hurricane Katrina were poor.

Eric Gay/ AP Photo

on the *Titanic* were given priority access to lifeboats, some reports claimed that the tsunami-affected areas that catered to well-off tourists received more assistance than the thousands of poor people who lived in the villages (Roberts 2005). Similarly, in the wake of Hurricane Katrina, some of the local poor residents waited at the back of an evacuation line while 700 guests and employees of the Hyatt Hotel were bused out first (Dowd 2005). Other victims of Hurricane Katrina were stranded for days without food, water, and medical supplies. Five days after the hurricane hit, 20,000 hungry, dehydrated, desperate people stranded in the rain-soaked and sweltering hot Louisiana Superdome in New Orleans, where overflowed toilets forced people to relieve themselves in hallways and stairwells, waited to be evacuated. For days, in some cases a week or more, some hurricane victims waited to be rescued from rooftops or from neck-high floodwater in their attics. For many, rescue came too late. Scores of media interviews and editorials suggested that the government's slow response was at least partly due to the fact that Katrina's victims were predominantly poor. A church pastor lamented, "I think a lot of it has to do with race and class. The people affected were largely poor people. Poor, black people" (quoted by Gonzalez 2005).

What Do You Think? Do you think that the federal response to the disaster left in the wake of Hurricane Katrina would have been different if Katrina had devastated an area of the country where wealthier people resided? Do you think, for example, that residents of Hollywood, California, or Long Island, New York, would have been stranded for days on their rooftops with signs saying, "HELP ME"?

Educational Problems and Poverty

Economic inequality is linked to lower average levels of educational attainment (International Labour Organization 2008a). In many countries, most children from the poorest households have no schooling, and enter their adult lives never having completed the first grade (United Nations Population Fund 2002).

In the United States, children living in poverty are more likely to suffer academically than are children who are not poor. "Overall, poor children receive lower grades, receive lower scores on standardized tests, are less likely to finish high school, and are less likely to attend or graduate from college than are non-poor youth" (Seccombe 2001, p. 323). Health problems associated with childhood poverty, including poorer vision, lead poisoning, asthma, and inadequate nutrition, contribute to poor academic performance (Rothstein 2004). The poor often attend schools that are characterized by lower quality facilities, overcrowded classrooms, and a higher teacher turnover rate (see also Chapter 8). Because poor parents have less schooling on average than do nonpoor parents, they may be less able to encourage and help their children succeed in school. Poor parents have fewer resources to provide their children with books, computers, travel, and other goods and experiences that promote cognitive development and educational achievement (Sobolewski & Amato 2005). With the skyrocketing costs of tuition and other fees, many poor parents cannot afford to send their children to college. Poor adults who want to escape poverty by furthering their education may have to work while attending school or may be unable to attend school because of unaffordable child care, transportation, and/or tuition, fees, and books.

Family Stress and Parenting Problems Associated with Poverty

The stresses associated with low income contribute to substance abuse, domestic violence, child abuse and neglect, divorce, and questionable parenting practices. For example, economic stress is associated with greater marital discord (Sobolewski & Amato 2005), and couples with incomes less than $25,000 are more likely to divorce than couples with incomes greater than $50,000 (Popenoe 2008). Child neglect is more likely to be found with poor parents who are unable to afford child care or medical expenses and leave children at home without adult supervision or fail to provide needed medical care. Poor parents are more likely than other parents to use harsh physical disciplinary techniques, and they are less likely to be nurturing and supportive of their children (Mayer 1997; Seccombe 2001).

Another family problem associated with poverty is teenage pregnancy. Poor adolescent teenagers are at higher risk of having babies than their nonpoor peers. Early childbearing is associated with increased risk of premature babies or babies with low birth weight, dropping out of school, and lower future earning potential as a result of lack of academic achievement. Luker (1996) noted that "the high rate of early childbearing is a measure of how bleak life is for young people who are living in poor communities and who have no obvious arenas for success" (p. 189). For poor teenage women who have been excluded from the American dream and disillusioned with education, "childbearing . . . is one of the few ways . . . such women feel they can make a change in their lives" (p. 182).

Having a baby is a lottery ticket for many teenagers: it brings with it at least the dream of something better, and if the dream fails, not much is lost. . . . In a few cases it leads to marriage or a stable relationship; in many others it motivates a woman to push herself for her baby's sake; and in still other cases it enhances the woman's self-esteem, since it enables her to do something productive, something nurturing and socially responsible. . . . (Luker 1996, p. 182)

Housing Problems

Housing problems include substandard housing, homelessness, and "the housing crisis."

Ric Feld/ AP Photo

Many Americans would be shocked to see the conditions under which many poor people in this country live.

Substandard Housing. Having a roof over one's head is considered a basic necessity. However, for the poor, that roof may be literally caving in. In addition to having leaky roofs, housing units of the poor often have holes in the floor and open cracks in the walls or ceiling. Low-income housing units often lack central heating and air conditioning, sewer or septic systems, and electric outlets in one or more rooms. Housing for the poor is also often located in areas with high crime rates and high levels of pollution.

Homelessness. Due to the recent rise in unemployment and foreclosures, more women, men, and children have been pushed into homelessness. The number of homeless nationwide is estimated to be more than one million (Lichtblau 2009). Over the course of a lifetime, an estimated 9 percent to 15 percent of the U.S. population becomes homeless (Hoback & Anderson 2007). Many factors contribute to homelessness: mental illness and the lack of mental health services, substance abuse and the lack of substance abuse services, low-paying jobs, unemployment, domestic violence, poverty, and prison release. However, the primary cause of homelessness is lack of affordable housing (Cunningham 2009). Housing is considered affordable when a household pays no more than 30 percent of its income on rent.

The primary cause of homelessness is lack of affordable housing.

Homeless individuals live on the street or outdoors; others live in homeless shelters or makeshift dwellings made of a variety of discarded materials such as pieces of wood and boards, cardboard, mattresses, fabric, and plastic tarps. For homeless people living on the street, every day is a struggle for survival:

Living on the streets makes you do a lot of things that you wouldn't normally do. . . . Comin' into this environment I've done a lot of things I said I wouldn't do. . . . There was some people that came along in a van and just threw sandwiches on the street and I picked them up and ate them. . . . The guilt almost killed me . . . but my stomach said, "Hey, listen, you better eat this food. (Dordick 1997, pp. 5–6)

Individuals who do not have a home of their own and who stay at the home of family or friends are known as **couch-homeless,** or "couch-surfers" (Hoback & Anderson 2007).

In recent years, there has been a surge in unprovoked violent attacks against homeless individuals. In a decade, at least 880 homeless men, women, and children have been attacked; 244 have been killed (National Coalition for the Homeless 2009). In most cases, the attacks are by teenage and young adult males. In Toms River, New Jersey, five high school students were charged with beating a 50-year-old homeless man nearly to death with pipes and baseball bats as he slept in the woods. In Spokane, Washington, a 22-year-old man was charged with first-degree murder in the case of a one-legged 50-year-old homeless man who was set on fire in his wheelchair on a downtown street; he died of his burns. In Nashville, Tennessee, two young men were charged with shoving a 32-year-old homeless woman who was sleeping on a boat ramp into the river and leaving her to drown (Lewan 2007).

Thousands of people in the United States are homeless on any given day.

There is also a new fascination with "bum bashing" or "bum fight" videos on YouTube—videos shot by young men and boys who are seen beating the homeless or who pay homeless people a few dollars to fight each other. Thousands of videos on YouTube have "bum fight" in the title, and over 6.8 million copies of bum fight DVDs and videos have been sold (Bowhay 2009).

Many acts of violence toward the homeless are not reported to the police, so documented cases may be just the tip of the iceberg. During the years he lived homeless on the street, David Pirtle was attacked five times and he did not report the attacks to police. "I was struck on the back, kicked, urinated on, spray-painted. . . . A lot of people who are homeless go through it, and it's just the way it is" (quoted in Dvorak 2009, p. DZ01).

What Do You Think? Under hate crime laws, violators are subject to harsher legal penalties if their crime is motivated by the victim's race, religion, national origin, or sexual orientation. In 2009, Maryland became the first state to add homeless status to its hate crime law. California, Maine, Alaska, and Puerto Rico, as well as several cities and counties, have also taken measures to recognize homeless status in their laws or procedures (Levin 2009). A bill in Congress (HR2216) would add homelessness to the federal hate crime law. Do you think that violent acts toward homeless individuals should be categorized as hate crimes and subject to harsher penalties? Why or why not?

couch-homeless Individuals who do not have a home of their own and who stay at the home of family or friends.

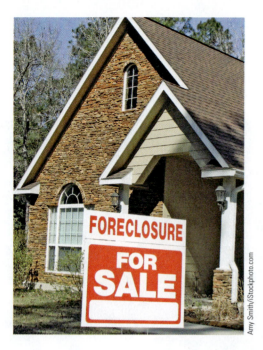

Amy Smith/iStockphoto.com

The "Housing Crisis." The recent housing crisis has caused financial ruin for millions of people in the United States, and has contributed to a global financial crisis that has exacerbated poverty around the world. In the 1990s and early 2000, inflated housing values enabled middle-class homeowners with maxed-out credit cards to keep spending by refinancing their mortgages. At the same time, there was an increase in **subprime mortgages**—high-interest or adjustable-rate mortgages that require little money down and are issued to borrowers with poor credit ratings or limited credit history. Subprime mortgage lending enabled low wage earners to buy a house, and the increased housing demand raised house values further. When the housing bubble burst and house values fell in 2007/08, millions of people were stuck with "upside-down mortgages," in which the amount owed on a mortgage is more than the value of the property. Many homeowners with subprime mortgages could not make their payments, and foreclosures skyrocketed. In this housing crisis, millions of people have lost their homes and their credit, and renters in foreclosed dwellings lost their lease with little notice. In response to this crisis, President Obama implemented a program to stabilize the housing market called Making Home Affordable. This program includes a refinancing program that offers refinancing at lower rates for eligible homeowners. Obama also signed the Protecting Tenants at Foreclosure Act of 2009, which gives tenants the right to stay in a foreclosed property at least until the end of the lease, and entitles month-to-month tenants to a 90-day notice before having to move out.

Intergenerational Poverty

Research finds that nearly half of U.S. children born to low-income parents became low-income adults (Corak 2006). Problems associated with poverty, such as health and educational problems, create a cycle of poverty from one generation to the next. Poverty that is transmitted from one generation to the next is called **intergenerational poverty.**

Intergenerational poverty creates a persistently poor and socially disadvantaged population, referred to as the underclass. Although the underclass is stereotyped as being composed of minorities living in inner-city or ghetto communities, the underclass is a heterogeneous population that includes poor whites living in urban and nonurban communities (Alex-Assensoh 1995).

William Julius Wilson attributes intergenerational poverty and the underclass to a variety of social factors, including the decline in well-paid jobs and their movement out of urban areas, the resultant decline in the availability of marriageable males able to support a family, declining marriage rates and an increase in out-of-wedlock births, the migration of the middle class to the suburbs, and the effect of deteriorating neighborhoods on children and youth (Wilson 1987, 1996).

War and Social Conflict

Poverty and economic inequality are often root causes of conflict and war within and between nations. Poorer countries are more likely than wealthier countries to be involved in civil war, and countries that experience civil war tend to become

subprime mortgages High-interest or adjustable-rate mortgages that require little money down and are issued to borrowers with poor credit ratings or limited credit history.

intergenerational poverty Poverty that is transmitted from one generation to the next.

and/or remain poor. Armed conflict and civil war are generally more likely to occur in countries with extreme and growing inequalities between ethnic groups (United Nations 2005).

> When inequalities become persistent and some groups are systematically barred from the benefits of growth . . . those at the bottom claim their share of the national income by any means possible (International Labour Organization 2008b)

In the developing world, most of the people recruited for armed conflict are unemployed. "They don't have education opportunities and they don't really see what the future holds for them other than war and misery" (World Population News Service 2003, p. 4). Tanzania president Benjamin Mkapa said that "countries with impoverished, disadvantaged and desperate populations are breeding grounds for present and future terrorists" (quoted by Schifferes 2004). A United Nations report suggested that "the most effective conflict prevention strategies . . . are those aimed at achieving reductions in poverty and inequality, full and decent employment for all, and complete social integration" (United Nations 2005, p. 94).

Not only does poverty breed conflict and war, but war also contributes to poverty. War devastates infrastructures, homes, businesses, and transportation systems. In the wake of war, populations often experience hunger and homelessness.

In the United States, the widening gap between the rich and the poor may lead to class warfare (hooks 2000). Briggs (1998) asked how long the United States can maintain social order "when increasing numbers of persons are left out of the banquet while a few are allowed to gorge?" (p. 474). Although Karl Marx predicted that the have-nots would revolt against the haves, Briggs did not foresee a revival of Marxism. "The means of surveillance and the methods of suppression by the governments of industrialized states are far too great to offer any prospect of success for such endeavors" (p. 476). Instead, Briggs predicted that American capitalism and its resulting economic inequalities would lead to social anarchy—a state of political disorder and weakening of political authority.

Strategies for Action: Alleviating Poverty

Government programs designed to alleviate poverty include various types of welfare and public assistance, the earned income tax credit, and policies and proposals that involve increasing wages. We also look at faith-based initiatives and international responses to poverty.

Government Public Assistance and Welfare Programs in the United States

Many public assistance programs stipulate that households are not eligible for benefits unless their income and/or assets fall below a specified guideline. Programs that have eligibility requirements based on income are called **means-tested programs.** Government public assistance programs designed to help the poor include Supplemental Security Income, Temporary Assistance for Needy Families, food programs, housing assistance, medical care, educational assistance, child care, child support enforcement, and the earned income tax credit (EITC).

means-tested programs
Assistance programs that have eligibility requirements based on income.

Supplemental Security Income. Supplemental Security Income (federal SSI), administered by the Social Security Administration, provides a minimum income to poor people who are age 65 or older, blind, or disabled. Under the 1996 welfare reforms, the definition of disability has been sharply restricted, and the eligibility standards have been tightened.

Temporary Assistance for Needy Families. Before 1996, a cash assistance program called Aid to Families with Dependent Children (AFDC) provided single parents (primarily women) and their children with a minimum monthly income. In 1996, Congress passed the Personal Responsibility and Work Opportunity Reconciliation Act (PRWORA), commonly referred to as "welfare reform," which replaced AFDC with a program called **Temporary Assistance for Needy Families (TANF)**. Within two years of receiving benefits, adult TANF recipients must be either employed or involved in work-related activities, such as on-the-job training, job search, and vocational education. A federal lifetime limit of five years is set for families receiving benefits, and able-bodied recipients ages 18 to 50 without dependents have a two-year lifetime limit. Some exceptions to these rules are made for individuals with disabilities, victims of domestic violence, residents in high unemployment areas, and those caring for young children. To qualify for TANF benefits, unwed mothers younger than age 18 are required to live in an adult-supervised environment (e.g., with their parents) and to receive education and job training. Legal immigrants who entered the United States before August 22, 1996, can receive TANF, but those who entered after this date can receive services only after they have been in the country for five years.

Food Assistance. The largest food assistance program in the United States is the **Supplemental Nutrition Assistance Program (SNAP)** (formerly known as the Food Stamp Program), followed by school meals and the Special Supplemental Food Program for Women, Infants, and Children (WIC). SNAP issues monthly benefits through coupons or a plastic card similar to a credit card. In 2008, the typical household receiving SNAP received a monthly benefit of $227; the benefit for an individual was $102.

To supplement SNAP, school meals, and WIC, many communities have food pantries (which distribute food to poor households), "soup kitchens" (which provide cooked meals on-site), and food assistance programs for the elderly population (such as Meals on Wheels). Despite the various forms of food assistance, a significant share of poor U.S. children (18 percent of young children and 12 percent of school-age children) receives no food assistance (Zedlewski & Rader 2005).

Housing Assistance. Housing costs are a major burden for the poor. Half of the working poor spend at least 50 percent of their income on housing (Grunwald 2006). Federal housing assistance programs include public housing, Section 8 housing, and other private project-based housing.

The **public housing** program, initiated in 1937, provides federally subsidized housing that is owned and operated by local public housing authorities (PHAs). To save costs and avoid public opposition, high-rise public housing units were built in inner-city projects. These have been plagued by poor construction, managerial neglect, inadequate maintenance, and rampant vandalism. As noted in the following, poor quality public housing has serious costs for its residents and for society:

Temporary Assistance for Needy Families (TANF) A federal cash welfare program that involves work requirements and a five-year lifetime limit.

Supplemental Nutrition Assistance Program (SNAP) The largest U.S. food assistance program.

public housing Federally subsidized housing that is owned and operated by local public housing authorities (PHAs).

Distressed public housing subjects families and children to dangerous and damaging living environments that raise the risks of ill health, school failure, teen parenting, delinquency, and crime—all of which generate long-term costs that taxpayers ultimately bear. . . . These severely distressed developments are not just old, outmoded, or run down. Rather, many have become virtually uninhabitable for all but the most vulnerable and desperate families. (Turner et al. 2005, pp. 1–2)

The HOPE VI Urban Revitalization Demonstration Program was established in 1992, to transform the nation's most distressed public housing projects by rebuilding public housing developments, expanding the opportunities of its residents, and building a sense of community among residents. This federal housing initiative has demolished 80,000 units of the worst public housing and built mixed-income housing developments in their place (Grunwald 2006).

Rather than building new housing units for low-income families, **Section 8 housing** involves federal rent subsidies provided either to tenants (in the form of certificates and vouchers) or to private landlords. Unlike public housing that confines low-income families to high-poverty neighborhoods, the aim with Section 8 housing is to disperse low-income families throughout the community. However, because of opposition by residents in middle-class neighborhoods, most Section 8 housing units remain in low-income areas.

The level of housing assistance available is sorely inadequate to meet the housing needs of low-income Americans. For every low-income family that receives federal housing assistance, three eligible families are without it (Grunwald 2006).

Lack of affordable housing is not just a problem for the poor living in urban areas. "The problem has climbed the income ladder and moved to the suburbs, where service workers cram their families into overcrowded apartments, college graduates have to crash with their parents, and firefighters, police officers, and teachers can't afford to live in the communities they serve" (Grunwald 2006).

A major barrier to building affordable housing is zoning regulations that set minimum lot size requirements, density restrictions, and other controls. Such zoning regulations serve the interests of upper-middle-class suburbanites who want to maintain their property values and keep out the "riffraff"—the lower-income segment of society who would presumably hurt the character of the community. Thus, one answer to the housing problem is to change zoning regulations that exclude affordable housing. Fairfax County, Virginia, is one of more than 100 communities that have adopted "inclusionary zoning," which requires developers to reserve a percentage of units for affordable housing (Grunwald 2006).

Alleviating Homelessness. Programs to alleviate homelessness include "homeless shelters" that provide emergency shelter beds, and transitional housing programs, which provide time-limited (usually two years) housing and services designed to help individuals gain employment, increase their income, and resolve substance abuse and other health problems. In 2000, the National Alliance to End Homelessness developed a plan to end homelessness in the United States by 2010. The plan, which included establishing permanent supportive housing, resulted in declines in homelessness. However, due to the recent housing crisis, recession, and budget cuts, cities that had reported declines in homelessness are now reporting increases, and the numbers of homeless are expected to grow as the U.S. economy continues to struggle (Cunningham 2009).

Section 8 housing A housing assistance program in which federal rent subsidies are provided either to tenants (in the form of certificates and vouchers) or to private landlords.

Foreclosure Protections. Since the housing crisis that started in 2006, millions of homes have been foreclosed, forcing homeowners and renters to find alternative housing. Before May 2009, most renters lost their leases upon foreclosure. The Protecting Tenants at Foreclosure Act of 2009 grants renters the right to stay until the end of the lease, and entitles month-to-month tenants to a 90-day notice before having to move out.

Medicaid. The largest U.S. public medical care assistance program is Medicaid, which provides medical services and hospital care for the poor through reimbursements to physicians and hospitals. However, many low-income individuals and families do not qualify for Medicaid and either cannot afford health insurance or cannot pay the deductible and co-payments under their insurance plan. Out-of-pocket medical expenses for poor adult Medicaid beneficiaries have grown twice as fast as their incomes in recent years, causing low-income beneficiaries to cut back on essential medical care (Center on Budget and Policy Priorities 2005).

In the earlier AFDC welfare program, all recipients were automatically entitled to Medicaid. Under the TANF program, states decide who is eligible for Medicaid; eligibility for cash assistance does not automatically convey eligibility for Medicaid. A provision of the 1996 welfare reform legislation guarantees welfare recipients at least one year of transitional Medicaid when leaving welfare for work (see Chapter 2 also).

Educational Assistance. Educational assistance includes Head Start and Early Head Start programs and college assistance programs (see also Chapter 8). Head Start and Early Head Start programs provide educational services for disadvantaged infants, toddlers, and preschool-age children and their parents. Evaluations of Head Start and Early Head Start programs indicate that they improve children's cognitive, language, and social-emotional development and strengthen parenting skills (Administration for Children and Families 2002). According to the Children's Defense Fund (2003), every $1 invested in high-quality early childhood care and education saves as much as $7, by increasing the likelihood that children will be literate, go to college, and be employed and by decreasing the likelihood that they will drop out of schools, be dependent on welfare, or be arrested for criminal activity.

To alleviate economic barriers for low-income individuals wanting to attend college, the federal government offers grants, loans, and work opportunities. The Pell Grant program aids students from low-income families. The guaranteed student loan program enables college students and their families to obtain low-interest loans with deferred interest payments. The federal college work-study program provides jobs for students with "demonstrated need."

Child Care Assistance. In the United States, lack of affordable, good child care is a major obstacle to employment for single parents and a tremendous burden on dual-income families and employed single parents. The cost of child care for a 4-year-old child ranges from about $4,000 to nearly $12,000 per year. Child care fees for an infant are even higher, ranging from about $5,000 to more than $15,000 a year (National Association of Child Care Resource and Referral Agencies 2008).

Some public- and private-sector programs and policies provide limited assistance with child care. The Dependent Care Assistance Plan provisions of the 1981 Economic Recovery Tax Act permit individuals to exclude the value of employer-provided child care services from their gross income. However, few employers provide on-site child care or subsidies for child care. At the same time, Congress increased the amount of the child care tax credit and modified the federal tax code to allow taxpayers to shelter pretax dollars for child care in "flexible spending plans." The Family Support Act of 1988 offered additional funding for child care services for the poor (in conjunction with mandatory work requirements). The Child Care and Development Block Grant, which became law in 1990, targeted child care funds to low-income groups, and the PRWORA appropriated funds for child care. However, child care assistance is inadequate; in 2005, 17 states had waiting lists for child care assistance, and many families earn more than the eligibility limit, but not enough to afford child care expenses (National Association of Child Care Resource and Referral Agencies 2006).

Child Support Enforcement. To encourage child support from absent parents, the PRWORA requires states to set up child support enforcement programs, and single parents who receive TANF are required to cooperate with child support enforcement efforts. The welfare reform law established a Federal Case Registry and National Directory of New Hires to track delinquent parents across state lines, increased the use of wage withholding to collect child support, and allowed states to seize assets and to revoke driving licenses, professional licenses, and recreational licenses of parents who fall behind in their child support.

Earned Income Tax Credit. The federal **earned income tax credit (EITC)**, created in 1975, is a refundable tax credit based on a working family's income and number of children. The EITC is designed to offset Social Security and Medicare payroll taxes on working poor families and to strengthen work incentives. The federal EITC lifts more children out of poverty than any other program (Llobrera & Zahradnik 2004).

In 2009, 21 states and the District of Columbia offered a state EITC that supplements the federal credit and works as a rebate for state taxes paid by low-income working people. In addition, local governments in Montgomery County, Maryland, and Denver, Colorado, San Francisco, California, and New York City, New York, offer their own version of EITCs.

Welfare in the United States: Myths and Realities

The majority of U.S. adults—63 percent in 2009—agreed that the government has a responsibility "to take care of people who can't take care of themselves" (Pew Research Center 2009). Nevertheless, negative attitudes toward welfare assistance and welfare recipients are not uncommon (Epstein 2004). For example, 72 percent of U.S. adults believe that "poor people have become too dependent on government assistance programs" (Pew Research Center 2009). What are some

earned income tax credit (EITC) A refundable tax credit based on a working family's income and number of children.

of the common myths about welfare that perpetuate negative images of welfare and welfare recipients?

Myth 1. People who receive welfare are lazy, have no work ethic, and prefer to have a "free ride" on welfare rather than work.

Reality. Most recipients of TANF and SNAP benefits are children and therefore are not expected to work. In 2006, one in five adult TANF recipients worked, earning an average monthly salary amount of $703 (Office of Family Assistance 2009). Unemployed adult welfare recipients experience a number of barriers that prevent them from working, including poor health, job scarcity, lack of transportation, lack of education, and/or the desire to stay home and care for their children (which often stems from the inability to pay for child care or the lack of trust in child care providers) (Zedlewski 2003). Welfare recipients who stay home to care for children *are* doing very important work: parenting. "Raising children is work. It requires time, skills, and commitment. While we as a society don't place a monetary value on it, it is work that is invaluable—and indeed, essential to the survival of our society" (Albelda & Tilly 1997, p. 111).

It is also important to note that many adults receiving public assistance do earn an income or are participating in work activities, including job training or education, job searches, and employment. However, many do not work because there are not enough jobs available. In May 2009, there were nearly six unemployed workers for every available job (Shierholz 2009).

Finally, most adult welfare recipients would rather be able to support themselves and their families than rely on public assistance. The image of a welfare "freeloader" lounging around enjoying life is far from the reality of the day-to-day struggles and challenges of supporting a household on a monthly TANF check of $372, which was the average monthly cash and cash-equivalent assistance to TANF families in 2006 (Office of Family Assistance 2009).

Myth 2. Most welfare mothers have large families with many children.

Reality. The average number of individuals in TANF families is 2.3, including an average of only 1.8 children (Office of Family Assistance 2009).

Myth 3. Welfare benefits are granted to many people who are not really poor or eligible to receive them.

Reality. Although some people obtain welfare benefits through fraudulent means, it is much more common for people who are eligible to receive welfare not to receive benefits (see Table 6.7). Only about half of families poor enough to qualify for TANF receive monthly cash assistance, and two-thirds of individuals eligible for SNAP receive benefits (Leftin & Wolkwitz 2009).

A main reason for not receiving benefits is lack of information; people do not know they are eligible. Many people who are eligible for public assistance do not apply for it because they do not want to be stigmatized as lazy people who just want a "free ride" at the taxpayers' expense—their sense of personal pride prevents them from receiving public assistance. Others have difficulty navigating the administrative process of applying for assistance. One homeless person explained,

TABLE 6.7 Percentage of Individuals Living Below the Poverty Level in Households That Receive Means-Tested Assistance, 2005

TYPE OF ASSISTANCE	PERCENTAGE
Any type of assistance	67.8
Cash assistance	20.9
Food stamps	37.9
Medicaid	54.7
Public or subsidized housing	17.3

Source: U.S. Census Bureau 2006.

I don't get welfare. I just can't . . . do it. I hate those people in there. They make you . . . sit and ask you questions that don't make any sense. . . . You're homeless but you have to have an address. What kind of shit is that? Give me a break. They want you to get so . . . upset that you do get up and walk out. (Dordick 1997, p. 58)

Finally, some individuals who are eligible for public assistance do not receive it because it is not available. In cities across the United States, thousands of eligible low-income households are on waiting lists for public housing assistance because there are not enough public housing units available, and some cities have even stopped accepting housing applications (U.S. Conference of Mayors 2008).

Myth 4. Immigrants place a huge burden on our welfare system.

Reality. In 2006, only 5.9 percent of adult recipients of TANF and 1.2 percent of child TANF recipients were qualified immigrants (Office of Family Assistance 2009).

Minimum Wage Increase and "Living Wage" Laws

In 2009, the federal minimum wage increased to $7.25 an hour. Some states have established a minimum wage that is higher than the federal minimum wage. As of 2009, 27 states and the District of Columbia have mandated a minimum wage that is higher than the federal $7.25.

What Do You Think? Just as there is a federal minimum wage, do you think that there should be a federal maximum wage? Why or why not? If you favor the idea of a maximum wage, what should that wage be?

Many cities and counties throughout the United States have **living wage laws** that require state or municipal contractors, recipients of public subsidies or tax breaks, or, in some cases, all businesses to pay employees wages that are significantly above the federal minimum, enabling families to live above the poverty line. Research findings show that businesses that pay their employees a living wage have lower worker turnover and absenteeism, reduced training costs, higher morale and productivity, and a stronger consumer market (Kraut et al. 2000).

As more individuals receiving welfare (TANF) reach their time limits and are forced to enter the job market, it is increasingly important to provide jobs that pay a living wage. As shown in this chapter's Social Problems Research Up Close feature, single mothers who work in low-wage jobs often have more hardships than those who are dependent on welfare.

Faith-Based Services for the Poor

In 2001, President G. W. Bush established the Faith-Based and Community Initiative, which provided federal funding for faith-based programs that serve the needy, such as homeless services and food aid programs. President Obama renamed the program **Faith-Based and Neighborhood Partnerships**, expanding the program to include non-faith-based neighborhood programs.

living wage laws Laws that require state or municipal contractors, recipients of public subsidies or tax breaks, or, in some cases, all businesses to pay employees wages that are significantly above the federal minimum, enabling families to live above the poverty line.

Faith-Based and Neighborhood Partnerships A program in which faith-based and other neighborhood organizations receive federal funding for programs that serve the needy, such as homeless services and food aid programs.

Making Ends Meet: Survival Strategies among Low-Income and Welfare Single Mothers

How do individuals in low-income jobs compare with those dependent on welfare in terms of their well-being? And how do both low-wage earners and welfare recipients survive on income that does not meet their basic needs? Researchers Kathryn Edin and Laura Lein (1997) conducted research to answer these questions.

Sample and Methods

The sample consisted of 379 Black American, white, and Mexican American single mothers from four cities (Chicago, San Antonio, Boston, and Charleston, South Carolina).

The mothers either received welfare cash assistance (N = 214) or were nonrecipients who held low-wage jobs earning $5 to $7 an hour between 1988 and 1992 (N = 165). Edin and Lein (1997) used a "snowball sampling" technique in which each mother who was interviewed was asked to refer researchers to one or two friends who might also participate in interviews. Nearly 90 percent of the mothers contacted agreed to be interviewed. Edin and Lein conducted multiple semistructured in-depth interviews with women in the sample on topics including the mothers' income and job experience, types and amount of welfare benefits they received, spending behavior, housing situation, use of medical care and child care, and hardships the women and their children experienced because of lack of financial resources.

Findings and Conclusions

Single mothers earning low wages had a higher monthly reported income than welfare-reliant mothers. However, the expenses of wage-earning mothers were also higher. This is because employed mothers usually have to pay for child care, transportation to work, and additional clothing to wear to work. If newly employed mothers have a federal housing subsidy, some of their new income is spent on the increase in the rent they must pay. Employed mothers are also usually not eligible for Medicaid, which means that they have more out-of-pocket medical expenses and often go uninsured.

The monthly expenses of both groups of women exceeded their reported monthly income, forcing women to use various strategies to make ends meet. Cash welfare and food stamps covered only three-fifths of welfare-reliant mothers' expenses. The main job of mothers earning low wages covered only 63 percent of their expenses. Edin and Lein (1997) found that women relied on three basic strategies to make ends meet: work in the formal, informal, or underground economy; cash assistance from absent fathers, boyfriends, relatives, and friends; and cash assistance or help from agencies, community groups, or charities in paying overdue bills. Welfare recipients had to keep their income-generating activities hidden from their welfare caseworkers and other government officials. Otherwise, their welfare checks would be reduced by nearly the same amount as their earnings. Many of the wage-earning mothers also concealed income generated "on the side" to maintain eligibility for food stamps, housing subsidies, or other benefits that would have been reduced or eliminated if they had reported this additional income.

Most of the single mothers in the study described experiencing serious material hardship during the previous 12 months. Material hardships included not having enough food and clothes, not receiving needed medical care, not having health insurance, having the utilities or phone cut off, not having a phone, and being evicted and/or homeless. An important finding was that wage-reliant mothers experienced more hardship than welfare-reliant mothers. In addition to the increased financial pressures of child care costs, transportation, health care, and work clothing, employed mothers worried about not providing adequate supervision of their children and struggled with balancing work and parenting responsibilities, especially when their children were sick. Nevertheless, almost all the mothers said they would rather work than rely on welfare. They believed that work provided important psychological benefits and increased self-esteem, avoided the stigma of welfare, and enabled them to be good role models for their children.

Source: Based on Edin & Lein 1997.

Critics are concerned about the degree to which the faith-based initiative violates the separation of church and state and affects the rights of clients seeking services. "How are the lives of the jobless improved when they are told they won't get work until they first get right with God?" (Boston 2005). Although religious groups are prohibited from using government funding to promote religion, this policy is difficult to enforce.

Millennium Development Goals Eight goals that comprise an international agenda for reducing poverty and improving lives.

International Responses to Poverty

In 2000, leaders from 191 United Nations member countries pledged to achieve eight **Millennium Development Goals**—an international agenda for reducing poverty and improving lives. One of the Millennium Development Goals (MDGs) is to halve, between 1990 and 2015, the proportion of people who live in severe poverty and who suffer from hunger. As can be seen in Table 6.8, several other MDGs involve alleviating problems related to poverty, such as disease, child and maternal mortality, and lack of access to education. From 1990 to 2005, the severe poverty rate in developing countries decreased from 42 percent of the population to 25 percent (World Bank 2009). Much of this progress is due to China, where the poverty rate dropped from 60 percent to 16 percent. As of 2005, all developing countries were "on track" to meet their poverty reduction goal, or had already exceeded it, except for countries in sub-Saharan Africa. With the crisis in soaring food prices in 2007, and the economic downturn in 2008/09, progress toward reducing poverty has certainly been stalled, and in some cases, reversed. Approaches for achieving poverty reduction throughout the world include promoting economic growth, investing in "human capital," providing financial aid and debt cancellation, and providing microcredit programs that provide loans to poor people.

TABLE 6.8 The Millennium Development Goals
1. Eradicate extreme poverty and hunger.
2. Achieve universal primary education.
3. Promote gender equality and empower women.
4. Reduce child mortality.
5. Improve maternal health.
6. Combat HIV/AIDS, malaria, and other diseases.
7. Ensure environmental sustainability.
8. Develop a global partnership for development.

Source: U.S. Census Bureau 2006.

Promoting Economic Growth and Development. In low-income countries, increases in gross domestic product (GDP) are associated with reductions in poverty. An expanding economy creates new employment opportunities and increased goods and services. As employment prospects improve, individuals are able to buy more goods and services. The increased demand for goods and services, in turn, stimulates economic growth. In 2009, world gross domestic product declined for the first time since World War II (World Bank 2009), impeding, and in some countries reversing, earlier gains in poverty reduction.

As economic and political leaders strategize ways to stimulate economic development, environmentalists urge them to consider the importance of controlling population growth and protecting the environment and natural resources, which are often destroyed and depleted in the process of economic growth. Advocates for the poor point out that economic growth does not always reduce poverty; in some cases, it increases it. Policies that involve cutting government spending, privatizing basic services, liberalizing trade, and producing goods primarily for export may increase economic growth at the national level, but the wealth ends up in the hands of the political and corporate elite at the expense of the poor. Growth does not help poverty reduction when public spending is diverted away from meeting the needs of the poor and instead is used to pay international debt, finance military operations, and support corporations that do not pay workers fair wages and that are hostile to unionization. The World Bank lends billions of dollars a year to developing nations to pay primarily for roads, bridges, and industrialized agriculture that mostly benefit corporations. "Relatively little attention or money has been given to developing basic social services, building schools and clinics, and building decent public sanitation and clean water systems in some of the world's poorest countries" (Mann 2000, p. 2). Thus, "economic growth, though essential for poverty reduction, is not enough. Growth must be pro-poor, expanding the opportunities and life choices of poor people" (UNDP 1997, pp. 72–73).

Economic development also threatens the lives and cultures of the 370 million indigenous people who live in 70 countries around the world. Indigenous people who live on land that is rich in natural resources are displaced by corporations that want access to the land and its natural resources, and by government forces that help the corporations expand their activities (Ramos et al. 2009). As remote areas are "developed," many indigenous people are forced to give up their traditional ways of life and become assimilated into the dominant culture. Conflict between indigenous people who want to continue their traditional ways of life and governments and multinationals who want to pursue economic development and the profits it produces can become violent. In June 2009, 600 Peruvian police opened fired on thousands of peaceful indigenous protesters who were blocking a road to protest new laws that allow the expansion of mining, oil drilling, logging, and destructive agriculture in their territories in the Amazon rain forest (Robinson 2009).

In June 2009, 600 Peruvian police opened fire on peaceful indigenous protestors who were blocking a road in an effort to stop the "economic development" in the Amazon rain forest—their home.

Investing in Human Capital. In many poor countries, large segments of the population are illiterate and without job skills and/or are malnourished and in poor health. A key feature of poverty reduction strategies involves investing in human capital—the skills, knowledge, and capabilities of individuals. Investments in **human capital** involve programs and policies that provide adequate nutrition, sanitation, housing, health care (including reproductive health care and family planning), and educational and job training.

Poor health is both a consequence and a cause of poverty; improving the health status of a population is a significant step toward breaking the cycle of poverty. Investments in education are also critical for poverty reduction. Increasing the educational levels of a population better prepares individuals for paid employment and for participation in political affairs that affect poverty and other economic and political issues. Improving the educational level and overall status of women in developing countries is also associated with lower birthrates, which in turn fosters economic development.

Providing Financial Aid and Debt Cancellation. To pay for investments in human capital, poor countries depend on financial aid from wealthier countries. To meet the MDG of halving poverty by 2015, the United Nations recommends that wealthier countries allocate just 0.7 percent of gross national income (GNI) to aid for poor countries. In 2008, five countries exceeded the target of 0.7 percent, but most donor countries failed to meet the target, including the United States, which gave only 0.18 percent of GNI (OECD 2009).

Another way to help poor countries invest in human capital and reduce poverty is to provide debt relief. In poor countries with large debts, money needed for health and education is instead spent on debt repayment. Canceling the debts of 32 of the poorest countries would cost citizens of the richest countries of the world just $2.10 a year for each citizen for 10 years (Oxfam 2005).

human capital The skills, knowledge, and capabilities of individuals.

Muhammad Yunus won the 2006 Nobel Peace Prize for his pioneering work in starting the Grameen Bank, which has more than 2,000 branches in more than 65,000 villages. It has provided loans to more than 6 million poor people, 96 percent of whom are women.

Microcredit Programs. The old saying "It takes money to make money" explains why many poor people are stuck in poverty: they have no access to financial resources and services. **Microcredit programs** refer to the provision of loans to people who are generally excluded from traditional credit services because of their low socioeconomic status. Microcredit programs give poor people the financial resources they need to become self-sufficient and to contribute to their local economies.

The Grameen Bank in Bangladesh, started in 1976, has become a model for the more than 3,000 microcredit programs that have served millions of poor clients (Roseland & Soots 2007). To get a loan from the Grameen Bank, borrowers must form small groups of five people "to provide mutual, morally binding group guarantees in lieu of the collateral required by conventional banks" (Roseland & Soots 2007, p. 160). Initially, only two of the five group members are allowed to apply for a loan. When the initial loans are repaid, the other group members may apply for loans.

Understanding Poverty and Economic Inequality

As we have seen in this chapter, economic prosperity has not been evenly distributed; the rich have become richer while the poor have become poorer. A common belief among U.S. adults is that the rich are deserving and the poor are failures. Blaming poverty on the individual rather than on structural and cultural factors implies not only that poor individuals are responsible for their plight but also that they are responsible for improving their condition. If we hold individuals accountable for their poverty, we fail to make society accountable

microcredit programs The provision of loans to people who are generally excluded from traditional credit services because of their low socioeconomic status.

for making investments in human capital that are necessary to alleviate poverty. Such human capital investments include providing health care, adequate food and housing, education, child care, and job training. Economist Lewis Hill (1998) believes that "the fundamental cause of perpetual poverty is the failure of the American people to invest adequately in the human capital represented by impoverished children" (p. 299). By blaming the poor for their plight, we also fail to recognize that there are not enough jobs for those who want to work and that many jobs do not pay wages that enable families to escape poverty. Lastly, blaming the poor for their condition diverts attention away from the recognition that the wealthy—individuals and corporations—receive far more benefits in the form of wealthfare or corporate welfare, without the stigma of welfare.

Ending or reducing poverty begins with the recognition that doing so is a worthy ideal and an attainable goal. Imagine a world where everyone had comfortable shelter, plentiful food, clean water and sanitation, adequate medical care, and education. If this imaginary world were achieved and if absolute poverty were effectively eliminated, what would be the effects on social problems such as crime, drug abuse, family problems (e.g., domestic violence, child abuse, and divorce), health problems, prejudice and racism, and international conflict? In the current global climate of conflict and terrorism, we might consider that "reducing poverty and the hopelessness that comes with human deprivation is perhaps the most effective way of promoting long-term peace and security" (World Bank 2005).

According to one source, the cost of eradicating poverty worldwide would be only about 1 percent of global income—and no more than 2 percent to 3 percent of national income in all but the poorest countries (UNDP 1997). Certainly the costs of allowing poverty to continue are much greater than that.

CHAPTER REVIEW

■ **What is the difference between absolute poverty and relative poverty?**

Absolute poverty refers to a lack of basic necessities for life, such as food, clean water, shelter, and medical care. In contrast, relative poverty refers to a deficiency in material and economic resources compared with some other population.

■ **How is poverty measured?**

The World Bank sets a "poverty threshold" of $1.25 per day for the least developed countries, and $2 per day for middle-income countries and regions, including Latin America and eastern Europe. Another measure of poverty is the proportion of people who consume less than a minimum amount of calories to maintain health—who are chronically hungry or undernourished. According to measures of relative poverty, members of a household are considered poor if their household income is less than 50 percent of the median household income in that country. Each year, the U.S. federal government establishes "poverty thresholds" that differ by the number of adults and children in a family and by the age of the family head of household. Anyone living in a household with pretax income below the official poverty line is considered "poor." To capture the multidimensional nature of poverty, the United Nations Development Programme developed the human poverty index that combines three measures of deprivation: (1) deprivation of a long, healthy life; (2) deprivation of knowledge; and (3) deprivation in decent living standards.

■ **Which sociological perspective criticizes wealthy corporations for using financial political contributions to influence politicians to enact policies that benefit corporations and the wealthy?**

The conflict perspective is critical of wealthy corporations that use financial political contributions to influence laws and policies that favor corporations and the rich. Such laws and policies, sometimes referred to as wealthfare or corporate welfare, include low-interest government loans to failing businesses and special subsidies and tax breaks to corporations.

- **In the United States what age group has the highest rate of poverty?**
U.S. children are more likely than adults to live in poverty. More than one-third of the U.S. poor population is children. Child poverty rates are much higher in the United States than in any other industrialized country.

- **What are some of the consequences of poverty and economic inequality for individuals, families, and societies?**
Poverty is associated with health problems and hunger, increased vulnerability from natural disasters, problems in education, problems in families and parenting, and housing problems. These various problems are interrelated and contribute to the perpetuation of poverty across generations, feeding a cycle of intergenerational poverty. In addition, poverty and economic inequality breed social conflict and war.

- **What are some of the U.S. government public assistance programs designed to help the poor?**
Government public assistance programs designed to help the poor include Supplemental Security Income, Temporary Assistance for Needy Families (TANF), food programs (such as school meal programs and SNAP), housing assistance, Medicaid, educational assistance (such as Pell Grants), child care, child support enforcement, and the earned income tax credit (EITC).

- **What are four common myths about welfare and welfare recipients?**
Common myths about welfare and welfare recipients are (1) that welfare recipients are lazy, have no work ethic, and prefer to have a "free ride" on welfare rather than work; (2) that most welfare mothers have large families with many children; (3) that welfare benefits are granted to many people who are not really poor or eligible to receive them; and (4) that immigrants place an enormous burden on our welfare system.

- **What are four general approaches for achieving poverty reduction throughout the world?**
Approaches for achieving poverty reduction throughout the world include promoting economic growth, investing in "human capital," providing financial aid and debt cancellation to nations, and providing microcredit programs that provide loans to poor people.

TEST YOURSELF

1. According to the text, which of the following countries has the lowest level of poverty as measured by the human poverty index?
   a. Sweden
   b. Italy
   c. United States
   d. France
2. According to the official U.S. poverty threshold guidelines, a single adult earning $12,000 a year is considered "poor."
   a. True
   b. False
3. In 2007, a CEO of a large U.S. corporation earned more in one workday than the typical worker earned
   a. in a week
   b. in a month
   c. in six months
   d. in a year
4. Corporate welfare refers to which of the following?
   a. Taxes corporations pay that provide most of the funding for federal welfare programs for the poor
   b. Tax-deductible contributions that corporations make to charitable organizations
   c. Laws and policies that benefit corporations
   d. Employee assistance programs offered by corporations to help employees who are struggling with debt
5. What age group in the United States has the highest rate of poverty?
   a. Younger than 18
   b. 18–29
   c. 30–55
   d. Older than 55
6. According to the text, the wealthy are hardest hit by natural disasters because they have more to lose than do the poor.
   a. True
   b. False
7. Subprime mortgages are mortgages that charge very low interest rates and are available only to people with excellent credit and ample collateral.
   a. True
   b. False
8. Which federal program lifts more children out of poverty than any other program?
   a. Public and Section 8 housing
   b. TANF
   c. EITC
   d. SNAP

9. Nearly half of recipients of welfare in the United States are immigrants.
   a. True
   b. False

10. In 2008, the United States was one of the few countries in the world that met the United Nations target (0.7 percent of GNI) of aid to poor countries.
    a. True
    b. False

Answers: 1: a; 2: b; 3: d; 4: c; 5: a; 6: b; 7: b; 8: c; 9: c; 10: b.

KEY TERMS

absolute poverty   189
corporate welfare   197
couch-homeless   209
culture of poverty   195
earned income tax credit (EITC)   215
extreme poverty   190
Faith-Based and Neighborhood Partnerships   217
feminization of poverty   199
human capital   220
human poverty index (HPI)   190
intergenerational poverty   210
living wage laws   217
means-tested programs   211

microcredit programs   221
Millennium Development Goals   218
public housing   212
relative poverty   189
Section 8 housing   213
subprime mortgages   210
Supplemental Nutrition Assistance Program (SANP)   212
Temporary Assistance for Needy Families (TANF)   212
underclass   195
wealth   194
wealthfare   197
working poor   201

MEDIA RESOURCES

Understanding Social Problems,
Seventh Edition Companion Website
www.cengage.com/sociology/mooney
Visit your book companion website, where you will find flash cards, practice quizzes, Internet links, and more to help you study.

Just what you need to know NOW! Spend time on what you need to master rather than on information you already have learned. Take a pretest for this chapter, and CengageNOW will generate a personalized study plan based on your results. The study plan will identify the topics you need to review and direct you to online resources to help you master those topics. You can then take a posttest to help you determine the concepts you have mastered and what you will need to work on. Try it out! Go to www.cengage.com/login to sign in with an access code or to purchase access to this product.

© AP Photo/The Herald, Michael V. Martina

7

> "When a man tells you that he got rich through hard work, ask him whose."
>
> **Don Marquis, Journalist**

Work and Unemployment

The Global Context: The New Global Economy | Sociological Theories of Work and the Economy | **Social Problems Research Up Close: Job Loss at Midlife** | Problems of Work and Unemployment | **The Human Side: Down but Not Out: From Hedge Funds to Pizza Delivery** | **Self and Society: How Do Your Spending Habits Change in Hard Economic Times?** | Strategies for Action: Responses to Problems of Work and Unemployment | Understanding Work and Unemployment | Chapter Review

225

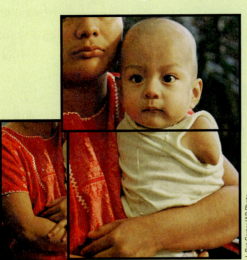

Francisca Herrera worked on a North Carolina tomato farm run by Ag-Mart, where she was repeatedly exposed to pesticides. "It happened morning, noon, and evening," she explained. Sprayers "would pass by close to where we were working. They didn't care if we were eating" (Collins 2008, p. 1A). Herrera said she was often told to work in fields that were still wet with pesticides, and that her supervisors ignored her complaints of frequent headaches and stomach pains. Francisca was pregnant during the time she worked on the tomato farm. When her son Carlitos was born, he had no arms and no legs. His parents blame pesticide exposure for their son's missing limbs.

J. Pat Carter/AP Photo

Francisca Herrera's son was born without limbs. During her pregnancy, she was exposed to pesticides while working on a tomato farm.

Health and safety hazards in the workplace, often due to employers' willful violations of health and safety regulations, are among the work-related problems discussed in this chapter. Other problems we examine in this chapter include unemployment, employment and retirement concerns of older Americans, slavery and forced labor, child labor, sweatshop labor, job dissatisfaction and alienation, work-family concerns, and declining labor strength and representation. We set the stage with a brief look at the global economy.

The Global Context: The New Global Economy

In recent decades, innovations in communication and information technology have spawned the emergence of a **global economy**—an interconnected network of economic activity that transcends national borders and spans the world. The globalization of economic activity means that our jobs, the products and services we buy, and our nation's economic policies and agendas are influenced by economic activities occurring around the world.

Beginning in 2007, the economic situation around the world took a downward turn, as banks faltered, credit froze, businesses closed, unemployment rates soared, and investments plummeted. This global financial crisis, which originated in the United States and spread throughout the world, illustrates the globalization of the **economic institution**. Some say that the cause of the global economic crisis was the lack of U.S. financial regulatory oversight that enabled

global economy An interconnected network of economic activity that transcends national borders.

economic institution The structure and means by which a society produces, distributes, and consumes goods and services.

financial institutions to engage in predatory and subprime lending during the housing boom in the early 2000s. As the housing boom turned to bust, and adjustable rate mortgages were reset to higher rates, millions of homeowners were unable to keep up with mortgage payments and foreclosures skyrocketed. Homeowners lost their homes, renters lost their leases, and banks suffered because the foreclosed homes they now owned were often worth less than the mortgages owed on them. As banks lost revenue, they had less money to lend so credit froze, consumer spending plummeted, businesses went bust, and stockholders watched their investments and retirement accounts take a nosedive. The whole banking system was faltering; some got bailed out, others (e.g., Bear Stearns) went bust. All this happened in the United States, but in this new global economy, what happens in Vegas does not stay in Vegas; the crisis spread around the world. This is because those risky subprime and adjustable rate mortgages were packaged and resold as "mortgage-backed securities" to financial institutions around the world.

The U.S. economic crisis also triggered a huge drop in world trade. Between 2000 and 2007, U.S. consumption accounted for more than a third of the growth in global consumption, and much of that consumption was based on borrowed money (Baily & Elliott 2009). When the United States went into recession and consumer spending declined, all the countries that depended on U.S. consumers to buy their goods and services lost a major source of revenue.

> In this new global economy, what happens in Vegas does not stay in Vegas.

The global economic crisis reignited debate between those who view U.S. capitalism as the cause of economic problems in the world, and those who hail capitalism as "the greatest engine of economic progress and prosperity known to mankind" (Ebeling 2009). After summarizing capitalism and socialism—the two main economic systems in the world—we describe how industrialization and postindustrialization have changed the nature of work, and look at the emergence of free trade agreements and transnational corporations.

What Do You Think? The economic health of a country is commonly measured by how much the country is producing (the total value of goods and services) and how much money consumers are spending on the purchase of goods and services. In what ways might high levels of production and consumption contribute to individual and social ills rather than to health and well-being?

Capitalism and Socialism

Socialism is an economic system characterized by state ownership of the means of production and distribution of goods and services. In a socialist economy, the government controls income-producing property. Theoretically, goods and services are equitably distributed according to the needs of the citizens. Socialist economic systems emphasize collective well-being rather than individualistic pursuit of profit. Critics of socialism argue that socialism creates excessive government control, reduces work incentives and technological development, and lowers the standard of living.

Under **capitalism,** private individuals or groups invest capital (money, technology, machines) to produce goods and services to sell for a profit in a competitive market. Whereas socialism emphasizes social equality, capitalism emphasizes

socialism An economic system characterized by state ownership of the means of production and distribution of goods and services.

capitalism An economic system characterized by private ownership of the means of production and distribution of goods and services for profit in a competitive market.

TABLE 7.1 Attitudes Toward Big Business in the United States	
PERCENT OF U.S. ADULTS AGREEING THAT. . .	
There's too much power concentrated in the hands of a few big companies.	77%
Businesses make too much profit.	62%
Free market needs regulation to best serve public interest needs.	62%

Source: Pew Research Center 2009.

individual freedom. Capitalism is characterized by economic motivation through profit, the determination of prices and wages primarily through supply and demand, and the absence of government intervention in the economy. Critics of capitalism argue that it creates too many social evils, including alienated workers, poor working conditions, near-poverty wages, unemployment, a polluted and depleted environment, and world conflict over resources. Table 7.1 reveals some of the attitudes that U.S. adults have toward big business in the United States—a largely capitalistic society.

In reality, there are no pure socialist or capitalistic economies. Rather, most countries have mixed economies, incorporating elements of both capitalism and socialism. Most developed countries, for example, have both private-owned and state-owned enterprises, as well as a social welfare system. The U.S. economy is dominated by capitalism, but there are elements of socialism in our welfare system and in government subsidies and low-interest loans to industry, fiscal stimulus money, and bailout money.

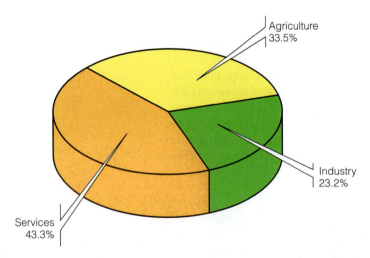

Agriculture 33.5%

Industry 23.2%

Services 43.3%

Figure 7.1 Global Employment, by Sector
Source: International Labor Organization 2009.

industrialization The replacement of hand tools, human labor, and animal labor with machines run by steam, gasoline, and electric power.

postindustrialization The shift from an industrial economy dominated by manufacturing jobs to an economy dominated by service-oriented, information-intensive occupations.

Industrialization, Postindustrialization, and the Changing Nature of Work

The nature of work has been shaped by the Industrial Revolution, the period between the mid-18th century and the early 19th century when the factory system was introduced in England. **Industrialization** dramatically altered the nature of work: Machines replaced hand tools, and steam, gasoline, and electric power replaced human or animal power. Industrialization also led to the development of the assembly line and an increased division of labor as goods began to be mass-produced. The development of factories contributed to the emergence of large cities. Instead of the family-centered economy characteristic of an agricultural society, people began to work outside the home for wages.

Postindustrialization refers to the shift from an industrial economy dominated by manufacturing jobs to an economy dominated by service and information technology jobs. In the global economy, jobs in the service sector outnumber jobs in both agriculture and industry (see Figure 7.1). In developed countries and the European Union, the majority of jobs (71 percent) are in services, followed by jobs in industry (25 percent) and agriculture (4 percent) (ILO 2009).

What Do You Think? Virtually all of the products and services produced in today's global economy depend on the use of petroleum. Indeed, most jobs in the world today would not exist without oil. Thus, professors Charles Hall and John Day (2009) note, "We do not live in an information age, or a postindustrial age . . . but a petroleum age" (p. 237). Do you think the world will ever achieve a "post-petroleum" economy? Why or why not?

McDonaldization of the Workplace

Sociologist George Ritzer (1995) coined the term **McDonaldization** to refer to the process by which the principles of the fast-food industry are being applied to more and more sectors of society, particularly the workplace. McDonaldization involves four principles:

1. *Efficiency.* Tasks are completed in the most efficient way possible by following prescribed steps in a process overseen by managers.
2. *Calculability.* Quantitative aspects of products and services (such as portion size, cost, and the time it takes to serve the product) are emphasized over quality.
3. *Predictability.* Products and services are uniform and standardized. A Big Mac in Albany is the same as a Big Mac in Tucson. Workers behave in predictable ways. For example, servers at McDonald's learn to follow a script when interacting with customers.
4. *Control through technology.* Automation and mechanization are used in the workplace to replace human labor.

What are the effects of McDonaldization on workers? In a McDonaldized workplace, employees are not permitted to use their full capabilities, be creative, or engage in genuine human interaction. Workers are not paid to think, just to follow a predetermined set of procedures. Because human interactions are unpredictable and inefficient (they waste time), "we're left with either no interaction at all, such as at ATMs, or a 'false fraternization.' Rule number 17 for Burger King workers is to smile at all times" (Ritzer, quoted by Jensen 2002, p. 41). Workers may also feel that they are merely extensions of the machines they operate. The alienation that workers feel—the powerlessness and meaninglessness that characterize a "McJob"— may lead to dissatisfaction with one's job and, more generally, with one's life.

McFlexible

We have shift patterns to suit the lifestyles of all our Crew.
Not bad for a McJob

We know it is important to have a life outside of work, so many of our Crew work shift patterns that allow them to pursue their hobbies and interests.
They also enjoy some great discounts as well as thorough training, in addition to access to our free private healthcare scheme after they've been with us for 3 years.
www.mcdonalds.co.uk

© 2006 McDonald's. The Golden Arches and I'm lovin' it logos are trademarks of McDonald's Corporation and affiliates.

McDonalds/PA Photos/Landov

This poster is part of McDonalds' advertising campaign to discredit the negative image implied by the term *McJob*.

McDonaldization The process by which principles of the fast food industry (efficiency, calculability, predictability, and control through technology) are being applied to more sectors of society, particularly the workplace.

The Globalization of Trade and Free Trade Agreements

Just as industrialization and postindustrialization changed the nature of economic life, so has the globalization of trade—the expansion of trade of raw materials, manufactured goods, and agricultural products across national and hemispheric borders. The first set of global trade rules were adopted through the General Agreement on Tariffs and Trade (GATT) in 1947. GATT members met periodically to revise trade agreements in negotiations called "rounds." In 1995, the World Trade Organization (WTO) replaced GATT as the organization overseeing the multilateral trading system.

In the 1980s and early 1990s, U.S. officials began negotiating regional free trade agreements that would open doors to U.S. goods in neighboring countries and reduce the massive U.S. trade deficit, which had grown from $25.3 billion in 1980, to $122 billion in 1985 (Schaeffer 2003). **Free trade agreements** are pacts between two countries or among a group of countries that make it easier to trade goods across national boundaries. Free trade agreements reduce or eliminate foreign restrictions on exports, reduce or eliminate tariffs (or taxes) on imported goods, and prevent U.S. technology from being copied and used by competitors through protection of "intellectual property rights." Treaties such as the Canada–U.S. Free Trade Agreement, the North American Free Trade Agreement (NAFTA), the Free Trade Area of the Americas (FTAA), and the Central American Free Trade Agreement (CAFTA) are designed to accomplish these trade goals.

U.S. officials have also used Section 301 of the Trade Acts of 1984 and 1988 to force trade negotiations with individual countries. If U.S. trade officials determine that other countries have denied U.S. corporations "reasonable" access to domestic markets, sold their goods in the United States at below-market prices, or failed to protect the patents and copyrights of U.S. companies, Section 301 allows the United States to impose retaliatory sanctions and tariffs on goods from these countries (Schaeffer 2003).

Through GATT and the WTO, free trade agreements, and Section 301, U.S. trade officials have expanded trading opportunities, benefiting large export manufacturing and service industries in the global north, specifically aircraft, auto, computer, pharmaceutical, and entertainment industries in western Europe, the United States, and Japan. But trade globalization also hurt the U.S. steel and textile-apparel industries and the workers employed in them, small businesses that cannot compete with large retail chain stores, supermarkets, franchises, and small farmers (Schaeffer 2003). Since NAFTA was signed in 1993, the growth of U.S. exports supported 1 million U.S. jobs, but the growth of imports from Mexico and Canada displaced production that would have supported 2 million jobs. Thus, NAFTA has resulted in the loss of 1 million U.S. jobs, two-thirds of which are in the manufacturing industries (Scott & Ratner 2005).

Foreign workers have also been hurt by trade agreements. Since NAFTA took effect, imports of highly subsidized U.S. and Canadian grain and other agricultural products undercut Mexico's agricultural economy and put over 2 million family farmers out of business, many of whom have come to the United States to find work. After NAFTA, annual illegal immigration from Mexico doubled (Faux 2008). Free trade agreements also undermine the ability of national, state, and local governments to implement environmental and food or product safety policies.

free trade agreements Pacts between two countries or among a group of countries that make it easier to trade goods across national boundaries by reducing or eliminating restrictions on exports and tariffs (or taxes) on imported goods and protecting intellectual property rights.

Transnational Corporations

Although free trade agreements have increased business competition around the world, resulting in lower prices for consumers for some goods, they have also opened markets to monopolies (and higher prices) because they have facilitated the development of large-scale transnational corporations. **Transnational corporations,** also known as *multinational corporations,* are corporations that have their home base in one country and branches, or affiliates, in other countries. The number of transnational corporations more than doubled from about 35,000 in 1990, to about 75,000 in 2005 (Roach 2007). Among the world's largest economies, 29 are companies, rather than countries (Roach 2007).

Transnational corporations provide jobs for U.S. managers, secure profits for U.S. investors, and help the United States compete in the global economy. Transnational corporations benefit from increased access to raw materials, cheap foreign labor, and the avoidance of government regulations.

> By moving production plants abroad, business managers may be able to work foreign employees for long hours under dangerous conditions at low pay, pollute the environment with impunity, and pretty much have their way with local communities. Then the business may be able to ship its goods back to its home country at lower costs and bigger profits. (Caston 1998, pp. 274–275)

Transnational companies can also avoid or reduce tax liabilities by moving their headquarters to a "tax haven." When Halliburton, a U.S.-based multinational corporation with operations in over 120 countries and the Pentagon's largest private contractor operating in Iraq, announced that it was moving its corporate headquarters from Texas to Dubai in 2007, critics accused the company of avoiding U.S. taxes. Indeed, Dubai's tax-free zones have lured about one-quarter of Fortune 500 companies to establish corporate headquarters there (Buncombe 2007).

The savings that big business reaps from cheap labor abroad are not passed on to consumers. "Corporations do not outsource to far-off regions so that U.S. consumers can save money. They outsource in order to increase their margin of profit" (Parenti 2007). For example, shoes made by Indonesian children working 12-hour days for 13 cents an hour cost only $2.60 but still sold for $100 or more in the United States. In 2006, the share of U.S. national income going to wages and salaries was at its *lowest level on record,* with data going back to 1929. In contrast, the share of national income reflected in corporate profits was at its *highest level on record* (Aron-Dine & Shapiro 2007).

Transnational corporations contribute to the trade deficit in that more goods are produced and exported from outside the United States than from within. Transnational corporations also contribute to the budget deficit, because the United States does not get tax income from U.S. corporations abroad, yet transnational corporations pressure the government to protect their foreign interests; as a result, military spending increases. Third, transnational corporations contribute to U.S. unemployment by letting workers in other countries perform labor that U.S. employees could perform. Finally, transnational corporations are implicated in an array of other social problems, such as poverty resulting from fewer jobs, urban decline resulting from factories moving away, and racial and ethnic tensions resulting from competition for jobs.

Sociological Theories of Work and the Economy

In sociology, structural functionalism, conflict theory, and symbolic interactionism serve as theoretical lenses through which we may better understand work and economic issues and activities.

transnational corporations
Also known as multinational corporations, corporations that have their home base in one country and branches, or affiliates, in other countries.

Structural-Functionalist Perspective

According to the structural-functionalist perspective, the economic institution is one of the most important of all social institutions. By providing the basic necessities common to all human societies, including food, clothing, and shelter, the economic institution contributes to social stability. After the basic survival needs of a society are met, surplus materials and wealth may be allocated to other social uses, such as maintaining military protection from enemies, supporting political and religious leaders, providing formal education, supporting an expanding population, and providing entertainment and recreational activities. Societal development is dependent on an economic surplus in a society (Lenski & Lenski 1987).

In light of the growing numbers of people needing food assistance, Mayor Rosemarie Vasquez of Montebello, California canceled the city's Fourth of July fireworks show, and instead, donated the money to a nonprofit food distribution organization.

The economic institution can also be dysfunctional when it fails to provide members with the goods and services they need, when the distribution of goods and services is grossly unequal, and when the production, distribution, and consumption of goods and services depletes and pollutes the environment.

The structural-functionalist perspective is also concerned with how changes in one aspect of society affect other aspects. For example, when unemployment rates rise, college enrollments go up, crime increases, and tax revenues decrease (unemployed people pay less in income tax and sales tax), which hurts the government's ability to pay for services such as education, garbage pickup, police and fire services, and road repairs.

What Do You Think? In 2009, nearly 50 U.S. cash-strapped cities canceled their annual Fourth of July fireworks. Mayor Bill Cervenik, of Euclid, Ohio said, "It came down to this: Did we want to spend $150,000 on something that would be over in a few hours? Or did we want to use that money to keep city workers employed?" In the Los Angeles suburb of Montebello, the City Council voted to use its $39,000 fireworks budget on donations to local food banks. Mayor Rosemarie Vasquez explained, "We figured that, instead of burning the money in the air, why not give it to people who need it?" (Huffstutter 2009). Protestors against the cancellation of fireworks shows argued that they are an important American tradition. If you were on the City Council, and the issue of whether or not to cancel the Fourth of July fireworks was being considered, what would your vote be?

Conflict Perspective

According to the conflict perspective, the ruling class controls the economic system for its own benefit and exploits and oppresses the working masses. The conflict perspective is critical of ways that the government caters to the interests of big business at the expense of workers, consumers, and the public interest.

This system of government that serves the interests of corporations—known as **corporatocracy**—involves ties between government and business. For example, in Chapter 2, we discussed how the pharmaceutical and health insurance industries influence politicians on matters related to health care.

Corporate interests find their way into politics through large political contributions and "soft money," which is money that flows through a loophole to provide political parties, candidates, and contributors a means to evade federal limits on political contributions. Critics of this system of campaign financing argue that corporations and interest groups purchase political influence through financial contributions and lobbying. For example, in 2008, Congress passed the Emergency Economic Stabilization Act, which created a $700 billion Troubled Assets Relief Program (TARP), allowing the government to purchase failing bank assets that resulted largely from the subprime mortgage crisis. Companies that received taxpayers' money from the TARP bailout program had spent $77 million on lobbying and $37 million on federal campaign contributions in 2008. Companies that spent the most on lobbying and political campaign contributions, including General Motors, Bank of America, and American International Group (AIG) are also the ones that received the most bailout money (Center for Responsive Politics 2009).

A survey of business leaders' views on political fund-raising found that the main reasons U.S. corporations make political contributions are fear of retribution and to buy access to lawmakers (*Multinational Monitor* 2000). Although 75 percent of the surveyed business leaders said that political donations give them an advantage in shaping legislation, nearly three-quarters (74 percent) said that business leaders are pressured to make large political donations. Half of the executives said that their colleagues "fear adverse consequences for themselves or their industry if they turn down requests" for contributions.

The pervasive influence of corporate power in government exists worldwide. The policies of the International Monetary Fund (IMF) and the World Bank pressure developing countries to open their economies to foreign corporations, promoting export production at the expense of local consumption, encouraging the exploitation of labor as a means of attracting foreign investment, and hastening the degradation of natural resources as countries sell their forests and minerals to earn money to pay back loans. In his book *Confessions of an Economic Hit Man,* John Perkins (2004) described his prior job as an "economic hit man"—a highly paid professional who would convince leaders of poor countries to accept huge loans (primarily from the World Bank) that were much bigger than the country could possibly repay. The loans would be used to help develop the country by paying for needed infrastructure, such as roads, electrical plants, airports, shipping ports, and industrial plants. One of the conditions of the loan was that the borrowing country had to give 90 percent of the loan back to U.S. companies (such as Halliburton or Bechtel) to build the infrastructure. The result: The wealthiest families in the country benefit from additional infrastructure and the poor masses are stuck with a debt they cannot repay. The United States uses the debt as leverage to ask for "favors," such as land for a military base or access to natural resources such as oil. According to Perkins, large corporations want "control over the entire world and its resources, along with a military that enforces that control" (quoted by MacEnulty 2005, p. 10).

Symbolic Interactionist Perspective

According to symbolic interactionism, the work role is a central part of a person's self-concept and social identity. When making a new social acquaintance,

corporatocracy A system of government that serves the interests of corporations and that involves ties between government and business.

During much of the 1990s, the U.S. economy flourished—incomes rose, unemployment and poverty rates fell, and stock market values soared. In the early 2000s, these economic trends changed and unemployment rates went back up. But unlike earlier job loss patterns of the 1980s that largely affected blue-collar workers, more recent job losses have increasingly involved middle-class professionals. Research presented here investigates how job loss affected the self-concepts and views about employment among a sample of unemployed professionals in midlife (Mendenhall et al. 2008).

Sample and Methods

The sample consisted of 77 men and women who were recruited through two Chicago-area networking groups that help unemployed managers and executives find employment, and also through announcements at local churches and posted flyers at cafés and public libraries. Participants had to meet four eligibility criteria: they had to (1) have been unemployed for at least three months during the past year; (2) have been married at the time of their job loss; (3) have children between the ages of 12 and 18 living at home; and (4) live in the Chicago area. Most participants were male (83 percent) and white (80 percent). Respondents had been unemployed for an average of 15 months at the time of their first interview. More than half had earned annual salaries of $100,000 or more; no one earned less than $50,000, and most had jobs with generous benefits.

The researchers interviewed participants at libraries, cafés, or on the University of Chicago campus. Interview topics included the job loss event, how the job loss affected family relationships, and educational plans for their children. Participants also completed a survey about their job loss experience, family economic circumstances, family relationships, and their own health and well-being. About a year later, participants were interviewed again, and completed another questionnaire.

Selected Findings

This study revealed some interesting patterns in how job loss among professionals in midlife (1) affected their self-concept, (2) influenced their job-seeking strategies, and (3) shaped the messages about employment they conveyed to their children. The participants viewed their job termination as evidence of a lack of employer loyalty and a change from a lifetime employment contract to one in which even high-level employees can be terminated without warning—a shift in which participants came to view themselves as free agents who effectively "rent" their services to employers. As free agents, there is no expectation of permanent employment or loyalty in the employee-employer relationship.

More than half of the participants 50 years and older said they perceived age discrimination in the job market, in that employers signaled that they were looking for entry-level employees, which is code for "young." In response, participants often de-emphasized their age and experience by omitting graduation dates and some of their work history on their resumes—a phenomenon known as "deprofessionalization." A former CEO explained that, in the first several months of job searching, he listed the date of his college graduation on his resume and got no responses. Then he deleted the date, and got half a dozen responses.

Many participants viewed their job loss experience as providing an opportunity to teach real-life lessons to their children about the world of work. A 50-year-old project manager said,

"I think [that my son getting a good glimpse into the reality of life is] a positive because . . . if he goes out there with rose-tinted glasses, he's going to get smacked upside the head real hard someday." (p. 203)

One 47-year-old information technology consultant used his job loss experience to teach his children to prepare for a job market in which neither employee nor employer expects a lifetime employment contract. He told his children,

"For most jobs now, you need to view [it] like the movie industry where . . . you're rented . . . you're contracted to do that movie, whether you're the lights or the cameraman or whatever, and then you're out of work again. And that's more the way most jobs are now where you should view a job as just your job until you're out of work again." (p. 204)

Respondents advised their children to take specific steps to prevent being overly dependent on employers. Some advised their children to own their own businesses; others urged their children to develop skills that could be "transferable" to a new employer. Others encouraged their children to choose careers where their potential client base was diversified:

"I've started talking to them about how . . . they should start thinking about [whether] they want to work for big companies or . . . go into a field where their client base is very much diversified, which is lawyers, doctors, CPAs, therapists. . . . Because then if someone fires you, you don't care because you have another 100 [clients] whereas when you work for a company, and . . . your boss . . . fires you . . . you're out of a job." (p. 204)

Source: Mendenhall et al. 2008.

one of the first questions we usually ask is, "What do you do?" The answer largely defines for us who that person is. An individual's occupation is one of the person's most important statuses; for many, it represents a "master status," that is, the most significant status in a person's social identity. This chapter's Social Problems Research Up Close feature describes a study that looks at how job loss of white-collar professionals in midlife affects their self-concepts and attitudes about work and unemployment.

Symbolic interactionism emphasizes the fact that attitudes and behavior are influenced by interaction with others. The applications of symbolic interactionism in the workplace are numerous: Employers and managers use interpersonal interaction techniques to elicit the attitudes and behaviors they want from their employees; union organizers use interpersonal interaction techniques to persuade workers to unionize. And, as noted in the Social Problems Research Up Close feature, parents teach their young adult children important lessons about work and unemployment through interaction with them.

Problems of Work and Unemployment

In this section, we examine unemployment and other problems associated with work. Poverty, minimum wage and living wage issues, and workplace discrimination are discussed in other chapters. Here, we discuss problems concerning unemployment and underemployment, employment and retirement concerns of older Americans, child labor, forced labor, sweatshop labor, health and safety hazards in the workplace, job dissatisfaction and alienation, work-family concerns, and labor unions and the struggle for workers' rights.

Figure 7.2 U.S. Unemployment Rates by Race and Hispanic Origin: August 2009
Source: Bureau of Labor Statistics 2009a.

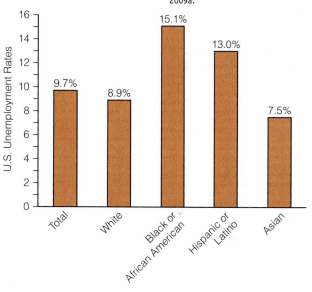

Unemployment and Underemployment

In 2008, an estimated 190 million people worldwide—6 percent of the global labor force—were unemployed. The highest rates of unemployment are in the Middle East and North Africa (ILO 2009).

Measures of **unemployment** in the United States consider individuals to be unemployed if they are currently without employment, are actively seeking employment, and are available for employment. In 2000, the U.S. unemployment rate dipped to a 31-year low of 4 percent. During the economic recession that began in 2007, mass layoffs pushed the unemployment rate to 9.7 in mid-2009 as companies went out of business and plants closed. A **recession** refers to a significant decline in economic activity spread across the economy and lasting for at least 6 months. Economists predicted that nearly one in 10 U.S. workers would be unemployed before the job market improved (Hagenbaugh 2009). In communities hardest hit by the recession, unemployment rates were over 20 percent. Rates of unemployment are higher among racial and ethnic minorities (see Figure 7.2) and among those with lower levels of education (see Chapter 8).

unemployment To be currently without employment, actively seeking employment, and available for employment, according to U.S. measures of unemployment.

recession A significant decline in economic activity spread across the economy and lasting for at least 6 months.

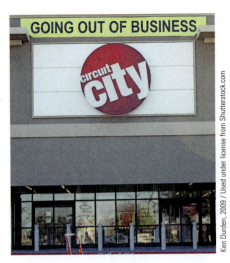

GOING OUT OF BUSINESS

When Circuit City closed its 567 stores nationwide, some 30,000 employees lost their jobs.

job exportation The relocation of jobs to other countries where products can be produced more cheaply.

outsourcing A practice in which a business subcontracts with a third party to provide business services.

automation The replacement of human labor with machinery and equipment.

long-term unemployment rate The share of the unemployed who have been out of work for 27 weeks or more.

underemployment Unemployed workers as well as (1) those working part-time but who wish to work full-time, (2) those who want to work but have been discouraged from searching by their lack of success, and (3) others who are neither working nor seeking work but who want and are available to work and have looked for employment in the last year. Also refers to the employment of workers with high skills and/or educational attainment working in low-skill or low-wage jobs.

The causes of unemployment are varied and complex. One simple but primary cause is lack of available jobs. In May 2009, there were nearly six unemployed U.S. workers for every available job (Shierholz 2009). Another cause of U.S. unemployment is **job exportation,** or the relocation of jobs to other countries where products can be produced more cheaply. **Outsourcing**, which involves a business subcontracting with a third party to provide business services, saves companies money as they pay lower salaries and no benefits to those who provide outsourced services. Many commonly outsourced jobs, including accounting, web development, information technology, telemarketing, and customer support, are outsourced to non-U.S. workers. **Automation,** or the replacement of human labor with machinery and equipment, also contributes to unemployment. Another cause of unemployment is increased global and domestic competition. Mass layoffs in the U.S. automobile industry have occurred, in part, due to competition from makers of foreign cars who, unlike U.S. automakers, do not have the burden of providing health insurance for their employees. Finally, unemployment results from mass layoffs that occur when plants close and companies downsize or go out of business. In 2009, Circuit City went out of business after revenues steadily declined due, in part, to competition from Best Buy and Wal-Mart. Circuit City was also hurt by the recession that started in late 2007, as consumers cut back on spending.

Long-Term Unemployment. The **long-term unemployment rate** refers to the share of the unemployed who have been out of work for 27 weeks or more. In mid-2009, 4.4 million Americans had been jobless for 27 weeks or more, which translates into 3 in 10 unemployed individuals experiencing long-term unemployment (Bureau of Labor Statistics 2009a).

Underemployment. Unemployment figures do not include "discouraged" workers, who have given up on finding a job and are no longer looking for employment. **Underemployment** is a broader term that includes unemployed workers as well as (1) those working part-time but who wish to work full-time ("involuntary" part-timers), (2) those who want to work but have been discouraged from searching by their lack of success ("discouraged" workers), and (3) others who are neither working nor seeking work but who indicate that they want and are available to work and have looked for employment in the last 12 months. The underemployment rate is always higher than the unemployment rate, which means that official unemployment rates—frequently reported in the media—undercount those whose employment needs are not being met. For example, although the official unemployment rate in June 2009 was 9.5, the underemployment rate was 16.5 (Bureau of Labor Statistics 2009b).

Underemployment also refers to the employment of workers with high skills and/or educational attainment working in low-skill or low-wage jobs, such as a college graduate mowing lawns, a physician driving a taxi, or an electrician working at a fast-food restaurant. This chapter's Human Side feature portrays a man who went from being a successful hedge fund manager to being a pizza delivery man.

Effects of Unemployment on Individuals, Families, and Societies. Unemployment has been linked to depression, low-self esteem, and increased mortality rates (Turner & Irons 2009). One study found that, among workers with no preexisting health conditions, losing a job due to a business

For the first 45 years of Ken Karpman's life, everything was close to perfect. He graduated from UCLA with a bachelor's degree and MBA, then got a high-paying job as an institutional equity sales trader. He married his dream girl, had two children, and traveled the world on expensive vacations. Over the span of Karpman's impressive 20-year career as a trader, he climbed the company ladder, reaching a salary of $750,000 a year.

"Life was good, we were making a lot of money—and why wouldn't this just continue on?" Karpman said. From all appearances, Ken and Stephanie Karpman were living the American dream in Tampa, Florida, nestled in their 4,000-square-foot home that sits on a golf course. "I had no idea what anything cost in a store," he said. "I'd just put it in the cart and buy." Karpman was so confident in his good fortune and the strong economy that he left his job in 2005, to start his own hedge fund. To pay for the new business and their standard of living, Karpman quickly burned through $500,000 in savings and, like so many Americans, took a line of credit against his house. But in the reversal of fortune that followed, Karpman was unable to attract investors and was forced to dissolve his hedge fund. He found himself jobless in a job market that had collapsed.

In the past, Karpman had found it easy to get a job. It wasn't so this time around. "When I used to go into a job interview, I probably came across as a jerk because I was like interviewing him to see whether this firm was worthy of me," he said. "Now it's kind of like you almost feel like you're coming in with your hat in your hand."

After a lengthy and fruitless job search, the Karpmans were shocked to find themselves in financial dire straits, with zero savings, hundreds of thousands of dollars in debt, and their home in foreclosure. Desperate for quick cash, Karpman tried to find a job bartending but came up empty. Finally, he drove his Mercedes to Mike's Pizza & Deli Station in Clearwater and applied for a job. Mike Dodaro, the owner of the pizza shop, said he was shocked when he read his application, but he offered him the job despite some reluctance to hire an overqualified candidate. Stephanie Karpman said she was more than a little surprised when he came home with his new job, initially saying, "You're kidding me, right? Delivering pizzas," she said. "Never in my wildest dreams did I think he'd be doing that."

Karpman's salary plummeted from six figures to $7.29 an hour—plus tips—but it's money that he's grateful to earn, even when it means delivering to neighbors or his old office building. "This whole progression down, it's amazing how many things you say, 'I can't do' and a week later you say, 'Yeah, I could do that,'" he said. "I'm not going to make a career out of this but, until I get something that pays more, this is what I'll do to keep food on the table."

The stress has also taken a toll on their marriage. Stephanie Karpman said she didn't want her husband to leave his trader job in the first place, and wishes he would have put more in savings. "There's no question of where the fault lies," Ken Karpman said. And when it comes to finger-pointing, "I point it in my direction."

Each day has brought new lows and new lessons in living with a little less "stuff" and a lot more humility. "The worst thing for me, for both of us probably, was, you know, to go to just friends around here, and say, 'Can I borrow some money?'" he said. "Pizza was a step up." The Karpmans are now on food stamps and a tight budget that doesn't nearly cover their children's $30,000 private school tuition. But thanks to an anonymous donor, the Karpmans' children's tuition has been covered through next year and they are deeply appreciative. "It's just something that kind of makes you a little misty every time you think about it, that somebody would do this for our kids," he said. "But we'll have a chance at some point to do that for another family."

The family's jet skis now collect dust in the garage near the Mercedes, with its broken transmission they cannot fix. The home they will soon lose has fallen into disrepair. Stephanie Karpman has closets full of clothes and handbags she likely won't be able to take with her and is eyeing consignment shops as a place to unload them. She said she has found herself going through her closet and wearing clothes she hasn't touched in years.

As Karpman counts every penny he earns, he still hopes he can come back from the financial brink and reclaim a lifestyle he, like so many Americans, never imagined he could lose.

Source: Hunter & Deutsch 2009.

closure increased the risk for a new health problem by 85 percent, with the most common health problems including hypertension, heart disease, and arthritis (Strully 2009). For workers who receive health insurance through their employer, losing a job can mean losing health insurance coverage (see also Chapter 2).

Long-term unemployment can have lasting effects, such as increased debt, diminished retirement and savings accounts (which are depleted to meet living expenses), home foreclosure, and/or relocation from secure housing and communities to unfamiliar places to find a job. But, even when individuals who are laid off find another job in one or two weeks, they still suffer damage to their self-esteem from having been told they are no longer wanted or needed at their workplace. And being fired affects worker trust and loyalty in future jobs. Employees who are not fired during a mass layoff are also affected, as they worry that "their job could be next" (Uchitelle 2006).

Unemployment is a risk factor for homelessness, substance abuse, and crime, as some unemployed individuals turn to illegitimate, criminal sources of income, such as theft, drug dealing, and prostitution. Globally, the high numbers of young adults without jobs create a risk for crime, violence, and political conflict (United Nations 2005). As discussed in other chapters, unemployed young adults are targeted for recruitment into terrorist groups.

In families, unemployment is also a risk factor for child and spousal abuse and marital instability. When an adult is unemployed, other family members are often compelled to work more hours to keep the family afloat. And unemployed noncustodial parents, usually fathers, fall behind on their child support payments.

Plant closings and large-scale layoffs affect communities by lowering property values and depressing community living standards. High numbers of unemployed adults create a drain on societies that provide support to those without jobs. And early and mid-career job displacement is linked to lower rates of social participation in church groups, charitable organizations, youth and community groups, and civic and neighborhood groups (Brand & Burgard 2008).

Unemployment and underemployment create a vicious cycle: The un- and underemployed (as well as those who fear job loss) cut back on spending, which hurts businesses that then must cut jobs to stay afloat. During times of economic stress or uncertainty, have you cut back on everyday spending? See this chapter's Self and Society feature.

Employment and Retirement Concerns of Older Americans

In the recent economic crisis, the stock market took a huge hit, losing more than half its value between September 30, 2007, and March 6, 2009 (Soto 2009). These losses decimated the retirement savings of older Americans. Many workers who were planning to retire could no longer afford to stop working. Others who had recently retired and then lost a lot of money in the market felt compelled to reenter the labor force. And others who had no retirement savings and could not get by on Social Security alone stayed in the work force. But, with more older people needing employment during a time of job scarcity, the unemployment rate for adults age 65 and older reached a record 6.8 percent in February 2009, the highest level recorded since the federal government began computing unemployment rates in 1948 (Johnson & Mommaerts 2009). Aside from how the recent economic recession affected older Americans, two ongoing concerns for older Americans are age discrimination and Social Security.

Age Discrimination in the Workplace. One of the obstacles older workers face in finding and keeping employment is age discrimination. Older workers may be more vulnerable to being "let go" because, although they have seniority on the job, they may also have higher salaries, and businesses that need to cut payroll expenses can save more money by letting higher-salaried personnel go. Prospective employers may view older job applicants as "too qualified" for

During the recent economic crisis, Americans bought fewer big-ticket items such as computers and new cars. In addition, many Americans cut back on their everyday spending. Difficult economic times make people reconsider what is important for them to have in their daily lives, and what they can live without or change to save money. How do your spending habits change in hard economic times? For each of the following spending categories, indicate how your spending has changed in response to tough economic times.

	Have Done	Have Considered	Have Not Done or Considered	n/a
1. Buy more generic brands	___	___	___	___
2. Take lunch from home instead of buying lunch out	___	___	___	___
3. Go to hairdresser/barber less often	___	___	___	___
4. Stopped purchasing bottled water; switched to refillable water bottle	___	___	___	___
5. Canceled one or more magazine subscriptions	___	___	___	___
6. Cut down on dry cleaning	___	___	___	___
7. Stopped buying coffee in the morning	___	___	___	___
8. Canceled or cut back on cable service	___	___	___	___
9. Canceled a newspaper subscription	___	___	___	___
10. Begun carpooling or using mass transit	___	___	___	___
11. Changed or canceled cell phone service	___	___	___	___
12. Canceled landline phone service and only use cell phone	___	___	___	___

Comparison Data: Table 1 presents the percentages of responses to each of the 12 previous items from a 2009 online Harris Poll of 2,177 U.S. adults. The most common changes in spending behavior include purchasing more generic brands and taking lunch from home instead of buying lunch out.

Table 1 Spending/Savings Over Past Six Months, in Percentages

"Have you done or considered doing any of the following over the past six months in order to save money?"

	HAVE DONE	HAVE CONSIDERED	HAVE NOT DONE OR CONSIDERED	N/A
1.	62	14	18	6
2.	47	8	13	32
3.	36	9	31	24
4.	33	11	23	33
5.	29	7	24	40
6.	20	4	18	59
7.	19	5	20	56
8.	17	23	46	15
9.	15	9	31	45
10.	14	8	34	44
11.	13	17	54	16
12	11	21	50	17

Source: Harris Poll 2009.

entry-level positions, less productive than younger workers, and/or more likely to have health problems that could affect not only their productivity, but the cost of employer-based group insurance premiums. In 1967, Congress passed the Age Discrimination in Employment Act (ADEA), which was designed to ensure continued employment for people between the ages of 40 and 65. In 1986, the upper age limit was removed, making mandatory retirement illegal in most occupations. Under ADEA, it is illegal to discriminate against people because of their age with respect to hiring, firing, promotion, layoff, compensation, benefits, job assignments, and training. Nevertheless, thousands of age discrimination cases are filed with the Equal Employment Opportunity Commission (EEOC) annually.

Social Security. Social Security, actually titled "Old Age, Survivors, Disability, and Health Insurance," protects against loss of income due to retirement, disability, or death. The current retirement age for collecting Social Security is 66, but workers can claim reduced benefits beginning at age 62.

When Social Security was established in 1935, it was not intended to be a person's sole economic support in old age; rather, it was meant to supplement other savings and assets. But Social Security is the major source of family income for nearly half of older Americans, and keeps many Americans out of poverty. Social Security may be even more important for future retirees, as fewer workers are receiving pensions from their employers. In 2008, 29 percent of employers (with 50 or more employees) offered pension plans, down from 48 percent in 1998 (Galinsky et al. 2008).

Because Social Security payments are based on the number of years of paid work and preretirement earnings, women and minorities, who often earn less during their employment years, receive less in retirement benefits. As the population ages, more older retired Americans are receiving Social Security and fewer younger workers are paying taxes to fund Social Security. Concerns about Social Security running out of money have led to a number of proposals, including privatizing Social Security, cutting benefits, raising the age at which a person can collect Social Security, and raising tax revenues to fund Social Security.

Employment Concerns of Recent College Grads

Most college students seek a college degree to improve their job opportunities. But having a college degree is no guarantee of employment in a struggling economy. The unemployment rate for young college graduates (adults under age 27 with a bachelor's degree or higher) was 5.9 percent in 2009—the highest since 1983 (Edwards 2009).

Two-thirds of college seniors worry about finding a job after graduation, and 40 percent of seniors say they expect to need financial help from their parents after college (Pugh 2009). Due to the high unemployment rate, high cost of living, and college debts averaging $20,000 per graduate, many college graduates end up living with their parents instead of on their own. There are 5 million "boomerang" kids in the United States—young adults aged between 25 and 35 who live with their parents (Brown 2009). Some new college graduates are glad that jobs are scarce. After graduation, some new grads are taking time off to relax and to travel instead of seeking employment (Weiss 2009). Other college grads question whether going to college was worth the expense and the effort. One survey of 4 million college graduates found that less than half were working in jobs that required a college degree (Pugh 2009).

Forced Labor and Slavery

Forced labor, also known as *slavery,* refers to any work that is performed under the threat of punishment and is undertaken involuntarily. There are more slaves in the world today than at any other time in history—27 million (Skinner 2008). A resurgence of slavery around the world is linked to three main factors: (1) rapid growth in population, especially in the developing world; (2) social and economic changes that have displaced many rural dwellers to urban centers and their outskirts, where people are powerless and jobless and are vulnerable to exploitation and slavery; and (3) government corruption that allows slavery to go unpunished, even though it is illegal in every country (Bales 1999).

Forced labor exists all over the world but is most prevalent in India, Pakistan, Bangladesh, and Nepal. Most forced laborers work in agriculture, mining, prostitution, and factories. Forced laborers produce goods we use every day, including sugar from the Dominican Republic, chocolate from the Ivory Coast, paper clips from China, carpets from Nepal, toys from China, and clothing from India.

> There are more slaves in the world today than at any other time in history.

Forms of Slavery and Forced Labor. The form of slavery most people are familiar with is **chattel slavery,** in which slaves are considered property that can be bought and sold. In the past, the high cost of purchasing a slave (about $40,000 in today's money) gave the master incentive to provide a minimum standard of care to ensure that the slave would be healthy enough to work and generate profit for the long term. Today, slaves are cheap, costing an average of $100 or less. In Port-au-Prince, Haiti, a 10-year-old girl can be bought for $50.00 (Skinner 2008). Because they are so cheap and abundant, slaves are no longer a major investment worth maintaining. If slaves become ill or injured, too old to work, or troublesome to the slaveholder, they are dumped or killed and replaced with another slave (Cernasky 2003).

Although chattel slavery still exists in some areas, most forced laborers today are not "owned" but are rather controlled by violence, the threat of violence, and/or debt. The most common form of forced labor today is called *bonded labor.* Bonded laborers are usually illiterate, landless, rural, poor individuals who take out a loan simply to survive or to pay for a wedding, funeral, medicines, fertilizer, or other necessities. Debtors must work for the creditor to pay back the loan, but often they are unable to repay it. Creditors can keep debtors in bondage indefinitely by charging the debtors illegal fines (for workplace "violations" or for poorly performed work) or charge laborers for food, tools, and transportation to the work site while keeping wages too low for the debt to ever be repaid. Alternatively, creditors can claim that all the labor the debtor performs is collateral for the debt and cannot be used to reduce it (Miers 2003).

Another common form of forced labor involves luring individuals with the promise of a good job and instead holding them captive and forcing them to work. Migrant workers are particularly vulnerable because, if they try to escape and report their abuse, they risk deportation. Organized crime rings are sometimes involved in the international trafficking of human beings, which often flows from developing nations to the West. A form of forced labor most common in South Asia is sex slavery, in which girls are forced into prostitution by their own husbands, fathers, and brothers to earn money to pay family debts. Other girls are lured by offers of good jobs and then are forced to work in brothels under the threat of violence.

forced labor Also known as slavery, any work that is performed under the threat of punishment and is undertaken involuntarily.

chattel slavery A form of slavery in which slaves are considered property that can be bought and sold.

Forced prison labor is a type of forced labor that is controlled by the state. Forced prison labor is particularly widespread in China.

Mark Peterson/CORBIS

Another type of forced labor is conducted by the state or military. Forced military service, which has been reported in parts of Africa, and forced prison labor (common in China) are examples of state and military forced labor.

Forced Labor in the United States. Each year, 14,000 to 17,000 people are trafficked into the United States and forced into slavery, most commonly in domestic work, farm labor, and the sex industry (Skinner 2008). Migrant workers are tricked into working for little or no pay as a means of repaying debts from their transport across the U.S. border, similar to debt bondage in South Asia. Traffickers posing as employment agents lure women into the United States with the promise of good jobs and education but then place them in "jobs" where they are forced to do domestic or sex work.

Sweatshop Labor

A U.S. Department of Labor investigation of the Daewoosa Samoa garment factory in American Samoa—a factory that produces men's sportswear for JCPenney—found that garment factory workers lived and worked under conditions of poor sanitation, malnutrition, electrical hazards, fire hazards, machinery hazards, illegally low wages, sexual harassment and invasion of privacy, workplace violence and corporal punishment, and overcrowded barracks in which two workers were forced to share each bed (National Labor Committee 2001). Female workers reported that the company owner routinely entered their barracks to watch them shower and dress. Workers reported incidents in which security guards slapped and kicked workers. The food provided to the workers at the Daewoosa Samoa garment factory consisted of a watery broth of rice and cabbage.

The workers at the Daewoosa Samoa garment factory are among the millions of people worldwide who work in **sweatshops**—work environments that are characterized by less-than-minimum wage pay, excessively long hours of work (often

sweatshops Work environments that are characterized by less-than-minimum wage pay, excessively long hours of work (often without overtime pay), unsafe or inhumane working conditions, abusive treatment of workers by employers, and/or the lack of worker organizations aimed to negotiate better working conditions.

Sweatshop labor commonly occurs in the garment industry.

Guenter Stand/ VISUM/ The Image Works

without overtime pay), unsafe or inhumane working conditions, abusive treatment of workers by employers, and/or the lack of worker organizations aimed at negotiating better working conditions. Sweatshop labor conditions occur in a wide variety of industries, including garment production, manufacturing, mining, and agriculture.

At a U.S. Senate panel hearing, members of a Senate subcommittee heard testimony from those who witnessed sweatshop conditions at overseas factories that produce goods for U.S. companies (Tate 2007). A worker in Colombia's flower industry, who was employed at a plant owned by the U.S. company Dole, described workers being exposed to hazardous pesticides, the firing of sick workers, forced pregnancy testing for women, and strong-arm union-busting tactics by companies.

Many products in the U.S. consumer market are made under sweatshop conditions. An investigative report on working conditions in five Chinese factories that produce products for Disney, Wal-Mart, Kmart, Mattel, and McDonald's revealed sweatshop conditions that violate Chinese labor laws (Students and Scholars Against Corporate Misbehavior 2005). Workers are forced to work grueling 12- to 15-hour days, earning just 33 cents to 41 cents an hour. In some factories, women are denied their legal maternity rights. Workers are housed in overcrowded dorm rooms and fed horrible food at the factory canteen. Workers are charged for the housing and food provided at the factory (even if they live and eat elsewhere), which often costs them one-fifth to one-third of their monthly wages. Some factories have no fans and become oppressively hot. Workers often faint from exhaustion and the unbearably stifling heat. Some workers are exposed to strong-smelling gases from working with glue, with no protective masks or ventilation system. Crushed fingers and other injuries are common in some factory departments. Workers have no health insurance, no pension, and no right to freedom of association or to organize.

Sweatshop Labor in the United States. Sweatshop conditions in overseas industries have been widely publicized. However, many Americans do not realize the extent to which sweatshops exist in the United States. The Department of Labor estimates that more than half of the country's 22,000 sewing shops violate minimum wage and overtime laws, and that 75 percent violate safety and health laws ("The Garment Industry" 2001). Most garment workers in the United States are immigrant women who typically work 60 to 80 hours a week, often earning less than minimum wage with no overtime, and many face verbal and physical abuse.

Immigrant farm workers, who process 85 percent of the fruits and vegetables grown in the United States, also work under sweatshop conditions. Many live in substandard and crowded housing provided by their employer and lack access to safe drinking water as well as bathing and sanitary toilet facilities. Farm workers commonly suffer from heat exhaustion, back and muscle strains, injuries resulting from the use of sharp and heavy farm equipment, and illness resulting from pesticide exposure (Austin 2002). Working 12-hour days under hazardous conditions, farm workers have the lowest annual family incomes of any U.S. wage and salary workers, and more than 60 percent of them live in poverty (Thompson 2002). Problems associated with immigrant labor are discussed further in Chapter 9.

Child Labor

Child labor involves a child performing work that is hazardous, that interferes with a child's education, or that harms a child's health or physical, mental, social, or moral

This eleven year-old child laborer in Bangladesh works in a metal parts factory earning the equivalent of $17.00 per month.

development. Even though virtually every country in the world has laws that prohibit or limit the extent to which children can be employed, child labor persists throughout the world. According to the most recent data available, there are an estimated 218 million school-age child laborers throughout the world—a number that may have increased since the global economic crisis that began in 2007 (ILO 2006).

Child laborers work in factories, workshops, construction sites, mines, quarries, fields, and on fishing boats. Tens of thousands of children in at least 24 countries and territories are recruited by armed forces where children are used in combat or for sexual exploitation (UNICEF 2009). Child laborers make bricks, shoes, soccer balls, fireworks and matches, furniture, toys, rugs, and clothing. They work in the manufacturing of brass, leather goods, and glass. They tend livestock and pick crops. They work long hours with no breaks and no days off, often in unsafe conditions where they are exposed to toxic chemicals and/or excessive heat, and they endure beatings and other forms of mistreatment from their employers, all for as little as a dollar a day. A former textile worker in Bangladesh described conditions at a company called Harvest Rich, where clothing is sewn for the U.S. firms including Wal-Mart and JCPenney. She testified that hundreds of children, some as young as 11 years old, were illegally working at Harvest Rich, sometimes for up to 20 hours a day. "Before clothing shipments had to leave for the United States, there are often mandatory

child labor Involves a child performing work that is hazardous, that interferes with a child's education, or that harms a child's health or physical, mental, social, or moral development.

19- to 20-hour shifts from 8:00 a.m. to 3:00 or 4:00 a.m. . . . The workers would sleep on the factory floor for a few hours before getting up for their next shift in the morning. If they did anything wrong, they were beaten every day." Workers had two days off a month and were paid $3.20 a week (Tate 2007). At another garment factory that made fleece jackets for Wal-Mart, 14- or 15-year-old kids worked 18- or 20-hour shifts, from 8:00 a.m. to midnight or 4:00 a.m., seven days a week. When they passed out, rulers struck them to wake them up. Some of the girls were raped by management (Tate 2007).

Illegal and oppressive employment of children also occurs in the United States in restaurants, grocery stores, meatpacking plants, garment factories, and agriculture. Between 300,000 and 800,000 U.S. child workers labor on commercial farms, often working 12-hour days and suffering from exposure to dangerous pesticides. They frequently work in 100 degree temperatures without adequate access to drinking water, toilets, or hand-washing facilities (Human Rights Watch 2003). Despite federal prohibitions, even youth employed in service and retail jobs are exposed to harmful conditions and dangerous equipment (e.g., paper balers, box crushers, and dough mixers) (Runyan et al. 2007).

Health and Safety Hazards in the U.S. Workplace

Many workplaces are safer today than in generations past. Nevertheless, fatal and disabling occupational injuries and illnesses still occur in troubling numbers. In 2008, 5,071 U.S. workers—most of whom were men—died of fatal work-related injuries (Bureau of Labor Statistics 2009c). The most common type of job-related fatality involves transportation accidents (see Figure 7.3). Industries with the highest rates of fatal injuries include agriculture/forestry/fishing and hunting, mining, transportation, and construction. Nonfatal occupational injuries and illnesses in U.S. workplaces are not uncommon—about 4 million cases were recorded in 2007 (Bureau of Labor Statistics 2009d). Sprains and strains are the most common nonfatal occupational injury or illness involving days away from work. Occupations that have the highest rates of nonfatal injuries and illnesses include nursing aides, orderlies, and attendants, who commonly experience back strains from lifting and moving patients.

The incidence of illnesses resulting from hazardous working conditions is probably much higher than the reported statistics show, because long-term latent

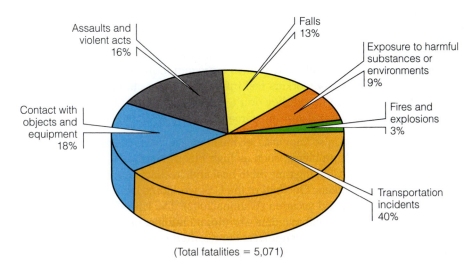

Assaults and violent acts 16%

Contact with objects and equipment 18%

Falls 13%

Exposure to harmful substances or environments 9%

Fires and explosions 3%

Transportation incidents 40%

(Total fatalities = 5,071)

Figure 7.3 Causes of Workplace Fatalities, 2008
Source: Bureau of Labor Statistics 2009c.

illnesses caused by, for example, exposure to carcinogens often are difficult to relate to the workplace and are not adequately recognized and reported. In addition, employees don't always report workplace injuries or illnesses for fear of losing their jobs, or in the case of undocumented immigrants, being deported. And companies don't always maintain accurate records of workplace injuries and illnesses, to avoid scrutiny and fines for possible violations of health and safety regulations.

In many cases, workplace injuries, illnesses, and deaths are due to employers' willful violations of health and safety regulations. The Sago Mine in West Virginia was cited 273 times for safety violations in the two years before an explosion there killed 12 miners in 2006 (Greenhouse 2008). Many of the fines were only $60, and none were more than $460.

What Do You Think? Suppose that a corporation is guilty of a serious violation of health and safety laws, in which "serious violation" is defined as one that poses a substantial probability of death or serious physical harm to workers. What penalty do you think that corporations should pay for such a violation? According to a report by the AFL-CIO (2005), serious violations of workplace health and safety laws carry an average penalty of only $873. And under federal law, causing the death of a worker by willfully violating safety rules—a misdemeanor with a 6-month maximum prison term—is a less serious crime than harassing a wild burro on federal lands, which is punishable by 1 year in prison (Barstow & Bergman 2003). Why do you think penalties for violating workplace health and safety laws are so weak?

Job Stress. Another work-related health problem is job stress. In a national sample of U.S. employees, more than one-fourth (26 percent) felt "overworked," and 27 percent felt "overwhelmed" by how much work they had to do often or very often in the past month (Galinsky et al. 2005). A 2008 national poll found that more than one in four respondents (28 percent) said they are somewhat or completely dissatisfied with the amount of on-the-job stress in their jobs (Gallup Organization 2009). Workers can be stressed by the number of hours they work, as the 40-hour workweek has become, for many workers, a 60-hour (and more) workweek. As the following excerpt reveals, employers are asking more of their employees, expecting them to work harder and faster. Employees also fear that they could lose their jobs if they don't work harder.

> With many companies shedding one group of workers after another, the employees who survive often have more and more responsibilities heaped on them. Many workers feel internal pressures to perform above expectations, fearing that if they don't, they might be next to lose their jobs to downsizing or offshoring. (Greenhouse 2008, p. 186)

Prolonged job stress, also known as **job burnout,** can cause or contribute to physical and mental health problems, such as high blood pressure, ulcers, headaches, anxiety, and depression. Taking time off to heal and "recharge one's batteries" is not an option for many workers: One-half of the U.S. workforce has no paid sick leave, and one-fourth has no paid vacation (Watkins 2002). The United States is the only advanced nation that does not mandate a minimum number of vacation days; in the 27 European Union countries, employers are required to give workers at least four weeks of vacation each year (some countries mandate five or six weeks of vacation)

job burnout Prolonged job stress that can cause or contribute to high blood pressure, ulcers, headaches, anxiety, depression, and other health problems.

(Greenhouse 2008). And, even when the workday is officially over, workers are connected to their jobs by their Blackberry, cell phones, e-mail accounts, and laptops. One in five employees works on a laptop computer while on vacation, 40 percent check office e-mail, and 50 percent check voicemail (Fram 2007).

Work-Family Concerns

Another source of stress for U.S. workers is the day-to-day struggle to meet the demands of family and work simultaneously. In about two-thirds of married couples with children under age 18, both parents are employed. And 71 percent of women in female-headed single-parent households and 83 percent of men in male-headed

This man, vacationing on the Greek island of Santorini, is among the one in five U.S. workers who works while on vacation.

single-parent households are employed (Bureau of Labor Statistics 2009e). Two-thirds of working parents feel they do not have enough time to spend with their families (Galinsky et al. 2004). U.S. workers clock more hours than workers in any other country participating in the Organization for Economic Co-operation and Development (OECD) (Mishel et al. 2009). More time at work means less time to care for and be with one's family.

Employed spouses strategize ways to coordinate their work schedules to have vacation time together, or simply to have meals together. Employed parents with young children must find and manage arrangements for child care and negotiate with their employers about taking time off to care for a sick child, or to attend a child's school event

The United States is the only advanced nation that does not mandate a minimum number of vacation days.

or extracurricular activity. Some employers post their workers' weekly schedules only a few days in advance, making it difficult to arrange child care. In some two-parent households, spouses or partners work different shifts so that one adult can be home with the children. However, working different shifts strains marriage relationships, because the partners rarely have time off together. Some employed parents who cannot find or afford child care leave their children with no adult supervision. More than 3 million children younger than age 13 are left without adult supervision for some period of time each week (Vandivere et al. 2003).

Employees with elderly and/or ill parents worry about how they will provide care for their parents, or arrange for and monitor their care, while putting in a 40-hour (or more) workweek. More than one-third (35 percent) of workers say that they have provided care for a relative or in-law age 65 or older in the past year (Bond et al. 2002).

Women tend to experience their jobs as interfering with family more so than do men; however, the level of work-family conflict women experience has remained stable over the past three decades, whereas men's reported level of work-family conflict has increased, probably due to the increased involvement of men in child care and household responsibilities (Galinsky et al. 2009).

Job Dissatisfaction and Alienation

A 2008 national poll found that 10 percent of employed U.S. adults are somewhat or very dissatisfied with their job or the work they do (Gallup Organization 2009). One source of dissatisfaction is stagnant, unfair, and/or declining wages.

Between 1979 and 2006, the top 10 percent of U.S. income earners received 91 percent of income growth, leaving only 9 percent to be shared among the remaining 90 percent of workers (Mishel et al. 2009). Workers in low-wage, low-status jobs with few or no benefits and little job security are vulnerable to feelings of dissatisfaction not only with their jobs but also with their limited housing and lifestyle options.

One form of job dissatisfaction is a feeling of alienation. Work in industrialized societies is characterized by a high degree of division of labor and specialization of work roles. As a result, workers' tasks are repetitive and monotonous and often involve little or no creativity. Limited to specific tasks by their work roles, workers are unable to express and utilize their full potential—intellectual, emotional, and physical. According to Marx, when workers are merely cogs in a machine, they become estranged from their work, the product they create, other human beings, and themselves. Marx called this estrangement "alienation." As we discussed earlier, the McDonaldization of the workplace also contributes to alienation.

Alienation usually has four components: powerlessness, meaninglessness, normlessness, and self-estrangement. Powerlessness results from working in an environment in which one has little or no control over the decisions that affect one's work. Meaninglessness results when workers do not find fulfillment in their work. Workers may experience normlessness if workplace norms are unclear or conflicting. For example, many companies that have family leave policies informally discourage workers from using them, or workplaces that officially promote nondiscrimination in reality practice discrimination. Alienation also involves a feeling of self-estrangement, which stems from the workers' inability to realize their full human potential in their work roles and lack of connections to others.

Labor Unions and the Struggle for Workers' Rights

In times of high unemployment, many people with jobs are thankful that they are employed. But having a job is no guarantee of having favorable working conditions and receiving decent pay and benefits. **Labor unions** are worker advocacy organizations that developed to protect workers and represent them at negotiations between management and labor.

Benefits and Disadvantages of Labor Unions to Workers. Labor unions have played an important role in fighting for fair wages and benefits, healthy and safe work environments, and other forms of worker advocacy. In 2008, the median weekly earnings of full-time wage and salary workers who were union members were $886, compared with a median of $691 for nonunion workers (Bureau of Labor Statistics 2009f). Unionized workers also received insurance and pension benefits worth more than those of nonunion employees, and union workers also got more paid time off than nonunion workers.

Labor unions are also influential in achieving better working conditions. For example, the United Food and Commercial Workers (UFCW), the country's largest union representing poultry-processing workers, was instrumental in the formation of an Occupational Safety and Health Administration (OSHA) rule that established a federal workplace "potty" policy governing when employees can use the bathroom while on the job. According to UFCW international president Doug H. Dority, "For years, workers in food processing industries have had to suffer the indignity of being denied the right to go to the bathroom when needed, just to maintain ever-increasing assembly-line speeds" ("New OSHA

Labor unions Worker advocacy organizations that developed to protect workers and represent them at negotiations between management and labor.

Policy Relieves Employees" 1998, p. 8). Dority claimed that "poultry processors often have no other choice than to relieve themselves where they stand on the assembly line because their floor boss will not let them leave their workstation" (p. 8). Now, OSHA mandates that employers must make toilet facilities available so that employees can use them when they need to.

The increasing number of women in labor unions, which nearly doubled from 20 percent to 39 percent between 1960 and 1998, has helped to strengthen labor unions' advocacy for women ("Labor's 'Female Friendly' Agenda" 1998). For example, a number of unions have been successful in bargaining for expanded family leave benefits, subsidized child care, elder care, and pay equity.

One of the disadvantages of unions is that members must pay dues and other fees, and these dues have been rising in recent years. Average annual dues for the 15 largest unions in 2004 ranged from $210 to $830, not including additional fees known as "special assessments." High union dues are problematic in light of the high salaries many union leaders make. In 2004, Donald Doser, president of the International Union of Operating Engineers, was the highest paid union leader, earning more than $800,000 (Brenner 2007).

Another disadvantage for unionized workers is the loss of individuality. Unionized workers are members of an overall bargaining unit in which the majority rules. Decisions made by the majority may conflict with individual employees' specific employment needs.

Declining Union Density. The strength and membership of unions in the United States have declined over the last several decades. **Union density**—the percentage of workers who belong to unions—grew in the 1930s and peaked in the 1940s and 1950s, when 35 percent of U.S. workers were unionized. In 2008, the percentage of U.S. workers belonging to unions had fallen to 12 percent (Bureau of Labor Statistics 2009f).

One reason for the decline in union representation is the loss of manufacturing jobs, which tend to have higher rates of unionization than other industries. Job growth has occurred in high technology and financial services, where unions have little presence. In addition, globalization has led to layoffs and plant closings at many unionized work sites as a result of companies moving to other countries to find cheaper labor. A major reason why union representation has declined is that corporations take active measures to keep workers from unionizing, and weak U.S. labor laws fail to support and protect unionization.

Corporate Antiunion Activities. In the 1960s and 1970s, U.S. corporations mounted an offensive attack on labor unions, "aiming to tame them or maim them" (Gordon 1996, p. 207). Corporations hired management consultants to help them develop and implement antiunion campaigns. They threatened unions with decertification, fired union leaders and organizers, and threatened to relocate their plants unless the unions and their members "behaved."

> One management consultant firm . . . was unusually blunt in broadcasting its methods. A late-1970s blurb promoting its manual promised: "We will show you how to screw your employees (before they screw you)— how to keep them smiling on low pay—how to maneuver them into low-pay jobs they are afraid to walk away from." (Gordon 1996, p. 208)

Human Rights Watch (2007) found that Wal-Mart engages in a variety of antiunion tactics. One antiunion policy requires managers to report any organizing

union density The percentage of workers who belong to unions.

activity at their stores to Wal-Mart's Union Hotline. Almost immediately, Wal-Mart sends one or more members of its Labor Relations Team to the store to implement an aggressive antiunion campaign in conjunction with store management. At group meetings with employees, management describes the negative consequences of organizing and often shows videos to reinforce this message. Managers and the videos tell workers that union dues and other union fees are expensive and that workers may lose wages and benefits as a result of unionization. They also warn workers that Wal-Mart can permanently replace workers who go on strike. "The videos dramatize the antiunion message by showing an example of a picket line that turns violent, characterizing unions as antiquated organizations, and portraying union organizers as harassing and bothersome people" (Human Rights Watch 2007, p. 7).

At least 23,000 workers each year are fired or discriminated against at their workplace because of involvement in union-related activity (Bonior 2006). A study of 62 union campaigns in the Chicago metropolitan area showed that significant numbers of employers faced with organizing campaigns engaged in the following antiunion strategies: (1) firing pro-union workers, (2) threatening to close a work site when workers try to form a union, (3) coercing workers into opposing unions with bribery or favoritism, (4) hiring high-priced union-busting consultants to fight union organizing drives, and (5) forcing employees to attend one-on-one antiunion meetings with their supervisors (Mehta & Theodore 2005).

Weak U.S. Labor Laws. The 1935 National Labor Relations Act (NLRA) is the primary federal labor law in the United States. The NLRA guarantees the right to unionize, bargain collectively, and to strike against private-sector employees. However, in addition to excluding public-sector workers, the law excludes agricultural and domestic workers, supervisors, railroad and airline employees, and independent contractors. As a result, millions of workers do not have the right under U.S. law to negotiate their wages, hours, or employment terms.

In addition, changes in U.S. labor law over the years have eroded workers' rights to freedom of association. Originally, labor law required employers to grant a demand for union recognition if a majority of workers signed a card indicating they wanted a union. But since 1947, employers can reject workers' demand for unionization and force a National Labor Relations Board (NLRB) election, which requires about one-third of workers to petition for the board to hold the election. "The company then uses the time leading up to the election to focus its campaign against union formation, while disallowing opportunities for opposing views" (Human Rights Watch 2007, p. 18).

Penalties for violating U.S. labor law are so weak that many employers consider them as a cost of doing business and a small price to pay for defeating workers' attempts to organize.

In the United States, the NLRB and the courts play an important role in upholding workers' rights to unionize and sanctioning employers who violate these rights. The NLRB has the authority to issue job reinstatement and "back pay" orders or other remedial orders to workers wrongfully fired or demoted for participating in union-related activities. Although it is illegal to fire workers for engaging in union activities, there are few consequences for employers that do so. Penalties for violating U.S. labor law are so weak that many employers consider them as a cost of doing business and a small price to pay for defeating workers' attempts to organize (Human Rights Watch 2009).

If you get fired for trying to organize, for example, you can apply to the NLRB. If the NLRB finds that you were illegally fired, the employer has to give you back pay for the time you were fired—minus any money that you may have earned at another job. As you can imagine, most people who are fired for trying to organize will, in fact, get another job somewhere, so there's no compensation for them at all. Then all the employer has to do is post on a bulletin board at the work site that they won't do it again. So there are effectively no sanctions, and it is in the employers' interest to fire people. They really don't suffer many consequences for doing so, and firing a leading union supporter sends a very powerful message to the rest of the employees. The message is: If you too try to lead an organizing campaign, you are going to lose your job; and if you vote for a union, you could lose your job. (Bonior 2006)

In addition, there is a backlog of thousands of cases of unfair labor practices by employers, and workers often wait years from the filing of a charge until the NLRB resolves a case, discouraging many workers from filing charges (Greenhouse 2008).

Labor Union Struggles Around the World. International norms established by the United Nations and the International Labour Organization declare the rights of workers to organize, negotiate with management, and strike (Human Rights Watch 2000). In European countries, labor unions are generally strong. However, in many less-developed countries and in countries undergoing economic transition, workers and labor unions struggle to have a voice in matters of wages and working conditions.

The International Confederation of Free Trade Unions (ICFTU) publishes an *Annual Survey of Violations of Trade Union Rights* that describes severe abuses of workers' rights in countries around the world. This annual survey is frightening documentation of the lack of workers' rights around the world and of the abuses of governments and employers in the continued suppression of workers' rights. The survey results in the last few years showed that between 75 and several hundred trade unionists are killed each year. Several thousands more are imprisoned, beaten in demonstrations, tortured by security forces or others, and often sentenced to long prison terms. And each year, hundreds of thousands of workers lose their jobs merely for attempting to organize a trade union.

Colombia remains the most dangerous place in the world to be a trade unionist, with 49 people killed for their trade union activity in 2008 (International Confederation of Free Trade Unions 2009). In some cases, the families of the unionist murder victims were also killed. Most reported cases of trade unionist assassinations are not properly investigated, and the murderers are not caught or punished.

Both employers and governments violate trade union rights. Governments in numerous countries hamper trade union activity or strike action and fail to enforce existing national and international laws that protect workers' rights. Some countries, such as Oman, Saudi Arabia, and Burma, do not recognize the right to form trade unions. Other countries, such as China, Egypt, and Syria, impose a trade union monopoly. And some governments, such as those in Belarus and Moldova (countries in eastern Europe), try to coerce workers into joining the government-supported union. Eager to secure financial benefits from participating in the global market, governments see trade unions as an obstacle to their economic development.

Strategies for Action: Responses to Problems of Work and Unemployment

Government, private business, human rights organizations, labor organizations, college student activists, and consumers play important roles in responding to problems of work and unemployment.

Reducing Unemployment

Efforts to reduce unemployment include (1) workforce development programs, and (2) programs and policies that create and save jobs.

Workforce Development. The 1998 **Workforce Investment Act** (WIA) provides a wide array of programs and services designed to assist individuals to prepare for and find employment. These include the following: assessment of skills and abilities; access to job vacancy listings; job search and placement assistance; individual career planning and counseling; resume preparation; English as a second language instruction; computer literacy; wage subsidies for on-the-job training; and support services such as transportation and child care to enable individuals to participate in WIA programs. Some workforce development programs focus on strategies to improve the employability of hard-to-employ individuals through providing targeted interventions such as substance abuse treatment, domestic violence services, prison release reintegration assistance, mental health services, and homelessness services, in combination with employment services (Martinson & Holcomb 2007).

Efforts to prepare high school students for work include the establishment of technical and vocational high schools and high school programs and school-to-work programs. School-to-work programs involve partnerships between business, labor, government, education, and community organizations that help prepare high school students for jobs (Leonard 1996). Although school-to-work programs vary, in general, they allow high school students to explore different careers, and they provide job skill training and work-based learning experiences, often with pay (Bassi & Ludwig 2000).

Educational attainment is often touted as the path to employment and economic security. Although education and skill development are important for individuals to realize their potential and to become better-informed and more productive citizens, education alone is not the answer to unemployment and economic insecurity. More than two-thirds of jobs in the United States require little education (Martinson & Holcomb 2007); thus, it is important to ensure that all jobs pay a minimum "living wage" (see also Chapter 6). But, as long as our economy allows people who work full-time to earn poverty-level wages, having a job is not necessarily the answer to economic self-sufficiency. As David Shipler (2005) explained in his book *The Working Poor:*

> A job alone is not enough. Medical insurance alone is not enough. Good housing alone is not enough. Reliable transportation, careful family budgeting, effective parenting, effective schooling are not enough when each is achieved in isolation from the rest. There is no single variable that can be altered to help working people move away from the edge of poverty. Only where the full array of factors is attacked can America fulfill its promise. (p. 11)

Workforce Investment Act
Legislation passed in 1998 that provides a wide array of programs and services designed to assist individuals to prepare for and find employment.

Job Creation and Preservation. In response to spiraling unemployment rates, President Obama signed into law the **American Recovery and Reinvestment Act of 2009** in February 2009, which provided an economic stimulus package of $787 billion to create and save jobs and to reinvigorate the U.S. economy. About $280 billion of these monies were administered by states and localities, largely for health, education, and transportation programs.

This economic stimulus package included measures to modernize the nation's infrastructure (roads, bridges, transit, and waterways), increase energy independence, expand educational opportunities, increase access to health care, provide tax relief, and save public sector jobs. The Congressional Budget Office estimates that the stimulus package will create 1 million to 3 million jobs. The stimulus package also included measures to increase or extend benefits under Medicaid, **unemployment insurance**, and nutrition assistance programs to help those most affected by the economic downturn. One analysis of stimulus package spending found that funds were not necessarily targeted to areas hardest hit by the recession (Grabell & La Fleur 2009).

Les Leopold, director of the Labor Institute and Public Health Institute in New York, proposed an interesting way to create jobs in the nonprofit sector. He proposed offering each of the 534,000 nonprofit U.S. charities $100,000 if they hire two new employees at $40,000 a year, plus benefits. This would generate more than a million jobs at a cost of $53.6 billion—only 3.6 percent of what the U.S. government has spent on TARP and the economic stimulus package (Leopold 2009).

Efforts to End Slavery and Child Labor

More than 50 years ago, the United Nations stated in Article 4 of its Universal Declaration of Human Rights, that "no one shall be held in slavery or servitude; slavery and the slave trade shall be prohibited in all their forms." Yet slavery persists throughout the world. The international community has drafted treaties on slavery, but many countries have yet to ratify and implement the different treaties.

One strategy to fight slavery is punishment. In at least 25 countries, slave trafficking is actively prosecuted and treated as a serious crime. However, slave traffickers often avoid punishment because, as a former official of the U.S. Agency for International Development explained, "government officials in dozens of countries assist, overlook, or actively collude with traffickers" (quoted by Cockburn 2003, p. 16). In many countries, the justice system is more likely to jail or expel sex slaves than to punish traffickers ("Sex Trade Enslaves Millions of Women, Youth" 2003).

In the United States, the Victims of Trafficking and Violence Protection Act, passed by Congress in 2000, protects slaves against deportation if they testify against their former owners. Convicted slave traffickers in the United States are subject to prison sentences, as shown in the following examples:

- Louisa Satia and Kevin Waton Nanji each received 9 years for luring a 14-year-old girl from Cameroon with promises of schooling and then isolating her in their Maryland home, raping her, and forcing her to work as their domestic servant for 3 years.
- Sardar and Nadira Gasanov were sentenced to 5 years each for recruiting women from Uzbekistan with promises of jobs, taking their passports, and forcing them to work in strip clubs and bars in Texas.
- Juan, Ramiro, and Jose Ramos each received 10 to 12 years for transporting Mexicans to Florida and forcing them to work as fruit pickers. (Cockburn 2003)

American Recovery and Reinvestment Act of 2009
Legislation that provided an economic stimulus package of $787 billion to create and save jobs and to reinvigorate the U.S. economy.

Unemployment insurance
A federal-state program that temporarily provides laid-off workers with a portion of their paychecks.

U.S. corporations are also being held accountable for enterprises that involve forced labor and other human rights and labor violations. In 2003, the Unocal oil company became the first corporation in history to stand trial in the United States for human rights violations abroad (George 2003). Unocal was accused of involvement in a pipeline project that used Myanmar (formerly Burma) military personnel to provide "security" for a natural gas pipeline project in the remote Yadana region near the Thai border. According to the Ninth U.S. Circuit Court of Appeals, the soldiers' true role was to force villagers in the pipeline region to work without pay. The military also forced villagers living along the pipeline route to relocate without compensation, raped and assaulted villagers, and imprisoned and/or executed those who opposed them. In 2004, Unocal announced that it had reached a settlement with the parties who alleged that Unocal was complicit in the human rights violations committed by the Myanmar military. In 2005, ChevronTexaco purchased Unocal.

In 1989, the General Assembly of the United Nations adopted the Convention on the Rights of the Child, which asserts the right that children should not be engaged in work deemed to be "hazardous or to interfere with the child's education, or to be harmful to the child's health." The International Labour Organization has taken a leading role in enforcing these rights, leading efforts to prevent and eliminate child labor. Although almost every country has laws prohibiting the employment of children below a certain age, some countries exempt certain sectors—often the very sectors where the highest numbers of child laborers are found. Penalties for violating child labor laws are also weak and poorly enforced. For example, in 2006, India strengthened its laws by extending the definition of hazardous work to include domestic labor and catering establishments but there is little evidence of enforcement.

Education is a primary means to combat child labor (International Programme on the Elimination of Child Labour 2006). Children with no access to education have little choice but to enter the labor market. In most countries, primary education is not free and parents must pay for costs such as uniforms and books. Parents who cannot afford these fees are also likely to view their children's labor as a necessary source of income to the household. So programs and strategies to combat poverty (see Chapter 6) are also central to the fight against child labor.

Responses to Sweatshop Labor

The Fair Labor Association (FLA), established in 1996, is a coalition of companies, universities, and nongovernmental organizations (NGOs) that works to promote adherence to international labor standards and improve working conditions worldwide. In 2009, 30 leading companies that sell brand-name apparel, footwear, and other goods voluntarily participated in FLA's monitoring system, which inspects their overseas factories and requires them to meet minimum labor standards, such as not requiring workers to work more than 60 hours a week. In addition, more than 200 colleges and universities require their collegiate licensees (companies that manufacture logo-carrying goods for colleges and universities) to participate in FLA's monitoring system.

In its first few years of operation, the FLA was criticized for allowing firms to select and directly pay their own monitors and to have a say in which factories were audited. In 2002, the FLA responded to these criticisms by taking much more control over external monitoring, with the FLA staff selecting factories for evaluation, choosing the monitoring organization, and requiring that inspections be unannounced (O'Rourke 2003). The FLA continues to be criticized, however, for having low standards in allowing below-poverty wages and excessive overtime and for requiring that only a small percentage of a manufacturer's supplier factories be inspected each year. Critics also suggest that companies use their participation in the FLA as a marketing tool. Once "certified" by the FLA, companies can sew a label into their products saying that the products were made under fair working conditions (Benjamin 1998). In 2006, United Students Against Sweatshops created "FLA Watch" to "expose the truth about the Fair Labor Association . . . and . . . the FLA's ongoing failure to defend the rights of workers" (FLA Watch 2007a). The home page of the FLA Watch website explains, "The FLA purports to be an 'independent' monitor of working conditions in the apparel industry. But the organization is funded and controlled by the very corporations that have been repeatedly found to be sweatshop violators." FLA Watch further accuses the Fair Labor Association of being "nothing more than a public relations mouthpiece for the apparel industry. Created, funded, and controlled by Nike, Adidas, and other leading sweatshop abusers, the FLA is a classic case of the 'fox guarding the hen house'" (FLA Watch 2007b).

Student Activism. United Students Against Sweatshops is a student activism group that is affiliated with the Worker Rights Consortium (WRC)—a nonprofit organization working to ensure that factories that produce clothing and other goods bearing school logos respect basic rights of workers, such as the right to unionize and to receive living wages and overtime pay. More than 180 colleges and universities are affiliated with the WRC and participate in a Designated Suppliers Program. This program requires brands that are licensed to make university apparel to be produced in factories in which employees are paid a living wage and are represented by a union or other form of employee representation.

What Do You Think? Does your college or university participate in either the Fair Labor Association or the Designated Suppliers Program? Do you think most students care if the college or university logo clothing or products they buy are made under sweatshop conditions?

Legislation. Perhaps the most effective strategy against sweatshop work conditions is legislation. In the United States, as of 2009, 8 states, 39 cities, 15 counties, 4 dioceses, and 118 school districts have passed "sweatfree" procurement laws that prohibit public entities (such as schools, police, and fire departments) from purchasing uniforms and apparel made under sweatshop conditions (SweatFree Communities n.d.).

In 2006, the Decent Working Conditions and Fair Competition Act was introduced in the U.S. Congress—the first proposed federal anti-sweatshop legislation in the United States. If passed, this legislation will prohibit the import, export, or sale of sweatshop goods in the United States.

Establishing and enforcing labor laws to protect workers from sweatshop labor conditions is difficult in a political climate that offers more protections to corporations than it does to workers.

Up to this point, it has been the companies that have demanded and won all sorts of enforceable laws—intellectual property and copyright laws backed up by sanctions—to defend their corporate trademarks, labels, and products. Yet, the corporations have long said that extending similar laws to protect the human rights of the 16-year-old girl in Bangladesh who sews the garment would be "an impediment to free trade." Under this distorted sense of values, the label is protected, but not the human being, the worker who makes the product. (National Labor Committee 2007)

International Anti-Sweatshop Efforts. The international community is also involved in efforts to improve working conditions and end sweatshop labor. The Clean Clothes Campaign is an international campaign in 12 European countries, focused on improving working conditions in the global garment and sportswear industries. The goals of the Clean Clothes Campaign are to:

- Put pressure on companies to take responsibility for ensuring decent working conditions.
- Support workers, trade unions, and NGOs in producer countries.
- Raise awareness among consumers by providing information about working conditions in the global garment and sportswear industry, so that citizens can use their power as consumers.
- Explore legal possibilities for improving working conditions, and lobby for legislation to promote good working conditions and for laws that would compel governments and companies to become ethical consumers.

Pressure from opponents of sweatshop labor and consumer boycotts of products made by sweatshop labor have resulted in some improvements in factories that make goods for companies such as Nike and Gap, which have cut back on child labor, use less dangerous chemicals, and require fewer employees to work 80-hour weeks. At many factories, supervisors have stopped hitting employees, have improved ventilation, and have stopped requiring workers to obtain permission to use the toilet. But improvements are not widespread, and oppressive forms of labor continue throughout the world. According to the National Labor Committee, two areas where "progress seems to grind to a halt" are efforts to form unions and efforts to achieve wage increases (Greenhouse 2000).

Responses to Worker Health and Safety Concerns

Over the past few decades, health and safety conditions in the U.S. workplace have improved as a result of media attention, demands by unions for change, more white-collar jobs, and regulations by OSHA. Through OSHA, the government develops, monitors, and enforces health and safety regulations in the workplace. But much work remains to be done to improve worker safety and health. Inadequate funding leaves OSHA unable to do its job effectively. The International Labour Organization recommends one labor inspector per 10,000 workers in an industrial economy. To meet this benchmark, OSHA would have needed 13,537 inspectors in 2007, but it had only 2,043 inspectors (AFL-CIO 2009).

Because "the task of monitoring and enforcement simply cannot be effectively carried out by a government administrative agency," Kenworthy (1995) suggested that the United States follow the example of many other industrialized countries: Turn over the bulk of responsibility for health and safety monitoring to the workforce (p. 114). Worker health and safety committees are a standard feature of companies in many other industrialized countries and are mandatory in most of Europe. These committees are authorized to inspect workplaces and cite employers for violations of health and safety regulations.

Research suggests that working beyond 50 hours a week jeopardizes health and safety in the workplace. As a result, most countries limit work hours to less than 48 hours, and about half of countries have a 40-hour limit. However, more than one in five (22 percent) people of the global workforce works more than 48 hours per week (Lee et al. 2007). Thus, efforts to improve health and safety of workers worldwide include establishing and enforcing laws that limit work hours.

In developing countries, governments fear that strict enforcement of workplace regulations will discourage foreign investment. Investment in workplace safety in developing countries, whether by domestic firms or foreign multinationals, is far below that in the rich countries. Unless global standards of worker safety are implemented and enforced in *all* countries, millions of workers throughout the world will continue to suffer under hazardous work conditions. Low unionization rates and workers' fears of losing their jobs—or their lives—if they demand health and safety protections leave most workers powerless to improve their working conditions.

Behavior-Based Safety Programs. A controversial health and safety strategy used by business management is behavior-based safety programs. Instead of examining how work processes and conditions compromise health and safety on the job, **behavior-based safety programs** direct attention to workers themselves as the problem. Behavior-based safety programs claim that 80 percent to 96 percent of job injuries and illnesses are caused by workers' own carelessness and unsafe acts (Frederick & Lessin 2000). These programs focus on teaching employees and managers to identify, "discipline," and change unsafe worker behaviors that cause accidents and encourage a work culture that recognizes and rewards safe behaviors.

Critics contend that behavior-based safety programs divert attention away from the employers' failures to provide safe working conditions. They also say that the real goal of behavior-based safety programs is to discourage workers from reporting illness and injuries. Workers whose employers have implemented behavior-based safety programs describe an atmosphere of fear in the workplace, such that workers are reluctant to report injuries and illnesses for fear of being labeled "unsafe workers." At one factory that had implemented a behavior-based safety program, when a union representative asked workers during shift meetings to raise their hands if they were afraid to report injuries, about half of 150 workers raised their hands (Frederick & Lessin 2000). Worried that some workers feared even raising their hand in response to the question, the union representative asked a subsequent group to write yes on a piece of paper if they were afraid to report injuries. Seventy percent indicated they were afraid to report injuries. Asked why they would not report injuries, workers said, "We know that we will face an inquisition," "We would be humiliated," and "We might be blamed for the injury."

A rule that OSHA issued protects workers by prohibiting discrimination against an employee for reporting a work-related fatality, injury, or illness (*Multinational Monitor* 2003). This rule also prohibits discrimination against an employee for filing a safety and health complaint or asking for health and safety records.

behavior-based safety programs A strategy used by business management that attributes health and safety problems in the workplace to workers' behavior, rather than to work processes and conditions.

TABLE 7.2 **How Do U.S. Work Policies Compare with Other Countries?**		

POLICY	UNITED STATES	OTHER COUNTRIES
Paid childbirth leave	No federal policy	168 countries offer paid leave to women; 98 countries offer 14 or more weeks of paid leave
Right to breast-feed at work	No federal policy	107 countries protect working women's right to breast-feed
Paid sick leave	No federal policy	145 countries provide paid sick leave
Paid annual leave	No federal policy	137 countries require employers to provide paid annual leave; 121 countries guarantee 2 weeks or more
Guaranteed leave for major family events (e.g., weddings, funerals)	No federal policy	49 countries guarantee leave for major family events (leave is paid in 40 countries)

Source: Based on Heymann et al. 2009.

Work-Family Policies and Programs

Policies that help women and men balance their work and family responsibilities are referred to by a number of terms, including *work-family, work-life,* and *family-friendly* policies. As shown in Table 7.2, the United States lags far behind many other countries in national work-family provisions.

Federal and State Family and Medical Leave Initiatives. In 1993, President Clinton signed into law the first national policy designed to help workers meet the dual demands of work and family. The **Family and Medical Leave Act (FMLA)** requires all public agencies and private-sector employers (with 50 or more employees who worked at least 1,250 hours in the preceding year) to provide up to 12 weeks of job-protected, *unpaid* leave so that they can care for a seriously ill child, spouse, or parent; stay home to care for their newborn, newly adopted, or newly placed foster child; or take time off when they are seriously ill. A 2008 amendment to the FMLA requires employers to provide up to 26 weeks of unpaid leave to employees to care for a seriously ill or injured family member who is in the armed forces, including the National Guard or Reserves. However, nearly 40 percent of employees are not eligible for the FMLA benefit because they work for companies with fewer than 50 employees or they work part-time (National Partnership for Women and Families 2008). Some employers do not comply with the FMLA either because they are unaware of their responsibilities under FMLA, or because they are deliberately violating the law (Galinsky et al. 2008). And some eligible employees do not take advantage of the FMLA benefit because of lack of awareness: In a survey of employees covered by the FMLA, only 38 percent correctly reported that the FMLA applied to them and about half said they did not know whether it did (Cantor et al. 2001). Finally, many eligible workers do not use their FMLA benefit because they cannot afford to take leave without pay, and/or they fear they will lose their job if they take time off. One survey found that 88 percent of workers who needed time off but did not take it said they would have taken leave if they could have received pay

Family and Medical Leave Act (FMLA) A federal law that requires public agencies and companies with 50 or more employees to provide eligible workers with up to 12 weeks of job-protected, unpaid leave so that they can care for an ill child, spouse, or parent; stay home to care for their newborn, newly adopted, or newly placed child; or take time off when they are seriously ill, and up to 26 weeks of unpaid leave to care for a seriously ill or injured family member who is in the Armed Forces, including the National Guard or Reserves.

during their absence; nearly one-third of workers who needed leave but did not take it cited worry about losing their job as a reason for not taking leave (Cantor et al. 2001).

Only three states—California, Washington, and New Jersey—offer paid leave programs that provide eligible workers a family leave benefit with a portion of their salary for up to six weeks. The proposed federal Family Leave Insurance Act would provide 8 to 12 weeks of partially paid leave for FMLA purposes, paid for by employer and worker contributions.

As shown in Table 7.2, U.S. employers are not required to provide workers with any paid sick leave, and nearly half (47 percent) of the private-sector workforce has no paid sick leave (Galinsky et al. 2008). The proposed Healthy Families Act would require employers with at least 15 employees to provide 7 days of paid sick leave annually for full-time employees who work at least 30 hours a week. Employees could take the benefit if they or a family member is sick. Supporters of the Healthy Families Act point out that, when sick employees go to work because they cannot afford to take unpaid sick days, or fear losing their job by taking a sick day, they risk spreading infectious diseases at the workplace. And sending sick children to school because their parents cannot afford to take unpaid sick days to stay home with them risks spreading infectious diseases at school. Recent concerns about the spread of swine flu have reinvigorated the initiative to establish a federally mandated paid sick day policy. President Obama supports paid sick leave for U.S. workers, giving supporters of the Health Families Act optimism that Congress will approve it.

Paul Richards/ AFP/ Getty Images

To control the spread of swine flu, the Centers for Disease Control and Prevention tell the public to stay home if they are sick. But many workers without paid sick leave cannot afford to stay home.

Employer-Based Work-Family Policies. Aside from government-mandated work-family policies, some corporations and employers have "family-friendly" work policies and programs, including unpaid or paid family and medical leave, child care assistance, assistance with elderly parent care, and flexible work options such as **flextime, compressed workweek**, and **telecommuting**. Organizations with higher proportions of full-time female employees are more likely to offer flexible scheduling, unpaid parental leave, and dependent care assistance (Davis & Kalleberg 2006). National surveys of more than 1,000 employers (with 100 or more employees) found that, between 1998 and 2008, some employer-provided work-family benefits increased, whereas others decreased (see Table 7.3).

> **What Do You Think?** A national survey found that more than half of employers (52 percent) that offer maternity leave provide at least partial pay to employees on maternity leave, whereas only 16 percent provide any pay for paternity leave (Galinsky et al. 2008). Do you think this is fair to fathers? Is this discrepancy a form of discrimination against men?

Efforts to Strengthen Labor

More than half (61 percent) of U.S. adults in 2009 agreed that "labor unions are necessary to protect the working person"—down from 74 percent in 2003 (Pew Research Center 2009). Although efforts to strengthen labor are viewed as problematic to corporations, employers, and some governments, such efforts have the potential to remedy many of the problems facing workers.

flextime A work arrangement that allows employees to begin and end the workday at different times so long as 40 hours per week are maintained.

compressed workweek A work arrangement that allows employees to condense their work into fewer days (e.g., four 10-hour days each week).

telecommuting A work arrangement involving the use of information technology that allows employees to work part- or full-time at home or at a satellite office.

TABLE 7.3 Employer-Based Work-Family Benefits and Policies, U.S., 1998–2008*

BENEFIT OR POLICY	PERCENTAGE OF EMPLOYERS* THAT PROVIDE BENEFIT/POLICY	
	1998	**2008**
Full pay for maternity leave	27%	16%
Partial pay for maternity leave	60%	67%
Private space for breast-feeding	37%	53%
Private space and storage facilities for women who are nursing to express milk	49%	66%
Some paid time off for paternity leave	13%	16%
Health insurance for unmarried partners	14%	31%
Allow some employees to periodically change their arrival and departure time	68%	79%
Provide information about elder care services	23%	39%
Child care at or near worksite	9%**	9%
Dependent Care Assistance Plans that help employees pay for child care with pretax dollars	46%**	46%

Source: Galinsky et al. 2008.

*Employers with 100 or more employees

**Report did not specify 1998 value, but did report that the 1998 value was not statistically different from the 2008 value.

In an effort to strengthen their power, some labor unions have merged with one another. Labor union mergers result in higher membership numbers, thereby increasing the unions' financial resources, which are needed to recruit new members and to withstand long strikes. Because workers must fight for labor protections within a globalized economic system, their unions must cross national boundaries to build international cooperation and solidarity. Otherwise, employers can play working and poor people in different countries against each other. An example of international union cooperation occurred when leaders from 21 unions in 11 countries on 5 continents resolved to form a global union network at International Paper Company (IP), the largest paper company in the world (CorpWatch 2002).

Strengthening labor unions requires combating the threats and violence against workers who attempt to organize or who join unions. One way to do this is to pressure governments to apprehend and punish the perpetrators of such violence. Although about 3,500 trade unionists were murdered between 1990 and 2005, only 600 cases were investigated, resulting in just 6 convictions (Moloney 2005). Another tactic is to stop doing business with countries where government-sponsored violations of free trade union rights occur.

Proposed legislation called the Employee Free Choice Act would allow workers to sign a card stating that they want to be represented by a union. If a majority of the employees in any workplace sign such a card, the company would then have to recognize the union and bargain over terms and conditions of employment. If the company does not negotiate a first contract in a timely manner after workers unionize, the Employee Free Choice Act requires binding arbitration. This legislation would also strengthen U.S. labor law enforcement by increasing penalties for violations. In 2007, the U.S. House of Representatives passed the Employee Free Choice Act and, as of this writing, the bill is pending in the U.S. Senate.

The late Senator Edward Kennedy supported the Employee Free Choice Act.

Challenges to Corporate Power and Globalization

Challenges to corporate power and globalization have taken root in the United States and throughout the world. The antiglobalization movement, also known as the global justice movement, is based on opposition to unregulated power of multinational corporations and financial markets and to the conditions of trade agreements that favor corporate interests over human rights and environmental protection. Antiglobalization activists have targeted the World Trade Organization, the International Monetary Fund (IMF), and the World Bank as forces that advance corporate-led globalization at the expense of social goals such as justice, community, national sovereignty, cultural diversity, ecological sustainability, and workers' rights. Some of the more notable actions of the antiglobalization movement include the 1999 protest in Seattle against the policies of the World Trade Organization, the protest against the 2000 meeting of the International Monetary Fund and the World Bank in Washington, DC, and the 2003 demonstration in Miami to protest the Free Trade Area of the Americas. In 2005, organized opposition to the Central American Free Trade Agreement occurred in countries such as Guatemala, Costa Rica, and Honduras. And in 2007, hundreds of South Koreans protested against a trade agreement between the United States and South Korea. Media attention to these events contributes to the growing worldwide awareness of the forces of corporate globalization and its social, environmental, and economic effects.

College student groups across the country have participated in boycotts against Coca-Cola in protest of the violence against union leaders at Colombian Coca-Cola plants.

Many global justice-minded critics of corporations believe that corporations pursue profit at the expense of the public good because corporations and their executives are greedy. But, as former corporate attorney Robert Hinkley points out, corporations behave as they do because they are bound by corporate law to try to make a profit for shareholders (Cooper 2004). Hinkley suggests redesigning corporations by adding what he calls a "Code for Corporate Citizenship" to corporate law. This code would consist of the following 28 words: "The duty of directors henceforth shall be to make money for shareholders *but not at the expense of the environment, human rights, public health and safety, dignity of employees, and the welfare of the communities in which the company operate*" (Cooper 2004, p. 6).

What Do You Think? Do you think that Hinkley's "Code for Corporate Citizenship" will ever make it to the legislature? Why or why not? If you were a legislator, how would you vote on Hinkley's corporate code?

Understanding Work and Unemployment

On December 10, 1948, the General Assembly of the United Nations adopted and proclaimed the Universal Declaration of Human Rights. Among the articles of that declaration are the following:

Article 23. Everyone has the right to work, to free choice of employment, to just and favourable conditions of work and to protection against unemployment.

Everyone, without any discrimination, has the right to equal pay for equal work.

Everyone who works has the right to just and favourable remuneration ensuring for himself and his family an existence worthy of human dignity, and supplemented, if necessary, by other means of social protection.

Everyone has the right to form and to join trade unions for the protection of his interests.

Article 24. Everyone has the right to rest and leisure, including reasonable limitation of working hours and periodic holidays with pay.

More than half a century later, workers around the world are still fighting for these basic rights as proclaimed in the Universal Declaration of Human Rights.

To understand the social problems associated with work and unemployment, we must first recognize that corporatocracy—the ties between government and corporations—serves the interests of corporations over the needs of workers. We must also be aware of the roles that technological developments and postindustrialization play on what we produce, how we produce it, where we produce it, and who does the producing. With regard to what we produce, the United States has moved away from producing manufactured goods to producing services. With regard to production methods, the labor-intensive blue-collar assembly line has declined in importance, and information-intensive white-collar occupations have increased. Although some people argue that the growth of multinational corporations brings economic growth, jobs, lower prices, and quality products to consumers throughout the world, others view global corporations as exploiting workers, harming the environment, dominating public policy, and degrading cultural values. "One thing is for certain—global corporations are an inescapable presence in the modern world and will be so for the foreseeable future" (Roach 2007, p. 2).

Decisions made by U.S. corporations about what and where to invest influence the quantity and quality of jobs available in the United States. As conflict theorists argue, such investment decisions are motivated by profit, which is part of a capitalist system. Profit is also a driving factor in deciding how and when technological devices will be used to replace workers and increase productivity. If goods and services are produced too efficiently, however, workers are laid off and high unemployment results. When people have no money to buy products, sales slump, recession ensues, and social welfare programs are needed to support the unemployed. When the government increases spending to pay for its social programs, it expands the deficit and increases the national debt. Deficit spending and a large national debt make it difficult to recover from the recession, and the cycle continues.

What can be done to break the cycle? Those adhering to the classic view of capitalism argue for limited government intervention on the premise that business will regulate itself by means of an "invisible hand" or "market forces." For example, if corporations produce a desired product at a low price, people will buy it, which means workers will be hired to produce the product, and so on.

Ironically, those who support limited government intervention also sometimes advocate government intervention to bail out failed banks and to lend money to troubled businesses. Such government help benefits the powerful segments of our society. Yet, when economic policies hurt less powerful groups, such as minorities, there has been a collective hesitance to support or provide social welfare programs. It is also ironic that such bailout programs, which contradict the ideals of capitalism, are needed because of capitalism. For example, the profit motive leads to multinationalization, which leads to unemployment, which leads to the need for government programs. The answers are as complex as the problems.

CHAPTER REVIEW

■ **According to a Pew Research survey, what do the majority of U.S. adults believe about big business?**
The majority of U.S. adults believe that (1) there's too much power concentrated in the hands of a few big companies; (2) businesses make too much profit; and (3) free market needs regulation to best serve public interest needs.

■ **The United States is described as a "postindustrialized" society. What does that mean?**
Postindustrialization refers to the shift from an industrial economy dominated by manufacturing jobs to an economy dominated by service-oriented, information-intensive occupations.

■ **What are transnational corporations?**
Transnational corporations are corporations that have their home base in one country and branches, or affiliates, in other countries.

■ **What are the four principles of McDonaldization?**
The four principles of McDonaldization are (1) efficiency, (2) predictability, (3) calculability, and (4) control through technology.

■ **What are some of the causes of unemployment?**
Causes of unemployment include not enough jobs, job exportation (the relocation of jobs to other countries), automation (the replacement of human labor with

machinery and equipment), increased global competition, and mass layoffs as plants close and companies downsize or go out of business.

■ **Does slavery still exist today? If so, where?**
Forced labor, commonly known as slavery, exists today all over the world, including the United States, but it is most prevalent in India, Pakistan, Bangladesh, and Nepal. Most forced laborers work in agriculture, mining, prostitution, and factories.

■ **What is the most common cause of job-related fatality and nonfatal job-related illness or injury?**
The most common type of job-related fatality involves transportation accidents. Sprains and strains are the most common nonfatal occupational injury or illness involving days away from work.

■ **What are some of the challenges that workers face in balancing work and family?**
Workers often struggle to find the time and energy to care for elderly parents and children and to have time with their families.

■ **How does unionization benefit employees?**
Compared with nonunion workers, union workers have higher average wages, receive more insurance and pension benefits, and get more paid time off.

■ **What is the American Recovery and Reinvestment Act of 2009?**
In response to spiraling unemployment rates, President Obama signed into law the American Recovery and Reinvestment Act of 2009, which provided an economic stimulus package of $787 billion to create and save jobs and to reinvigorate the U.S. economy.

■ **What is the federal Family and Medical Leave Act?**
In 1993, President Clinton signed into law the Family and Medical Leave Act (FMLA), which requires all companies with 50 or more employees to provide eligible workers with up to 12 weeks of job-protected, unpaid leave so that they can care for a seriously ill child, spouse, or parent; stay home to care for their newborn, newly adopted, or newly placed child; or take time off when they are seriously ill. The United States is the only advanced nation that does not mandate paid leave for family and medical reasons.

■ **What does the antiglobalization movement oppose?**
The antiglobalization movement, also known as the global justice movement, is based on opposition to unregulated power of multinational corporations and financial markets and to the conditions of trade agreements that favor corporate interests over human rights and environmental protection.

TEST YOURSELF

1. The global financial crisis that began in 2007 began in what country?
   a. China
   b. India
   c. United States
   d. Mexico
2. According to your text, corporations outsource labor to other countries so that U.S. consumers can save money on cheaper products.
   a. True
   b. False
3. According to a 2005 report by the AFL-CIO (2005), serious violations of workplace health and safety laws carry an average penalty of
   a. $873
   b. $8,773
   c. $78,337
   d. $788,333
4. The most common type of job-related fatality involves which of the following?
   a. Mining accidents
   b. Construction accidents
   c. Transportation accidents
   d. Homicide
5. The United States is the only advanced nation that does not mandate a minimum number of vacation days.
   a. True
   b. False
6. The 1935 National Labor Relations Act (NLRA) gives all U.S. workers the right to unionize, bargain collectively, and to strike.
   a. True
   b. False
7. Which is the most dangerous country in the world to be involved in trade union activities?
   a. China
   b. Ireland
   c. Egypt
   d. Colombia
8. The Decent Working Conditions and Fair Competition Act is the first proposed federal legislation in the United States aimed at which of the following?
   a. Combating sweatshop labor
   b. Preventing outsourcing of jobs
   c. Strengthening unions
   d. Prohibiting age discrimination in the workplace

9. More than one in five people of the global workforce works more than 48 hours per week.
   a. True
   b. False

10. More than half of U.S. adults say that "labor unions are necessary to protect the working person."
    a. True
    b. False

Answers: 1: c; 2: b; 3: a; 4: c; 5: a; 6: b; 7: d; 8: a; 9: a; 10: a.

KEY TERMS

American Recovery and Reinvestment Act of 2009   253
automation   236
behavior-based safety programs   257
capitalism   227
chattel slavery   241
child labor   244
compressed workweek   259
corporatocracy   233
economic institution   226
Family and Medical Leave Act (FMLA)   258
flextime   259
forced labor   241
free trade agreements   230
global economy   226
industrialization   228
job burnout   246

job exportation   236
labor unions   248
long-term unemployment rate   236
McDonaldization   229
outsourcing   236
post-industrialization   228
recession   235
socialism   227
sweatshops   242
telecommuting   259
transnational corporations   231
underemployment   236
unemployment   235
unemployment insurance   253
union density   249
Workforce Investment Act   252

MEDIA RESOURCES

Understanding Social Problems,
Seventh Edition Companion Website
www.cengage.com/sociology/mooney
Visit your book companion website, where you will find flash cards, practice quizzes, Internet links, and more to help you study.

Just what you need to know NOW! Spend time on what you need to master rather than on information you already have learned. Take a pretest for this chapter, and CengageNOW will generate a personalized study plan based on your results. The study plan will identify the topics you need to review and direct you to online resources to help you master those topics. You can then take a posttest to help you determine the concepts you have mastered and what you will need to work on. Try it out! Go to www.cengage.com/login to sign in with an access code or to purchase access to this product.

8

Problems in Education

> "Let us think of education as the means of developing our greatest abilities, because in each of us there is a private hope and dream which, fulfilled, can be translated into benefit for everyone and greater strength for our nation."
>
> *John F. Kennedy, Thirty-fifth president of the United States*

The Global Context: Cross-Cultural Variations in Education | Sociological Theories of Education | Who Succeeds? The Inequality of Educational Attainment | Problems in the American Educational System | **Self and Society: The Student Alienation Scale** | **Social Problems Research Up Close: Bullying and Victimization among Black and Hispanic Adolescents** | **The Human Side: You Want Heroes?** | Strategies for Action: Trends and Innovations in American Education | Understanding Problems in Education | Chapter Review

A freshman at one of the cutting-edge high schools in Baltimore, Maryland, Mariya "has enough talent to fill stadiums" (Toppo 2006, p. 1). At 15, she is already considered a gifted poet. She is the type of student every teacher wants in the classroom—bright and inquisitive, a "superstar" with a promising future.

Yet, Mariya is at home in bed—she didn't go to school today, or yesterday, or the day before. She has missed 35 of 62 days this semester and has already failed the ninth grade. Perhaps it's the morning sickness. Or perhaps, like her mother before her, she has just given up. At 15 and pregnant, Mariya's mother dropped out of school. This time it will be different, she insists. Mariya will graduate from high school—baby and all.

But day after day, with her grandmother and little sister by her side, Mariya remains at home. She studies her growing belly, wishing she could finish the year and go on to tenth grade. Realizing how much time she has already missed, she just sighs. "If I fail, I fail." She's tired. "I just don't feel like going to school."

Dropping out is just one of the many issues that must be addressed in today's schools. Students continue to graduate from high school unable to read, work simple mathematics problems, or write grammatically correct sentences. Graduates discover that they are ill prepared for corporations that demand literate, articulate, informed employees. Teachers leave the profession because of uncontrollable discipline problems, inadequate pay, and overcrowded classrooms. Students and teachers alike are "dumbing down"—lowering their standards, expectations, and role performances to fit increasingly undemanding and unresponsive systems of learning.

Yet education is often claimed as a panacea—the cure-all for poverty and prejudice, drugs and violence, war and hatred, and the like. Can one institution, riddled with problems, be a solution to other social problems? In this chapter, we focus on this question and on what is being called an educational crisis. We begin with a look at education around the world.

The Global Context: Cross-Cultural Variations in Education

Looking only at the American educational system might lead one to conclude that most societies have developed some method of formal instruction for their members. After all, the United States has more than 140,000 schools, 4.6 million primary and secondary school teachers and college faculty, 5.2 million administrators and support staff, and 74.1 million students (NCES 2009a). In reality, many societies have no formal mechanism for educating the masses. One in five adults cannot read or write—776 million people worldwide—the majority of them women (UNESCO 2009).

Education at a Glance, a publication of the Organization for Economic Cooperation and Development (OECD), reports education statistics on 30 countries (OECD 2009). Some interesting findings are revealed. First, in general, educational levels are rising. For example, many countries saw an increase in the proportion of young people attending colleges and universities. However, large numbers of people do not graduate from tertiary institutions. On the average, 27 percent of all adults in OECD-participating countries have completed postsecondary education;

One in five adults cannot read or write—776 million people worldwide—the majority of them women.

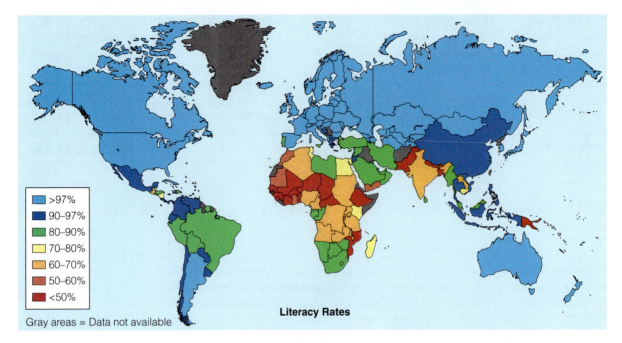

Literacy Rates	
>97%	
90–97%	
80–90%	
70–80%	
60–70%	
50–60%	
<50%	

Gray areas = Data not available

Figure 8.1 Literacy Percentages Around the World, 15 Years Old and Older, 2008
Source: Human Development Report 2008.

university and college completion rates range from a high of 47 percent in Canada to a low of 10 percent in Turkey.

Second, there is a clear link between education and income, and between education and employment. In general, across the OECD-participating countries, the more education people have, the higher their income and the greater the likelihood they will be employed. Across the 19 OECD-participating countries for which data were available, investment in a college degree, on the average, has a return rate of 11.5 percent (OECD 2009).

Third, across all OECD-participating countries, an average of $8,553 is spent per student each year they are in school from elementary school to college. Spending, although increasing in general, varies widely by country. In Mexico, Poland, and the Slovak Republic, just $4,000 or less is spent per student per year. However, in Norway, Austria, and the United States, on the average, more than $10,000 per student per year is invested in education costs (OECD 2009).

Fourth, in reference to teachers, the average student-teacher ratio in elementary schools in OECD-participating countries is 21:1, but increases as grade level increases. Teachers' salaries in almost all countries have increased over the last decade, with the greatest increases in Finland, Hungary, and Mexico (OECD 2009). Despite the fact that public school teachers' salaries in the United States averaged $51,009 in 2007 (AFT 2008a), approximately the OECD average, teachers in the United States work more hours than in any of the other 29 countries.

Fifth, educational attainment is increasing. In 18 of the 30 OECD-participating countries, the percentage of 25- to 34-year-olds who have completed high school is between 80 and 97 percent (OECD 2009). Ranging from a high of 47 percent in Canada to less than 10 percent in Brazil, the average percentage of 25- to 34-year-olds who have completed college across OECD-participating countries is 33 percent (OECD 2009).

Finally, the *Programme for International Student Assessment (PISA)* measures "how well students who are nearing the end of compulsory education are prepared to meet the challenges of today's knowledge societies—what PISA refers to as 'literacy'" (OECD 2009, p. 80). A random sample of over 400,000 15-year-old students in 60 countries participates in the PISA. The results indicate that just 1.3 percent of students earned the highest levels of science proficiency and, with the exception of Finland, Estonia, and Hong Kong/China, every country had at least 10 percent who scored at the lowest level of science proficiency.

Of all OECD-participating countries, 8.6 percent of students read at the highest level of reading literacy. Korea had the largest percentage of students at this level, whereas Mexico had the lowest, with less than 1 percent of its students achieving this level. Mathematics proficiency, as science, was quite low, with just 3.3 percent of students achieving the highest proficiency level and 12 percent performing at or below the lowest level. Of the 57 countries for which mathematics proficiency scores are available, students from Finland and Korea topped the list, U.S. students ranked thirty-fifth, and students from Kyrgyzstan had the lowest mathematics literacy scores as measured by PISA (OECD 2009).

The New York Times/Redux Pictures

Globally, few countries have the quality of schools or accessibility to education as enjoyed in the United States. Here, students in Lahtora, India, have outdoor lessons because their classroom is too crowded and too cold to have classes inside.

Sociological Theories of Education

The three major sociological perspectives—structural functionalism, conflict theory, and symbolic interactionism—are important in explaining different aspects of American education.

Structural-Functionalist Perspective

According to structural functionalism, the educational institution serves important tasks for society, including instruction, socialization, the sorting of individuals into various statuses, and the provision of custodial care (Sadovnik 2004). Many social problems, such as unemployment, crime and delinquency, and poverty, can be linked to the failure of the educational institution to fulfill these basic functions (see Chapters 4, 6, and 7). Structural functionalists also examine the reciprocal influences of the educational institution and other social institutions, including the family, political institution, and economic institution.

Instruction. A major function of education is to teach students the knowledge and skills that are necessary for future occupational roles, self-development, and social functioning. Although some parents teach their children basic knowledge and skills at home, most parents rely on schools to teach their children to read, spell, write, tell time, count money, and use computers. As discussed later, many U.S. students display a low level of academic achievement. The failure of schools to instruct students in basic knowledge and skills both causes and results from many other social problems.

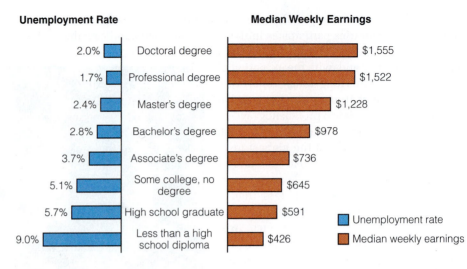

Figure 8.2 Unemployment Rate and Median Weekly Earnings for Individuals 25 and Over by Highest Level of Education, 2008
Source: BLS 2009.

Socialization. The socialization function of education involves teaching students to respect authority—behavior that is essential for social organization (Merton 1968). Students learn to respond to authority by asking permission to leave the classroom, sitting quietly at their desks, and raising their hands before asking a question. Students who do not learn to respect and obey teachers may later disrespect and disobey employers, police officers, and judges.

The educational institution also socializes youth into the dominant culture. Schools attempt to instill and maintain the norms, values, traditions, and symbols of the culture in a variety of ways, such as celebrating holidays (e.g., Martin Luther King Jr. Day and Thanksgiving); requiring students to speak and write in standard English; displaying the American flag; and discouraging violence, drug use, and cheating.

As the number and size of racial and ethnic minority groups have increased, American schools are faced with a dilemma: Should public schools promote only one common culture, or should they emphasize the cultural diversity reflected in the U.S. population? Some evidence suggests that most Americans believe that schools should do both—they should promote one common culture and emphasize diverse cultural traditions.

Multicultural education—that is, education that includes all racial and ethnic groups in the school curriculum—promotes awareness and appreciation for cultural diversity (also see Chapter 9). Two-thirds of a sample of U.S. adults responded that it was "very important" that colleges and universities teach students to participate in a diverse society (Orfield & Lee 2006).

Sorting Individuals into Statuses. Schools sort individuals into statuses by providing credentials for individuals who achieve various levels of education at various schools within the system. These credentials sort people into different statuses—for example, "high school graduate," "Harvard alumna," and "English major." In addition, schools sort individuals into professional statuses by awarding degrees in fields such as medicine, engineering, and law. The significance of such statuses lies in their association with occupational prestige and income—the higher one's education, the higher one's income. Furthermore, unemployment rates and earnings are tied to educational status as seen in Figure 8.2.

multicultural education
Education that includes all racial and ethnic groups in the school curriculum, thereby promoting awareness and appreciation for cultural diversity.

Custodial Care. The educational system also serves the function of providing custodial care by providing supervision and care for children and adolescents until they are 18 years old (Merton 1968). Despite 12 years and almost 13,000 hours of instruction, some school districts are increasing the number of school hours and/or days beyond the "traditional" schedule. In 2006, Massachusetts began the Expanded Learning Time Initiative, which adds an additional 300 hours of learning over the course of a school year by having longer and/or more school days (ELT 2009). Concerns over mandatory year-round schedules led a parents' group to sue local school district. In 2009, the North Carolina Supreme Court held that a school system does not need parental permission to assign students to year-round schools (Bowens 2009).

Conflict Perspective

Conflict theorists emphasize that the educational institution solidifies the class positions of groups and allows the elite to control the masses. Although the official goal of education in society is to provide a universal mechanism for achievement, in reality, educational opportunities and the quality of education are not equally distributed.

Conflict theorists point out that the socialization function of education is really indoctrination into a capitalist ideology (Sadovnik 2004). In essence, students are socialized to value the interests of the state and to function to sustain it. Such indoctrination begins in kindergarten. Rosabeth Moss Kanter (1972) coined the term the *organization child* to refer to the child in nursery school who is most comfortable with supervision, guidance, and adult control. Teachers cultivate the organization child by providing daily routines and rewarding those who conform. In essence, teachers train future bureaucrats to be obedient to authority.

In addition, to conflict theorists, education serves as a mechanism for **cultural imperialism,** or the indoctrination into the dominant culture of a society. When cultural imperialism exists, the norms, values, traditions, and languages of minorities are systematically ignored. A Mexican American student recalls his feelings about being required to speak English (Rodriguez 1990):

To cover financial costs associated with the increasing needs of educational programs, school systems often find it necessary to contract with major corporations such as Coca Cola.

> When I became a student, I was literally "remade"; neither I nor my teachers considered anything I had known before as relevant. I had to forget most of what my culture had provided, because to remember it was a disadvantage. The past and its cultural values became detachable, like a piece of clothing grown heavy on a warm day and finally put away. (p. 203)

Conflict theorists are also quick to note that learning is increasingly a commercial enterprise as corporations anxious to bombard students with advertising and other pro-capitalist messages fund necessary financial support for

cultural imperialism The indoctrination into the dominant culture of a society.

equipment, laboratories, and technological upgrades. Approximately 75 percent of high schools, 65 percent of middle schools, and 30 percent of elementary schools have exclusive contracts with soft drink companies (Shin 2007).

Finally, the conflict perspective focuses on what Kozol (1991) called the "savage inequalities" in education that perpetuate racial disparities. Kozol documented gross inequities in the quality of education in poorer districts, largely composed of minorities, compared with districts that serve predominantly white middle-class and upper-middle-class families. Kozol revealed that schools in poor districts tend to receive less funding and to have inadequate facilities, books, materials, equipment, and personnel. For example, disadvantaged children are more likely to attend schools with fewer certified or experienced teachers and to have fewer advanced placement courses than schools attended by more advantaged children (Viadero 2006).

Symbolic Interactionist Perspective

Whereas structural functionalism and conflict theory focus on macro-level issues, such as institutional influences and power relations, symbolic interactionism examines education from a micro-level perspective. This perspective is concerned with individual and small-group issues, such as teacher-student interactions and the self-fulfilling prophecy.

Teacher-Student Interactions. Symbolic interactionists have examined the ways that students and teachers view and relate to one another. For example, children from economically advantaged homes may be more likely to bring to the classroom social and verbal skills that elicit approval from teachers. From the teachers' point of view, middle-class children are easy and fun to teach. They grasp the material quickly, do their homework, and are more likely to "value" the educational process. Children from economically disadvantaged homes often bring fewer social and verbal skills to those same middle-class teachers, who may, inadvertently, hold up social mirrors of disapproval. Teacher disapproval contributes to lower self-esteem among disadvantaged youth.

Self-Fulfilling Prophecy. The **self-fulfilling prophecy** occurs when people act in a manner consistent with the expectations of others. For example, a teacher who defines a student as a slow learner may be less likely to call on that student or to encourage the student to pursue difficult subjects. The teacher may also be more likely to assign the student to lower ability groups or curriculum tracks (Riehl 2004). As a consequence of the teacher's behavior, the student is more likely to perform at a lower level.

A classic study by Rosenthal and Jacobson (1968) provided empirical evidence of the self-fulfilling prophecy in the public school system. Five elementary school students in a San Francisco school were selected at random and identified for their teachers as "spurters." Such a label implied that they had superior intelligence and academic ability. In reality, they were no different from the other students in their classes. At the end of the school year, however, these five students scored higher on their intelligence quotient (IQ) tests and made higher grades than their classmates who were not labeled as spurters. In addition, the teachers rated the spurters as more curious, interesting, and happy and more likely to succeed than the non-spurters. Because the teachers expected the spurters to do well, they treated the students in a way that encouraged better school performance.

self-fulfilling prophecy A concept referring to the tendency for people to act in a manner consistent with the expectations of others.

Who Succeeds? The Inequality of Educational Attainment

Figure 8.3 shows the extent of the variation in highest level of education attained by individuals 25 years of age and older in the United States. As noted earlier, conflict theory focuses on such variations in discussions of educational inequalities. Educational inequality is based on social class and family background, race and ethnicity, and gender. Each of these factors influences who succeeds in school.

Social Class and Family Background

One of the best predictors of educational success and attainment is socioeconomic status. Children whose families are in middle and upper socioeconomic brackets are more likely to perform better in school and to complete more years of education than children from lower socioeconomic class families. Using father's education as a proxy for socioeconomic status, global statistics indicate that students are significantly more likely to attend college if their fathers graduated from college—more than twice as likely in Austria, Germany, France, and England (OECD 2009).

Socioeconomic status also predicts academic achievement. On standardized tests such as the SAT and the ACT, "children from the lowest-income families have the lowest average test scores, with an incremental rise in family income associated with a rise in test scores" (Corbett et al. 2008, p. 3). Muller and Schiller (2000) reported that students from higher socioeconomic backgrounds are more likely to enroll in advanced courses for mathematics credit and to graduate from high school—two indicators of future educational and occupational success. In addition, compared with low-income students, high-income students are six times more likely to graduate with a bachelor's degree in five years (Toppo 2004). Moreover, of top-tier colleges and universities, less than 3 percent of enrolled students come from the lowest socioeconomic group (Viadero 2006).

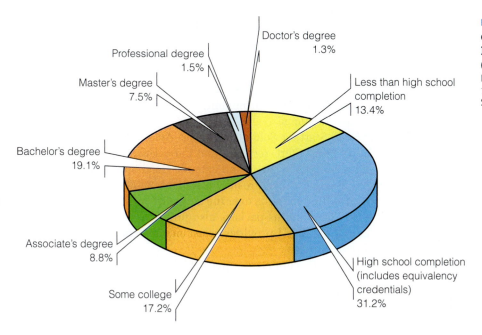

Figure 8.3 **Highest Level of Education Attained by Individuals Age 25 Years Old and Older, 2008**
NOTE: Percentages do not sum to 100 percent because of rounding.
Source: NCES 2009a.

Doctor's degree 1.3%
Professional degree 1.5%
Master's degree 7.5%
Bachelor's degree 19.1%
Associate's degree 8.8%
Some college 17.2%
Less than high school completion 13.4%
High school completion (includes equivalency credentials) 31.2%

What Do You Think? Research suggests that poor children who attend high-quality preschools are "less likely to drop out of school, repeat grades, or need special education, [and] as adults, are less likely to commit crimes, more likely to be employed, and likely to have higher earnings" (Olson 2007, p. 30). However, access to preschools varies by family income—the wealthier the family, the higher the probability a child will attend preschool. Further, although 38 states have state-financed preschools, yearly funding levels vary from a high of $10,989 per child in New Jersey to less than $2,000 per child in South Carolina and Maine (NIEER 2008). Given the above, should the federal government establish a nationalized program of preschools for all 3- and 4-year-olds? If so, should all preschool teachers be required to have a bachelor's degree?

Families with low incomes have fewer resources to commit to educational purposes—less money to buy books or computers or to pay for tutors. Disadvantaged parents are less involved in learning activities. As parental education and income levels increase, the likelihood of a parent taking a child to a library, play, concert or other live show, art gallery, museum, or historical site also increases. Further, parents who have less education and lower income levels are less likely to be involved in school-related activities, for example, attending a school event or volunteering at a school (NCES 2009a).

Head Start and Early Head Start. In 1965, Project **Head Start** began to help preschool children from the most disadvantaged homes. Head Start provides an integrated program of health care, parental involvement, education, and social services. Today, over 900,000 preschoolers (3- and 4-year-olds) in fifty states participate in the program which, in 2007, had an average cost of $7,326 per child (Office of Head Start 2008). Graduates of Head Start "score better on intelligence and achievement tests, their health status is better, and they have the socio-emotional traits to help them adjust to school" (Zigler et al. 2004, p. 341).

Early Head Start, a program for infants and toddlers from low-income families, has also been evaluated. A seven-year national evaluation of Early Head Start programs culminated in a 2004 report to Congress. In summarizing the report, the authors conclude that "participating children perform significantly better in cognitive, language, and social-emotional development than their peers who do not participate. The program also had important impacts on many aspects of parenting and the home environment, and supported parents' progress toward economic self-sufficiency" (MPR 2007, p. 1).

Head Start Begun in 1965 to help preschool children from the most disadvantaged homes, Head Start provides an integrated program of health care, parental involvement, education, and social services for qualifying children.

Title I Part of the *Elementary and Secondary Education Act* of 1965, Title I provides assistance to school districts with a high proportion of poor children.

The Improving Head Start for School Readiness Act of 2007 authorized new funds to expand Early Head Start, requires that Early Head Start teachers have child development credentials by 2010, and extends Head Start funding through 2012 (Head Start Act 2008). Further, President Obama's 2009 economic recovery package provides $2.1 billion in funds over the course of two years for Head Start and Early Head Start. The stimulus package also provides $13 billion for Title I initiatives (see Figure 8.4). **Title I** provides assistance to school districts with a high proportion of poor children.

Although helping ailing state economies, federal stimulus dollars cannot entirely compensate for the loss of state expenditures on education (Goodnough 2009). Thirty-six of the 47 states presently facing deficits have or are proposing cuts to education (Johnson et al. 2009). Economic shortfalls in California are so severe that schools districts have been forced "to lay off thousands of teachers, expand class sizes, close schools, eliminate bus service,

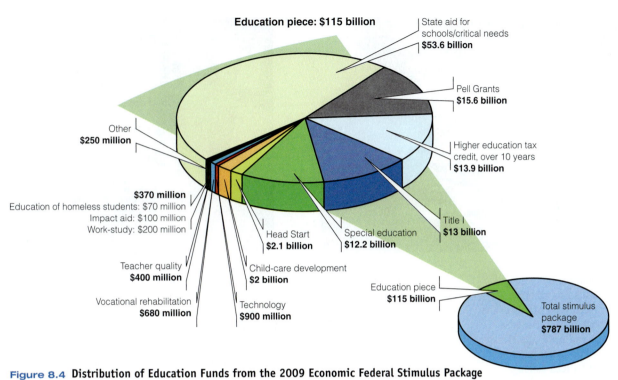

Figure 8.4 Distribution of Education Funds from the 2009 Economic Federal Stimulus Package
Source: Klein 2009.

cancel summer school programs, and possibly shorten the academic year" (Associated Press 2009, p. 1).

Further, local dollars, which make up 45 percent of school funding, varies by socioeconomic status of the district (NCES 2009b). For example, local expenditures on schools come from taxes, usually property taxes, and as housing prices decline, property taxes decline. The U.S. system of decentralized funding for schools has several additional consequences:

- Low socioeconomic status school districts are more likely to be in urban areas and in inner cities where the value of older and dilapidated houses have depreciated; less desirable neighborhoods are hurt by "white flight," with the result that the tax base for local schools is lower in deprived areas.
- Low socioeconomic status school districts are less likely to have businesses or retail outlets where revenues are generated; such businesses have closed or moved away.
- Because of their proximity to the downtown area, low socioeconomic status school districts are more likely to include hospitals, museums, and art galleries, all of which are tax-free facilities. These properties do not generate revenues.
- Low socioeconomic status neighborhoods are often in need of the greatest share of city services; fire and police protection, sanitation, and public housing consume the bulk of the available revenues. Precious little is left over for education in these districts.
- In low socioeconomic status school districts, a disproportionate amount of the money has to be spent on maintaining the school facilities, which are old and in need of repair, and on free or reduced-priced lunches, so less is available for the children themselves.

Although states provide additional funding to supplement local taxes, this funding is not always enough to lift schools in poorer districts to a level that even approximates the funding available to schools in wealthier districts. For

Rose A. Young Elementary School at Three Mile Creek designed by
VCBO Architecture, Salt Lake City, Utah

Dan Habib

As shown above, the expenditure per pupil varies dramatically by socioeconomic status of the school district. Between 2007 and 2008, the average expenditure per pupil was $9,361 (NCES 2009a).

example, in *Leandro v. State* (1997), the North Carolina Supreme Court held that the state constitution required that all schools must provide adequate resources to fully educate disadvantaged students (i.e., those who are poor, in special education, and have limited English proficiency). Yet, over a decade later, at least one county in North Carolina is doing such a poor job of educating its disadvantaged students that a judge has called it "academic genocide" (Waggoner 2009).

Race and Ethnicity

Between 1987 and 2007, the percentage of racial and ethnic minorities in public schools increased from 32 percent to 44 percent (NCES 2009b). To a large extent, this demographic shift in the racial and ethnic makeup of public school students is a result of the growth in the number of Hispanic students—the fastest-growing minority in the United States.

In comparison to whites, Hispanics and blacks are less likely to succeed in school at almost every level. As early as the start of kindergarten, Hispanics and blacks have lower mean achievement scores than other racial or ethnic groups in reading and mathematics (NCES 2009b). By fourth grade, over 80 percent of blacks and Hispanics are reading below grade level compared to 58 percent of whites. Similarly, by eighth grade, over 80 percent of black and Hispanic students compared to 59 percent of white students are below grade level in mathematics (CDF 2008). Further, as Table 8.1 indicates, although educational attainment has increased over time, racial and ethnic disparities remain. Note that, in general, the high school graduation gap between racial and ethnic groups is narrowing; the college graduation gap, however, is getting wider—whites and Asians on one side and Hispanics and Black Americans on the other.

It is important to note, however, that socioeconomic status interacts with race and ethnicity. Because race and ethnicity are so closely tied to socioeconomic status, i.e., a disproportionate number of racial and ethnic minorities are poor, it *appears* that race or ethnicity determines school success. Although race and ethnicity may have an independent effect on educational achievement, their relationship is largely a result of the association between race and ethnicity and socioeconomic status.

Although race and ethnicity may have an independent effect on educational achievement, their relationship is largely a result of the association between race and ethnicity and socioeconomic status.

TABLE 8.1 Educational Attainments by Race, Ethnicity, and Sex, 1970 and 2007

	1970		2007	
	MALES	FEMALES	MALES	FEMALES
Four years of high school or more				
White	54.0	55.0	85.3	89.1
Black	30.1	32.5	81.9	82.6
Hispanic	37.9	34.2	58.2	62.5
Asian and Pacific Islander	NA	NA	89.8	85.9
Total	51.9	52.8	85.0	86.4
Four years of college or more				
White	14.4	8.4	29.9	28.3
Black	4.2	4.6	18.0	19.0
Hispanic	7.8	4.3	11.8	13.7
Asian and Pacific Islander	NA	NA	55.2	49.3
Total	13.5	8.1	29.5	28.0

Note: NA = Not available
Source: U.S. Census Bureau 2009.

In addition to the socioeconomic variables, there are several reasons why minority students have academic difficulty. First, minority children may be English language learners (ELL). Although ELL students come from over 400 different language backgrounds, over 75 percent are Spanish speakers (Goldenberg 2008). Now imagine, as Goldenberg (2008) suggests, you are in second grade and don't speak English very well and are expected to learn in one year

> . . . irregular spelling patterns, diphthongs, syllabication rules, regular and irregular plurals, common prefixes and suffixes, antonyms and synonyms; how to follow written instructions, interpret words with multiple meanings, locate information in expository texts . . . read fluently and correctly at least 80 words per minute, add approximately 3,000 words to your vocabulary . . . and write narratives and friendly letters using appropriate forms, organization, critical elements, capitalization, and punctuation, revising as needed. (p. 8)

Not surprisingly, ELL students score significantly below non-ELL students on standardized tests in both reading and mathematics. To help ELL students, some educators advocate **bilingual education,** teaching children in both English and their non-English native language.

Advocates claim that bilingual education results in better academic performance of minority students, enriches all students by exposing them to different languages and cultures, and enhances the self-esteem of minority students. Critics argue that bilingual education limits minority students and places them at a disadvantage when they compete outside the classroom, reduces the English skills of minorities, costs money, and leads to hostility with other minorities who are also competing for scarce resources.

bilingual education In the United States, teaching children in both English and their non-English native language.

What Do You Think? In 2002, voters in Boston's 56,000-student school district voted to end bilingual education in favor of English-only instruction (Vaznis 2009). According to a recent report, since the enactment of the law, dropout rates of ELL students have nearly doubled, and test scores have not improved and in some cases have gone down. Do you think bilingual education or English-only instruction better serves the needs of non-native English-speaking students? Of society as a whole?

The second reason why racial and ethnic minorities don't perform well in school, and compounding the difficulty ELL students have, is that many of the tests used to assess academic achievement and ability are biased against minorities. Questions on standardized tests often require students to have knowledge that is specific to the white middle-class majority culture, knowledge that racial and ethnic minorities may not have.

A third factor that hinders minority students' academic achievement is overt racism and discrimination. Discrimination against minority students takes the form of unequal funding, as discussed earlier, as well as racial profiling and school segregation.

Increasingly, school classrooms will be characterized by racial and ethnic diversity. By 2040, less than half of all school-age children will be non-Hispanic whites.

Studies indicate that minority students, and specifically black students, may be the victims of what is being called "learning while black" (Morse 2002). The allegation is often supported by an examination of school discipline statistics. For example, in 2006, 15 percent of Black American students were suspended and expelled from school, compared with between 3 percent to 8 percent of all other racial and minority groups (NCES 2009b). Although the debate is likely to continue, it must be noted that differences in discipline patterns are not necessarily a consequence of racism. They may reflect differences in behavior.

School Desegregation. In 1954, the U.S. Supreme Court ruled in *Brown v. Board of Education* that segregated education was unconstitutional because it was inherently unequal. In 1966, a landmark study titled *Equality of Educational Opportunity* (Coleman et al. 1966) revealed that almost 80 percent of all U.S. schools attended by whites contained 10 percent or fewer blacks and that, with the exception of Asian Americans, whites outperformed minorities on standardized tests. Coleman and colleagues emphasized that the only way to achieve quality education for all racial groups was to desegregate the schools. This recommendation, known as the **integration hypothesis,** advocated busing to achieve racial balance.

Despite the Coleman report, court-ordered busing, and a societal emphasis on the equality of education, public schools remain largely segregated. Most black and Hispanic U.S. students attend schools that are predominantly minority in enrollment. For example, in 2007, over half of all Black American and Hispanic public school students attended schools that had a combined minority enrollment of 75 percent or more (NCES 2009b). Commenting on the "restoration of apartheid schooling in America" Jonathan Kozol (2005) notes that "[s]chools that were already deeply segregated 25 or 30 years ago . . . are no less segregated now, while thousands of other schools that had been integrated either voluntarily or by force of law have since been rapidly re-segregating . . ." (p. 18). Attending a desegregated school, while having no negative achievement effect on white students, tends to have positive effects on black and Hispanic students in terms of learning and graduation rates (Orfield & Lee 2006).

integration hypothesis A theory that the only way to achieve quality education for all racial and ethnic groups is to desegregate the schools.

In 2007, the U.S. Supreme Court held that public school systems "can not seek to achieve or maintain integration through measures that take explicit account of a student's race" (Greenhouse 2007, p. 1). The court's decision reflects a general trend toward using socioeconomic or income-based integration rather than race-based integration variables. Kahlenberg (2006) advocated this approach for several reasons. First, "socioeconomic integration more directly and effectively achieves the first aim of racial integration: raising the achievement of students" (Kahlenberg 2006, p. 10). Second, socioeconomic integration, because of the relationship between race and income, achieves racial integration, and racial integration, in turn, fosters racial tolerance and social cohesion. Lastly, unlike race-based integration that is subject to "strict scrutiny" by the government, school assignments based on socioeconomic status are perfectly legal. Nonetheless, several school districts throughout the United States remain under federal oversight as required by decades-old racial desegregation cases.

Gender

Worldwide, women receive less education than men. An estimated 776 million adults in the world are illiterate, and two-thirds of them are women (OECD 2009). Further, girls comprise more than 70 percent of the 125 million children who don't attend school (Save the Children 2009). Although progress in reducing the education gender gap has been made, gender parity in primary and secondary schools has not been achieved. Globally, of the 181 countries with data available, one-third have not achieved gender parity in primary schools; of the 177 countries with data available, two-thirds have not achieved gender parity in secondary schools (EFA Global Monitoring Report 2007).

Historically, U.S. schools have discriminated against women. Before the 1830s, U.S. colleges accepted only male students. In 1833, Oberlin College in Ohio became the first college to admit women. Even so, in 1833, female students at Oberlin were required to wash male students' clothes, clean their rooms, and serve their meals and were forbidden to speak at public assemblies (Fletcher 1943; Flexner 1972).

In the 1960s, the women's movement sought to end sexism in education. Title IX of the Education Amendments of 1972 states that no person shall be discriminated against on the basis of sex in any educational program receiving federal funds. These guidelines were designed to end sexism in the hiring and promoting of teachers and administrators. Title IX also sought to end sex discrimination in granting admission to college and awarding financial aid. Finally, the guidelines called for an increase in opportunities for female athletes by making more funds available to their programs.

What Do You Think? In 2006, the U.S. Department of Education gave "public school districts broad new latitude to expand the number of single-sex classes and schools . . ." (Schemo 2006, p. 1). Despite mixed research results, i.e., under some conditions single-sex classes (SSC) benefit some males and some females some of the time, by 2008, there were over 300 single-sex classrooms and schools in the United States (Spielhagen 2007; Sadker & Zittleman 2009). Proponents argue that SSC allow schools to cater to the needs of both girls and boys by providing students an equal opportunity for success in all subjects. Critics argue that SSC are discriminatory and cater to gender stereotypes. What do you think about SSC? Given the option, would you have preferred to be in SSC?

Although gender inequality in education continues to be a problem worldwide, the push toward equality has had considerable effect in the United States. For example, in 1970, nearly twice as many men as women had four years of college or more—8.1 percent compared with 13.5 percent. By 2007, 29.5 percent of men and 28 percent of women had four years of college or more (see Table 8.1). Further, scores on the National Assessment of Educational Progress (NAEP) exam indicate that over the last few decades the gender gap in both mathematics and reading scores has decreased. Where differences exist, for example, in a specific grade level, they generally follow academic stereotypes—boys outscore girls in mathematics, girls outscore boys in reading. Similarly, on the SAT college admissions test, males, on the average, outperform females, with the largest achievement gap on the mathematics portion of the exam (Corbett et al. 2008).

Why does the gender gap, however slight, persist and in such a predictable direction? Traditional gender roles may account for some of the differences in educational achievement between women and men. As noted in Chapter 10, schools, teachers, and educational materials reinforce traditional gender roles. Further, a body of evidence suggests that the tests themselves are biased (Sadker & Zittleman 2009). For example, standardized tests are timed and generally contain multiple-choice questions—a format that benefits boys.

Most of the research on gender inequality in the schools focuses on how female students are disadvantaged in the educational system. But what about male students? Although often achieving higher scores on standardized tests (Corbett et al. 2008), boys are more likely to lag behind girls in the classroom, be diagnosed with attention deficit/hyperactivity disorder (ADHD), have learning disabilities, feel alienated from the learning process, and drop out or be expelled from school (see this chapter's Self and Society feature and assess your own school alienation rate) (Dobbs 2005; Mead 2006; Tyre 2008). Further, Black American males compared to white males score lower on the NAEP, are less likely to be in advanced placement (AP) classes and more likely to be in special education classes, and are less likely to graduate from high school or college (Schott Report 2008). Thus, the problems boys have in school may indeed require schools to devote more resources and attention to them (e.g., recruit male teachers). Unfortunately, in a time of economic crisis, that may not be possible.

Problems in the American Educational System

When a random sample of adults were asked the average grade they would give local public schools and public schools throughout the nation, the average grade given was 2.40 and 2.02, respectively, on a five-point scale (NCES 2009a). The implication that American schools are just "average" is strengthened by the Educational Research Center's (ERC) annual assessment of education in the United States. Analyzing data across states in several substantive areas (e.g., learning, teachers) the Center assigned an overall score to the nation's schools of 75.9 percent—a C (ERC 2009). Not surprisingly, the Obama administration hopes to address many of the problems—low academic achievement, high dropout rates, questionable teacher training, school violence, and the challenges of higher education—that the ERC and other experts identified.

Indicate your agreement with each statement by selecting one of the responses provided:

1. It is hard to know what is right and wrong because the world is changing so fast.
__ Strongly agree __ Agree __ Disagree __ Strongly disagree

2. I am pretty sure my life will work out the way I want it to.
__ Strongly agree __ Agree __ Disagree __ Strongly disagree

3. I like the rules of my school because I know what to expect.
__ Strongly agree __ Agree __ Disagree __ Strongly disagree

4. School is important in building social relationships.
__ Strongly agree __ Agree __ Disagree __ Strongly disagree

5. School will get me a good job.
__ Strongly agree __ Agree __ Disagree __ Strongly disagree

6. It is all right to break the law as long as you do not get caught.
__ Strongly agree __ Agree __ Disagree __ Strongly disagree

7. I go to ball games and other sports activities at school.
__ Always __ Most of the time __ Some of the time __ Never

8. School is teaching me what I want to learn.
__ Always __ Most of the time __ Some of the time __ Never

9. I go to school parties, dances, and other school activities.
__ Always __ Most of the time __ Some of the time __ Never

10. A student has the right to cheat if it will keep him or her from failing.
__ Always __ Most of the time __ Some of the time __ Never

11. I feel like I do not have anyone to reach out to.
__ Always __ Most of the time __ Some of the time __ Never

12. I feel that I am wasting my time in school.
__ Always __ Most of the time __ Some of the time __ Never

13. I do not know anyone that I can confide in.
__ Strongly agree __ Agree __ Disagree __ Strongly disagree

14. It is important to act and dress for the occasion.
__ Always __ Most of the time __ Some of the time __ Never

15. It is no use to vote because one vote does not count very much.
__ Strongly agree __ Agree __ Disagree __ Strongly disagree

16. When I am unhappy, there are people I can turn to for support.
__ Always __ Most of the time __ Some of the time __ Never

17. School is helping me get ready for what I want to do after college.
__ Strongly agree __ Agree __ Disagree __ Strongly disagree

18. When I am troubled, I keep things to myself.
__ Always __ Most of the time __ Some of the time __ Never

19. I am not interested in adjusting to American society.
__ Strongly agree __ Agree __ Disagree __ Strongly disagree

20. I feel close to my family.
__ Always __ Most of the time __ Some of the time __ Never

21. Everything is relative and there just aren't any rules to live by.
__ Strongly agree __ Agree __ Disagree __ Strongly disagree

22. The problems of life are sometimes too big for me.
__ Always __ Most of the time __ Some of the time __ Never

23. I have lots of friends.
__ Always __ Most of the time __ Some of the time __ Never

24. I belong to different social groups.
__ Strongly agree __ Agree __ Disagree __ Strongly disagree

Interpretation

This scale measures four aspects of alienation: powerlessness, or the sense that high goals (e.g., straight As) are unattainable; meaninglessness, or lack of connectedness between the present (e.g., school) and the future (e.g., job); normlessness, or the feeling that socially disapproved behavior (e.g., cheating) is necessary to achieve goals (e.g., high grades); and social estrangement, or lack of connectedness to others (e.g., being a "loner"). For items 1, 6, 10, 11, 12, 13, 15, 18, 19, 21, and 22, the response indicating the greatest degree of alienation is "strongly agree" or "always." For all other items, the response indicating the greatest degree of alienation is "strongly disagree" or "never."

Source: Mau 1992. Used by permission of Libra Publishers, Inc., 3089 Clairemont Drive, Suite 383, San Diego, California 92117.

Low Levels of Academic Achievement

The Educational Research Center uses three indicators to measure achievement in public elementary and secondary schools: current levels of performance, improvement over time, and the achievement gap between poor and nonpoor learners (ERC 2009). Based on a 100-point scale, the achievement average for the nation was 69 (D+), ranging from a high of 85.2 (B) for Massachusetts and a low of 55.9 (F) for Mississippi (ERC 2009).

One way to measure performance is to look at the results of the National Assessment of Education Progress, the "nation's report card" on public and private school student performances. Although varying by race, ethnicity, and gender, in general, mathematics scores of 9-year-olds (fourth graders) and 13-year-olds (eighth graders) has significantly increased since 2004; the average mathematics score for 17-year-olds (twelfth graders) over the same time period is unchanged (NCES 2009b). Reading scores, however, have increased for each of the three age groups.

It is important to note that even statistically significant increases, although a step in the right direction, may mask poor performances. For example, as noted, reading scores increased significantly for 9-, 13-, and 17-year-olds between 2004 and 2008. However, 33 percent of 9-year-olds and 26 percent of 13-year-olds scored below the "basic" reading level in 2007. Similarly, 18 percent of 9-year-olds and 29 percent of 13-year-olds scored below the "basic" mathematics level in 2007 (NCES 2009b).

U.S. students are also outperformed by many of their foreign counterparts—something particularly troubling in a knowledge-based global economy. Based upon the Program for International Assessment (PISA), U.S. 15-year-olds perform below the OECD country averages in both science and mathematics. The U.S. average high school graduation rate is also below the average high school graduation rate for OECD-participating countries (OECD 2009). Perhaps most troubling, the international achievement gap "widens the longer children are in school" and is "not merely an issue for poor children attending schools in poor neighborhoods; instead, it affects most children in most schools" (McKinsey and Company 2009, p. 7). The international achievement gap impacts individuals and communities and costs the economy millions of dollars in lost human potential (McKinsey and Company 2009).

School Dropouts

The *status dropout rate* is the percentage of an age group that is not in school and has not earned a high school degree or its equivalent. In 2008, the status dropout rate for 16- to 24-year-olds was 9 percent, a reduction of 14 points since 1980 (NCES 2009b). As Figure 8.5 indicates, dropout rates vary considerably by race and ethnicity, and by nativity. They also vary by gender, with males having higher dropout rates than females, 11 percent and 8 percent, respectively.

Students from lower socioeconomic backgrounds who have only one parent in the home and who have changed schools frequently are more likely to drop out (Barton 2005; Tyler & Lofstrom 2009). Further, a study funded by the Bill and Melinda Gates Foundation, using focus groups and a survey of more than 500 dropouts, reports that respondents identified five major reasons for dropping

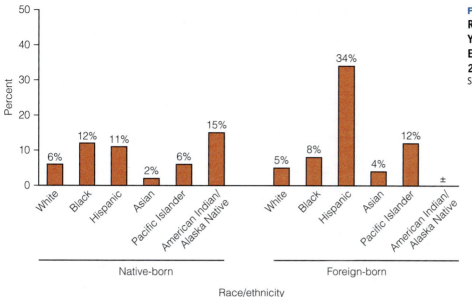

Figure 8.5 Status Dropout Rates of 16- to 24- Year-Olds, by Race/ Ethnicity and Nativity, 2007
Source: NCES 2009b.

± Reporting standards not met (too few cases).

out of high school: (1) "classes were not interesting"; (2) "missed too many days and can't catch up"; (3) "spent time with people who were not interested in school"; (4) "had too much freedom and not enough rules"; and (5) "was failing in school" (Bridgeland et al. 2006, p. 3).

The former students were also asked what could be done to improve a student's chances of remaining in school. Eighty-one percent responded "opportunities for real-world learning (internships, service-learning, etc.) to make classrooms more relevant" and "better teachers who keep classes interesting" (Bridgeland et al. 2006, p. 13). Other top responses included smaller classes and more individual attention, better communication between parents and schools, and increased supervision at home and at school to ensure students attend classes.

The economic and social consequences of dropping out of school are significant. A million children drop out of high school every year, reducing tax revenues and increasing societal costs for public assistance, crime, and health care (Tyler & Lofstrom 2009). Some programs are successful in reducing the dropout rate. A review of dropout prevention programs by Tyler and Lofstrom (2009) suggests that successful programs share five common elements: (1) management of students, (2) mentoring, (3) family involvement, (4) curricular reform, and (5) contending with out-of-school problems.

> A million children drop out of high school every year, reducing tax revenues and increasing societal costs for public assistance, crime, and health care.

Second-chance initiatives such as General Educational Development (GED) certification allow students to complete their high school requirements. Alternatively, early or middle school college programs allow dropouts to enroll in community colleges or, in some cases, four-year degree programs. There, they receive a secondary school education, earn a high school diploma, and often accrue college credits (Manzo 2005).

What Do You Think? California, as roughly half of all states, requires an exit exam for high school graduation. Researchers have found that, after the exam was put into place in California, (1) lower-achieving students had significantly higher dropout rates, and (2) the negative effect of the exit exam was stronger for minority and female students than for nonminority and male students, even when levels of academic achievement were held constant (Reardon et al. 2009). These findings support the "**stereotype threat**" hypothesis, i.e., the tendency for minorities and women to perform poorly on high-stakes tests because of the anxiety created by fear that a poor performance will validate negative societal stereotypes. Do you think high schools should or should not have exit exams as a requirement for graduation? Do you think exit exams unfairly target minorities and females?

Crime, Violence, and School Discipline

Despite the horrors of high-publicity school killings such as those at Columbine and Virginia Tech, the chance of a student dying at school is quite rare—about one homicide or suicide per 1.6 million students (NCES 2009c). The unlikelihood of such an event is reflected in students perceptions of safety. In 2008, 93 percent of students said they felt very or somewhat safe while at school, a number that has changed little over the last 15 years (MetLife 2008). In 2007, less than 5 percent of students reported being the victim of school-related crime in the previous six months.

In the 2005–2006 school year, the latest term for which data are available, over 75 percent of schools reported at least one violent crime. As in the nation as a whole, the number of school-related nonviolent offenses exceeded the number of school-related violent offenses. Nearly half of all schools reported that a theft had taken place during the school year—nearly one million thefts in all (NCES 2009c). Schools that experienced higher-than-average rates of school-related violent crime, whether against a student or teacher, were disproportionately in public versus private schools, in poor, urban neighborhoods, and were more likely to have gang-related activity (NCES 2009c; CDC 2008).

Discipline problems such as verbal abuse of teachers, disorder in classrooms, disrespect for teachers, fighting, insubordination, and the use of drugs or alcohol are also a concern. Of the disciplinary actions available to school personnel, suspension and expulsion are the most serious. In 2006, one in 14 public school students was suspended and one in 476 students expelled (NCES 2009b). Of late, one particular type of school-related problem, bullying, has become the focus of research and legislative action. For example, the Safe Schools Improvement Act of 2009, if passed, would require states and school districts to clearly articulate policies that prohibit bullying, develop bullying prevention programs, and maintain a bullying database of recorded bullying incidents.

Bullying is characterized by an "imbalance of power that exists over a long period of time between two individuals, two groups, or a group and an individual in which the more powerful intimidate or belittle others" (Hurst 2005, p. 1) (see this chapter's Social Problems Research Up Close feature). Bullying may be direct (e.g., hitting someone) or indirect (e.g., spreading rumors) and may be considered a type of aggression (Wong 2009). Sometimes called cyberbullying, bullying can also take place remotely, i.e., through electronic communication devices (e.g., cell phone) (see Chapter 14).

stereotype threat The tendency of minorities and women to perform poorly on high-stakes tests because of the anxiety created by the fear that a negative performance will validate societal stereotypes about one's member group.

Bullying Bullying "entails an imbalance of power that exists over a long period of time in which the more powerful intimidate or belittle others" (Hurst 2005, p. 1).

Researchers often study the variables associated with being a student bully or victim. The hope is that, if we can identify the characteristics of a student bully or victim early on, we can develop successful interventions that will reduce the prevalence of the behavior. To that end, Peskin et al. (2006) investigate variables associated with being a bully, a victim, or both, in a sample of middle and high school students.

Sample and Methods

Students from eight predominantly black and Hispanic secondary schools located in a large urban school district in Texas were selected for study. Classes were sampled by grade, resulting in a sample size of 1,413 respondents and a response rate of 52 percent. Nearly 60 percent of the sample was females. Middle school students (sixth, seventh, and eighth graders) comprised 56 percent of the sample, ninth graders comprised 11 percent of the sample, and tenth through twelfth graders comprised 32 percent of the sample. Sixty-four percent of the sample described themselves as Hispanic, with the remainder self-identifying as African American.

Students were asked about their participation in bullying and their rates of victimization in the last 30 days. Response options included two categories: 0 to 2 times, and 3 or more times. Students were asked the frequency of their bullying behavior: (1) upsetting other students for the fun of it, (2) group teasing, (3) harassing, (4) teasing, (5) rumor spreading, (6) starting arguments, (7) getting others to fight, and (8) excluding others. There were four student victimization variables: (1) called names, (2) picked on by others, (3) made fun of, and (4) got hit and pushed.

In addition to measuring the frequency of bullying and victimization events, students were classified into one of four categories:

> . . . a student was classified as a bully if he/she participated in at least two of the "bullying" behaviors at least

three times in the last 30 days. Victims were classified as those students who reported that at least one of the "victim" behaviors happened to them at least three times in the last 30 days. Four mutually exclusive categories were constructed: (1) bullies; (2) victims; (3) those who reported both bullying and being a victim (bully-victim); and (4) students reporting neither behavior (p. 471)

Findings and Conclusions

Seven percent of the sample was defined as bullies, 12 percent as victims, and 5 percent as bully-victims. Grade level was found to be significantly related to bully or victim status. The prevalence of bullying is highest in the ninth grade (11.5 percent) and lowest in the sixth grade (4.9 percent) and tenth grade (6.2 percent). The highest rate of victimization is in the sixth grade, with 20.8 percent of sixth graders meeting the criteria for being a victim. The lowest rate of being a victim was among eleventh and twelfth graders (7.5 percent). Bully-victims also varied by grade level, with the highest levels occurring in the eleventh and twelfth grades (7.9 percent) and lowest levels in the sixth grade (3.8 percent) and ninth grade (3.8 percent).

There were no significant relationships between the dependent variable and gender. Males and females were equally likely to report being bullies, victims, and bully-victims. However, race and ethnicity were significantly associated with the dependent variable. Blacks, compared with Hispanics, were more likely to be bullies, victims, and bully-victims. For example, although only 3.7 percent of Hispanics reported being bully-victims, 8.6 percent of blacks reported being bully-victims.

When specific bullying behaviors are examined, "upsetting students for the fun of it" was the most common kind of bullying activity, followed by teasing, group teasing, and starting arguments. The lowest prevalence of bullying behaviors

was getting others to fight. Males were significantly more likely to participate in harassing and teasing behaviors than females and, with the exception of rumor spreading, African Americans were more likely to be involved in bullying behaviors than Hispanics.

Victims report that the most common form of bullying is name calling, followed by being made fun of. Males were significantly more likely to be hit and pushed than females, and blacks were significantly more likely to be made fun of, called names, or be hit and pushed than their Hispanic counterparts. Further, eleventh and twelfth graders were statistically more likely to report being picked on or made fun of than students in other class levels.

The authors conclude that the results of the study suggest the need for early intervention and the direction it should take:

> While steps to decrease physical types of bullying may be targeted largely at males, steps to reduce verbal and relational types should be targeted at all students. . . . Interventions should be developed in middle school as the prevalence of these behaviors seems to peak as students begin high school. . . . In our study, teasing and name calling were most prevalent; thus, targeted actions for the reduction of these behaviors may be a focus for intervention activities. (p. 479)

Finally, the authors note that future research should concentrate on the role of race and ethnicity in the prevalence of bullying and victimization. Rather than simply noting racial and ethnic differences, however, research must begin to articulate the way in which racial and ethnic differences impact "the content" of bullying behavior. Only then can we fully appreciate the role of social factors "in the development of bullying and victimization problems" (p. 480).

Source: Peskin et al. 2006.

Research by Wong (2009), NCES (2009c), and AASA (2009) indicates that:

- In 2007, about a third of students between 12 and 18 reported being bullied in the last year.
- Students who bully often perform poorly academically, have high dropout rates, are more likely to get into fights, drink alcohol, vandalize property, and be truant.
- Victims of bullying report, in order of frequency, being the subject of rumors, being pushed, shoved, tripped, spit on, being threatened with harm, and being intentionally excluded from activities.
- Most bullying takes place inside the school building and is more often directed at females than males.
- Being bullied is associated with low self-esteem, anxiety, depression, alcohol and drug use, running away, and suicide.

Curtis Compton/ AP Photo

In 2009, Jaheem Herrera (pictured) and Carl Walker-Hoover, both 11 years old, committed suicide. Despite countless complaints to school personnel by both their mothers, antigay bullying by classmates continued. Neither child self-identified as gay but each was subjected to a daily barrage of slurs including "queer," "faggot," and "homo."

In 2009, 11-year-olds Jaheem Herrera and Carl Walker-Hoover, in different schools in different states, came home after school, went to their rooms, and hung themselves. Each had been tormented by classmates for being gay. According to a report by the Gay, Lesbian, and Straight Education Network (GLSEN), perceived sexual orientation and gender expression follow physical appearance as the most common reasons for bullying (Robbins & Byard 2009).

In response to crime, violence, and other disciplinary problems, schools throughout the country have police officers patrolling the halls, require students to pass through metal detectors before entering school, and conduct random locker searches. Video cameras set up in classrooms, cafeterias, halls, and buses purportedly deter some school violence. A relatively new way to deal with school "troublemakers" is the *alternative* school. In theory, such schools deter disciplinary problems while providing low-achieving students with the remedial help they need. A recent evaluation of Mississippi alternative schools, however, concluded that alternative schools may contribute to the "school-to-prison pipeline" by "pushing out and criminalizing students who misbehave" (ACLU 2009, p. 5).

Inadequate School Facilities and Programs

A report by the American Federation of Teachers (AFT) has documented many of the troubling conditions that exist in U.S. schools:

... rodent infestation, mice droppings, fallen ceiling tiles, poor lighting, mold that has caused mushrooms to grow, crumbling exterior walls, asbestos, severely overcrowded classrooms and hallways, freezing rooms in the winter and extreme heat in the summer, old carpeting, clogged bathroom toilets and no stall doors, inadequate circuit breakers causing frequent outages, and poor ventilation. (AFT 2006, p. 4)

Presently, it is estimated that state governments need over $254 billion for school infrastructure repairs and maintenance, renovations and new construction, retrofitting, major improvements, and existing building additions (AFT 2009). Older schools have greater needs than newer ones, and are more often located in disadvantaged neighborhoods. Courts have consistently held that the

More school buildings and facilities are in need of repair. Mold, defective ventilation systems, faulty plumbing, and the like are not uncommon. Still, quality education is expected to continue in the classrooms despite such deplorable conditions.

Mark Richards / PhotoEdit

quality of building facilities is part and parcel of equal educational opportunities resulting in government monitoring of state spending on infrastructure needs to ensure equitable distributions of funds (AFT 2009).

There is a considerable body of evidence that documents the relationship between school environment and academic achievement. For example, in a survey of school principals, nearly half indicated that harmful environmental factors negatively impact the quality of instruction (Chaney & Lewis 2007). Tanner (2008) found a significant relationship between school environment (e.g., space, movement patterns, light, etc.) and academic achievement of third graders even when controlling for socioeconomic status of the school. Air quality, noise, overcrowding, inadequate space and lighting all affect a child's ability to learn, a teacher's ability to teach, and a staff member's ability to be effective.

In evaluating research on the impact of school environment on academic achievement, four variables emerged as the most significant—lighting, air quality, humidity and temperature, and acoustics (AFT 2009). These four variables are not only linked to poor school performance, they impact the health of students and teachers. For example, poor air quality is associated with higher incidences of asthma, and mold spores that result from damp environments (e.g., leaky roof) can trigger allergies and exacerbate respiratory diseases (AFT 2009).

Green schools are "education buildings that operate in harmony with the natural environment" (AFT 2009, p. 4). Green schools reduce energy use by using natural and solar light. They conserve water and use recycled materials, reduce air pollution and excessive noise, and enhance indoor climate control. Collectively, such innovations reduce absenteeism, increase student performance and teacher retention, and, ultimately, save money.

Special education programs pose yet another problem for school systems. Before 1997, when the Individuals with Disabilities Education Act (IDEA) was adopted, more than 1 million students with disabilities were excluded from public schools; 3.5 million were not receiving appropriate educational services (NCD 2000). With the implementation of IDEA, which was reaffirmed in 2004, more

green schools Schools that are "in harmony with the natural environment."

than 6.7 million children and young people with disabilities, ages 3 through 21, qualified for educational interventions under IDEA in 2007 (NCES 2009b). Pursuant to IDEA mandates, public school special education programs are now required to provide guaranteed access to education for disabled children. These specially designed instructional programs are structured to meet the unique needs of each child at no cost to the parent. As part of the program, if the school district cannot meet the unique needs of the child, by law, it becomes the school system's fiscal responsibility to pay for the child's education in an appropriate private school placement even if the child never received "special education or related services" through the public schools (Walsh & Robelen 2009).

Although the number of special education students in private settings is estimated to be just over 1 percent, financially strapped school districts may find it difficult to cover the high costs of private school placements for these students. There is also concern that Black American students, particularly Black American males, are disproportionately placed in special education programs that contribute to their comparatively low high school graduation rates (Schott Report 2008). Finally, many teachers are opting to leave special education classrooms for traditional classrooms or to pursue different fields entirely. With a lack of certified teachers in this area, a critical shortage has occurred nationwide, affecting the quality of education taking place in special education classrooms.

What Do You Think? All students, whether in regular classrooms or special education classes, are required by law to have a free and appropriate education. However, because of the increased costs associated with educating a child with special needs, this often means that students of special education are receiving a higher-quality, more individualized education than mainstream students who are in overcrowded classrooms in poor school districts (Berger 2007). How do you balance the unique needs of the students in special education with the basic needs of the mainstream students who are poor? How would you allocate funds?

Recruitment and Retention of Quality Teachers

School districts with inadequate funding and facilities, low salaries, lack of community support, and minimal professional development have difficulty attracting and retaining qualified school personnel. According to the U.S. Department of Education, nearly 25 percent of new teachers quit teaching within the first 3 years (Boyd et al. 2009.) High teacher turnover is problematic in a number of ways (Boyd et al. 2009). First, newer teachers are less experienced and often less effective. Second, teacher turnover contributes to a lack of continuity in programs and educational reforms. Finally, recruiting and training expenses in addition to the time and effort devoted to replacing teachers who have left the profession is considerable.

A recent report by the Bill and Melinda Gates Foundation concludes that the key to a good education hinges on the quality of the teachers in the classroom (Blankinship 2009) (see this chapter's The Human Side). Unfortunately, in the United States, qualified and experienced teachers are not equally distributed across all school districts. Because teacher salaries are the largest component of a school district's costs, poor school districts have less money to compete for qualified teachers

School districts with inadequate funding and facilities, low salaries, lack of community support, and minimal professional development have difficulty attracting and retaining qualified school personnel.

(McKinsey and Company 2009). Thus, they are more likely to employ beginning teachers with less than three years' experience, and are more likely to assign these teachers to areas outside their specialty. Knowledge of subject taught is one of the key characteristics of an effective teacher (Stotsky 2009). Further, students in poorer school districts are twice as likely to be taught by substitute teachers when compared to students in more affluent districts (Barton 2004).

Recruiting and retaining quality teachers in poverty-level schools is critical to the success of its students. For example, Hanushek et al. (2005) report that, if a child from a poor family has a good teacher for five consecutive years, the achievement gap between that child and a child from a higher-income family would be closed. Because minority students disproportionately populate poor school districts, it is also important to recruit and retain teachers who meet the needs of children from diverse backgrounds and of varying abilities. The number of minority teachers who can serve as role models, have similar life experiences, and have similar language and cultural backgrounds is far too few for the number of minority students.

Even seasoned teachers are not necessarily competent or effective. There is evidence that those who choose teaching as a career, on average, have lower college entrance exam scores than the average college student. Further, there is a documented relationship between college entrance scores and the likelihood of continued teaching—the *lower* college entrance scores, the *higher* the probability of still teaching 10 years post-baccalaureate (NCES 2007). Nonetheless, there is some evidence that the quality of teachers is improving. Both teachers and principals, when asked about the qualifications and competence of new teachers, were more likely to rate them as excellent compared with teachers' and principals' ratings 20 years ago (MetLife 2008).

In an effort to place quality teachers in the classroom, many states have implemented mandatory competency testing (e.g., Praxis Series). The need for teachers who are officially classified as "highly qualified" is tied to federal mandates that place an emphasis on the importance of having licensed teachers in the classroom. Additionally, teachers who have a bachelor's degree and have been in the classroom for three or more years are also eligible for national board certification. Some studies indicate that students of "highly qualified" teachers and/or board-certified teachers perform better on standardized tests and have shown greater testing gains than students of teachers who are not "highly qualified" and/or board-certified (Viadero 2005; NBPTS 2007).

In an effort to meet the demands of placing teachers in classrooms while facing teacher shortages due to baby boomer retirements, states are now allowing skilled professionals who have an interest in teaching but did not receive a teaching degree to enter the teaching profession. Called lateral entry by some states, the program allows the person to obtain a lateral entry teaching license while actually teaching in the classroom. In addition, more than half of the states have adopted **alternative certification programs,** whereby college graduates with degrees in fields other than education can become certified if they have "life experience" in industry, the military, or other relevant jobs. Teach for America (TFA), a program originally conceived by a Princeton University student in an honors thesis, is an alternative teacher education program with the aim of recruiting liberal arts graduates into teaching positions in economically deprived and socially disadvantaged schools. Critics argue that the program may place unprepared personnel in schools. However, an analysis of TFA teachers versus traditional teachers concludes that TFA teachers are more effective in the classroom than traditional teachers as measured by student achievement (Xu et al. 2007).

alternative certification program A program whereby college graduates with degrees in fields other than education can become certified if they have "life experience" in industry, the military, or other relevant jobs.

Forrest "Frosty" Troy, editor and founder of the Oklahoma Observer *has long been an advocate for teachers. In this essay, he writes about America's "unsung heroes."*

"Where are the heroes of today?" a radio talk show host thundered. He blames society's shortcomings on public education.

Too many people are looking for heroes in all the wrong places. Movie stars and rock musicians, athletes and models aren't heroes. They're celebrities.

Heroes abound in public schools, a fact that doesn't make the news. There is no precedent for the level of violence, drugs, broken homes, child abuse, and crime in today's America. Public education didn't create these problems but deals with them every day.

You want heroes?

Consider Dave Sanders, the schoolteacher shot to death while trying to shield his students from two neo-Nazi youth on a bombing and shooting rampage at Columbine High School in Littleton, Colorado. Sanders gave his life, along with 12 students, but other less-heralded heroes survived the Colorado blood bath.

You want heroes?

Columbine special education teacher, Robin Ortiz, braved gunfire moving from classroom to classroom, shouting at students and teachers to get out of the building. His action alone cleared the east side of the high school. No one will ever know how many lives he saved.

You want heroes?

For Ronnie Holuby, a Fort Gibson, Oklahoma, middle school teacher, it was a routine school day until gunfire erupted. He opened a door to the schoolyard and two students fled past him. A 13-year-old student had shot five other students when Holuby stepped outside, walking deliberately toward the boy, telling him to hand over the gun. He kept walking. Finally the boy handed him the gun. Holuby walked the boy to the side of the building, then sought to help a wounded girl.

You want heroes?

Jane Smith, a Fayetteville, North Carolina, teacher, was moved by the plight of one of her students, a boy dying for want of a kidney transplant. So this pretty white woman told the family of this handsome 14-year-old black boy that she would give him one of her kidneys. And she did. When

they subsequently appeared together hugging on the *Today Show*, even tough little Katie Couric was near tears.

You want heroes?

Doris Dillon dreamed all her life of being a teacher. She not only made it, she was one of those wondrous teachers who could wring the best out of every single child. One of her fellow teachers in San Jose, California, said, "She could teach a rock to read." Suddenly she was stricken with Lou Gehrig's disease, which is always fatal, usually within five years. She asked to stay on the job—and did. When her voice was affected, she communicated by computer. Did she go home? She is running two elementary school libraries. When the disease was diagnosed, she wrote the staff and all the families that she had one last lesson to teach—that dying is part of living. Her colleagues named her Teacher of the Year.

You want heroes?

Bob House, a teacher in Gay, Georgia, tried out for *Who Wants to Be a Millionaire*. After he won the million dollars, a network film crew wanted to follow up to see how it had impacted his life. New cars? Big new house? Instead, they found both Bob House and his wife still teaching. They explained that it was what they had always wanted to do with their lives and that would not change. The community was both stunned and gratified.

You want heroes?

Last year, the average public school teacher spent $468 of their own money for student necessities—workbooks, pencils—supplies kids had to have but could not afford. That's a lot of money from the pockets of the most poorly paid teachers in the industrial world. Public schools don't teach values? The critics are dead wrong. Public education provides more Sunday school teachers than any other profession. The average teacher works more hours in nine months than the average 40-hour employee does in a year.

You want heroes?

For millions of kids, the hug they get from a teacher is the only hug they will get that day because the nation is living through the worst parenting in history. Many have never been taken to church or synagogue in their lives.

A Michigan principal moved me to tears with the story of her attempt to rescue a badly abused little boy who doted on a stuffed animal on her desk—one that said, "I love you!" He said he'd never been told that at home.

This is a constant in today's society—two million unwanted, unloved, abused children in the public schools, the only institution that takes them all in.

You want heroes?

Visit any special education class and watch the miracle of personal interaction, a job so difficult that fellow teachers are awed by the dedication they witness. There is a sentence from an unnamed source that says, "We have been so anxious to give our children what we didn't have that we have neglected to give them what we did have."

What is it that our kids really need? What do they really want? Mathematics, science, history, and social studies are important, but children need love, confidence, encouragement, someone to talk to, someone to listen, standards to live by. Teachers provide upright examples, the faith and assurance of responsible people. Kids need to be accountable to caring parents who send well-disciplined children to school. These human values are essential in a democracy—anything that threatens them makes our whole society a little less free, our nation a little less strong. These values can be neither created nor preserved without continuous effort, and that effort must come from more than teachers who have students only six hours of the day.

Despite the problems, public school teachers laugh often and much. They have the respect of intelligent people and the affection of students who care. Teachers strive to find the best in their students, even where some see little hope. No other American bestows a finer gift than teaching—reaching out to the brilliant and the retarded, the gifted and the average.

Teachers leave the world a little bit better than they found it, knowing if they have redeemed just one life, they have done God's work. They are America's unsung heroes.

Source: Troy 2001.

The Challenges of Higher Education in America

Although there are many types of postsecondary education, higher education usually refers to two- or four-year, public or private, degree-granting institutions. In the last decade, enrollment in degree-granting institutions has increased by 26 percent. Full-time students, women, younger students, minorities, and students enrolled at four-year schools disproportionately contributed to this growth (NCES 2009a).

Higher education employs an estimated 3.6 million people, 2.6 million in professional positions (e.g., administrators, faculty, nonteaching professional staff, etc.) and 0.9 million in nonprofessional staff (e.g., clerical, service, maintenance, etc.) (NCES 2009a). In recent years, there has been a significant decrease in full-time tenured or tenure track faculty, once the "core" of academia, and significant increases in noninstructional staff (e.g., administrators) and non-tenure track, part- or full-time instructors (AFT 2008b).

As the number of students has increased, so have the costs associated with getting a college degree. Annual expenses for tuition, room, and board, estimated to be $11,164 and $28,846 at public and private undergraduate institutions, respectively, have increased 30 percent over the last decade (NCES 2009a). In light of such increases, in 2008, the U.S. Congress approved the Higher Education Act, which requires that institutions with the largest tuition increases submit a report to the U.S. Department of Education, justifying the increases and outlining cost-saving initiatives (Klein & Sawchuk 2008). The act also simplifies the federal student aid process.

About 65 percent of all undergraduates receive some type of student aid; in 2008, the average total amount of financial aid was $9,100 (Wei et al. 2009). Full-time students attending four-year private doctoral-granting institutions are the most likely to receive student aid, and federal and state grants and student loans comprise the largest proportion of student aid. However, budget deficits and cutbacks to education are likely to reduce the student aid "pot." Even in California, the most affordable college education in the United States (NCPPHE 2009), proposed budget cuts to the state student grant program would impact hundreds of thousands of students, cutting community college enrollment in half and disproportionately hurting low-income would-be students (ICAS 2009).

Access to and completion of higher education, particularly among minority and/or low-income students, is also problematic. For example, racial and ethnic minorities, although varying by state, are less likely to be enrolled in college and to receive a degree (see Table 8.2). College completion rates at all degree-granting institutions have decreased over time, leading to concerns over U.S. competitiveness in global markets, particularly labor markets (NCPPHE 2009). Of particular concern is whether the United States is training a sufficient number of graduates in science, technology, engineering, and mathematics occupations (see Chapter 14).

To address some of the issues in higher education, a number of policy changes have been adopted. For example, to increase enrollment and, consequently, revenues, 17 states now

TABLE 8.2 First-Time, Full-Time Students Completing a Bachelor's Degree within Six Years of College Entrance, 2007

	WHITES	BLACKS
Delaware	73%	41%
Ilinois	65%	34%
Maryland	73%	42%
Michigan	58%	32%
	WHITES	HISPANICS
Ilinois	65%	45%
New Jersy	66%	49%
New York	63%	43%
Texas	56%	38%
	WHITES	NATIVE AMERICANS
New Mexico	47%	25%
North Daktoa	48%	17%
Washington	65%	41%

Source: NCPPHE 2009.

allow community colleges to award bachelor's degrees (Lewin 2009). Some bachelor degree granting institutions, on the other hand, are now offering three-year degree programs to attract students. Critics argue that economic considerations should not drive educational policy; proponents argue that an economic crisis must be met with innovation (Strauss 2009).

College completion rates at all degree-granting institutions have decreased over time, leading to concerns over U.S. competitiveness in global markets, particularly labor markets.

Finally, the Obama administration has made a commitment to several higher education programs. The federal stimulus package includes $15.6 billion in Pell Grant student aid, and $13.9 billion over ten years in higher education tax cuts (see Figure 8.4). Additionally, the President's 2010 budget includes billions of dollars to (1) assist degree-granting institutions that primarily serve minority and economically disadvantaged students; (2) help states develop programs that focus on increasing graduation rates among low-income students; and (3) develop local projects that are "models for innovative reform and improvement in postsecondary education" (Office of Management and Budget 2009, p. 69).

What Do You Think? In June 2009, hundreds of students and young immigrants wearing graduation gowns walked in front of the U.S. Capitol in support of the DREAM Act (Zehr 2009a). The act, if passed, would "provide a path to legalization for undocumented youths who graduated from U.S. high schools and attend college or serve in the military for two years" (p. 1). Critics of the act argue that it would grant amnesty to millions of illegal immigrants and further strain federal and state budgets. It is estimated that about 65,000 undocumented students who have lived in the United States for five or more years graduate from U.S. high schools every year (Gonzales 2009). Because of their illegal status, they are ineligible for financial aid and barred from legal employment. Do you think the DREAM Act should be passed?

Strategies for Action: Trends and Innovations in American Education

Americans consistently rank improving education as one of their top priorities. Recent attempts to improve schools include raising graduation requirements, barring students from participating in extracurricular activities if they are failing academic subjects, lengthening the school year, prohibiting dropouts from obtaining driver's licenses, and extending the number of years permitted to complete a high school degree. However, with a new president and secretary of education, educational reformers are calling for changes that go beyond these get-tough policies that maintain the status quo.

National Educational Policy

The challenges facing national educational policies are considerable. In a 2009 address to the U.S. Hispanic Chamber of Commerce, President Obama stated that "[W]e've let our grades slip, our schools crumble, our teacher quality fall short, and other nations outpace us" (Robelen 2009). Although it may be difficult so early in his administration to predict the totality of his educational policy, there is evidence of both significant changes and "things as they were" politics.

No Child Left Behind. The No Child Left Behind (NCLB) Act of 2001 was signed into law in January 2002. The federally funded plan was organized around four principles: *accountability* for learning outcomes, *flexibility* in funding, *expanding school options for parents,* and the use of sound *teaching methods,* including the use of only "highly qualified" teachers in the classroom by 2006.

However, soon after the act was signed, it became clear that the implementation of NCLB was problematic. For example, accountability efforts required that, to make adequate yearly progress (AYP), the factor by which a school is measured, all student groups (e.g., students who are black, white, Native American, of limited English proficiency, disabled, or others) must attain a set level of achievement in both reading and mathematics. According to the act, if one student group does not reach the set levels, the entire school receives a failing grade and is in danger of being severely sanctioned (NCDPI 2007). Because of the tremendous disparities between student groups, the original NCLB regulations were changed to allow for alternative testing of students who were disabled or had limited English proficiency; deadlines for making AYP in reading and mathematics were also extended to 2014 for all states (Olson & Hoff 2006).

In addition to concerns about accountability, critics of the law also argue that it unfairly burdens the states which must absorb the financial cost of its provisions. The National Education Association, the largest organization of teachers in the nation, joined schools in several districts to bring the first federal lawsuit against the U.S. Department of Education for failing to provide funding for NCLB initiatives. *Pontiac v. Spelling* is now in the hands of the 6th circuit U.S. Court of Appeals (Walsh 2009).

Further, empirical tests of NCLB are mixed. For example, a study by Hall and Kennedy (2006) indicates that, overall, achievement gains were found in elementary schools where reading scores showed gains in 27 of 31 states and mathematics scores improved in 29 of 32 states. However, the picture in middle schools and high schools was less clear, with a combination of gains and losses in both reading and mathematics. Although minority and low-income students showed achievement gains, whites and higher-income students, in general, showed more dramatic gains, thus increasing the achievement gap between the two groups.

Despite the former administration's claim that all 50 states have successfully implemented NCLB and that achievement gaps are narrowing (Spelling 2008), critics continue to argue that the act's effectiveness is questionable. Reauthorization attempts have stalled, and some commentators have suggested that the Obama administration is considering changing the name of the act as a means of distancing itself from the previous administration's educational policies. Other commentators have noted that U.S. Education Secretary Arne Duncan's emphasis on accountability and testing are reminiscent of the Bush administration (Ravitch 2009).

Guiding Principles. For every child to get a "world-class education," national educational policy, as presently conceived, has three guiding principles (White House 2009). The first is a focus on *early childhood education.* Arguing that the developmental years prior to entering the school system are important determinants of learning outcomes, the stimulus package of 2009 allocated just over $4 billion for early child care development and education in such programs as Early Head Start (see Figure 8.4). Further, the 2010 federal budget provides $8.5 billion over 10 years for the creation of a home visitation program to, among other things, help with school preparation (Office of Management and Budget 2009).

The second guiding principle is *reform and investment in elementary and secondary education*. In a global economy, the relatively low levels of academic achievement of many U.S. students necessitate rigorous standards, quality methods of assessment, and increased support for and accountability of classroom teachers. The new administration will "invest in innovative strategies to help teachers to improve student outcomes, and use rewards and incentives to keep talented teachers in the schools that need them the most" (White House 2009, p. 1). Recruitment of outstanding teachers and removal of ineffective ones are also part of this principle.

The final guiding principle is *restoring the nation's leadership position in higher education*. The new administration is committed to reestablishing the United States as an educational leader and, by 2020, having the highest proportion of college graduates in the world. To accomplish that goal, federal aid programs will be expanded and simplified, and substantial investments will be made in both two-year and four-year degree granting institutions.

National Standards. The National Governors Association and the Council of Chief State School Officers have led the move to adopt a common set of academic standards. Presently, 46 states have joined the National Common Core State Standards Initiative. The initiative was motivated by a patchwork quilt of state standards that allowed some states to excel under NCLB by setting "proficiency" levels lower than other states. Supported by $350 million in stimulus funds, member states of the National Common Core State Standards Initiative are drafting English and mathematics standards with the hope of voluntary adoption. Ultimately, the initiative hopes to develop standardized assessment measures to be used across all school districts (McNeal 2009).

Stabilization and Innovation. Budget shortfalls threatened state and local school districts with education budgets so deep that thousands of teachers would be laid off, class sizes increased, capital improvement projects cancelled, and needed programs eliminated. The over $100 billion earmarked for education as part of the American Recovery and Reinvestment Act of 2009 was intended to avert massive teacher layoffs but, according to the secretary of education, had a larger goal—"to drive a set of reforms that we believe will transform public education in America" (Duncan 2009, p. 1).

Almost half of the education stimulus dollars, $54 billion, has been set aside for "fiscal stabilization" in state aid for critical education needs. However, for states to receive these additional funds, governors must pledge to improve teacher quality, raise academic standards, intervene in failing schools, and develop "robust data systems that allow districts to better track the growth of individual students" (Duncan 2009, p. 1). These **four assurances** must be met and documented for states to continue receiving stimulus funds (Dillon 2009, p. 1). An additional $5 billion in what has been called "Race to the Top Funds" is also available to states that have used innovative programs to address education-related problems.

four assurances The four assurances refer to a state's commitment to improving teacher quality, raising academic standards, intervening in failing schools, and developing assessment databases in return for federal stimulus dollars.

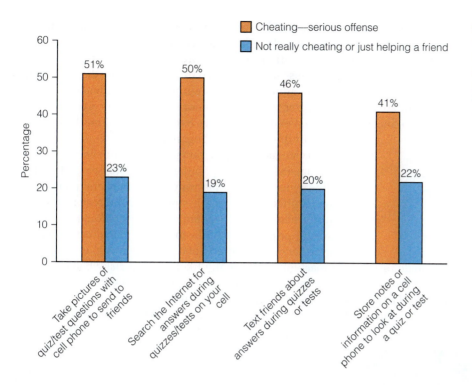

Figure 8.6 Percent of 7th Through 12th Grade Students Indicating If Act Is Cheating, 2009 (*N* = 1013)
Source: Benenson Strategy Group 2009.

Legend:
- Cheating—serious offense
- Not really cheating or just helping a friend

Categories (bar values):
- Take pictures of quiz/test questions with cell phone to send to friends: 51%, 23%
- Search the Internet for answers during quizzes/tests on your cell: 50%, 19%
- Text friends about answers during quizzes or tests: 46%, 20%
- Store notes or information on a cell phone to look at during a quiz or test: 41%, 22%

Y-axis: Percentage (0 to 60)

Character Education and Service Learning

Research indicates that cheating is a fairly common event among students in the United States. One type of cheating, electronic cheating, is on the rise. In a survey of over 1,000 U.S. students enrolled in seventh through twelfth grade, over half admitted using the Internet to cheat, including 38 percent who copied entire passages of text from the Internet and submitted it as their own work (Benenson Strategy Group 2009). Further, 35 percent of cell phone owners admitted to using their cell phone to cheat at least once. Perhaps more disturbing is the number of students who didn't consider anything wrong with these activities (see Figure 8.6). To many educators, results like these signify the need for **character education.**

Character education entails teaching students to act morally and ethically, including the ability to "develop just and caring relationships, contribute to community, and assume the responsibilities of democratic citizenship" (Lickona & Davidson 2005). Despite most schools' emphasis on academic achievement, knowledge without character is potentially devastating. Sanford McDonnell (2009), former CEO of McDonnell Aircraft Corporation and chairman emeritus of the Character Education Partnership, recounts a letter written by a principal and former concentration camp survivor to his teachers at the start of a new school year:

> My eyes saw what no person should witness: gas chambers built by learned engineers, children poisoned by educated physicians, infants killed by trained nurses, women and babies shot and burned by high school and college graduates. Your efforts must never produce learned monsters and skilled psychopaths. Reading, writing, and arithmetic are important only if they serve to make our children more humane. (p. 1)

character education
Education that emphasizes the moral and ethical aspects of an individual.

President Obama's educational reform policy includes additional dollars for expanding and redefining character education. Government analyses of research on established character education programs indicate that classroom activities designed to teach "core values" are ineffective in producing the desired student outcomes. Keeping with the administration's emphasis on evidence-based programs, the 2010 budget redirects funds to a more broadly conceived notion of character education administered through the Safe and Drug-Free Schools and Communities National Programs.

> Despite most schools' emphasis on academic achievement, knowledge without character is potentially devastating.

Service learning programs, one type of character education, are increasingly popular at universities and colleges nationwide. Service learning programs are community-based initiatives in which students volunteer in the community and receive academic credit for doing so. Service learning programs have been found to increase academic performance, promote "hands-on" learning, enhance civic engagement and moral reasoning, reduce the likelihood of risky behaviors, and increase the self-esteem of program participants (Jacobs 1999; Independent Sector 2002; Ramierz-Valles & Brown 2003; CNCS 2008).

The Edward M. Kennedy Serve America Act, signed in 2009, creates several programs designed to increase student participation in service activities, and is hoped to increase the participation of volunteers in AmeriCorps and similar programs (Zehr 2009b). A President Obama appointee and Nike, Inc. vice president, Maria Eitel, who will head the Corporation for National and Community Service, will administer the funding for the act. Examples of new programs include Summer of Service, which pays middle and high school students $500 toward college tuition for 100 hours of community service, and Youth Engagement Zones, which partners schools and community organizations to facilitate community service by secondary school students (Zehr 2009b).

Charles Dharapak / AP Photo

After signing the Edward M. Kennedy Serve America Act, President Obama, First Lady Michelle Obama, and former president Bill Clinton joined the Student Conservation Association and AmeriCorps for tree planting at Kenilworth Park and Aquatic Gardens in Washington, DC.

Use of Computer Technology and E-Learning

Computers in the classroom allow students to access large amounts of information. The proliferation of computers both in school and at home may mean that teachers will become facilitators and coaches rather than the sole providers of information. Not only do computers enable students to access enormous amounts of information, they also allow students to progress at their own pace. However, computer technology is not equally accessible to all students. For example, although most students have access to computers at school, access to computers at home varies dramatically by parents' education, parents' income, and race and ethnicity. In that respect, schools "bridge the digital divide" between poor minority students and their more advantaged counterparts (DeBell & Chapman 2006).

The Enhancing Education Through Technology Program is part of the NCLB Act. The goals of the program are threefold: (1) to improve student achievement

Students work on their laptops at the Philadelphia School of the Future. Sponsored as part of a public/private partnership between the City of Philadelphia and Microsoft, this model for future schools opened on September 6, 2006.

Tim Shaffer, Microsoft/ AP Photo

through the use of technology resources; (2) to ensure that teachers integrate technology into the curriculum in such a way as to improve student achievement; and (3) to help students become technically literate by the eighth grade (NCES 2003). Thus, two of the three goals of the program are to increase academic achievement through the use of technological resources.

Recently, a study mandated by Congress assessed the use of educational technology in improving student achievement. Reading and math scores on standardized tests of students using educational technology selected for their presumed effectiveness were compared with reading and mathematics scores on standardized tests of students not using the educational software. The performances of 9,424 students in 132 school districts were evaluated. The results indicated that there were no statistically significant achievement differences between students in classes where the software was used and students in classrooms where the software was not used (NCES 2007).

E-learning separates the student and the teacher by time and/or place. They are, however, connected by some communication technology (e.g., videoconferencing, e-mail, real-time chat room, or closed-circuit television). Some classes have blended learning, a mix of online and traditional face-to-face learning. Based on sampled school districts, a report examining e-learning in the nation's elementary and secondary schools found the following (Picciano & Seaman 2009):

- In 2008, there were over a million K–12 students taking online courses.
- Seventy-five percent of the school districts offered online or blended courses.

e-learning Learning in which, by time or place, the learner is separated from the teacher.

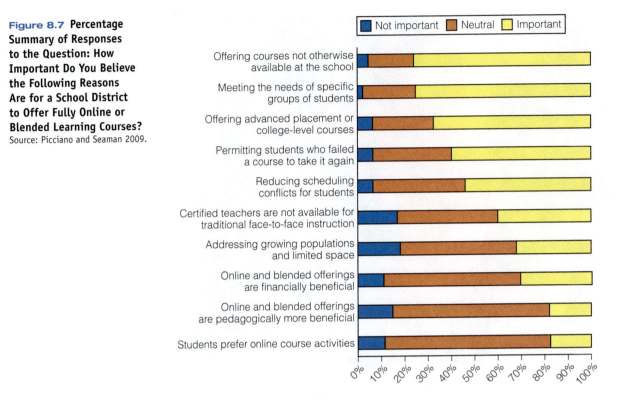

Figure 8.7 Percentage Summary of Responses to the Question: How Important Do You Believe the Following Reasons Are for a School District to Offer Fully Online or Blended Learning Courses?
Source: Picciano and Seaman 2009.

- School districts depended on multiple online service providers (e.g., private, postsecondary institutions).
- Online learning meets specific needs of students (e.g., additional help).
- Online learning provides courses that simply would not be available to students otherwise.

Figure 8.7 graphically portrays additional benefits of online education and their relative importance. Although low in rank, financial considerations were identified as a benefit of online education and may increasingly be so in the face of funding cuts.

Online education often serves a segment of the population that would not otherwise be able to attend school—older, married, full-time employees, and people from remote areas. Thus, for example, the number of public school on-line course offerings is higher in rural areas than in towns, cities, or urban areas (Bausell & Klemick 2007). Further, some research suggests that online learning benefits those who have historically been disadvantaged in the classroom. A study by DeNeui and Dodge (2006) indicates that females in a blended course were more likely to use a learning management system (e.g., Blackboard) than males, and to significantly outperform them as measured by final grades in an introductory psychology class.

The Debate over School Choice

school vouchers Tax credits that are transferred to the public or private school that parents select for their child.

Traditionally, children have gone to school in the district where they live. School vouchers, charter schools, private schools, and home schooling provide parents with alternative school choices for their children. **School vouchers** are tax credits that are transferred to the public or private school that parents select for their child. For example, the Washington, DC, voucher program provides up to $7,500 to students from disadvantaged families who can then use the funds to send their child to the school of their choice. The program, however, as vouchers in general, has been a contentious issue.

Proponents of the voucher system argue that it increases the quality of schools by creating competition for students. Those who oppose the voucher system argue that it will drain needed funds and the best students away from public schools. Forster (2009), after examining 17 empirical studies on the impact of vouchers on public schools, concludes that 16 of the studies "find that vouchers improve public schools and one—the only one examining a program that insulates public schools from voucher competition—finds no visible difference. No empirical studies find that vouchers harm public schools" (Forster 2009, p. 34).

Opponents also argue that vouchers increase segregation because white parents use the vouchers to send their children to private schools with few minorities, and that the use of vouchers for religious schools violates the constitutional guarantee of separation of church and state. However, the U.S. Supreme Court, in reviewing the voucher program in Cleveland, Ohio, held that the use of tax dollars for enrollment in religious schools is not unconstitutional (NSBA 2007).

Vouchers can be used for charter schools. **Charter schools** originate in contracts, or charters, which articulate a plan of instruction that local or state authorities must approve. Although foundations, universities, private benefactors, and entrepreneurs can fund charter schools, many are supported by tax dollars. Today, there are more than 4,000 charter schools in 40 states with enrollments of over 1 million students (Rand 2009).

Charter schools, like school vouchers, were designed to expand schooling options and to increase the quality of education through competition. Like vouchers, charter schools have come under heavy criticism for increasing school segregation, reducing public school resources, and "stealing away" top students. Proponents, including President Obama and Secretary of Education Duncan, argue that charter schools encourage innovation and reform, and increase student learning outcomes.

A recent evaluation of charter schools in six cities and three states addresses these issues. Using longitudinal data to assess changes over time, the researchers conclude that (1) there are few differences between achievement test scores of middle and high school charter students when compared with their public school counterparts; (2) charter schools do not lure top students away from traditional schools; (3) charter schools do not impact school achievement at neighboring traditional schools; and (4) students who attended charter high schools are more likely to graduate and to attend college (Rand 2009).

Another school choice parents can make is to send their children to a private school. In 2007, private school enrollment in elementary and secondary schools numbered nearly 6 million—about 11 percent of all kindergarten through twelfth grade students (NCES 2009b). The primary reason parents send their children to private schools is for religious instruction. The second most common reason is the belief that private schools are superior to public schools in terms of academic achievement. Contrary to expectations, however, there is evidence that public school students fair as well or better academically as private school students. For example, Lubienski and Lubienski (2006), using National Assessment

charter schools Schools that originate in contracts, or charters, which articulate a plan of instruction that must be approved by local or state authorities.

of Educational Progress (NAEP) data, report that mathematics scores for public school students were higher than mathematics scores for private school students. Parents also choose private schools for their children to have greater control over school policy, to avoid busing, or to obtain a specific course of instruction, such as dance or music.

Some parents choose not to send their children to school at all but to teach them at home. In 2007, 1.5 million were homeschooled, the majority of whom, 77 percent, were white (NCES 2009b). Reasons for homeschooling children vary. The three most common reasons given by parents for **homeschooling** include the need for religious and moral instruction, avoidance of the negative environment of public schools (e.g., drugs), and concerns over the quality of instruction in public schools (NCES 2009b). How does being schooled at home instead of attending public school affect children? Some evidence suggests that homeschooled children perform as well as or better than their institutionally schooled counterparts (Winters 2001).

Understanding Problems in Education

What can we conclude about the educational crisis in the United States? Any criticism of education must take into account the fact that just over a century ago, the United States had no systematic public education system at all. Many American children did not receive even a primary school education. Instead, they worked in factories and on farms to help support their families. Whatever education they received came from the family or the religious institution. In the mid-1800s, educational reformer Horace Mann advocated mandatory education for all U.S. children. In 1852, the first compulsory education laws in the United States were passed, requiring school-aged children to attend 12 weeks of school each year. By World War I, every state mandated primary school education and, by World War II, secondary education was compulsory as well.

Today, the United States spends more money per student, at every level of education, than any other industrialized nation in the world (OECD 2009; NCES 2009b), and federal spending on education continues to escalate (see Figure 8.8). Nonetheless, U.S. students are outperformed at almost every level by their international peers. Significant educational reform is needed to meet the needs of a global economy in the 21st century and, perhaps more importantly, to fulfill Horace Mann's dream of education as the "balanced wheel of social machinery," equalizing social differences among members of an immigrant nation.

First, we must invest in teacher education and in teaching practices that have been empirically documented to work in raising student outcomes. President Obama's emphasis on "best practices" is a step in the right direction. Teacher's salaries also need to better reflect the priority Americans place on children, education, and the education of children. In a 2007 Gallup Poll, when a representative sample of parents was asked, "In your opinion, what would be the best way to improve the education your oldest child receives at his or her school?" the most common responses were teacher-related (e.g., better educated) (Gallup Poll 2007).

Second, the "savage inequalities" in education, primarily based on race, ethnicity, and socioeconomic status, must be addressed. Segregation, rather than

homeschooling The education of children at home instead of in a public or private school.

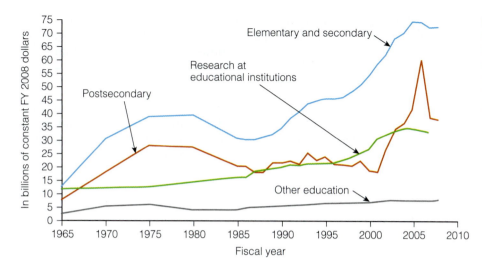

Figure 8.8 Federal On-Budget Funds for Education, by Level or Other Educational Purpose: Selected Years, 1965 Through 2008

Source: NCES 2009a.

decreasing is increasing, a reflection of housing patterns, local school districts' heavy reliance on property taxes, and immigration patterns. Public schools should provide all U.S. children with the academic and social foundations necessary to participate in society in a productive and meaningful way but, for many children, schools perpetuate an endless downward cycle of failure, alienation, and hopelessness.

Legally required protections can help. Settled in 2004, *Williams v. California* was a class-action suit against the state of California. Attorneys representing over 1 million children argued that low-income students were permanently disadvantaged as a result of the schools they attended. The settlement held that (1) all students must have needed textbooks and materials; (2) every classroom must be clean, safe, and in good condition; and (3) every teacher must meet standards set down by the state of California and federal law. Implementation of what is now called the "Williams Legislation" prompted one California official to remark that schools are now "prepared, making sure that every student has all the text materials that are required and that the teaching environment is safe and clean. Some of the classes even get upset if I haven't chosen their classroom for visitation" (Allen 2009, p. 18).

However, implementation of the Williams settlement cost over $1 billion at a time when California, as well as many other states, is faced with one of the worst budget deficits in decades. Swelling student enrollments and an increasingly diverse student population compound the problem. There are, however, some reasons to be optimistic. McKinsey and Company (2009), for example, report that the "wide variation in performance among schools and school systems serving similar students suggests that the opportunity and output gaps related to today's achievement gap can be substantially closed. Many teachers and schools across the country are proving that race and poverty are not destiny (McKinsey and Company 2009, p. 6).

Finally, as a society, we must attend to and be cognizant of the importance of early childhood development. Poliakoff (2006) notes

Children's physical, emotional, and cognitive development are profoundly shaped by the circumstances of their preschool years. Before some children are even born, birth weight, lead poisoning, and nutrition have taken a toll on their capacity for academic achievement. Other factors—excessive television watching, little exposure to conversation or books, parents who are absent or distracted, inadequate nutrition—further compromise their early development. (p. 10)

We must provide support to families so that children grow up in healthy, safe, and nurturing environments. Children are the future of our nation and of the world. Whatever resources we provide to improve the lives and education of children are sure to be wise investments in our collective future.

CHAPTER REVIEW

- **Do all countries educate their citizens?**
No. Many societies have no formal mechanism for educating the masses. As a result, worldwide, there are millions of illiterate adults around the world. The problem of illiteracy is greater in developing countries than in developed nations and, worldwide, disproportionately affects women rather than men (see Figure 8.1).

- **According to the structural-functionalist perspective, what are the functions of education?**
Education has four major functions. The first is instruction—that is, teaching students knowledge and skills. The second is socialization that, for example, teaches students to respect authority. The third is sorting individuals into statuses by providing them with credentials. The fourth function is custodial care—a babysitting agency of sorts.

- **What is a self-fulfilling prophecy?**
A self-fulfilling prophecy occurs when people act in a manner consistent with the expectations of others.

- **What variables predict school success?**
Three variables tend to predict school success. Socioeconomic status predicts school success: the higher the socioeconomic status, the higher the likelihood of school success. Race predicts school success, with nonwhites and Hispanics having more academic difficulty than whites and non-Hispanics. Gender also predicts success, although it varies by grade level.

- **What are the three reasons given why Black and Hispanic Americans, in general, do not perform as well in school as their white and Asian counterparts?**
First, because race and ethnicity are so closely tied to socioeconomic status, it appears that race or ethnicity

alone can determine school success when, in fact, it may be socioeconomic status. Second, many minorities are not native English speakers, making academic achievement significantly more difficult. And third, racial and ethnic minorities may be the victim of racism and discrimination.

- **What are some of the conclusions of the study summarized in the Social Problems Research Up Close feature?**
The results of the study indicate that there are more victims of bullying than bullies, and that bullying behavior is highest in the ninth grade and lowest in the sixth grade and tenth grade. The highest rate of victimization is in the sixth grade. Males and females were equally likely to report being bullies, victims, and bully-victims. Blacks, compared to Hispanics, were more likely to be bullies, victims, and bully-victims. "Upsetting students for the fun of it" was the most common kind of bullying activity, followed by teasing, group teasing, and starting arguments.

- **What are some of the problems associated with the American school system?**
One of the main problems is the lack of student achievement in our schools—particularly when U.S. data are compared with data from other industrialized countries. Minority dropout rates are high, and school violence, crime, and discipline problems continue to be a threat. School facilities are in need of repair and renovations, and personnel, including teachers, have been found to be deficient. Higher education must also address several challenges.

- **What are the "four assurances"?**
For states to receive stimulus funds, they must agree to the four assurances—a commitment to improve teacher quality, raise academic standards, intervene in failing schools, and develop accountability databases.

What are the arguments for and against school choice?
Proponents of school choice programs argue that they reduce segregation and that schools that have to compete with one another will be of a higher quality. Opponents argue that school choice programs increase segregation and treat disadvantaged students unfairly. Low-income students cannot afford to go to private schools, even with vouchers. Furthermore, those opposed to school choice are quick to note that using government vouchers to help pay for religious schools is unconstitutional.

TEST YOURSELF

1. All societies have some formal mechanism to educate their citizenry.
   a. True
   b. False
2. According to structural functionalists, which of the following is not a major function of education?
   a. Teach students knowledge and skills
   b. Socialize students into the dominant culture
   c. Indoctrinate students into the capitalist ideology
   d. Provide custodial care for children
3. The number of children speaking a language other than English in the home has decreased in recent years.
   a. True
   b. False
4. In a study supported by the Bill and Melinda Gates Foundation, researchers identified five major reasons why students drop out of high school. Which of the following is a reason given for dropping out of school?
   a. Classes were not interesting
   b. Had to drop out to work
   c. Pregnancy
   d. Hated school
5. Which of the following statements about bullying is not true?
   a. Student bullies tend to be marginal students
   b. More females are bullied than males
   c. There are serious consequences for the student victims of bullying
   d. Most bullying takes place outside so it is undetected

6. Public school special education programs are required to provide guaranteed access to education for disabled children.
   a. True
   b. False
7. Over the next decade, the need for teachers is likely to
   a. decrease because of the baby boom retirements
   b. increase because of immigration patterns
   c. remain stable
   d. decrease because of the aging population
8. Which of the following is not one of the four assurances?
   a. Improve teacher quality
   b. Reduce school segregation
   c. Raise academic standards
   d. Intervene in failing schools
9. Computer software in the classroom has been shown to significantly increase academic achievement.
   a. True
   b. False
10. School vouchers are tax credits that are transferred to the public or private school that parents select for their child.
   a. True
   b. False

Answers: 1: b; 2: c; 3: b; 4: a; 5: d; 6: a; 7: b; 8: b; 9: b; 10: a.

KEY TERMS

alternative certification program 289
bilingual education 277
bullying 284
character education 295
charter schools 299

cultural imperialism 271
e-learning 297
four assurances 294
green schools 287
Head Start 274
home-schooling 300

integration hypothesis 278
multicultural education 270
school vouchers 298
self-fulfilling prophecy 272
stereotype threat 284
Title I 274

MEDIA RESOURCES

Understanding Social Problems,
Seventh Edition Companion Website
www.cengage.com/sociology/mooney
Visit your book companion website, where you will find flash cards, practice quizzes, Internet links, and more to help you study.

Just what you need to know NOW! Spend time on what you need to master rather than on information you already have learned. Take a pretest for this chapter, and CengageNOW will generate a personalized study plan based on your results. The study plan will identify the topics you need to review and direct you to online resources to help you master those topics. You can then take a posttest to help you determine the concepts you have mastered and what you will need to work on. Try it out! Go to www.cengage.com/login to sign in with an access code or to purchase access to this product.

Race, Ethnicity, and Immigration

"The 21st century will be the century in which we redefine ourselves as the first country in world history which is literally made up of every part of the world."

**Kenneth Prewitt,
Census Bureau director**

The Global Context: Diversity Worldwide | Racial and Ethnic Group Diversity and Relations in the United States | Immigrants in the United States | **Self and Society: Attitudes Toward U.S. Immigrants and Immigration** | Sociological Theories of Race and Ethnic Relations | Prejudice and Racism | **Social Problems Research Up Close: Two-Faced Racism** | Discrimination Against Racial and Ethnic Minorities | **The Human Side: Anti-Immigrant Hate: One Immigrant's Experience** | Strategies for Action: Responding to Prejudice, Racism, and Discrimination | Understanding Race, Ethnicity, and Immigration | Chapter Review

305

Following Obama's election in the 2008 Presidential campaign, racially charged incidents occurred around the country.

Mario Tama/Getty Images

Kaylon Johnson, age 33, was a campaign volunteer for Barack Obama in the 2008 presidential election, and was among the millions of people around the world who celebrated the historic election of the first African American U.S. president. About a month after the election, Johnson drove into a gas station in Shreveport, Louisiana, in his GMC Yukon, adorned with Obama bumper stickers. Wearing an Obama T-shirt, Johnson went inside the station to buy a soft drink and was returning to his vehicle when three white men jumped him, shouting "Fuck Obama!" and "Nigger president!" The beating left Johnson with a broken nose and a fractured eye socket that required surgery (Keller 2009). The attack on Johnson was one of hundreds of racially charged incidents around the country related to Obama's campaign and ultimate victory, such as the following:

- A life-sized likeness of Obama was found hanging from a noose in a tree at the University of Kentucky.
- A sign inside a general store in Standish, Maine, read, "Osama Obama Shotgun Pool." Customers could place $1 bets on the date they thought the new president would be assassinated. At the bottom of the sign someone had written, "Let's hope someone wins."
- A University of Alabama professor reported that an Obama poster was torn off her office door. When she put up another one, someone wrote a racial slur and a death threat on it.

The election of Barack Obama represents a significant milestone in U.S. history, and has ushered in a new era of hope for racial and ethnic minorities. Another milestone for minorities was achieved when Sonia Sotomayor was sworn in as the first Hispanic Supreme Court Justice in 2009. But just one year earlier, in 2008, Mississippi's Charleston High School held its first racially integrated prom. And in the same year that Obama took office and Sotomayor took a seat on the Supreme Court, a group of largely African American children from Creative Steps Day Care were kicked out of the Valley Swim Club in Huntingdon Valley, Pennsylvania. The day care center's director had contracted for the children to use the swim club once a week. When the children came for their first weekly swim, several of the children reported hearing racist comments from white club members. One of the boys, for example, reported that a woman at the club said she feared the children "might do something" to her child. A few days later, the day care center's check for $1,950 was returned and the kid's swimming privileges were canceled. When actor Tyler Perry heard about what happened, he offered to send 65 children from the largely minority Creative Steps Day Care to Disney World! As we shall see in this chapter, prejudice and discrimination against minority groups typically have negative outcomes for minority groups and for society in general.

A **minority group** is a category of people who have unequal access to positions of power, prestige, and wealth in a society and who tend to be targets of prejudice and discrimination. Minority status is not based on numerical representation in society but rather on social status. For example, although Hispanic individuals outnumber non-Hispanic whites in California, Texas, and New Mexico, they are considered a "minority" because they are underrepresented in positions of power prestige, and wealth, and because they are targets of prejudice and discrimination.

In this chapter, we focus on prejudice and discrimination, their consequences for racial and ethnic minorities, and the strategies designed to reduce these problems. We also examine issues related to U.S. immigration, because immigrants often bear the double burden of being minorities *and* foreigners who are not welcomed by many native-born Americans. We begin by examining racial and ethnic diversity worldwide and in the United States, emphasizing first that the concept of race is based on social rather than biological definitions.

Althea Wright, director of Creative Steps Day Care, speaks to the media about the incident at the Valley Swim Club.

The Global Context: Diversity Worldwide

A first grade teacher asked the class, "What is the color of apples?" Most of the children answered red. A few said green. One boy raised his hand and said, "white." The teacher tried to explain that apples could be red, green, or sometimes golden, but never white. The boy insisted his answer was right and finally said, referring to the apple, "Look inside" (Goldstein 1999). Like apples, human beings may be similar on the "inside," but they are often classified into categories according to external appearance. After examining the social construction of racial and ethnic categories, we review patterns of interaction among racial and ethnic groups and examine racial and ethnic diversity in the United States.

The Social Construction of Race and Ethnicity

The concept of **race** refers to a category of people who are believed to share distinct physical characteristics that are deemed socially significant. Cultural definitions of race have taught us to view race as a scientific categorization of people based on biological differences between groups of individuals. However, "races are not biologically real but are cultural and social inventions created in specific cultural, historical, and political contexts. . . . Races are not scientifically valid because there are no objective, reliable, meaningful criteria scientists can use to construct or identify racial groupings" (Mukhopadhyay et al. 2007, pp. 1, 5).

Historically in the United States, racial classification has been primarily based on skin color and secondarily on hair texture as well as the size and shape of the eyes, lips, and nose. But distinctions among human populations are graded, not abrupt. Skin color is not black or white but rather ranges from dark

minority group A category of people who have unequal access to positions of power, prestige, and wealth in a society and who tend to be targets of prejudice and discrimination.

race A category of people who are believed to share distinct physical characteristics that are deemed socially significant.

to light with many gradations of shades. Noses are not either broad or narrow but come in a range of shapes. Physical traits such as these, as well as hair color and other characteristics, come in an infinite number of combinations. For example, a person with dark skin can have any blood type and can have a broad nose (a common combination in West Africa), a narrow nose (a common combination in East Africa), or even blond hair (a combination found in Australia and New Guinea) (Cohen 1998).

Further, skin color, hair texture, and facial features are only a few of the many traits that vary among human beings. What if we classified people into racial categories based on eye color instead of skin color? Or hair color? Or blood type? (There are no racial blood types; blood types cut across races.) Is there any scientific reason for selecting certain traits over others in determining racial categories? The answer is "No."

> What if we classified people into racial categories based on eye color instead of skin color?

Some physical variations among people are the result of living for thousands of years in different geographic regions. According to anthropologists, modern humans evolved in Africa about 100,000 years ago and developed dark skin from the natural skin pigment, melanin, to protect against the sun's ultraviolet radiation, levels of which are high near the equator. When humans began migrating to regions farther from the equator, lighter skin developed because of the reduced exposure to ultraviolet radiation. But more importantly, lighter skin was necessary to allow enough UV rays to penetrate the skin to produce vitamin D, which is important for human health (Mukhopadhyay et al. 2007).

The science of genetics also challenges the notion of race. Geneticists have discovered that the genes of any two unrelated people, chosen at random from around the globe, are 99.9 percent alike (Ossorio & Duster 2005). Furthermore, "most human genetic variation—approximately 85 percent—can be found between any two individuals from the same group (racial, ethnic, religious, etc.). Thus, the vast majority of variation is within-group variation" (Ossorio & Duster 2005, p. 117). Classifying people into different races fails to recognize that, over the course of human history, migration and intermarriage have resulted in the blending of genetically transmitted traits. Thus there are no "pure" races; people in virtually all societies have genetically mixed backgrounds (Keita & Kittles 1997). And contrary to what some people believe, all humans belong to the same species.

The American Anthropological Association has passed a resolution stating that "differentiating species into biologically defined 'races' has proven meaningless and unscientific" (Etzioni 1997, p. 39).

> To summarize, races are unstable, unreliable, arbitrary, culturally created divisions of humanity. This is why scientists . . . have concluded that race, as scientifically valid biological divisions of the human species, is fiction not fact. (Mukhopadhyay et al. 2007, p. 14)

As clear evidence that race is a social rather than a biological concept, different societies construct different systems of racial classification, and these systems change over time. For example, "at one time in the not-too-distant past in the United States, Italians, Greeks, Jews, the Irish, and other 'white' ethnic groups were not considered to be white. Over time . . . the category of 'white' was reshaped to include them" (Rothenberg 2002, p. 3).

The significance of race is not biological but social and political, because race is used to separate "us" from "them" and becomes a basis for unequal treatment

TABLE 9.1 Who's Hispanic?

THE U.S. CENSUS BUREAU APPROACH TO DEFINING WHO IS HISPANIC:

Q. *I immigrated to Phoenix from Mexico. Am I Hispanic?*

A. You are if you say so.

Q. *My parents moved to New York from Puerto Rico. Am I Hispanic?*

A. You are if you say so.

Q. *My grandparents were born in Spain but I grew up in California. Am I Hispanic?*

A. You are if you say so.

Q. *I was born in Maryland and married an immigrant from El Salvador. Am I Hispanic?*

A. You are if you say so.

Q. *My mom is from Chile and my dad is from Iowa. I was born in Des Moines. Am I Hispanic?*

A. You are if you say so.

Q. *I was born in Argentina but grew up in Texas. I don't consider myself Hispanic. Does the Census count me as Hispanic?*

A. Not if you say you aren't.

Source: Passel & Taylor 2009.

of one group by another. Despite the increasing acceptance that "there is no biological justification for the concept of 'race'" (Brace 2005, p. 4), its social significance continues to be evident throughout the world.

> **What Do You Think?** Do you think the time will ever come when a racial classification system will no longer be used? Why or why not? What arguments can be made for discontinuing racial classification? What arguments can be made for continuing it?

Ethnicity, which refers to a shared cultural heritage or nationality, is also socially constructed in part. Consider that the U.S. government has two approaches to defining who is Hispanic. One approach defines a Hispanic or Latino as someone who can trace their roots to one (or more) of 20 Spanish-speaking countries (Latin America or Spain). The other approach, which the Census Bureau uses, is based on a person's definition of themselves. In other words, a person is Hispanic if they say they are Hispanic (see Table 9.1).

Patterns of Racial and Ethnic Group Interaction

When two or more racial or ethnic groups come into contact, one of several patterns of interaction occurs; these include genocide, expulsion, colonialism, segregation, acculturation, pluralism, and assimilation. These patterns of interaction occur when the groups exist in the same society or when different groups from different societies come into contact. Although not all patterns of interaction between racial and ethnic groups are destructive, author and Mayan shaman

Ethnicity Shared cultural heritage or nationality.

Martin Prechtel reminded us, "Every human on this earth, whether from Africa, Asia, Europe, or the Americas, has ancestors whose stories, rituals, ingenuity, language, and life ways were taken away, enslaved, banned, exploited, twisted, or destroyed" (quoted by Jensen 2001, p. 13).

Genocide refers to the deliberate, systematic annihilation of an entire nation or people. The European invasion of the Americas, beginning in the 16th century, resulted in the decimation of most of the original inhabitants of North and South America. Some native groups were intentionally killed, whereas others fell victim to diseases brought by the Europeans. In the 20th century, Hitler led the Nazi extermination of 12 million people, including 6 million Jews, in what is known as the Holocaust. In the early 1990s, ethnic Serbs attempted to eliminate Muslims from parts of Bosnia—a process they called "ethnic cleansing." In 1994, genocide took place in Rwanda when Hutus slaughtered hundreds of thousands of Tutsis (called "cockroaches" by the Hutus)—an event depicted in the 2004 film *Hotel Rwanda*. And as this book goes to press, genocide is occurring in the Darfur region of Sudan, where the Sudanese government, using Arab *Janjaweed* militias, its air force, and organized starvation, is systematically killing the African Muslim communities because some among them have challenged the authoritarian rule of the Sudanese government (see also Chapter 15).

In 1994, Hutus in Rwanda committed genocide against the Tutsis, resulting in 800,000 deaths.

genocide The deliberate, systematic annihilation of an entire nation or people.

expulsion Occurs when a dominant group forces a subordinate group to leave the country or to live only in designated areas of the country.

colonialism Occurs when a racial or ethnic group from one society takes over and dominates the racial or ethnic group(s) of another society.

segregation The physical separation of two groups in residence, workplace, and social functions.

Expulsion occurs when a dominant group forces a subordinate group to leave the country or to live only in designated areas of the country. The 1830 Indian Removal Act called for the relocation of eastern tribes to land west of the Mississippi River. The movement, lasting more than a decade, has been called the Trail of Tears because tribes were forced to leave their ancestral lands and endure harsh conditions of inadequate supplies and epidemics that caused illness and death. After Japan's attack on Pearl Harbor in 1941, President Franklin Roosevelt authorized the removal of any people considered threats to national security. All people on the West Coast with at least one-eighth Japanese ancestry were transferred to evacuation camps surrounded by barbed wire, where 120,000 Japanese Americans experienced economic and psychological devastation. In 1979, Vietnam expelled nearly 1 million Chinese from the country as a result of long-standing hostilities between China and Vietnam.

Colonialism occurs when a racial or ethnic group from one society takes over and dominates the racial or ethnic group(s) of another society. The European invasion of North America, the British occupation of India, and the Dutch presence in South Africa before the end of apartheid are examples of outsiders taking over a country and controlling the native population. As a territory of the United States, Puerto Rico is essentially a colony whose residents are U.S. citizens, but they cannot vote in presidential elections unless they move to the mainland.

Segregation refers to the physical separation of two groups in residence, workplace, and social functions. Segregation can be *de jure* (Latin meaning "by law") or *de facto* ("in fact"). Between 1890 and 1910, a series of U.S. laws, which came to be known as **Jim Crow laws,** were enacted to separate blacks from whites by prohibiting blacks from using "white" buses, hotels, restaurants, and drinking fountains. In 1896, the U.S. Supreme Court (in *Plessy v. Ferguson*) supported de jure segregation of blacks and whites by declaring that "separate but equal" facilities were constitutional. Blacks were forced to live in separate neighborhoods and attend separate schools. Beginning in the 1950s, various rulings overturned these Jim Crow laws, making it illegal to enforce racial segregation. Although de jure segregation is illegal in the United States, de facto segregation still exists in the tendency for racial and ethnic groups to live and go to school in segregated neighborhoods.

Acculturation refers to adopting the culture of a group different from the one in which a person was originally raised. Acculturation may involve learning the dominant language, adopting new values and behaviors, and changing the spelling of the family name. In some instances, acculturation may be forced. For decades, the Australian government removed aboriginal children from their families and placed them in missions or foster families, forcing them to abandon their language and traditional aboriginal culture. Authorities targeted children of mixed descent—what they referred to as "half-caste"—because they thought these children could be more easily acculturated into white society. Today, these individuals are known as the "stolen generations" because they had been stolen from their families and their culture.

Pluralism refers to a state in which racial and ethnic groups maintain their distinctness but respect each other and have equal access to social resources. In Switzerland, for example, four ethnic groups—French, Italians, Swiss Germans, and Romansch—maintain their distinct cultural heritage and group identity in an atmosphere of mutual respect and social equality. In the United States, the political and educational recognition of multiculturalism reflects efforts to promote pluralism.

Assimilation is the process by which formerly distinct and separate groups merge and become integrated as one. Assimilation is sometimes referred to as the "melting pot," whereby different groups come together and contribute equally to a new, common culture. Although the United States has been referred to as a melting pot, in reality, many minorities have been excluded or limited in their cultural contributions to the predominant white Anglo-Saxon Protestant tradition.

Assimilation can be of two types: secondary and primary. *Secondary assimilation* occurs when different groups become integrated in public areas and in social institutions, such as neighborhoods, schools, the workplace, and in government. *Primary assimilation* occurs when members of different racial or ethnic groups are integrated in personal, intimate associations, as with friends, family, and spouses. Nineteen states had **antimiscegenation laws** banning interracial marriage until 1967, when the Supreme Court (in *Loving v. Virginia*) declared these laws unconstitutional. (The term *miscegenation* comes from the Latin *miscere,* meaning "to mix" and *genus,* "kind.")

Between 1980 and 2000, the proportion of interracial or interethnic marriages more than doubled, from 4 percent in 1980, to 9 percent in 2000 (Amato et al. 2007). Since 1960, the number of black-white married couples has increased

This is a scene from the 2002 Australian film *Rabbit-Proof Fence*, which tells the story of three aboriginal girls who were taken from their family by the Australian government as part of a program to force Aborigines to adopt the white culture.

Penny Tweedie/Marimax/Dimension Films/The Kobal Collection/Picture Desk

Jim Crow laws Between 1890 and 1910, a series of U.S. laws enacted to separate blacks from whites by prohibiting blacks from using "white" buses, hotels, restaurants, and drinking fountains.

acculturation The process of adopting the culture of a group different from the one in which a person was originally raised.

pluralism A state in which racial and ethnic groups maintain their distinctness but respect each other and have equal access to social resources.

assimilation The process by which formerly distinct and separate groups merge and become integrated as one.

antimiscegenation laws Laws banning interracial marriage until 1967, when the Supreme Court (in *Loving v. Virginia*) declared these laws unconstitutional.

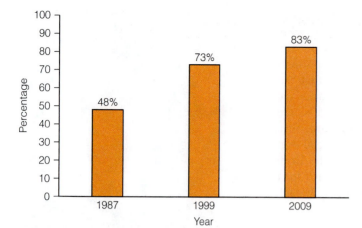

five-fold, the number of Asian-white married couples has increased more than tenfold, and the number of Hispanics married to non-Hispanics has tripled. This rise in nontraditional unions is related to the increased independence that adult children have from their parents and communities of origin as well as to increased societal acceptance of interracial dating and mixed marriages (Rosenfeld & Kim 2005) (see Figure 9.1).

Nevertheless, 15 percent of U.S. adults disapprove of marriage between blacks and whites (Gallup 2009a).

Figure 9.1 Percentage of U.S. Adults Who Agree That "It's All Right for Blacks and Whites to Date." Source: Pew Research Center 2009.

Racial and Ethnic Group Diversity and Relations in the United States

In the same year that Barack Obama was elected president, the U.S. Census Bureau projected that minorities, who comprised roughly one-third of the U.S. population in 2008, would become the majority in number in 2042, reaching 54 percent of the U.S. population in 2050 (U.S. Census Bureau 2008a). Currently, minority populations outnumber non-Hispanic whites in four states—California, New Mexico, Hawaii, and Texas. In these states, Hispanics are the largest minority group, except for Hawaii, where the largest minority group is Asian Americans (*New York Times* 2005).

Racial Diversity in the United States

The first census in 1790 divided the U.S. population into four groups: free white males, free white females, slaves, and other people (including free blacks and Indians). To increase the size of the slave population, the *one-drop rule* appeared, which specified that even one drop of "negroid" blood defined a person as black and therefore eligible for slavery. The "one-drop rule" is still operative today: biracial individuals are typically seen as a member of whichever group has the lowest status (Wise 2009). So, individuals who are part white and part black, such as Barack Obama, are socially considered black by most Americans.

Currently, minority populations outnumber non-Hispanic whites in four states—California, New Mexico, Hawaii, and Texas.

In 1960, the census recognized only two categories: white and nonwhite. In 1970, the census categories consisted of white, black, and "other" (Hodgkinson 1995). In 1990, the U.S. Census Bureau recognized four racial classifications: (1) white, (2) black, (3) American Indian, Aleut, or Eskimo, and (4) Asian or Pacific Islander. The 1990 census also included the category of "other." Beginning with the 2000 census, the Office of Management and Budget required federal agencies to use a minimum of five race categories: (1) white, (2) black or African American, (3) American Indian or Alaska Native, (4) Asian, and (5) Native Hawaiian or other Pacific Islander. In addition, respondents to federal surveys and the census now have the option of officially identifying themselves as being more than one race rather than checking only one racial category (see Figure 9.2).

Figure 9.2 **2010 Census Questions**
Source: U.S. Census Bureau 2009a.

➡ **NOTE: Please answer BOTH Question 8 about Hispanic origin and Question 9 about race. For this census, Hispanic origins are not races.**

8. **Is Person 1 of Hispanic, Latino, or Spanish origin?**

   ☐ **No,** not of Hispanic, Latino, or Spanish origin
   ☐ Yes, Mexican, Mexican Am., Chicano
   ☐ Yes, Puerto Rican
   ☐ Yes, Cuban
   ☐ Yes, another Hispanic, Latino, or Spanish origin — *Print origin, for example, Argentinean, Colombian, Dominican, Nicaraguan, Salvadoran, Spaniard, and so on.* ↘

   ☐☐☐☐☐☐☐☐☐☐☐☐☐☐☐☐☐☐☐☐

9. **What is Person 1's race?** Mark ☒ one or more boxes.

   ☐ White
   ☐ Black, African Am., or Negro
   ☐ American Indian or Alaska Native — *Print name of enrolled or principal tribe.* ↘

   ☐☐☐☐☐☐☐☐☐☐☐☐☐☐☐☐☐☐☐☐

   ☐ Asian Indian   ☐ Japanese   ☐ Native Hawaiian
   ☐ Chinese   ☐ Korean   ☐ Guamanian or Chamorro
   ☐ Filipino   ☐ Vietnamese   ☐ Samoan
   ☐ Other Asian — *Print race, for example, Hmong, Laotian, Thai, Pakistani, Cambodian, and so on.* ↘   ☐ Other Pacific Islander — *Print race, for example, Fijian, Tongan, and so on.* ↘

   ☐☐☐☐☐☐☐☐☐☐☐☐☐☐☐☐☐☐☐☐

   ☐ Some other race — *Print race.* ↘

   ☐☐☐☐☐☐☐☐☐☐☐☐☐☐☐☐☐☐☐☐

Figure 9.3 shows the racial and Hispanic composition of the United States in 2008, and projected to the year 2050.

Mixed-Race Identity. The number of Americans who self-identify as members of two or more races is expected to triple from 5.2 million in 2005, to 16.2 million in 2050 (U.S. Census Bureau 2008a). The census option for identifying as "mixed race" avoids putting children of mixed-race parents in the difficult position of choosing the race of one parent over the other when filling out race data on school and other forms. It also avoids impairment of children's self-esteem and social functioning that comes from choosing the racial category of "other." Such a category implies that the society does not recognize and respect mixed-race individuals, and thus "children growing up within mixed families may feel ashamed of their 'irregular' racial makeup and may experience rejection and alienation in the wider social community" (Zack 1998, p. 23).

Some critics of the new mixed-race option are concerned that the wide-scale recognition of mixed-race identity will decrease the numbers within minority groups and disrupt the solidarity and loyalty based on racial identification. For example, what will happen to organizations and movements devoted to equal rights for blacks if much of the "black" population acquires a new mixed-race identity? However, census data suggest that the mixed-race option will not have the large national impact that critics fear. As shown in Figure 9.3, only a small

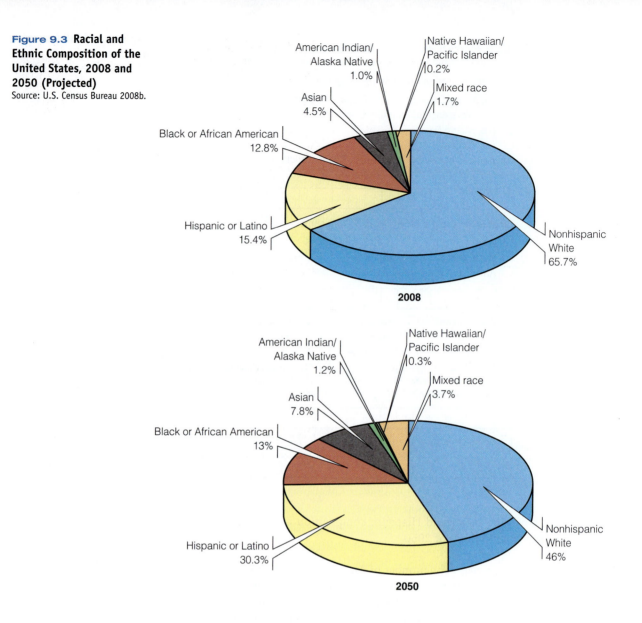

Figure 9.3 Racial and Ethnic Composition of the United States, 2008 and 2050 (Projected)
Source: U.S. Census Bureau 2008b.

2008

- Native Hawaiian/Pacific Islander 0.2%
- American Indian/Alaska Native 1.0%
- Mixed race 1.7%
- Asian 4.5%
- Black or African American 12.8%
- Hispanic or Latino 15.4%
- Nonhispanic White 65.7%

2050

- Native Hawaiian/Pacific Islander 0.3%
- American Indian/Alaska Native 1.2%
- Mixed race 3.7%
- Asian 7.8%
- Black or African American 13%
- Hispanic or Latino 30.3%
- Nonhispanic White 46%

percentage of the U.S. population identify themselves as being of more than one race (U.S. Census Bureau 2008b).

Ethnic Diversity in the United States

Ethnicity—one's membership in a group with a shared cultural heritage or nationality—can be distinguished on the basis of language, forms of family structures and roles of family members, religious beliefs and practices, dietary customs, forms of artistic expression such as music and dance, and national origin. Two individuals with the same ethnic background may identify with different races. For example, although most Hispanics are white, a small percentage identifies themselves as black (Navarro 2003). Conversely, two individuals with the same racial identity may have different ethnicities. For example, a black American and a black Jamaican have different cultural, or ethnic, backgrounds.

U.S. citizens come from a variety of ethnic backgrounds. The largest ethnic population in the United States is of Hispanic origin. (The terms *Hispanic* and *Latino* are used interchangeably here.) In 2008, 15 percent of the U.S. population was Hispanic, with the majority being Mexican (see Figure 9.4). The current Census Bureau classification system does not allow people of mixed Hispanic or Latino ethnicity to identify themselves as such. Individuals with one Hispanic and one non-Hispanic parent still must say that they are either Hispanic or not Hispanic. In addition, Hispanics must select one country of origin, even if their parents are from different countries.

The use of racial and ethnic labels is often misleading and imprecise. The ethnic classification of "Hispanic/Latino," for example, lumps together disparate groups such as Puerto Ricans, Mexicans, Cubans, Venezuelans, Colombians, and others from Latin American countries. The racial term *American Indian* includes more than 300 separate tribal groups that differ enormously in language, tradition, and social structure. The racial label *Asian* includes individuals from China, Japan, Korea, India, the Philippines, or one of the countries of Southeast Asia. And what about people who are Asian but who live in the United States? The term *Asian American* is used to describe people with Asian racial features who are born in the United States, as well as those who immigrate to the United States. Columbia University professor Derald Wing Sue, an Asian American who was born in the United States, said, "When I get out of a cab after having a conversation

Although Barack Obama's background is biracial—he is the son of a black Kenyan immigrant father and a white Kansas native mother—he identifies as a black African American.

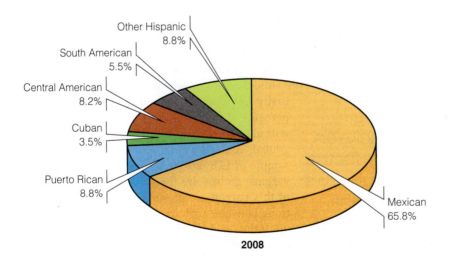

Figure 9.4 Percent Distribution of U.S. Hispanics by Type, 2008
Source: U.S. Census Bureau 2009d.

Other Hispanic 8.8%
South American 5.5%
Central American 8.2%
Cuban 3.5%
Puerto Rican 8.8%
Mexican 65.8%

2008

AP/Wide World Photos

Following Hurricane Katrina in 2005, a national survey found that the majority of blacks believe that the government's response to the crisis would have been faster if most of Katrina's victims had been white.

with a white cab driver, they'll say something like, 'Boy, you speak excellent English. . . . From their perspective, that's meant as a compliment, but another hidden meaning is being communicated, and that is that I am a perpetual foreigner in my own land" (quoted in Tanneeru 2007).

Race and Ethnic Group Relations in the United States

Despite significant improvements over the last two centuries, race and ethnic group relations continue to be problematic. The racial divide in the United States sharpened in 2005, in the wake of Hurricane Katrina, which left victims—who were predominantly black and poor—waiting for days to be rescued from their flooded attics or rooftops or to be evacuated from overcrowded "shelters" with no food, water, medical supplies, or working toilets. Following this crisis, a national survey found that most blacks (77 percent), compared with only 17 percent of whites, believe that the government's response to the disaster would have been faster if most of Katrina's victims had been white (Pew Research Center 2005).

Racial tensions concerning the treatment of African Americans spiked again in 2007, after a series of racially charged events occurred in Jena, Louisiana. The events began with the hanging of nooses on the "white tree" (where only white students gathered) on the Jena High School campus the day after black students sat under the tree. Three white students who were responsible for hanging the nooses received a punishment of only three days of in-school suspension for their "prank." But to blacks, this prank was a serious act of racial hate that recalled the history of lynching of blacks in the United States. The black/white racial tensions stemming from this prank and the failure of the school board and superintendent to take it seriously led to fights between black and white students. Six black students, known as the *Jena Six,* were arrested and charged with crimes related to their alleged involvement in the assault of a white student. Five of the six students were charged with attempted murder. Critics accused District Attorney Reed Walters, who is white, of prosecuting blacks more harshly than whites. After massive protests in Jena and nationwide, charges were reduced to aggravated second-degree battery in four of the cases.

In response to a question asking whether relations between blacks and whites will always be a problem for the United States or whether a solution will eventually be worked out, more than a third (38 percent) of U.S. adults said that race relations will always be a problem (Gallup Organization 2009a). A Gallup Poll survey asked a national sample of U.S. adults to rate relations between various groups in the United States. Results of this survey are presented in Table 9.2. Relations between various racial and ethnic groups are influenced by prejudice and discrimination (discussed later in this chapter). Race and ethnic

In response to a question asking whether relations between blacks and whites will always be a problem for the United States or whether a solution will eventually be worked out, more than a third . . . of U.S. adults said that race relations will always be a problem.

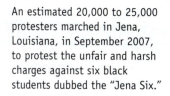

An estimated 20,000 to 25,000 protesters marched in Jena, Louisiana, in September 2007, to protest the unfair and harsh charges against six black students dubbed the "Jena Six."

© AP Photo/The Plain Dealer, Lisa DeJong

TABLE 9.2 Perceptions of Race and Ethnic Relations in the United States, 2008

A national sample of U.S. adults was asked to rate relations between various groups in the United States. The results are depicted in this table.

	VERY OR SOMEWHAT GOOD	VERY OR SOMEWHAT BAD
Whites and blacks	68%	30%
Whites and Hispanics	65%	33%
Blacks and Hispanics	49%	40%
Whites and Asians	81%	12%

Note: Percentages do not add to 100 because of "no opinion" responses.
Source: Gallup Organization 2009a.

relations are also complicated by issues concerning immigration—the topic we turn to next.

What Do You Think? A Gallup Organization (2009a) poll asked U.S. adults, "If blacks and whites honestly expressed their true feelings about race relations, do you think this would do more to bring races together or cause greater racial division?" More than half (59 percent) replied "bring races together," and about a third (34 percent) said "cause greater division" (7 percent had no opinion). What would your answer be? Why?

Immigrants in the United States

The growing racial and ethnic diversity of the United States is largely the result of immigration as well as the higher average birthrates among many minority groups. Immigration generally results from a combination of "push" and "pull" factors.

For each of the following questions, choose the answer that best reflects your attitudes. After answering all the questions, compare your answers with the results of a national Gallup Poll sample of U.S. adults.

1. On the whole, do you think immigration is a good thing or a bad thing for this country?
   a. Good thing
   b. Bad thing

2. In your view, should immigration be kept at its present level? Increased? Or decreased?
   a. Present level
   b. Increased
   c. Decreased

3. Which comes closer to your point of view: (a) Illegal immigrants in the long run become productive citizens and pay their fair share of taxes, or (b) illegal immigrants cost the taxpayers too much by using government services like public education and medical services?
   a. Pay fair share
   b. Cost too much

4. Which comes closer to your point of view: (a) illegal immigrants mostly take jobs that American workers want, or (b) illegal immigrants mostly take low-paying jobs Americans don't want?
   a. Take jobs that American workers want
   b. Take jobs that American workers do not want

Comparison Data

Gallup Poll data collected in 2008 revealed the following attitudes toward U.S. immigrants and immigration:

1. Bad thing: 30 percent; Good thing: 30 percent; Mixed or no opinion: 6 percent

2. Present level: 39 percent; Increased: 18 percent; Decreased: 39 percent; No opinion: 3 percent

3. Pay fair share of taxes: 31 percent; Cost taxpayers too much: 63 percent; Neither/both/no opinion: 6 percent

4. Take jobs Americans want: 15 percent; Take low-paying jobs Americans don't want: 79 percent; Neither/both/no opinion: 6 percent.

Source: Gallup Organization 2009b.

Adverse social, economic, and/or political conditions in a given country "push" some individuals to leave that country; whereas favorable social, economic, and/or political conditions in other countries "pull" some individuals to those countries. Before reading further, you may want to assess your attitudes toward U.S. immigrants and immigration in this chapter's Self and Society feature.

U.S. Immigration: A Historical Perspective

For the first 100 years of U.S. history, all immigrants were allowed to enter and become permanent residents. The continuing influx of immigrants, especially those coming from nonwhite, non-European countries, created fear and resentment among native-born Americans, who competed with immigrants for jobs and who held racist views toward some racial and ethnic immigrant populations. Increasing pressures from U.S. citizens to restrict or entirely halt the immigration of various national groups led to legislation that did just that. America's open-door policy on immigration ended in 1882 with the Chinese Exclusion Act, which suspended the entrance of the Chinese to the United States for 10 years and declared Chinese ineligible for U.S. citizenship. The Immigration Act of 1917 required all immigrants to pass a literacy test before entering the United States. And in 1921, the Johnson Act introduced a limit on the number of immigrants who could enter the country in a single year, with stricter limitations for certain countries (including those in Africa and the Near East). The 1924 Immigration Act further limited the number of immigrants allowed into the United States and completely excluded the Japanese. Other federal immigration laws

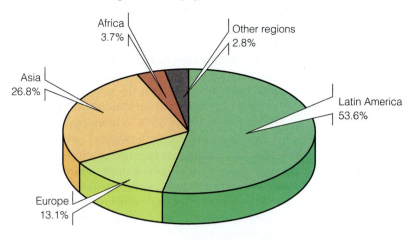

Total foreign-born U.S. population in 2007: 38.1 million

Africa 3.7%

Other regions 2.8%

Asia 26.8%

Latin America 53.6%

Europe 13.1%

Figure 9.5 U.S. Foreign-Born Residents by Region of Birth, 2007
Source: U.S. Census Bureau 2009b.

TABLE 9.3 Foreign-Born Population and Percentage of Total Population for the United States, 1890–2007

YEAR	NUMBER (IN MILLIONS)	PERCENTAGE OF TOTAL
2007	38.1	12.2
1990	19.8	7.9
1970	9.6	4.7
1950	10.3	6.9
1930	14.2	11.6
1890	9.2	14.8

Source: Lollock 2001; and U.S. Census Bureau 2009b.

include the 1943 Repeal of the Chinese Exclusion Act, the 1948 Displaced Persons Act (which permitted refugees from Europe), and the 1952 Immigration and Nationality Act (which permitted a quota of Japanese immigrants).

In the 1960s, most immigrants were from Europe, but most immigrants today are from Latin America (predominantly Mexico) or Asia (see Figure 9.5). In 2007, more than 1 in 10 U.S. residents (about 12 percent) were born in a foreign country (U.S. Census Bureau 2009b) (see Table 9.3).

Guest Worker Program

The United States has two guest worker programs that allow employers to import unskilled labor for temporary or seasonal work: the H-2A program for agricultural work and the H-2B program for nonagricultural work. In his final days in the White House, former President G. W. Bush changed regulations for the H-2B program by expanding the types of jobs considered to be temporary from jobs lasting no more than one year to jobs that last up to three years, making more jobs eligible for the program. The revised regulations also allow employers wanting to use the guest worker program to simply say they searched for U.S. workers and were unable to find enough. Previously, employers were required to obtain certification from the Department of Labor that there is a shortage of U.S. workers. H-2 visas generally do not permit guest workers to bring their families to the United States. Most guest workers are from Mexico, followed by Jamaica and Guatemala (SPLC 2007a).

A Southern Poverty Law Center report revealed that the guest worker program constitutes a "modern-day system of indentured servitude" (SPLC 2007a, p. 2). Guest workers often incur debts ranging from $500 to more than $10,000 to pay for visas, travel costs, and recruiter fees. When they arrive at their jobs, their employers often take their identity documents (passports and Social Security cards) to ensure that workers do not leave before their contract is fulfilled. Employers often use threats to control their workers. "In one case where workers refused to work until they received their pay after not having been paid in several weeks, the employer responded by threatening to call immigration and declare that the workers had 'abandoned' their work and were thus 'illegal' workers" (SPLC 2007a, p. 16). Guest workers are often paid substantially less than the minimum wage and are rarely paid overtime pay, despite working well over 40 hours a week. A Guatemalan guest worker in the forestry industry said, "Our pay would come out to approximately $25 for a 12-hour workday" (SPLC 2007a, p. 19). Although guest workers perform some of the most difficult and dangerous jobs in America, many who are injured on the job are unable to obtain medical treatment and workers' compensation benefits. Although employers hiring H-2A workers are required to provide them with free housing, the quality of housing is often substandard, even dangerous, and located in isolated rural areas where workers are dependent on their employers for transportation to work, grocery stores, and banks (for which employers often charge exorbitant fees). In some cases, guest workers are literally locked up in their living quarters. Immigrant women working at low-wage jobs are often targets of sexual violence; one survey found that three out of four Latinas in the South view sexual harassment as a major workplace problem (SPLC 2009a). Unable to obtain legal assistance, guest workers who are abused and denied their legal rights must endure the abuse or try to escape in a foreign land without passports, money, or tickets home.

Illegal Immigration

Many immigrants come to the United States illegally, without going through legal channels such as the H-2 visa program. An estimated 12 million undocumented immigrants are in the United States (the terms *undocumented immigrants* and *unauthorized immigrants* are used interchangeably). More than half of undocumented immigrants (57 percent) are Mexican (see Table 9.4).

Border Crossing. The U.S. Customs and Border Protection is charged with deterring people from illegally crossing the border into the United States and apprehending those who do. In addition, some groups of U.S. citizens have taken action to try to prevent illegal border crossing. For example, in 2003, a Texas rancher invited

TABLE 9.4 Undocumented U.S. Immigrants by Country or Region of Origin

COUNTRY OR REGION OF ORIGIN	PERCENTAGE OF UNDOCUMENTED IMMIGRANTS
Mexico	59
Asia	11
Central America	11
South America	7
Caribbean	4
Other parts of the world	8

Source: Passel & Cohn 2009.

a vigilante group called Ranch Rescue to his property to stop undocumented Latinos from crossing his land. Fatima Leiva and Edwin Mancia, Salvadorans, were among a group of immigrants traveling on foot when members of Ranch Rescue detained them. During the detention, Mancia was struck on the back of the head and attacked by a rottweiler dog owned by a Ranch Rescue member. Although Ranch Rescue purports to be protecting the United States from illegal immigration, the underlying motive for their anti-immigrant actions is hate. The president and spokesperson for Ranch Rescue has described Mexicans as "dog turds" who are "ignorant, uneducated and desperate for a life in a decent nation because the one [they] live in is nothing but a pile of dog [excrement] made up of millions of little dog turds" (quoted in SPLC 2005a, p. 3). Between 2005 and 2007, 250 new organizations that target immigrants—called "nativist extremist" organizations—have sprung up across the country, but largely in states that border Mexico. Members of these organizations, some of them armed, engage in vigilante roundups of undocumented Latino immigrants (Beirich et al. 2007).

Despite efforts to seal the U.S.-Mexican border, illegal border crossings occur every day. Some people cross (or attempt to cross) the U.S.-Mexican border with the help of a *coyote*—a hired guide who typically charges $2,000 to $3,000 to lead people across the border. Crossing the border illegally involves a number of risks, including death from drowning (e.g., while trying to cross the Rio Grande) or dehydration.

Undocumented Immigrants in the Workforce. More than 8 million undocumented immigrants are in the U.S. labor force, comprising 5.4 percent of the U.S. workforce (Passel & Cohn 2009). Sociologist Robert Maril (2004) noted, "The vast majority of illegal immigrants leave their home countries to work hard, save their money, then return to their homeland. . . . These individuals do not travel their difficult and dangerous journeys searching for a welfare handout; they immigrate to work" (pp. 11–12). Virtually all undocumented men are in the labor force. Their labor force participation exceeds that of men who are legal immigrants or who are U.S. citizens because undocumented men are less likely

On May 1, 2006, thousands of immigrant workers did not show up for work to show the nation how valuable immigrant labor is for the U.S. economy. Across the country, many businesses closed for the day; others experienced high rates of employee absenteeism.

to be disabled, retired, or in school. Undocumented women are less likely to be in the labor force than undocumented men because they are more likely to be stay-at-home mothers.

Undocumented workers often do work that U.S. workers are unwilling to do. Workers routinely work 60 or more hours per week and earn less than the minimum wage. They are not paid overtime and have no benefits. Illegal workers who cleaned Wal-Mart stores generally worked 7 nights a week, 364 days a year, and were often locked in stores (Greenhouse 2005). Immigrant workers who are legally admitted to the United States to work under the temporary foreign worker visa program also endure abusive labor conditions. "Because of language barriers and their vulnerable status under immigration laws, these workers may be the most exploited in the nation" (Bauer, quoted in SPLC 2005b).

Policies Regarding Illegal Immigration. In 1986, Congress approved the Immigration Reform and Control Act, which made hiring illegal immigrants an illegal act punishable by fines and even prison sentences. In 2005, Wal-Mart agreed to pay a record $11 million to settle charges that it used hundreds of illegal immigrants to clean its stores (Greenhouse 2005).

Recent concerns about terrorism and drug trafficking have led to stepped-up efforts to apprehend "illegal aliens." U.S. Customs and Border Protection, an agency within the Department of Homeland Security, had 18,000 border patrol agents in 2008, more than double the number in 2001. In 2008, these agents apprehended more than 948,000 people attempting to enter the United States illegally (U.S. Customs and Border Protection 2009). In the Secure Fence Act of 2006, President G. W. Bush authorized the construction of a 700-mile fence along the 2,100-mile U.S. border with Mexico. Public opposition to the building of a border fence has come from landowners who do not want to sell their land to the government; environmentalists who are concerned about the damage the fence project will do to endangered plants and animals in the Lower Rio Grande Valley National Wildlife Refuge; business owners who are concerned about the border fence's impact on the economy; and others who say that a fence will not effectively stop illegal immigration, will cost taxpayers more than $2 billion, and will damage diplomatic relations with Mexico.

Several states, cities, and counties have passed various laws and ordinances related to illegal immigration (Bazar 2007). In 2006, Hazelton, Pennsylvania, became the first city to take business licenses away from employers who hire undocumented workers and to fine landlords $1,000 a day for renting to illegal immigrants. In 2007, Arizona passed one of the toughest illegal immigration bills in the country, stipulating that employers who knowingly hire illegal immigrants would have their business license suspended or revoked.

In contrast to states that have taken a hard-line approach to illegal immigration, more than 30 major cities, including San Francisco, New York City, and Newark, have declared themselves "sanctuary cities" for undocumented immigrants. These cities have "sanctuary policies" that prohibit police from asking about immigrant status without criminal cause and instruct city employees not to notify the federal government of the presence of undocumented immigrants living or working in their communities.

Some churches have also played a role in providing a sanctuary to undocumented immigrants. In August 2006, Elvira Arellano, a Mexican citizen who lived illegally in the United States, took refuge in the Adalberto United Methodist Church in an attempt to avoid deportation by U.S. Immigration and Customs

After taking refuge in a Chicago church for one year to avoid deportation to Mexico, Elvira Arellano was arrested in California and deported to Mexico, leaving behind her 8-year-old son, a U.S. citizen.

© AP Photo/Julio Cortez

Enforcement. Arellano, whose young son is a U.S. citizen, said she should not have to choose between leaving her son in the states and taking him to Mexico. Arellano was arrested in August 2007, when she left the church and traveled to California and was deported to Mexico. Elvira Arellano's deportation has ignited a debate concerning immigration policies involving illegal immigrants whose children are U.S. citizens. Worksite raids and home raids by immigration officials have resulted in the arrest, detention, and deportation of thousands of immigrants, leaving tens of thousands of children, including children who are U.S. citizens, separated from their parents, or effectively deported with their families (Kremer et al. 2009). A report documenting the harm these children suffer states that "In a country that emphasizes the importance of family unity in the socialization and upbringing of its children, an immigration system that promotes family separation is a broken system" (Kremer et al. 2009, p. 5). Some parents even lose legal custody of their children. In 2007, Encarnación Bail Romero, a Guatemalan immigrant, was arrested along with 136 undocumented immigrants at a Missouri processing plant. While serving time in jail for presenting false identification to the police, a county court terminated her rights to her 2-year-old son Carlos on grounds of abandonment, and the child was adopted by a local couple (Thompson 2009).

Policies and proposals regarding illegal immigration range from increasing border security and imposing sanctions on employers who hire undocumented workers to providing a path to legalization of workers already here and implementing humane policies that protect family unity. Most Americans (63 percent) say they favor providing a way for illegal immigrants to gain legal citizenship if they (1) pass a background check, (2) pay fines, and (3) have a job (Pew Research Center 2009). The Development, Relief and Education for Alien Minors Act (DREAM Act), introduced in Congress in 2009, would provide a path to legal status for undocumented immigrants who were brought to the United States as children. If passed, the DREAM Act would permit certain immigrant students who have grown up

Immigrants in the United States **323**

in the United States to apply for temporary legal status and to eventually obtain permanent status and become eligible for U.S. citizenship if they go to college or serve in the U.S. military. The law would also eliminate a federal provision that penalizes states that provide in-state tuition to undocumented immigrants.

What Do You Think? Although the Supreme Court ruled in 1982 that states must educate undocumented immigrants through the twelfth grade, states are divided in their laws regarding supporting undocumented immigrants through in-state college tuition. Since 2001, 10 states have passed laws permitting certain undocumented students to receive in-state tuition (National Immigration Law Center 2009). Immigrants' rights advocates argue that it is in the best interest of the United States to help undocumented immigrant youth achieve a college education. But opponents worry that granting in-state tuition to undocumented immigrants would (1) be a financial burden to taxpayers, (2) make it harder for legal citizens to get accepted into state universities, and (3) open the door to granting other benefits now reserved for state residents. Do you think that undocumented immigrants should qualify for in-state college tuition? Why or why not?

Becoming a U.S. Citizen

In 2007, of all foreign-born U.S. residents, about half were **naturalized citizens** (immigrants who applied and met the requirements for U.S. citizenship) (U.S. Census Bureau 2009b). To become a U.S. citizen, immigrants must have been lawfully admitted for permanent residence; must have resided continuously as a lawful permanent U.S. resident for at least five years; must be able to read, write, speak, and understand basic English (certain exemptions apply); and must show that they have "good moral character" (U.S. Citizenship and Immigration Services 2009). Applicants who have been convicted of murder or an aggravated felony are permanently denied U.S. citizenship. In addition, applicants are denied citizenship if, in the last five years, they have engaged in any one of a variety of offenses, including prostitution, illegal gambling, controlled substance law violation (except for a single offense of possession of 30 grams or less of marijuana), habitual drunkenness, willful failure or refusal to support dependents, and criminal behavior involving "moral turpitude." To become a U.S. citizen, one must take the oath of allegiance and swear to support the Constitution and obey U.S. laws, renounce any foreign allegiance, and bear arms for the U.S. military or perform services for the U.S. government when required. Finally, applicants for U.S. citizenship must pass an examination on English (speaking, reading, and writing) and U.S. government and history administered by the U.S. Citizenship and Immigration Services.

Myths about Immigration and Immigrants

Despite the prejudice, discrimination, lack of social support, and language barriers they experience, many foreign-born U.S. residents work hard to succeed educationally and occupationally. Although nearly a third of foreign-born U.S. adults age 25 and older have not completed high school, more than one-quarter have a bachelor's degree or higher (U.S. Census Bureau 2009b). Despite the achievements and contributions of immigrants, a number of myths about immigration and immigrants persist, largely perpetuated by anti-immigrant groups and campaigns:

naturalized citizens
Immigrants who apply for and meet the requirements for U.S. citizenship.

Myth 1. Immigrants take jobs away from native workers.

Reality: Immigrant employment is relatively concentrated in a small number of sectors, including building and grounds maintenance, food preparation, and construction (Holzer 2005). Although immigrant labor displaces some U.S. workers, over the long term, immigrants have modest negative effects on the employment of less educated U.S. workers. Immigrants provide labor in sectors where shortages might occur otherwise, and immigrant labor helps reduce the prices of some products and services, such as housing and food (Holzer 2005). In 2007, former President G. W. Bush's economic advisers reported that immigrants tend to complement (not substitute for) natives, raising natives' productivity and income by more than $30 billion a year (Bauer & Reynolds 2009).

Myth 2. Immigrants hurt the U.S. economy.

Reality: The United States was built on the labor of immigrants, and the benefits of immigrants to the U.S. economy continue today. One report concluded that "the immediate effect of eliminating the undocumented workforce would include an estimated $1.757 trillion in annual lost spending, $651.511 billion in annual lost output, and 8.1 million job losses" (Perryman 2008, p. 40).

Myth 3. Immigrants drain the public welfare system and our public schools.

Reality: Although immigrants are more likely to live in poverty compared to the native U.S. population, they are less likely than low-income natives to use benefits such as Medicaid, Temporary Assistance for Needy Families (TANF), and food stamps. In part, this is because many noncitizen immigrants are ineligible for federal public benefit programs (although some states provide assistance to immigrants who are not eligible for federally funded services). In addition, many immigrants do not know that they—and/or their children—may be eligible to receive welfare benefits, or they are afraid that seeking benefits will have a negative effect on their legal status (Capps et al. 2005). Children of unauthorized immigrants, 73 percent of whom are U.S. citizens, comprise only 6.8 percent of students in elementary and secondary schools (Passel & Cohn 2009). The states bear the cost of social services, education, and medical services for the immigrant population. However, research suggests that the economic benefits that immigrants provide for the states outweigh the costs associated with supporting them. For example, a study of immigrants in North Carolina found that, over the prior 10 years, Latino immigrants had cost the state $61 million in a variety of benefits—but were responsible for more than $9 billion in state economic growth (Beirich 2007). Further, many immigrants pay Social Security taxes, even though the workers will not collect Social Security payments because they are not citizens.

Myth 4. Latino immigrants do not want to learn English.

Reality: The demand for English classes for immigrants exceeds their availability; out of 176 providers of English as a Second Language (ESL) classes, more than half reported waiting lists ranging from a few weeks to more than 3 years (Bauer & Reynolds 2009). Further, 96 percent of foreign-born Latinos in the United States said it is "very important" to teach English to children of immigrants.

Myth 5. Undocumented immigrants have children in the United States as a means of gaining legal status.

Reality: Of the approximately 5 million children of undocumented immigrants in the United States, more than 3 million are born in and therefore are citizens of the United States. But having children who are U.S. citizens does not provide immigrants with a means of gaining legal status in the United States. Children under 21 are not allowed to petition for their parents' U.S. citizenship.

Myth 6. Immigrants have high rates of criminal behavior.

Reality: Immigrants are less likely than natives to commit crimes. Because they risk deportation, undocumented immigrants have a strong motivation to avoid involvement with the law. In 2000, the U.S. incarceration rate for native-born men aged between 18 and 39 was 3.5 percent—five times greater than that of their foreign-born counterparts (Bauer & Reynolds 2009). El Paso, Texas, a city with a high immigrant population, is among the safest big cities in America. Criminologist Jack Levin said, "If you want to find a safe city, first determine the size of the immigrant population. If the immigrant community represents a large proportion of the population, you're likely in one of the country's safer cities" (quoted in Balko 2009).

Myth 7. Undocumented immigrants should have just "gotten in line" to gain legal entry into the United States.

Reality: Despite the demand of U.S. business for immigrant labor, only 5,000 permanent visas for lawful entry of less-skilled workers are available per year. Temporary work visas are also limited, which means that "the avenues for lawful entry into the U.S. by the lower-skilled, lower educated immigrant that makes up the vast majority of the undocumented population are virtually nonexistent" (Kremer et al. 2009, pp. 2–3).

Sociological Theories of Race and Ethnic Relations

Some theories of race and ethnic relations suggest that individuals with certain personality types are more likely to be prejudiced or to direct hostility toward minority group members. Sociologists, however, concentrate on the impact of the structure and culture of society on race and ethnic relations. Three major sociological theories lend insight into the continued subordination of minorities.

Structural-Functionalist Perspective

Structural functionalists emphasize that each component of society affects the stability of the whole. From this perspective, racial and ethnic inequality is dysfunctional for society (Williams & Morris 1993; Schaefer 1998). A society that practices discrimination fails to develop and utilize the resources of minority members. Prejudice and discrimination aggravate social problems, such as crime and violence, war, unemployment and poverty, health problems, family problems, urban decay, and drug use—problems that cause human suffering as well as impose financial burdens on individuals and society.

Picca and Feagin (2007) explained how "the system of racial oppression in the United States affects not only Americans of color but white Americans and society as a whole" (p. 271):

Whites lose when they have to pay huge taxes to keep people of color in prisons because they are not willing to remedy patterns of unjust enrichment and . . . to pay to expand education, jobs, or drug-treatment programs that would be less costly. They lose by driving long commutes so they do not have to live next to people of color in cities. . . . They lose when white politicians use racist ideas and arguments to keep from passing legislation that would improve the social welfare of all Americans. Most of all, whites lose . . . by not having in practice the democracy that they often celebrate to the world in their personal and public rhetoric. (p. 271)

The structural-functionalist analysis of manifest and latent functions also sheds light on issues of race and ethnic relations. For example, the manifest function of the civil rights legislation in the 1960s was to improve conditions for racial minorities. However, civil rights legislation produced an unexpected negative consequence, or latent dysfunction. Because civil rights legislation supposedly ended racial discrimination, whites were more likely to blame blacks for their social disadvantages and thus perpetuate negative stereotypes such as "blacks lack motivation" and "blacks have less ability" (Schuman & Krysan 1999).

Conflict Perspective

The conflict perspective examines how competition over wealth, power, and prestige contributes to racial and ethnic group tensions. Consistent with this perspective, the "racial threat" hypothesis views white racism as a response to perceived or actual threats to whites' economic well-being or cultural dominance by minorities.

For example, between 1840 and 1870, large numbers of Chinese immigrants came to the United States to work in mining (the California Gold Rush of 1848), railroads (the transcontinental railroad, completed in 1869), and construction. As Chinese workers displaced whites, anti-Chinese sentiment rose, resulting in increased prejudice and discrimination and the eventual passage of the Chinese Exclusion Act of 1882, which restricted Chinese immigration until 1924. More recently, white support for Proposition 209—a 1996 resolution passed in California that ended state affirmative action programs—was higher in areas with larger Latino, African American, or Asian American populations, even after controlling for other factors (Tolbert & Grummel 2003). In other words, opposition to affirmative action programs that help minorities was higher in areas with greater racial and ethnic diversity, suggesting that whites living in diverse areas felt more threatened by the minorities.

In another study, researchers interviewed individuals in white racist Internet chat rooms to examine the extent to which people would advocate interracial violence in response to alleged economic and cultural threats (Glaser et al. 2002). The researchers posed three scenarios that might be perceived as threatening: interracial marriage, minority in-migration (i.e., blacks moving into one's neighborhood), and job competition (i.e., competing with a black person for a job). Respondents' reactions to interracial marriage were the most volatile, followed by in-migration. The researchers concluded that violent ideation among white racists stems from perceived threats to white cultural dominance and separateness rather than from perceived economic threats.

Furthermore, conflict theorists suggest that capitalists profit by maintaining a surplus labor force, that is, by having more workers than are needed. A surplus labor force ensures that wages will remain low because someone is always available to take a disgruntled worker's place. Minorities who are disproportionately unemployed serve the interests of the business owners by providing surplus labor, keeping wages low, and, consequently, enabling them to maximize profits.

Conflict theorists also argue that the wealthy and powerful elite fosters negative attitudes toward minorities to maintain racial and ethnic tensions among workers. So long as workers are divided along racial and ethnic lines, they are less likely to join forces to advance their own interests at the expense of the capitalists. In addition, the "haves" perpetuate racial and ethnic tensions among the "have-nots" to deflect attention away from their own greed and exploitation of workers.

Symbolic Interactionist Perspective

The symbolic interactionist perspective focuses on the social construction of race and ethnicity—how we learn conceptions and meanings of racial and ethnic distinctions through interaction with others—and how meanings, labels, and definitions affect racial and ethnic groups. We have already explained that contemporary race scholars agree that there is no scientific, biological basis for racial categorizations. However, people have learned to think of racial categories as real, and, as the *Thomas Theorem* suggests, if things are defined as real, they are real in their consequences. Ossorio and Duster (2005) explain:

> People often interact with each other on the basis of their beliefs that race reflects physical, intellectual, moral, or spiritual superiority or inferiority. . . . By acting on their beliefs about race, people create a society in which individuals of one group have greater access to the goods of society—such as high-status jobs, good schooling, good housing, and good medical care—than do individuals of another group. (p. 119)

The labeling perspective directs us to consider the role that negative stereotypes play in race and ethnicity. **Stereotypes** are exaggerations or generalizations about the characteristics and behavior of a particular group. When Americans in a 1990 National Opinion Research Center poll were asked to evaluate various racial and ethnic groups, blacks were rated least favorably (Shipler 1998). Most of the respondents labeled blacks as less intelligent than whites (53 percent), lazier than whites (62 percent), and more likely than whites to prefer being on welfare to being self-supporting (78 percent). Negative stereotyping of minorities leads to a self-fulfilling prophecy. As Schaefer (1998) explained:

> Self-fulfilling prophecies can be devastating for minority groups. Such groups often find that they are allowed to hold only low-paying jobs with little prestige or opportunity for advancement. The rationale of the dominant society is that these minority individuals lack the ability to perform in more important and lucrative positions. Training to become scientists, executives, or physicians is denied to many subordinate group individuals, who are then locked into society's inferior jobs. As a result, the false definition becomes real. The subordinate group has become inferior because it was defined at the start as inferior and was therefore prevented from achieving the levels attained by the majority. (p. 17)

Even stereotypes that appear to be positive can have negative effects. The view of Asian Americans as a "model minority" involves the stereotypes of Asian Americans as excelling in academics and occupational success. These stereotypes mask the struggles and discrimination that many Asian Americans experience and also put enormous pressure on Asian American youth to live up to the social expectation of being a high academic achiever (Tanneeru 2007).

stereotypes Exaggerations or generalizations about the characteristics and behavior of a particular group.

The symbolic interactionist perspective is concerned with how individuals learn negative stereotypes and prejudicial attitudes through language. Different connotations of the colors white and black, for example, may contribute to negative attitudes toward people of color. The white knight is good, and the black knight is evil; angel food cake is white, and devil's food cake is black. Other negative terms associated with black include *black sheep, black plague, black magic, black mass, blackballed,* and *blacklisted*. The continued use of derogatory terms such as *Jap, gook, spic, frog, kraut, coon, chink, wop, towel-head, kike,* and *mick* also confirms the power of language in perpetuating negative attitudes toward minority group members. Similarly, advocates for immigrant rights suggest that the term *illegal aliens* is derogatory; they prefer the term *undocumented* or *unauthorized immigrant* as a more neutral term.

What Do You Think? Are racial and ethnic jokes insulting and harmful to minority group members? Or are such jokes innocent and playful forms of fun? Have you ever felt offended by hearing a racial or ethnic joke? How did you respond to the person telling the joke?

In the next section, we explore the concepts of prejudice and racism in more depth and discuss ways in which socialization and the media perpetuate negative stereotypes and prejudicial attitudes toward racial and ethnic groups.

Prejudice and Racism

Prejudice refers to negative attitudes and feelings toward or about an entire category of people. Prejudice can be directed toward individuals of a particular religion, sexual orientation, political affiliation, age, social class, sex, race, or ethnicity. **Racism** is the belief that race accounts for differences in human character and ability and that a particular race is superior to others. In a national sample of U.S. adults, twice as many blacks as whites said that racism is a "big problem" in our society today (see Figure 9.6) (*Washington Post* 2009). This chapter's Social

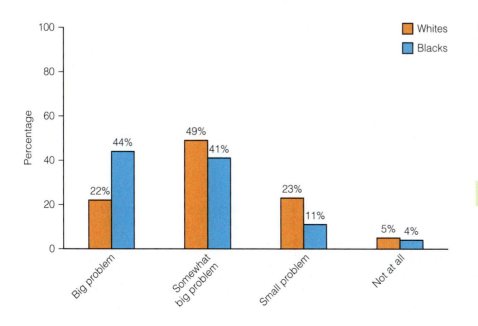

Figure 9.6 How Big a Problem Is Racism in Our Society Today?
Source: Washington Post 2009.

prejudice Negative attitudes and feelings toward or about an entire category of people.

racism The belief that race accounts for differences in human character and ability and that a particular race is superior to others.

In a book titled *Two-Faced Racism*, sociologists Leslie Picca and Joe Feagin examined accounts of racism in the everyday life of college students. The following is their research briefly described.

Sample and Methods

Between 2002 and 2003, Picca and Feagin, with the help of numerous college and university faculty members, recruited 934 college students to participate in a research project that required students to keep a diary or journal of "everyday events and conversations that deal with racial issues, images, and understandings" (p. 31). Students did not write their own name on the journals, but did indicate their gender, racial group, and age. Instructors who recruited students for this research decided whether to use the journal writing as a class assignment or as extra credit. The majority of students (63 percent) were from colleges and universities in the South, 19 percent were from the Midwest, 14 percent were from the West, and 4 percent were from the Northeast. Most of the participating students were between 18 and 25 years of age, although many were in their late 20s and a few were older. Rates of participation reported by the 23 instructors varied from 0 percent to 80 percent (15 percent to 30 percent average). The present study is based on journals written by 626 white students (68 percent women and 32 percent men).

Selected Findings

About three-quarters of the roughly 9,000 journal entries submitted by white college students described racist events. In relatively few of these accounts, students described their own racist actions or thoughts.

> **Kristi:** When I went to pick up the laundry, I saw a young black man sitting in the driver's side of a minivan with the engine running. My first thought was that he was waiting for a friend to rob the store and he was the getaway driver. . . . I am so embarrassed and saddened by my thinking. . . . (p. 1)

More frequently, students described the racist comments or actions of friends, family members, acquaintances, and strangers. A few journal accounts described antiracist actions by whites.

Students wrote about racist accounts targeted at African Americans, Latinos, Asian Americans, Native Americans, Jewish Americans, and Middle Easterners. The most frequently targeted minority group was black Americans.

In analyzing the racist accounts described in the journals, the researchers drew on the work of sociologist Erving Goffman, who used theatrical metaphors to describe and understand human behavior. Goffman suggested that when people are "backstage" in private situations, they act differently than they do when they are in "frontstage" public situations. Picca and

Feagin's research reveals that the expression of racist attitudes has gone backstage to private settings where whites are among other whites, especially family and friends because "many . . . whites realize it is no longer socially acceptable to be blatantly racist in frontstage areas" (p. xi). Consider the following diary entry:

> **Hannah:** Three of my friends (a white girl and two white boys) and I went back to my house to drink a little more before we ended the night. My one friend, Dylan started telling jokes. . . . Dylan said: "What's the most confusing day of the year in Harlem?" "Father's Day . . . who's your daddy?" Dylan also referred to black people as "porch monkeys." Everyone laughed a little, but it was obvious that we all felt a little less comfortable when he was telling jokes like that. My friend Dylan is not a racist person. He has more black friends than I do, that's why I was surprised he so freely said something like that. Dylan would never have said something like that around anyone who was a minority. . . . It is this sort of "joking" that helps to keep racism alive today. People know the places they have to be politically correct and most people will be. However, until this sort of "behind-the-scenes" racism comes to an end, people will always harbor those stereotypical views that are so prevalent in our country. This kind of joking really does

Problems Research Up Close feature suggests that racism in the United States is more common than many people realize.

Forms of Racism

Compared with traditional, "old-fashioned" prejudice, which is blatant, direct, and conscious, contemporary forms of prejudice are often subtle, indirect, and unconscious. Three variants of these more subtle forms of prejudice are aversive racism, modern racism, and "Racism 2.0."

aversive racism A subtle form of prejudice that involves feelings of discomfort, uneasiness, disgust, fear, and pro-white attitudes.

Aversive Racism. **Aversive racism** represents a subtle, often unintentional, form of prejudice exhibited by many well-intentioned white Americans who

bother me, but I don't know what to do about it. I know that I should probably stand up and say I feel uncomfortable when my friends tell jokes like that, but I know my friends would just get annoyed with me and say that they obviously don't mean anything by it. (pp. 17–18)

Hannah's journal entry illustrates the social norm that defines expressions of racism as acceptable in private "backstage" settings, and inappropriate in public "frontstage" settings. Hannah's journal entry also exemplifies a theme of "white innocence," whereby white students offer excuses for and minimize the significance of racist events. Whites often view racist comments not as "real racism," but as harmless remarks that are not to be taken seriously.

Although white women also display racist behavior, the journal entries in this research suggest that "white men disproportionately make racist jokes and similar racially barbed comments" (p. 133):

Carissa: The white men got on the subject of so-called nicknames for black people. Some mentioned were porch monkeys, jigaboos, tree swingers, etc. The one thing I took notice of was that not one girl made a comment. (p. 133)

Some students made comments in their journals or on the cover sheet that point to a general lack of awareness of racism.

One student, for example, wrote, "This assignment made me realize how many racial remarks are said every single day and I usually never catch any of them or pay close attention" (p. 39).

Many students were offended by the racist remarks they recorded in their journals, but did not express their disapproval and, so, participated in racism as a "bystander." Some students indicated that, after participating in the journal writing assignment, they are more aware of the need to intervene in social situations where racism is being expressed:

Kyle: As my last entry in this journal, I would like to express what I have gained out of this assignment. I watched my friends and companions with open eyes. I was seeing things that I didn't realize were actually there. By having a reason to pick out the racial comments and actions, I was made aware of what is really out there. Although I noticed that I wasn't partaking in any of the racist actions or comments, I did notice that I wasn't stopping them either. I am now in a position to where I can take a stand and try to intervene in many of the situations. (p. 275)

Conclusion

Picca and Feagin's study of college students' journals reveals that racism is still a pervasive feature of U.S. society, especially in "backstage" social contexts. The researchers also conclude: "Our data probably underestimate the severity and extent of the racist performances that whites engage in behind closed doors. Numerous students indicated that they suppressed, or thought about suppressing, racial events that they were involved in or observed" (pp. 33–34). For example, one student wrote:

As I am getting ready to turn in this assignment, I looked back over some of the entries . . . and thought that I should change a lot of them for fear of whoever is reading them might be offended even though I was very reserved in some of the accounts of what happened. (p. 34)

Based on the research presented in *Two-Faced Racism*, Picca and Feagin suggest that "the majority of whites still participate in openly racist performances in the backstage arena . . . and do not define such performances as problematic and deserving of action aimed at eradication" (p. 22). The researchers further note that "most of our college student diarists did not, or could not, see the connection between the everyday racial performances they recorded and the unjust discrimination and widespread suffering endured by people of color in society generally" (p. 28).

Source: Picca & Feagin 2007.

possess strong egalitarian values and who view themselves as unprejudiced. The negative feelings that aversive racists have toward blacks and other minority groups are not feelings of hostility or hate but rather feelings of discomfort, uneasiness, disgust, and sometimes fear (Gaertner & Dovidio 2000). Aversive racists may not be fully aware that they harbor these negative racial feelings; indeed, they disapprove of individuals who are prejudiced and would feel falsely accused if they were labeled prejudiced. "Aversive racists find blacks 'aversive,' while at the same time find any suggestion that they might be prejudiced 'aversive' as well" (Gaertner & Dovidio 2000, p. 14).

Another aspect of aversive racism is the presence of pro-white attitudes, as opposed to anti-black attitudes. In several studies, respondents did not indicate that blacks were worse than whites, only that whites were better than blacks (Gaertner & Dovidio 2000). For example, blacks were not rated as being lazier than whites, but whites were rated as being more ambitious than blacks. Gaertner and Dovidio (2000) explain that "aversive racists would not characterize blacks more negatively than whites because that response could readily be interpreted by others or oneself to reflect racial prejudice" (p. 27). Compared with anti-black attitudes, pro-white attitudes reflect a more subtle prejudice that, although less overtly negative, is still racial bias.

Modern Racism. Like aversive racism, **modern racism** involves the rejection of traditional racist beliefs, but a modern racist displaces negative racial feelings onto more abstract social and political issues. The modern racist believes that serious discrimination in the United States no longer exists, that any continuing racial inequality is the fault of minority group members, and that demands for affirmative action for minorities are unfair and unjustified. "Modern racism tends to 'blame the victim' and places the responsibility for change and improvements on the minority groups, not on the larger society" (Healey 1997, p. 55). Like aversive racists, modern racists tend to be unaware of their negative racial feelings and do not view themselves as prejudiced.

Racism 2.0. The election of Barack Obama has unveiled a new form of racism: Racism 2.0. According to Tim Wise (2009), **Racism 2.0** is a type of racism that "allows for and even celebrates the achievements of individuals of color, but only because those individuals generally are seen as different from a less appealing, even pathological black or brown rule" (p. 9). The fact that Barack Obama's supporters and admirers include a huge segment of the white population does not necessarily imply that these white Obama supporters are nonracist.

> If whites come to like, respect, and even vote for persons of color like Barack Obama, but only because they view them as having "transcended" their blackness in some way, to claim that the success of such candidates proves the demise of racism makes no sense at all. (Wise 2009, p. 9)

What Do You Think? A national poll (*Washington Post* 2009) asked a sample of U.S. adults: "If you honestly assessed yourself, would you say that you have at least some feelings of racial prejudice?" How would you answer that question? Would it surprise you to learn that 34 percent of whites and 38 percent of blacks answered "Yes"?

modern racism A subtle form of racism that involves the belief that serious discrimination in America no longer exists, that any continuing racial inequality is the fault of minority group members, and that the demands for affirmative action for minorities are unfair and unjustified.

Racism 2.0 A form of racism that allows for and celebrates the achievements of individuals of color who are viewed as having "transcended" their minority status.

Learning to Be Prejudiced: The Role of Socialization and the Media

Psychological theories of prejudice focus on forces within the individual that give rise to prejudice. For example, the frustration-aggression theory of prejudice (also known as the scapegoating theory) suggests that prejudice is a form of hostility that results from frustration. According to this theory, minority groups serve as convenient targets of displaced aggression. The authoritarian-personality theory of prejudice suggests that prejudice arises in people with a certain personality type. According to this theory, people with an authoritarian personality—who are highly conformist, intolerant, cynical, and preoccupied with power—are prone to being prejudiced.

Rather than focus on the individual, sociologists focus on social forces that contribute to prejudice. Earlier we explained how intergroup conflict over wealth, power, and prestige gives rise to negative feelings and attitudes that serve to protect and enhance dominant group interests. Prejudice is also learned through socialization and the media.

Learning Prejudice Through Socialization. Although most researchers agree that the majority of children learn conceptions of racial and ethnic distinctions by the time they are about 6 years old, Van Ausdale and Feagin (2001) suggest that children as young as 3 years old have acquired prejudicial attitudes:

> Well before they can speak clearly, children are exposed to racial and ethnic ideas through their immersion in and observation of the large social world. Since racism exists at all levels of society and is interwoven in all aspects of American social life, it is virtually impossible for alert young children either to miss or ignore it. . . . Children are inundated with it from the moment they enter society. (pp. 189–190)

In the socialization process, individuals adopt the values, beliefs, and perceptions of their family, peers, culture, and social groups. Prejudice is taught and learned through socialization, although it need not be taught directly and intentionally. Parents who teach their children to not be prejudiced yet live in an all-white neighborhood, attend an all-white church, and have only white friends may be indirectly teaching negative racial attitudes to their children. Socialization can also be direct, as in the case of parents who use racial slurs in the presence of their children or who forbid their children from playing with children from a certain racial or ethnic background. Children can also learn prejudicial attitudes from their peers. The telling of racial and ethnic jokes among friends, for example, perpetuates stereotypes that foster negative racial and ethnic attitudes.

What Do You Think? In Canada, a 7-year-old girl went to school with a swastika drawn on her arm. Alerted by the school, Child and Family Services investigated the child's home and, after finding neo-Nazi symbols and flags, they removed the girl and her 2-year-old brother from the home and placed them in temporary custody with their aunt. In a custody hearing, a social worker that interviewed the girl said that the girl spoke about how people of other races should be dead because this is a white man's world, and she provided graphic descriptions of how to kill people. The girl told the social worker that she watched skinhead videos with her parents and that her parents belonged to a skinhead website (CBCNews.ca 2009). This case has drawn international attention and sparked debate over whether the government has the right to protect children from their parents' racist views. What do you think?

Prejudice and the Media. The media contribute to prejudice by portraying minorities and immigrants in negative and stereotypical ways. An analysis of character portrayals in the 2003–2004 prime-time television season revealed that Latinos were more likely than any other group to be cast in low-status occupations, such as domestic worker, and as criminals. Nearly half (46 percent) of Arabs and Middle Easterners were cast as criminals (Glaubke & Heintz-Krowles 2004).

Byron Calvert operates Panzerfaust Records, one of the nation's largest "white power" music labels.

© AP Photo/The Minnesota Public Radio, Jeff Horwich

On his CNN show *Lou Dobbs Tonight* Dobbs has vilified immigrants as invaders who bring crime and disease to our communities. Dobbs proclaimed that "illegal aliens are criminals," even though illegal immigrants are not considered criminals under current U.S. law. Dobbs also referred to U.S.-born children of undocumented immigrants as "anchor babies," suggesting inaccurately that having a child who is a U.S. citizen helps immigrants gain legal status and protects them from deportation (Leadership Council on Civil Rights Education Fund 2009). The 2009 swine flu outbreak provided more fuel for anti-immigrant rants. Radio show host Michael Savage suggested that Mexican immigrants were being used as biological weapons against the United States:

> Could this be a terrorist attack through Mexico? Could our dear friends in the radical Islamic countries have concocted this virus and planted it in Mexico knowing that . . . [Homeland Security Secretary] Janet Napolitano would do nothing to stop the flow of human traffic from Mexico? And they are a perfect mule—perfect for bringing this virus into America. (quoted in SPLC 2009b).

Another media form that is used to promote hatred toward minorities and to recruit young people to the white power movement is "white power music," which contains anti-Semitic, racist, and homophobic lyrics. A CD that was distributed to thousands of middle and high school students by the neo-Nazi record label Panzerfaust Records contains the following lyrics from the group the Bully Boys: "Whiskey bottles/baseball bats/pickup trucks/and rebel flags/we're going on the town tonight/hit and run/let's have some fun/we've got jigaboos on the run." And in a song called "Wrecking Ball," the band H8Machine advises kids to "destroy all your enemies," promising that "the best things come to those who hate" (SPLC 2004). The Southern Poverty Law Center has identified 123 domestic and 227 international white power bands that promote hate and intolerance through their music (SPLC 2002).

The Internet also spreads messages of hate toward minority groups through the websites of various white supremacist and hate group organizations. After Hurricane Katrina in 2005, white supremacists posted hundreds of messages on the Internet expressing hopes that blacks in New Orleans would be wiped out. One message suggested that "they pile up all the niggers and use them as human sand bags against the rising storm surge" (Potok 2005).

The Southern Poverty Law Center, a nonprofit organization that combats hate, intolerance, and discrimination, identified 566 hate websites in 2006 (SPLC 2007b). And in 2007, about 12,000 white supremacist propaganda videos, hate music concert videos, and Holocaust denial videos were posted on video-sharing websites, such as YouTube (Mock 2007). Hate websites use sophisticated graphics, music, and entertaining games to lure children, teenagers, and adults (Nemes 2002). Once individuals are hooked on racist ideology, they use Internet discussion groups or chat rooms to connect with other like-minded individuals with whom they can share their racist views and have those views reinforced.

Although a number of countries, including Germany and France, have passed laws prohibiting hate speech on the Internet, free speech in the United States is protected under the First Amendment. Thus, many German neo-Nazi white supremacist groups use U.S.-based Internet servers to avoid prosecution under German law (Kaplan & Kim 2000).

Discrimination Against Racial and Ethnic Minorities

Whereas prejudice refers to attitudes, **discrimination** refers to actions or practices that result in differential treatment of categories of individuals. Although prejudicial attitudes often accompany discriminatory behavior or practices, one can be evident without the other.

Individual versus Institutional Discrimination

Individual discrimination occurs when individuals treat other individuals unfairly or unequally because of their group membership. Individual discrimination can be overt or adaptive. In **overt discrimination,** individuals discriminate because of their own prejudicial attitudes. For example, a white landlord may refuse to rent to a Mexican American family because of a prejudice against Mexican Americans. Or a Taiwanese American college student who shares a dorm room with an African American student may request a roommate reassignment from the student housing office because of prejudice against blacks.

Suppose that a Cuban American family wants to rent an apartment in a predominantly non-Hispanic neighborhood. If the landlord is prejudiced against Cubans and does not allow the family to rent the apartment, that landlord has engaged in overt discrimination. But what if the landlord is not prejudiced against Cubans but still refuses to rent to a Cuban family? Perhaps that landlord is engaging in **adaptive discrimination,** or discrimination that is based on the prejudice of others. In this example, the landlord may fear that other renters who are prejudiced against Cubans may move out of the building or neighborhood and leave the landlord with unrented apartments. Overt and adaptive individual discrimination can coexist. For example, a landlord may not rent an apartment to a Cuban family because of a prejudice and the fear that other tenants may move out.

discrimination Actions or practices that result in differential treatment of categories of individuals.

individual discrimination The unfair or unequal treatment of individuals because of their group membership.

overt discrimination Discrimination that occurs because of an individual's own prejudicial attitudes.

adaptive discrimination Discrimination that is based on the prejudice of others.

Institutional discrimination occurs when normal operations and procedures of social institutions result in unequal treatment of and opportunities for minorities. Institutional discrimination is covert and insidious and maintains the subordinate position of minorities in society. For example, the practice of businesses moving out of inner-city areas results in reduced employment opportunities for America's highly urbanized minority groups. In the retail industry, traditional warehouses designed to house inventory are being replaced by large distribution centers that require more space, thus encouraging location away from central cities to lower-priced land in less populated areas. A study by the U.S. Equal Employment Opportunity Commission (EEOC) found that, as areas become less populated, the percentage of minority workers declines (EEOC 2004).

Institutional discrimination also occurs in education. When schools use standard intelligence tests to decide which children will be placed in college preparatory tracks, they are limiting the educational advancement of minorities whose intelligence is not fairly measured by culturally biased tests developed from white middle-class experiences. And the funding of public schools through local tax dollars results in less funding for schools in poor and largely minority school districts.

Institutional discrimination is also found in the criminal justice system, which more heavily penalizes crimes that minorities are more likely to commit. For example, the penalties for crack cocaine, more often used by minorities, have traditionally been higher than those for other forms of cocaine use, even though the same prohibited chemical substance is involved. As conflict theorists emphasize, majority group members make rules that favor their own group.

Slightly more than a third (36 percent) of U.S. adults believe that discrimination against blacks is rare (Pew Research Center 2009). Yet, as the following sections reveal, discrimination against U.S. racial and ethnic minorities is more common than we may wish to believe.

Employment Discrimination

When a national sample of U.S. adults was asked, "Do you feel that racial minorities in this country have equal job opportunities as whites?" nearly half (46 percent) said "No" (Gallup Organization 2009a). Despite laws against it, discrimination against minorities occurs today in all phases of the employment process, from recruitment to interview, job offer, salary, promotion, and firing decisions.

A sociologist at Northwestern University studied employers' treatment of job applicants in Milwaukee, Wisconsin, by dividing job applicant "testers" into four groups: blacks with a criminal record, blacks without a criminal record, whites with a criminal record, and whites without a criminal record (Pager 2003). Applicant testers, none of whom actually had a criminal record, were trained to behave similarly in the application process and were sent with comparable résumés to the same set of employers. The study found that white applicants with no criminal record were the most likely to be called back for an interview (34 percent) and that black applicants with a criminal record were the least likely to be called back (5 percent). But surprisingly, white applicants with a criminal record (17 percent) were more likely to be called back for a job interview than were black applicants *without* a criminal record (14 percent). The researcher concluded that "the powerful effects of race thus continue to direct employment decisions in ways that contribute to persisting racial inequality" (Pager 2003, p. 960).

institutional discrimination Discrimination in which the normal operations and procedures of social institutions result in unequal treatment of minorities.

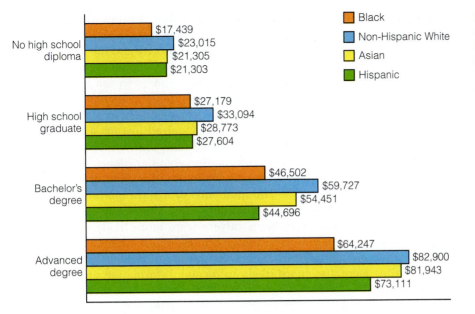

Legend:
- Black
- Non-Hispanic White
- Asian
- Hispanic

No high school diploma
- $17,439
- $23,015
- $21,305
- $21,303

High school graduate
- $27,179
- $33,094
- $28,773
- $27,604

Bachelor's degree
- $46,502
- $59,727
- $54,451
- $44,696

Advanced degree
- $64,247
- $82,900
- $81,943
- $73,111

Discrimination in hiring may be unintended. For example, many businesses rely on their existing employees to refer new recruits when a position opens up. Word-of-mouth recruitment is inexpensive and efficient; some companies offer bonuses to employees who bring in new recruits. But this traditional recruitment practice tends to exclude minority workers, because they often do not have a network of friends and family members in higher positions of employment who can recruit them (Schiller 2004).

Employment discrimination contributes to the higher rates of unemployment and lower incomes of blacks and Hispanics compared with those of whites (see Chapters 6 and 7). Lower levels of educational attainment among minority groups account for some, but not all, of the disadvantages they experience in employment and income. As shown in Figure 9.7, mean earnings of whites are higher than those for blacks, Hispanics, and Asians at the same level of educational attainment.

Workplace discrimination also includes unfair treatment. In one workplace with a large Hispanic workforce, Hispanic workers were selected each week to clean the lunchroom without being paid for that work (Greenhouse 2003). One Hispanic worker said that this treatment was a matter of dignity and that it made the workers feel humiliated.

Housing Discrimination and Segregation

Before the 1968 Federal Fair Housing Act and the 1974 Equal Credit Opportunity Act, discrimination against minorities in housing and mortgage lending was as rampant as it was blatant. Banks and mortgage companies commonly engaged in "redlining"—the practice of denying mortgage loans in minority neighborhoods on the premise that the financial risk was too great, and the ethical standards of the National Association of Real Estate Boards prohibited its members from introducing minorities into white neighborhoods. Instead, realtors practiced "geographic steering," whereby they discouraged minorities from moving into certain areas by showing them homes only in minority neighborhoods.

Although housing discrimination is illegal today, it is not uncommon. To assess discrimination in housing, researchers use a method called "paired testing." In a paired test, two individuals—one minority and the other nonminority—are trained to pose as home seekers, and they interact with real estate agents, landlords, rental agents, and mortgage lenders to see how they are treated. The testers are assigned comparable or identical income, assets, and debt as well as comparable or identical housing preferences, family circumstances, education, and job characteristics. A paired testing study of housing discrimination in 23 metropolitan areas found that whites in the rental market were more likely to receive information about available housing units and had more opportunities to inspect available units than did blacks and Hispanics (Turner et al. 2002). The incidence of discrimination was greater for Hispanic renters than for black renters. The same study found that, in the home sales market, white home buyers were more likely to be able to inspect available homes and to be shown homes in more predominantly non-Hispanic white neighborhoods than were comparable black and Hispanic buyers. Whites were also more likely to receive information and assistance with financing.

In a study of housing discrimination in the Philadelphia area, Massey and Lundy (2001) found that, compared with whites, African Americans were less likely to have a rental agent return their calls, less likely to be told that a unit was available, more likely to pay application fees, and more likely to have credit mentioned as a potential problem in qualifying for a lease. Sex and class exacerbated these racial effects. Lower-class blacks experienced less access to rental housing than middle-class blacks, and black females experienced less access than black males. Lower-class black females were the most disadvantaged group. They experienced the lowest probability of contacting and speaking to a rental agent and, even if they did make contact, they faced the lowest probability of being told of a housing unit's availability. Lower-class black females also faced the highest chance of paying an application fee. On average, lower-class black females were assessed $32 more per application than white middle-class males.

In other research on mortgage lending discrimination, minorities were less likely to receive information about loan products, they received less time and information from loan officers, they were often quoted higher interest rates, and they had higher loan denial rates than whites, other things being equal (Turner & Skidmore 1999).

Despite continued housing discrimination, homeownership rates among minorities and low-income groups increased substantially in the 1990s, reaching record rates in many central cities. However, minority and low-income homeowner rates still lag behind the overall homeownership rate. Also, many of the gains in minority and low-income homeownership rates are due to increases in *subprime lending*—loans with higher fees and higher interest rates that are offered to borrowers who have poor (or nonexistent) credit records (Williams et al. 2005).

Residential segregation of racial and ethnic groups also persists. Almost a quarter of all census tracts within the largest U.S. metropolitan areas are more than 90 percent white, and 12 percent are more than 90 percent minority (Turner & Fortuny 2009). For years, sociologists have known that U.S. minorities, who are disproportionately represented among the poor, tend to be segregated in concentrated areas of low-income housing, often in inner-city areas of concentrated poverty (Massey & Denton 1993). After Hurricane Katrina in 2005, citizens across the United States and the world were shocked by the degree of poor, segregated communities such as the Lower Ninth Ward of New Orleans, where 98 percent of residents were black and 36 percent lived below the poverty line.

Educational Discrimination and Segregation

Both institutional discrimination and individual discrimination in education negatively affect racial and ethnic minorities and help to explain why minorities (with the exception of Asian Americans) tend to achieve lower levels of academic attainment and success (see also Chapter 8). Institutional discrimination is evidenced by inequalities in school funding—a practice that disproportionately hurts minority students (Kozol 1991). Nearly half of school funding comes from local taxes. In 2002, the federal government supplied 8 percent of educational expenditures, and state government contributed 48 percent; local governments provided the remainder (Schiller 2004). Because minorities are more likely than whites to live in economically disadvantaged areas, they are more likely to go to schools that receive inadequate funding. Inner-city schools, which serve primarily minority students, receive less funding per student than do schools in more affluent, primarily white areas.

Another institutional education policy that is advantageous to whites is the policy that gives preference to college applicants whose parents or grandparents are alumni. The overwhelming majority of alumni at the highest-ranked universities and colleges are white. Thus, white college applicants are the primary beneficiaries of these so-called legacy admissions policies. About 10 percent to 15 percent of students in most Ivy League colleges and universities are children of alumni. Harvard University accepts about 11 percent of its overall applicant pool, but the admission rate is 40 percent for legacy applicants (Schmidt 2004). As a result of pressure from state lawmakers and minority rights activists, in 2004, Texas A&M University became the first public college to abandon its legacy admittance policy.

Minorities also experience individual discrimination in the schools as a result of continuing prejudice among teachers. One college student completing a teaching practicum reported that some of the teachers in her school often spoke about children of color as "wild kids who slam doors in your face" (Lawrence 1997, p. 111). In a survey conducted by the Southern Poverty Law Center, 1,100 educators were asked whether they had heard racist comments from their colleagues in the past year. More than one-fourth of survey respondents answered yes ("Hear and Now" 2000). It is likely that teachers who are prejudiced against minorities discriminate against them, giving them less teaching attention and less encouragement.

Racial and ethnic minorities are also treated unfairly in educational materials, such as textbooks, which often distort the history and heritages of people of color (King 2000). For example, Zinn (1993) observed, "To emphasize the heroism of Columbus and his successors as navigators and discoverers, and to deemphasize their genocide, is not a technical necessity but an ideological choice. It serves, unwittingly—to justify what was done" (p. 355).

Finally, racial and ethnic minorities are largely isolated from whites in a largely segregated school system. U.S. schools in the 2000–2001 school year were more segregated than they were in 1970 (Orfield 2001). School segregation is largely due to the persistence of housing segregation and the termination of court-ordered desegregation plans. Court-mandated busing became a means to achieve equality of education and school integration in the early 1970s, after the Supreme Court (in *Swann v. Charlotte-Mecklenberg*) endorsed busing to desegregate schools. But in the 1990s, lower courts lifted desegregation orders in dozens of school districts (Winter 2003a). And in 2007, the United States Supreme Court issued a landmark ruling, in a bitterly divided 5-to-4 vote, that race cannot be a factor in the assignment of children to public schools. The decision jeopardizes similar plans in hundreds of districts nationwide, and it further restricts how public school systems may achieve racial diversity. Recent data suggest that racial and ethnic segregation in U.S. suburban schools has declined slightly, as the percentage of minority students in suburban school districts has increased from 28 percent in the 1993–94 school year to 41 percent in the 2006–07 school year (Fry 2009).

Hate Crimes

In June 1998, James Byrd, Jr., a 49-year-old father of three, was walking home from a niece's bridal shower in the small town of Jasper, Texas. According to police reports, three white men riding in a gray pickup truck saw Byrd, a black man, walking down the road and offered him a ride. The men reportedly drove down a dirt lane and, after beating Byrd, chained him to the back of the pickup truck and dragged him for two miles down a winding, narrow road. The next day, police found Byrd's mangled and dismembered body. The three men who were arrested had ties to white supremacist groups. James Byrd had been brutally murdered simply because he was black.

The murder of James Byrd exemplifies a **hate crime**—an unlawful act of violence motivated by prejudice or bias. Examples of hate crimes, also known as "bias-motivated crimes" and "ethnoviolence," include intimidation (e.g., threats), destruction of or damage to property, physical assault, and murder.

The Federal Bureau of Investigation (FBI) reported that, in 2007, there were 7,624 hate crimes in the United States. However, FBI hate crime data undercount the actual number of hate crimes because (1) not all U.S. jurisdictions report hate crimes to the FBI (reporting is voluntary; more than 4,000 police agencies did not participate in the FBI's hate crime data collection); (2) it is difficult to prove that crimes are motivated by hate or prejudice; (3) law enforcement agencies shy away from classifying crimes as hate crimes because it makes their community "look bad"; and (4) victims are often reluctant to report hate crimes to the authorities.

From the first year that FBI hate crime data were published in 1992, most hate crimes have been based on racial bias (see Figure 9.8). In 2007, most of the reported race-based hate crimes were directed against blacks (69 percent), followed by whites (19 percent). FBI data on hate crimes reveal that more than twice as many hate crimes are reported against African Americans as any other group (Leadership Council on Civil Rights Education Fund 2009). "From lynching, to burning crosses and churches, to murdering a man by chaining him to a truck and dragging him down a road for three miles, anti-black violence has been and still remains the prototypical hate crime, intended not only to injure and kill individuals but to terrorize an entire group of people" (p. 25).

hate crime An unlawful act of violence motivated by prejudice or bias.

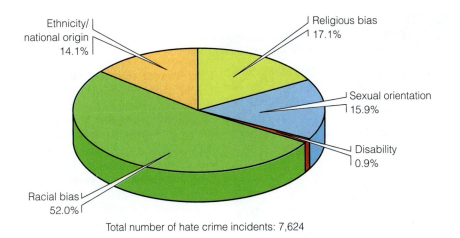

Ethnicity/
national origin
14.1%

Religious bias
17.1%

Figure 9.8 **Hate Crime Incidence by Category of Bias, 2007**
Source: FBI 2008a.

Sexual orientation
15.9%

Disability
0.9%

Racial bias
52.0%

Total number of hate crime incidents: 7,624

After the terrorist attacks of September 11, 2001, hate crimes against individuals perceived to be Muslim and/or Middle Eastern increased significantly. A poll by the Council on American-Islamic Relations showed that more than half of America's 7 million Muslims (57 percent) said they experienced bias or discrimination after September 11 (Morrison 2002). On the positive side, the same poll found that most American Muslims (79 percent) also experienced special kindness or support from friends or colleagues of other faiths. Acts of kindness include verbal assurances, support, and even offers to help guard local mosques and Islamic schools.

More recently, violence against U.S. immigrants has become a growing national problem. This chapter's *The Human Side* feature describes the hate crime victimization of a Hispanic immigrant in Georgia—a state with one of the fastest-growing immigrant populations.

Motivations for Hate Crimes. Levin and McDevitt (1995) found that the motivations for hate crimes were of three distinct types: thrill, defensive, and mission. Thrill hate crimes are committed by offenders who are looking for excitement and attack victims for the "fun of it." Defensive hate crimes involve offenders who view their attacks as necessary to protect their community, workplace, or college campus from "outsiders" or to protect their racial and cultural purity from being "contaminated" by interracial marriage and childbearing. A study of white racists in Internet chat rooms found that the topic of interracial marriage was more likely than other topics (such as blacks moving into one's neighborhood and competing for one's job) to elicit advocacy of violence (Glaser et al. 2002). For example, one respondent said, "better kill her. kill him and her. pull a oj. . . . i'm not kidding. I would do it if it was my sister. i would gladly go to prison then live a free life knowing some mud babies were calling me uncle whitey" (p. 184).

Mission hate crimes are perpetrated by white supremacist group members or other offenders who have dedicated their lives to bigotry. The Ku Klux Klan, the first major racist white supremacist group in the United States, began in Tennessee shortly after the Civil War. Klansmen have threatened, beaten, mutilated, and lynched blacks as well as whites who dared to oppose them. Other hate groups known to engage in violent crimes are the Identity Church Movement, the neo-Nazis, and the skinheads. The Southern Poverty Law Center identified

Domingo Lopez Vargas left his dirt-poor Guatemalan farm village to come to the United States, where he hoped to earn decent money for his wife and nine children. After picking oranges in Florida, he moved to Georgia, where the booming construction business lured immigrant workers. Unlike many of his compadres, Lopez had legal status, which helped him find steady work hanging doors and windows. When work dried up, Lopez joined the more than 100,000 jornaleros—day laborers—who wait for landscaping and construction jobs on street corners and in front of convenience stores all across Georgia. Usually there are plenty of pickup trucks that swing by, offering $8 to $12 an hour for digging, planting, painting, or hammering. But, this day, nada. By late afternoon, Lopez had tired of waiting in the cold, so he walked up the street to pick up a few things at a grocery store.

"I got milk, shampoo, and toothpaste," Lopez recalled. "When I was leaving the store, this truck stopped right in front of me and said, 'Do you want to work?' . . . I said, yes, how much? They said nine dollars an hour. I didn't ask what kind of job. I just wanted to work, so I said yes."

Until that afternoon, Lopez said, "Americans had always been very nice to me"—which might explain why he wasn't concerned that the four guys in the pickup truck looked awfully young to be contractors. Or why he didn't think twice about being picked up so close to sunset. "I took

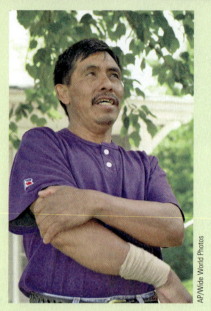

Domingo Lopez Vargas, an immigrant day laborer, was brutally beaten in a hate crime in Canton, Georgia.

the offer because I know sometimes people don't stop working until 9 at night," he says.

The four young men, all high school students, drove Lopez to a remote spot strewn with trash. "They told me to pick up some plastic bags that were on the ground. I thought that was my job, to clean up the trash. But when I bent over to pick it up, I felt somebody hit me from behind with

a piece of wood, on my back." It was just the start of a 30-minute pummeling that left Lopez bruised and bloody from his thighs to his neck. "I thought I was dying," he said. "I tried to stand up but I couldn't." Finally, after he handed over all the cash in his wallet, $260, along with his Virgin Mary pendant, the teenagers sped away.

As a result of injuries incurred by the beating, Lopez could not work for four months, and he was left with $4,500 in medical bills. Sometimes he still puzzles over his attackers' motives. "They were young," he speculates, "and maybe they didn't have enough education. Or maybe their families . . . taught them to kill people, and that is what they have learned."

Having sworn off day labor, Lopez works night shifts now, cutting up chickens at the nearby Tyson plant, wincing through the pain that shoots up his right arm when he lowers the boom on a bird. But it's only temporary, he says. "I called my wife and told her what happened. She told me to move back to Guatemala. I wanted to, but I didn't have enough money to go back and the police officers told me not to move out of the country because they will still need me to work on the case. After the case is finished, I want to go back to my family."

Source: Adapted from Moser 2004.

926 hate groups in the United States in 2008—a more than 50 percent increase since 2000, when there were 602 hate groups in the United States (Holthouse 2009). Although some members of hate groups have visible features, such as tattoos and armbands, that identify their hate group membership, many hate group members are not so easily identifiable. Criminologist Jack Levin, who studies hate crimes, noted that "many white supremacist groups are going more mainstream . . . they are eliminating the sheets and armbands. . . . The groups realize if they want to be attractive to middle-class types, they need to look middle class" (quoted in Leadership Council on Civil Rights Education Fund 2009, p. 19).

Some anti-immigrant hate groups, such as Federation for American Immigration Reform (FAIR), the Center for Immigration Studies (CIS), and NumbersUSA portray themselves as legitimate mainstream advocates against illegal immigration, but "some of these organizations have disturbing links to or relationships with extremists in the anti-immigrant movement" (p. 14).

Hate on Campus. Karl Nichols, a white residence hall director at the University of Mississippi, learned about racism in college, but not in the classroom and not from a textbook. Two chunks of asphalt were hurled through Nichols's dormitory room window along with a note warning, "You're going to get it, you Godforsaken nigger-lover" (SPLC 2000, p. 10). The next night, someone attempted to set Nichols's door on fire. According to the university's investigation of these incidents, Nichols "may have violated racist taboos . . . by openly displaying his affinity for African American individuals and black culture, by dating black women, by playing black music . . . and by promoting diversity" in his dormitory (p. 10).

Karl Nichols is not alone. At Brown University in Rhode Island, a black student was beaten by three white students who told her she was a "quota" who did not belong at a university. At the State University of New York in Binghamton, an Asian American student was left with a fractured skull after a racially motivated assault by three students. Two students at the University of Kentucky—one white, one black—were crossing the street just off campus when 10 white men attacked them. The attackers yelled racist slurs at the black student and choked him until he could not speak or move. The assailants called the white student a "nigger lover" as they broke his hand and nose. "I definitely thought I was going to lose my life," the black student said later. The white student was shocked by the incident, commenting, "I didn't know that much hate existed" (SPLC 2000, p. 7).

According to the FBI, more than 1 in 10 hate crimes (11.3 percent in 2007) occur at schools or colleges (FBI 2008a). Far more common than hate crimes are "bias incidents," which are events that are not crimes but still can have the same negative and divisive effects. Howard J. Ehrlich, director of the Prejudice Institute in Baltimore, estimates that one-fourth of racial and ethnic minority college students and up to 5 percent of white college students are targets of bias-motivated name-calling, e-mails, telephone calls, verbal aggression, and other forms of psychological intimidation each year (Willoughby 2003).

Hate Group Members in the Military. In December 1995, two members of the 82nd Airborne at Fort Bragg, North Carolina, who belonged to a white supremacist skinhead gang, shot and killed a black couple in a random, racially motivated double murder that shocked the nation. These hate crime murders led to congressional hearings and a major investigation of extremism in the military. The killers were sentenced to life in prison, and 19 other members of the 82nd Airborne were dishonorably discharged for neo-Nazi gang activities.

The Fort Bragg murders were not the first instance of military personnel involvement in hate groups. In the late 1980s and early 1990s, a number of cases came to light in which extremists in the military were caught diverting stolen firearms and explosives to neo-Nazi and white supremacist organizations, conducting guerrilla training for paramilitary racist militias, and murdering non-white civilians. According to investigations by the Southern Poverty Law Center, the military's tough stance against hate group affiliations among military personnel has relaxed since the recent war in Iraq and Afghanistan and the pressure to maintain enlistment numbers. An FBI report titled "White Supremacist Recruitment of Military Personnel since 9/11" revealed that racist extremists were taking advantage of lowered wartime recruiting standards to enlist in the military (FBI 2008b). Department of Defense investigator Scott Barfield said,

> Recruiters are knowingly allowing neo-Nazis and white supremacists to join the armed forces, and commanders don't remove them from the military even after we positively identify them as extremists or gang members. . . . Last year, for the first time, they didn't make their recruiting goals. They don't want to start making a big deal again about neo-Nazis in the military, because then parents who are already worried about their kids signing up and dying in Iraq are going to be even more reluctant about their kids enlisting if they feel they'll be exposed to gangs and white supremacists. (quoted in Holthouse 2006)

In one year, Barfield, who is based at Fort Lewis, identified and submitted evidence on 320 extremists there in the past year, but only two were discharged.

Strategies for Action: Responding to Prejudice, Racism, and Discrimination

Because racial and ethnic tensions exist worldwide, strategies for combating prejudice, racism, and discrimination globally require international cooperation and commitment. The World Conference Against Racism, Racial Discrimination, Xenophobia, and Related Intolerance, held in Durban, South Africa, in 2001, exemplifies international efforts to reduce racial and ethnic tensions and inequalities and to increase harmony among the various racial and ethnic populations of the world. Unfortunately, the U.S. delegation to this conference withdrew because of the expectation that hateful language would be used against Israel (because of the Israeli-Palestinian conflict).

In the following sections, we discuss the Equal Employment Opportunity Commission's role in responding to employment discrimination and examine the issue of affirmative action in the United States. We also discuss educational strategies to promote diversity and multicultural awareness and appreciation in schools. Finally, we look at apologies and reparations as a means of achieving racial reconciliation.

The Equal Employment Opportunity Commission

Equal Employment Opportunity Commission (EEOC) A U.S. federal agency charged with ending employment discrimination in the United States that is responsible for enforcing laws against discrimination, including Title VII of the 1964 Civil Rights Act that prohibits employment discrimination on the basis of race, color, religion, sex, or national origin.

The **Equal Employment Opportunity Commission (EEOC),** a U.S. federal agency charged with ending employment discrimination in the United States, is responsible for enforcing laws against discrimination, including Title VII of the 1964 Civil Rights Act that prohibits employment discrimination on the basis of race, color, religion, sex, or national origin. The EEOC investigates, mediates, and may file lawsuits against private employers on behalf of alleged victims of discrimination. For example, after receiving and investigating complaints from across the country that Walgreens discriminated against African American retail management and pharmacy employees in promotion, compensation, and assignment, the EEOC filed a discrimination suit against Walgreens that was settled for $20 million in monetary relief for an estimated 10,000 class members (EEOC 2007b).

The most frequently filed claims with the EEOC are allegations of race discrimination, racial harassment, or retaliation from opposition to racial discrimination. In fiscal year 2008, the EEOC received 33,937 charges alleging race-based discrimination, accounting for 36 percent of the agency's private-sector caseload (EEOC 2009). Because of budget shortfalls and staffing declines, the EEOC has a backlog of cases in the tens of thousands (Lee 2006).

In 2007, the EEOC launched a national initiative to combat racial discrimination in the workplace. The goals of this initiative, called E-RACE (Eradicating Racism And Colorism from Employment), are to (1) identify factors that contribute to race and color discrimination, (2) explore strategies to improve the administrative processing and litigation of race and color discrimination cases, and (3) increase public awareness of race and color discrimination in employment.

Affirmative Action

Affirmative action refers to a broad range of policies and practices in the workplace and educational institutions to promote equal opportunity as well as diversity. Affirmative action is an attempt to compensate for the effects of past discrimination and prevent current discrimination against women and racial and ethnic minorities. Vietnam veterans and people with disabilities may also qualify under affirmative action policies. Although the largest category of affirmative action beneficiaries is women, the majority of students in two sociology classes did not know that women were covered by affirmative action (Beeman et al. 2000).

What Do You Think? Although women are the largest category designated to benefit from affirmative action, a survey of 35 introductory sociology texts published in the 1990s revealed that nearly 90 percent of the texts did not mention affirmative action in their sections on gender inequality, and only 20 percent of texts included women in their definitions of affirmative action (Beeman et al. 2000). Why do you think many textbooks overlook or minimize the benefits women may receive from affirmative action?

Federal Affirmative Action. Affirmative action policies developed in the 1960s, from federal legislation that required any employer (universities as well as businesses) who received contracts from the federal government to make "good faith efforts" to increase the pool of qualified minorities and women (U.S. Department of Labor 2002). Such efforts can be made by expanding recruitment and training programs. Hiring decisions are to be made on a nondiscriminatory basis.

Affirmative Action in Higher Education. The Supreme Court's 1974 ruling in *Regents of the University of California v. Bakke* marked the beginning of the decline of affirmative action. Allan Bakke, a white male, had applied to the University of California at Davis medical school and was rejected, even though his grade point average and score on the medical school admissions test were higher than those of several minority applicants who had been admitted. The medical school had established fixed racial quotas, guaranteeing admission to 16 minority applicants regardless of their qualifications. Bakke claimed that such quotas discriminated against him as a white male and that the University

affirmative action A broad range of policies and practices in the workplace and educational institutions to promote equal opportunity as well as diversity.

of California had violated his Fourteenth Amendment right to equal protection under the law. The Supreme Court ruled in Bakke's favor by a 5-to-4 vote (showing how split the court was), concluding that the University of California unwittingly engaged in "reverse discrimination," which was unconstitutional. Affirmative action programs, the court ruled, could not use fixed quotas in admission, hiring, or promotion policies. However, the court affirmed the right for universities and employers to consider race as a factor in admission, hiring, and promotion to achieve diversity.

Since the *Bakke* case, numerous legal battles have challenged affirmative action (Olson 2003). In *Hopwood v. Texas,* Cheryl Hopwood and three other white individuals who had been denied admission to the University of Texas School of Law claimed that their Fourteenth Amendment rights to equal protection under the law had been violated by the university's affirmative action admission policies. In 1997, the Fifth Circuit Court of Appeals decided that the University of Texas could no longer use race as a factor in awarding financial aid, admitting students, and hiring and promoting faculty. Higher courts refused to review the decision. But in 2003, the U.S. Supreme Court, after hearing the appeals from two white applicants who applied but were not accepted to the University of Michigan, affirmed the right of colleges to consider race in admissions, but the court rejected Michigan's use of a point system to do so. According to the University of Michigan undergraduate admissions procedure, minority status provided 20 points on a 150-point scale for admission. The court found that the point system was problematic in that it turned race into the decisive factor for some applicants instead of just one of many factors (Winter 2003b). In response, the University of Michigan abandoned its point system and created an undergraduate admissions policy similar to its law school admissions policy, which may serve as a model for how other universities can achieve a diverse student body while following court guidelines. In the new "holistic review" approach, the university considers the unique circumstances of each student, prioritizing academics and treating all other factors, including race, equally. Applicants are now required to write more essays, including one on cultural diversity. Other universities, including the University of Wisconsin, University of Washington, and University of California, have also adopted "holistic admissions" procedures to increase their minority student populations.

Some universities, such as the University of Texas at Austin, use the "10 percent plan" as a way to maintain minority enrollment. Graduates in the top 10 percent of their high school classes are admitted automatically to the public college or university of their choice; standardized test scores and other factors are not considered.

In 2006, Michigan voters decided to ban affirmative action in university admissions, becoming the fifth state to do so (following California, Florida, Texas, and Washington). The continued weakening of affirmative action in higher education will necessitate creative admissions policies, such as holistic admissions and the 10 percent plan, to ensure access of minorities to higher education.

Attitudes Toward Affirmative Action. Affirmative action remains a divisive issue among Americans. Among first-year college students, nearly half (48 percent) agreed that "affirmative action in college admissions should be abolished" (Pryor et al. 2008). In a 2009 NBC News/*Wall Street Journal* Poll, a sample of U.S. adults was asked whether (a) "Affirmative action programs are still needed to counteract the effects of discrimination against minorities, and are a good idea as long as there are no rigid quotas"; or (b) "Affirmative action programs have gone too far in favoring minorities, and should be ended because they unfairly discriminate against whites." Nearly two-thirds (63 percent) said affirmative action programs are still needed, 28 percent said they should be ended, and 9 percent were unsure (Polling Report, Inc. 2009).

Public opinion poll results are influenced by how survey questions are worded and framed. Survey questions that ask whether respondents favor "affirmative action programs for women and minorities" elicit more favorable responses than questions that ask about affirmative action for minorities only (Paul 2003). In addition, terms such as *affirmative action, equal,* and *opportunity* in survey questions yield more support for affirmative action policies, whereas terms such as *special preferences, preferential treatment,* and *quotas* tend to lessen support.

Supporters of affirmative action suggest that such policies have many social benefits. In a review of more than 200 scientific studies of affirmative action, Holzer and Neumark (2000) concluded that these policies produce benefits for women, minorities, and the overall economy. Holzer and Neumark (2000) found that employers who adopt affirmative action increase the relative number of women and minorities in the workplace by an average of 10 percent to 15 percent. Since the early 1960s, affirmative action in education has contributed to an increase in the percentage of blacks attending college by a factor of 3 and the percentage of blacks enrolled in medical school by a factor of 4. Black doctors choose more often to practice medicine in inner cities and rural areas serving poor or minority patients than their white medical school classmates do (Holzer & Neumark 2000). Increasing the numbers of minorities in educational and professional positions also provides positive role models for other, especially younger, minorities "who can identify with them and form realistic goals to occupy the same roles themselves" (Zack 1998, p. 51).

Opponents of affirmative action suggest that such programs constitute reverse discrimination, which hurts whites. In 2004, 20 firefighters who are white (one is also Hispanic) sued the city of New Haven, Connecticut, after the city threw out the results of an examination given to determine promotion to positions of captain and lieutenant. Of the 118 candidates who took the test, 27 were black. None of the black candidates scored high enough to qualify for promotion, whereas all 20 of the plaintiffs qualified. The city decided to scrap the test results and promote no one, leading the 20 white firefighters who passed the test to sue the city for reverse discrimination. The hotly debated case reached the Supreme Court; in June 2009, the court ruled that the 20 white New Haven firefighters who were denied promotion were victims of illegal racial discrimination (Mahony & Kovner 2009).

During times of high unemployment rates, it is tempting for some whites to blame unemployment on affirmative action policies (as well as on immigrant labor). However, the main causes of unemployment among the white population are corporate downsizing, computerization and automation, factory relocations outside the United States, and other macroeconomic conditions, not affirmative action.

Attorney Karen Torre, center, stands with the firefighters she represented in a reverse discrimination case that made it all the way to the Supreme Court.

Some critics of affirmative action argue that it undermines the self-esteem of women and minorities. Although affirmative action may have this effect in rare cases, affirmative action can raise the self-esteem of women and minorities in many cases, by providing them with opportunities for educational advancement and employment (Plous 2003). Another criticism of affirmative action is that it fails to help the most impoverished of minorities—those whose deep and persistent poverty impairs their ability to compete not only with whites but also with other more advantaged minorities (Wilson 1987).

Opposition to affirmative action threatens the future of such policies and programs and the future educational and occupational opportunities of minorities. After California passed Proposition 209—an initiative to abolish affirmative action—black student enrollment in the UCLA law school declined from 10.3 percent in 1996 to 1.4 percent in 2000 (Greenberg 2003). Black and Latino enrollment has also significantly declined at the undergraduate program and law school at the University of California in Berkeley; the University of California, San Diego School of Medicine; and the University of Texas Law School.

Educational Strategies

Schools and universities play an important role in whether minorities succeed in school and in the job market. One way to improve minorities' chances of academic success is to reduce or eliminate disparities in school funding. As noted earlier, schools in poor districts—which predominantly serve minority students—have traditionally received less funding per pupil than do schools in middle- and upper-class districts (which predominantly serve white students). More than two dozen states have been forced by the courts to come up with a new system of financing schools to increase inadequate funding of schools in poor districts (Goodnough 2001). Other educational strategies focus on reducing prejudice, racism, and discrimination and fostering awareness and appreciation of racial and ethnic diversity. These strategies include multicultural education, "whiteness studies," and efforts to increase diversity among student populations.

Positive outcomes for students who take college diversity courses include increased racial understanding and cultural awareness, increased social interaction with students who have backgrounds different from their own, improved cognitive development, increased support for efforts to achieve educational equity, and higher satisfaction with their college experience.

Multicultural Education in Schools and Communities. In schools across the nation, multicultural education, which encompasses a broad range of programs and strategies, works to dispel myths, stereotypes, and ignorance about minorities; to promote tolerance and appreciation of diversity; and to include minority groups in the school curriculum (see also Chapter 8). With multicultural education, the school curriculum reflects the diversity of U.S. society and fosters an awareness and appreciation of the contributions of different racial and ethnic groups to U.S. culture. The Southern Poverty Law Center's program Teaching Tolerance publishes and distributes materials and videos designed to promote better human relations among diverse groups. These materials are sent to schools, colleges, religious organizations, and a variety of community groups across the nation.

Many colleges and universities have made efforts to promote awareness and appreciation of diversity by

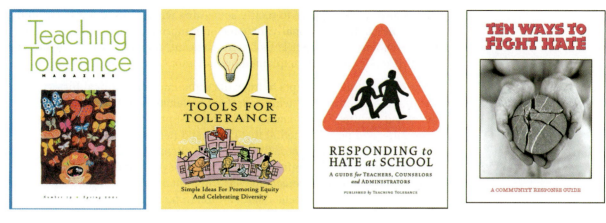

More than 600,000 teachers receive *Teaching Tolerance* magazine—a free resource for tolerance education in the classroom (Dees 2000). Other free materials available from the Southern Poverty Law Center (http://www.spl-center.org) include *101 Tools for Tolerance, Responding to Hate at School,* and *Ten Ways to Fight Hate*.

offering courses and degree programs in racial and ethnic studies and by sponsoring multicultural events and student organizations. A national survey by the Association of American Colleges and Universities revealed that 54 percent of colleges and universities required students to take at least one course that emphasizes diversity, and another 8 percent were in the process of developing such a requirement (Humphreys 2000). Positive outcomes for students who take college diversity courses include increased racial understanding and cultural awareness, increased social interaction with students who have backgrounds different from their own, improved cognitive development, increased support for efforts to achieve educational equity, and higher satisfaction with their college experience (Humphreys 1999).

Whiteness Studies. Traditionally, studies of and courses on race have focused on the social disadvantages that racial minorities experience, while ignoring or minimizing the other side of the racial inequality equation—the social advantages conferred upon whites. Courses in Whiteness Studies, which are being offered in many colleges and universities, focus on increasing awareness of white privilege—an awareness that is limited among white students. "White privilege is so fundamental as to be largely invisible, expected, and normalized" (Picca & Feagin 2007, p. 243). In an often-cited paper called "White Privilege: Unpacking the Invisible Knapsack," Peggy McIntosh (1990) likened white privilege to an "invisible weightless knapsack" that whites carry with them without awareness of the many benefits inside the knapsack. Just a few of the many benefits McIntosh associates with being white include being able to

- Go shopping without being followed or harassed.
- Be assured that my skin color will not convey that I am financially unreliable when I use checks or credit cards.
- Swear, or dress in secondhand clothes, or not answer letters without having people attribute these choices to the bad morals, poverty, or the illiteracy of my race.
- Avoid ever being asked to speak for all the people of my racial group.
- Be sure that, if a traffic cop pulls me over or if the IRS audits my tax return, it is not because I have been singled out because of my race.
- Take a job with an affirmative action employer without having coworkers on the job suspect that I got it because of race.

Whiteness studies "serve to rectify something wrong with the way we study race in America: By traditionally focusing on minority groups, the implicit message that scholarship projects is that nonwhites are 'deviant,' that's why they are studied" (Conley, 2002). As explained by Professor Gregory Jay (2005):

> Whiteness Studies attempts to trace the economic and political history behind the invention of "whiteness," to challenge the privileges given to so-called "whites," and to analyze the cultural practices (in art, music, literature, and popular media) that create and perpetuate the fiction of "whiteness." . . . "Whiteness Studies" is an attempt to think critically about how white skin preference has operated systematically, structurally, and sometimes unconsciously as a dominant force in American— and indeed in global—society and culture. (n.p.)

Diversification of College Student Populations. Recruiting and admitting racial and ethnic minorities in institutions of higher education can foster positive relationships among diverse groups and enrich the educational experience of all students—minority and nonminority alike (American Council on Education & American Association of University Professors 2000). Psychologist Gordon Allport's (1954) "contact hypothesis" suggested that contact between groups is necessary for the reduction of prejudice between group members. For example, native-born Americans with higher levels of contact with immigrants tend to have more positive views of immigrants and immigration than those with less contact (NPR et al. 2005). Ensuring a diverse student population can provide students with opportunities for contact with different groups and can thereby reduce prejudice. One study found that students with the most exposure to diverse populations during college had the most cross-racial interactions five years after leaving college (Gurin 1999).

Retrospective Justice Initiatives: Apologies and Reparations

In 2003, Brown University President Ruth J. Simmons appointed a Steering Committee on Slavery and Justice to investigate and issue a public report on the university's historical relationship to slavery and the trans-Atlantic slave trade. The steering committee's research concluded that "there is no question that many of the assets that underwrote the University's creation and growth derived, directly and indirectly, from slavery and the slave trade" (Brown University Steering Committee on Slavery and Justice 2007, p. 13). The committee's final report recommended that Brown University should acknowledge and make amends for its past ties to the slave trade.

Various governments around the world have issued official apologies for racial and ethnic oppression. After World War II, West Germany signed a reparations agreement with Israel in which West Germany agreed to pay Israel for the enslavement and persecution of Jews during the Holocaust and to compensate for Jewish property that the Nazis stole. In 2008, the Australian government issued a formal apology for the treatment of the country's aboriginal people, specifically, for the decades during which the government removed aboriginal children from their families in a forced acculturation program.

In the United States, President Gerald Ford and Congress apologized to Japanese Americans for their internment during World War II, and reparations of $20,000 were granted to each surviving internee who was a U.S. citizen or legal resident alien at time of internment. In 1993, President Bill Clinton apologized to native Hawaiians for overthrowing the government of their nation. In 1994, the state of Florida offered monetary compensation to the survivors and descendents of the 1923 "Rosewood massacre," in which a white mob attacked and murdered black residents in Rosewood, Florida, and set fire to the town. And in 1997, the U.S. government offered monetary reparations to surviving victims of the Tuskegee syphilis study, in which blacks suffering from syphilis were denied medical treatment.

Although various forms of reparations were offered to Native American tribes to compensate for land that had been taken by force or deception, it wasn't until 2008 that the Senate Committee on Indian Affairs passed a resolution apologizing to all Indian tribes for the mistreatment and violence committed against them. In 2008, the U.S. House of Representatives issued a formal apology to African Americans for slavery, and in 2009, the U.S. Senate passed a resolution apologizing for slavery, adding the stipulation that the official apology cannot be used to support claims for restitution. The resolution "acknowledges the fundamental injustice, cruelty, brutality, and inhumanity of slavery and Jim Crow Laws," and "apologizes to African Americans . . . for the wrongs committed against them and their ancestors who suffered under slavery and Jim Crow laws" (CNN 2009).

What Do You Think? Some scholars argue that achieving black/white racial reconciliation in the United States requires that the U.S. government not only issue an official apology but also provide substantial monetary and other reparations to African Americans (Brooks 2004). Do you think African Americans should receive monetary or other reparations? Why or why not?

The growing movement to redress past large-scale violations of human rights is based on the moral principles of taking responsibility for and attempting to rectify past wrongdoings. Supporters of the reparative justice movement believe that the granting of apologies and reparations to groups that have been mistreated promotes dialogue and healing, increases awareness of present inequalities, and stimulates political action to remedy current injustices. Some who are opposed to this movement argue that "the quest for historical redress, and for monetary reparations in particular, is just one more symptom of the 'culture of complaint,' of the elevation of victimhood and group grievance over self-reliance and common nationality," whereas others claim that "preoccupation with past injustice is a distraction from the challenge of present injustice" (Brown University Steering Committee on Slavery and Justice 2007, p. 39). Advocates of restitution programs face opposition from those who question whether restitution programs are morally justified. Many Americans view the enslavement of African Americans as a closed chapter in U.S. history that has nothing to do with the current status of black Americans. Consider the following letter sent to the Brown University Steering Committee on Slavery and Justice (2007):

> You disgust me, as you disgust many other Americans. Slavery was wrong, but at that time it was a legal enterprise. It ended, case closed. You cite slavery's effects as being the reason that black people are so far behind, but that just illustrates your ignorance. Black people, here and now, are behind because some can't keep their hands off drugs, or guns, or can't move forward, can't get off welfare, can't do the simple things to improve their life. . . . They don't deserve money, they deserve a boot in the backside over and over until they can find their own way. . . . (p. 9)

In addition, proponents of proposals for restitution face difficult questions about what form reparations should take, who the beneficiaries of the reparations should be, who should bear the costs of payment, and whether a reparative payment should be a single-shot event or a continuing program (Posner & Vermeule 2003).

The Brown University Steering Committee on Slavery and Justice (2007) examined "retrospective justice initiatives" from around the world and concluded that the most successful generally combined three elements: (1) formal acknowledgment of an offense; (2) a commitment to truth telling, to ensure that the relevant facts are uncovered, discussed, and properly memorialized; and (3) the making of some form of amends in the present to give material substance to expressions of regret and responsibility. In the committee's view, "reparative justice is not an invitation to 'wallow in the past' but a way for societies to come to terms with painful histories and move forward" (Brown University Steering Committee on Slavery and Justice 2007, p. 39).

Understanding Race, Ethnicity, and Immigration

After considering the material presented in this chapter, what understanding about race and ethnic relations are we left with? First, we have seen that racial and ethnic categories are socially constructed; they are largely arbitrary, imprecise, and misleading. Although some scholars suggest that we abandon racial and ethnic labels, others advocate adding new categories—multiethnic and multiracial—to reflect the identities of a growing segment of the U.S. and world population.

Conflict theorists and structural functionalists agree that prejudice, discrimination, and racism have benefited certain groups in society. But racial and ethnic disharmony has created tensions that disrupt social equilibrium. Symbolic interactionists note that negative labeling of minority group members, which is learned through interaction with others, contributes to the subordinate position of minorities.

Prejudice, racism, and discrimination are debilitating forces in the lives of minorities and immigrants. Despite these negative forces, many minority group members succeed in living productive, meaningful, and prosperous lives. But many others cannot overcome the social disadvantages associated with their minority status and become victims of a cycle of poverty (see Chapter 6). Minorities are disproportionately poor, receive inferior education and health care, and, with continued discrimination in the workplace, have difficulty improving their standard of living.

Achieving racial and ethnic equality requires alterations in the structure of society that increase opportunities for minorities—in education, employment and income, and political participation. In addition, policy makers concerned with racial and ethnic equality must find ways to reduce the racial and ethnic wealth gap and to foster wealth accumulation among minorities (Conley 1999). Social class is a central issue in race and ethnic relations. Professor and activist bell hooks (2000) (who spells her name in all lowercase) warned that focusing on issues of race and gender can deflect attention away from the larger issue of class division that increasingly separates the haves from the have-nots. Addressing class inequality must, suggests hooks, be part of any meaningful strategy to reduce inequalities that minority groups suffer. Civil rights activist Lani Guinier, in an interview with Paula Zahn on the *CBS Evening News* on July 18, 1998, suggested that "the real challenge is to . . . use race as a window on issues of class, issues of gender, and issues of fundamental fairness, not just to talk about race as if it's a question of individual bigotry or individual prejudice. The issue is more than about making friends—it's about making change." But, as Shipler (1998) noted, making change requires members of society to recognize that change is necessary, that there is a problem that needs rectifying:

Rev. Joseph Lowery, giving the benediction at President Obama's inauguration in January 2009.

> One has to perceive the problem to embrace the solutions. If you think racism isn't harmful unless it wears sheets or burns crosses or bars blacks from motels and restaurants, you will support only the crudest antidiscrimination laws and not the more refined methods of affirmative action and diversity training. (p. 2)

The historic election of the first African American U.S. president and the appointment of the first Hispanic woman to the U.S. Supreme Court have raised hopes for a new era in race and ethnic relations in the United States. We end this chapter with an excerpt from the benediction that Dr. Joseph Lowery delivered at President Obama's inauguration:

> Lord . . . we ask you to help us work for that day when black will not be asked to get back, when brown can stick around—and when yellow will be mellow—when the red man can get ahead, man—and when white will embrace what's right. (quoted in Kovac 2009).

CHAPTER REVIEW

■ **What is a minority group?**

A minority group is a category of people who have unequal access to positions of power, prestige, and wealth in a society and who tend to be targets of prejudice and discrimination. Minority status is not based on numerical representation in society but rather on social status.

■ **What is meant by the idea that race is socially constructed?**

The concept of race refers to a category of people who are believed to share distinct physical characteristics that are deemed socially significant. Races are cultural and social inventions; they are not scientifically valid because there are no objective, reliable, meaningful criteria scientists can use to identify racial groupings. Different societies construct different systems of racial classification, and these systems change over time. The significance of race is not biological but social and political, because race is used to separate "us" from "them" and becomes a basis for unequal treatment of one group by another.

■ **What are the various patterns of interaction that may occur when two or more racial or ethnic groups come into contact?**

When two or more racial or ethnic groups come into contact, one of several patterns of interaction occurs, including genocide, expulsion, colonialism, segregation, acculturation, pluralism, and assimilation.

- **Beginning with the 2000 census, what are the five race categories used to identify the race composition of the United States?**
Beginning with the 2000 census, the five race categories are (1) white, (2) black or African American, (3) American Indian or Alaska Native, (4) Asian, and (5) Native Hawaiian or other Pacific Islander. In addition, respondents to federal surveys and the census now have the option of officially identifying themselves as being of more than one race, rather than checking only one racial category.

- **What is an ethnic group?**
An ethnic group is a population that has a shared cultural heritage or nationality. Ethnic groups can be distinguished on the basis of language, forms of family structures and roles of family members, religious beliefs and practices, dietary customs, forms of artistic expression such as music and dance, and national origin. The largest ethnic population in the United States is Hispanics or Latinos.

- **What percentage of the U.S. population (in 2007) was born outside the United States?**
More than 1 in 10 U.S. residents (12.2 percent) were born in a foreign country.

- **What were the manifest function and latent dysfunction of the civil rights movement?**
The manifest function of the civil rights legislation in the 1960s was to improve conditions for racial minorities. However, civil rights legislation produced an unexpected consequence, or latent dysfunction. Because civil rights legislation supposedly ended racial discrimination, whites were more likely to blame blacks for their social disadvantages and thus perpetuate negative stereotypes such as "blacks lack motivation" and "blacks have less ability."

- **How does contemporary prejudice differ from more traditional, "old-fashioned" prejudice?**
Traditional, old-fashioned prejudice is easy to recognize, because it is blatant, direct, and conscious. More contemporary forms of prejudice are often subtle, indirect, and unconscious. In addition, racist expressions have gone "backstage" to private social settings.

- **Is it possible for an individual to discriminate without being prejudiced?**
Yes. In overt discrimination, individuals discriminate because of their own prejudicial attitudes. But sometimes individuals who are not prejudiced discriminate because of someone else's prejudice. For example, a store clerk may watch black customers more closely because the store manager is prejudiced against blacks and has instructed the employee to follow black customers in the store closely. Discrimination based on someone else's prejudice is called adaptive discrimination.

- **Are U.S. schools segregated?**
Racial and ethnic minorities are largely isolated from whites in an increasingly segregated school system. One study found that U.S. schools in the 2000–2001 school year were more segregated than they were in 1970. The upward trend in school segregation is due to large increases in minority student enrollment, continuing white flight from urban areas, the persistence of housing segregation, and the termination of court-ordered desegregation plans.

- **According to FBI data, the majority of hate crimes are motivated by what kind of bias?**
Since the FBI began publishing hate crime data in 1992, the majority of hate crimes have been based on racial bias.

- **What is the role of the Equal Employment Opportunity Commission (EEOC) in combating employment discrimination?**
The Equal Employment Opportunity Commission (EEOC) is responsible for enforcing laws against discrimination, including Title VII of the 1964 Civil Rights Act that prohibits employment discrimination on the basis of race, color, religion, sex, or national origin. The EEOC investigates, mediates, and may file lawsuits against private employers on behalf of alleged victims of discrimination. The most frequently filed claims with the EEOC are allegations of race discrimination, racial harassment, or retaliation from opposition to racial discrimination.

- **What group constitutes the largest beneficiary of affirmative action policies?**
Affirmative action policies are designed to benefit racial and ethnic minorities, women, and, in some cases, Vietnam veterans and people with disabilities. The largest category of affirmative action beneficiaries is women.

- **What are Whiteness Studies?**
Courses in Whiteness Studies, which are being offered in many colleges and universities, focus on increasing awareness of white privilege—an awareness that is limited among white students.

- **According to the Brown University Steering Committee on Slavery and Justice, successful retrospective justice initiatives contain what three elements?**
The Brown University Steering Committee on Slavery and Justice examined retrospective justice initiatives from around the world and concluded that the most successful generally combined three elements: (1) formal acknowledgment of an offense; (2) a commitment to truth telling, to ensure that the relevant facts are uncovered, discussed, and properly memorialized; and (3) the making of some form of amends in the present to give material substance to expressions of regret and responsibility.

TEST YOURSELF

1. All humans belong to the same species.
   a. True
   b. False
2. Which of the following occurs when a racial or ethnic group from one society takes over and dominates the racial or ethnic group(s) of another society?
   a. Secondary assimilation
   b. Colonialism
   c. Genocide
   d. Pluralism
3. According to projections by the U.S. Census Bureau, in what year will racial and ethnic minorities outnumber non-Hispanic whites in the United States?
   a. 2012
   b. 2042
   c. 2099
   d. Never
4. A Southern Poverty Law Center report reveals that the guest worker program constitutes a "modern-day system of _____."
   a. amnesty
   b. colonialism
   c. paid apprenticeship
   d. indentured servitude
5. Which minority group in the United States is considered a "model minority"?
   a. Women
   b. Black women
   c. Scandinavian immigrants
   d. Asian Americans
6. Which of the following has/have moved "backstage"?
   a. Racist behavior
   b. Undocumented immigrants
   c. Multicultural education
   d. Affirmative action
7. Under U.S. law, hate groups are not allowed to express racial or ethnic hatred on Internet websites.
   a. True
   b. False
8. Which of the following is responsible for enforcing laws against discrimination?
   a. Local police departments
   b. Equal Employment Opportunity Commission
   c. U.S. Department of Labor
   d. The Supreme Court
9. A national survey by the Association of American Colleges and Universities found that more than half of colleges and universities require students to take at least one course that emphasizes diversity.
   a. True
   b. False
10. The U.S. Congress has issued an official apology to African Americans for slavery and Jim Crow laws.
   a. True
   b. False

Answers: 1: a; 2: b; 3: b; 4: d; 5: d; 6: a; 7: b; 8: b; 9: a; 10: a.

KEY TERMS

acculturation   311
adaptive discrimination   330
affirmative action   345
antimiscegenation laws   311
assimilation   311
aversive racism   330
colonialism   310
discrimination   335
ethnicity   309, 313
Equal Employment Opportunity Commission (EEOC)   344

expulsion   310
genocide   310
hate crime   340
individual discrimination   335
institutional discrimination   336
Jim Crow laws   311
minority group   307
modern racism   330
naturalized citizens   324
overt discrimination   335
pluralism   311

prejudice   329
race   307
racism   329
Racism 2.0   332
segregation   310
stereotypes   328

MEDIA RESOURCES

Understanding Social Problems,
Seventh Edition Companion Website
www.cengage.com/sociology/mooney
Visit your book companion website, where you will find flash cards, practice quizzes, Internet links, and more to help you study.

 Just what you need to know NOW! Spend time on what you need to master rather than on information you already have learned. Take a pretest for this chapter, and CengageNOW will generate a personalized study plan based on your results. The study plan will identify the topics you need to review and direct you to online resources to help you master those topics. You can then take a posttest to help you determine the concepts you have mastered and what you will need to work on. Try it out! Go to www.cengage.com/login to sign in with an access code or to purchase access to this product.

Meredith Mullins/iStockphoto.com

Gender Inequality

"Only a radical transformation of the relationship between women and men to one of full and equal partnership will enable the world to meet the challenges of the 21st century."

Beijing Declaration and Platform for Action

357

Captain of the high school track team. Champion sprinter. Fluent in German. Played the clarinet. High school class valedictorian. Wing leader, Reserve Officers' Training Corps. Accepted at West Point. Graduated second in class. Third-generation Army. Commissioned officer. Platoon leader. Decorated soldier.

Male or female? In the 1950s, these characteristics would have described a male. Today, it is more difficult to tell, and in this case, this description is of a female. But she is no ordinary female.

Emily Tatum Perez was born in Germany, the daughter of Black American and Hispanic parents (Partlow & Parker 2006; Portraits of Sacrifice 2006; Thornburgh 2006). Raised in an area known for gangs, drugs, and violence, she still excelled— leading her track team, graduating at the top of her class, and starting a successful AIDS ministry at her local church. The American Red Cross later honored her for her volunteering efforts.

Since childhood, Emily had wanted to join the Army. In September 2001, she entered West Point just two weeks before the tragic events of 9/11. She rose to the top, holding the second highest rank in her senior class. And, when Emily was named Brigade Command Sergeant Major, she became the highest-ranking minority woman in the history of West Point.

Emily graduated in "The Class of 9/11," with a bachelor of science in sociology. She left West Point and was commissioned as a 2nd Lieutenant in the U.S. Army and, in December 2005, was sent to Iraq as a Medical Service Corps Officer. On September 26, 2006, Emily Tatum Perez, 23, an extraordinary soldier, an extraordinary woman, returned to her beloved West Point where she was buried, the first female graduate to die in the war in Iraq.

Emily Tatum Perez, a West Point graduate, was killed in Iraq in 2006. Since her death, thousands of mourners—family, friends, and strangers— have left comments on her memorial web pages and her MySpace page. Her roommate wrote, "Where you used to be, there is a hole in the world, which I find myself constantly walking around in the daytime, and falling in at night. I miss you. . . ."

Clearly, Emily Perez could do it all. But not long ago, she would not have had the opportunities she had to excel. West Point did not admit women to the academy until 1976, one indication of the gender inequality that existed then—and now as women continue to be barred from direct ground combat. The term *gender inequality*, however, raises the question: unequal in what way? Depending on the issue, both women and men are victims of inequality. When income, career advancement, household work, and sexual harassment are the focus, women are most often disadvantaged. But when life expectancy, emotional freedom, and death and disease are considered, often men are disadvantaged. In this chapter, we seek to understand inequalities for *both* genders.

We also look at **sexism**—the belief that innate psychological, behavioral, and/ or intellectual differences exist between women and men and that these differences connote the superiority of one group and the inferiority of the other. As with race and ethnicity, such attitudes often result in prejudice and discrimination at both the individual and institutional levels. Individual discrimination is

sexism The belief that innate psychological, behavioral, and/ or intellectual differences exist between women and men and that these differences connote the superiority of one group and the inferiority of the other.

reflected by physicians who will not hire a male nurse because they believe that women are more nurturing and empathetic and are therefore better nurses. Institutional discrimination—that is, discrimination built into the fabric of society—is exemplified by the difficulty many women experience in finding employment; they may have no work history and few job skills as a consequence of living in traditionally defined marriages.

Discerning the basis for discrimination is often difficult because the different types of minority statuses may *intersect*. For example, elderly Black American and Hispanic women are more likely to receive lower wages and to work in less prestigious jobs than younger white women. They may also experience discrimination if they are "out" as homosexuals. Such **double or triple (multiple) jeopardy** occurs when a person is a member of two or more minority groups. In this chapter, however, we emphasize the impact of gender inequality.

Gender refers to the social definitions and expectations associated with being female or male and should be distinguished from **sex,** which refers to one's biological classification. In most Western cultures, we take for granted that there are two categories of gender. However, in many other societies, three and four genders have been recognized. For example, in parts of Mexico *muxes* (pronounced MOO-shays) live "between two genders" (Lacey 2008, p.1), and many Polynesian cultures recognize the *mahū (*pronounced MAH-ho*)*—individuals who take on the work roles of members of the opposite sex (Nanda 2000).

Recognition that gender is not binary (i.e., either female or male) is increasing. In 2009, President Obama proposed guidelines barring discrimination against federal transgender employees (Rutenberg 2009). A **transgendered individual** (sometimes called "trans") is a person whose sense of gender identity—masculine or feminine—is inconsistent with their birth (sometimes called chromosomal) sex (male or female). Transgender is not a sexual orientation and transgendered individuals may have any sexual orientation—heterosexual, homosexual, or bisexual. Transsexuals are transgendered individuals "who have changed or are in the process of changing [their] . . . physical sex to conform to [their] . . . internal sense of gender identity" (HRC 2005, p. 7).

What Do You Think? Physically, men and women are different from birth. Men, in general, are stronger, taller, and heavier and have more facial hair. Women develop breasts, have higher-pitched voices, menstruate, and bear children. But are these physical characteristics related to behavioral differences? Are women innately nurturers? Are men innately aggressive? Noting, for example, that most societies are (were) patriarchal, most would answer yes. Others, however, would be quick to point out the role of socialization in traditional **gender role** assignment. Cross-cultural evidence is mixed. Some societies are characterized by traditional gender roles; in other societies, androgyny dominates; and in still other societies, traditional gender roles are reversed. What do you think? Are gender roles a matter of nature or nurture?

The Global Context: The Status of Women and Men

There is no country in the world in which women and men have equal status. Although much progress has been made in closing the gender gap in areas such as education, health care, employment, and government, gender inequality is still prevalent throughout the world.

double or triple (multiple) jeopardy The disadvantages associated with being a member of two or more minority groups.

gender The social definitions and expectations associated with being female or male.

sex A person's biological classification as male or female.

transgendered individual A transgender individual is a person whose sense of gender identity—masculine or feminine—is inconsistent with their birth (sometimes called chromosomal) sex (male or female).

gender roles A pattern of socially defined behaviors and expectations associated with being female or male.

A professor of English and creative writing at Colby College in Maine, Jennifer Finney Boylan is an award-winning author and nationally recognized speaker. One of her books, She's Not There: A Life in Two Genders, *is a memoir of the author's transition from James to Jennifer (Boylan 2009). Her life experiences as both man and woman make her uniquely qualified to address such issues as the meaning of gender, commitment, and marriage.*

As many Americans know . . . Governor John Baldacci of Maine signed a law that made this state the fifth in the nation to legalize gay marriage. It's worth pointing out, however, that there were some legal same-sex marriages in Maine already, just as there probably are in all 50 states. These are marriages in which at least one member of the couple has changed genders since the wedding.

I'm in such a marriage myself and, quite frankly, my spouse and I forget most of the time that there is anything particularly unique about our family, even if we are—what is the phrase?—"differently married."

Deirdre Finney and I were wed in 1988 at the National Cathedral in Washington. In 2000, I started the long and complex process of changing from male to female. Deedie stood by me, deciding that her life was better with me than without me. Maybe she was crazy for doing so; lots of people have generously offered her this unsolicited opinion over the years. But what she would tell you, were you to ask, is that the things that she loved in me have mostly remained the same, and that our marriage, in the end, is about a lot more than what genders we are, or were.

Deirdre is far from the only spouse to find herself in this situation; each week we hear from wives and husbands going through similar experiences together. Reliable statistics on transgendered people always prove elusive, but just judging from my e-mail, it seems as if there are a whole lot more transsexuals—and people who love them—in New England than say, Republicans. Or Yankees fans.

I've been legally female since 2002, although the definition of what makes someone "legally" male or female is part of what makes this issue so unwieldy. How do we define legal gender? By chromosomes? By genitalia? By spirit? By whether one asks directions when lost?

We accept as a basic truth the idea that everyone has the right to marry somebody. Just as fundamental is the belief that no couple should be divorced against their will.

For our part, Deirdre and I remain legally married, even though we're both legally female. If we had divorced last month, before Governor Baldacci's signature, I would have been allowed on the following day to marry a man only. There are states, however, that do not recognize sex

---

The World Economic Forum assessed the gender gap in 128 countries by measuring the extent to which women have achieved equality with men in four areas: economic participation and opportunity, educational attainment, health and survival, and political empowerment (Hausmann et al. 2009). Table 10.1 presents the overall scores and rankings of (1) the 10 countries with the smallest gender gap (i.e., the least gender inequality); (2) the 10 countries with the largest gender gap (i.e., the most gender inequality); and (3) the United States, which did not rank in the top 10 or the bottom 10 but ranked number 27 of the 128 countries studied. Ties were possible. For example, Finland, the Philippines, and Latvia tied for first place on the composite measure of health and survival. Further, note that the overall score approximates the proportion of the gender gap a country has *closed*—in the United States, 0.7179 or 71.79 percent.

Gender inequality varies across cultures, not only in its extent or degree but also in its forms. For example, in the United States, gender inequality in family roles commonly takes the form of an unequal division of household labor and child care, with women bearing the heavier responsibility for these tasks. In other countries, forms of gender inequality in the family include the expectation that wives ask their husbands for permission to use birth control (see Chapter 12), the practice of aborting female fetuses in cultures that value male children over female children, and unequal penalties for spouses who commit adultery, with wives receiving harsher punishment. For example, the 2009 movie, *The*

changes. If I were to attempt to remarry in Ohio, for instance, I would be allowed to wed a woman only.

Gender involves a lot of gray area. And efforts to legislate a binary truth upon the wide spectrum of gender have proven only how elusive sexual identity can be. The case of J'Noel Gardiner, in Kansas, provides a telling example. Ms. Gardiner, a postoperative transsexual woman, married her husband, Marshall Gardiner, in 1998. When he died in 1999, she was denied her half of his $2.5 million estate by the Kansas Supreme Court on the ground that her marriage was invalid. Thus in Kansas, any transgendered person who is anatomically female is now allowed to marry only another woman.

Similar rulings have left couples in similar situations in Florida, Ohio, and Texas. A 1999 ruling in San Antonio, in *Littleton v. Prange*, determined that marriage could be only between people with different chromosomes. The result, of course, was that lesbian couples in that jurisdiction were then allowed to wed as long as one member of the couple had a Y chromosome, which is the case with both transgendered male-to-females and people born with conditions like androgen insensitivity syndrome. This ruling made Texas, paradoxically, one of the first states in which gay marriage was legal.

A lawyer for the transgendered plaintiff in the *Littleton* case noted the absurdity of the country's gender laws as they pertain to marriage: "Taking this situation to its logical conclusion, Mrs. Littleton, while in San Antonio, Texas, is a male and has a void marriage; as she travels to Houston, Texas, and enters federal property, she is female and a widow; upon traveling to Kentucky she is female and a widow; but, upon entering Ohio, she is once again male and prohibited from marriage; entering Connecticut, she is again female and may marry; if her travel takes her north to Vermont, she is male and may marry a female; if instead she travels south to New Jersey, she may marry a male."

Legal scholars can (and have) devoted themselves to the ultimately frustrating task of defining "male" and "female" as entities fixed and unmoving. A better use of their time, however, might be to focus on accepting the elusiveness of gender— and to celebrate it. Whether a marriage like mine is a same-sex marriage or some other kind is hardly the point. What matters is that my spouse and I love each other, and that our legal union has been a good thing—for us, for our children and for our community. . . .

Source: Boylan 2009.

Stoning of Soraya M., tells the true story of an Iranian woman, although falsely accused, stoned to death for committing adultery.

A global perspective on gender inequality must also take into account the different ways in which such inequality is viewed. For example, many non-Muslims view the practice of Muslim women wearing a headscarf in public as a symbol of female subordination and oppression. To Muslims who embrace this practice (and not all Muslims do), wearing a headscarf reflects the high status of women and represents the view that women should be respected and not treated as sexual objects.

Similarly, cultures differ in how they view the practice of female genital mutilation (FGM), also known as female genital cutting or female circumcision. There are several forms of FGM, ranging from a symbolic nicking of the clitoris to removal of the clitoris and labia and partial closure of the vaginal opening by stitching the two sides of the vulva together, leaving only a small opening for the passage of urine and menstrual blood. After marriage, the sealed opening is reopened to permit intercourse and childbearing. After childbirth, the woman's vulva is often stitched back together.

Nonmedical personnel perform most FGM procedures using unsterilized blades or string. Health risks associated with FGM include pain, hemorrhage, infection, shock, scarring, and infertility. Worldwide, between 100 million and 140 million girls and young women are estimated to have experienced FGM, and an additional 3 million are at risk each year (WHO 2008).

COUNTRY	OVERALL SCORE*	RANKINGS				
		OVERALL	ECONOMIC PARTICIPATION AND OPPORTUNITY	EDUCATIONAL ATTAINMENT	HEALTH AND SURVIVAL	POLITICAL EMPOWERMENT
TOP 10 COUNTRIES						
Norway	0.8239	1	6	1	53	2
Finland	0.8195	2	19	1	1	1
Sweden	0.8139	3	5	33	75	4
Iceland	0.7999	4	20	61	96	3
New Zealand	0.7859	5	7	1	69	6
Philippines	0.7568	6	8	1	1	22
Denmark	0.7538	7	28	1	97	10
Ireland	0.7518	8	48	1	81	8
Netherlands	0.7399	9	51	59	72	12
Latvia	0.7397	10	13	1	1	31
United States	**0.7179**	**27**	**12**	**1**	**37**	**56**
BOTTOM 10 COUNTRIES						
Bahrain	0.5927	121	126	66	112	127
Ethiopia	0.5867	122	96	126	101	70
Turkey	0.5853	123	124	108	88	106
Egypt	0.5832	124	120	105	84	124
Morocco	0.5757	125	127	117	85	86
Benin	0.5582	126	105	128	67	75
Pakistan	0.5549	127	128	123	123	50
Saudi Arabia	0.5537	128	129	85	62	130
Chad	0.5290	129	81	130	62	105
Yemen	0.4664	130	130	129	1	129

Source: Hausmann et al. 2009.

*All overall scores are reported on a scale of 0 to 1, with 1 representing maximum gender equality.

People from countries in which FGM is not the norm generally view this practice as a barbaric form of violence against women. For example, an Ethiopian immigrant in the United States was convicted of aggravated battery and cruelty to children for using a pair of scissors to remove his daughter's clitoris. He was sentenced to 10 years in prison (Haines 2006). In countries where it commonly occurs, FGM is viewed as an important and useful practice. In some countries, it is considered a rite of passage that enhances a woman's status. In other countries, it is aesthetically pleasing. For others, FGM is a moral imperative based on religious beliefs (Yoder et al. 2004; WHO 2008).

Inequality in the United States

Although attitudes toward gender equality are becoming increasingly liberal, the United States has a long history of gender inequality. Women have had to fight for equality: the right to vote, equal pay for comparable work, quality education, entrance into male-dominated occupations, and legal equality. As shown in Table 10.1, the World Economic Forum (Hausmann et al. 2009)—based on its assessment of women's economic participation and opportunities, political empowerment, educational attainment, and health and survival—ranks the United States only 27th in the world in terms of gender equality. Most U.S. citizens agree that American society does not treat women and men equally: Women have lower incomes, hold fewer prestigious jobs, earn fewer graduate degrees, and are more likely than men to live in poverty.

Men are also victims of gender inequality. In 1963, sociologist Erving Goffman wrote that in the United States there is only

> one complete unblushing male . . . a young, married, white, urban, northern heterosexual, Protestant father of college education, fully employed, of good complexion, weight and height, and a recent record in sports. . . . Any male who fails to qualify in one of these ways is likely to view himself . . . as unworthy, incomplete, and inferior. (p. 128)

Although standards of masculinity have relaxed, Williams (2000) argues that masculinity is still based on "success"—at work, on the athletic field, on the streets, and at home. Similarly, Vandello et al. (2008) conclude the notion that ". . . manhood is tenuous, and therefore requires public proof, is consistent with research across multiple areas" (p. 1326).

When U.S. college students were asked to list the best and worst things about being the opposite sex, the same qualities, although in opposite categories, emerged (Cohen 2001). For example, what males list as the best thing about being female (e.g., free to be emotional), females list as the worst thing about being male (e.g., not free to be emotional). Similarly, what females list as the best thing about being male (e.g., higher pay), males listed as the worst thing about being female (e.g., lower pay). As Cohen (2001) noted, although "some differences are exaggerated or oversimplified . . . we identif[ied] a host of ways in which we 'win' or 'lose' simply because we are male or female" (p. 3).

In traditional Muslim societies, women are forbidden to show their faces or other parts of their bodies when in public. Muslim women wear a veil to cover their faces and a chador, a floor-length loose-fitting garment, to cover themselves from head to toe. Although some women adhere to this norm out of fear of repercussions, many others believe veiling was first imposed on Muhammad's wives out of respect for women and the desire to protect them from unwanted advances. More than half a million Muslims are living in the United States.

Sociological Theories of Gender Inequality

Both structural functionalism and conflict theory concentrate on how the structure of society and, specifically, its institutions contribute to gender inequality. However, these two theoretical perspectives offer opposing views of the development and maintenance of gender inequality. Symbolic interactionism, on the other hand, focuses on the culture of society and how gender roles are learned through the socialization process.

Structural-Functionalist Perspective

Structural functionalists argue that preindustrial society required a division of labor based on gender. Women, out of biological necessity, remained in the home performing functions such as bearing, nursing, and caring for children. Men, who were physically stronger and could be away from home for long periods of time, were responsible for providing food, clothing, and shelter for their families. This division of labor was functional for society and became defined as both normal and natural over time.

Industrialization rendered the traditional division of labor less functional, although remnants of the supporting belief system still persist. With increased control over reproduction (e.g., contraception), declining birth rates, and fewer jobs dependent upon physical size and strength, women's opportunities for education and workforce participation increased (Wood & Eagly 2002). Thus, modern conceptions of the family have, to some extent, replaced traditional ones—families have evolved from extended to nuclear, authority is more egalitarian, more women work outside the home, and greater role variation exists in the division of labor. Structural functionalists argue, therefore, that, as the needs of society change, the associated institutional arrangements also change.

Conflict Perspective

Many conflict theorists hold that male dominance and female subordination are shaped by the relationships men and women have to the production process. During the hunting and gathering stage of development, males and females were economic equals, both controlling their own labor and producing needed subsistence. As society evolved to agricultural and industrial modes of production, private property developed and men gained control of the modes of production, whereas women remained in the home to bear and care for children. Inheritance laws that ensured that ownership would remain in their hands furthered male domination. Laws that regarded women as property ensured that women would remain confined to the home.

> Thus, unlike structural functionalists, conflict theorists hold that the subordinate position of women in society is a consequence of social inducement rather than biological differences that led to the traditional division of labor.

As industrialization continued and the production of goods and services moved away from the home, the gaps between females and males continued to grow—women had less education, lower incomes, and fewer occupational skills and were rarely owners. World War II necessitated the entry of a large number of women into the labor force, but, in contrast with previous periods, many of them did not return to the home at the end of the war. They had established their own place in the workforce and, facilitated by the changing nature of work and technological advances, now competed directly with men for jobs and wages.

Conflict theorists also argue that continued domination by males requires a belief system that supports gender inequality. Two such beliefs are (1) that women are inferior outside the home (e.g., they are less intelligent, less reliable, and less rational); and (2) that women are more valuable in the home (e.g., they have maternal instincts and are naturally nurturing). Thus, unlike structural functionalists, conflict theorists hold that the subordinate position of women in society is a consequence of social inducement rather than biological differences that led to the traditional division of labor.

Symbolic Interactionist Perspective

Although some scientists argue that gender differences are innate, symbolic interactionists emphasize that, through the socialization process, both females and males are taught the meanings associated with being feminine and masculine. Gender assignment begins at birth as a child is classified as either female or male. However, the learning of gender roles is a lifelong process whereby individuals acquire society's definitions of appropriate and inappropriate gender behavior.

Gender roles are taught by the family, in the school, in peer groups, and by media presentations of girls and boys and women and men (see the discussion on the social construction of gender roles later in this chapter). Most importantly, however, gender roles are learned through symbolic interaction as the messages that others send us reaffirm or challenge our gender performances. Tenenbaum (2009), in an examination of parent-child interactions, found that discussions regarding course selections for high school followed gender-stereotyped patterns. Here, a father talks to his fifth grade daughter (how the conversation was coded by the researchers is in brackets):

> But you know, spelling is English, right? That's what English is. For the most part generally speaking, girls do better with those kind of skills, they have a harder time with math, generally, you know? [Code: Lack of ability.] Now, of course, you know it's nice to do the things you're really good at too, and you like to do. But, sometimes when you're trying to be, when you grow up to be someone in life, you also gotta take classes that are, not really, how do I say? You're not really good at, you have to put more practice in, right? [Code: Lack of ability.] So, I picked, I selected algebra, cause that's kinda, high school, mathematics. (pp. 458–459)

Although the father encouraged his daughter to take mathematics, he twice conveyed to his daughter that she (and girls in general) was not very good in math.

Feminist theory, although also consistent with a conflict perspective, incorporates many aspects of symbolic interactionism. Feminists argue that conceptions of gender are socially constructed as societal expectations dictate what it means to be female or what it means to be male. Thus, women are generally socialized into **expressive roles** (i.e., nurturing and emotionally supportive roles), and males are more often socialized into **instrumental roles** (i.e., task-oriented roles). These roles are then acted out in countless daily interactions as boss and secretary, doctor and nurse, football player and cheerleader "do gender."

JIMMY CHOO

Image courtesy of The Advertising Archives

As conflict theorists are quick to note, women are often portrayed provocatively as a means of selling a product or service. However, one of the most disturbing uses of women in advertising is their portrayal in violent scenarios. To a symbolic interactionist, what kind of meaning about the nature of female-male relationships does this advertisement convey?

expressive roles Roles into which women are traditionally socialized (i.e., nurturing and emotionally supportive roles).

instrumental roles Roles into which men are traditionally socialized (i.e., task-oriented roles).

Feminists also hold that gender "is a central organizing factor in the social world and so must be included as a fundamental category of analysis in sociological research" (Renzetti & Curran 2003, p. 8). Noting that the impact of the structure and culture of society is not the same for different groups of women and men, feminists encourage research on gender that takes into consideration the differential effects of age, race and ethnicity, and sexual orientation.

Gender Stratification: Structural Sexism

As structural functionalists and conflict theorists agree, the social structure underlies and perpetuates much of the sexism in society. **Structural sexism,** also known as institutional sexism, refers to the ways the organization of society and specifically its institutions subordinate individuals and groups based on their sex classification. Structural sexism has resulted in significant differences in the education and income levels, occupational and political involvement, and civil rights of women and men.

Education and Structural Sexism

Literacy rates worldwide indicate that women are less likely than men to be able to read and write, with millions of women being denied access to even the most basic education (UNESCO 2009). For example, on average, women in South Asia have only half as many years of education as their male counterparts. In addition, worldwide, for every 48 boys who do not attend elementary school, there are 53 girls who do not attend elementary school (UNICEF 2009).

In 2007, few differences existed between men and women in their completion rates of high school and college degrees (NCES 2009a). In fact, in recent years, most U.S. colleges and universities have had a higher percentage of women than men enrolling directly from high school (see Chapter 8). This trend is causing some concern that many young American men may not have the education they need to compete in today's global economy. Although differences are narrowing, men are still more likely to go on to complete a graduate or professional degree than are women. As Figure 10.1 indicates, dramatic differences also appear in the types of advanced degrees that women and men earn. For example, women earn 81 percent of master's degrees in library science but only 23 percent of master's degrees in engineering and engineering technologies. Concern over the lack of women in STEM (science, technology, engineering, and mathematics) degree programs is fueled by the relationship between education and income. As York (2008) documents in a sample of high school valedictorians, the median annual income associated with careers female graduates selected ($74,608) is significantly lower than the median annual income associated with careers male graduates selected ($97,734).

One explanation for why women earn fewer advanced degrees than men is that women are socialized to choose marriage and motherhood over long-term career preparation. From an early age, women are exposed to images and models of femininity that stress the importance of domestic family life. Ceci et al. (2009) report that math-proficient women compared to math-proficient men are less likely to enter math-intensive fields such as computer science,

structural sexism The ways in which the organization of society, and specifically its institutions, subordinate individuals and groups based on their sex classification.

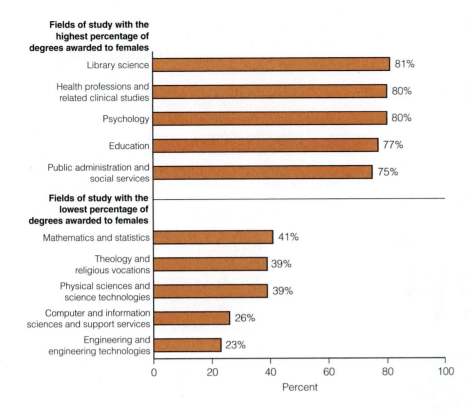

Figure 10.1 **Percentage of Master's Degrees Awarded to Females by Degree-Granting Institutions in Selected Fields of Study: Academic Year 2006–07**
Source: NCES 2009a.

engineering, and physics and, if they do, are more likely to leave such careers than their male counterparts. Although the reasons are complex, the greater priority women place on home over career explains much of the gender variation.

There are also structural limitations that discourage women from advancing in higher education. For example, women seeking academic careers may find that securing a tenure-track position is more difficult for them than it is for men, and that having children pre-tenure negatively impacts the likelihood of getting tenure, that is, there is a "pregnancy penalty" (Ceci et al. 2009). Similarly, in an analysis of data from the *National Study of Postsecondary Faculty*, Leslie (2007) reports that, as the number of children increases, the number of hours a female faculty member works decreases, and the number of hours a male faculty member works increases (see the discussion on household division of labor later in this chapter). Finally, McBrier (2003) found that women faculty in law schools progress through the professorial ranks at a slower pace than men. She attributes the differences to a variety of variables, including family and geographic constraints, social capital differences, degree prestige, and prior work history.

Work and Structural Sexism

According to the International Labour Organization (ILO), in 2008, women made up 40.4 percent of the world's *total* labor force (ILO 2009). Globally, women are disproportionately employed in the agricultural and service sectors, and in vulnerable employment (e.g., unpaid family workers). Women are also more likely

> Women work 66 percent of the total number of labor hours worldwide, yet receive just 10 percent of the world's income.

Societal definitions of the appropriateness of gender roles have traditionally restricted women and men in terms of educational, occupational, and leisure-time pursuits. Fifty years ago, little boys playing with dolls and displaying nurturing behaviors would have been unheard of.

occupational sex segregation The concentration of women in certain occupations and men in other occupations.

glass escalator effect The tendency for men seeking or working in traditionally female occupations to benefit from their minority status.

pink-collar jobs Jobs that offer few benefits, often have low prestige, and are disproportionately held by women.

to be unemployed when compared to men. Increasing rates of female unemployment are, in part, a consequence of the global economic crisis, particularly in developing countries. For example, because women are disproportionately employed in the agricultural sector, as food prices increase as a result of the recession, food exports drop, leading to higher rates of female unemployment (ITUC 2009).

Worldwide, women tend to work in jobs that have little prestige and low or no pay, where no product is produced, and where they are the facilitators for others. Women are also more likely to hold positions of little or no authority within the work environment and to receive lower wages than men (WHO 2006; ILO 2009; IPC 2008). Women work 66 percent of the total number of labor hours worldwide, yet receive just 10 percent of the world's income (Foerstel 2008).

No matter what the job, if a woman does it, it is likely to be valued less than if a man does it. For example, in the early 1800s, 90 percent of all clerks were men and being a clerk was a prestigious profession. As the job became more routine, in part because of the advent of the typewriter, the pay and prestige of the job declined and the number of female clerks increased. Today, 91 percent of clerks are female (U.S. Census Bureau 2009), and the position is one of relatively low pay and prestige.

The concentration of women in certain occupations and men in other occupations is referred to as **occupational sex segregation**. Although occupational sex segregation remains high, as indicated by the third column in Table 10.2, it has decreased in recent years for some occupations. For example, between 1983 and 2007, the percentage of female physicians nearly doubled from 16 percent to 30 percent, female dentists increased from 7 percent to 28 percent, and female clergy increased from 6 percent to 15 percent (U.S. Census Bureau 2009).

Although the pace is slower, increasingly men are applying for jobs that women traditionally held. Spurred by the loss of jobs in the manufacturing sector and the present economic crisis, many jobs traditionally defined as male (e.g., auto worker, construction worker) have been lost. Thus, over the last 20 years, there has been a significant increase in the number of men in traditionally held female jobs: for example, a 50 percent increase in male telephone operators, 45 percent increase in male tellers, and 40 percent increase in male preschool and kindergarten teachers (Bourin & Blakemore 2008). Some evidence suggests that men in traditionally held female jobs have an advantage in hiring, promotion, and salaries called the **glass escalator effect** (Williams 2007). Table 10.2, for example, indicates that elementary and middle school teachers are overwhelmingly female—81 percent. Yet the average median weekly salary for male teachers is $994 compared to $871 for female teachers. Despite the increase of men into traditionally-held female occupations, women are still heavily represented in low-prestige, low-wage, **pink-collar jobs** that offer few benefits.

TABLE 10.2 The Wage Gap in the Ten Most Common Occupations* for Women and Men (Full-Time Workers Only), 2008

	PERCENT OF WORKERS IN OCCUPATION THAT ARE FEMALE	MEDIAN WEEKLY EARNINGS FOR WOMEN	MEDIAN WEEKLY EARNINGS FOR MEN	WOMEN'S EARNINGS AS % OF MEN'S
All female full-time workers (47,209,000)	44.3%	$638	$798	79.94%
10 most common occupations for women				
Secretaries and administrative assistants	96.5%	$614	$736	83.4%
Elementary and middle school teachers	81.0%	$871	$994	87.6%
Registered nurses	90.1%	$1,011	$1,168	86.6%
Nursing, psychiatric, and home health aides	87.7%	$424	$485	87.4%
First-line supervisors/managers of retail sales workers	43.2%	$556	$781	71.2%
First-line supervisors/managers of office and administrative support workers	68.6%	$688	$848	81.1%
Cashiers	74.0%	$349	$399	87.5%
Customer service representatives	67.2%	$568	$607	93.6%
Accountants and auditors	60.5%	$908	$1,178	77.1%
Receptionists and information clerks	93.3%	$502	$537	93.5%
10 most common occupations for men				
Driver/sales workers and truck drivers	4.3%	$709	$542	76.5%
Managers, all other	37.5%	$1,359	$1,010	74.3%
First-line supervisors/managers of retail sales workers	43.2%	$781	$556	71.2%
Laborers and freight, stock, and material movers, hand	14.2%	$508	$417	82.1%
Construction laborers	14.2%	$558	*	*
Retail salespersons	42.5%	$623	$440	70.6%
Janitors and building cleaners	27.4%	$493	$397	80.5%
Carpenters	1.1%	$655	*	*
Sales representatives, wholesale and manufacturing	26.2%	$1,064	$846	79.5%
Cooks	36.6%	$404	$363	89.9%

Source: IWPR 2009b.

*Data are only made available where there is an estimated minimum of 50,000 workers in an occupation.

Sex segregation in occupations continues for several reasons. First, cultural beliefs about what is an "appropriate" job for a man or a woman still exist. Snyder and Green's (2008) analysis of nurses in the United States is a case in point. Using survey data and in-depth interviews, the researchers identified patterns of sex segregation. Over 88 percent of all patient-care nurses were in sex-specific specialties (e.g., intensive care and psychiatry for male nurses, and labor or delivery and outpatient services for female nurses). Interestingly, although women rarely mentioned gender as a reason for their choice of specialty, male nurses frequently did so, acknowledging the "process of gender affirmation that led them to seek out 'masculine' positions within what was otherwise construed to be a women's profession" (p. 291).

Second, opportunity structures for men and women differ. For example, women and men, upon career entry, are often channeled by employers into gender-specific jobs that carry different wages and promotion opportunities. However, even women in higher-paying jobs may be victimized by a **glass ceiling**—an often invisible barrier that prevents women and other minorities from moving into top corporate positions. For example, women and minorities have different social networks than do white men, which contribute to this barrier. White men in high-paying jobs are more likely to have interpersonal connections with individuals in positions of authority (Padavic & Reskin 2002). In addition, women often find that their opportunities for career advancement are adversely affected after returning from family leave. Female lawyers returning from maternity leave found their career mobility stalled after being reassigned to less prestigious cases (Williams 2000).

Finally, there is evidence that working mothers pay a price for motherhood. Using an experimental design, Correll et al. (2007) report that, even when qualifications, background, and work experience were held constant, "evaluators rated mothers as less competent and committed to paid work than non-mothers" (p. 1332). Other examples of the "**motherhood penalty**" include women who feel pressured to choose professions that permit flexible hours and career paths, sometimes known as mommy tracks (Moen & Yu 2000). Thus, women dominate the field of elementary education, which permits them to be home when their children are not in school. Nursing, also dominated by women, often offers flexible hours. Although the type of career pursued may be the woman's choice, it is a **structured choice**—a choice among limited options as a result of the structure of society.

Income and Structural Sexism

Worldwide, on the average, women earn just half of what men earn (Social Watch 2009). In 2008, full-time working women in the United States earned, on

glass ceiling An invisible barrier that prevents women and other minorities from moving into top corporate positions.

motherhood penalty The tendency for women with children, particularly young children, to be disadvantaged in hiring, wages, and the like compared to women without children.

structured choice Choices that are limited by the structure of society.

the average, 79.9 percent of the weekly median earnings of full-time working men, a decrease of 0.3 percent from the previous year (IWPR 2009a). The gender pay gap not only varies over time, it also varies by state. In Vermont, a female college graduate over the age of 25 working full-time earns, on the average, 87 percent of what her male counterpart earns. In Wyoming, she earns just 62 percent of what a full-time working man earns (AAUW 2007a).

Racial differences also exist. Although women in general earn 79.9 percent as much as men, Black American and Hispanic American women earn just 71 percent and 58 percent, respectively, of men's salaries (NOW 2007). Even among celebrities, a significant income gap exists. According to *Forbes* magazine's Celebrity 100 issue, the top female athlete (Maria Sharapova) earned 20 percent of what the top male athlete (Tiger Woods) earned in 2008 (*Forbes* 2009).

There are several arguments as to why the gender pay gap exists. One, the **human capital hypothesis,** holds that pay differences between females and males are a function of differences in women's and men's levels of education, skills, training, work experience, and the like. For example, Rose and Hartman (2008), using a longitudinal data set, found that over a 15-year period, women worked fewer years than men and, when they worked, they worked fewer hours per year. Rose and Hartman conclude that over ". . . the 15 years, the more likely a woman is to have dependent children and be married, the more likely she is to be a low earner and have fewer hours in the labor market" (Rose & Hartman 2008, p. 1). Bertrand et al. (2009) reported similar findings, concluding that the "presence of children is associated with less accumulation of job experience, more career interruptions, and shorter work hours for female MBAs but not for male MBAs" (p. 24). Based on their analysis, the two years following the birth of her first child, a woman can expect her annual salary to decrease by $45,000, and to decrease by $80,000 a year in subsequent years.

One variation of the human capital hypothesis is called the *life-cycle human capital hypothesis.* Here it is argued that women have less incentive to invest in education and marketable skills because they know that they will be working less than their male counterparts as wives and mothers, and that their careers will be interrupted by family responsibilities. Alternatively, men's incentives to acquire marketable skills increase with greater family responsibilities and it is, or so it is argued, this human capital difference that is responsible for the female-male pay gap (Polachek 2006).

Human capital theorists also argue that women make educational choices (e.g., school attended, major, etc.) that limit their occupational opportunities and future earnings. Women, for example, are more likely to major in the humanities, education, or the social sciences rather than science and engineering, which results in reduced incomes (York 2008). Ironically, the relatively low rate of women majoring in science and engineering is a function of the lack of female faculty in science and engineering departments (Sonnert et al. 2007).

Female-male human capital differences are, however, a result of structural constraints (e.g., no national system of child care) as well as expectations that women should remain in the home. The results of a survey indicate that few working mothers (11 percent) or stay-at-home mothers (10 percent) believe that a full-time working mother is the "ideal situation for a child" (Pew Research Center 2007). Further, as Table 10.3 indicates, 72 percent of men with children under 18 years of age responded that working full-time was the ideal situation for them compared to only 20 percent of women with children under the age of 18.

human capital hypothesis
The hypothesis that pay differences between females and males are a function of differences in women's and men's levels of education, skills, training, and work experience.

TABLE 10.3 What Working Situation Would Be Ideal for You?

Considering everything, what would be the ideal situation for you—working full-time, working part-time, or not working at all outside the home?

	IDEAL SITUATION WOULD BE				N
	NOT WORKING	PART-TIME WORK	FULL-TIME WORK	DON'T KNOW	
	%	%	%	%	
Have children under 18					
Fathers	16	12	72	0	343
Mothers	29	50	20	1	414
Mothers with children under 18					
Employed full-time	21	49	29	1	184
Employed part-time	15	80	5	0	75
Not employed	48	33	16	3	153

Source: Pew Research Center 2007.

The second explanation for the gender gap is called the **devaluation hypothesis.** It argues that women are paid less because the work they perform is socially defined as less valuable than the work men perform. Guy and Newman (2004) argued that these jobs are undervalued in part because they include a significant amount of **emotion work**—that is, work that involves caring, negotiating, and empathizing with people, which is rarely specified in job descriptions or performance evaluations.

Finally, there is evidence that, even when women and men have equal education and experience (and, therefore, not a matter of human capital differences) and are in the same occupations (and, therefore, not a matter of women's work being devalued), pay differentials remain. Table 10.2 indicates that men make more than women in each of the 10 most sex-segregated occupations for females and males. Even among secretaries and administrative assistants, a profession that is 96.5 percent female, women earn a weekly median salary of $614, and men, on the average, $736—a difference of $6,344 annually. An examination of high- and low-wage professions is also telling. For example, the profession with the highest median weekly earning for women is pharmacist, and about half of all pharmacists are women. Yet female pharmacists earn 86.1 percent of what male pharmacists earn (IWPR 2009b).

Clearly the human capital differences and the devaluation hypotheses may explain some of the variation in women's and men's earnings. But women's and men's earnings vary for many different reasons including discrimination in education, and in hiring, promotions, and salaries in the workplace. For example, the American Association of University Women (2007b) report that:

devaluation hypothesis The hypothesis that women are paid less because the work they perform is socially defined as less valuable than the work men perform.

emotion work Work that involves caring for, negotiating, and empathizing with people.

. . . just one year after college graduation, women earn only 80 percent of what their male counterparts earn. Ten years after graduation, women fall further behind, earning only 69 percent of what men earn. Even when controlling for hours, occupation, parenthood, and other factors known to affect earnings, the research indicates that one-quarter of the pay gap remains unexplained and is likely due to sex discrimination. (p. 1)

Similarly, when the Government Accountability Office examined the gender pay gap between female and male federal employees, some of it was explained by female-male differences in occupation, education, and work experience. However, that portion of the pay gap that could not be explained by such variables grew over the time period studied—1988 to 2007. The report concludes that the portion of the pay gap that cannot be explained "neither confirms nor refutes the presence of discriminatory practices" in government wages (Sherrill 2009, p. 2).

Comparable worth refers to the belief that individuals in occupations, even in different occupations, should be paid equally if the job requires "comparable" levels of education, training, and responsibility. Is it fair that "maids and housecleaners," who are 87 percent women, earn $3,000 a year less than "janitors and building cleaners" (Billitteri 2008)? Does the pay differential reflect differences in the requirements of the job or the traditionally low esteem women's work has held? Advocates of comparable worth, consistent with the devaluation hypothesis, argue that comparable worth policies are the only way to ensure pay equity. Critics argue that comparable worth policies undermine our system of market-based wages and are demeaning to women (Billitterri 2008).

Bernat Armangue/AP Photo

Shortly after Prime Minister José Luis Rodríguez Zapatero appointed Spain's first female Minister of Defense, Carmé Chacón, she gave birth to a son. Chacón's appointment is part of Spain's "commitment to gender parity" in a "traditionally macho society whose new equality laws are among the most progressive in Europe" (Burnett 2008, p. 1).

Politics and Structural Sexism

Women received the right to vote in the United States in 1920, with the passage of the Nineteenth Amendment. Even though this amendment went into effect more than 85 years ago, women still play a rather minor role in the political arena. In general, the more important the political office is, the lower the probability that a woman will hold it. Although women constitute 52 percent of the population, the United States has never had a woman president or vice president and, until 2009, when Justice Sonia Sotomayor was appointed, the United States had only two female Supreme Court Justices. The highest-ranking women ever to serve in the U.S. government have been former Secretaries of State Madeleine Albright and Condoleezza Rice, current Speaker of the House Nancy Pelosi, and current Secretary of State Hillary Clinton. In 2009, women represented only 14 percent of all governors and held only 16.8 percent of all U.S. Congressional seats. Worldwide, the percentage of women in legislative posts varies by region of the world (see Figure 10.2).

comparable worth The belief that individuals in occupations, even in different occupations, should be paid equally if the job requires "comparable" levels of education, training, and responsibility.

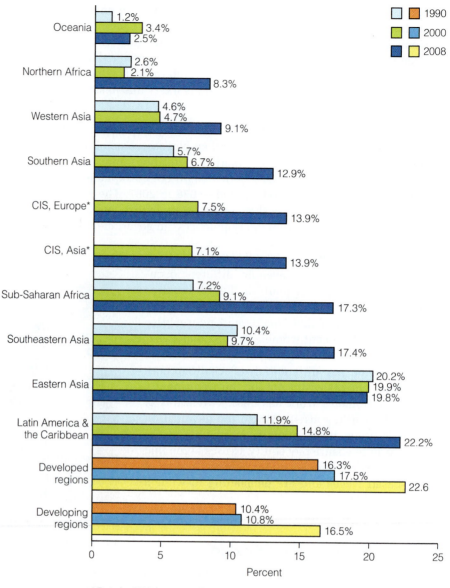

Figure 10.2 Proportion of Seats Held by Women in National Legislatures: 1990, 2000, and 2008
Source: U.N. 2008.

Legend: 1990, 2000, 2008

Region	1990	2000	2008
Oceania	1.2%	3.4%	2.5%
Northern Africa	2.6%	2.1%	8.3%
Western Asia	4.6%	4.7%	9.1%
Southern Asia	5.7%	6.7%	12.9%
CIS, Europe*		7.5%	13.9%
CIS, Asia*		7.1%	13.9%
Sub-Saharan Africa	7.2%	9.1%	17.3%
Southeastern Asia	10.4%	9.7%	17.4%
Eastern Asia	20.2%	19.9%	19.8%
Latin America & the Caribbean	11.9%	14.8%	22.2%
Developed regions	16.3%	17.5%	22.6
Developing regions	10.4%	10.8%	16.5%

Percent

* Data for 1990 are not available.

In response to the underrepresentation of women in the political arena, some countries have instituted electoral quotas. Today, there are 74 countries "where quotas have been implemented in the constitution, regulations, and laws, or where political parties have implemented their own internal quotas" (IDEA 2009, p. 1). Quotas are particularly useful in countries where women are underrepresented and their candidacy is threatened by longstanding patriarchal traditions. In Iraq, for example, women running for office are often faced with death threats. Although a proposed law required that 25 percent of the seats in the 18 provincial councils in Iraq be allocated to women, the final version included only a vague reference to quotas, stating that there had to be "a woman at the end of every three winners" (Rubin & Dagher 2009, p. 1).

The relative absence of women in politics, as in higher education and in high-paying, high-prestige jobs in general, is a consequence of structural limitations. Running for office requires large sums of money, the political backing of powerful individuals and interest groups, and a willingness of the voting public to elect women. Disproportionately lacking these resources, minority women have even greater structural barriers to election and, not surprisingly, represent an even smaller percentage of elected officials. Of the 535 U.S. congressional representatives, women of color represent 3.7 percent, and no minority woman is presently a member of the U.S. Senate (CAWP 2009).

Running for office requires large sums of money, the political backing of powerful individuals and interest groups, and a willingness of the voting public to elect women. Disproportionately lacking these resources, minority women have even greater structural barriers to election and, not surprisingly, represent an even smaller percentage of elected officials.

There is also evidence of gender discrimination against female candidates. In an experiment where two congressional candidates' credentials were presented to a sample of respondents, in one case as Ann Clark and in the other as Andrew Clark, Republican respondents were significantly more likely to say they would vote for a father with young children rather than a mother with young children. They were also more likely to vote for women without small children than women with small children—additional evidence of the "motherhood penalty." The opposite pattern was detected for Democrat respondents (Morin & Taylor 2008).

Finally, gender bias and sexist commentaries dominated the most recent presidential election. In a gender analysis of the 2008 election, Pozner (2009) notes that Sarah Palin's vice presidential candidacy was characterized by sexual innuendos and trivializing taunts. Palin was described as "by the far the best-looking woman ever to rise to such heights" and "the first indisputably fertile female to dare dance with the big dogs." Donny Deutsch of MSNBC stated that Palin was divisive because "this is the first woman in power with sexual appeal. . . . We're used to seeing a woman in power as nonthreatening." The logical conclusion of such a statement is not lost on Pozner, who asks, "So, Palin's polarizing because she's . . . hot?" (p. 1).

Finally, arguing that part of the American ethic of politics is whether or not you are "man enough to serve," Faludi (2008) contends that John McCain, who survived years in a prisoner of war camp, typifies the American ideal of what it means to be a man. On the other hand, Tucker Carlson of MSNBC described Obama as "kind of a wuss." A TV talk show host described President Obama's bowling style as "prissy" and continued that "Americans want their president, if it's a man, to be a real man." Finally, Don Imus, as Faludi notes, "never to be outdone in the sexual slur department, dubbed Mr. Obama as a 'sissy boy'" (pp. 1–2). Thus, to some, a candidate's gender performance, or at least the perception of it, should play a role in an individual's voting behavior.

Civil Rights, the Law, and Structural Sexism

In many countries, victims of gender discrimination cannot bring their cases to court. This is not true in the United States. The 1963 Equal Pay Act and Title VII of the 1964 Civil Rights Act make it illegal for employers to discriminate in wages or employment on the basis of sex. Nevertheless, such discrimination still occurs as evidenced by the thousands of grievances filed each year with the Equal Employment Opportunity Commission (EEOC)—28,372 grievances in 2008 (EEOC 2009a).

In one of the largest employment discrimination suits ever filed, Wal-Mart Stores, Inc. has been named in a class-action suit in which 2 million former and current female employees allege that management discriminated against them by (1) paying male employees more, and (2) denying promotions to women. When faced with lawsuits of this nature, employers use various techniques to justify their employment practices. Wal-Mart, for example, stated that the stores were independently owned, thus supporting their contention that there was no company-wide policy or practice of discrimination in place (Kravets 2007). Another technique employers use to justify differences in pay is the use of different job titles for the same type of work. Repeatedly the courts have ruled, however, that jobs that are "substantially equal," regardless of title, must result in equal pay. In the Wal-Mart case, if the courts ultimately rule in favor of the plaintiffs, Wal-Mart Stores, Inc. faces the possibility of having to pay billions of dollars in damages. Recently, Wal-Mart requested a review of a lower court's decision that all female employees who worked at Wal-Mart since December 26, 1998, constitute a "class" and can, therefore, continue in their class-action suit (Egelko 2009). The decision is forthcoming.

Discrimination, although illegal, takes place at both the institutional and the individual levels (see Chapter 9). Institutional discrimination includes screening devices designed for men, hiring preferences for veterans, the practice of promoting from within an organization based on seniority, and male-dominated recruiting networks (Reskin & McBrier 2000). For example, the Augusta National Golf Club, home of the Masters Golf Tournament and a virtual "who's who of the corporate world," has refused to change its policy, forbidding women from joining, despite years of political pressure from women's groups and negative publicity (Goldman 2009). One of the most blatant forms of individual discrimination is sexual harassment, discussed later in this chapter.

In the United States, women often have difficulty obtaining home mortgages or rental property because they have lower incomes, shorter work histories, and less collateral. Until fairly recently, husbands who raped their wives were exempt from prosecution. Even today, some states require a legal separation agreement and/or separate residences for a wife who has been raped to receive full protection under the law. Women in the military have traditionally been restricted in the duties they can perform, and, finally, since the U.S. Supreme Court's 1973 *Roe v. Wade* decision, which made abortion legal, the right of a woman to obtain an abortion has steadily been limited and narrowed by subsequent legislative acts and judicial decisions (Connolly 2005). The debate has continued with several recent court decisions (see Chapter 14).

What Do You Think? The arguments are as old as the "battle of the sexes": (1) women who are working, particularly married women, are simply making money for the "extras," whereas men who are working are working to support their families; (2) women who are working will get pregnant and quit so there is no need to invest in their professional development; and (3) a woman's primary concern is her family, not her career—hiring or promoting women is unfair to her career-minded male counterpart. Each of these arguments justifies treating men and women differently. Assuming equal qualifications, is preferential treatment in the workplace ever justified? If so, under what conditions?

The Social Construction of Gender Roles: Cultural Sexism

As social constructionists note, structural sexism is supported by a system of cultural sexism that perpetuates beliefs about the differences between women and men. **Cultural sexism** refers to the ways the culture of society—its norms, values, beliefs, and symbols—perpetuate the subordination of an individual or group because of the sex classification of that individual or group.

For example, the *belief* that females are less valuable than males has serious consequences. In India, technology and a preference for male babies has led to the abortion of over 10 million girls in the past 20 years (Ramesh 2007). Contrary to what is often assumed, educated Indian women are more likely to abort unwanted female fetuses than uneducated, poor Indian women. Educated, wealthier women have access to the ultrasound technology necessary to determine the sex of their baby (Rao 2006). In 2009, Swedish health authorities ruled that gender-based abortion is not against the law.

Cultural sexism takes place in a variety of settings, including the family, the school, and the media, as well as in everyday interactions.

Family Relations and Cultural Sexism

From birth, males and females are treated differently. The toys that male and female children receive convey different messages about appropriate gender behavior. For example, a study by Professor Becky Francis of Roehampton University in England, examined the impact of educational toys on 3- to 5-year-olds' learning. When parents were asked their child's favorite toys, boys' toys "involved action, construction, and machinery," whereas girls' toys were more often "dolls and perceived 'feminine' interests, such as hairdressing" (Lepkowska 2008, p. 1) After purchasing and analyzing the toys parents selected, Francis concluded that girls' toys had limited "learning potential" whereas boys' toys "were far more diverse" and propelled boys "into a world of action as well as technology" designed to "be exciting and stimulating" (quoted in Lepkowska 2008, p. 1).

Household Division of Labor. Globally, women and girls continue to be responsible for household maintenance including cooking, gathering firewood and fetching water, and taking care of younger siblings (IPC 2008). In a study of household labor in 10 Western countries, Bittman and Wajcman (2000)

Claire Deprez/Reporters/Redux Pictures

According to Hochschild (1989), women are expected to work "second shifts" by having gainful outside employment as well as performing household chores and child care once they arrive home after a day's work. Research indicates that women contribute significantly more hours to home and child care and have less access to leisure time than their male counterparts.

report that "women continue to be responsible for the majority of hours of unpaid labor," ranging from a low of 70 percent in Sweden to a high of 88 percent in Italy (p. 173). A study in Mexico found that women in the paid labor force performed household chores for an additional 33 hours of work each week; men contributed 6 hours a week to domestic chores (UNICEF 2007).

In the United States, girls and boys work within the home in approximately equal amounts until the age of 18 when girls' household labor begins to increase.

cultural sexism The ways in which the culture of society perpetuates the subordination of an individual or group based on the sex classification of that individual or group.

Although men's share of the household labor has more than doubled in the last 25 years, men's share of inside chores is still only between 25 percent and 33 percent of the total (Coltrane & Adams 2008). The fact that women, even when working full-time, contribute significantly more hours to home care than men is known as the "second shift" (Hochschild 1989). According to Coltrane and Adams (2008), husbands of working wives are most likely to do child care, meal preparation, cleanup after meals (e.g., washing dishes), and house cleaning, and least likely to do laundry. The two researchers predict that, as the hours worked outside the home by employed husbands and wives converge, there will be a continued trend toward egalitarian households.

Although some changes have occurred, the traditional division of labor remains to a large extent. Three explanations emerge from the literature. The first explanation is the *time-availability approach.* Consistent with the structural-functionalist perspective, this position claims that role performance is a function of who has the time to accomplish certain tasks. Because women are more likely to be at home, they are more likely to perform domestic chores.

A second explanation is the *relative resources approach.* This explanation, consistent with a conflict perspective, suggests that the spouse with the least power is relegated the most unrewarding tasks. Because men, on average, have more education, higher incomes, and more prestigious occupations, they are less responsible for domestic labor. Thus, for example, because women earn less money than men do, on average, they turn down overtime and other work opportunities to take care of children and household responsibilities, which subsequently reduces their earnings potential even further (Williams 2000).

Gender role ideology, the final explanation, is consistent with a symbolic interactionist perspective. It argues that the division of labor is a consequence of traditional socialization and the accompanying attitudes and beliefs. Women and men have been socialized to perform various roles and to expect their partners to perform other complementary roles. Women typically take care of the house, and men take care of the yard. These patterns begin with household chores assigned to girls and boys and are learned through the media, schools, books, and toys. A test of the three positions found that, although all three had some support, gender role ideology was the weakest of the three in predicting work allocation (Bianchi et al. 2000). It should be noted that the three positions are not mutually exclusive. For example, Mannino and Deutsch (2007), in their longitudinal study of 81 married women, found support for the relative resources and gender role ideology perspectives.

The School Experience and Cultural Sexism

Sexism is also evident in schools. The bulk of research on gender images in books and other instructional materials document the way males and females are portrayed stereotypically. For example, in a study of 200 "top-selling" children's picture books, women and girls were significantly underrepresented, with twice as many male title and main characters. Males were also more likely to be in the illustrations, to be pictured in the outdoors and, if an adult, to be visibly portrayed as employed outside of the home. Both men and women were over nine times more likely to be pictured in traditional rather than in nontraditional occupations, and "female main characters . . . were more than three times more likely than were male main characters . . . to perform nurturing or caring behaviors" (Hamilton et al. 2006, p. 761). Comparisons to award-winning children's books of the 1980s and 1990s revealed no significant reduction in gender stereotypes.

The differing expectations and/or encouragement that females and males receive contribute to their varying abilities, as measured by standardized tests, in disciplines such as reading, math, and science. Are such differences a matter of aptitude? Social science research would indicate otherwise. For example, in an experiment at the University of Waterloo, male and female college students, all of whom were good in math, were shown either gender-stereotyped or gender-neutral advertisements. When, subsequently, female students who had seen the female-stereotyped advertisements took a math test, they performed significantly lower than women who had seen the gender-neutral advertisements (Begley 2000).

Further, the results of a recent study at the University of Virginia document the pervasiveness of gender stereotypes and their impact on science achievement (Nosek et al. 2009). Over 500,000 respondents from 34 countries participated in the research designed to measure *implicit bias*—bias that we are unaware of. About 70 percent of the respondents were found to have stereotypical views of the relationship between gender and science in the predictable direction. In countries where gender stereotypes were the strongest, that is, being male was associated with science but being female was not, males performed better and females performed worse on standardized math and science tests. The authors suggest that "implicit stereotypes and sex differences in science participation and performance are mutually reinforcing, contributing to the persistent gender gap in science engagement" (Nosek 2009, p. 10593).

As discussed in Chapter 8, Title IX of the 1972 Educational Amendments Act prohibits sex discrimination in educational programs and activities that receive federal financial assistance (NCES 2009b). An evaluation of Title IX, 35 years after its adoption, suggests that sex discrimination continues at many levels. Women are underrepresented in administrative positions at all levels of education; women's and girls' athletic programs lag behind boys' and men's programs in terms of participation, resources, and coaching; despite significant gains, women and girls continue to perform behind boys and men in science and math; and career and technical education remain sex-segregated, in part, because of stereotyped career counseling in high school (NCWGE 2008).

Additionally, in 2006, the U.S. Department of Education granted school districts permission to expand single-sex education. This change does *not* "require that sex segregation be used only if there is adequate justification to show that it will be better than co-education in accomplishing the desired objectives, such as increasing gender equity in education" (NCWGE 2008, p. 2).

Sexism is also reflected in the way that teachers treat their students. Millions of young girls are subjected to sexual harassment by male teachers, who then fail them when they refuse the teachers' sexual advances (Quist-Areton 2003). There is also convincing evidence that elementary and secondary school teachers pay more attention to boys than to girls—talking to them more, asking them more questions, listening to them more, counseling them more, giving them more extended directions, and criticizing and rewarding them more frequently. In *Still Failing at Fairness,* Sadker and Zittleman (2009) recount a fifth grade teacher's instructions to her students. "There are too many of us here to all shout out at once. I want you to raise your hands, and then I'll call on you. If you shout out,

I'll pick somebody else." The discussion on presidents continues, with Stephen calling out (pp. 65–66):

> *Stephen:* I think Lincoln was the best president. He held the country together during a war.
> *Teacher:* A lot of historians would agree with you.
> *Kelvin* (seeing that nothing happened to Stephen, calls out): I don't. Lincoln was OK but my Dad liked Reagan. He always said Reagan was a great president.
> *David* (calling out): Reagan? Are you kidding?
> *Teacher:* Who do you think our best president was, David?
> *David:* FDR. He saved us from the Depression.
> *Max* (calling out): I don't think it's right to pick one best president. There were a lot of good ones.
> *Teacher:* That's interesting.
> *Rebecca* (calling out): I don't think the presidents today are as good as the ones we used to have.
> *Teacher:* Ok, Rebecca. But you forgot the rule. You're supposed to raise your hand.

Gender-based interactions are not the only things that affect student outcomes. Dee (2006) tested the effect of teachers' gender on school performance by using the National Education Longitudinal Survey. Analyzing data from a nationally representative sample of eighth grade boys and girls and their teachers, he found that (1) when the teacher is a female, boys are more likely to be seen as disruptive; (2) when the teacher is a female, girls are less likely to be seen as either disruptive or inattentive; (3) when the teacher is male, female students are less likely to perceive a class as useful, to ask questions in class, or to look forward to the class; and (4) when the teacher is female, regardless of the subject, boys are less likely to report looking forward to the class. In conclusion, the author noted that, "simply put, girls have better educational outcomes when taught by women, and boys are better off when taught by men." (p. 71) (see also Chapter 8)

What Do You Think? In 1970, Whitney Darrow published a book entitled *I'm Glad I'm a Boy! I'm Glad I'm a Girl!* Although some argue that it was intended to be a satire of gender roles, it was a popular adoption in school libraries, regardless of intent. The book distinguishes between boys and girls by what they like to do and who they are: Boys are doctors, girls are nurses; boys have trucks, girls have dolls; boys can eat, girls can cook; boys are policemen, girls are meter maids; boys are football players, girls are cheerleaders; boys are pilots, girls are stewardesses; boys build houses, girls keep houses. Given the information in this chapter thus far, have things really changed over the last 40 years?

Media, Language, and Cultural Sexism

Another concern social scientists voice is the extent to which the media portray females and males in a limited and stereotypical fashion and the impact of such portrayals. For example, Levin and Kilbourne (2009), in *So Sexy So Soon,* document the sexualizing of young girls and boys. Advertising, books, cartoons, songs, toys, and television shows create:

[a] narrow definition of femininity and sexuality [that] encourages girls to focus heavily on appearance and sex appeal. They learn at a very young age that their value is determined by how beautiful, thin, "hot," and sexy they are. And boys, who get a very narrow definition of masculinity that promotes insensitivity and macho behavior, are taught to judge girls based on how close they come to an artificial, impossible, and shallow ideal. (p. 2)

Girls and women are often invisible. A study of six months of Sunday morning television talk shows in the United States found that women were almost nonexistent as panelists (Common Dreams 2005).

Men are also victimized by media images. A study of 1,000 adults found that two-thirds of the respondents thought that women in television advertisements were pictured as "intelligent, assertive, and caring," whereas men were portrayed as "pathetic and silly" (Abernathy 2003). In a study of beer and liquor ads, Messner and Montez de Oca (2005) concluded that, although beer ads of the 1950s and 1960s focused on men in their work roles and only depicted women as a reflection of men's home lives, present-day ads portray young men as "bumblers" and "losers" and women as "hotties" and "bitches."

One of the largest studies of women and men in the media ever conducted is the Global Media Monitoring Project (Gallagher 2006). Nearly 13,000 news stories on television, on the radio, and in newspapers were monitored by 40,000 news personnel in 76 countries. The results indicate the following:

- Women are dramatically underrepresented as news subjects, and, when they are the subject of news, they are most often "stars" or ordinary women rather than figures of authority or experts.
- Although the number of female reporters has increased in recent years, female reporters are still a minority in electronic (37 percent) and print media (29 percent).
- The older a woman is, the less likely she is to be on television—for example, by the age of 50, only 7 percent of newscasters are women.
- Male journalists are more likely to work on the "hard" news (e.g., politics) whereas women are more often assigned "soft stories" (e.g., social issues).
- Just 10 percent of all news stories focus specifically on women; news on gender equality or inequality is almost nonexistent.

As with media images, both the words we use and the way we use them can reflect gender inequality. Radio talk show host Rush Limbaugh, referring to female reporters as "infobabes" and stating that they have "chickified the news," is demeaning to women and men (Media Matters 2009). The term *nurse* carries the meaning of "a woman who . . ." and the term *engineer* suggests "a man who. . . ." Terms such as *broad, old maid,* and *spinster* have no male counterparts. Sexually active teenage females are described by terms carrying negative connotations, whereas terms for equally sexually active male teenagers are considered complimentary. Language is so gender-stereotyped that the placement of male or female before titles is sometimes necessary, as in the case of "female police officer" or "male prostitute." Furthermore, as symbolic interactionists note, the embedded meanings of words carry expectations of behavior. This chapter's Self and Society assesses your attitudes toward language and gender.

> Language is so gender-stereotyped that the placement of female or male before titles is sometimes necessary, as in the case of "female police officer" or "male prostitute."

Please use the following definition in completing this exercise: Sexist language includes words, phrases, and expressions that unnecessarily differentiate between females and males or exclude, trivialize, or diminish either gender.

For each of the statements below, choose the descriptor (1 = strongly disagree; 2 = tend to disagree; 3 = undecided; 4 = tend to agree; 5 = strongly agree) that most closely corresponds with your beliefs about language and enter the number in the blank following each statement.

1. Women who think that being called a "chairman" is sexist are misinterpreting the word "chairman." ____

2. We should not change the way the English language has traditionally been written and spoken. ____

3. Worrying about sexist language is a trivial activity. ____

4. If the original meaning of the word "he" was "person," we should continue to use "he" to refer to both males and females today. ____

5. When people use the term "man and wife," the expression is not sexist if the users don't mean it to be. ____

6. The English language will never be changed because it is too deeply ingrained in the culture. ____

7. The elimination of sexist language is an important goal. ____

8. Most publication guidelines require newspaper writers to avoid using ethnic and racial slurs. So, these guidelines should also require writers to avoid sexist language. ____

9. Sexist language is related to sexist treatment of people in society. ____

10. When teachers talk about the history of the United States, they should change expressions, such as "our forefathers," to expressions that include women. ____

11. Teachers who require students to use nonsexist language are unfairly forcing their political views upon their students. ____

12. Although change is difficult, we still should try to eliminate sexist language. ____

Scoring

All items are scored on a 5-point Likert-type scale. Items 1, 2, 3, 4, 5, 6 and 11 are reverse-scored. High scores (4–5) indicate a positive attitude toward inclusive language; low scores (1–2) indicate a negative attitude toward inclusive language. A score of 3 indicates neutrality or uncertainty. Each respondent's score on the instrument is the total of all the items.

Interpretation

The range of possible total scores on the 12-item IASNL is 12 to 60. Across the 12 items, total scores between 42.1 and 60 reflect a supportive attitude toward nonsexist language; total scores between 12 and 30 reflect a negative attitude toward nonsexist language; and total scores between 30.1 and 42 reflect a neutral attitude.

Sources: Parks & Roberton 2000; Parks & Roberton 2001.

Religion and Cultural Sexism

Most Americans claim membership in a church, synagogue, or mosque. Research indicates that women attend religious services more often, rate religion as more important to their lives, and are more likely to believe in an afterlife than are men (Davis et al. 2002; Smith 2006; Adams 2007). In general, religious teachings have tended to promote traditional conceptions of gender. For example, in 2009, the Vatican began investigating the religious and secular lives of U.S. nuns (Goodstein 2009). Many nuns fear that the investigation is motivated by the church's patriarchal need to control them, some of whom have stopped wearing the traditional habit, live outside of convents, joined advocacy groups, worked in academia, and advocated ordination of female priests.

Gallagher (2004) found that most evangelical Christians believe that marriage should be considered an equal partnership, but also believe that the male is the head of the household. Female Orthodox Jews are not counted as part of the *minyan* (i.e., *a* quorum required at prayer services), are not allowed to read from the Torah, and are required to sit separately from men at religious services. In addition, Roman Catholic Church doctrine forbids the use of artificial forms of contraception, and Muslim women are required to be veiled in public at all times (Renzetti & Curran 2003). Women cannot serve as ordained religious leaders in the Roman Catholic Church, in Orthodox Jewish synagogues, or in Islamic temples.

Despite this "stained glass ceiling" (Adams 2007), women are increasingly active in leadership positions in churches and temples across the nation. For example, in 2009, the Lutheran Church of Great Britain ordained a woman bishop—the first female bishop in Great Britain (Lutheran World Federation 2009), and Alysa Stanton became the first U.S. Black American female rabbi (Kaufman 2009). Nonetheless, today in the United States, 85 percent of clergy are male (U.S. Census Bureau 2009), and even in denominations that allow for the ordination of women, female clergy often do not hold the same status as their male counterparts and are often limited in their duties (Renzetti & Curran 2003).

However, religious teachings are not all traditional in their beliefs about women and men. Quaker women have been referred to as the "mothers of feminism" because of their active role in the early feminist movement. Reform Judaism has allowed ordination of women as rabbis for more than 25 years, and gays and lesbians for more than 15 years. In addition, the women-church movement—a coalition of feminist faith-sharing groups composed primarily of Roman Catholic women—offers feminist interpretations of Christian teachings. Within many other religious denominations, individual congregations choose to interpret their religious teachings from an inclusive perspective by replacing masculine pronouns in hymns, the Bible, and other religious readings (Anderson 1997; Renzetti & Curran 2003).

Social Problems and Traditional Gender Role Socialization

Cultural sexism, transmitted through the family, school, media, and language, perpetuates traditional gender role socialization. Although the recent trend toward **gender tourism** suggests that traditional gender roles may be changing, to a large extent the social definitions of what it means to be a woman and what it means to be a man have varied little over the decades. These definitions, in turn, are associated with several social problems including the feminization of poverty, social-psychological costs, death and illness, conflict in relationships, and gendered violence.

The Feminization of Poverty

Today, women and girls comprise 70 percent of the poorest people in the world (Foerstel 2008). Further, women are more likely to be unemployed than men. In 2008, the world unemployment rate for men was 5.9 percent; for women, 6.3 percent (*ILO News* 2009). Women are also disproportionately affected by the global economic crisis, leading labor market specialists to predict that up to an additional 22 million women may become unemployed over the next several years.

gender tourism The recent tendency for definitions of masculinity and femininity to become less clear, resulting in individual exploration of the gender continuum.

Jane Hodges, Bureau Director for Gender Equality at the International Labour Organization explained this continued trend toward the feminization of poverty: "Women's lower employment rates, weaker control over property and resources, concentration in informal and vulnerable forms of employment with lower earnings, and less social protection, all place women in a weaker position than men to weather crises" (*ILO News* 2009, p. 1).

Women in the United States make up the majority of minimum wage workers and are significantly more likely to live in poverty than men (ITUC 2009). Two groups of women are the most likely to be poor in the United States—heads of households with dependent children and women over the age of 65 who have outlived their spouses. For example, 13.2 percent of male-headed households are below the poverty level compare to 28.3 percent of female-headed households. Hispanic and black female-headed households are the poorest of all families, headed by a single woman (U.S. Census Bureau 2009).

It is often assumed that antipoverty programs designed to address overall economic inequality will reduce the feminization of poverty. However, Brady and Kall's (2008) analysis of poverty in 18 affluent countries suggests that not only is the feminization of poverty universal, it is unique in its origins, tied to the percent of children in single-mother families and the male-to-female ratio of the elderly in a country.

The Social Psychological Costs of Gender Socialization

How we feel about ourselves begins in early childhood. Significant others, through the socialization process, expect certain behaviors and prohibit others based upon our birth sex. Both girls and boys, often as a consequence of these expectations, feel varying degrees of self-esteem, autonomy, depression, and life dissatisfaction. For example, Jose and Brown (2009) studied depression, stress, and rumination (i.e., worrying) in a co-ed sample of 10- to 17-year-olds. Each of the three dependent variables was significantly higher in females when compared to males.

Adolescent girls are also more likely to be dissatisfied with their looks, including physical attractiveness, appearance, and body weight than adolescent boys. Canadian nonprofit group Media Awareness Network (Mnet) summarizes the present state of research on the "cult of thinness" (Mnet 2009). Research indicates that 25 percent of college-aged women have tried to lose weight using unhealthy means (e.g., vomiting, laxative abuse, fasting). Women's magazines have 10 times the advertisements and articles promoting weight loss than men's magazines, and 75 percent of the covers of women's magazines contain at least one message about women changing their appearance (e.g., cosmetic surgery, diet).

According to Mnet, media messages that perpetuate insecurities in women are simply a matter of economics. If you liked the way you looked, you wouldn't buy anything. The diet industry alone makes billions of dollars a year selling products that don't work. The social-psychological costs are high. Research on the impact of media images "indicates that exposure to images of thin, young, air-brushed female bodies is linked to depression, loss of self-esteem, and the development of unhealthy eating habits in women and girls" (Mnet 2009, p. 1).

Boys too, from an early age, are concerned about body image, and as adults, their self-esteem is also linked to body shape and weight (Grogan 2008). Men are also affected by traditional gender socialization, which places enormous cultural pressure to be successful in their work and to earn high incomes. Sanchez and Crocker (2005) found that, among college-age women *and* men, the more participants were invested in traditional ideals of gender, the lower their self-concept and psychological well-being. Traditional male socialization also discourages males from expressing emotion and asking for help—part of what William Pollack (2000a) calls the **boy code.** Much of the boy code continues into adulthood as described in journalist Norah Vincent's 2006 book *Self-Made Man* (see this chapter's Social Problems Research Up Close feature).

In a recent book, sociologist Michael Kimmel (2008) describes the lives of nearly 400 predominantly white, college-educated males between 16 and 21 years of age—the residents of *Guyland.* The young men of Guyland live with male friends (e.g., in dormitories, apartments, fraternity houses) or with parents who "just don't get it." They are unmarried, and work in low-paying entry-level jobs, leaving plenty of free time to be "guys." What do the guys of Guyland do? They hook up with women; play video games; listen to, watch, and talk about sports; drink, sometimes nightly; gamble online and watch pornography; live by the adage "Bros Over Hos;" and are misogynistic, homophobic, and, sometimes, violent.

The boys and men of Guyland, as Kimmel describes them, are in "suspended animation," somewhere between boyhood and adulthood, lost in a world they don't understand and have difficulty maneuvering, frustrated by "thwarted entitlement"— the belief that minorities, women, and gays are encroaching on their territory. With compassion and concern for today's young men, Kimmel warns that bullying and hazing, binge drinking and self-medication, sexual violence, gay bashing, and suicides will continue unless, as one reviewer states, "we enable young men to chart their own paths, to stay true to themselves, and to travel safely through *Guyland*, emerging as responsible and fully formed men of integrity and honor" (Harper Collins 2008, p. 1).

According to sociologist Michael Kimmel, the residents of Guyland (2008), representing over 22 million young men between the ages of 16 and 26, are a major consumer group. They are targeted by Hollywood, magazines such as *Stuff, Men's Health,* and *Maxim,* by television sports channels, and by a variety of technology-related products such as computers, video games, software, and CDs.

The Impact of Gender Socialization on Death and Illness

Men are less likely to go to a doctor than women for a variety of structural and cultural reasons. Men, for example, work longer hours than women and are more likely to be working full-time, making it difficult to see a doctor or attend preventive medicine programs that are often only available during the day. Men are also less likely to have a regular physician or to go to the hospital, even if time permits (White & Witty 2009).

At every stage of life, "American males have poorer health and a higher risk of mortality than females" (Gupta 2003, p. 84). On average, men in the United States die about five years earlier than women, although gender differences in life expectancy have been shrinking (U.S. Census Bureau 2009). Traditionally defined gender roles for men are linked to high rates of cirrhosis of the liver (e.g., alcohol consumption), many cancers (e.g., tobacco use), and cardiovascular diseases (e.g., stress). Men also engage in self-destructive behaviors more often than women—poor diets, lack of exercise, higher drug use, refusal to ask for help or

boy code A set of societal expectations that discourages males from expressing emotion, weakness, or vulnerability, or asking for help.

Journalist Norah Vincent wanted to know what it is like to be a man. So, for 18 months, she observed and participated in the world of men, not as Norah Vincent, but as "Ned"—the male persona she created for the purpose of her research. Norah Vincent describes her journey into manhood in the book *Self-Made Man* (2006).

Sample and Methods

Norah Vincent used a method of research known as participant observation—a method in which the researcher participates in the phenomenon being studied to obtain an insider's perspective of the people and/or behavior being observed. She describes her book as "a travelogue as much as anything else . . . a six-city tour of an entire continent, a woman's-eye view of one guy's approximated life, not an authoritative guide to the whole vast and variegated terrain of manhood in America" (p. 17).

To participate in and observe the world of men, Norah sought the help of a makeup artist to transform herself into Ned. She learned how to apply artificial beard stubble, wore masculine-looking glasses, had her hair cut into a man's hairstyle, wore men's clothing, and stuffed her pants with a prosthetic penis known as a "packable softie." Norah bulked up her upper body muscle through six months of weight lifting and increased protein consumption, and she flattened her breasts by wearing

a flat-front sports bra two sizes too small. Norah, who already had a deep voice, hired a voice coach to help her speak like a man.

Having achieved a male identity, Ned spent 18 months exploring what it means to be a man by (1) joining a men's bowling team, (2) going to men's strip clubs, (3) dating women, (4) living at a monastery, (5) working at a sales job, and (6) attending a men's movement group.

Findings and Conclusion

Examples of Ned's experiences and observations include the following:

1. *Bowling with the guys.* One of Ned's first outings was to the local bowling alley, where he played in a Monday night men's league. It is here that he was introduced to male interaction cues:

   Our evenings together always started out slowly with a few grunted hellos that among women would have been interpreted as rude. This made my female antennae twitch a little. Were they pissed off at me about something? But among these guys no interpretation was necessary. . . . If they were pissed at you, you'd know it. These gruff greetings . . . were coming from long, wearing workdays, usually filled with hard physical labor and the slow, soul-deadening deprecation that comes of being told what to do all day. (pp. 29–30)

2. *Strip clubs.* Ned went to strip clubs to get a glimpse of "a substratum of the male sexual psyche that most women either don't know about, don't want to know about, or both" (p. 65). After weeks of observing and interacting with various strippers and other patrons, Ned offers the following reflections:

   I'd been inside a part of the male world that most women and even a lot of men never see, and I'd seen it as just another one of the boys. In those places male sexuality felt like something you weren't supposed to feel but did, like something heavy you were carrying around and had nowhere to unload except in the lap of some damaged stranger, and then only for five minutes. . . . It wasn't nearly so simple as men objectifying women. . . . Nobody won . . . nobody was more or less victimized than anyone else. The girls got money. The men got an approximation of sex and flirtation. But in the end everyone was equally debased by the experience. (pp. 90–91)

3. *Dating.* Ned found that most of the women he dated "were carrying the baggage of previous hurts at the hands of men, which in many cases had prejudiced them unfairly against the male sex" (p. 100). Ned described that, as a "man" dating women,

wear a seatbelt, and stress-related activities. Being married improves men's health more than it does women's (Williams & Umberson 2004), in large part because wives encourage their husbands to take better care of themselves.

Women's health is also gendered. Although men have higher rates of HIV/AIDS worldwide, the disease disproportionately affects women in many areas of the world (sees Chapter 2). For example, in sub-Saharan Africa, 61 percent of those infected are women (WHO 2009). Women's inequality contributes to the spread of the disease. First, in many of these societies, "women lack the power in relationships to refuse sex or negotiate protected sex" (Heyzer 2003, p. 1). Second, women are often the victims of rape and sexual assault, with little social or legal

. . . I often felt attacked, judged, on the defensive. Whereas with the men I met and befriended as Ned there was a presumption of innocence—that is, you're a good guy until you prove otherwise—with women there was quite often a presumption of guilt: you're a cad like every other guy until you prove otherwise. (p. 101)

Ned experienced the difficulty that heterosexual men have in living up to the ideals of women, observing that "women wanted a take-control man, at the same time, they wanted a man who was vulnerable to them, a man who would show his colors and open his doors, someone expressive, intuitive, attuned" (p. 111).

4. *A Catholic monastery.* In the weeks Ned spent at a Catholic monastery, he observed men in a context in which women and sexual expression were, at least in principle, excluded from the male experience. Even in this context, however, male gender role expectations were evident. Ned was at the dinner table with the monks and he told Father Richard that he looked very good for his age:

As soon as the remark came out of my mouth, everyone at the table stopped eating mid-forkful and looked at me as if I had three heads. Father Richard . . . said a very suspicious, squint-eyed "thank you," and looked away, clearly

embarrassed. But the implication from other quarters was clear: "What the hell's wrong with you, kid? Don't you know that properly socialized males don't behave that way with each other?" (pp. 144–45)

5. *Work.* Ned interviewed for several jobs and landed a job as a salesman. One of his coworkers advised him, "You're a man. . . . You gotta pitch like a man." Ned learned that, in sales work, women and men had different strategies:

Girls pitched differently. They flirted. They cajoled and smiled and eased their way into the sales underhandedly, which was exactly how I'd started out trying to do it. I'd tried initially to ask for the sale the way I asked for food in a restaurant as a woman, or the way I asked for help at a gas station— pleadingly. But coming from a man this was off-color. It didn't work. It bred contempt in both men and women. . . . People see weakness in a woman and they want to help. They see weakness in a man and they want to stamp it out. (p. 213)

6. *Men's movement group.* Ned attended a men's movement group in which 25 to 30 guys met once a month. He noted that hugging was a central part of the group:

Most men don't tend to share much physical affection with their male

friends, so here the guys made a point of hugging each other long and hard at every possible opportunity as a way of offsetting what the world had long deprived them of, and what they in turn had been socialized to disallow themselves. (p. 232)

In the concluding chapter of *Self-Made Man,* Norah explains that the experience of being a guy was not what she had expected:

I had thought that by being a guy I would get to do all the things I didn't get to do as a woman, things I'd always envied about boyhood when I was a child: the perceived freedoms of being unafraid in the world. . . . But when it actually came to the business of being Ned I rarely felt free at all. (pp. 275–76)

As Ned, Norah found that "somebody is always evaluating your manhood. Whether it's other men, other women, even children. And everybody is always on the look-out for your weakness or your inadequacy" (p. 276). Norah described the male role as a "straightjacket . . . that is no less constrictive than its feminine counterpart. You're not allowed to be a complete human being. Instead you get to be a coached jumble of stoic poses. You get to be what's expected of you" (p. 276).

Source: Vincent 2006.

recourse. Third, gender norms often dictate that men have more sexual partners than women, putting women at greater risk.

The World Health Organization (2009) has identified additional ways traditional definitions of gender impact the health and well-being of women and girls. For example, every day, over 1,600 women die from preventable complications during pregnancy and childbirth. Primarily responsible for household duties, women are exposed to hundreds of pollutants as they cook that contribute to their disproportionately high rates of death from chronic obstructive pulmonary disease (COPD). Deaths from lung cancer and other tobacco-related illnesses are expected to rise as the tobacco industry targets women in developing countries.

Finally, many women and girls throughout the world have a higher probability of suffering or dying from a variety of diseases because of gender: they are more likely to be poor, less likely to be seen as worthy of care when resources are short, and in many countries they are forbidden to travel unaccompanied by a male, making access to a hospital difficult.

Gender Based Violence

Men are more likely than women to be involved in violence—to kill and be killed; to wage war and die both as combatants and noncombatants; to take their own lives, usually with the use of a firearm; to engage in violent crimes of all types; to bully, harass, and abuse. Although the most serious of violent acts are exceptions rather than the norm, Pollack argues that male violence is a consequence of gender socialization and definitions of masculinity that hold "as long as nobody is seriously hurt, no lethal weapons are employed, and especially within the framework of sports and games—football, soccer, boxing, wrestling—aggression and violence are widely accepted and even encouraged in boys" (Pollack 2000b, p. 40).

Women and girls are often the victims of male violence. Worldwide, as many as 71 percent of women will be physically or sexually abused in their lifetime (see Chapter 5) (WHO 2009). Attacks on women's and girls' bodies routinely take place as they are beaten, raped, and killed in the name of religion, war, and honor. Over 5,000 women and girls are killed each year in **honor killings**—murders, often public, as a result of a female dishonoring, or being perceived to have dishonored, her family or community (Foerstel 2008). In 2007, Du'a Khalil Aswad, 17, was brutally beaten, kicked and stoned, and hit from behind with a cement block, as she struggled to sit up. Her crime? She had fallen in love with someone outside her religious sect. The murder, posted on YouTube, became an international outrage, yet no one has been prosecuted to date.

Violence against women is "rising every day. . . . [B]eheadings, rapes, beatings, suicides through self-immolation, genital mutilation, trafficking, and child abuse masquerading as marriage of girls as young as nine are all on the increase" (Judd 2008, p. 1). Violence against women and girls, a problem of pandemic proportions, is rooted in gender inequality and the lingering notion that women and girls are property—a belief rooted in ancient law and many of the world's religions.

Some of the most significant costs of traditional gender roles for men and boys are shortened life expectancy, increased likelihood of diseases associated with stress and risky behaviors (e.g., smoking), limited expression of emotions, and higher rates of suicide and violence-related deaths.

honor killings Murders, often public, as a result of a female dishonoring, or being perceived to have dishonored, her family or community.

Yet, in the United States, being female is not part of the federal hate crime statutes despite the fact that, as one advocate testified before Congress, "women and girls . . . are exposed to terror, brutality, serious injury, and even death because of their sex" (LCCREF 2009, p. 33).

Attacks on women's and girls' bodies routinely take place as they are beaten, raped, and killed in the name of religion, war, and honor.

What Do You Think? Worldwide, much of the violence against women is steeped in "harmful traditional practices," including honor killings, female genital mutilation, forced marriage, and dowry killings. Dowry killings involve "a woman being killed by her husband or in-laws because her family is unable to meet their demands for her dowry—a payment made to a woman's in-laws upon her engagement or marriage as a gift to her new family" (UNIFEM 2007, p. 1). Bride burning, acid attacks, and many deaths ruled as suicides are linked to dowry disputes. In India, courts have ruled that "customary payments" to a groom's family are not illegal (Mahapatra 2008), but in 2009, the India Supreme Court expressed outrage at such "barbaric practices" (Venkatesan 2009). Do you think dowry killings can be prevented through law? Why or why not?

Strategies for Action: Toward Gender Equality

In recent decades, there has been a growing awareness of the need to increase gender equality throughout the world. Strategies to achieve this end have focused on empowering women in social, educational, economic, and political spheres and improving women's access to education, nutrition, health care, and basic human rights. But as we will see in the following section on grassroots movements, there is also a men's movement that is concerned with gender inequities and the issues facing men.

Grassroots Movements

Efforts to achieve gender equality in the United States have been largely fueled by the feminist movement. Despite a conservative backlash, feminists, and to a lesser extent men's activist groups, have made some gains in reducing structural and cultural sexism in the workplace and in the political arena.

Feminism and the Women's Movement. **Feminism** is the belief that women and men should have equal rights and responsibilities. The U.S. feminist movement began in Seneca Falls, New York, in 1848, when a group of women wrote and adopted a women's rights manifesto modeled after the Declaration of Independence. Although many of the early feminists were primarily concerned with suffrage, feminism has its "political origins . . . in the abolitionist movement of the 1830s," when women learned to question the assumption of "natural superiority" (Anderson 1997, p. 305). Early feminists were also involved in the temperance movement, which advocated restricting the sale and consumption of alcohol, although their greatest success was the passing of the Nineteenth Amendment in 1920, which recognized women's right to vote.

feminism The belief that men and women should have equal rights and responsibilities.

The rebirth of feminism almost 50 years later was facilitated by a number of interacting forces: an increase in the number of women in the labor force, the publication of Betty Friedan's book *The Feminine Mystique,* an escalating divorce rate, the socially and politically liberal climate of the 1960s, student activism, and the establishment of the Commission on the Status of Women by John F. Kennedy. The National Organization for Women (NOW) was established in 1966, and remains one of the largest feminist organizations in the United States, with more than 500,000 members in 550 chapters across the country.

One of NOW's hardest fought battles is the struggle to win ratification of the **equal rights amendment** (ERA), which states that "equality of rights under the law shall not be denied or abridged by the United States, or by any state, on account of sex." The proposed 28th amendment to the Constitution passed both the House of Representatives and the Senate in 1972, but failed to be ratified by the required 38 states by the 1979 deadline, which was later extended to 1982. With the exception of 2008, when presidential politics took precedence, the bill has been reintroduced into Congress every year since 1982 (ERA 2009). Six states are presently considering ratification. If three of the six ratify the amendment, it could become law within two years of reaching the 38-state requirement (Cook 2009; ERA 2009).

Proponents of the ERA argue that its opponents used scare tactics—saying that the ERA would lead to unisex bathrooms, mothers losing custody of their children, and mandatory military service for women—to create a conservative backlash. However, Susan Faludi, in *Backlash: The Undeclared War against American Women* (1991), contends that contemporary arguments against feminism are the same as those levied against the movement 100 years ago and that the negative consequences predicted by opponents of feminism (e.g., women unfulfilled and children suffering) have no empirical support. Proponents also argue that "without the explicit wording and intention of women's rights documented in the principles of our government, women remain second-class citizens" (Cook 2009, p. 1).

Today, a new wave of feminism is being led by young women and men who grew up with the benefits their mothers won, but who are shocked by the stoning to death of a woman accused of adultery in Afghanistan, the continuing practice of female genital mutilation in at least 25 countries, the fact that millions of women throughout the world lack access to modern contraception, and the fact that, even in one of the "freest" countries in the world, U.S. women face increasing restrictions on abortion, can be denied prescription contraception by a pharmacist who morally objects, earn less than men earn for doing the same job, and experience alarming rates of date rape and intimate partner abuse. These young feminists are more inclusive than their predecessors, welcoming all who champion the cause of global equality. Not surprisingly, the new feminists are likely to attract a more diverse group of supporters than their predecessors because future feminist efforts focus on gender equality rather than "gender sameness" (Parker 2000). Some observers, however, note that the new diversity of the women's movement may contribute to "a tension within feminism between the felt urgency to present concerns and grievances as a single unified group of women, and the need to give voice to the variations in concerns and grievances that exist among feminists on the basis of race and ethnicity, social class, sexual orientation, age, physical ability/disability, and a host of other factors" (Renzetti & Curran 2003, p. 25).

equal rights amendment (ERA) The proposed 28th amendment to the Constitution, which states that "equality of rights under the law shall not be denied or abridged by the United States, or by any state, on account of sex."

The Men's Movement. As a consequence of the women's rights movement, men began to reevaluate their own gender status. As with any grassroots movement, the men's movement has a variety of factions. One of the early branches of the men's movement is known as the mythopoetic men's movement, which began after the publication of Robert Bly's (1990) *Iron John*—a fairy tale about men's wounded masculinity that was on the *New York Times* bestseller list for more than 60 weeks (Zakrzewski 2005). Participants in the men's mythopoetic movement met in men-only workshops and retreats to explore their internal masculine nature, male identity, and emotional experiences through the use of stories, drumming, dance, music, and discussion.

The ManKind Project (MKP), founded in 1985, grew out of the men's mythopoetic movement. It is an international organization with over 40,000 members (Clothier 2009). Their motto, "changing the world one man at a time," reflects the stated values of the organization: connection to feelings, leadership, fatherhood and respect for the elderly (MKP 2009a). MKP conducts New Warrior Training Adventure weekends in which, through "group discussions, games, guided visualizations, journaling, and individual process work" each man can get "in touch with the truth about himself—not his job, not his possessions, not his roles in life—himself" (MKP 2009b, p. 1).

In addition to enthusiastic supporters, there are also critics. Some experts argue that MKP "practices therapy without a license; targets vulnerable members of 12-step recovery groups; purposefully withholds the details of the program, thus keeping potential participants from making a fully informed decision whether or not to attend; and does not screen applicants who may be too emotionally frail for the rigors of the program" (Vogel 2008, p. 1).

In 2005, 15 days after attending a MKP-Houston New Warrior Training Adventure weekend, Michael Scinto committed suicide, but not before writing a letter describing his experience. According to Scinto, he had to sign a confidentiality waiver as a condition of attending the workshop; he had to engage in such activities as sitting nude among the 40 or so other participants as each described their sexual histories; he had his life threatened; and, when he became uncomfortable with the activities and wanted to leave, he was forbidden to do so. His parents filed a wrongful death suit that was settled in 2008, and included monetary damages and court-imposed changes in screening procedures of applicants, full disclosure of MKP-Houston information, and an open-door policy at weekend retreats.

The men's movement also includes men's organizations that advocate gender equality and work to make men more accountable for sexism, violence, and homophobia. For example, the National Organization for Men Against Sexism (NOMAS) was founded in 1975, and "advocates a perspective that is pro-feminist, gay affirmative, antiracist, dedicated to enhancing men's lives, and committed to justice on a broad range of social issues including class, age, religion, and physical abilities" (NOMAS 2009, p. 1).

Other men's groups oppose feminism and view the feminist agenda as an organized form of male bashing. The agenda of antifeminist men's groups is to maintain and promote traditional gender ideology and roles. For example, the Promise Keepers, part of a Christian men's movement, and Louis Farrakhan's Nation of Islam, have often been criticized as patriarchal and antifeminist (Renzetti & Curran 2003).

Some men's groups focus on issues concerning children and fathers' rights. Groups such as the American Coalition for Fathers and Children, Dads Against

Other concerns on the agenda of men's rights groups include the domestic violence committed against men by women, false allegations of child sexual abuse, wrongful paternity suits, and the oppressive nature of restrictive masculine gender norms.

Discrimination, and Fathers4Justice are attempting to change the social and legal bias against men in divorce and child custody decisions, which tend to favor women. Members of the National Coalition of Free Men (NCFM) are also concerned with issues surrounding fatherhood, concentrating on "how gender-based expectations limit men legally, socially, and psychologically" (NCFM 2009, p. 1).

Other concerns on the agenda of men's rights groups include the domestic violence committed against men by women, false allegations of child sexual abuse, wrongful paternity suits, and the oppressive nature of restrictive masculine gender norms. Just as women have fought against being oppressed by expectations to conform to traditional gender stereotypes, men are beginning to want the same freedom from traditional gender expectations. For example, men who enter nontraditional work roles, such as nurse and primary school teacher, are often stigmatized for participating in "feminine" work. A study of men in nontraditional work roles found that these men commonly experience embarrassment, discomfort, shame, and disapproval from friends and peers (Simpson 2005).

National Public Policy

A number of important federal statutes have been passed to help reduce gender inequality. They include the Equal Pay Act of 1963, Title VII of the Civil Rights Act of 1964, Title IX of the Education Amendments of 1972, the Victims of Trafficking and Violence Protection Act of 2000, and the Violence against Women Reauthorization Act of 2005. In 2009, President Obama signed the Ledbetter Fair Pay Act, which reversed the 2007 U.S. Supreme Court decision that held that victims of pay discrimination had 180 days to file a grievance after the act of discrimination. The act now defines each paycheck as a separate act of discrimination (Mehmood 2009). Finally, the Paycheck Fairness Act awaits legislative approval. If passed, it would close loopholes in the Equal Pay Act of 1963, provide the same protections against pay discrimination as racial and ethnic minorities currently enjoy, and prohibit retaliation by employers against employees who disclose their wages (NWLC 2009; Current Legislation 2009).

Presently, two hotly debated public policies concern sexual harassment and affirmative action.

Sexual Harassment. Sexual harassment is a form of sex discrimination that violates Title VII of the 1964 Civil Rights Act. The U.S. EEOC (2009b) defines **sexual harassment** in the workplace as "unwelcome sexual advances, requests for sexual favors, and other verbal or physical conduct of a sexual nature . . . when this conduct explicitly or implicitly affects an individual's employment, unreasonably interferes with an individual's work performance, or creates an intimidating, hostile, or offensive work environment" (p. 1). Sexual harassment can be of two types: (1) *quid pro quo,* in which an employer requires sexual favors in exchange for a promotion, salary increase, or any other employee benefit, and (2) the existence of a hostile environment that unreasonably interferes with job performance, as in the case of sexually explicit comments or insults being made to an employee. Common examples of sexual harassment include

sexual harassment In reference to workplace harassment, when an employer requires sexual favors in exchange for a promotion, salary increase, or any other employee benefit and/or the existence of a hostile environment that unreasonably interferes with job performance.

unwanted touching, the invasion of personal space, making sexual comments about a person's body or attire, and telling sexual jokes (Uggen & Blackstone 2004).

The victim as well as the harasser may be a woman or a man, although adult women are the most frequent targets of sexual harassment (Uggen & Blackstone 2004). Women who work in male-dominated occupations and blue-collar jobs are the most likely to experience sexual harassment (Jackson & Newman 2004). In 1998, the U.S. Supreme Court extended protection to victims of same-sex harassment. Victims of sexual harassment are not limited to those who are harassed but include anyone affected by the offensive conduct.

Sexual harassment occurs in a variety of settings, including the workplace, public schools, military academies, and college campuses. In 2009, Big Vanilla Pasadena and Big Vanilla Athletic were ordered to pay over $100,000 in damages to four women who were sexually harassed by employees of the facilities and then fired when they complained about the harassment. The employer will also be subject to EEOC oversight of employee training on sexual harassment regulations and the display of antidiscrimination notices in the workplace (EEOC 2009c).

What Do You Think? A report by the American Association of University Women entitled "Drawing the Line: Sexual Harassment on Campus" (AAUW 2006) surveyed more than 2,000 U.S. undergraduate college students. The results indicate that (1) two-thirds of undergraduates report having been the victim of sexual harassment; (2) females were more likely to say they were upset, ashamed, embarrassed, or humiliated as a result of the harassment than males; (3) a majority of the harassers admitted to thinking that sexual harassment is funny; and (4) although few of the students reported the harassment, nearly half wish there was an office or person to contact about the unwanted advances. Have you ever been the victim of sexual harassment? Did you report it and, if so, to whom? What would you do to reduce the incidence of sexual harassment on your campus?

Any person who experiences sexual harassment can file a charge of discrimination at any EEOC office. In the fiscal year 2008, the EEOC received 13,867 charges of sexual harassment; males filed 15 percent of those claims (EEOC 2009b).

Affirmative Action. As discussed in Chapter 9, **affirmative action** refers to a broad range of policies and practices to promote equal opportunity as well as diversity in the workplace and on campuses. Affirmative action policies, developed in the 1960s from federal legislation, require that any employer (universities as well as businesses) that receives contracts from the federal government must make "good faith efforts" to increase the number of female and other minority applicants. Such efforts can be made through expanding recruitment and training programs and by making hiring decisions on a nondiscriminatory basis.

However, a 1996 California ballot initiative, the first of its kind, abolished race and sex preferences in government programs, which included state colleges and universities. Over the next three years, several other states followed suit. In 2003, the U.S. Supreme Court held that universities have a "compelling interest" in a diverse student population and therefore may take minority status into consideration when making admissions decisions. Since that time, several states have addressed the issue of affirmative action in government programs with varying

affirmative action A broad range of policies and practices to promote equal opportunity as well as diversity in the workplace and on campuses.

In developing nations where suitable housing, clean water, food, health care, and sanitary living conditions are scarce, women and girls are particularly likely to suffer. They are also often unable to make "quality of life" decisions, including where to live, who to marry, whether to go to school or work outside the home, and how to raise their children.

results. For example, in Nebraska, voters outlawed affirmative action programs; in Colorado, voters turned down a ban of affirmative action programs; and in Missouri and Oklahoma, anti-affirmative action initiatives were stopped before they reached the ballot. Over the next several years, Arizona and a number of other states will decide the fate of equal opportunity legislation (ACLU 2009).

International Efforts

International efforts to address problems of gender inequality date back to the 1979 Convention on the Elimination of All Forms of Discrimination Against Women (CEDAW), often referred to as the international women's bill of rights, adopted by the United Nations in 1979. Although 185 countries have ratified the women's bill of rights, including every country in Europe and South and Central America, 70 have filed "reservations," meaning they have been exempt from portions of the agreement (Foerstel 2008). The United States is the only industrialized country in the world that has not ratified the document.

Another significant international effort occurred in 1995, when representatives from 189 countries adopted the Beijing Declaration and Platform for Action at the Fourth World Conference on Women sponsored by the United Nations. The platform reflects an international commitment to the goals of equality, development, and peace for women everywhere. The platform identifies strategies to address critical areas of concern related to women and girls, including poverty, education, health, violence, armed conflict, and human rights. Today, over a decade after the platform was adopted, there are concerns that many of the stated goals have not been met (Foerstal 2008).

In addition to the CEDAW and the Beijing Platform, in 2000, all of the members of the United Nations adopted the Millennium Declaration. One of the eight Millennium Development Goals, as stated in the Millennium Declaration, is the promotion of gender equality and women's empowerment by 2015. Progress has been slow. In an evaluation of this goal, a United Nations' report (2008) concludes the following:

> Ensuring gender equality and empowering women in all respects—desirable objectives in themselves—are required to combat poverty, hunger and disease, and to ensure sustainable development. The limited progress in empowering women and achieving gender equality is a pervasive shortcoming that extends beyond the goal itself. Relative neglect of, and de facto bias against women and girls continues to prevail in most countries. (p. 5)

In addition to international efforts, individual countries have instituted programs or policies designed to combat sexism and gender inequality. For example, Japan implemented the Basic Law for a Gender-Equal Society, a "blueprint for gender equality in the home and workplace" (Yumiko 2000, p. 41). However, a United Nations report ranked Japan behind all other industrialized countries in women's empowerment. For example, Japanese women earn just 44 percent of what men earn and are more than 90 percent of the 8 million part-time workers (Faiola 2007).

For the first time in history, men and women have come together to address gender issues in Brazil at the Global Symposium on Engaging Men and Boys in Achieving Gender Equality (Moreno 2009). The government of the United Kingdom has proposed legislation that would require "gender pay audits" of businesses to ensure equality of treatment in wages between women and men (Bennett 2009). However, Afghanistan, a country known for its patriarchal mores, recently passed legislation that applies to the Shiite minority—a religious denomination of Islam. The objectionable provisions state that a woman has no legal right to refuse her husband's sexual advances, that she must get her husband's permission to go to school or work outside the home, and she must "make herself up" or "dress up" if her husband so desires (Filkins 2009, p. 1). Finally, in 2009, reversing an earlier government stance, the Obama administration has paved the way for women, under some circumstances, to be granted asylum in the United States. In the case of a Mexican woman who sought asylum after her common-law husband "repeatedly raped her at gunpoint, held her captive, stole from her, and at one point tried to burn her alive when he learned she was pregnant," the government response stated that "it is possible [that she] and other applicants who have experienced domestic violence could qualify for asylum" (Preston 2009, p. 1).

Understanding Gender Inequality

Gender roles and the social inequality they create are ingrained in our social and cultural ideologies and institutions and are therefore difficult to alter as indicated by the persistence of discriminatory attitudes displayed in Figure 10.3. Nevertheless, as we have seen in this chapter, growing attention to gender issues in social life has spurred some change. Women who have traditionally been expected to give domestic life first priority are now finding it more acceptable to seek a career outside the home; the gender pay gap is narrowing; and significant improvements in educational disparities between men and women, boys and girls, have been made. Most of these improvements, however, are in developed countries. Globally, millions of women and girls continue to be victimized by poverty, gendered violence, illiteracy, and limited legal rights and political representation.

Eliminating gender stereotypes and redefining gender in terms of equality does not simply mean liberating women but liberating men as well as society. Men are also victimized by discrimination and gender stereotypes that define what he "should" do rather than what he is capable, interested, and willing to do. The National Coalition of Free Men (2009) has incorporated this view into their position statement:

> We have heard in some detail from the women's movement how such sex-stereotyping has limited the potential of women. More recently, men have become increasingly aware that they too are assigned limiting roles which they are expected to fulfill regardless of their individual abilities,

Figure 10.3 Men's Discriminatory Attitudes Toward Women by Region of the World
Source: UNICEF 2007.

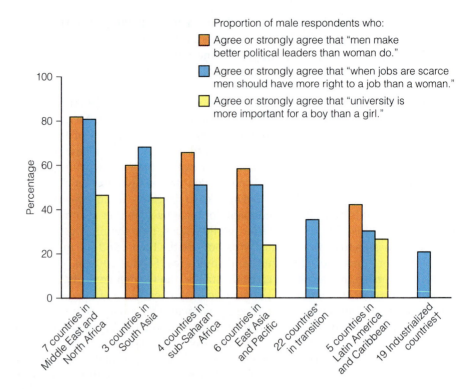

Proportion of male respondents who:

■ Agree or strongly agree that "men make better political leaders than woman do."

■ Agree or strongly agree that "when jobs are scarce men should have more right to a job than a woman."

□ Agree or strongly agree that "university is more important for a boy than a girl."

*Countries such as Croatia, Albania, Latvia, etc.(incomplete data).
†Countries such as the United States, England, France, etc.(incomplete data).

interests, physical/emotional constitutions, or needs. Men have few or no effective choices in many critical areas of life. They face injustices under the law. And typically they have been handicapped by socially defined "shoulds" in expressing themselves in other than stereotypical ways.

Increasingly, both women and men are embracing **androgyny**—the blending of both masculine and feminine characteristics. The concept of androgyny implies that both masculine and feminine characteristics and roles are *equally valued.* However, "achieving gender equality . . . is a grindingly slow process, since it challenges one of the most deeply entrenched of all human attitudes" (Lopez-Claros & Zahidi 2005, p. 1).

An international strategy for achieving gender equality, *gender mainstreaming* is "the process of assessing the implications for women and men of any planned action, including legislation, policies or programmes, in all areas and at all levels" (IASC 2009, p. 7). Difficult? Yes. But, regardless of whether traditional gender roles emerged out of biological necessity as the structural functionalists argue, or out of economic oppression as the conflict theorists hold, or both, it is clear that, today, gender inequality carries a high price: poverty, loss of human capital, feelings of worthlessness, violence, physical and mental illness, and death. Perhaps we have reached a time when we need to ask ourselves, are the costs of gender inequality too high to continue to pay?

androgyny Having both feminine and masculine characteristics.

CHAPTER REVIEW

■ Does gender inequality exist worldwide?

There is no country in the world in which men and women are treated equally. Although women suffer in terms of income, education, and occupational prestige, men are more likely to suffer in terms of mental and physical health, mortality, and the quality of their relationships.

■ How do the three major sociological theories view gender inequality?

Structural functionalists argue that the traditional division of labor was functional for preindustrial society and has become defined as both normal and natural over time. Today, however, modern conceptions of the family have replaced traditional ones to some extent. Conflict theorists hold that male dominance and female subordination evolved in relation to the means of production—from hunting and gathering societies in which females and males were economic equals to industrial societies in which females were subordinate to males. Symbolic interactionists emphasize that, through the socialization process, both females and males are taught the meanings associated with being feminine and masculine.

■ What is meant by the terms *structural sexism* and *cultural sexism*?

Structural sexism refers to the ways in which the organization of society, and specifically its institutions, subordinate individuals and groups based on their sex classification. Structural sexism has resulted in significant differences between education and income levels, occupational and political involvement, and civil rights of women and men. Structural sexism is supported by a system of cultural sexism that perpetuates beliefs about the differences between women and men. Cultural sexism refers to the ways the culture of society—its norms, values, beliefs, and symbols—perpetuate the subordination of an individual or group because of the sex classification of that individual or group.

■ What is the difference between the glass ceiling and the glass escalator?

The glass ceiling is an often invisible barrier that prevents women and other minorities from moving into top corporate positions. The glass escalator, on the other hand, refers to the tendency for men seeking traditionally female jobs to have an edge in hiring and promotion practices.

■ What are some of the problems caused by traditional gender roles?

First is the feminization of poverty. Women are socialized to put family ahead of education and careers, a belief that is reflected in their less prestigious occupations and lower incomes. Second are social-psychological costs. Women tend to have lower self-esteem and higher rates of depression and eating disorders than men. Men and boys, on the other hand, are often subject to the emotional restrictions of the "boy code." Third, traditional gender roles carry health costs in terms of death and illness. For example, many traditionally defined male behaviors shorten life expectancy, and women in many regions of the world disproportionately suffer from HIV/AIDS. Finally, gendered violence is responsible for the deaths and suffering of both men and women.

■ What strategies can be used to end gender inequality?

Grassroots movements, such as feminism and the women's rights movement and the men's rights movement, have made significant inroads in the fight against gender inequality. Their accomplishments, in part, have been the result of successful lobbying for passage of laws concerning sex discrimination, sexual harassment, and affirmative action. Besides these national efforts, international efforts continue as well. One of the most important is the Convention on the Elimination of All Forms of Discrimination Against Women (CEDAW), also known as the international women's bill of rights, which was adopted by the United Nations in 1979.

TEST YOURSELF

1. The United States is the most gender-equal nation in the world.
   a. True
   b. False
2. Symbolic interactionists argue that
   a. male domination is a consequence of men's relationship to the production process
   b. gender inequality is functional for society
   c. gender roles are learned in the family, in the school, in peer groups, and in the media
   d. women are more valuable in the home than in the workplace
3. Recent trends indicate that more males enter college from high school than females.
   a. True
   b. False

4. The devaluation hypothesis argues that female and male pay differences are a function of women's and men's different levels of education, skills, training, and work experience.
   a. True
   b. False

5. The glass ceiling is an often invisible barrier that prevents women and other minorities from moving into top corporate positions.
   a. True
   b. False

6. Which of the following statements is true about women and U.S. politics?
   a. proportionately there are more female governors than U.S. Congresswomen
   b. in the history of the U.S. Supreme Court there have only been three female justices
   c. the highest ranking woman in the U.S. government was vice-president Hillary Clinton
   d. women received the right to vote with the passage of the 21st amendment

7. The Global Media Monitoring Project found that
   a. the number of female reporters has decreased in recent years
   b. women are dramatically underrepresented as news subjects
   c. older women were more likely to appear on television than younger women
   d. male journalists are more likely to work on stories dealing with social issues than women journalists

8. The feminization of poverty refers to
   a. the tendency for women to be caretakers of the poor
   b. feminists' criticism of public policy on poverty
   c. the disproportionate number of women who are poor
   d. gender role socialization of poor women

9. *Quid pro quo* sexual harassment refers to the existence of a hostile working environment.
   a. True
   b. False

10. Feminists would argue which of the following?
    a. The ERA should be part of the Constitution
    b. The passage of the Nineteenth Amendment was a mistake
    c. Occupational sex segregation benefits everyone
    d. NOMAS is a radical, sexist organization

Answers: 1: b; 2: c; 3: b; 4: b; 5: a; 6: b; 7: b; 8: c; 9: b; 10: a.

KEY TERMS

affirmative action 393
androgyny 396
boy code 385
comparable worth 373
cultural sexism 377
devaluation hypothesis 372
double or triple (multiple)
  jeopardy 359
emotion work 372
equal rights amendment 390

expressive roles 365
feminism 389
gender 359
gender role 359
gender tourism 383
glass ceiling 370
glass escalator effect 368
honor killing 388
human capital hypothesis 371
instrumental roles 365

motherhood penalty 370
occupational sex
  segregation 368
pink-collar jobs 368
sex 359
sexism 358
sexual harassment 392
structural sexism 366
structured choice 370
transgendered individual 359

MEDIA RESOURCES

Understanding Social Problems,
Seventh Edition Companion Website
www.cengage.com/sociology/mooney
Visit your book companion website, where you will find flash cards, practice quizzes, Internet links, and more to help you study.

CENGAGENOW™

Just what you need to know NOW! Spend time on what you need to master rather than on information you already have learned. Take a pretest for this chapter, and CengageNOW will generate a personalized study plan based on your results. The study plan will identify the topics you need to review and direct you to online resources to help you master those topics. You can then take a posttest to help you determine the concepts you have mastered and what you will need to work on. Try it out! Go to www.cengage.com/login to sign in with an access code or to purchase access to this product.

Aaron Harris/Canadian Press/AP Photo

11

"Homophobia alienates mothers and fathers from sons and daughters, friend from friend, neighbor from neighbor, Americans from one another. So long as it is legitimated by society, religion, and politics, homophobia will spawn hatred, contempt, and violence, and it will remain our last acceptable prejudice."

Byrne Fone

Sexual Orientation and the Struggle for Equality

The Global Context: A World View of the Status of Homosexuality | Homosexuality and Bisexuality in the United States: A Demographic Overview | The Origins of Sexual Orientation Diversity | Sociological Theories of Sexual Orientation Inequality | Heterosexism, Homophobia,

399

At the Homewood-Flossmoor High School in the suburbs of Chicago, Myka Held was involved in a campaign to promote gay tolerance by selling T-shirts that say, "gay? fine by me" and having students and teachers wear them on the same day at school. The T-shirt campaign was started by students at Duke University.

As a junior and senior at Homewood-Flossmoor High School in the suburbs of Chicago, Myka Held played a key role in leading a campaign to promote acceptance of gay and lesbian students. The campaign involved selling gay-friendly T-shirts to students and teachers and having as many people as possible wear the T-shirts to school on a designated day. The T-shirts say, "gay? fine by me." The shirts were purchased through a nonprofit organization (http://www.finebyme.org) that students at Duke University started. "I think it's really important for gay people out there to know that there are straight people who support them," Ms. Held said (quoted in Puccinelli 2005). "I have always supported equal rights for every person and have been disgusted by discrimination and prejudice. As a young Jewish woman, I believe it is my duty to stand up and support minority groups. . . . In my mind, fighting for gay rights is a proxy for fighting for every person's rights" (Held 2005).

sexual orientation The classification of individuals as heterosexual, bisexual, or homosexual, based on their emotional and sexual attractions, relationships, self-identity, and behavior.

heterosexuality The predominance of emotional and sexual attraction to individuals of the other sex.

homosexuality The predominance of emotional and sexual attraction to individuals of the same sex.

bisexuality Emotional and sexual attraction to members of both sexes.

As the opening vignette illustrates, fighting for sexual orientation equality is an issue not just for lesbians, gays, and bisexuals but also for all those who value fairness and respect for humans in all their diversity. It is beyond the scope of this chapter to explore how sexual diversity and its cultural meanings vary throughout the world. Rather, in this chapter, we focus on Western conceptions of diversity in sexual orientation. The term **sexual orientation** refers to the classification of individuals as heterosexual, bisexual, or homosexual, based on their emotional and sexual attractions, relationships, self-identity, and behavior. **Heterosexuality** refers to the predominance of emotional and sexual attraction to individuals of the other sex. **Homosexuality** refers to the predominance of emotional and sexual attraction to individuals of the same sex, and **bisexuality** is emotional and sexual attraction to members of both sexes. Much of the current literature on the treatment and political and social agendas of individuals who are gay, lesbian, and bisexual includes transgender individuals. As discussed in

Chapter 10, *transgender* individuals are people whose sense of gender identity (masculine or feminine) is inconsistent with their birth sex (male or female). Transgender individuals who have changed or are in the process of changing their physical sex to match their gender identity are known as *transsexuals*. The terms **LGBT** or **GLBT** are often used to refer collectively to individuals who are lesbian, gay, bisexual, and transgender.

After summarizing the legal status of lesbians and gay men around the world, we discuss the prevalence of homosexuality, heterosexuality, and bisexuality in the United States, discuss beliefs about the origins of sexual orientation diversity, and apply sociological theories to better understand societal reactions to sexual diversity. Then, after detailing the ways in which nonheterosexuals are victimized by prejudice and discrimination, we end the chapter with a discussion of strategies to reduce antigay prejudice and discrimination.

The Global Context: A World View of the Status of Homosexuality

Homosexual behavior has existed throughout human history and in most, perhaps all, human societies (Kirkpatrick 2000). A global perspective on laws and social attitudes regarding homosexuality reveals that countries vary tremendously in their treatment of same-sex sexual behavior—from intolerance and criminalization to acceptance and legal protection. At least 86 member states of the United Nations criminalize consensual same-sex behavior among adults (Ottosson 2008). In the majority of these countries, homosexuality among both males and females is illegal; in 37 of these countries, only homosexuality among males is illegal. Legal penalties vary for violating laws that prohibit homosexual sexual acts. In many countries, homosexuality is punishable by prison sentences and/or corporal punishment, such as whipping or lashing. In some countries, people found guilty of engaging in same-sex sexual behavior may receive the death penalty (see Table 11.1).

TABLE 11.1 Countries in Which Homosexual Acts Are Subject to the Death Penalty

- Iran
- Mauritania
- Saudi Arabia
- Sudan
- United Arab Emirates
- Yemen
- Nigeria

Source: Ottosson 2008.

What Do You Think? Why do you think that male homosexuality is illegal in many countries, but female homosexuality is not?

One factor that affects the status of homosexuality in a society is religion. In a survey of 129 countries, researchers found that, in countries where a majority of people report that religion is important in their daily lives, people are also likely to report that their society is *not* a good place for individuals who are gay and lesbian to live (Pelham & Crabtree 2009).

In general, countries throughout the world are moving toward increased legal protection of sexual orientation minorities, as discrimination on the basis of sexual orientation has become part of a broad international human rights agenda. In 45 countries throughout the world, laws prohibit employment discrimination based on sexual orientation. In six countries (including the United States), some parts of the country prohibit sexual orientation discrimination in

LGBT or GLBT Terms often used to refer collectively to lesbians, gays, bisexuals, and transgender individuals.

A global perspective on laws and social attitudes regarding homosexuality reveals that countries vary tremendously in their treatment of same-sex sexual behavior—from intolerance and criminalization to acceptance and legal protection.

the workplace (Ottosson 2008). In 1996, South Africa became the first country in the world to include in its constitution a clause banning discrimination based on sexual orientation. Canada, Ecuador, Fiji, Portugal, Sweden, and Switzerland also have constitutions that ban discrimination based on sexual orientation.

In 2008, 66 nations at the General Assembly of the United Nations signed a resolution decriminalizing homosexuality and confirming that international human rights protections include sexual orientation and gender identity. The resolution includes language prohibiting harassment, discrimination, exclusion, stigmatization, and prejudice against members of the GLBT population. G. W. Bush had refused to sign the resolution while he was in office. President Obama signed the resolution in 2009.

The growing legal recognition of same-sex relationships provides evidence of the changing status of homosexuality throughout the world. In 2001, the Netherlands became the first country in the world to offer full legal marriage to same-sex couples. In 2003, Belgium legalized same-sex marriages. In 2005, Spain, Canada, and South Africa became the third, fourth, and fifth countries, respectively, to legalize same-sex marriage.

Other countries recognize same-sex **registered partnerships** or "civil unions," which are federally recognized relationships that convey most but not all the rights of marriage (some countries also offer registered partnerships or civil unions to opposite-sex couples). Legally recognized registered partnerships or civil unions for same-sex couples are available in Australia, Belgium, Brazil, Canada, Denmark, Finland, France, Germany, Great Britain, Greenland, Hungary, Iceland, Israel, Italy, New Zealand, Norway, Portugal, Spain, Sweden, in Mexico City, and the northeastern state of Coahuila, Mexico. As we discuss later in this chapter, a number of U.S. states give legal recognition to same-sex couples.

Homosexuality and Bisexuality in the United States: A Demographic Overview

Before looking at demographic data concerning homosexuality and bisexuality in the United States, it is important to understand the ways in which identifying or classifying individuals as homosexual, gay, lesbian, and bisexual is problematic.

Sexual Orientation: Problems Associated with Identification and Classification

The classification of individuals into sexual orientation categories (e.g., gay, straight, bisexual, lesbian, homosexual, or heterosexual) is problematic for a number of reasons (Savin-Williams 2006). First, distinctions among sexual orientation categories are simply not as clear-cut as many people would believe. Consider the early research on sexual behavior by Kinsey and his colleagues

registered partnerships Federally recognized relationships that convey most but not all the rights of marriage.

(1948; 1953), who found that, although 37 percent of men and 13 percent of women had had at least one same-sex sexual experience since adolescence, few of the individuals reported exclusive homosexual behavior. These data led Kinsey to conclude that heterosexuality and homosexuality represent two ends of a sexual orientation continuum and that most individuals are neither entirely homosexual nor entirely heterosexual but fall somewhere along this continuum. In other words, most people are, to some degree, bisexual.

Sexual orientation classification is complicated by the fact that many people who are sexually attracted to or have had sexual relations with individuals of the same sex do not view themselves as homosexual or bisexual. Consider the findings of a national survey of U.S. adults that investigated (1) sexual attraction to individuals of the same sex, (2) sexual behavior with people of the same sex, and (3) homosexual self-identification (Michael et al. 1994). This survey found that 4 percent of women and 6 percent of men said that they are sexually attracted to individuals of the same sex, and 4 percent of women and 5 percent of men reported that they had had sexual relations with a same-sex partner after age 18. Yet less than 3 percent of the men and less than 2 percent of women in this study identified themselves as homosexual or bisexual (Michael et al. 1994). A more recent survey of more than 4,000 men in New York City found that 4 percent reported a gay identity but 12 percent reported same-sex sexual behavior in the past year; almost 10 percent who self-identified as straight and 10 percent of married men had at least one sexual encounter with another man during the previous year (Pathela et al. 2006).

> . . . Distinctions among sexual orientation categories are simply not as clear-cut as many people would believe.

Finally, because of the social stigma associated with nonheterosexual identities, many individuals conceal or falsely portray their sexual orientation identities to protect themselves against prejudice and discrimination.

The Prevalence of LGBT Individuals and Same-Sex Couple Households in the United States

In the national survey by Michael et al. (1994), less than 3 percent of the men and less than 2 percent of women in this study identified themselves as homosexual or bisexual. Smith and Gates (2001) estimated that between 4 percent and 5 percent of the total U.S. adult population is gay. A 2004 national poll found that about 5 percent of U.S. high school students identify as lesbian or gay (Curtis 2004). And a national study of U.S. college students found that 4.1 percent identified either as gay, lesbian, or bisexual (see Table 11.2). An estimated 1 million to 3 million Americans older than age 65 are gay, lesbian, bisexual, or transgendered—a number that is expected to double over the next quarter century (Cahill 2007).

According to U.S. Census data, there are an estimated 564,743 same-sex couple households in the United States (Gates 2009). By the end of 2008, approximately 35,000 U.S. same-sex couples were legally married. But many more same-sex couples view their relationship as a marriage, as nearly 150,000 same-sex couples identified one partner as "husband" or "wife" (Gates 2009). About one-fifth (22.3 percent) of gay male couple households and one-third (34.3 percent) of lesbian couple households have children present in the home (Simmons & O'Connell 2003).

TABLE 11.2 Sexual Identity of U.S. College Students*	
Heterosexual	94.0%
Gay/lesbian	1.6%
Bisexual	2.5%
Transgender	0.2%
Unsure	1.8%

Source: American College Health Association 2008.
*Based on a sample of 20,507 students at 39 campuses.

Residents of America's first gay and lesbian retirement community, Palms of Manasota, in Palmetto, Florida.

Palms of Manasota

Why are data on the numbers of U.S. adults and couples identifying as GLBT relevant? Primarily these data are important because census numbers on the prevalence of adults and couples identifying as GLBT can influence laws and policies that affect gay individuals and their families. In anticipation of the 2000 census, the Policy Institute of the National Gay and Lesbian Task Force and the Institute for Gay and Lesbian Strategic Studies conducted a public education campaign urging people to "out" themselves on the 2000 census. The slogan was, "The more we are counted, the more we count" (Bradford et al. 2002, p. 3). "The fact that the Census documents the actual presence of same-sex couples in nearly every state legislative and U.S. congressional district means antigay legislators can no longer assert that they have no gay and lesbian constituents" (Bradford et al. 2002, p. 8).

The Origins of Sexual Orientation Diversity

One of the prevailing questions regarding sexual orientation centers on its origin or "cause." Questions about the causes of sexual orientation are typically concerned with the origins of homosexuality and bisexuality. Because heterosexuality is considered normative and "natural," causes of heterosexuality are rarely considered.

In much of the biomedical and psychological research on sexual orientation, an attempt is made to identify one or more causes of sexual orientation diversity. The driving question behind this research is: Is sexual orientation inborn or is it learned or acquired from environmental influences? Although a number of factors have been correlated with sexual orientation, including genetic factors, gender role behavior in childhood, and fraternal birth order, there is no single theory that can explain diversity in sexual orientation.

Beliefs about What "Causes" Homosexuality

Aside from what "causes" homosexuality, sociologists are interested in what people *believe* about the "causes" of homosexuality. Most individuals who are gay believe that homosexuality is an inherited, inborn trait. In a national study of homosexual men, 90 percent believe that they were born with their homosexual orientation; only 4 percent believe that environmental factors are the sole cause (Lever 1994). The percentage of Americans who believe that homosexuality is something a person is born with increased from 19 percent in 1989, to 41 percent in 2008 (Gallup Organization 2009).

Individuals who believe that homosexuality is biologically based tend to be more accepting of homosexuality (see Figure 11.1). In contrast, "those who believe homosexuals choose their sexual orientation are far less tolerant of gays and lesbians and more likely to conclude homosexuality should be illegal than those who think sexual orientation is not a matter of personal choice" (Rosin & Morin 1999, p. 8).

Although a number of factors have been correlated with sexual orientation, including genetic factors, gender role behavior in childhood, and fraternal birth order, no single theory can explain diversity in sexual orientation.

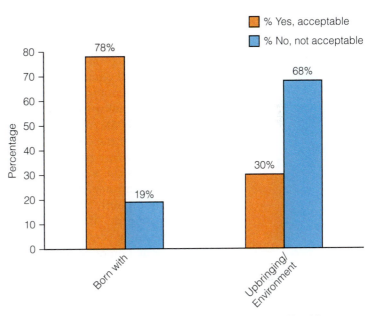

Figure 11.1 Should Homosexuality Be Considered an Acceptable Lifestyle? Based on Views about Origin of Homosexuality
Source: Saad 2007.

Can Homosexuals Change Their Sexual Orientation?

Those who believe that individuals who are homosexual choose their sexual orientation tend to think that these individuals can and should change their sexual orientation. Various forms of **reparative therapy** or **conversion therapy** are dedicated to changing the sexual orientation of individuals who are homosexual. Some religious organizations sponsor "ex-gay ministries," which claim to "cure" homosexual individuals and transform them into heterosexual individuals through prayer and other forms of "therapy." Most reparative therapy programs allegedly achieve "conversion" to heterosexuality through embracing evangelical Christianity and being "born again" (Cianciotto & Cahill 2007).

Consider the following examples:

- A 21-year-old gay man reported that his strict religious family, upon learning of his orientation, threatened to disown him and have him excommunicated unless he changed. He went into an ex-gay program voluntarily because he could not bear to lose his family, but other young men, he said, were forced into the program after being kidnapped. The program "counselors" strapped him to a chair with electrodes and sensors and showed him pictures of nude men, shocking him when he became aroused; they continued this treatment until he did not respond. He finally fled the program after being sexually abused by a male orderly.

reparative therapy or conversion therapy Various therapies that are aimed at changing homosexuals' sexual orientation.

- Sixteen-year-old Zach was enrolled by his parents in a Christian camplike program to change him into a heterosexual. The program, called Refuge, discourages homosexual behavior by imposing the following rules on its participants: no secular music, no more than 15 minutes per day behind a closed bathroom door, no contact with any practicing homosexual, no masturbation, and (no joke) no Calvin Klein underwear (Buhl 2005).

Critics of reparative therapy and ex-gay ministries take a different approach: "It is not gay men and lesbians who need to change . . . but negative attitudes and discrimination against gay people that need to be abolished" (Besen 2000, p. 7). The American Psychiatric Association, the American Psychological Association, the American Academy of Pediatrics, the American Counseling Association, the National Association of School Psychologists, the National Association of Social Workers, and the American Medical Association agree that homosexuality is not a mental disorder and needs no cure—that sexual orientation *cannot* be changed and that efforts to change sexual orientation do not work and may, in fact, be harmful (Human Rights Campaign 2000; Potok 2005). According to Lambda Legal (the nation's largest organization working for the civil rights of the LGBT population), ex-gay programs and conversion therapy practitioners can be held liable for the harm they cause to clients who can sue for malpractice, consumer fraud, false advertising, or child abuse and neglect laws for minors forced to attend an ex-gay program (Cianciotto & Cahill 2007).

Close scrutiny of reports of "successful" reparative therapy reveal that (1) many claims come from organizations with an ideological perspective on sexual orientation rather than from unbiased researchers, (2) the treatments and their outcomes are poorly documented, and (3) the length of time that clients are followed after treatment is too short for definitive claims to be made about treatment success (Human Rights Campaign 2000). Indeed, at least 13 ministries of Exodus International—the largest ex-gay ministry network that claims to include over 170 ex-gay programs in 17 countries—have closed because their directors reverted to homosexuality (Fone 2000).

Sociological Theories of Sexual Orientation Inequality

Sociological theories do not explain the origin or "cause" of sexual orientation diversity; rather, they help to explain societal reactions to homosexuality and bisexuality and ways in which sexual identities are socially constructed.

Structural-Functionalist Perspective

Structural functionalists, consistent with their emphasis on institutions and the functions they fulfill, emphasize the importance of monogamous heterosexual relationships for the reproduction, nurturance, and socialization of children. From a structural-functionalist perspective, homosexual relations, as well as heterosexual nonmarital relations, are "deviant" because they do not fulfill the main function of the family institution—producing and rearing children. Clearly, however, this argument is less salient in a society in which (1) other institutions,

Many same-sex couples rear children in a stable, nurturing environment, which negates the argument that homosexual relations do not fulfill the family institution's main function of producing and rearing children.

Douglas C. Pizac/ AP Photo

most notably schools, have supplemented the traditional functions of the family, (2) reducing (rather than increasing) population is a societal goal, and (3) same-sex couples can and do raise children.

Some structural functionalists argue that antagonisms between individuals who are heterosexual and homosexual disrupt the natural state, or equilibrium, of society. Durkheim (1993), however, recognized that deviation from society's norms can also be functional. Specifically, the gay rights movement has motivated many people to reexamine their treatment of sexual orientation minorities and has produced a sense of cohesion and solidarity among members of the gay population. Gay activism has also been instrumental in advocating HIV/AIDS prevention strategies and health services that benefit the society as a whole.

The structural-functionalist perspective is concerned with how changes in one part of society affect other aspects. For example, research has shown that the worldwide increase in legal and social support of sexual orientation equality has been influenced by three cultural changes: the rise of individualism, increasing gender equality, and the emergence of a global society in which nations are influenced by international pressures (Frank & McEneaney 1999).

According to Frank and McEneaney (1999), individualism "appears to loosen the tie between sex and procreation, allowing more personal modes of sexual expression" (p. 930). They add:

> Whereas once sex was approved strictly for the purpose of family reproduction, sex increasingly serves to pleasure individualized men and women in society. This shift has involved the casting off of many traditional regulations on sexual behavior, including prohibitions of male-male and female-female sex. (p. 936)

This child is protesting Proposition 8—the voter referendum that reversed the California Supreme Court ruling in 2008 that legalized same-sex marriage.

David McNew/Getty Images

Gender equality involves the breakdown of sharply differentiated sex roles, thereby supporting the varied expressions of male and female sexuality. Globalization permits the international community to influence individual nations. For example, when Zimbabwe president Robert Mugabe pursued antihomosexual policies in 1995, 70 U.S. Congress members as well as several international human rights organizations pressured Mugabe to halt his antihomosexual campaign.

The structural-functionalist perspective is also concerned with latent functions, or unintended consequences. A latent function of the gay rights movement is increased opposition to gay rights. For example, after California granted same-sex couples the right to be legally married, a fierce campaign was launched to pass Proposition 8—a voter referendum that, in 2008, amended the state constitution to prohibit same-sex marriage.

Conflict Perspective

The conflict perspective frames the gay rights movement and the opposition to it as a struggle over power, prestige, and economic resources. This struggle is largely over prestige, or social respect. Sexual orientation minorities want to be recognized as full and decent human beings who are deserving of all the rights and protections entitled to individuals who are heterosexual. A major achievement in gaining prestige for gay and lesbian individuals occurred in 1973, when the American Psychiatric Association (APA) removed homosexuality from its official list of mental disorders (Bayer 1987).

More recently, gay and lesbian individuals have been waging a political battle to win civil rights protections against employment discrimination and to be allowed to marry a same-sex partner. Conflict theory helps to explain why many business owners and corporate leaders support nondiscrimination policies: It is good for the "bottom line." The majority (85 percent) of Fortune 500 companies have included sexual orientation in their nondiscrimination policies, and employers are increasingly offering benefits to domestic partners of LGBT employees (Human Rights Campaign 2009a). Gay-friendly work policies help employers maintain a competitive edge in recruiting and maintaining a talented and productive workforce. Companies are also competing for the gay and lesbian dollar. Many GLBT as well as heterosexual consumers prefer to purchase products and services from businesses that provide workplace protections for GLBT employees. Recent trends toward increased social acceptance of homosexuality may, in part, reflect the corporate world's competition for the gay and lesbian consumer dollar.

Many business leaders also see an economic advantage to supporting same-sex marriage. In support of marriage equality in California, Levi Strauss & Co. stated that, "Ending marriage discrimination will improve businesses' ability to attract the best and the brightest to California and enhance California's reputation as a diverse, inclusive and innovative community, both of which are key factors to continued economic growth and prosperity in this state . . ." (quoted in Human Rights Campaign 2009a).

Symbolic Interactionist Perspective

Symbolic interactionism focuses on the meanings of heterosexuality, homosexuality, and bisexuality; how these meanings are socially constructed; and how they influence the social status, self-concepts, and well-being of nonheterosexual individuals. The meanings we associate with same-sex relations are learned from society—from family, peers, religion, and the media. The negative meanings associated with homosexuality are reflected in the current slang use of the phrase "That's so gay" or "You're so gay," which is meant to convey that something or someone is stupid or worthless (Kosciw & Diaz 2005).

The symbolic interactionist perspective also points to the effects of labeling on individuals. Once individuals become identified or labeled as lesbian, gay, or bisexual, that label tends to become their **master status.** In other words, the dominant heterosexual community tends to view "gay," "lesbian," and "bisexual" as the most socially significant statuses of individuals who are identified as such. Esterberg (1997) noted that, "unlike heterosexuals, who are defined by their family structures, communities, occupations, or other aspects of their lives, lesbians, gay men, and bisexuals are often defined primarily by what they do in bed" (p. 377).

Symbolic interactionism draws attention to how social interaction affects our self-concept, behavior, and well-being. When gay and lesbian individuals interact with people who express antigay attitudes, they may develop what is known as **internalized homophobia**—a sense of personal failure and self-hatred among lesbians and gay men resulting from social rejection and stigmatization. Internalized homophobia has been linked to increased risk for depression, substance abuse and addiction, anxiety, and suicidal thoughts (Gilman et al. 2001; Bobbe 2002). Family members have a powerful effect on the self-concepts, behavior, and well-being of lesbian, gay, and bisexual youth. A study of young adults who are lesbian, gay, and bisexual found that higher rates of family rejection during adolescence were significantly associated with poorer health outcomes (Ryan et al., 2009). Young LGBT adults who reported higher levels of family rejection were more likely to have attempted suicide, have high levels of depression, use illegal drugs, and engage in unprotected intercourse compared with peers from families that reported no or low levels of family rejection.

Labels also affect the attitudes of individuals who are heterosexual. When Gallup tested alternative terms for referring to gay Americans, it found that using the label *gays and lesbians* results in somewhat more favorable, pro-gay responses than does using the term *homosexual* (Saad 2005). The negative meanings associated with homosexuality are explored in the following section.

> **What Do You Think?** How might the experience of being gay, lesbian, bisexual, or transgendered be similar to the experience of being a racial or ethnic minority? How might it be different?

Heterosexism, Homophobia, and Biphobia

The United States, along with many other countries throughout the world, is predominantly heterosexist. **Heterosexism** refers to "the institutional and societal reinforcement of heterosexuality as the privileged and powerful norm" (SIECUS 2000).

master status The status that is considered the most significant in a person's social identity.

internalized homophobia A sense of personal failure and self-hatred among some lesbians and gay men due to social rejection and stigmatization.

heterosexism The institutional and societal reinforcement of heterosexuality as the privileged and powerful norm, based on the belief that heterosexuality is superior to homosexuality.

This questionnaire is designed to examine which of the following statements most closely describes your behavior during past encounters with people you thought were homosexual. Rate each of the following self-statements as honestly as possible by choosing the frequency that best describes your behavior: (1) never, (2) rarely, (3) occasionally, (4) frequently, or (5) always.

1. I have spread negative talk about someone because I suspected that he or she was gay. _____

2. I have participated in playing jokes on someone because I suspected that he or she was gay. _____

3. I have changed roommates and/or rooms because I suspected my roommate to be gay. _____

4. I have warned people who I thought were gay and who were a little too friendly with me to keep away from me. _____

5. I have attended antigay protests. _____

6. I have been rude to someone because I thought that he or she was gay. _____

7. I have changed seat locations because I suspected the person sitting next to me to be gay. _____

8. I have had to force myself to stop from hitting someone because he or she was gay and very near me. _____

9. When someone I thought to be gay has walked toward me as if to start a conversation, I have deliberately changed directions and walked away to avoid the person. _____

10. I have stared at a gay person in such a manner as to convey my disapproval of his or her being too close to me. _____

11. I have been with a group in which one (or more) person(s) yelled insulting comments to a gay person or group of gay people. _____

12. I have changed my normal behavior in a restroom because a person I believed to be gay was in there at the same time. _____

13. When a gay person has "checked" me out, I have verbally threatened him or her. _____

14. I have participated in damaging someone's property because the person was gay. _____

15. I have physically hit or pushed someone I thought was gay because the person brushed against me when passing by. _____

16. Within the past few months, I have told a joke that made fun of gay people. _____

17. I have gotten into a physical fight with a gay person because I thought he or she had been making moves on me. _____

18. I have refused to work on school and/or work projects with a partner I thought was gay. _____

19. I have written graffiti about gay people or homosexuality. _____

20. When a gay person has been near me, I have moved away to put more distance between us. _____

Scoring

The revised Self-Report of Behavior Scale is scored by totaling the number of points endorsed on all items (never, 1; rarely, 2; occasionally, 3; frequently, 4; always, 5), yielding a range from 20 to 100 total points. The higher the score, the more negative are the attitudes toward homosexuals.

Comparison Data

Sunita Patel (1989) originally developed the Self-Report of Behavior Scale in her psychology master's thesis research at East Carolina University. Shartra Sylivant (1992) and by Tristan Roderick (1994) revised the scale.

In the study of Roderick et al. (1998), the mean score for 182 college women was 24.76. The mean score for 84 men was significantly higher, 31.60. In Sylivant's (1992) high school sample, the mean scores were 33.74 for young women and 44.40 for young men.

The revised Self-Report of Behavior Scale is reprinted by the permission of the students and faculty who participated in its development: S. Patel, S. L. McCammon, T. E. Long, L. J. Allred, K. Wuensch, T. Roderick, and S. Sylivant.

Source: Patel et al. 1995.

Heterosexism is based on the belief that heterosexuality is superior to homosexuality; it results in prejudice and discrimination against homosexuals and bisexuals. Prejudice refers to negative attitudes, whereas discrimination refers to behavior that denies individuals or groups equality of treatment. Before reading further, you may wish to complete this chapter's Self and Society feature, which assesses your behaviors toward individuals you perceive to be homosexual.

Homophobia

The term **homophobia** is commonly used to refer to negative attitudes and emotions toward homosexuality and those who engage in it. Homophobia is not necessarily a clinical phobia (i.e., one involving a compelling desire to avoid the feared object despite recognizing that the fear is unreasonable). Other terms that refer to negative attitudes and emotions toward homosexuality include *homonegativity* and *antigay bias*.

As shown in Table 11.3, more than a third of U.S. adults have negative attitudes toward homosexuality, as they disagreed that "homosexuality is an acceptable alternative lifestyle." In general, individuals who are more likely to have negative attitudes toward homosexuality are older, are Republican, attend religious services regularly, are less educated, live in the South or Midwest, and reside in small rural towns (Loftus 2001; Curtis 2003; Page 2003; Saad 2007). Public opinion surveys also indicate that men are more likely than women to have negative attitudes toward gay individuals (Saad 2007). But many studies on attitudes toward homosexuality do not distinguish between attitudes toward gay men and attitudes toward lesbians. Research that has assessed attitudes toward male versus female homosexuality has found that heterosexual women and men hold similar attitudes toward lesbians, but that men are more negative toward gay men (Louderback & Whitley 1997; Price & Dalecki 1998).

What Do You Think? Some public opinion surveys, such as the one referred to in Table 11.3, use the term *homosexual lifestyle*. Do all homosexual individuals have the same lifestyle? Do all heterosexual individuals have the same lifestyle? What does the expression *homosexual lifestyle* imply?

TABLE 11.3 Homosexuality as an Acceptable Alternative Lifestyle

	YES (%)	NO (%)
Men	53	44
Women	61	35
18–34 years	75	23
35–54 years	58	39
55+ years	45	51
Republican	36	58
Independent	60	36
Democrat	72	27
Worship services		
Attend weekly	33	64
Attend nearly weekly/monthly	57	40
Attend less often/never	74	22

Source: Saad 2007.

homophobia Negative attitudes and emotions toward homosexuality and those who engage in same-sex sexual behavior.

Cultural Origins of Homophobia. Why do many Americans disapprove of homosexuality? Antigay bias has its roots in various aspects of U.S. culture.

Religion. Most Americans who view homosexuality as unacceptable say that they object on religious grounds (Rosin & Morin 1999). Indeed, conservative Christian ideology has been identified as the best predictor of homophobia (Plugge-Foust & Strickland 2000). Many religious leaders teach that homosexuality is sinful and prohibited by God. The Roman Catholic Church rejects all homosexual expression and resists any attempt to validate or sanction the homosexual orientation. Some fundamentalist churches have endorsed the death penalty for homosexual people and teach the view that AIDS is God's punishment for engaging in homosexual sex. An organization of Christian fundamentalists claimed that the destruction brought on by Hurricane Katrina in 2005 was God's judgment against New Orleans for holding the annual gay Southern Decadence party (Curtis 2005).

The Westboro Baptist Church (Topeka, Kansas), headed by the antigay Reverend Fred Phelps, maintains a website called godhatesfags.com. Members of this church have held antigay demonstrations near the funerals of people who have died from AIDS, carrying signs reading "Gays Deserve to Die." Phelps and his followers have also picketed funerals of American servicemen and servicewomen who have died in the Iraq war, saying that U.S. military casualties are God's way of punishing the United States for being nice to lesbians and gays (Intelligence Briefs 2006).

Theologians and religious scholars have different viewpoints on the Bible's position on homosexuality. Many scholars believe that key passages actually are denouncing orgies and prostitution—or in the case of the town of Sodom, inhospitality—and not homosexuality. Two Old Testament passages that do condemn homosexual acts are found amid a long list of religious prohibitions, including eating pork and wearing mixed fabrics—rules that have been abandoned by most contemporary Christians (Potok 2005).

Some religious groups, such as the Quakers and the United Church of Christ (UCC), are accepting of homosexuality, and other groups have made reforms toward increased acceptance of lesbians and gays. In the early 1970s, the UCC became the first major Christian church to ordain an openly gay minister, and in 2005, the UCC became the largest Christian denomination to endorse same-sex marriages. Some Episcopal priests perform "ceremonies of union" between same-sex couples; some Reform Jewish groups sponsor gay synagogues (Fone 2000). In June 2003, the Reverend V. Gene Robinson was elected bishop of the Episcopal diocese of New Hampshire, becoming the first openly gay bishop in the church's history. Although the official position of the United Methodist Church is one that condemns homosexuality, some Methodist ministers advocate acceptance of and equal rights for lesbians and gay men (Ontario Consultants on Religious Tolerance 1999). Acceptance of gays and lesbians is even found among Southern Baptists, one of the more conservative Christian religions. The Association of Welcoming and Affirming Baptists, organized in 1993, is a network of nearly 50 congregations and other church groups that advocate for the inclusion of LGBT individuals within the Baptist community of faith.

Concern about HIV and AIDS. Although most cases of HIV and AIDS worldwide are attributed to heterosexual transmission, the rates of HIV and AIDS in the United States are much higher among gay and bisexual men than among other groups. Lesbians, incidentally, have a very low risk for sexually transmitted HIV—a lower risk than heterosexual women. Nevertheless, many people associate HIV and AIDS with homosexuality and bisexuality. In 2007, 23 percent of U.S. adults agreed with the statement, "AIDS might be God's punishment for immoral sexual behavior," down from 43 percent in 1987 (Pew Research Center 2007). This association between AIDS and homosexuality has fueled antigay sentiments.

Rigid Gender Roles. Disapproval of homosexuality also stems from rigid gender roles. When Cooper Thompson (1995) was asked to give a guest presentation on male roles at a suburban high school, male students told him that the most humiliating put-down was being called a fag. The boys in this school gave Thompson the impression that they were expected to conform to rigid, narrow standards of masculinity to avoid being labeled as a fag.

From a conflict perspective, heterosexual men's subordination and devaluation of gay men reinforces gender inequality. "By devaluing gay men . . . heterosexual men devalue the feminine and anything associated with it" (Price & Dalecki 1998, pp. 155–156). Negative views toward lesbians also reinforce the patriarchal system of male dominance. Social disapproval of lesbians is a form of punishment for women who relinquish traditional female sexual and economic dependence on men. Not surprisingly, research findings suggest that individuals with traditional gender role attitudes tend to hold more negative views toward homosexuality (Louderback & Whitley 1997).

Myths and Negative Stereotypes. Prejudice toward homosexuals can also stem from some of the myths and negative stereotypes regarding homosexuality. One negative myth about homosexuals is that they are sexually promiscuous and lack "family values," such as monogamy and commitment to relationships. Although some homosexuals do engage in casual sex, as do some heterosexuals, many homosexual couples develop and maintain long-term committed relationships. Between 64 percent and 80 percent of lesbians report that they are in a committed relationship at any given time, and between 46 percent and 60 percent of gay men report being in a committed relationship (Cahill et al. 2002). According to research, about one-third of gay male couples are monogamous, and most non-monogamous gay couples are open with each other about their outside sexual activities (*The Advocate* 2002; LaSala 2004; Shernoff 2006). Although a majority of male couples are not sexually exclusive, they are emotionally monogamous—a concept one researcher coined as "monogamy of the heart" (LaSala 2005).

What Do You Think? According to research on the topic, nonmonogamy generally is more accepted in the gay male subculture than in the heterosexual society or in the lesbian subculture. Why do you think this is so? Do you think the higher acceptance of nonmonogamy among gay males is explained by their sexual orientation? Or their sex and gender?

Another myth is that homosexuals, as a group, are child molesters. In fact, 95 percent of all reported incidents of child sexual abuse are committed by heterosexual men (SIECUS 2000). Most often, the abuser is a father, stepfather, or heterosexual relative of the family. Research cited by Cahill and Jones (2003) indicated that a child's risk of being molested by a relative's heterosexual partner is more than 100 times greater than the risk of being molested by someone who is homosexual or bisexual. Furthermore, "when a man abuses a young girl, the problem is not heterosexuality. . . . Similarly, when a priest sexually abuses a boy or underage teen, the problem is not homosexuality. The problem is child abuse" (Cahill & Jones 2003, p. 1).

Biphobia

Just as the term *homophobia* is used to refer to negative attitudes toward gay men and lesbians, **biphobia** refers to "the parallel set of negative beliefs about and stigmatization of bisexuality and those identified as bisexual" (Paul 1996, p. 449). Both heterosexuals and homosexuals often reject bisexuals; thus, bisexuals experience "double discrimination."

Biphobia includes negative stereotyping of bisexuals, the exclusion of bisexuals from social and political organizations of lesbians and gay men, and fear and distrust of, as well as anger and hostility toward, people who identify themselves as bisexual (Firestein 1996). Individuals with negative attitudes toward bisexual individuals often believe that bisexuals are actually homosexuals who are in denial about their true sexual orientation or are trying to maintain heterosexual privilege (Israel & Mohr 2004). Bisexual individuals are sometimes viewed as heterosexuals who are looking for exotic sexual experiences. One negative stereotype that encourages biphobia is the belief that bisexuals are, by definition, nonmonogamous. However, many bisexual women and men prefer and have long-term committed monogamous relationships.

Discrimination Against Sexual Orientation Minorities

In June 2003, a Supreme Court decision in *Lawrence v. Texas* invalidated state laws that criminalize **sodomy**—oral and anal sexual acts. This historic decision overruled a 1986 Supreme Court case (*Bowers v. Hardwick*), which upheld a Georgia sodomy law as constitutional. The 2003 ruling, which found that sodomy laws were discriminatory and unconstitutional, removes the stigma and criminal branding that sodomy laws have long placed on GLBT individuals. Before this historic ruling was made, sodomy was illegal in 13 states (Alabama, Florida, Idaho, Kansas, Louisiana, Mississippi, Missouri, North Carolina, Oklahoma, South Carolina, Texas, Utah, and Virginia) and four states (Kansas, Missouri, Oklahoma, and Texas) targeted only same-sex acts. Penalties for engaging in sodomy ranged from a $200 fine to 20 years in prison. In states that criminalized both same- and opposite-sex sodomy, sodomy laws were usually not used against heterosexuals but were used primarily against gay men and lesbians (ACLU 1999).

Like other minority groups in U.S. society, homosexuals and bisexuals experience various forms of discrimination. Next, we look at sexual orientation discrimination in the workplace, in the military, in marriage and parenting, in violent expressions of hate, and in treatment by police.

Discrimination and Harassment in the Workplace

Most U.S. adults (89 percent) agree that homosexuals should have equal job opportunities (Gallup Organization 2009). Yet, as of Fall 2009, it was still legal in 29 states to fire, decline to hire or promote, or otherwise discriminate against

biphobia Negative attitudes and emotions toward bisexuality and people who identify as bisexual.

sodomy Oral and anal sexual acts.

employees because of their sexual orientation, and in 37 states, it was legal to fire someone for being transgender (National Gay and Lesbian Task Force 2009a).

As we discuss later, many workplaces have nondiscrimination policies that cover LGBT employees. But gay-friendly policies don't ensure friendly attitudes and behaviors from coworkers. In the largest-ever national poll of LGBT employees, just over half of LGBT workers "hide themselves at work"—they are either out to only a few people or to no one at all—because of the workplace environment (see Figure 11.2) (Fidas 2009). Nearly half of LGBT employees (48 percent) say that they hear coworkers express negative views concerning LGBT issues at least once in a while, and 61 percent report hearing coworkers tell jokes about LGBT people at least once in a while. More than one in five respondents in this study said they looked for a new job in the last year because of the uncomfortable working environment in their current job. A gay police officer who felt compelled to hide his sexual orientation from his coworkers describes his experience (Carney 2007):

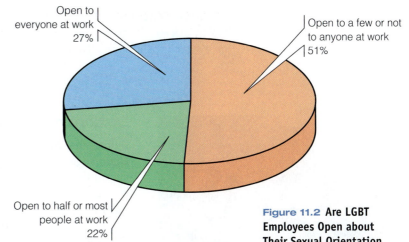

Figure 11.2 Are LGBT Employees Open about Their Sexual Orientation at Work?
Source: Fidas 2009.

> Can you imagine going to work every day and avoiding any conversations about with whom you had a date . . . or a great weekend . . . or an argument—basically not sharing any part of your personal life for fear of reprisal or being ostracized. I did this in a career that prides itself on integrity, honesty and professionalism—and where a bond with one's colleagues and partner is critical in dangerous and potentially deadly situations.

Discrimination in the Military

In 1993, President Clinton compromised on the issue of whether or not gays should be allowed to serve in the military by instituting a "Don't ask, don't tell" (DADT) policy. Although the DADT policy does not prohibit gays from serving in the military, it prohibits gays from revealing their sexual orientation (and it also prohibits recruiting officers from asking recruits about their sexual orientation). Service members who reveal their sexual orientation, or who are investigated for being suspected of being gay, can lose their jobs.

More than 12,000 service members have been discharged under DADT since it took effect, and countless others have decided not to join the military due to this discriminatory law. The "Don't ask, don't tell" policy also provides an additional means for servicemen to harass lesbian service members by threatening to "out" those who refused their advances or threatening to report them, thus ending their careers (Human Rights Watch 2001). A Government Accountability Office report found that the DADT policy cost nearly $200 million for the replacement and training of personnel who were discharged as a result of the policy (Human Rights Campaign 2005).

To commemorate the 14th anniversary of "Don't ask, don't tell," the Human Rights Campaign displayed 12,000 flags on the National Mall to pay tribute to the 12,000 men and women discharged from the military under DADT.

Discrimination in Marriage

Before the 2003 Massachusetts Supreme Court ruling in *Goodridge v. Department of Public Health,* no state had declared that same-sex couples have a constitutional right to be legally married. In response to growing efforts to secure legal recognition of same-sex couples, opponents of same-sex marriage have prompted antigay marriage legislation. In 1996, Congress passed and President Clinton signed the **Defense of Marriage Act (DOMA),** which (1) states that marriage is a "legal union between one man and one woman"; (2) denies federal recognition of same-sex marriage; and (3) allows states to either recognize or not recognize same-sex marriages performed in other states. As of this writing, 29 states had banned gay marriage through a constitutional amendment, and 11 states had laws banning same-sex marriage (Human Rights Campaign 2009b). Many of these states ban not only same-sex marriage, but also other forms of partner recognition, such as domestic partnerships and civil unions. Some states ban domestic partnerships for opposite-sex couples as well.

The Federal Marriage Amendment, also called the Marriage Protection Amendment, would amend the U.S. Constitution to define marriage as being a union between a man and a woman. It was first introduced in Congress in 2003, and has been defeated each time it has come up for a vote. Public opinion is evenly split on the issue of a constitutional amendment banning same-sex marriage: In 2008, 49 percent favored such an amendment, 48 percent opposed it (Gallup Organization 2009).

Arguments in Favor of Same-Sex Marriage. Advocates of same-sex marriage argue that banning same-sex marriages or refusing to recognize same-sex marriages granted in other states is a violation of civil rights that denies same-sex couples the many legal and financial benefits that are granted to heterosexual married couples. For example, married couples have the right to inherit from a spouse who dies without a will, to avoid inheritance taxes between spouses, to make crucial medical decisions for a partner and to take family leave to care for a partner in the event of the partner's critical injury or illness, to receive Social Security survivor benefits, and to include a partner in health insurance coverage. Other rights bestowed on married (or once-married) partners include assumption of a spouse's pension, bereavement leave, burial determination, domestic violence protection, reduced rate memberships, divorce protections (such as equitable division of assets and visitation of partner's children), automatic housing lease transfer, and immunity from testifying against a spouse. Finally, unlike 17 other countries that recognize same-sex couples for immigration purposes, the

Defense of Marriage Act Federal legislation that states that marriage is a legal union between one man and one woman and denies federal recognition of same-sex marriage.

United States does not recognize same-sex couples in granting immigration status because such couples are not considered "spouses."

Another argument for same-sex marriage is that it would promote relationship stability among gay and lesbian couples. "To the extent that marriage provides status, institutional support, and legitimacy, gay and lesbian couples, if allowed to marry, would likely experience greater relationship stability" (Amato 2004, p. 963).

Greater relationship stability benefits not only same-sex couples, but their children as well. Children in same-sex families would also benefit by legalizing same-sex marriage because they would be granted a range of securities and benefits, including the right to get health insurance coverage and Social Security survivor benefits from a nonbiological parent and the right to continue living with a nonbiological parent should their biological mother or father die (Tobias & Cahill 2003). Ironically, the same pro-marriage groups that stress that children are better off in married-couple families disregard the benefits of same-sex marriage to children.

There are also religious-based arguments in support of same-sex marriage. In a sermon titled "The Christian Case for Gay Marriage," Jack McKinney (2004) interprets Luke 4: "Jesus is saying that one of the most fundamental religious tasks is to stand with those who have been excluded and marginalized. . . . [Jesus] is determined to stand with them, to name them beloved of God, and to dedicate his life to seeing them empowered." McKinney goes on to ask, "Since when has it been immoral for two people to commit themselves to a relationship of mutual love and caring? No, the true immorality around gay marriage rests with the heterosexual majority that denies gays and lesbians more than 1,000 federal rights that come with marriage." As noted earlier, in 2005, the United Church of Christ became the largest Christian denomination to endorse same-sex marriages.

Finally, a cross-cultural and historical view of marriage and family suggests that marriage is a social construction that comes in many forms. In response to supporters of a constitutional amendment banning gay marriage as a threat to civilization, the American Anthropological Association released the following statement:

> The results of more than a century of anthropological research on households, kinship relationships, and families, across cultures and through time, provide no support whatsoever for the view that either civilization or viable social orders depend upon marriage as an exclusively heterosexual institution. Rather, anthropological research supports the conclusion that a vast array of family types, including families built upon same-sex partnerships, can contribute to stable and humane societies. (n.d.)

Arguments Against Same-Sex Marriage. Some opponents of same-sex marriage who view homosexuality as sick and/or immoral argue that granting legal status to same-sex unions would convey social acceptance of homosexuality and would teach youth to view homosexuality as an accepted "normal" lifestyle.

Opponents of same-sex marriage commonly argue that such marriages would subvert the stability and integrity of the heterosexual family. However, Sullivan (1997) suggested that homosexuals are already part of heterosexual families:

> [Homosexuals] are sons and daughters, brothers and sisters, even mothers and fathers, of heterosexuals. The distinction between "families" and "homosexuals" is, to begin with, empirically false; and the stability of existing families is closely linked to how homosexuals are treated within them. (p. 147)

Many opponents of same-sex marriage base their opposition on their religious views. In a Pew Research Center national poll, the majority of Catholics and Protestants oppose legalizing same-sex marriage, whereas the majority of secular respondents favor it (Green 2004). However, marriage is a *civil* option that does not require religious sanctioning; legalizing same-sex marriage does not take away churches' rights to refuse to perform marriage ceremonies for same-sex couples.

Finally, opponents of gay marriage point to public opinion polls that suggest that the majority of Americans are against same-sex marriage. A 2008 Gallup Poll found that more than half (56 percent) of Americans oppose same-sex marriage—down from 68 percent in 1996 (Gallup Organization 2009). However, a national survey of college freshmen found that two-thirds agree that "same-sex couples should have the right to legal marital status" (Pryor et al. 2008).

What Do You Think? Should public opinion polls on same-sex marriage determine government's legislative decisions regarding same-sex marriage? Why or why not?

Discrimination in Child Custody and Visitation

Several respected national organizations—including the Child Welfare League of America, the American Psychological Association, the American Psychiatric Association, and the National Association of Social Workers—have taken the position that a parent's sexual orientation is irrelevant in determining child custody (Landis 1999). In a review of research on family relationships of lesbians and gay men, Patterson (2001) concluded that "the greater majority of children with lesbian or gay parents grow up to identify themselves as heterosexual" and that "the home environments provided by lesbian and gay parents are just as likely as those provided by heterosexual parents to enable psychosocial growth among family members" (pp. 279, 283). Other scholars have concluded that:

> The sexual orientation of a parent does not negatively affect child adjustment. Children raised by lesbian mothers show similar adjustment (as compared to children raised by heterosexual parents) in terms of intellectual and behavioral functioning, psychological and socioemotional development, and attachment to family and friends. (Chamberlain et al. 2008, p. 111)

Children raised in lesbian families actually experience lower rates of physical and sexual abuse than the national norms (Chamberlain et al. 2008). Nevertheless, some court judges are biased against lesbian and gay parents in custody and visitation disputes. For example, in 1999, the Mississippi Supreme Court denied custody of a teenage boy to his gay father and instead awarded custody to his heterosexual mother who remarried into a home "wracked with domestic violence and excessive drinking" (Human Rights Campaign 2000b, p. 1).

Discrimination in Adoption and Foster Care

Just under half of U.S. adults (46 percent) think that same-sex couples should be able to adopt children (Gallup Organization 2009). There are two types of adoption: joint adoption in which a couple jointly adopts a child and second-parent adoption (or stepparent adoption) in which an individual adopts the child of his or her partner. Most adoptions by LGBT individuals are second-parent adoptions in which an individual adopts the biological child of his or her partner. Second-parent adoptions ensure that the children have the security and benefits of having two legal parents, which is especially important if the biological parent dies or becomes incapacitated.

Laws that govern adoption vary widely among the states, and in many cases, adoption decisions are made on a case-by-case basis. Same-sex couples are prohibited from adopting in Florida, Mississippi, and Utah, and all unmarried couples—gay or straight—are not allowed to adopt in Michigan and Arkansas. Nebraska, Utah, and Arkansas have policies prohibiting gays from being foster parents.

Hate Crimes Against Sexual Orientation Minorities

On October 6, 1998, Matthew Shepard, a 21-year-old student at the University of Wyoming, was abducted and brutally beaten. Two motorcyclists who had initially thought he was a scarecrow found him tied to a wooden ranch fence. His skull had been smashed, and his head and face had been slashed. The only apparent reason for the attack: Matthew Shepard was gay. On October 12, Shepard died of his injuries. Media coverage of his brutal attack and subsequent death focused nationwide attention on hate crimes against sexual orientation minorities.

Anti-LGBT hate crimes are crimes against individuals or their property that are based on bias against the victim because of perceived sexual orientation or gender identity. Such crimes include verbal threats and intimidation, vandalism, sexual assault and rape, physical assault, and murder.

According to the Federal Bureau of Investigation (FBI), 15.9 percent of hate crimes in 2007 were motivated by sexual orientation bias, involving 1,512 victims (FBI 2008). But as discussed in Chapter 9, FBI hate crime statistics underestimate the incidence of hate crimes. The National Coalition of Anti-Violence Programs (2008) documented 2,430 victims of antigay hate crimes in 2007, including 21 anti-LGBT murders. And because it is not uncommon for heterosexual men and women to be mistaken for gay men and lesbians, 6 percent of victims of anti-LGBT violence in 2007 identified as being heterosexual.

Antigay Hate and Harassment in Schools and on Campuses. In April 2009, 11-year-old Carl Joseph Walker-Hoover hanged himself with an extension cord in his family's home after being tormented by continuous antigay bullying and harassment at his middle school. Less than two weeks later, Jaheem Herrera, also 11 years old, hanged himself at home after being subjected to antigay taunts from his classmates. Neither Carl nor Jaheem identified as gay, yet they were targets of antigay harassment from their peers.

A national survey of students ages 13 through 18 and secondary schoolteachers found that actual or perceived sexual orientation is one of the most common reasons that students are harassed by their peers, second only to physical appearance (Harris Interactive & GLSEN 2005). The following other findings from this survey, published in a report titled *From Teasing to Torment: School Climate in America,* revealed the extent of antigay behavior in U.S. schools:

- Fifty-two percent of teens frequently hear students make homophobic remarks.
- Sixty-nine percent of teens frequently hear students say "that's so gay" or "you're so gay," expressions in which "gay" is meant to mean something bad or devalued.

- Ninety percent of LGBT teens (vs. 62 percent of non-LGBT teens) have been verbally or physically harassed or assaulted during the past year because of their perceived or actual appearance, gender, sexual orientation, gender expression, race/ethnicity, disability, or religion.

Another national survey of 1,732 LGBT students ages 13 to 20 found that:

- Nearly two-thirds (64 percent) reported feeling unsafe at school because of their sexual orientation.
- Nearly two-thirds (64 percent) reported that they had been verbally harassed at least some of the time in school in the past year because of their sexual orientation.
- Over one-third (38 percent) of students had experienced physical harassment, and nearly one-fifth (18 percent) had been physically assaulted at school because of their sexual orientation. (Kosciw & Diaz 2005)

Given the harsh treatment of LGBT youth in school settings, it is not surprising that 40 percent of gay youth report that their schoolwork is negatively affected by conflicts over their sexual orientation (GLSEN 2000). More than one-fourth of gay youth drop out of school—usually to escape the harassment, violence, and alienation they endure there (Chase 2000). Lesbian, gay, and bisexual youth who report high levels of victimization at school also have higher levels of substance use, suicidal thoughts, and sexual risk behaviors than heterosexual peers who report high levels of at-school victimization (Bontempo & D'Augelli 2002). A survey of youths' risk behavior, conducted by the Massachusetts Department of Education in 1999, found that 30 percent of gay teens attempted suicide in the previous year, compared with 7 percent of their straight peers (reported in Platt 2001).

Antigay hate is also common among college students (see this chapter's Social Problems Research Up Close feature). In 2007, more than one in ten hate crimes based on antigay bias occurred in schools or colleges (FBI 2008). In a survey of 484 young adults at six community colleges in California, 10 percent reported physically assaulting or threatening people whom they believed to be homosexual, and 24 percent reported calling homosexuals insulting names (Franklin 2000). From the findings of the study, the researcher concluded that some young adults believe that antigay harassment and violence is socially acceptable.

Police Mistreatment of Sexual Orientation Minorities

Most cases of antigay violence are *not* reported to the police (National Coalition of Anti-Violence Programs 2008). One reason that LGBT victims of hate crimes do not report these crimes to the police is the perception that the police will not be helpful and may even inflict further abuses. Among victims of antigay hate crimes who did report the crime to the police, about half of the victims described the police as "courteous"; the other half described police as either "indifferent," verbally abusive, and/or physically abusive (National Coalition of Anti-Violence Programs 2008). A report by Amnesty International (2005) found that police mistreatment and abuse of LGBT people are widespread in the United States. Types of police mistreatment revealed in the report include targeted and discriminatory enforcement of laws against LGBT people, verbal abuse, inappropriate pat-down and strip searches, failure to protect LGBT people in holding cells, inappropriate response or failure to respond to hate crimes or domestic violence calls, sexual harassment and abuse, and physical abuse. In one case, a

A survey by the Policy Institute of the National Gay and Lesbian Task Force (Rankin 2003) sheds light on the campus climate for GLBT students, staff, faculty, and administrators.

Sample and Methods

Because of the difficulty in identifying lesbian, gay, bisexual, and transgender individuals, Rankin (2003) used purposeful sampling of GLBT individuals and snowball sampling. Contacts were made with "out" GLBT individuals on campus, and they were asked to share the survey with other members of the GLBT community who were not so open about their sexual/gender identity.

Surveys, both paper and pencil and online versions, were administered to students, faculty, staff, and administrators at 14 colleges and universities from around the country. The surveys were designed to collect data about (1) respondents' personal campus experiences as members of the GLBT community, (2) their perception of the climate for GLBT members of the academic community, and (3) their perceptions of institutional actions, including administrative policies and academic initiatives regarding GLBT issues and concerns on campus. A total of 1,669 usable surveys were returned, representing the following groups:

- 1,000 students (undergraduate and graduate), 150 faculty, and 467 staff/administrators
- 720 men, 848 women
- 326 people of color
- 572 gay people (mostly male), 458 lesbians, 334 bisexuals, and 68 transgendered individuals
- 825 "closeted" people

Findings and Discussion

Some of the findings of the Campus Climate Assessment are discussed in the following sections.

Lived Oppressive Experiences

Nineteen percent of the respondents reported that they had feared for their physical safety within the past year because of their sexual orientation or gender identity, and 51 percent had concealed their sexual or gender identity to avoid intimidation. The study also found that 27 percent of faculty, staff, and administrators and 40 percent of students had concealed their sexual identity to avoid discrimination in the past year.

Respondents were also asked if they had experienced harassment in the past year. Harassment was defined as "conduct that has interfered unreasonably with your ability to work or learn on this campus or has created an offensive, hostile, intimidating working or learning environment." Undergraduate students were the most likely to have experienced harassment (36 percent), followed by faculty (27 percent), graduate students (23 percent), and staff (19 percent). Derogatory remarks were the most common form of harassment. Other types of harassment included verbal harassment or threats, anti-GLBT graffiti, threats of physical violence, denial of services, and physical assault.

Perceptions of Anti-GLBT Oppression on Campus

About three-fourths of faculty, students, administrators, and staff rated the campus climate as homophobic. In contrast, most respondents rated the campus generally (not specific to GLBT people) as friendly,

concerned, and respectful. Thus "even though respondents feel that the overall campus climate is hospitable, heterosexism and homophobia are still prevalent" (Rankin 2003, p. 31).

Institutional Actions

Less than half (37 percent) of participants agreed that "the college/university thoroughly addresses campus issues related to sexual orientation/gender identity." However, the majority of respondents agreed that their classrooms or their job sites were accepting of GLBT people (63 percent) and that the college or university provided resources on GLBT issues and concerns (71 percent).

These findings indicate that intolerance and harassment of GLBT students, staff, faculty, and administrators continue to be prevalent on U.S. campuses. It is important to note that the colleges and universities that participated in this study are not representative of most institutions of higher learning in the United States and "may be among the most gay-friendly campuses in the country" (Rankin 2003, p. 3). Rankin explains that "all of the institutions who participated in this survey had a visible GLBT presence on campus, including, in most cases, a GLBT campus center. Most had sexual orientation nondiscrimination policies. As only 100 of the 5,500 U.S. colleges and universities have GLBT student centers, the 14 universities surveyed here are not representative of most institutions of higher education in the U.S." (p. 3). Thus, Rankin suggests that the findings of this study "may significantly understate the problems facing GLBT students and staff at U.S. colleges and universities" (p. 3).

31-year-old gay African American man who was arrested by Chicago police after an altercation with his landlord alleged that officers handcuffed him and raped him with a billy club that had been dipped in cleaning fluid. The victim received an out-of-court settlement of $20,000.

Some police departments have taken initiatives to improve their practices regarding the treatment of LGBT individuals. For example, Washington, DC, created the Gay and Lesbian Liaison Unit, and some police departments provide training on hate crimes against LGBT individuals.

Effects of Antigay Bias and Discrimination on Heterosexuals

Lesbians, gays, bisexuals, and transgendered individuals are not the only victims of anti-LGBT prejudice and discrimination. In the following list, we explain how heterosexuals are also victims of anti-LGBT bias and discrimination.

1. *Restriction of gender expression.* Because of the antigay climate, heterosexuals, especially males, are hindered in their own self-expression and intimacy in same-sex relationships. "The threat of victimization (i.e., antigay violence) probably also causes many heterosexuals to conform to gender roles and to restrict their expressions of (nonsexual) physical affection for members of their own sex" (Garnets et al. 1990, p. 380). Fear of being labeled "gay" leads some youth to avoid activities that they might otherwise enjoy and benefit from (e.g., arts for boys and athletics for girls) (GLSEN 2000).

2. *Dysfunctional sexual behavior.* Some cases of rape and sexual assault are related to homophobia and compulsory heterosexuality. For example, college men who participate in gang rape, also known as "pulling train," entice each other into the act "by implying that those who do not participate are unmanly or homosexual" (Sanday 1995, p. 399). Homonegativity also encourages early sexual activity among adolescent men. Adolescent male virgins are often teased by their male peers, who say things like "You mean you don't do it with girls yet? What are you, a fag or something?" Not wanting to be labeled and stigmatized as a "fag," some adolescent boys "prove" their heterosexuality by having sex with girls.

3. *Loss of rights for individuals in unmarried relationships.* Social disapproval of same-sex relationships in the United States has contributed to the passage of state constitutional amendments that prohibit same-sex marriage, which can also result in denying rights and protections to opposite-sex unmarried couples. For example, some antigay marriage measures also threaten the provision of domestic partnership benefits to unmarried heterosexual couples.

4. *Heterosexual victims of hate crimes and harassment.* As we discussed earlier in this chapter, hate crimes are crimes of perception, meaning that victims of antigay hate crimes may not be homosexual; they may just be perceived as being homosexual. Recall that 6 percent of victims of antigay hate crimes in 2007 was heterosexual (National Coalition of Anti-Violence Programs 2008). And in a survey of students across the nation, 13 percent of heterosexual students reported that they have been called names, teased, bullied, or hurt at school because people thought they were gay (Harris Interactive & GLSEN 2005). Heterosexual youth with gay and lesbian family members are often taunted by their peers.

5. *Fear and grief.* Many heterosexual family members and friends of homosexuals live with the fear that their lesbian or gay friend or family member could be victimized by antigay prejudice and discrimination. They may also be afraid of being mistreated themselves simply for having a gay or lesbian friend or family member. And finally, imagine the grief that family and

friends of victims of antigay hate experience, such as the mother of Matthew Shepard.

6. *School shootings.* Antigay harassment has also been a factor in school shootings. In March 2001, 15-year-old Charles Andrew Williams fired more than 30 rounds in a San Diego suburban high school, killing 2 and injuring 13 others. A woman who knew Williams reported that the students had teased Williams and called him gay (Dozetos 2001). According to the Gay, Lesbian, and Straight Education Network (GLSEN), Williams's story is not unusual. Referring to a study of harassment of U.S. students, a GLSEN report concluded, "For boys, no other type of harassment provoked as strong a reaction on average; boys in this study would be less upset about physical abuse than they would be if someone called them gay" (Dozetos 2001).

7. *Loss of talented and dedicated professionals.* Employment discrimination against LGBT individuals deprives the public of vital services otherwise provided by talented and dedicated professionals. For example, the discriminatory military "Don't ask, don't tell" policy has resulted in the discharge of hundreds of specialists with critical military skills, including more than 50 linguists who specialize in Arabic (Human Rights Campaign 2005a). Or consider the case of Michael P. Carney, a gay police officer who kept his sexual orientation hidden from his coworkers and resigned due to the stress of working in a homophobic work environment. During his years as a police officer, Carney (1) helped save a man who jumped from a bridge in a suicide attempt; (2) was recognized for a youth mentorship program he cofounded; and (3) received a letter of commendation from the Police Commission for outstanding police work in capturing a bank robber. After Carney publicly revealed his sexual orientation, he tried to be reinstated in the police force, but was repeatedly denied, despite his remarkable record. Although Carney won his job back in an employment discrimination suit, for the two and a half years he fought the case, he could not help save lives, mentor youth, or catch bank robbers (Carney 2007).

Michael P. Carney (left) spent two and a half years in an employment discrimination legal battle to be reinstated as a police officer—two and a half years that he could not serve in the police force.

Strategies for Action: Reducing Antigay Prejudice and Discrimination

As shown in Table 11.4, over the past few decades, attitudes toward homosexuality in the United States have become more accepting, and support for protecting civil rights of gays and lesbians has increased (Gallup Organization 2009). Many of the efforts to change policies and attitudes regarding sexual orientation minorities have been spearheaded by organizations that specifically advocate for GLBT rights, such as the Human Rights Campaign (HRC), the National Gay and Lesbian Task Force (NGLTF), the Gay and Lesbian Alliance Against Defamation (GLAAD), the International Lesbian and Gay Association (ILGA), the International Gay and Lesbian Human Rights Commission, Amnesty International, the Lambda Legal Defense and Education Fund, the National Center for Lesbian Rights, and GLSEN. But the effort to achieve sexual orientation equality is not just a "gay agenda"; it is a human rights agenda that many heterosexuals and mainstream organizations support. For example, in 2007, the United Nations

TABLE 11.4 Changes in Attitudes Toward Homosexuality and Gay Rights

	1982	1996	2008
Percentage of U.S. adults who agree that			
homosexuality should be considered an acceptable lifestyle	34	44	57
homosexuals should have equal rights in terms of job opportunities	59	84	89
marriages between same-sex couples should be recognized by the law as valid, with the same rights as traditional marriages		27	40

Source: Gallup Organization 2009.

Humans Rights Council launched the *Yogyakarta Principles on the Application of International Law in Relation to Issues of Sexual Orientation and Gender Identity,* proposing international legal standards under which governments should end violence, abuse, and discrimination against lesbian, gay, bisexual, and transgender people and ensure full equality.

Next, we look the important roles that "coming out," media, and the political process play in the struggle for sexual orientation equality. Then we discuss efforts to reduce employment discrimination against sexual orientation minorities, provide recognition and support to lesbian and gay families, and include sexual orientation in hate crime legislation. We also provide an overview of educational policies and programs designed to increase acceptance and support of LGBT students in schools and on campuses.

The Role of "Coming Out" in the Struggle for Equality

The term **"coming out"** refers to the process of LGBT individuals accepting and being open with others about their sexual orientation identity. Coming out may involve revealing one's sexual orientation to family, friends, and coworkers through verbal disclosure and/or through behavior (such as displaying photos of one's same-sex partner or bringing one's partner to family or business functions).

> ...Coming out not only is an important step in the lives of LGBT individuals, but also a critical component of the gay rights movement.

National Coming Out Day (October 11) is recognized in many countries as a day to raise awareness of the LGBT population and foster discussion of gay rights issues. National Coming Out Day signifies the recognition that coming out is not only an important step in the lives of LGBT individuals, but also a critical component of the gay rights movement. As more openly gay and lesbian Americans come out to their family, friends, and coworkers, heterosexuals have more personal contact with LGBT individuals. A national poll of U.S. adults found that 4 in 10 Americans have close friends or relatives who are gay (Neidorf & Morin 2007). In another poll, more than half of the respondents (56 percent) said they have a friend or acquaintance who is gay or lesbian, nearly one-third (32 percent) said they work with someone who is gay or lesbian, and nearly one-fourth (23 percent) said that someone in their family is gay or lesbian (Newport 2002). A national poll of U.S. high school students found that 16 percent of students have a gay or lesbian person in their family, and 72 percent know someone who is gay or lesbian (Curtis 2004).

coming out The process of LGBT individuals accepting and being open with others about their sexual orientation identity.

National Coming Out Day Celebrated on October 11, this day is recognized in many countries as a day to raise awareness of the LGBT population and foster discussion of gay rights issues.

Psychologist Gordon Allport (1954) asserted that contact between groups is necessary for the reduction of prejudice—an idea known as the **contact hypothesis**. In general, heterosexuals have more favorable attitudes toward gay men and lesbian women if they have had prior contact with or know someone who is gay or lesbian (Mohipp & Morry 2004). Contact with openly gay individuals reduces negative stereotypes and ignorance and increases support for gay and lesbian equality (Wilcox & Wolpert 2000).

Harvey Milk, who in 1977 became the first openly gay person to be elected to public office as a member of the San Francisco Board of Supervisors, believed that the only way for homosexuals to combat homophobia was for lesbians and gays to make themselves visible—to come out of the closet:

> Gay people, we will not win our rights by staying quietly in our closets. . . . We are coming out to fight the lies, the myths, the distortions! We are coming out to tell the truth about gays! (quoted in Shilts 1982, p. 365)

Harvey Milk, the first openly gay person to be elected to public office in the United States, encouraged gays to come out of the closet.

When GLBT individuals come out, they risk rejection from friends, coworkers, and family. When Reverend Mel White, a closeted gay Christian man finally came out to his mother, her response was, "I'd rather see you at the bottom of that swimming pool, drowned, than to hear this" (White 2005, p. 28). Coming out can also have positive outcomes in terms of strengthening relationships (see this chapter's Human Side feature).

Gays and Lesbians in the Media

Positive depictions of gays and lesbians in the popular media also play a role in the struggle for sexual orientation equality. The 2008 movie *Milk*, starring Sean Penn, revived an important piece of history in the gay rights movement: the election of the first openly gay public official who encouraged gay individuals to come out of the closet and who helped pass landmark legislation banning antigay employment discrimination in California. Prior to the release of *Milk*, many Americans, including those active in the gay rights movement, had little or no awareness of Harvey Milk and his pivotal role in the gay rights movement. As revealed by many web blogs, seeing the movie *Milk* inspired some closeted individuals to come out and reinvigorated many gay rights activists.

In 1998, Ellen DeGeneres came out on her sitcom *Ellen,* and by 2010, many television viewers had seen gay and lesbian characters in television shows such as *Buffy the Vampire Slayer, Queer Eye for the Straight Guy, Six Feet Under, Will & Grace*, and *The L Word.* "Honest, nonstereotyped and diverse portrayals of gays and lesbians in prime time can offer youth a realistic representation of the gay community . . . [and] can offer positive role models for gay and lesbian youth" (Miller et al. 2002, p. 21).

Actor Sean Penn won an Oscar for best actor for his portrayal of Harvey Milk in the 2008 movie *Milk*.

One study found that college students reported lower levels of antigay prejudice after watching television shows with prominent homosexual characters (e.g., *Six Feet Under* and *Queer Eye for the Straight Guy*) (Schiappa et al. 2005). The researchers propose that, just as Allport's contact hypothesis suggests that intergroup contact may reduce prejudice, the "parasocial contact hypothesis" suggests that contact with individuals through the media has similar effects.

contact hypothesis The idea that contact between groups is necessary for the reduction of prejudice.

The following is a letter from a mother to her son, after her son came out to her and disclosed his homosexuality:

Being your mother has been an exquisite joy and a delight, well, for 99 percent of the time—as has being the mother of your siblings, but you, my son were somebody different.

I remember that even as a very small child we described you as being the one who marched to the tune of the different drum. . . .

Of course, we are a very normal family, whatever the definition of normal family might be, and we had our fights and arguments, and yelling sessions followed by slamming of doors, and the accusation that I was a pretty horrible sort of mother who never did nor ever would understand whatever it was that was the center of the disagreement. . . .

During your adolescence you became rather withdrawn, and would lock yourself away in your bedroom after you had returned home from school. . . . You did your homework, ate your dinner, and spent hours on the phone with your friend Steven. I was so worried about you. . . . Steven seemed to be the only friend you had made, and he was a very solitary introverted boy . . . precisely the kind of person I saw you becoming. . . .

All of these were among the images which I had on the October evening in 1992, when you felt that you could no longer hide your different sexuality from us.

That night is deeply etched in my mind, so deeply engraved that if I close my eyes now, I can see the events unfolding with such clarity. . . . I see you crying, I see you hugging yourself coiled into an almost fetal position, your father discarding his newspaper to hug you. . . . I know that I was crying. I know that I felt an almost overwhelming sense of relief that my suspicions were confirmed, and that I had answers for the many questions which had been plaguing me for a long, long time.

It was my sense of bewilderment and inadequacy that I found to be the most difficult with which to deal. I was the Mother, the one who always had the answers to your questions, the one to whom you had turned when you were sick or in pain, the one who had always "Kissed it better." . . . And I had failed you—for you had to battle all the hostilities vented against you during your school days all by yourself with no family support, and I found that very hard to accept. But basically, I who should have known so much actually knew very little about being gay, and the little which I did know was a mishmash of myths and stereotypes and misconceptions. . . . I certainly never thought a great deal about differing sexual orientations, nor did it ever occur to me that none of us sets out to design our sexuality, we simply are who we are. . . .

I think that in many ways our positions became reversed after your coming out. I took all my questions to you, and not only did you provide me with answers, but you gave me books to read and, most importantly, you introduced me to your friends.

For these things, I thank you from the bottom of my heart. Actually, there is much for which to thank you. For your honesty and your trust in us, your family, I am truly grateful. For leading me to PFLAG (Parents, Families, and Friends of Lesbians and Gays), well that has been a life changing experience. . . . Your coming out was the beginning of a different phase in both our lives, we became very open and I suppose unafraid with each other. . . .

And after all these years I have to agree with you that "it isn't fair!"—completely unfair that people like yourself, fine good people must face discrimination from people who do not know you and to whom you have done no harm, from religious leaders who preach a gospel of love, but who practice something entirely different, from politicians who rely on your taxes and contributions for the running of this country and who refuse to give legal status to long-term same-sex relationships; from people who call you sick and disgusting and who condemn you for pursuing your so-called lifestyle, who insist that this lifestyle should be lived out in complete silence and virtual invisibility. . . . It's now my turn to get a bit red in the face, stamp my foot, and yell IT SIMPLY IS NOT FAIR.

But now you have met your soulmate, the someone special to whom you have chosen to publicly commit your life. In the rings which you and Tim exchanged is an engraving which states the way in which your lives are now complete because of the love which you share. You are happy in the way I have always hoped and dreamed you would be, there is a serenity in your life, that I, ever the romantic, believe comes from loving and being loved in return, and because of this happiness in your life I, too, am happy, contented, and complete.

With all my love to you and to my new son, Tim.

Your mother, Nanette R. McGregor. Australia

Source: Adapted from McGregor n.d.

Reducing Employment Discrimination Against Sexual Orientation Minorities

Most Americans are in favor of equal rights for homosexuals in the workplace. Yet, as of this writing, federal law protects individuals from discrimination only on the basis of race, religion, national origin, sex, age, and disability. A bill called the **Employment Nondiscrimination Act (ENDA),** introduced in Congress in 1994, would make it illegal to discriminate in the workplace on the basis of sexual orientation. ENDA would not apply to religious organizations, businesses with fewer than 15 employees, or the military. ENDA failed to pass the Senate when it came up for a vote in 1996. In 2007, ENDA was amended to prohibit employment discrimination on the basis of gender expression and gender identity—in effect, expanding employment protections to transgender individuals. But, to get enough votes to pass the bill, these expanded protections were removed before ENDA came up for a vote in the House, which passed the bill in November 2007. Many advocates for equal rights were disappointed that the final ENDA bill did not include protections for all transgendered individuals. Others argue that the passing of the bill in the House was an important step forward and may pave the way for future progress toward protecting all transgendered individuals from employment discrimination. At the time of this writing, a transgender-inclusive version of ENDA is pending in Congress. With the support of President Obama, ENDA is expected to become law.

Regarding sexual orientation discrimination in the military, President Obama has pledged to end the "Don't ask, don't tell" policy. The Military Readiness Enhancement Act, if passed, would repeal DADT and replace it with a statute banning discrimination on the basis of sexual orientation. This legislation would also allow those discharged under "Don't ask, don't tell" to seek reentry into the military. Most European countries, including Canada, Australia, France, Germany, Spain, Switzerland, and Denmark, have lifted their bans on gay individuals in the military. Great Britain actively encourages gay individuals to enlist (Lyall 2005). In a national poll, the majority of U.S. adults (59 percent) favor allowing gays and lesbians to serve openly in the military (Pew Research Center for People & The Press 2009). Although some prominent retired military generals and admirals have supported a repeal of "Don't ask, don't tell," about 1,000 retired officers sent a letter to Obama saying they were "greatly concerned" about the impact a repeal could have on recruitment, morale, and unit cohesion (Bumiller 2009).

Lieutenant Dan Choi was discharged after he publicly revealed that he is gay on the *Rachel Maddow Show*. As part of an online campaign to repeal DADT, Choi explained:

> As an infantry officer, an Iraq combat veteran and a West Point graduate with a degree in Arabic, I refuse to lie to my commanders. I refuse to lie to my peers. I refuse to lie to my subordinates. As a result, the Army sent a letter discharging me. . . . I have served for a decade under "Don't Ask, Don't Tell"—an immoral policy that forces American soldiers to lie about their sexual orientation. Worse, it forces others to tolerate deception. As I learned at West Point, deception and lies poison a unit and cripple a fighting force. . . . In the ten years since I first raised my right hand at the United States Military Academy at West Point and committed to fighting for my country, I have learned many lessons. Courage, integrity, honesty, and selfless service are some of the most important. That's why my discharge from the Army is so painful. I am not accustomed to begging, but I am begging President Obama today: Do not fire me. My subordinates know I'm gay. They don't care. They are professionals. My soldiers are more than a unit or a fighting force—we are a family and we support each other. (Courage Campaign n.d.)

Employment Nondiscrimination Act (ENDA) A bill that would make it illegal to discriminate based on sexual orientation.

Lieutenant Dan Choi was discharged from the army after he publicly disclosed that he is gay on the *Rachel Maddow Show*. Despite his appeal and petition signed by almost 162,000 people, in June 2009, a panel of New York National Guard officers recommended that Choi be discharged. The final decision is pending, as of this writing.

Local and State Bans on Antigay Employment Discrimination. With the absence of federal legislation prohibiting antigay discrimination, some state and local governments have taken measures to prohibit employment discrimination based on sexual orientation. The scope of these measures varies, from prohibiting discrimination only in public employment to comprehensive protection against discrimination in public and private employment, education, housing, public accommodations, credit, and union practices.

In 1974, Minneapolis became the first municipality to ban antigay job discrimination. By 2008, 181 cities and counties prohibited sexual orientation discrimination in the private workplace (Human Rights Campaign 2009a). In 1982, Wisconsin became the first state to prohibit antigay employment discrimination. By 2009, 21 states (and the District of Columbia) had laws banning sexual orientation discrimination in the workplace (National Gay and Lesbian Task Force 2009a).

Nondiscrimination Policies in the Workplace. In 1975, AT&T became the first employer to add sexual orientation to its nondiscrimination policy. By 2008, 85 percent of Fortune 500 companies included sexual orientation in their equal employment opportunity or nondiscrimination policies.

Effects of Nondiscrimination Laws and Policies. A survey of 126 U.S. cities and counties that had implemented laws or policies prohibiting discrimination on the basis of sexual orientation found that such laws and policies sometimes had negative effects and "resulted in divisions in the community, greater controversy or tension, or the increased mobilization of those opposed to gay rights" (Button et al. 1997, p. 120). However, the more commonly cited effects were positive and included the following:

- Reduced discrimination against lesbians and gays in public employment and other institutions covered by the law
- Increased feelings of security, comfort, and acceptance by lesbians and gay men, resulting in lesbians and gay men being more likely to "come out of the closet"
- An increased sense of legitimacy among gays and lesbians that helped them overcome "internalized homophobia"

What Do You Think? Do you think that social acceptance of homosexuality leads to the creation of laws that protect lesbians and gays? Or does the enactment of laws that protect lesbians and gays help to create more social acceptance of gays?

Providing Legal Recognition and Support to Lesbian and Gay Couples

In 2004, Massachusetts became the first state to allow same-sex couples to be legally married. As of this writing, five states permit same-couples to marry. As noted earlier, the 1996 Defense of Marriage Act (DOMA), which was passed *before* same-sex marriage was legal in any U.S. states, denies federal recognition of same-sex marriage and allows states to either recognize or not recognize same-sex marriages performed in other states. Now that several states have legalized same-sex marriage, there are 35,000 legally married same-sex couples in the United States (as of 2009) who are denied federal benefits such as Social Security survivor's benefits and medical leave to care for an ill or injured spouse. In 2009, the Respect for Marriage Act (RMA) was introduced into Congress that, if passed, would overturn DOMA and grant federal recognition to same-sex marriages, regardless of the state laws in which they reside. The bill would *not* require states to recognize same-sex marriages performed in other states.

Several other states allow same-sex couples legal status that entitles them to many of the same rights and responsibilities as married heterosexual couples (see Table 11.5). For example, in some states, same-sex couples can apply for a civil union license. A **civil union** is a legal status parallel to civil marriage under state law, and entitles same-sex couples to almost all of the rights and responsibilities available under state law to married couples. Federal law does not recognize the rights of partners in same-sex civil unions, so they do not have the more than 1,000 federal protections that go along with civil marriage, nor is their legal status recognized in all other states.

Some states, counties, cities, and workplaces allow unmarried couples, including gay couples, to register as **domestic partners** (or "reciprocal beneficiaries" in Hawaii). The rights and responsibilities granted to domestic partners vary from place to place but may include coverage under a partner's health and pension plan, rights of inheritance and community property, tax benefits, access to married student housing, child custody and child and spousal support obligations, and mutual responsibility for debts.

TABLE 11.5 Legal Relationship Recognition for U.S. Same-Sex Couples

States with Full Marriage Equality*

Massachusetts (2004); Connecticut (2008); Iowa (2009); Vermont (2009); New Hampshire (2009)

States with Broad Relationship Recognition Laws**

Civil Unions: Vermont*** (2000); New Jersey (2006); New Hampshire (2007)
Domestic Partnerships: California (2005); Oregon (2007); Washington (2008); District of Columbia (2008); Nevada (2009)

States That Recognize Same-Sex Marriages Performed in Other States

New York (2008); Washington, DC (2009)

States with Limited Relationship Recognition Laws

Designated Beneficiaries: Colorado (2009)
Domestic Partnerships: Maryland (2008); Wisconsin (2009)
Reciprocal Beneficiaries: Hawaii (1997)

Source: National Gay and Lesbian Task Force 2009b.
*Same-sex marriage in Maine, originally scheduled to take effect in September 2009, was overturned by a voter referendum in November 2009.
**A broad relationship recognition law extends to same-sex couples nearly all the rights and responsibilities extended to married couples under state law. Other relationship recognition laws offer more limited rights and protections.
***In 2009, when the Vermont Legislature passed marriage recognition, it changed the civil union law so that new civil unions cannot be created but civil unions already in existence are still recognized.
****Date in parentheses refer to the date the state passed the legislation.

civil union A legal status that entitles same-sex couples who apply for and receive a civil union certificate to nearly all of the benefits available to married couples.

domestic partners A status granted to unmarried couples, including gay and lesbian couples, by some states, counties, cities, and workplaces that conveys various rights and responsibilities.

**"Same-sex marriage is nothing new.
We've been having the same sex for 25 years."**

Copyright 2004 by Randy Glasbergen, www.glasbergen.com

In 1991, the Lotus Development Corporation became the first major American firm to extend domestic partner recognition to gay and lesbian employees. More than half (57 percent) of Fortune 500 companies today offer same-sex domestic partner health benefits (Human Rights Campaign 2009a). Twenty states offer benefits to the domestic partners of state employees, and the Domestic Partnership Benefits and Obligations Act, currently pending in Congress, would provide domestic partnership benefits for federal employees and their partners (Badgett 2009). However, domestic partner benefits are usually taxed as income by the federal government, whereas spousal benefits are not.

Protecting Gay and Lesbian Parental Rights

One in three lesbian women has given birth, and one in six gay men has fathered or adopted a child (Gates et al. 2007). A number of policies and rulings reflect the increasing provision of parental rights and responsibilities to gay and lesbian parents and co-parents:

- A Massachusetts court ruled that two women may be listed as "mother" on a birth certificate when one of the women donated the egg for the child and the other carried the child (LAWbriefs 2000).
- A Pennsylvania court ordered a nonbiological, nonadoptive parent to pay child support for the five children she jointly parented with her former partner (LAWbriefs 2003).
- In 2005, the Supreme Court of Pennsylvania overturned a lower court decision that found it to be in the best interest of a child not to have any contact with her nonbiological lesbian mother because the child's biological mother had alienated the child from her. The court noted that a heterosexual parent who alienated the child would not be rewarded for such behavior and that "this scenario is equally applicable" (LAWbriefs 2005a).
- The West Virginia Supreme Court awarded a woman custody of the child she and her deceased lesbian partner raised together. The deceased partner's parents had fought for custody of the child, and, after being denied by a trial court, an appellate court reversed the decision and awarded custody to the grandparents. The Supreme Court overruled the appellate court decision, recognizing the parental rights of the woman who parented the child for his entire life (LAWbriefs 2005b).
- In Tennessee, Christy Berry had primary custody of her young son after her divorce from the child's father. The father sought a change in custody, claiming that Ms. Berry's sexual orientation would harm their child. A trial court granted the father's request for custody, but an appeals court reversed the decision, noting that the boy was doing well with his mother and that there was no evidence her sexual orientation had any adverse effect on him (LAWbriefs 2005b).
- In 2001, Guadalupe Benitez—a lesbian woman who wanted to have a child and raise it with her partner—sued two doctors who refused to artificially inseminate her for religious reasons. Benitez claims that the doctors violated California's antidiscrimination laws that protect gays and lesbians. In

2008, the California Supreme Court ruled that doctors cannot use their religious beliefs as a reason for discriminating against same-sex couples.

Antigay Hate Crimes Legislation

Hate crime laws call for tougher sentencing when prosecutors can prove that the crime committed was a hate crime. As of June 2009, 31 states and the District of Columbia had hate crime laws that include sexual orientation, 14 states had hate crime laws that did not include sexual orientation, and 5 states had no hate crime laws (Human Rights Campaign 2009c).

The 1969 federal hate crimes law covers hate crimes based on race, religion, and national origin. For years, gay rights advocates lobbied Congress to add federal hate crime law protections based on sexual orientation, gender, gender identity, and disability. In October 2009, President Obama signed into law the Matthew Shepard and James Byrd, Jr. Hate Crimes Prevention Act (HCPA). This new law expands federal hate crimes protection to cover hate crimes based on actual or perceived sexual orientation, gender, gender identity, and disability. The new law was named after Mathew Shepard, a gay Wyoming teenager who died after being severely beaten and James Byrd Jr., an African American man who was attacked, chained to a vehicle, and dragged to his death in Texas.

Educational Strategies: Policies and Programs in the Public Schools

If schools are to promote the health and well-being of all students, they must address the needs and promote acceptance of gay, lesbian, and bisexual youth. Three strategies for attaining these goals are to (1) include sexual orientation diversity in sex education programs, (2) create learning environments that promote tolerance of sexual orientation diversity, and (3) establish gay-straight alliances in middle and high schools.

Sex Education. As of 2007, only one state (New Jersey) and the District of Columbia mandated that homosexuality be discussed in public school sex education classes; one state (Louisiana) banned teachers from discussing homosexuality in the classroom (SIECUS 2008). And eight states required sex education teachers to teach negative messages about homosexuality and/or same-sex sexual activity. For example, some states require teachers to refer to homosexuality and/or same-sex sexual activity as a public health risk or must state that homosexuality is not acceptable to the general public. Only one state (California) required sex education teachers to convey a positive message about homosexuality (e.g., students should respect people of all sexual orientations).

Promotion of Tolerance in Learning Environments. Twenty-four states and the District of Columbia have laws, regulations, or policies that prohibit antigay discrimination, harassment, and/or bullying in schools, and 21 states prohibit bullying in school but list no categories of protection (Human Rights Campaign 2009d). Congresswoman Linda Sanchez from California introduced the Safe Schools Improvement Act (H.R. 3132) in 2007—a federal bill that, if passed, would require school districts to adopt policies prohibiting bullying and harassment based on race, gender, religion, sexual orientation, and gender identity/expression, among others.

A number of programs exist that aim to create a "harassment-free" climate and promote understanding and acceptance of sexual orientation and gender diversity in the school setting. GLSEN—a national organization fighting against harassment and discrimination in K–12 schools—conducts training for school staffs around the country and has developed the faculty training program of the Massachusetts Department of Education: the Safe Schools for Gay and Lesbian Students program, the first statewide effort with the aim of ending homophobia in schools (GLSEN 2000). The National Education Association has also implemented national training programs to educate teachers in every state about the role they can and must play to stop antigay harassment in their schools (Chase 2000). And the Human Rights Campaign (2009e) encourages K–5 schools to use a resource called *The Welcoming Schools Guide*—a tool to help K–5 educators create learning environments that are respectful of LGBT individuals and families. The guide focuses on topics related to family diversity, gender stereotyping, and bullying in K–5 learning environments.

Schools that do not protect students against harassment can face legal challenges. A number of court rulings have held school districts responsible for failing to protect LGBT students from discrimination, violence, and harassment and have ordered school districts to pay between $40,000 and $1 million in damages (Cianciotto & Cahill 2003).

Gay-Straight Alliances. **Gay-straight alliances (GSAs)** are school-sponsored clubs for gay teens and their straight peers. GSAs create safe, supportive environments for gay youth and their allies. Most clubs also organize school events to increase acceptance of sexual minorities and reduce antigay harassment. For example, GSAs organize a Day of Silence during which students do not speak for an entire day in recognition of the daily harassment that LBGT students endure. In an effort to combat antigay remarks commonly made in school, members of one GSA in Hamilton, Wisconsin, designed classroom posters that read, "I just heard you say 'That's so gay.' What I think you meant was . . . ," followed by a list of adjectives such as ridiculous, silly, absurd, and foolish (Kilman 2007). After the posters were displayed, students reported that "that's so gay" remarks decreased. The first GSA was established in 1988, and today there are more than 3,500 GSAs in U.S. schools (GLSEN 2007).

Campus Policies and Programs Dealing with Sexual Orientation

Student groups have been active in the gay liberation movement since the 1960s. Because of the activism of students and the faculty and administrators who support them, nearly 400 U.S. colleges and universities have nondiscrimination policies that include sexual orientation (Singh & Wathington 2003). Other measures to support the LGBT college student population include gay and lesbian studies programs, social centers, and support groups, as well as campus events and activities that celebrate diversity. Many campuses have Safe Zone or Ally programs designed to visibly identify students, staff, and faculty who support the LGBT population. The Safe Zone and Ally programs may require a training session that provides a foundation of knowledge needed to be an effective ally to LGBT students and those questioning their sexuality. Participants in Safe Zone or Ally programs display some type of sign or placard outside their office or residence hall room that identifies them as individuals who are willing to provide a safe

gay-straight alliances (GSAs) School-sponsored clubs for gay teens and their straight peers.

(from left to right) Courtesy of Pennsylvania State University, Courtesy of University of North Carolina at Chapel Hill, Courtesy of Purdue University.

Logos of Safe Zone programs at three universities and colleges. *From left to right:* Pennsylvania State University, University of North Carolina at Chapel Hill, and Purdue University.

haven, a listening ear, and support for LGBT people and those struggling with sexual orientation issues (Safe Zone Resources 2003).

Understanding Sexual Orientation and the Struggle for Equality

Recent years have witnessed a growing acceptance of lesbians, gay men, and bisexuals as well as increased legal protection and recognition of these marginalized populations. The advancements in gay rights are notable and include the Supreme Court's 2003 decriminalization of sodomy, the election of the first openly gay bishop in the New Hampshire Episcopal diocese, the growing adoption of nondiscrimination policies covering sexual orientation, and the increased legal recognition of same-sex couples and families. But the winning of these battles in no way signifies that the war is over. As evidenced by the passing of Proposition 8 in California in 2008, gay rights previously won can be taken away. LGBT individuals employed at workplaces with antidiscrimination policies still experience harassment and rejection from their coworkers, and students in schools with policies against bullying are still subjected to antigay taunts. Although many countries worldwide are increasing legal protections for LGBT individuals, homosexuality is formally condemned in some countries, with penalties ranging from fines to imprisonment and even death.

Many of the advancements in gay rights have been the result of political action and legislation. Barney Frank (1997), an openly gay U.S. representative, emphasized the importance of political participation in influencing social outcomes. He noted that demonstrative and cultural expressions of gay activism, such as "gay pride" celebrations, marches, demonstrations, or other cultural activities promoting gay rights, are important in organizing gay activists. However, he warned:

> Too many people have seen the cultural activity as a substitute for democratic political participation. In too many cases over the past decades we have left the political arena to our most dedicated opponents [of gay rights], whose letter writing, phone calling, and lobbying have easily triumphed over our marching, demonstrating, and dancing. The most important lesson . . . for people who want to make America a fairer place is that politics—conventional, boring, but essential politics—will

ultimately have a major impact on the extent to which we can rid our lives of prejudice. (p. xi)

As both structural functionalists and conflict theorists note, nonheterosexuality challenges traditional definitions of family, child rearing, and gender roles. Every victory in achieving legal protection and social recognition for sexual orientation minorities fuels the backlash against them by groups who are determined to maintain traditional notions of family and gender. Often, this determination is rooted in and derives its strength from religious ideology.

> Every victory in achieving legal protection and social recognition for sexual orientation minorities fuels the backlash against them by groups who are determined to maintain traditional notions of family and gender.

As symbolic interactionists note, meanings associated with homosexuality are learned. Powerful individuals and groups opposed to gay rights focus their efforts on maintaining the negative meanings of homosexuality to keep the gay, lesbian, and bisexual population marginalized.

But political efforts to undermine gay rights and recognition must contend with the fact that "gay is everywhere": 2000 census data reveal that 99.3 percent of U.S. counties reported same-sex cohabiting partners, compared with 52 percent of counties in 1990. With the gay population being more "out" than ever before, more heterosexual individuals are reporting having family, work-related, or friendship or acquaintance ties to lesbian and gay individuals. This increased contact, gay-straight alliances in schools, and more frequent portrayals of lesbians and gays in the media have inspired many heterosexual individuals to support the gay rights movement.

But, as one scholar noted, "The new confidence and social visibility of homosexuals in American life have by no means conquered homophobia. Indeed it stands as the last acceptable prejudice" (Fone 2000, p. 411). True, the American public is becoming increasingly supportive of gay rights. But as Yang (1999) pointed out, as the antigay minority diminishes in size, "it often becomes more dedicated and impassioned" (p. ii):

> Our task in the coming years is to get the heterosexual Americans who support our cause to feel as passionately outraged by the injustices we face and to be as strongly motivated to act in support of our rights as our adversaries are in their opposition to our rights. (p. iii)

CHAPTER REVIEW

■ **Are there any countries in which homosexuality is illegal?**
Yes, at least 86 member states of the United Nations criminalize consensual same-sex behavior among adults. In 37 of these countries, only male homosexuality is illegal; female homosexuality is not. Legal penalties for homosexuality vary from prison sentences and/or corporal punishment (such as whipping or lashing) to the death penalty.

■ **Is there any country where same-sex couples can be legally married?**
Yes. In 2001, the Netherlands became the first country in the world to offer legal marriage to same-sex

couples. Belgium, Spain, Canada, and South Africa as well as a few U.S. states have also legalized same-sex marriage.

■ **In what ways is the classification of individuals into sexual orientation categories problematic?**
Classifying individuals into sexual orientation categories is problematic for a number of reasons: (1) Some people conceal or falsely portray their sexual orientation identity; (2) attractions, love, behavior, and self-identity do not always match; and (3) sexual orientation can change over time.

- **What is the relationship between beliefs about what "causes" homosexuality and attitudes toward homosexuality?**
Individuals who believe that homosexuality is biologically based or inborn tend to be more accepting of homosexuality. In contrast, individuals who believe that homosexuals choose their sexual orientation are less tolerant of gays and lesbians.

- **What is the position of the American Psychiatric Association, the American Psychological Association, the American Academy of Pediatrics, and the American Medical Association on conversion or reparative therapy for gays and lesbians?**
These organizations agree that sexual orientation cannot be changed and that efforts to change sexual orientation (conversion or reparative therapy) do not work and may, in fact, be harmful.

- **What three cultural changes have influenced the worldwide increase in liberalized national policies on same-sex relations and the gay rights movement?**
The worldwide increase in liberalized national policies on same-sex relations and the gay rights social movement have been influenced by (1) the rise of individualism, (2) increasing gender equality, and (3) the emergence of a global society in which nations are influenced by international pressures.

- **What is internalized homophobia? What are the effects of internalized homophobia on lesbian and gay individuals?**
Internalized homophobia is a sense of personal failure and self-hatred among lesbian and gay individuals, resulting from social rejection and stigmatization. Internalized homophobia has been linked to increased risk for depression, substance abuse and addiction, anxiety, and suicidal thoughts.

- **Bisexuals experience "double discrimination." What does that mean?**
Bisexual individuals face rejection not only from heterosexuals but also from many homosexual individuals. Hence, bisexuals experience "double discrimination."

- **In the United States, what factors help to explain the increased acceptance of homosexuality and support for gay rights in the last decade?**
Factors that help explain the increased acceptance of homosexuality and support of gay rights include the increasing levels of education among U.S. adults, the positive depiction of gays and lesbians in the popular media, and increased contact between heterosexuals and openly gay and lesbian individuals.

- **In June 2003, what was the Supreme Court's decision in *Lawrence v. Texas*?**
In *Lawrence v. Texas*, the Supreme Court ruled that sodomy laws (laws that criminalized oral and anal sexual acts) were discriminatory and unconstitutional, removing the stigma and criminal branding that sodomy laws have long placed on GLBT individuals. Before this historic ruling, sodomy was illegal in 13 states; in 4 states, sodomy laws targeted only same-sex acts. In states that criminalized both same- and opposite-sex sodomy, sodomy laws were usually not used against heterosexuals but were used primarily against gay men and lesbians.

- **Is employment discrimination based on sexual orientation illegal in all 50 states?**
As of 2009, only 21 states ban employment discrimination based on sexual orientation. This means it is legal in 30 states to discriminate against someone on the basis of sexual orientation. As of this writing, a federal bill called the Employment Nondiscrimination Act (ENDA), which would ban employment discrimination against individuals on the basis of sexual orientation, is pending before Congress.

- **What is the "Don't ask, don't tell" policy?**
In 1993, President Bill Clinton instituted a "Don't ask, don't tell" policy in which recruiting officers are not allowed to ask about sexual orientation and homosexuals are encouraged not to volunteer such information. Thousands of service members have been discharged under "Don't ask, don't tell" since it took effect, and countless others have decided not to join the military because of this discriminatory law.

- **Same-sex couples want the same legal and financial benefits that are granted to heterosexual married couples. What are some of these benefits?**
Married couples have the right to inherit from a spouse who dies without a will, to avoid inheritance taxes between spouses, to make crucial medical decisions for a partner and to take family leave to care for a partner in the event of the partner's critical injury or illness, to receive Social Security survivor benefits, and to include a partner in health insurance coverage. Other rights bestowed on married (or once-married) partners include assumption of a spouse's pension, bereavement leave, burial determination, domestic violence protection, reduced rate memberships, divorce protections (such as equitable division of assets and visitation of partner's children), automatic housing lease transfer, immunity from testifying against a spouse, and spousal immigration eligibility.

- **How common is it for anti-LGBT hate crimes to involve heterosexual victims?**
Because it is not uncommon for heterosexual men and women to be mistaken for gay men and lesbians, 6 percent of anti-LGBT violence in 2007 identified as being heterosexual.

- **According to a national survey of students ages 13 through 18 and secondary school teachers, what are the two most common reasons that students are harassed by their peers?**
The most common reason students are harassed by their peers is physical appearance; the second most common reason for being harassed is actual or perceived sexual orientation.

- **A report by Amnesty International (2005) found that police mistreatment and abuse of LGBT people are widespread in the United States. What types of police mistreatment were revealed in this report?**
Police mistreatment of LGBT individuals reported by Amnesty International included targeted and discriminatory enforcement of laws against LGBT people, verbal abuse, inappropriate pat-down and strip searches, failure to protect LGBT people in holding cells, inappropriate response or failure to respond to hate crimes or domestic violence calls, sexual harassment and abuse, and physical abuse.

- **What are "civil unions"?**
A civil union is a legal status parallel to civil marriage under state law that entitles same-sex couples to almost all the rights and responsibilities available under state law to married couples.

- **What are three strategies that promote the health and well-being of all students and address the needs and promote acceptance of GLBT youth?**
Three strategies discussed in your text are to (1) include sexual orientation diversity in sex education programs, (2) create learning environments that promote tolerance of sexual orientation diversity, and (3) establish gay-straight alliances in middle and high schools.

- **What are Safe Zone or Ally programs?**
Safe Zone or Ally programs are designed to visibly identify students, staff, and faculty who support the LGBT population. Participants in Safe Zone or Ally programs display a sign or placard outside their office or residence hall room that identifies them as individuals who are willing to provide a safe haven and support for LGBT people and those struggling with sexual orientation issues.

TEST YOURSELF

1. In some countries, people found guilty of engaging in same-sex sexual behavior may receive the death penalty.
   a. True
   b. False
2. A national study of U.S. college students found that what percent identified as gay, lesbian, or bisexual?
   a. Less than one percent
   b. 2.5
   c. 4.1 percent
   d. 8.6 percent
3. In the United States, about one-fifth of gay male couple households and one-third of lesbian couple households have children present in the home.
   a True
   b. False
4. Most reparative therapy programs allegedly achieve "conversion" to heterosexuality through
   a. uncovering repressed childhood sexual trauma and confronting the perpetrator
   b. hormonal treatments
   c. intensive psychotherapy and family therapy
   d. embracing evangelical Christianity and being "born again"

5. Why has Reverend Fred Phelps of the Westboro Baptist Church (Topeka, Kansas) picketed funerals of American servicemen and servicewomen who have died in the Iraq war?
   a. To protest the "Don't ask, don't tell" military policy
   b. To protest efforts to repeal the "Don't ask, don't tell" military policy
   c. To publicize his view that U.S. military casualties are God's way of punishing the United States for being nice to lesbians and gays
   d. To show support of Iraq's harsh penalties for homosexuality
6. The majority of Americans believe that homosexual individuals should have equal job opportunities.
   a. True
   b. False
7. A national survey of college freshmen found that the majority agrees that same-sex couples should not have the right to legal marital status.
   a. True
   b. False

8. Who was Harvey Milk?
   a. The first openly gay person to be elected to public office as a member of the San Francisco Board of Supervisors
   b. The first openly gay person to be discharged from the military under the "Don't ask don't tell" policy
   c. A 12-year-old boy who committed suicide after being subjected to antigay harassment by his peers
   d. A parent who started the organization Parents, Families, and Friends of Lesbians and Gays (PFLAG) after his son committed suicide due to antigay harassment

9. The "Don't ask, don't tell" discriminates against gay and lesbian individuals who work
   a. in the public schools
   b. for city and state government
   c. for religious organizations
   d. in the military

10. More than half of Fortune 500 companies today offer same-sex domestic partner health benefits.
    a. True
    b. False

Answers: 1: a; 2: c; 3: a; 4: d; 5: c; 6: a; 7: b; 8: a; 9: d; 10: a.

KEY TERMS

biphobia 414
bisexuality 400
civil union 429
coming out 424
contact hypothesis 425
conversion therapy 405
Defense of Marriage Act 416
domestic partners 429

Employment Nondiscrimination Act (ENDA) 427
gay-straight alliances (GSAs) 432
GLBT 401
heterosexism 409
heterosexuality 400
homophobia 411
homosexim 410

internalized homophobia 409
LGBT 401
master status 409
National Coming Out Day 424
registered partnerships 402
reparative therapy 405
sexual orientation 400
sodomy 414

MEDIA RESOURCES

Understanding Social Problems,
Seventh Edition Companion Website
www.cengage.com/sociology/mooney

Visit your book companion website, where you will find flash cards, practice quizzes, Internet links, and more to help you study.

Just what you need to know NOW! Spend time on what you need to master rather than on information you already have learned. Take a pre-test for this chapter, and CengageNOW will generate a personalized study plan based on your results. The study plan will identify the topics you need to review and direct you to online resources to help you master those topics. You can then take a post-test to help you determine the concepts you have mastered and what you will need to work on. Try it out! Go to www.cengage.com/login to sign in with an access code or to purchase access to this product.

© AP Photo/Bikas Das

12

Population Growth and Urbanization

"Population may be the key to all the issues that will shape the future: economic growth; environmental security; and the health and wellbeing of countries, communities, and families."

Nafis Sadik, former executive director, United Nations Population Fund

The Global Context: A World View of Population Growth and Urbanization | Sociological Theories of Population Growth and Urbanization | Social Problems Related to Population Growth and Urbanization | **Social Problems Research Up Close: Relationship Between Urban Sprawl and Physical Activity, Obesity, and Health** | Strategies for Action: Responding to Problems of Population Growth,

In August 2007, a bridge on Interstate 35W in Minnesota collapsed into the Mississippi River, killing at least a dozen people and injuring many more. Two years before this tragic event, the Interstate 35W bridge was deemed "structurally deficient." Indeed, federal government reports indicated that, in 2006, nearly a quarter of the nation's 600,000 major bridges (longer than 20 feet) were "structurally deficient" or "functionally obsolete" (McLaughlin 2007). Some observers attributed the tragedy that occurred on Interstate 35W to the government's failure to allocate adequate funding for the repair and maintenance of our nation's bridges. But another contributing factor is our growing population, and the increasing demands such growth places on our nation's infrastructure, especially in urban areas where population growth is highest. Bridges that were built decades ago are now carrying more passengers, and more weight, than ever. In only 10 years, from 1995 to 2005, the weight load on urban highways increased by half (Lohn 2007).

Eric Miller/Reuters /Landov

The collapse of Interstate 35W could be blamed on the government's failure to provide adequate funding for the repair and maintenance of our nation's bridges. But another contributing factor is the increasing demands our growing population places on our nation's infrastructure, especially in urban areas where population growth is highest.

In this chapter, we focus on problems associated with population growth. Because the growth of the world's population is occurring largely in urban areas, we also explore the problems associated with the ever-increasing urbanization of our planet.

The Global Context: A World View of Population Growth and Urbanization

Although thousands of years passed before the world's population reached 1 billion, the population exploded from 1 billion to 6 billion in less than 300 years (see Figure 12.1). World population was 1.6 billion when we entered the 20th century, and 6.1 billion when we entered the 21st century.

World Population: History, Current Trends, and Future Projections

Humans have existed on this planet for at least 200,000 years. For 99 percent of human history, population growth was restricted by disease and limited food

Figure 12.1 **World Population Growth Through History**
Source: Mcfalls 2007.

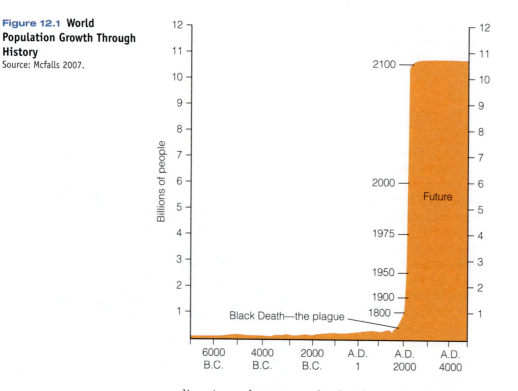

supplies. Around 8000 B.C., the development of agriculture and the domestication of animals led to increased food supplies and population growth, but even then, harsh living conditions and disease still put limits on the rate of growth. This pattern continued until the mid-18th century, when the Industrial Revolution improved the standard of living for much of the world's population. The improvements included better food, cleaner drinking water, improved housing and sanitation, and advances in medical technology, such as antibiotics and vaccinations against infectious diseases—all of which contributed to rapid increases in population.

Population Doubling Time. Population **doubling time** is the time required for a population to double from a given base year if the current rate of growth continues. It took several thousand years for the world's population to double to a size of 14 million, but then took only a thousand years to nearly double to 27 million and another thousand to reach 50 million. From there, it took only 500 years to double from 50 million to 100 million, and 400 years for the next doubling to occur. When the Industrial Revolution began around 1750, population growth exploded, taking only 100 years to double. The most recent doubling—from three billion in 1960, to six billion in 1999—took only about 40 years (Weeks 2008). Although world population will continue to grow in the coming decades, it will probably not double in size again.

Current Population Trends and Future Projections. World population is projected to grow from 6.8 billion in 2009, to 9.1 billion in 2050 (United Nations 2009). Most of the world's population live in less developed countries (see Figure 12.2). The most populated country in the world today is China, where nearly one in five people on this planet live. By 2050, India will become the most populated country (see Figure 12.3).

doubling time The time required for a population to double in size from a given base year if the current rate of growth continues.

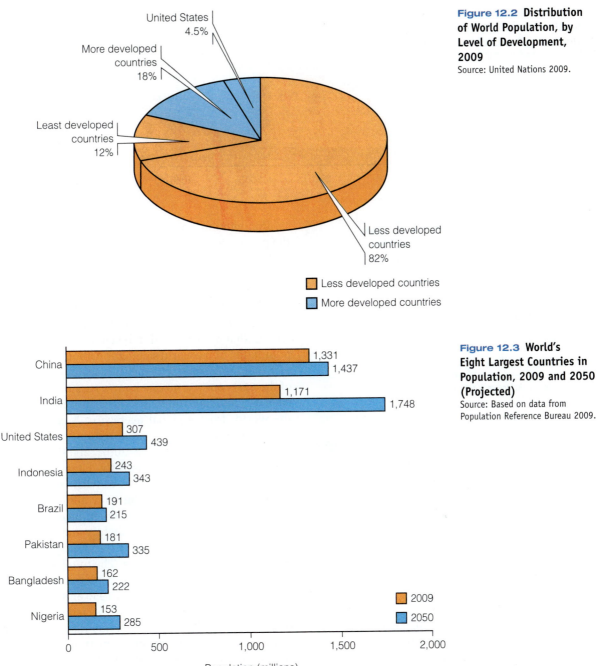

Figure 12.2 Distribution of World Population, by Level of Development, 2009
Source: United Nations 2009.

United States 4.5%

More developed countries 18%

Least developed countries 12%

Less developed countries 82%

■ Less developed countries
■ More developed countries

Figure 12.3 World's Eight Largest Countries in Population, 2009 and 2050 (Projected)
Source: Based on data from Population Reference Bureau 2009.

China 1,331 / 1,437
India 1,171 / 1,748
United States 307 / 439
Indonesia 243 / 343
Brazil 191 / 215
Pakistan 181 / 335
Bangladesh 162 / 222
Nigeria 153 / 285

■ 2009
■ 2050

Population (millions)

Nearly all of world population growth is in less developed countries, mostly in Africa and Asia (see Figure 12.4). As the size of a country's population grows, so does its **population density,** or the number of people per unit of land area. Population density is highest in less developed countries (see Table 12.1).

Higher population growth in developing countries is largely due to higher **total fertility rates**—the average lifetime number of births per woman in a population. As shown in Table 12.2, the least developed regions of the world have the highest rates of fertility.

population density The number of people per unit of land area.

total fertility rate The average lifetime number of births per woman in a population.

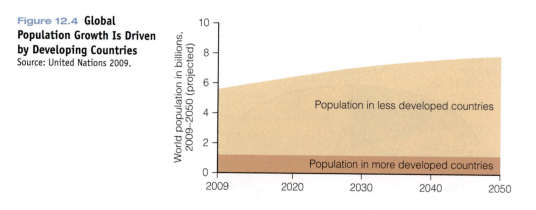

Figure 12.4 Global Population Growth Is Driven by Developing Countries
Source: United Nations 2009.

TABLE 12.1 Population Density

AREA	POPULATION DENSITY (PEOPLE PER SQUARE MILES/KILOMETERS)
World	19/50
More developed countries	10/27
Less developed countries	26/67
Least developed countries	15/40

Source: Population Reference Bureau 2009.

TABLE 12.2 Fertility Rates by Region

World	2.6
More developed	1.7
Less developed	2.7
Less developed (excluding China)	3.1
Least developed	4.6

Source: Population Reference Bureau 2009.

replacement-level fertility The level of fertility at which a population exactly replaces itself from one generation to the next; currently, the number is 2.1 births per woman (slightly more than 2 because not all female children will live long enough to reach their reproductive years).

Immigration also affects population size. Much of the population growth occurring in the United States is due to high immigration rates. Other factors that can affect the size of a population are armed conflict, economic stagnation, and high rates of disease such as HIV/AIDS.

Will there be an end to the rapid population growth that has occurred in recent decades? Will the population of the world stabilize? Although some predict that population will stabilize around the middle of the 21st century, no one knows for sure. There has been a significant reduction in fertility rates around the world, from a global average of 5 children per woman in the 1950s, to 2.6 children between 2005 and 2010 (United Nations 2007; 2009). To reach population stabilization, fertility rates throughout the world would need to achieve what is called "replacement level," whereby births would replace, but not outnumber, deaths. **Replacement-level fertility** is 2.1 births per woman, that is, slightly more than 2 because not all female children will live long enough to reach their reproductive years. The number of countries that have achieved below-replacement fertility rates has grown from 5 between 1950 and 1955, to 76 between 2005 and 2010, and is expected to reach 196 by 2050 (United Nations 2009). In some of these countries, population will continue to grow for several decades because of **population momentum**—continued population growth as a result of past high fertility rates that have resulted in a large number of young women who are currently entering their childbearing years. But there are 37 countries or areas in whose populations are projected to decrease between 2009 and 2050 (United Nations 2009). The U.S. population, however, will continue to increase through 2050 because of immigration.

In sum, there are two population trends occurring simultaneously that appear to be contradictory: (1) The total number of people on this planet is rising and is expected to continue to increase over the coming decades; and (2) fertility rates are so low in some countries that the countries' populations are likely to decline over the coming years. As we discuss later in this chapter, each of these trends presents a set of problems and challenges.

TABLE 12.3 The Percent of Population Ages 60+ and 80+, by Region: 2010 and 2050

	AGES 60+		AGES 80+	
	2010	2050	2010	2050
World	11%	22%	2%	4%
More developed regions	9%	20%	1%	3%
North America	18%	28%	4%	8%
Less developed regions	9%	20%	1%	3%

Source: United Nations 2009.

The Growing Elderly Population. Another demographic trend that presents its own set of challenges is the increasing number and proportion of older individuals in the total population. Between 2010 and 2050, the percentage of older individuals in the world population is expected to double (see Table 12.3). The number of people over age 60 is expected to reach almost 2 billion by 2050, representing 22 percent of the world's population. This growth in the size and share of the elderly population raises the question: How will societies provide housing, medical care, transportation, and other needs of the increasing elderly population? As the ratio of people in older dependent age groups increases relative to those in working-age groups, there are fewer workers to support the nonworking older population. However, declining fertility rates may counterbalance this shift by reducing the number of younger dependent people.

> Between 2010 and 2050, the percentage of older individuals in the world population is expected to double.

An Overview of Urbanization Worldwide and in the United States

As early as 5000 B.C., cities of between 7,000 and 20,000 people existed along the Nile, Tigris-Euphrates, and Indus River valleys. But not until the Industrial Revolution in the 19th century did **urbanization,** the transformation of a society from a rural to an urban one, spread rapidly.

As population has increased, so has the proportion of people living in urban areas. An **urban area** is a spatial concentration of people whose lives are centered around nonagricultural activities. Although countries differ in their definitions of "urban," most countries designate places with 2,000 people or more as being urbanized. According to the U.S. census definition, an "urban population" consists of individuals living in cities or towns of 2,500 or more inhabitants.

The number of cities, as well as the share of population that lives in cities, has grown at an incredible pace. Worldwide, the number of cities with more than 1 million inhabitants grew from 86 in 1950, to 400 in 2006, and by 2015, there will be at least 550 (Davis 2006). The number of **megacities**—urban areas with 10 million residents or more—is also increasing. Megacities are named after the major city at their core, but they encompass a large area and a number of urban settlements within it, and therefore constitute complex urban agglomerations. In 1950, there were only two cities with more than 10 million inhabitants: New

population momentum Continued population growth as a result of past high fertility rates that have resulted in a large number of young women who are currently entering their childbearing years.

urbanization The transformation of a society from a rural to an urban one.

urban area A spatial concentration of people whose lives are centered around nonagricultural activities.

megacities Urban areas with 10 million residents or more.

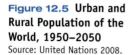

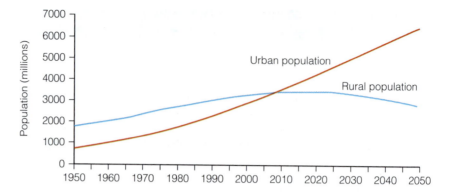

Figure 12.5 **Urban and Rural Population of the World, 1950–2050**
Source: United Nations 2008.

York and Tokyo. By 2007, there were 19 megacities, and in 2025, 27 cities are projected to have more than 10 inhabitants (United Nations 2008).

In 2008, the world population reached a historic landmark: For the first time in history, the urban population equaled the rural population of the world. The world's urban population now surpasses the rural population, and by 2050, 70 percent of the world population is expected to live in urban areas (see Figure 12.5). The urban population is growing faster than the overall population, which means that the urban areas of the world are expected to absorb all the population growth expected over the next few decades while at the same time drawing in some of the rural population. Nevertheless, in Africa and Asia, more than half of the population lives in rural areas (United Nations 2008).

In many countries, natural increase (the difference of births minus deaths) accounts for 60 percent or more of urban population, with the remaining 40 percent due to a combination of migration from rural to urban areas and the reclassification of areas from rural to urban (United Nations Population Fund 2007). Rural dwellers migrate to urban areas to flee war or natural disasters or to find employment. As foreign corporate-controlled commercial agriculture displaces traditional subsistence farming in poor rural areas, peasant farmers flock to the city looking for employment. Some rural dwellers migrate to urban areas in search of a better job—one that has higher wages and better working conditions. Governments have also stimulated urban growth by spending more to improve urban infrastructures and services while neglecting the needs of rural areas (Clark 1998).

History of Urbanization in the United States. Urbanization of the United States began as early as the 1700s, when most major industries were located in the most populated areas, including New York City, Philadelphia, and Boston. Unskilled laborers, seeking manufacturing jobs, moved into urban areas as industrialization accelerated in the 19th century. The "pull" of the city was not the only reason for urbanization, however. Technological advances were making it possible for fewer farmers to work the same amount of land. Thus "push" factors were also involved—making a living as a farmer became more difficult as technology replaced workers.

Urban populations continued to multiply as a large influx of European immigrants in the late 1800s and early 1900s settled in U.S. cities. This influx was followed by a major migration of southern rural blacks to northern urban areas. People were lured to the cities by the promise of employment and better wages and urban amenities such as museums, libraries, and entertainment. Immigrants are also attracted to cities with large ethnic communities that provide a familiar cultural environment.

Suburbanization. In the late 19th century, railroad and trolley lines enabled people to live outside the city and commute into the city to work. As more people moved to the **suburbs**—urban areas surrounding central cities—the United States underwent **suburbanization.** As city residents left the city to live in the suburbs, cities lost population and experienced **deconcentration,** or the redistribution of the population from cities to suburbs and surrounding areas.

Many factors contributed to suburbanization and deconcentration. After World War II, many U.S. city dwellers moved to the suburbs out of concern for the declining quality of city life and the desire to own a home on a spacious lot. Suburbanization was also spurred by racial and ethnic prejudice, as the white majority moved away from cities that were becoming increasingly diverse because of immigration. Mass movement into suburbia was encouraged by the federal interstate highway system (financed by the government under the guise of ensuring national defense), the affordability of the automobile, and the dismantling of metropolitan mass transit systems (Lindstrom & Bartling 2003). In the 1950s, Veterans Administration and Federal Housing Administration loans made housing more affordable, enabling many city dwellers to move to the suburbs. Suburb dwellers who worked in the central city could commute to work or work in a satellite branch of a company in suburbia that was connected to the main downtown office. As increasing numbers of people moved to the suburbs, so did businesses and jobs. Without a strong economic base, city services and the quality of city public schools declined, which further increased the exodus from the city.

> **What Do You Think?** In cities across the country, families with children have left the city to move to the suburbs, where they can afford a bigger house with more space. Consequently, many cities have experienced a decline in their population of children (Egan 2005). How might significant reductions in the youth population affect cities? What could cities do to attract more families with children?

U.S. Metropolitan Growth and Urban Sprawl. Simply defined, a **metropolitan area** is a densely populated core area together with adjacent communities. Metropolitan areas have grown rapidly in the United States, both in number and size. Most Americans (84 percent) live in one of the 366 metropolitan areas in the nation (Office of Management and Budget 2008). One U.S. state—New Jersey—is entirely occupied by metropolitan areas as designated by the U.S. census.

The ever-increasing outward growth of urban areas is known as **urban sprawl.** Urban sprawl results in the loss of green open spaces, the displacement and endangerment of wildlife, traffic congestion and noise, and pollution liabilities—problems that we discuss later in this chapter.

Those who enjoy the conveniences and amenities of urban life but who find large metropolitan areas undesirable may choose to live in a **micropolitan area**—a small city (between 10,000 and 50,000 people) located beyond congested metropolitan areas. These areas are large enough to attract jobs, restaurants,

suburbs Urban areas surrounding central cities.

suburbanization The development of urban areas outside central cities, and the movement of populations into these areas.

deconcentration The redistribution of the population from cities to suburbs and surrounding areas.

metropolitan area A densely populated core area together with adjacent communities.

urban sprawl The ever-increasing outward growth of urban areas that results in the loss of green open spaces, the displacement and endangerment of wildlife, traffic congestion and noise, and pollution.

micropolitan area A small city (between 10,000 and 50,000 people) located beyond congested metropolitan areas.

community organizations, and other benefits yet are small enough to elude traffic jams, high crime rates, and high costs of housing and other living expenses associated with larger cities.

Sociological Theories of Population Growth and Urbanization

The three main sociological perspectives—structural functionalism, conflict theory, and symbolic interactionism—can be applied to the study of population and urbanization.

Structural-Functionalist Perspective

Structural functionalism focuses on how changes in one aspect of the social system affect other aspects of society. For example, the **demographic transition theory** of population describes how industrialization has affected population growth. According to this theory, high fertility rates are necessary in traditional agricultural societies to offset high mortality and to ensure continued survival of the population. As a society becomes industrialized and urbanized, improved sanitation, health, and education lead to a decline in mortality. The increased survival rate of infants and children along with the declining economic value of children leads to a decline in fertility rates. About one-third of the world's countries have completed the demographic transition—the progression from a population with short lives and large families to one in which people live longer and have smaller families (Cincotta et al. 2003). Many countries with low fertility rates have entered what is known as a "second demographic transition," in which fertility falls below the two-child replacement level. This second demographic transition has been linked to greater educational and job opportunities for women, increased availability of effective contraception, and the rise of individualism and materialism (Population Reference Bureau 2004).

Urbanization plays a significant role in the demographic transition. Because health care delivery is more cost-effective in cities than in rural areas, governments prioritize urban health clinics. With greater access to health care, urban dwellers are the first to experience declines in infant mortality and fertility. A study of contraceptive use in Kenya revealed that women who lived in urban areas were more likely to have used contraception (Kimuna & Adamchak 2001). Recent declines in fertility throughout Africa are mostly an urban phenomenon (Cincotta et al. 2003).

Structural functionalists view the development of urban areas as functional for societal development. Although cities initially served as centers of production and distribution, today they are centers of finance, administration, education, health care, and information.

Urbanization is also dysfunctional, because it leads to increased rates of anomie, or normlessness, as the bonds between individuals and social groups become weak (see also Chapters 1 and 4). Whereas social cohesion in rural areas is based on shared values and beliefs, social cohesion in urban areas is based on interdependence created by the specialization and social diversity of the urban population. Anomie is linked to higher rates of deviant behavior, including crime, drug

demographic transition theory A theory that attributes population growth patterns to changes in birth rates and death rates associated with the process of industrialization.

addiction, and alcoholism. Overcrowding, poverty, the rapid spread of infectious disease, and environmental destruction are also considered dysfunctions associated with urbanization.

Conflict Perspective

The conflict perspective focuses on how wealth and power, or the lack thereof, affect population problems. In 1798, Thomas Malthus predicted that the population would grow faster than the food supply and that masses of people were destined to be poor and hungry. According to Malthusian theory, food shortages would lead to war, disease, and starvation, which would eventually slow population growth. However, conflict theorists argue that food shortages result primarily from inequitable distribution of power and resources (Livernash & Rodenburg 1998).

Conflict theorists also note that population growth results from pervasive poverty and the subordinate position of women in many less developed countries. Poor countries have high infant and child mortality rates. Hence, women in many poor countries feel compelled to have many children to increase the chances that some will survive into adulthood. Their subordinate position prevents many women from limiting their fertility. In many developing countries, a woman must get her husband's consent before she can receive any contraceptive services. Thus, according to conflict theorists, population problems result from continued economic and gender inequality.

Power and wealth also affect the development and operations of urban areas. The capitalistic pursuit of wealth contributed to the development of cities, because capitalism requires that the production and distribution of goods and services be centrally located, thus, at least initially, leading to urbanization. Today, global capitalism and corporate multinationalism, in search of new markets, cheap labor, and raw materials, have largely spurred urbanization of the developing world. Capitalism also contributes to migration from rural areas into cities because peasant farmers who have traditionally produced goods for local consumption are being displaced by commercial agriculture that is geared to producing fruits, flowers, and vegetables for export to the developed world. Displaced from their traditional occupations, peasant farmers are flocking to cities to find employment (Clark 1998).

The conflict perspective also focuses on how individuals and groups with wealth and power influence decisions that affect urban populations. For example, according to citizens' groups working to stop urban sprawl in central and eastern Europe, city officials may be bribed to approve a new shopping mall or other development project (Sheehan 2001). In addition, deteriorating conditions in U.S. inner cities are often ignored because the residents of inner cities lack the wealth, power, and status to solicit government spending on needed infrastructure and services, such as sidewalk and road repairs, street cleaning, and beautification projects.

Symbolic Interactionist Perspective

The symbolic interactionist perspective focuses on how meanings, labels, and definitions learned through interaction affect population problems. For example, many societies are characterized by **pronatalism**—a cultural value that promotes having children. Throughout history, many religions have

pronatalism A cultural value that promotes having children.

worshiped fertility and recognized it as being necessary for the continuation of the human race. In many countries, religions prohibit or discourage birth control, contraceptives, and abortion. Women in pronatalistic societies learn through interaction with others that deliberate control of fertility is socially unacceptable. Women who use contraception in communities in which family planning is not socially accepted face ostracism by their community, disdain from relatives and friends, and even divorce and abandonment by their husbands (Women's Studies Project 2003). However, once some women learn new definitions of fertility control, they become role models and influence the attitudes and behaviors of others in their personal networks (Bongaarts & Watkins 1996).

Social Problems Related to Population Growth and Urbanization

Social problems related to population growth and urbanization include environmental problems, poverty and unemployment, global insecurity, poor maternal and infant health, transportation and traffic problems, and the effects of sprawl on wildlife and human health. First, we consider the problems faced by countries where fertility rates are low and population is declining.

Problems Associated with Below-Replacement Fertility

In a growing number of countries—including China, Japan, and all European countries—fertility rates have fallen well below the 2.1 children replacement level (see Table 12.4). Because these low fertility rates will eventually lead to population decline, some reports have sounded an alarm about the possibility of a "birth dearth." Low fertility rates lead not only to a decline in population size but also to an increasing proportion of the population comprising elderly members. A birth dearth eventually results in fewer workers to support the pension, social security, and health care systems for the elderly. Below-replacement fertility rates also raise concern about the ability of a country to maintain a productive economy, because there may not be enough future workers to replace current workers as they age and retire.

Environmental Problems and Resource Scarcity

According to a survey of faculty at the State University of New York College of Environmental Science and Forestry (2009), overpopulation is the world's top environmental problem, followed closely by climate change and the need to replace fossil fuels with renewable energy sources. As we discuss in Chapter 13, population growth places increased demands on natural resources, such as

TABLE 12.4	Number of Countries with Below-Replacement Fertility				
1950–1955	1970–1975	1990–1995	2005–2010	2025–2030	2045–2050
5	18	53	76	119	147

Source: United Nations 2009.

forests, water, cropland, and oil and results in increased waste and pollution. Over the past 50 years, the earth's ecosystems have been degraded more rapidly and extensively than in any other comparable period of time in human history (Millennium Ecosystem Assessment 2005).

The countries that suffer most from shortages of water, farmland, and food are developing countries with the highest population growth rates. However, countries with the largest populations do not necessarily have the largest impact on the environment. This is because the demands that humanity makes on the earth's natural resources—each person's **environmental footprint**—is determined by the patterns of production and consumption in that person's culture. The environmental footprint of an average person in a high-income country is much larger than that of someone in a low-income country. Hence, although population growth is a contributing factor in environmental problems, patterns of production and consumption are at least as important in influencing the effects of population on the environment.

> **What Do You Think?** The growing world population raises questions about how we can provide enough food, water, energy, housing, etc., to meet everyone's needs. Gar Smith (2009) argues that "meeting the 'growing demands of a growing population' is not solving the problem: It's perpetuating it" (p. 15). What does he mean by this statement? Do you agree? Why or why not?

Poverty and Unemployment

Poverty and unemployment are problems that plague countries with high population growth as well as urban areas in both rich and poor countries. Nearly half (48 percent) of people worldwide live on less than $2 a day (Population Reference Bureau 2009). Less developed, poor countries with high birth rates do not have enough jobs for a rapidly growing population, and land for subsistence farming becomes increasingly scarce as populations grow. In some ways, poverty leads to high fertility, because poor women are less likely to have access to contraception and are more likely to have large families in the hope that some children will survive to adulthood and support them in old age. But high fertility also exacerbates poverty, because families have more children to support and national budgets for education and health care are stretched thin.

Until recently, poverty was deeper and more widespread in rural areas than in cities. However, poverty today is increasing more rapidly in urban areas than in rural areas (United Nations Population Fund 2007).

In the United States, the highest rates of poverty are in the central cities, in part because of higher unemployment rates in central cities. In the United States and other industrialized countries, urban unemployment and poverty are partly the results of deindustrialization, or the loss and/or relocation of manufacturing industries. Cities most affected by the recession that began in 2007 have seen unemployment and poverty rates soar (see Table 12.5).

TABLE 12.5 U.S. Cities with the Highest Poverty Rates, 2007

CITY	PERCENT OF POPULATION BELOW POVERTY LEVEL
Detroit, MI	33.8
Cleveland, OH	29.5
Buffalo, NY	28.7
El Paso, TX	27.4
Memphis, TN	26.2
Miami, FL	25.5
Milwaukee, WI	24.4
Newark, NJ	23.9
Philadelphia, PA	23.8
Cincinnati, OH	23.5

Source: Bishaw & Semega 2008.

environmental footprint The demands that humanity makes on the Earth's natural resources.

Nearly one in three city dwellers—almost 1 billion people—live in slums characterized by overcrowding, little employment, and poor water, sanitation, and health care services.

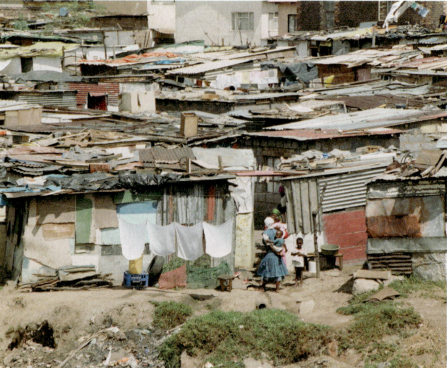

© David Turnley/Corbis

Urban Housing and Sanitation Problems

Many cities are experiencing a housing crisis, because the number of low-income renters has increased while the number of low-cost rental units has dropped. Housing that is available and affordable is often substandard, characterized by outdated plumbing and wiring, overcrowding, rat infestations, toxic lead paint, and fire hazards (see also Chapter 6).

Nearly a billion people, a sixth of the world's population, live in slums characterized by overcrowding, little employment, and poor water, sanitation, and health care services.

slums Concentrated areas of poverty and poor housing in urban areas.

ghettos In the United States, slums that are occupied primarily by African Americans.

barrios In the United States, slums that are occupied primarily by Latinos.

Concentrated areas of poverty and poor housing in urban areas are called **slums.** In the United States, slums that are occupied primarily by African Americans are known as **ghettos,** and those occupied primarily by Latinos are called **barrios.** Nearly a billion people, a sixth of the world's population, live in slums characterized by overcrowding, little employment, and poor water, sanitation, and health care services (United Nations Population Fund 2007).

More than one-third of the growing urban population in developing countries live in slum conditions (United Nations 2008). In sub-Saharan Africa, urbanization has become virtually synonymous with slum growth; nearly three-quarters of the urban population in sub-Saharan Africa live in slum conditions (Davis 2006). Worldwide, the growth of slums is outpacing the growth of urbanization:

Thus, the cities of the future, rather than being made out of glass and steel . . . are instead largely constructed out of crude brick, straw, recycled plastic, cement blocks, and scrap wood. Instead of cities of light soaring toward heaven, much of the twenty-first century urban world squats in squalor, surrounded by pollution, excrement, and decay. (Davis, 2006, p. 19)

Global Insecurity

A Population Institute report titled *Breeding Insecurity: Global Security Implications of Rapid Population Growth* warns that rapid population growth is a contributing factor to global insecurity, including civil unrest, war, and terrorism (Weiland 2005). The report points out that many developing countries are characterized by a "youth bulge"—a high proportion of 15- to 29-year-olds relative to the adult population. Youth bulges result from high fertility rates and declining infant mortality rates, a common pattern in developing countries today. Youth bulges are growing rapidly throughout Africa and the Middle East. For example, young adults in Iraq make up nearly half of the adult population. The combination of a youth bulge with other characteristics of rapidly growing populations, such as resource scarcity, high unemployment rates, poverty, and rapid urbanization, sets the stage for political unrest. "Large groups of unemployed young people, combined with overcrowded cities and lack of access to farmland and water create a population that is angry and frustrated with the status quo and thus is more likely to resort to violence to bring about change" (Weiland 2005, p. 3).

Poor Maternal, Infant, and Child Health

As noted in Chapter 2, maternal deaths (deaths related to pregnancy and childbirth) are the leading cause of mortality for reproductive-age women in the developing world. Having several children at short

© Peter Johnson/Corbis

Worldwide, pregnancy is the leading cause of death for young women ages 15 to 19. Most (95 percent) maternal deaths occur in Africa and Asia. This woman in sub-Saharan Africa has a 1 in 16 risk of dying in pregnancy or childbirth, compared to a 1 in 2,800 chance for a woman in a developed country.

intervals increases the chances of premature birth, infectious disease, and death for the mother or the baby. Childbearing at young ages (teens) also increases the risks of health problems and death for both women and infants (United Nations Population Division 2009). In developing countries, one in four children is born unwanted, increasing the risk of neglect and abuse. In addition, the more children a woman has, the fewer the parental resources (parental income and time and maternal nutrition) and social resources (health care and education) available to each child.

Transportation and Traffic Problems

Urban areas are often plagued with transportation and traffic problems. A study of U.S. urban areas found that, in 2005, drivers experienced 38 hours of delays—up from 14 hours in 1982 (Schrank & Lomax 2007). This study revealed that a total of 2.9 billion gallons of fuel were wasted in 2005 because of traffic congestion.

Congested traffic is linked to stress and respiratory problems due to vehicle exhaust. And more than 20 million people are severely injured or killed on the world's roads each year (World Health Organization 2003). Indeed, far more people are killed and injured in automobile accidents than by violent crime. Therefore, it has been argued that, despite higher crime rates in the inner city, the suburbs are the more dangerous place to live, because suburbanites "drive three times as much, and twice as fast, as urban dwellers" (Durning 1996, p. 24).

Effects of Sprawl on Wildlife and Human Health

The spread of urban and suburban areas is increasingly replacing natural habitats with pavement, buildings, and human communities. The loss of open green space, trees, and plant life affects animals whose homes are turned into parking lots, shopping centers, office buildings, and housing developments. According to the U.S. Fish and Wildlife Service, habitat loss resulting from urban and suburban sprawl is the number one reason that wildlife species are becoming increasingly endangered (Shevis 1999).

Displacement of wildlife due to urban sprawl endangers not only the wildlife, but humans as well. For example, as more wooded areas are cleared for development, deer are displaced. "With no place to go, they bound into suburban backyards in search of food and water and across highways, frequently injuring themselves and causing harm to drivers. . . . Deer cause an estimated half-million vehicle accidents a year, killing 100 people and injuring thousands more" (Shevis 1999, pp. 2–3).

Suburban sprawl has also brought humans into greater contact with ticks that carry Lyme disease (UNEP 2005). If bitten by an infected tick, humans can develop various symptoms, including skin rash, neurological problems, fatigue, abdominal and joint pain, headache, and heart damage. Other health problems associated with sprawl include lack of physical activity and obesity (see this chapter's Social Problems Research Up Close feature).

Traffic congestion wastes fuel, increases air pollution, and creates stress and frustration among drivers.

© EGDigital/iStockphoto

Health experts agree that most Americans are overweight and do not get enough exercise. As discussed in Chapter 2, obesity has reached epidemic proportions and is the second leading cause of preventable deaths in the United States. Although eating too much of the wrong kinds of foods is a major cause of obesity, lack of physical activity also plays a major role.

Here, we describe the first national study to investigate whether our physical activity levels, weight, and health are related to the type of place in which we live (Ewing et al. 2003). This study assesses how sprawl development—where homes are far from workplaces, shops, restaurants, and other destinations—affects physical activity, weight, and health of community residents.

Sample and Methods

Ewing and colleagues (2003) used the Centers for Disease Control and Prevention Behavioral Risk Factor Surveillance System (BRFSS) surveys (1998–2000) to obtain data on physical activity levels,

body mass index and obesity, hypertension, diabetes, and heart disease of respondents. The BRFSS is a random telephone survey administered to U.S. adults. More than 200,000 respondents from 448 counties and more than 175,000 respondents from 83 metropolitan areas were selected from the larger BRFSS samples because they lived in areas for which urban sprawl indexes were available.

Ewing and coworkers (2003) measured the degree of sprawl in 448 counties by using a "sprawl index" based on data available from the U.S. Census Bureau and other federal sources. The bigger the sprawl index, the more compact the metropolitan area; the smaller the index, the more sprawling the region.

Findings and Conclusions

Sprawling areas are not conducive to walking or biking, and people who live in sprawling communities have fewer opportunities to walk or bicycle as part of their daily routine. Rather, residents in sprawling areas depend on driving as the

most convenient form of transportation. Therefore, it is not surprising that Ewing and colleagues found that people who live in counties with sprawling development are likely to walk less and weigh more than people who live in less sprawling counties. The people living in the most sprawling areas were likely to weigh 6 pounds more than people in the most compact county. In addition, after controlling for factors such as age, education, sex, and race and ethnicity, people in more sprawling counties were more likely to suffer from hypertension (high blood pressure). The researchers did not find a statistically significant relationship between level of sprawl and diabetes or cardiovascular disease.

These results suggest that (1) there is a direct association between the level of sprawl development in a community and the health of the people who live there, and (2) we can improve public health by creating more compact communities that offer opportunities and design features (e.g., sidewalks) to include walking in our daily routines.

Strategies for Action: Responding to Problems of Population Growth, Population Decline, and Urbanization

Whereas some countries are struggling to slow population growth, others are challenged with maintaining or even increasing their populations. After describing efforts to maintain or increase population in countries with low fertility rates, we look at strategies to slow population growth, including providing access to family planning services and contraceptive methods, involving men in family planning, implementing a one-child policy as in China, improving the status of women through education and employment, and increasing economic development and improving health status. Because populations increasingly live in urban areas, strategies that address population problems also address, indirectly, problems of urbanization. Additional strategies to alleviate urban problems include those designed to lessen poverty and stimulate economic development in inner cities, implement "smart growth" and "new urbanism" in development, improve transportation and ease traffic congestion, and curb urban growth in developing countries.

Efforts to Maintain or Increase Population in Countries with Low Fertility Rates

In some countries with below-replacement fertility levels, population strategies have focused on *increasing* rather than decreasing the population. For example, Australia's total fertility rate hit a record low of 1.73 in 2001, prompting the government to begin paying a $3,000 bonus in 2004 (which increased to $4,000 in 2005) to families who have babies (Lalasz 2005). The town of Yamatsuri, Japan, offers a $9,200 monetary reward (over a 10-year period) to persuade women who have at least two children to have more (Wiseman 2005). Aside from monetary rewards, many countries encourage childbearing by implementing policies designed to help women combine child rearing with employment. For example, as noted in Chapters 5 and 7, many European countries have generous family leave policies and universal child care. Another way to increase population is to increase immigration. Spain, for example, has eased restrictions on immigration as a way to gain population.

What Do You Think? One strategy for encouraging childbearing in European countries with low fertility rates is to provide work-family supports to make it easier for women to combine childbearing with employment. If the United States offered more generous work-family benefits, such as paid parenting leave and government-supported child care, would the U.S. birth rate increase? Would such policies affect the number of children you would want to have?

Efforts to Curb Population Growth: Reducing Fertility

A number of strategies are associated with efforts to reduce the number of children women have. To some population experts, stopping population growth is a critical issue. "Zero population growth, which characterized human population for more than 99 percent of its history, must be achieved once again, at least as a long-term average, if the human species is to survive" (McFalls 2007, p. 27).

Provide Access to Family Planning Services. Since the 1950s, governments and nongovernmental organizations such as the International Planned Parenthood Federation have sought to lower fertility through family planning programs that provide reproductive health services and access to contraceptive information and methods. Such programs, along with developments in contraceptive technology, have been largely successful: Globally, the average number of children born to each woman has fallen from 5 in the 1950s, to 2.6 between 2005 and 2010, because more women today want to limit family size and are using modern methods of birth control to control the number and spacing of births (United Nations 2007; 2009). Worldwide, more than half of married women ages 15 to 49 use some form of modern contraception (68 percent of U.S.

The region of the world with the highest fertility rate is Africa, where women have an average of five children in their lifetime.

women use modern contraception). However, in 21 countries, less than 10 percent of women use modern contraception (Population Reference Bureau 2009). Most least developed countries have fertility levels above 5, and fertility rates are 6 or higher in seven countries (United Nations 2009).

Although access to contraceptives has increased worldwide since the 1950s, there is still a significant unmet need for contraception. Worldwide, about 200 million women want to delay or avoid pregnancy but are not using modern contraceptive methods (Cohen 2006).

Family planning programs are effective in increasing use of contraceptives. For example, in rural Pakistan, married women living in households served by a "lady health worker," who provides doorstep delivery of contraceptive supplies, are more likely than other married women to use modern contraceptives (Douthwaite & Ward 2005). The provision of doorstep services through female workers is especially valuable in Pakistan, where women's mobility is limited and female modesty is highly valued.

In some countries, family planning personnel refuse or are forbidden by law or policy to make referrals for contraceptive and abortion services for unmarried women. Furthermore, many women throughout the world do not have access to legal, safe abortion. Without access to contraceptives, many women who experience unwanted pregnancy resort to abortion—even under illegal and unsafe conditions. More than half of the nearly 80 million unintended pregnancies that occur worldwide every year end in abortion. Research in central Asia and eastern Europe has shown that increased contraception use has resulted in significant declines in the rate of abortion (Leahy 2003).

Since the mid-1990s, most least developed countries have experienced a per capita decrease in donor funding for family planning (United Nations Population Division 2009). In 1994, President Reagan passed the Global Gag Rule—a rule prohibiting the United States from funding international family planning groups that provide abortion or abortion referrals. President Clinton canceled the Global Gag Rule in 1993, and G. W. Bush reinstated it in 2001. One of President Obama's first actions after taking office was to strike down the Global Gag Rule.

China's One-Child Policy. In 1979, China initiated a national family planning policy that encourages families to have only one child by imposing a monetary fine on couples that have more than one child. The implementation and enforcement of this policy varies from one province to another, and there are a number of exemptions that allow some couples to have two or even more children. For example, most urban couples are limited to having only one child, but some rural couples are allowed to have a second child if their first child is a girl. Families that need children for labor, ethnic minority populations, and married couples who themselves are only children are also exempt from the one-child rule.

China has been criticized for using extreme measures to enforce its one-child policy, including steep fines, seizure of property, and forced sterilizations and abortions. In addition, because of a traditional preference for male heirs, many Chinese couples have aborted female fetuses with the hope of

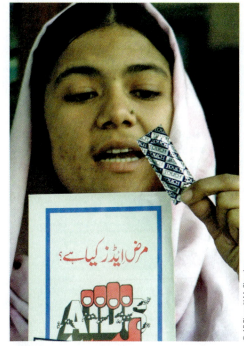

In Pakistan, where women's mobility is limited and female modesty is highly valued, "lady health workers" provide doorstep contraceptive services.

Some couples in China are eligible for an exemption from the one-child policy and are encouraged to have a second child.

having a boy in a subsequent pregnancy. This has led to an imbalanced sex ratio, with many more Chinese males than females. The Chinese government has defended the one-child policy, claiming that it has been effective in preventing at least 300 million births and has contributed to China's recent economic development ("China Is Softening Messages" 2007). China has the highest rate of modern contraceptive use in the world: Of married women ages 15 to 49 in China, 90 percent use modern contraceptives (Population Reference Bureau 2009).

Another problem with the one-child policy is that, as older Chinese retire, there are fewer workers to take their place and to support pensions for the elderly. The ratio of working-age adults to retiree was more than 7:1 in 1979; it will be 1.6:1 in 20 years (Mackinnon 2009). Concerned about how it will care for its growing older population, China has recently shifted its policy to not just allow, but encourage those couples who are eligible to have a second child.

Improve the Status of Women: The Importance of Education and Employment. Throughout the developing world, the primary status of women is that of wife and mother. Women in developing countries traditionally have not been encouraged to seek education or employment; rather, they are encouraged to marry early and have children.

Improving the status of women by providing educational and occupational opportunities is vital to curbing population growth. Educated women are more likely to marry later, want smaller families, and use contraception. Recent data from many countries have shown that women with at least a secondary-level education eventually give birth to between one-third and one-half as many children as women with no formal education (Population Reference Bureau 2007) (see Figure 12.6).

Figure 12.6 Lifetime Births per Woman by Highest Level of Education
Source: Population Reference Bureau 2007.

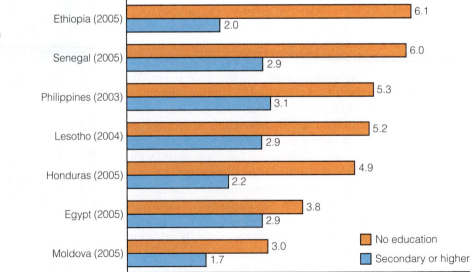

Better-educated women tend to delay marriage and exercise more control over their reproductive lives, including decisions about childbearing. In addition, "education can result in smaller family size when the education provides access to a job that offers a promising alternative to early marriage and child-bearing" (Population Reference Bureau 2004, p. 18). Providing employment opportunities for women is also important to slow population growth, because high levels of female labor force participation and higher wages for women are associated with smaller family size. Primary school enrollment of both young girls and boys is also related to declines in fertility rates.

In countries where primary school enrollment is widespread or nearly universal, fertility declines more rapidly because (1) schools help spread attitudes about the benefits of family planning, and (2) universal education increases the cost of having children, because parents sometimes are required to pay school fees for each child and because they lose potential labor that children could provide (Population Reference Bureau 2004). However, providing access to contraception and increasing girls' education are unlikely to slow population growth without changes in male attitudes toward family planning. Indeed, attempts to provide free primary education may increase African men's desire for more children because the costs are decreased (Frost & Dodoo 2009).

> Better-educated women tend to delay marriage and exercise more control over their reproductive lives, including decisions about childbearing.

In the United States, as in other countries of the world, the cultural norm is for women and couples to, sooner or later, want to have children. However, a small but growing segment of U.S. women and men does not want children and chooses to be childfree. In the United States, 7 percent of women ages 35 to 44 are voluntarily childless, making voluntary childlessness more common than involuntary childlessness (Hollander 2007). In general, childfree couples are more educated, live in urban areas, are less religious, and do not adhere to traditional gender ideology (Parks 2005). Although one of the motives some individuals who are childless by choice cite is concern for overpopulation and a deep caring for the health of the planet, voluntarily childless individuals are often criticized as being selfish and individualistic, as well as less well-adjusted and less nurturing (Parks 2005). In this chapter's The Human Side feature, one man explains why he chose to have a vasectomy and not have children.

Involvement of Men in Family Planning.
Although men play a central role in family planning decisions, they often do not have access to information and services that would empower them to make informed decisions about contraceptive use (Women's Studies Project 2003). Therefore, family planning programs need to direct educational programs and health services to men. For example, men need education about the health risks to women when pregnancies are spaced too closely or when pregnancies occur before age 20 and after age 40.

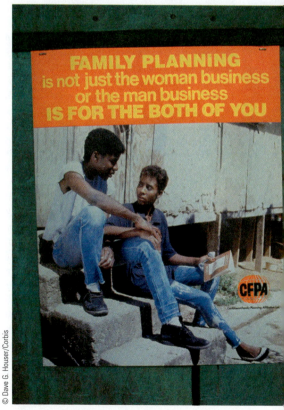

© Dave G. Houser/Corbis

Recognizing that men play a crucial role in family planning decisions, family planning programs are making efforts to include men in family planning education and services.

Matt Leonard lives in California, where he works on climate justice and energy issues with the organizations Greenpeace and Rising Tide North America. In this Human Side feature, Matt Leonard explains why he has chosen to have a vasectomy—a simple surgical procedure that makes men unable to cause pregnancy.

Courtesy of Matt Leonard

Last year, I . . . had a vasectomy. While it's actually a very common procedure (nearly 500,000 are performed every year in the United States), it raises eyebrows—and a lot of questions.

The first one is always simply: Why?

Although this was a very personal decision for me, it was also a choice I made out of larger societal, political, and environmental motivations. I consider the environmental ones paramount. In an economic system that demands infinite growth with finite resources, not doubling my own consumption is one small stone in a big river.

More importantly, I live in the United States, and any child I had would have been raised here and would consume (despite my best efforts) far more resources than I am comfortable accepting. Living even a modest lifestyle in the United States comes as a direct result of the oppression, domination, and deaths of many unseen people, not to mention the exploitation of natural resources at rates that threaten the ability of our planet to sustain life. These facts shouldn't be cause for guilt or shame; instead, they should spur us to organize to confront the systems and institutions that have created these problems. On a personal level, contributing another person to the system that I have spent my adult life fighting is just not something I'm willing to do. . . .

The next question is usually: *But what if you change your mind?*

I view my decision as permanent. As I see it, I already made the decision years ago not to have children, based on sound, rational reasons. If I change my mind in the future, I believe that change would be fundamentally selfish, and I am comfortable committing myself to rational reasons now.

People typically follow up with: *Aren't there other forms of birth control?*

Yes, of course, and most of us here in the United States are lucky to be able to choose the form that is best for our lifestyles, our preferences, and our relationships. A vasectomy fit my needs best.

I guess there's always abstinence, but that's no fun, right? I suppose the rhythm method is an option, but almost everyone knows how [in]effective that is. Condoms are fine and dandy in many situations, but they have their downsides as well, and can seem pointless if you are in a monogamous relationship.

All the other common birth control methods have one aspect in common: They place the onus on women. Not only does our society expect women to deal with the logistics of birth control, but these methods also have severe physiological drawbacks, from roller-coaster hormonal changes to intensifying menstruation cycles to weight and skin changes. Although these methods have come a long way in a few decades, they still burden women and their bodies. Is it any coincidence that in a male-dominated society, the medical establishment has thus far focused on birth control methods that leave the burden solely on women?

For men, vasectomies are simple. There are almost no side effects and no long-term impacts; it's a quick, low-cost, outpatient procedure. Having decided that I want to take an active role in birth control, a vasectomy is fair, easy, and it confronts my privilege on this issue.

What if you decide you want children in the future? people ask.

Many of my friends whom I deeply respect have chosen to have children or will do so in the future. Some people do feel that there is something special and important about having a blood-related child. I just don't share that feeling.

There are thousands of beautiful children all over the world who need parents, and if I ever decide that being a father is something I want in my life, I would be remiss to ignore the existing children needing support and love. For me, adoption is the best option. We need more parents in this world, not more kids.

Finally, *But don't we need the smart, progressive people to reproduce?*

I'm of the nurture-over-nature camp. I think the whole "passing on genes" obsession can sometimes border on eugenics. I'm fairly confident there is no gene that instructs your child to fight for justice, peace, and sustainability. That comes from living those values and instilling them in the communities we are a part of. That's what I want to prioritize in my life—and I feel I can share those things more effectively without a child.

And besides—I've got messed-up teeth, I'm legally blind, bald, and have a history of heart disease. Let Matt Damon pass on his genes instead.

Source: M. Leonard 2009. Reprinted by permission.

Another important component of family planning and reproductive health programs involves changing traditional male attitudes toward women. According to traditional male gender attitudes, (1) a woman's most important role is being a wife and mother, (2) it is a husband's right to have sex with his wife at his demand, and (3) it is a husband's right to refuse to use condoms and to forbid his wife to use any other form of contraception. In parts of Africa, men express their reluctance to use condoms by arguing that "you wouldn't eat a banana with the peel on or candy with the wrapper on" (Frost & Dodoo 2009). A number of programs around the world work with groups of boys and young men to change such traditional male gender attitudes (Schueller 2005).

Increase Economic Development and Improve Health. Although fertility reduction can be achieved without industrialization, economic development may play an important role in slowing population growth. Families in poor countries often rely on having many children to provide enough labor and income to support the family. Economic development decreases the economic value of children and is also associated with more education for women and greater gender equality.

Economic development tends to result in improved health status of populations. Reductions in infant and child mortality are important for fertility decline, because couples no longer need to have many pregnancies to ensure that some children survive into adulthood. Finally, the more developed a country is, the more likely women are to be exposed to meanings and values that promote fertility control through their interaction in educational settings and through media and information technologies (Bongaarts & Watkins 1996).

> Families in poor countries often rely on having many children to provide enough labor and income to support the family.

Efforts to Restore Urban Prosperity

Strategies to restore prosperity to U.S. cities and well-being to their residents, businesses, and workers include strategies to attract new businesses, create jobs, repopulate cities, and in some cases, shrink and deurbanize cities. The economic development and revitalization of cities also involves improving affordable housing options (see Chapter 6), alleviating urban problems related to HIV/AIDS, addiction, and crime (discussed in Chapters 2 through 4), and reducing problems of traffic and transportation, which we address later in this chapter.

Empowerment Zone/Enterprise Community Program. The federal Empowerment Zone/Enterprise Community Initiative, or EZ/EC program, provides tax incentives, grants, and loans to businesses to create jobs for residents living within various designated zones or communities, many of which are in urban areas. Federal money provided to empowerment zones and enterprise communities is also used to train and educate youth and families and to improve child care, health care, and transportation. The EZ/EC program provides grant funding so that communities can design local solutions that empower residents to participate in the revitalization of their neighborhoods.

Infrastructure Improvements. Urban revitalization often involves making improvements in the **infrastructure**—the underlying foundation that enables

infrastructure The underlying foundation that enables a city to function, including such things as water and sewer lines, phone lines, electricity cables, sidewalks, streets, bridges, curbs, lighting, and storm drainage systems.

a city to function. Infrastructure includes things such as water and sewer lines, phone lines, electricity cables, sidewalks, streets, bridges, curbs, lighting, and storm drainage systems. Improving infrastructure also helps to attract business to an area, increases property values, and renews residents' sense of pride in their neighborhood (Cowherd 2001).

Brownfield Redevelopment. **Brownfields** are abandoned or undeveloped sites that are located on contaminated land. There are more than 400,000 brownfields throughout the United States and an estimated 5 million acres of abandoned industrial sites in U.S. cities—roughly the same amount of land occupied by 60 of the nation's largest cities (U.S. Department of Housing and Urban Development 2004).

Cleaning up and redeveloping brownfields is not only an important environmental measure but also is a key component of urban revitalization because it provides jobs; increases tax revenues; attracts more businesses, residents, and tourists; and helps to curb urban sprawl. Hundreds of urban brownfield sites across the United States have been successfully redeveloped into residential, business, and recreational property. Recognizing that a major obstacle to brownfield redevelopment is lack of funding, the U.S. Department of Housing and Urban Development developed a funding program known as the Brownfields Economic Development Initiative, which has provided millions of dollars to communities across the country for the purpose of brownfield redevelopment (U.S. Department of Housing and Urban Development 2004).

Gentrification and Incumbent Upgrading. **Gentrification** is a type of neighborhood revitalization in which middle- and upper-income individuals buy and rehabilitate older homes in an economically depressed neighborhood. The city provides tax incentives for investing in old housing with the goal of attracting wealthier residents back into these neighborhoods and increasing the tax base. After the house is renovated, the owner may live there, rent, or sell the house. A downside to gentrification is that low-income city residents are often forced into substandard housing because less affordable housing is available. In effect, gentrification often displaces the poor and the elderly (Johnson 1997). However, a survey of residents in two gentrifying neighborhoods in Portland, Oregon, found that most residents—including owners and renters, whites and minorities, newcomers and long-time residents—liked how their neighborhood had changed and were optimistic that it would continue to improve (Sullivan 2007).

An alternative to gentrification is **incumbent upgrading,** in which aid programs help residents of depressed neighborhoods buy or improve their homes and stay in the community. Both gentrification and incumbent upgrading improve decaying neighborhoods, attracting residents as well as businesses.

Community Development Corporations. In many low- and moderate-income urban neighborhoods, **community development corporations (CDCs)**—nonprofit groups formed by residents, small business owners, congregations, and other local stakeholders—work to create jobs and affordable housing and create or renovate parks and other community facilities, such as child care centers, senior centers, arts and cultural centers, and health care centers. Most CDCs augment their housing and economic development projects with other community building activities, such as budget or credit counseling, immigration services,

brownfields Abandoned or undeveloped sites that are located on contaminated land.

gentrification A type of neighborhood revitalization in which middle- and upper-income individuals buy and rehabilitate older homes in an economically depressed neighborhood.

incumbent upgrading Aid programs that help residents of depressed neighborhoods buy or improve their homes and stay in the community.

community development corporations (CDCs) Nonprofit groups formed by residents, small business owners, congregations, and other local stakeholders that work to create jobs and affordable housing and renovate parks and other community facilities.

prisoner reentry services, community gardening, education and training, and homeless services. CDCs are funded by a variety of sources, including federal, state, and local governments, as well as banks, foundations, corporations, religious institutions, and individual donors. A major strength of CDCs is that they involve community residents in planning and implementing urban renewal projects, giving residents a sense of empowerment in their communities.

De-Urbanization. Flint, Michigan, one of the poorest U.S. cities, where one in five adults is jobless, has lost about half its population. The mass exodus of Flint residents, along with the collapse of property values, has left whole sections of the city almost completely abandoned. In response, Flint pioneered a "shrink to survive" urban renewal strategy known as **de-urbanization**, which involves completely leveling entire urban districts and returning the land to nature. Similar de-urbanization plans are being considered in other cities, including Detroit, Philadelphia, Pittsburgh, Baltimore, and Memphis (T. Leonard 2009).

In Flint, Michigan, a pioneering urban renewal strategy involves bulldozing entire districts and returning the land to nature.

Improve Transportation and Alleviate Traffic Congestion

An important strategy for reducing traffic congestion, as well as improving air quality and reducing greenhouse gas emissions, involves increasing the use of public transit, such as buses, trains, and subways. The majority of Americans supports building more rail systems serving cities, suburbs, and entire regions to give them the option of not driving their cars (U.S. Conference of Mayors 2001).

In Austin, Texas, a community bike program makes bikes available for anyone to use and then leave in a prominent place so someone else can use the bike.

de-urbanization An urban survival strategy that involves completely leveling entire urban districts and returning them to nature.

Some cities require motorists to pay a "congestion charge" for driving in a "congestion charge zone" during weekday high-traffic periods. Congestion charges encourage travelers to use public transport, bicycles, motorcycles, or alternative fuel vehicles, which are exempt from the charge. In 1998, Singapore became the first city to use a congestion charge. Other cities that levy congestion charges include Oslo, Bergen, and Trondheim in Norway, and London, England.

Finally, the development of communities that enable residents to walk or ride a bicycle to schools, shops, and other locations can help relieve traffic congestion, and alleviate air pollution associated with motor vehicles. In Denmark and the Netherlands, cities have created networks of streets for bicycle use where motor vehicles are banned (Newman & Kenworthy 2007).

These strategies are part of the "smart growth" and "new urbanism" movement discussed in the next section. You can assess your attitudes toward walking and proposals to create more walkable communities in this chapter's Self and Society feature.

Self and Society | Attitudes Toward Walking and Creating Better Walking Communities

For each of the following items, select the answer that best represents your attitudes. You can compare your answers with those of a national random sample of U.S. adults who participated in a 2002 telephone survey that Belden Russonello & Stewart Research and Communications (2003) conducted (*Americans' Attitudes Toward Walking*).

1. Which of the following statements describes you more? (a) If it were possible, I would like to walk more throughout the day either to get to specific places or for exercise, or (b) I prefer to drive my car wherever I go.

2. How much of a factor is each of the following in why you do not walk more right now? (a major reason; somewhat of a reason; not much of a reason; not a reason at all):

2a. Things are too far to get to, and it is not convenient to walk.

2b. Not enough time to walk.

2c. Laziness.

2d. It is hard to walk where I live because of traffic and lack of places to walk.

2e. It is hard to walk where I live because there are not enough sidewalks or crosswalks.

2f. Physically, I am unable to walk more.

2g. I do not like to walk.

2h. There is too much crime to walk where I live.

3. For each of the following proposals to create more walkable communities, indicate your views using the following key: strongly favor, somewhat favor, somewhat oppose, strongly oppose.

3a. Better enforce traffic laws such as speed limit.

3b. Use part of the transportation budget to design streets with sidewalks, safe crossings, and other devices to reduce speeding in residential areas and make it safer to walk, even if this means driving more slowly.

3c. Use part of the state transportation budget to create more sidewalks and stop signs in communities, to make it safer and easier for children to walk to school, even if this means less money to build new highways.

3d. Increase federal spending for making sure people can safely walk across the street, even if this means fewer tax dollars go to building roads.

3e. Have your state government use more of its transportation budget for improvements in public transportation, such as trains, buses, and light rail, even if this means less money to build new highways.

3f. Design communities so that more stores and other places are within walking distance of homes, even if this means building homes closer together.

Comparison Data

1. Walk more: 55 percent; drive: 41 percent; don't know: 5 percent

2. Reasons for not walking more:

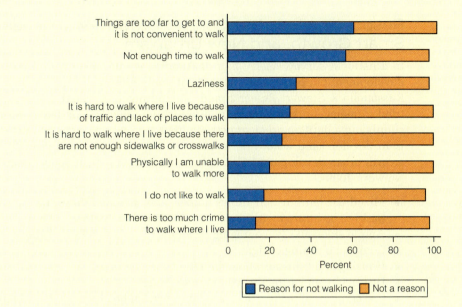

3. Proposals to create more walkable communities:

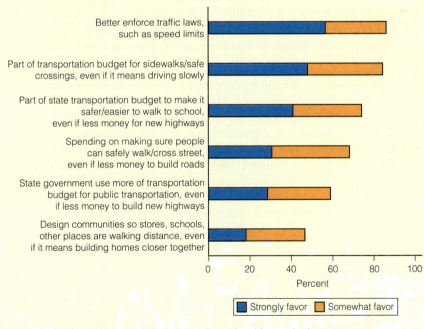

Source: Adapted from Belden Russonello & Stewart Research and Communications 2003.

Responding to Urban Sprawl: Growth Boundaries, Smart Growth, and New Urbanism

In a national survey, nearly half of Americans (46 percent) said that slowing development of open space was an "extremely high" or "high" priority (Belden Russonello & Stewart Research and Communications 2004). Some cities have tried to manage urban sprawl by establishing growth boundaries. Rather than simply put a limit on urban growth, another approach to managing urban sprawl is to develop land according to principles known as **smart growth.** A smart growth urban development plan entails the following principles (Smart Growth Network 2009):

- Mixed-use land, which allows homes, jobs, schools, shops, workplaces, and parks to be located within close proximity of each other
- Ample sidewalks, encouraging residents to walk to jobs and shops
- Compact building design
- Housing and transportation choices
- Distinctive and attractive community design
- Preservation of open space, farmland, natural beauty, and critical environmental areas
- Redevelopment of existing communities, rather than letting them decay and building new communities around them
- Regional planning and collaboration among businesses, private residents, community groups, and policy makers on development and redevelopment issues

smart growth A strategy for managing urban sprawl that serves the economic, environmental, and social needs of communities.

new urbanism A movement in urban planning that approaches the idea of sustainable urban communities with the goal of raising the quality of life for all those in the community by creating compact communities with a sustainable infrastructure.

regionalism A form of collaboration among central cities and suburbs that encourages local governments to share common responsibility for common problems.

Smart growth is similar to another movement in urban planning called **new urbanism.** The goals and methods of new urbanism are similar to those of smart growth, but the impetus for these movements is slightly different. Smart growth approaches the idea of sustainable urban communities with the primary goal of stopping sprawl. The new urbanism approach is to raise the quality of life for all those in the community by creating compact communities with a sustainable infrastructure. Smart growth and new urbanism are often impeded by local zoning codes that mandate large housing setbacks, wide streets, and separation of residential and commercial areas (Pelley 1999).

Regionalism

The various social problems that face urban areas may best be addressed through **regionalism**—a form of collaboration among central cities and suburbs that encourages local governments to share common responsibility for common problems. Central cities, declining inner suburbs, and developing suburbs are often in conflict over the distribution of government-funded resources, zoning and land use plans, transportation and transit reform, and development plans. Rather than compete with each other, regional government provides a mechanism for achieving the interests of an entire region. A metropolitan-wide government would handle the inequities and concerns of both suburban and urban areas. As might be expected, suburban officials resist regionalization because they believe it will hurt their neighborhoods economically by draining off money for the cities.

Strategies for Reducing Urban Growth in Developing Countries

In developing countries, limiting population growth is essential for alleviating social problems associated with rapidly growing urban populations. Another strategy for minimizing urban growth in less developed countries involves redistributing the population from urban to rural areas. Such redistribution strategies include the following: (1) promoting agricultural development in rural areas, (2) providing incentives to industries and businesses to relocate from urban to rural areas, (3) providing incentives to encourage new businesses and industries to develop in rural areas, and (4) developing the infrastructure of rural areas, including transportation and communication systems, clean water supplies, sanitary waste disposal systems, and social services. Of course, these strategies require economic and material resources, which are in short supply in less developed countries.

Understanding Problems of Population Growth and Urbanization

What can we conclude from our analysis of population growth and urbanization? First, although fertility rates have declined significantly in recent years and although many countries are experiencing a decline in their fertility rates, world population will continue to grow for several decades. This growth will occur largely in urban areas in developing regions. Finance columnist Paul Farrell (2009) says that population growth is "the key variable in every economic equation . . . impacting every other major issue facing world economies . . . from peak oil to global warming . . . from foreign policy to nuclear threats . . . from religion to science . . . everything." Given the problems associated with population growth, such as environmental problems and resource depletion, global insecurity, poverty and unemployment, and poor maternal and infant health, most governments recognize the value of controlling population size and support family planning programs. However, efforts to control population must go beyond providing safe, effective, and affordable methods of birth control. Slowing population growth necessitates interventions that change the cultural and structural bases for high fertility rates. Two of these interventions are increasing economic development and improving the status of women, which includes raising their levels of education, their economic position, and their (and their children's) health. Addressing problems associated with population growth also requires the willingness of wealthier countries to commit funds to providing reproductive health care to women, improving the health of populations, and providing universal education for people throughout the world (see Table 12.6).

Attention to urban problems and issues is increasingly important because the United States and the rest of the world are rapidly becoming urbanized. Aside from population concerns, problems affecting urban residents include poverty and unemployment, inferior or unaffordable housing, and traffic and transportation problems.

TABLE 12.6 Annual Expenditures on Luxury Items Compared with Funding Needed to Meet Selected Basic Needs

PRODUCT	ANNUAL EXPENDITURE
Makeup	$18 billion
Pet food in Europe and the United States	$17 billion
Perfume	$15 billion
Ocean cruises	$14 billion
Ice cream in Europe	$11 billion
SOCIAL OR ECONOMIC GOAL	ADDITIONAL ANNUAL INVESTMENT NEEDED TO ACHIEVE GOAL
Reproductive health care for all women	$12 billion
Elimination of hunger and malnutrition	$19 billion
Universal literacy	$5 billion
Clean drinking water for all	$10 billion
Immunizing every child	$1.3 billion

Source: World Watch Institute 2004.

The social forces affecting urbanization in industrialized countries are different from those in developing countries. Countries such as the United States have experienced urban decline as a result of deindustrialization, deconcentration, and the shift to a service economy in which jobs that pay well and come with full benefits are scarce. At the same time, developing countries have experienced rapid urban growth as a result of industrialization, fueled in part by a global economy in which transnational corporations locate industry in developing countries to gain access to cheap labor, raw materials, and new markets.

Like other social problems, slowing population growth and improving the cities of the world require political will and leadership. When Enrique Peñalosa became Mayor of Bogotá, Colombia, in 1998, his vision was to make Bogota a city designed for people, not cars. Under Peñalosa's leadership, the city created or renovated 1,200 parks, introduced a successful bus transit system, built extensive bicycle paths and pedestrian streets, reduced rush hour traffic by 40 percent, planted 100,000 trees, and involved local citizens directly in improving their neighborhoods (Brown 2006).

One of the shadows lingering over cities throughout the world is cast by environmental problems, because urban populations consume the largest share of the world's natural resources and contribute most of the pollution and waste that compromise the health of our planet. Global environmental problems, discussed in Chapter 13, are linked to both population growth and urban life. "Key global environmental problems have their roots in cities—from the vehicular exhaust that pollutes and warms the atmosphere, to the

As we ponder ways to improve cities and as we strive to meet the needs of growing populations, we must include the larger environment in the equation.

urban demand for timber that denudes forests and threatens biodiversity, to the municipal thirst that heightens tensions over water" (Sheehan 2003, p. 131). As we ponder ways to improve cities and as we strive to meet the needs of growing populations, we must include the larger environment in the equation.

In the current economic crisis that began in 2007, control of population and urban growth may not seem like priorities. But as finance columnist Paul Farrell (2009) warns,

> Population is the core problem that, unless confronted and dealt with, will render all solutions to all other problems irrelevant. Population is the one variable in an economic equation that impacts, aggravates, irritates, and accelerates all other problems.

CHAPTER REVIEW

- **How long did it take for the world's population to reach 1 billion? How long did it take for it to reach 6 billion?**
  It took thousands of years for the world's population to reach 1 billion, and just another 300 years for the population to grow from 1 billion to 6 billion.

- **Where is most of the world's population growth occurring?**
  Most world population growth is in developing countries, primarily in Africa and Asia. Most of this growth is occurring in urban areas.

- **In the history of urbanization, why is 2008 a historic milestone?**
  In the year 2008, the urban population equaled the rural population for the first time in history.

- **What is "urban sprawl," and why is it a problem?**
  Urban sprawl refers to the ever-increasing outward growth of urban areas. Urban sprawl results in the loss of green, open spaces, the displacement and endangerment of wildlife, traffic congestion and noise, and pollution liabilities. Sprawl is also linked to negative health effects in humans, because people in high-sprawl areas walk less and are more overweight than people who live in low-sprawl areas.

- **What is the demographic transition?**
  The demographic transition is the progression from a population with short lives and large families to one in which people live longer and have smaller families. About one-third of countries have completed the demographic transition.

- **Many countries are experiencing below-replacement fertility (fewer than 2.1 children born to each woman). Why are some countries concerned about a "birth dearth"?**
  In countries with below-replacement fertility, there are or will be fewer workers to support a growing number of elderly retirees and to maintain a productive economy.

- **What kinds of environmental problems are associated with population growth?**
  Population growth places increased demands on natural resources, such as forests, water, cropland, and oil, and results in increased waste and pollution. Although population growth is a contributing factor in environmental problems, patterns of production and consumption are at least as important in influencing the effects of population on the environment.

- **Why is population growth considered a threat to global security?**
  In developing countries, rapid population growth results in a "youth bulge"—a high proportion of 15- to 29-year-olds relative to the adult population. The combination of a youth bulge with other characteristics of rapidly growing populations, such as resource scarcity, high unemployment rates, poverty, and rapid urbanization, sets the stage for civil unrest, war, and terrorism, because large groups of unemployed young people resort to violence in an attempt to improve their living conditions.

- **Efforts to curb population growth include what strategies?**
  Efforts to curb population growth include strategies to reduce fertility by providing access to family planning services, involving men in family planning, implementing a one-child policy as in China, and improving the status of women by providing educational and employment opportunities. Achievements in economic development and health are also associated with reductions in fertility.

- **Globally, what was the average number of children born to each woman in the 1950s? Between 2005 and 2010?**
  Globally, the average number of children born to each woman has fallen from 5 in the 1950s to 2.6 between 2005 and 2010.

- **What are brownfields?**

  Brownfields are abandoned or undeveloped sites that are located on contaminated land. Cleaning up and redeveloping brownfields is a key component of urban revitalization because it provides jobs, increases tax revenues, and potentially attracts more businesses, residents, and tourists.

- **What does the term mixed-use land refer to? What are the benefits of mixed-use land?**

  The designation of *mixed-use land* is a strategy of the smart growth movement whereby homes, jobs, schools, shops, workplaces, and parks are located within close proximity of each other. Mixed-use land encourages walking as a means of transportation, which minimizes the use of cars.

TEST YOURSELF

1. In 2009, world population was 6.8 billion. In 2050, world population is projected to be _____ billion.
   a. 6.5
   b. 6.8
   c. 7.7
   d. 9.1

2. How many countries have achieved below-replacement fertility rates?
   a. None
   b. 5
   c. 28
   d. 76

3. What would happen if every country in the world achieved below-replacement fertility rates?
   a. Population growth would stop and world population would remain stable.
   b. World population would immediately begin to decline.
   c. World population would continue to grow for several decades.
   d. World population would decline, but then go up again.

4. A "megacity" is defined as a city that has at least one million residents.
   a. True
   b. False

5. Conflict theorists argue that food shortages result primarily from overpopulation of the planet.
   a. True
   b. False

6. Pronatalism is a cultural value that promotes which of the following?
   a. Car ownership
   b. Abstaining from sex until one is married
   c. Having children
   d. Urban living

7. Worldwide, the growth of slums is outpacing the growth of urbanization.
   a. True
   b. False

8. According to the U.S. Fish and Wildlife Service, what is the number one reason that wildlife species are becoming increasingly endangered?
   a. Air pollution from motor vehicles
   b. Habitat loss resulting from urban and suburban sprawl
   c. Hormones in the water supply that result from the widespread use of oral contraceptives
   d. Deaths of wild animals due to being hit by motor vehicles

9. Today, more than half of married women worldwide use some form of modern contraception.
   a. True
   b. False

10. U.S. women with advanced education are more likely than women with less education to voluntarily choose to have no children.
    a. True
    b. False

Answers: 1: d; 2: d; 3: c; 4: b; 5: b; 6: c; 7: a; 8: b; 9: a; 10: a.

KEY TERMS

barrios   450
brownfields   460
community development
   corporations (CDCs)   460
deconcentration   445
demographic transition theory   446
de-urbanization   461
doubling time   440
environmental footprint   449
gentrification   460

ghettos   450
incumbent upgrading   460
infrastructure   459
megacities   443
metropolitan area   445
micropolitan area   445
new urbanism   464
population density   441
population momentum   442
pronatalism   447

regionalism   464
replacement-level fertility   442
slums   450
smart growth   464
suburbanization   445
suburbs   445
total fertility rate   441
urban area   443
urbanization   443
urban sprawl   445

MEDIA RESOURCES

Understanding Social Problems,
Seventh Edition Companion Website
www.cengage.com/sociology/mooney
Visit your book companion website, where you will find flash cards, practice quizzes, Internet links, and more to help you study.

 Just what you need to know NOW! Spend time on what you need to master rather than on information you already have learned. Take a pretest for this chapter, and CengageNOW will generate a personalized study plan based on your results. The study plan will identify the topics you need to review and direct you to online resources to help you master those topics. You can then take a post-test to help you determine the concepts you have mastered and what you will need to work on. Try it out! Go to www.cengage.com/login to sign in with an access code or to purchase access to this product.

13

Environmental Problems

> "Just as we have the capacity to hasten the degradation and destruction of our planet, so, too, do we have the capacity to preserve, build, and improve the quality of life on our planet. The choice is ours."
>
> **Hal Burdett**
> **Population Institute**

This Greenpeace activist is climbing the 630–foot chimney at Kingsnorth coal power plant.

Will Rose/Greenpeace/PA Wire/ AP Photo

In 2007, six Greenpeace environmental activists climbed a 630-foot chimney at Kingsnorth coal power plant in England, intending to shut down the plant by occupying the chimney. They planned to write the words, "Gordon, bin it" on the chimney to pressure Prime Minister Gordon Brown to stop the building of new coal power plants, but after writing "Gordon," the six activists were served with an injunction and came down from the chimney. The "Kingsnorth Six," as they are called, were criminally charged for property damage; it cost 35,000 euros (US $53,000) to remove the graffiti. Jurors in the case found the Kingsnorth Six "not guilty," accepting the defense arguments that the six activists had a "lawful excuse" to damage property at the Kingsnorth power station to prevent even greater damage caused by global warming (McCarthy 2008). The Criminal Damage Act of 1971 allows individuals to damage property to prevent even greater damage—such as breaking down the door of a burning house to put out a fire. James Hansen, a top NASA climate scientist, testified for the defense, and told the jury that carbon dioxide emissions from the Kingsnorth power plant would contribute to climate change that would damage property. Hansen told the court that the 20,000 tons of carbon dioxide that Kingsnorth emitted daily could be responsible for the extinction of up to 400 species, and that humanity was in "grave peril." "Somebody needs to step forward and say there has to be a moratorium, draw a line in the sand, and say no more coal-fired power stations" (quoted in Vidal 2008).

In this chapter, we focus on environmental problems that threaten the lives and well-being of people, plants, and animals all over the world—today and in future generations. After examining how globalization affects environmental problems, we view environmental issues through the lens of structural functionalism, conflict theory, and symbolic interactionism. We then present an overview of major environmental problems, examining their social causes and exploring strategies that are used in the attempt to reduce or alleviate them.

The Global Context: Globalization and the Environment

Two aspects of globalization that have affected the environment are (1) the permeability of international borders to pollution and environmental problems, and (2) the growth of free trade and transnational corporations.

Permeability of International Borders

Environmental problems such as global warming and destruction of the ozone layer (discussed later in this chapter) extend far beyond their source to affect the entire planet. A striking example of the permeability of international borders to pollution is the spread of toxic chemicals (such as polychlorinated biphenyls [PCBs]) from the Southern Hemisphere into the Arctic. In as few as five days, chemicals from the tropics can evaporate from the soil, ride the winds thousands of miles north, condense in the cold air, and fall on the Arctic in the form of toxic snow or rain (French 2000). This phenomenon was discovered in the mid-1980s, when scientists found high levels of PCBs in the breast milk of Inuit women in the Canadian Arctic region.

Another environmental problem involving permeability of borders is **bioinvasion:** the intentional or accidental introduction of organisms in regions where they are not native. Bioinvasion is largely a product of the growth of global trade and tourism (Chafe 2005). Invasive species compete with native species for food, start an epidemic, or prey on native species.

Red fire ants are an example of a bioinvasion. In 1957, they traveled from Paraguay and Brazil on shiploads of lumber to Mobile, Alabama, and have since spread throughout the southern states, causing damage to gardens and yards, invading the food supplies (seeds, young plants, and insects) of other animals, and harming humans with their painful sting (Hilgenkamp 2005). You might be surprised to learn that the domestic cat is considered among the world's 100 worst invasive species. Cats, native to northeast Africa, have spread to every part of the world and are responsible for the decline and extinction of many species of birds (Global Invasive Species Database 2006).

The Growth of Transnational Corporations and Free Trade Agreements

As discussed in Chapter 7, the world's economy is dominated by transnational corporations, many of which have established factories and other operations in developing countries where labor and environmental laws are lax. Transnational corporations have been implicated in environmentally destructive activities—from mining and cutting timber to dumping toxic waste.

The World Trade Organization (WTO) and free trade agreements such as the North American Free Trade Agreement (NAFTA) and the Free Trade Area of the Americas (FTAA) allow transnational corporations to pursue profits, expand markets, use natural resources, and exploit cheap labor in developing countries while weakening the ability of governments to protect natural resources or to implement environmental legislation. Transnational corporations have influenced the world's most powerful nations to institutionalize an international system of governance that values commercialism, corporate rights, and "free" trade over the environment, human rights, worker rights, and human health (Bruno & Karliner 2002).

Under NAFTA's Chapter 11 provisions, corporations can challenge local and state environmental policies, federal controlled substances regulations, and

Toxic chemicals travel thousands of miles from the Southern Hemisphere to the Arctic, where they have been found in the breast milk of Inuit women.

Bryan & Cherry Alexander Photography/Alamy

bioinvasion The intentional or accidental introduction of plant, animal, insect, and other species in regions where they are not native.

court rulings if such regulatory measures and government actions negatively affect the corporation's profits. Any country that decides, for example, to ban the export of raw logs as a means of conserving its forests or, as another example, to ban the use of carcinogenic pesticides, can be charged under the WTO by member states on behalf of their corporations for obstructing the free flow of trade and investment. A secret tribunal of trade officials would then decide whether these laws were "trade-restrictive" under the WTO rules and should therefore be struck down. Once the secret tribunal issues its edict, no appeal is possible. The convicted country is obligated to change its laws or face the prospect of perpetual trade sanctions (Clarke 2002, p. 44). For example, in the late 1990s, Ethyl, a U.S. chemical company, used NAFTA rules to challenge Canada's decision to ban the gasoline additive methylcyclopentadienyl manganese tricarbonyl (MMT), which is believed to have harmful effects on human health. Ethyl won the suit, and Canada paid $13 million in damages and legal fees to Ethyl and reversed the ban on MMT (Public Citizen 2005).

Sociological Theories of Environmental Problems

The three main sociological theories—structural functionalism, conflict theory, and symbolic interactionism—provide insights into social causes of and responses to environmental problems.

Structural-Functionalist Perspective

Structural functionalism focuses on how changes in one aspect of the social system affect other aspects of society. For example, agriculture, forestry, and fishing provide 50 percent of all jobs worldwide and 70 percent of jobs in sub-Saharan Africa, East Asia, and the Pacific (World Resources Institute 2000). As croplands become scarce or degraded, as forests shrink, and as marine life dwindles, millions of people who make their living from these natural resources must find alternative livelihoods. Globally, there are an estimated 30 million **environmental refugees**—individuals who have migrated because they can no longer secure a livelihood as a result of environmental problems (Margesson 2005). As individuals lose their source of income, so do nations. In one-fourth of the world's nations, crops, timber, and fish contribute more to the nation's economy than industrial goods do (World Resources Institute 2000).

The structural-functionalist perspective raises our awareness of latent dysfunctions—negative consequences of social actions that are unintended and not widely recognized. For example, the more than 840,000 dams worldwide provide water to irrigate farmlands and supply some of the world's electricity. Yet dam building has had unintended negative consequences for the environment, including the loss of wetlands and wildlife habitat, the emission of methane (a gas that contributes to global warming) from rotting vegetation trapped in reservoirs, and the alteration of river flows downstream, which kills plant and animal life (Environmental Defense Fund 2001). Dams have also displaced millions of people from their homes.

The expanding production of biofuels also has unintended consequences. More than one-third of U.S. corn production in 2008 was used to produce

environmental refugees Individuals who have migrated because they can no longer secure a livelihood as a result of deforestation, desertification, soil erosion, and other environmental problems.

ethanol (an alternative fuel discussed later in this chapter), and about half of the vegetable oils produced in the European Union was used for biodiesel fuel. Although biofuels help reduce dependence on fossil fuels, they have caused significant increases in global food prices (World Water Assessment Program 2009). As philosopher Kathleen Moore pointed out, "Sometimes in maximizing the benefits in one place, you create a greater harm somewhere else" (quoted by Jensen 2001, p. 11).

Conflict Perspective

The conflict perspective focuses on how wealth, power, and the pursuit of profit underlie many environmental problems. Per capita, wealthy nations consume more natural resources and generate higher amounts of pollution and waste. Wealthy nations exploit less developed nations for raw materials, labor, and as a market to sell goods to (Barbosa 2009).

The capitalistic pursuit of profit encourages making money from industry regardless of the damage done to the environment. McDaniel (2005) noted that "our culture tolerates environmentalism only so long as it has minimal impact on big business. . . . In an economically centered culture, jobs come first, not the health of people or the environment" (pp. 22–23). To maximize sales, manufacturers design products intended to become obsolete. As a result of this **planned obsolescence,** consumers continually throw away used products and purchase replacements. Industry profits at the expense of the environment, which must sustain the constant production and absorb ever-increasing amounts of waste.

> The capitalistic pursuit of profit encourages making money from industry regardless of the damage done to the environment.

Industries also use their power and wealth to influence politicians' environmental and energy policies as well as the public's beliefs about environmental issues. ExxonMobil, the world's largest oil company, has spent millions of dollars on lobbying and has funded numerous organizations that have tried to discredit scientific findings that link fossil fuel burning to global climate change (Mooney 2005). Current government policies that support ethanol have been linked to political financial contributions by major players in the ethanol industry, such as Archer Daniels Midland, the nation's largest ethanol producer (Food & Water Watch and Network for New Energy Choices 2007). The American Coalition for Clean Coal Electricity (ACCCE), consisting of 48 mining, rail, manufacturing, and power-generation companies, spent nearly $10 million on lobbying in 2008, far more than any other group devoted to influencing climate change legislation. In the 2008 election cycle, most members of Congress received contributions from political action committees (PACS) and individuals employed by ACCCE member firms, with John McCain being the top recipient, followed by Barack Obama (Lavelle 2009).

Symbolic Interactionist Perspective

The symbolic interactionist perspective focuses on how meanings, labels, and definitions learned through interaction and through the media affect environmental problems. Whether an individual recycles, drives a sport-utility vehicle (SUV), or joins an environmental activist group is influenced by the meanings and definitions of these behaviors that the individual learns through interaction with others.

planned obsolescence The manufacturing of products that are intended to become inoperative or outdated in a fairly short period of time.

Large corporations and industries commonly use marketing and public relations strategies to construct favorable meanings of their corporation or industry. The term **greenwashing** refers to the way in which environmentally and socially damaging companies portray their corporate image, products, and services as being "environmentally friendly" or socially responsible. Greenwashing is commonly used by public relations firms that specialize in damage control for clients whose reputations and profits have been hurt by poor environmental practices. For example, coal is associated with the devastation of communities through the mining practice of mountaintop removal, and the burning of coal is the biggest contributor to pollution that causes global warming. The coal industry has spent enormous sums to convince the public that coal is clean. The "clean coal" campaign has invited widespread criticism from environmentalists: "Saying coal is clean is like talking about healthy cigarettes. There's no such thing as clean coal" (Beinecke 2009).

Although greenwashing involves manipulation of public perception to maximize profits, many corporations make genuine and legitimate efforts to improve their operations, packaging, or overall sense of corporate responsibility toward the environment. For example, in 1990, McDonald's announced that it was phasing out foam packaging and switching to a new, paper-based packaging that is partially degradable. But many environmentalists are not satisfied with what they see as token environmentalism, or as Peter Dykstra of Greenpeace suggested, 5 percent of environmental virtue to mask 95 percent of environmental vice (Hager & Burton 2000).

What Do You Think? Earth Day events have been sponsored by corporations such as Office Depot, Texas Instruments, Raytheon Missile Systems, and Waste Management (a Houston-based company responsible for numerous hazardous waste sites). Some environmentalists, including Earth Day's founder, former senator Gaylord Nelson, consider the participation of corporations in Earth Day evidence of the celebration's success. But other environmentalists accuse corporations of using their financial support of Earth Day as a public relations greenwashing strategy. Do you think that organizers of Earth Day events should accept sponsorship from corporations with poor environmental records? Why or why not?

Environmental Problems: An Overview

Over the past 50 years, humans have altered **ecosystems**—the complex and dynamic relationships between forms of life and the environments they inhabit—more rapidly and extensively than in any other comparable period of time in history (Millennium Ecosystem Assessment 2005). As a result, humans have created environmental problems, including depletion of natural resources; air, land, and water pollution; global warming and climate change; environmental illness; environmental injustice; threats to biodiversity; and light pollution. Because many of these environmental problems are related to the ways that humans produce and consume energy, we begin this section with an overview of global energy use.

greenwashing The way in which environmentally and socially damaging companies portray their corporate image and products as being "environmentally friendly" or socially responsible.

ecosystems The complex and dynamic relationships between forms of life and the environments they inhabit.

Go into any large grocery or "big box" store today and take notice of how many products are advertised as "green," "all natural," or "earth-friendly." Marketing companies know that "green" sells, as consumers are becoming more eco-minded. But how valid are the environmental claims made on the labels of the products we buy? Are the claims trustworthy? Or do marketers deliberately mislead consumers regarding the environmental practices of a company or the environmental benefits of a product or service—a practice known as greenwashing? In this Social Problems Research Up Close feature, we present a study that attempts to answer these questions.

Sample and Methods

In 2008 and 2009, TerraChoice, an environmental marketing company, sent researchers into leading "big box" retailers in the United States, Canada, the United Kingdom, and Australia with instructions to record every product making an environmental claim. For each product, the researchers recorded product details, claim(s) details, any supporting information, and any explanatory detail or offers of additional information or support. In the United States and Canada, a total of 2,219 products making 4,996 green claims were recorded. These claims were tested against guidelines provided by the U.S. Federal Trade Commission, Competition Bureau of Canada, Australian Competition and Consumer Commission, and the International Organization for Standardization (ISO) 14021 standard for environmental labeling. The researchers wanted to answer the following questions about each green claim they identified: (1) Is the claim truthful? (2) Does the company offer validation for its claim from an independent and trusted third party? (3) Is the claim specific, using terms that have agreed-upon definitions, not vague ones like "natural" or "nontoxic"? (4) Is the claim relevant to the product it accompanies? (5) Does the claim address the product's principal environmental impact(s) or does it distract consumers from the product's real problems?

Selected Findings and Conclusions

This study found that green claims are most common for products related to children (toys and baby products), cosmetics, and cleaning products. Of the 2,219 North American products surveyed, over 98 percent committed at least one of the following seven "sins" of greenwashing:

1. *Sin of the hidden trade-off*, committed by suggesting a product is "green" based on an unreasonably narrow set of attributes without attention to other important environmental issues. Paper, for example, is not necessarily environmentally preferable just because it comes from a sustainably harvested forest. Other important environmental issues in the papermaking process, including energy, greenhouse gas emissions, and water and air pollution, may be equally or more significant.
2. *Sin of no proof,* committed by an environmental claim that cannot be substantiated by easily accessible supporting information or by a reliable third-party certification. Common examples are facial or toilet tissue products that claim various percentages of postconsumer recycled content without providing any evidence.
3. *Sin of vagueness,* committed by every claim that is so poorly defined or broad that consumers are likely to misunderstand its real meaning. "All natural" is an example. Arsenic, uranium, mercury, and formaldehyde are all naturally occurring, and poisonous. "All natural" isn't necessarily "green."
4. *Sin of irrelevance,* committed by making an environmental claim that may be truthful but is unimportant or unhelpful for consumers seeking environmentally preferable products. "CFC-free" is a common example, because it is a frequent claim despite the fact that CFCs are banned by law.
5. *Sin of lesser of two evils,* committed by claims that may be true within the product category, but that may distract the consumer from the greater environmental impacts of the category as a whole. Organic cigarettes are an example of this category, as are fuel-efficient sport-utility vehicles.
6. *Sin of fibbing,* the least frequent sin, is committed by making environmental claims that are simply false. The most common examples were products falsely claiming to be Energy Star-certified or -registered.
7. *Sin of worshiping false labels* is committed by a product that, through either words or images, gives the impression of third-party endorsement where no such endorsement actually exists.

Although many companies are making meaningful efforts to minimize their impact on the environment, this study alerts consumers to the widespread practice of greenwashing and suggests that there is a need for more accountability and transparency in the marketing of products as "green."

Source: TerraChoice Group Inc. 2009.

Energy Use Worldwide: An Overview

Being mindful of environmental problems means seeing the connections between energy use and our daily lives. Most of us don't think about how dependent we are on energy:

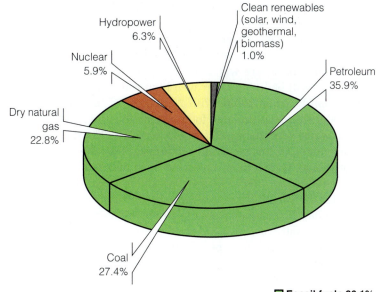

Figure 13.1 **World Energy Production by Source, 2006**
Source: Energy Information Administration (2008).

Hydropower 6.3%

Clean renewables (solar, wind, geothermal, biomass) 1.0%

Nuclear 5.9%

Petroleum 35.9%

Dry natural gas 22.8%

Coal 27.4%

■ Fossil fuels 86.1%

Everything we consume or use—our homes, their contents, our cars and the roads we travel, the clothes we wear, and the food we eat—requires energy to produce and package, to distribute to shops or front doors, to operate, and then to get rid of. We rarely consider where this energy comes from or how much of it we use—or how much we truly need. (Sawin 2004, p. 25)

Most of the world's energy—86 percent in 2006—comes from fossil fuels, which include petroleum (or oil), coal, and natural gas (Energy Information Administration 2008) (see Figure 13.1). As you continue reading this chapter, notice that the major environmental problems facing the world today—air, land, and water pollution, destruction of habitats, biodiversity loss, global warming, and environmental illness—are linked to the production and use of fossil fuels.

The next most common source of energy is hydroelectric power (6.3 percent), which involves generating electricity from moving water. As water passes through a dam into a river below, a turbine in the dam produces energy. Although hydroelectric power is nonpolluting and inexpensive, it is criticized for affecting natural habitats. For example, dams make certain fish unable to swim upstream to reproduce.

Nuclear power, accounting for 5.9 percent of world energy production, is associated with a number of problems related to radioactive nuclear waste—problems that are discussed later in this chapter. Only 1 percent of the world's energy comes from clean renewable resources, which include geothermal power (from the heat of the earth), solar power, wind power, and biomass (e.g., fuel wood, crops, and animal wastes), which are primarily used by poor populations in developing countries.

Depletion of Natural Resources: Our Growing Environmental Footprint

Humans have used more of the earth's natural resources since 1950 than in the million years preceding 1950 (Lamm 2006). Population growth, combined with consumption patterns, is depleting natural resources such as forests, water, minerals, and fossil fuels.

Water supplies around the world are dwindling, while the demand for water continues to increase because of industrialization, rising living standards, and changing diets that include more food products that require larger amounts of water to produce: milk, eggs, chicken, and beef. Currently, 50 countries are facing moderate or severe water stress; by the year 2030, nearly half the world's population will be living in areas of high water stress (WWF 2008; World Water Assessment Program 2009). With 70 percent of freshwater use going to agriculture, water shortages threaten food production and supply.

The world's forests are also being depleted. The demand for new land, fuel, and raw materials has resulted in **deforestation**—the conversion of forestland to nonforestland. Global forest cover has been reduced by half of what it was 8,000 years ago (Gardner 2005). Every year, about 13 million hectares of the world's forests (an area about the size of Greece) are cut down and converted to land use (UNEP 2008). The major causes of deforestation are the expansion of agricultural land, human settlements, wood harvesting, and road building.

Deforestation displaces people and wild species from their habitats; soil erosion caused by deforestation can cause severe flooding; and, as we explain later in this chapter, deforestation contributes to global warming. Deforestation also contributes to **desertification**—the degradation of semiarid land, which results in the expansion of desert land that is unusable for agriculture. Overgrazing by cattle and other herd animals also contributes to desertification. The problem of desertification is most severe in Africa (Reese 2001). As more land turns into desert, populations can no longer sustain a livelihood on the land, and so they migrate to urban areas or other countries, contributing to social and political instability.

The demands that humanity makes on the earth's natural resources are known as the **environmental footprint.** A person's environmental footprint is determined by the patterns of production and consumption in that person's culture. Environmental footprints are a measure of how much of the earth people require to provide the natural resources they consume, and are often expressed in terms of global hectares per person (1 hectare equals about 2.47 acres). Collectively, human demand on the world's natural resources exceeds the earth's capacity by nearly a third; 77 percent of us are living in ecological debt, using more natural resources than the earth can sustain. If we continue current patterns of consumption, we would need the equivalent of two planets to support us by 2030 (WWF 2008).

> If we continue current patterns of consumption, we would need the equivalent of two planets by 2030 to support us.

deforestation The conversion of forestland to nonforestland.

desertification The degradation of semiarid land, which results in the expansion of desert land that is unusable for agriculture.

environmental footprint The demand that each person makes on the Earth's natural resources, often measured in terms of global hectares (gha) per person.

Air Pollution

Transportation vehicles, fuel combustion, industrial processes (such as burning coal and processing minerals from mining), and solid waste disposal have contributed to the growing levels of air pollutants, including carbon monoxide, sulfur dioxide, nitrogen dioxide, mercury, and lead. Air pollution, which is linked to heart disease, lung cancer, emphysema, chronic bronchitis, and asthma, kills about 3 million people a year (Pimentel et al. 2007).

In the United States, six out of ten people live in counties where they are exposed to unhealthy levels of air pollution (ozone or particulate pollution) (American Lung Association 2009). Indoor air pollution from burning wood and biomass for heating and cooking, which we discuss next, is a significant cause of respiratory illness, lung cancer, and blindness in developing countries.

Indoor Air Pollution. When we hear the phrase *air pollution,* we typically think of industrial smokestacks and vehicle exhausts pouring gray streams of chemical matter into the air. But indoor air pollution is also a major problem, especially in poor countries.

More than half of the world's population cook food and generate heat by burning dung, wood, crop waste, or coal on open fires or stoves without chimneys (World Health Organization 2005). The resulting indoor smoke contains health-damaging pollutants including small soot or dust particles that are able to penetrate deep into the lungs. Exposure is particularly high among women and children, who spend the most time near the domestic hearth or stove. Exposure to indoor air pollution increases the risk of pneumonia, chronic respiratory disease, asthma, cataracts, tuberculosis, and lung cancer.

Even in affluent countries, much air pollution is invisible to the eye and exists where we least expect it—in our homes, schools, workplaces, and public buildings. Sources of indoor air pollution include lead dust (from old lead-based paint); secondhand tobacco smoke; by-products of combustion (e.g., carbon monoxide) from stoves, furnaces, fireplaces, heaters, and dryers; and other common household, personal, and commercial products (American Lung Association 2005). Some of the most common indoor pollutants include carpeting (which emits more than a dozen toxic chemicals); mattresses, sofas, and pillows (which emit formaldehyde and fire retardants); pressed wood found in kitchen cabinets and furniture (which emits formaldehyde); and wool blankets and dry-cleaned clothing (which emit perchloroethylene). Air fresheners, deodorizers, and disinfectants emit the pesticide paradichlorobenzene. Potentially harmful organic solvents are present in numerous office supplies, including glue, correction fluid, printing ink, carbonless paper, and felt-tip markers. Many homes today contain a cocktail of toxic chemicals: "Styrene (from plastics), benzene (from plastics and rubber), toluene and xylene, trichloroethylene, dichloromethane, trimethylbenzene, hexanes, phenols, pentanes, and much more outgas from our everyday furnishings, construction materials, and appliances" (Rogers 2002).

Destruction of the Ozone Layer. The ozone layer of the earth's atmosphere protects life on earth from the sun's harmful ultraviolet rays. Yet the ozone layer has been weakened by the use of certain chemicals, particularly chlorofluorocarbons (CFCs), used in refrigerators, air conditioners, spray cans, and other applications. The depletion of the ozone layer allows hazardous levels of ultraviolet rays to reach the earth's surface and is linked to increases in skin cancer and cataracts, weakened immune systems, reduced crop yields, damage to ocean ecosystems and reduced fishing yields, and adverse effects on animals.

Indoor pollution is a serious problem in developing countries. As this woman cooks food for her family, she is exposed to harmful air contaminants from the fumes.

The ozone hole in 2008 was 9.8 million square miles—just larger than the size of North America (NASA 2009). The ozone hole was largest in 2006, when it reached a record-breaking area of 10.3 million square miles. Despite measures that have ended production of CFCs, the ozone hole is not expected to shrink significantly for about another decade. This is because CFCs already in the atmosphere remain for 40 to 100 years.

Acid Rain. Air pollutants, such as sulfur dioxide and nitrogen oxide, mix with precipitation to form **acid rain.** Polluted rain, snow, and fog contaminate crops, forests, lakes, and rivers. As a result of the effects of acid rain, all the fish have died in a third of the lakes in New York's Adirondack Mountains (Blatt 2005). Because winds carry pollutants in the air, industrial pollution in the Midwest falls back to earth as acid rain on southeast Canada and the northeast New England states. Acid rain is not just a problem in North America; it decimates plant and animal species around the globe. In China, most of the electricity comes from burning coal, which creates sulfur dioxide pollution and acid rain that falls on one-third of China, damaging lakes, forests, and crops (Woodward 2007). Acid rain also deteriorates the surfaces of buildings and statues. "The Parthenon, Taj Mahal, and Michelangelo's statues are dissolving under the onslaught of the acid pouring out of the skies" (Blatt 2005, p. 161).

Global Warming and Climate Change

Global warming refers to the increasing average global air temperature, caused mainly by the accumulation of various gases (greenhouse gases) that collect in the atmosphere. According to the Intergovernmental Panel on Climate Change (IPCC)—a team of more than 1,000 scientists from 113 countries—"Warming of the climate system is unequivocal, as is now evident from observations of increases in global average air and ocean temperatures, widespread melting of snow and ice, and rising global average sea level" (2007a, p. 5). Average global surface temperatures have increased by about 0.74°C over the past century (between 1906 and 2005) (Intergovernmental Panel on Climate Change 2007a). If current greenhouse gas emissions trends continue, average global temperatures may rise another 2°C by 2035. Although 2°C may not seem significant, the effects of a global temperature increase of 2 degrees are expected to be catastrophic (Global Humanitarian Forum 2009).

Causes of Global Warming. The prevailing scientific view is that **greenhouse gases**—primarily carbon dioxide (CO_2), methane, and nitrous oxide—accumulate in the atmosphere and act like the glass in a greenhouse, holding heat from the sun close to the earth. Most scientists believe that global warming has resulted from the marked increase in global atmospheric concentrations of greenhouse gases since industrialization began. Global increases in carbon dioxide concentration are due primarily to the use of fossil fuels. Deforestation also contributes to increasing levels of carbon dioxide in the atmosphere. Trees and other plant life use carbon dioxide and release oxygen into the air. As forests are cut down or are burned, fewer trees are available to absorb the carbon dioxide. The greenhouse gases methane and nitrous oxide are primarily due to agriculture (Intergovernmental Panel on Climate Change 2007a). Despite the scientific evidence that human activity causes global warming, one-third of U.S. adults believe that natural changes in the environment cause global warming (Maibach et al. 2009).

acid rain The mixture of precipitation with air pollutants, such as sulfur dioxide and nitrogen oxide.

global warming The increasing average global air temperature, caused mainly by the accumulation of various gases (greenhouse gases) that collect in the atmosphere.

greenhouse gases Gases (primarily carbon dioxide, methane, and nitrous oxide) that accumulate in the atmosphere and act like the glass in a greenhouse, holding heat from the sun close to the earth.

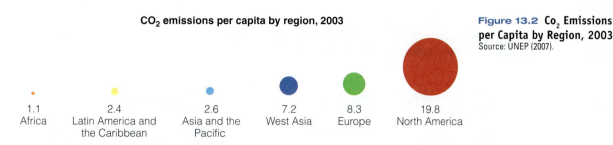

CO₂ emissions per capita by region, 2003

Figure 13.2 Co₂ Emissions per Capita by Region, 2003
Source: UNEP (2007).

1.1	2.4	2.6	7.2	8.3	19.8
Africa	Latin America and the Caribbean	Asia and the Pacific	West Asia	Europe	North America

Units of measurement: Tons per capita

But the growth of emissions is strongest in developing countries, particularly China. With less than 5 percent of the world's population, the United States produces 20 percent of the world's fossil fuel CO_2 emissions (Flavin & Engelman 2009). In North America, carbon emissions per person far exceed that of other regions of the world (see Figure 13.2).

Even if greenhouse gases are stabilized, global air temperature and sea level are expected to continue to rise for hundreds of years. That is because global warming that has already occurred contributes to further warming of the planet—a process known as a *positive feedback loop.* For example, the melting of Siberia's frozen peat bog could release billions of tons of methane, a potent greenhouse gas, into the atmosphere (Pearce 2005). And the melting of ice and snow—another result of global warming—exposes more land and ocean area, which absorbs more heat than ice and snow, further warming the planet.

> Even if greenhouse gases are stabilized, global air temperature and sea level are expected to continue to rise for hundreds of years.

Effects of Global Warming and Climate Change. Numerous effects of global warming and climate change have been observed and are anticipated in the future (Intergovernmental Panel on Climate Change 2007b; National Assessment Synthesis Team 2000; UNEP 2007) (see this chapter's Photo Essay). Global warming and climate change are projected to affect regions in different ways. As temperature increases, some areas will experience heavier rain, whereas other regions will get drier, although global warming has produced an overall 2.2 percent increase in the air's humidity over a 30-year period (from 1973 to 2002) (Willett et al. 2007). Some regions will experience increased water availability and crop yields, but other regions, particularly tropical and subtropical regions, are expected to experience decreased water availability and a reduction in crop yields. Global warming results in shifts of plant and animal habitats and the increased risk of extinction of some species. Regions that experience increased rainfall as a result of increasing temperatures may face increases in waterborne diseases and diseases transmitted by insects. Over time, climate change may produce opposite effects on the same resource. For example, in the short term, forest productivity is likely to increase in response to higher levels of carbon dioxide in the air, but over the long term, forest productivity is likely to decrease because of drought, fire, insects, and disease.

As a result of global warming, sea ice in the Arctic is melting earlier and forming later each year. Arctic sea ice during the 2007 melt season was at the lowest level since satellite measures began in 1979. In just two years, from 2005 to 2007, sea ice plummeted from 2.14 million square miles to 1.65 million square

▲ The golden toad of Monteverde may be the first species to become extinct due to global warming.

Michael Patricia Fogden/Minden Pictures/National Geographic Stock

Eric Mille/ AP Photo

▲ Global warming is expected to produce increases in draught, threatening agricultural production in areas of the world where hunger is already epidemic.

Hans Strand/CORBIS

▲ Shrinking Arctic ice threatens the survival of the polar bear.

Global warming and climate change may be the most significant threats facing the world today. In this photo essay, we present just a few examples of the many current and projected effects of global warming and climate change.

Threat of Species Extinction. Scientists have predicted that, in certain areas of the world, global warming will lead to the extinction of up to 43 percent of plant and animal species, representing the potential loss of 56,000 plant species and 3,700 vertebrate species (Malcolm et al. 2006). The U.S. Geological Survey (2007) predicts that the entire polar bear population of Alaska may be extinct in the next 43 years.

Sea-Level Rise. As higher global temperatures melt ice caps, sea level rises. Scientists predict that, by the end of the century, sea level will rise up to 2.5 feet (Siddall et al. 2009). Rising sea levels pose a threat to coastal populations around the world.

Extreme Weather: Hurricanes, Droughts, and Heat Waves. Rising temperatures are causing drought in some parts of the world, and too much rain in other parts. Globally, the proportion of land surface in extreme drought is predicted to increase from 1 percent to 3 percent (present day) to 30 percent by 2090 (Intergovernmental

Panel on Climate Change 2007). Warmer tropical ocean temperatures can cause more intense hurricanes (Chafe 2006). With rising temperatures, an increase in the number, intensity, and duration of heat waves is expected, with the accompanying adverse health effects (Intergovernmental Panel on Climate Change 2007).

Spread of Disease. Increased heavy rains and flooding caused by global warming contribute to the spread of disease. Flooding, for example, provides fertile breeding grounds for mosquitoes that carry a variety of diseases including encephalitis, dengue fever, yellow fever, West Nile virus, and malaria (Knoell 2007). With the warming of the planet, mosquitoes are now living in areas in which they previously were not found, placing more people at risk of acquiring one of the diseases carried by the insect.

Another effect of global warming is an increase in the number and size of forest fires. In the Western region of the United States, wildfire activity increased suddenly and markedly in the mid-1980s, with higher frequency of large wildfires, longer wildfire durations, and longer wildfire seasons (Westerling et al. 2006). Warmer temperatures dry out brush and trees, creating ideal conditions for fires to spread. Global warming also means that spring comes earlier, making the fire season longer.

▲ Warmer tropical ocean temperatures are linked to increased hurricane intensity.

This man is being ▶ treated for dengue fever at a military hospital in Brazil.

Global warming has ▶ contributed to an increase in megafires in the U.S. West.

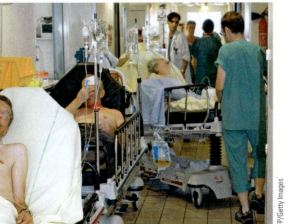

◀ Nearly 15,000 people died in France during a heat wave in August 2003. Hospitals were so overcrowded with individuals who had heat-related illness cases that they were treated in hospital corridors.

miles—a 23 percent drop. Scientists say the Arctic Ocean in summer could be ice-free by 2030 (National Snow and Ice Data Center 2007). As global warming melts glaciers and permafrost (soil at or below freezing temperature for two or more years), the sea level will continue to rise. As sea levels rise, some island countries, as well as some barrier islands off the U.S. coast, are likely to disappear, and low-lying coastal areas will become increasingly vulnerable to storm surges and flooding.

In urban areas, flooding can be a problem where storm drains and waste management systems are inadequate. Increased flooding associated with global warming is expected to result in increases in drownings and in diarrheal and respiratory diseases. Increases in the number of people exposed to insect- and water-related diseases, such as malaria and cholera, are also expected. Climate change kills an estimated 30,000 people per year, mostly in the developing world (Global Humanitarian Forum 2009). The majority of these deaths are attributed to crop failure leading to malnutrition and water problems such as flooding and drought.

Land Pollution

About 30 percent of the world's surface is land, which provides soil to grow the food we eat. Increasingly, humans are polluting the land with nuclear waste, solid waste, and pesticides. In 2009, 1,264 hazardous waste sites (also called Superfund sites) were on the National Priority List (EPA 2009).

Nuclear Waste. Nuclear waste, resulting from both nuclear weapons production and nuclear reactors or power plants, contains radioactive plutonium, a substance linked to cancer and genetic defects. Radioactive plutonium has a half-life of 24,000 years, meaning that it takes 24,000 years for the radioactivity to be reduced by half (Mead 1998). Thus, nuclear waste in the environment remains potentially harmful to human and other life for thousands of years.

In the United States, nuclear waste is being stored temporarily in 121 aboveground sites in 39 states. The first planned U.S. repository for nuclear waste was in Yucca Mountain, 100 miles northwest of Las Vegas. However, President Obama has rejected this plan, saying there are too many questions about whether nuclear waste storage at Yucca Mountain would be safe. The question remains about how to safely dispose of nuclear waste.

Nuclear plants have about 52,000 tons of radioactive spent fuel, with about 10,000 tons of that amount sealed in casks (Vedantam 2005b). Because of inadequate oversight and gaps in safety procedures, radioactive spent fuel is missing or unaccounted for at some U.S. nuclear power plants, which raises serious safety concerns (Vedantam 2005a). Accidents at nuclear power plants, such as the 1986 accident at Chernobyl, and the potential for nuclear reactors to be targeted by terrorists add to the actual and potential dangers of nuclear power plants.

Recognizing the hazards of nuclear power plants and their waste, Germany became the first country to order all of its 19 nuclear power plants shut down by 2020 ("Nukes Rebuked" 2000). Belgium is also phasing out nuclear reactors, and Austria, Denmark, Italy, and Iceland have prohibitions against nuclear energy. Nevertheless, at the end of 2007, there were 439 operating nuclear reactors worldwide, with 13 more under construction (International Atomic Energy Agency 2008). The United States had the most operating reactors (104) followed by France (59), Japan (55), and Russia (31).

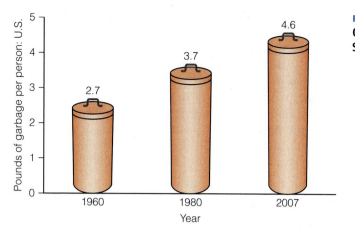

Figure 13.3 **Pounds of Garbage per Person, United States, 2008**

Solid Waste. In 1960, each U.S. citizen generated 2.7 pounds of garbage on average every day. This figure increased to 3.7 pounds in 1980, and to 4.6 pounds in 2007 (see Figure 13.3) (EPA 2008). This figure does not include mining, agricultural, and industrial waste; demolition and construction wastes; junked autos; or obsolete equipment wastes. About one-third of solid waste is recycled; most ends up in landfills. The availability of landfill space is limited, however. Some states have passed laws that limit the amount of solid waste that can be disposed of; instead, they require that bottles and cans be returned for a deposit or that lawn clippings be used in a community composting program.

What Do You Think? More than 380 billion plastic shopping bags are used in the United States annually, and only a fraction of the plastic bags are recycled (0.6 percent) (Cheeseman 2007). Most plastic bags, which contain toxic chemicals, end up in landfills where it takes 1,000 years for them to degrade. After the Republic of Ireland enacted a 15-cent tax on plastic shopping bags in 2002, the use of plastic bags declined by 90 percent. Several other countries, including China, have also banned plastic shopping bags. In spring 2007, San Francisco became the first U.S. city to ban plastic bags from supermarkets and chain pharmacies. Would you support a ban, or a tax, on plastic bags in your community?

Solid waste includes discarded electrical appliances and electronic equipment, known as **e-waste.** Ever think about where your discarded computer, cell phone, CD player, television, or other electronic product ends up when you replace it with a newer model? Most discarded electronics end up in landfills, incinerators, or hazardous waste exports; only about 11 percent are recycled (Electronics TakeBack Coalition 2009). With about 400 million units of e-waste discarded each year, e-waste is the fastest growing type of municipal waste in the United States. The main concern about dumping e-waste in landfills is that hazardous substances, such as lead, cadmium, barium, mercury, PCBs, and polyvinyl chloride, can leach out of e-waste and contaminate the soil and groundwater. Recycling electronics or incinerating them can release toxic emissions into the air and/or expose recycling workers to hazardous chemicals (Silicon Valley Toxics Coalition 2001).

e-waste Discarded electrical appliances and electronic equipment.

Pesticides. Pesticides are used worldwide for crops and gardens; outdoor mosquito control; the care of lawns, parks, and golf courses; and indoor pest control. Pesticides contaminate food, water, and air and can be absorbed through the skin, swallowed, or inhaled. Many common pesticides are considered potential carcinogens and neurotoxins (Blatt 2005). Even when a pesticide is found to be hazardous and is banned in the United States, other countries from which we import food may continue to use it. In an analysis of 1,260 domestic and 4,252 imported food samples, pesticide residues were detected in 31.2 percent of the domestic samples and 33.4 percent of the imported samples (Food and Drug Administration 2008). Pesticides also contaminate our groundwater supplies.

Water Pollution

Our water is being polluted by a number of harmful substances, including pesticides, vehicle exhaust, acid rain, oil spills, and industrial, military, agricultural waste, and sewage. Fertilizer runoff from agricultural lands in the Mississippi River Basin is the main cause of the annual "Dead Zone"—a seasonal phenomenon in which oxygen depletion causes an area of the Gulf of Mexico the size of Massachusetts to become uninhabitable to marine organisms.

Water pollution is most severe in developing countries, where more than 1 billion people lack access to clean water. In developing nations, more than 80 percent of untreated sewage is dumped directly into rivers, lakes, and seas that are also used for drinking and bathing (World Water Assessment Program 2009). Mining operations, located primarily in developing countries, are notoriously damaging to the environment. Modern gold-mining techniques use cyanide to extract gold from low-grade ore. In 2000, a dam holding cyanide-laced waste at a Romanian gold mine broke, dumping 22 million gallons of cyanide-laced waste into the Tisza River, which flowed into Hungary and Serbia. Some have called this event the worst environmental disaster since the 1986 Chernobyl nuclear explosion (Sampat 2003).

In the United States, one indicator of water pollution is the thousands of fish advisories issued by the U.S. Environmental Protection Agency (EPA) that warn against the consumption of certain fish caught in local waters because of contamination with pollutants such as mercury and dioxin. The EPA advises women who may become pregnant, pregnant women, nursing mothers, and young children to avoid eating certain fish altogether (swordfish, shark, king mackerel, and tilefish) because of the high levels of mercury (EPA 2004).

Pollutants also find their way into the water we drink. At Camp Lejeune—a Marine Corps base in Onslow County, North Carolina—as many as 1 million people were exposed to water contaminated with trichloroethylene (TCE), an industrial degreasing solvent, and perchloroethylene (PCE), a dry-cleaning agent from 1957 until 1987 (Sinks 2007). Exposure to these chemicals has been linked to a number of health problems, including kidney, liver, and lung damage, as well as cancer, childhood leukemia, and birth defects. In this chapter's The Human Side feature, a retired Marine tells the story of his daughter's illness and death that he believes resulted from the contaminated water at Camp Lejeune.

Chemicals, Carcinogens, and Health Problems

About 3 million tons of toxic chemicals are released into the environment each year (Pimentel et al. 2007). Chemicals in the environment enter our bodies via the food and water we consume, the air we breathe, and the substances with which we come in contact. During a 2004 World Health Organization convention in Budapest, 44 different hazardous chemicals were found in the bloodstreams of top European Union officials (Schapiro 2004). And in a study of umbilical cord blood of 10 newborns, researchers found an average of 200 industrial chemicals, pesticides, and other pollutants (Environmental Working Group 2005).

In the United States, the EPA has required testing on only about 200 of the more than 80,000 chemicals that have been on the market since 1976 (Gearhart 2009).

The *11th Report on Carcinogens* (U.S. Department of Health and Human Services 2004) lists 246 chemical substances that are "known to be human carcinogens" or "reasonably anticipated to be human carcinogens," meaning that they are linked to cancer. However, the report suggests that these 246 chemical substances may constitute only a fraction of actual human carcinogens. In a

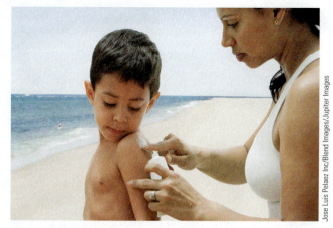

This mother puts sunscreen on her child to protect against sunburn. But sunscreen, like other personal care products, may contain harmful chemicals. Learn what chemicals are in the personal care products you use at the Environmental Working Group's website Skin Deep at http://www.cosmeticsdatabase.com.

review of 152 research studies of environmental pollution and breast cancer, researchers concluded that the evidence supports a link between breast cancer and a number of environmental pollutants (Brody et al. 2007). Many of the chemicals we are exposed to in our daily lives can cause not only cancer but also other health problems, such as infertility, birth defects, and a number of childhood developmental and learning problems (Fisher 1999; Kaplan & Morris 2000; McGinn 2000; Schapiro 2007). Chemicals found in common household, personal, and commercial products can result in a variety of temporary acute symptoms, such as drowsiness, disorientation, headache, dizziness, nausea, fatigue, shortness of breath, cramps, diarrhea, and irritation of the eyes, nose, throat, and lungs. Long-term exposure can affect the nervous system, reproductive system, liver, kidneys, heart, and blood. Fragrances, which are found in many consumer products, may produce sensory irritation, respiratory problems, and possible neurotoxic effects (Fisher 1998). The average adult uses nine personal care products a day, with roughly 120 chemicals among them, including those that have known or suspected adverse health effects on users (Zandonella 2007).

What Do You Think? Some businesses and local governments are voluntarily limiting fragrances in the workplace or banning them altogether to accommodate employees who experience ill effects from them. What do you think about banning fragrances in the workplace or other public places?

Vulnerability of Children. Half of U.S. children 1 year to 5 years old nationwide are estimated to have blood lead levels that are linked to adverse effects on cognitive abilities (Miranda et al. 2007). Asthma, the number one childhood illness in the United States, is linked to air pollution. In California and other agricultural states, hundreds, perhaps thousands, of children in the past decade have been exposed to pesticides and other farm chemicals that are linked to illness, brain damage, birth defects, and death (Associated Press 2007).

Jerry Ensminger holds a portrait of his daughter Janey, whose death from leukemia is believed to have been caused by contaminated water at Camp Lejeune.

Gerry Broome/AP Photo

Retired Marine Master Sergeant Jerry Ensminger is one of at least 850 individuals who have filed administrative claims worth $4 billion against Camp Lejeune in Onslow County, North Carolina, for damages that allegedly resulted from drinking and bathing in water contaminated with the dry-cleaning agent PCE and the industrial degreasing solvent TCE (Hefling 2007).

Here, we present Jerry Ensminger's account of his daughter, Janey, who died in 1985 of leukemia at age 9. Ensminger's story is excerpted from the book *Poisoned Nation* (2007). Sergeant Ensminger has gathered evidence pointing to the cause of Janey's illness and death: contaminated water at Camp Lejeune that Janey's mother was exposed to during her pregnancy with

Janey. As this text goes to print, Jerry Ensminger and other individuals who filed claims against Camp Lejeune are awaiting the completion of a government scientific study designed to determine whether there is an association between exposure to contaminated water and certain birth defects and cancers among children born between 1968 and 1985, to women who lived at Camp Lejeune during some portion of their pregnancy.

My little girl died in my arms and fifteen years later I found out that the people I had faithfully served for almost 25 years knew she was being poisoned by the water all along. . . . It was 1997 before I finally found out why my daughter died. Even then, it was just by chance. A local TV station had picked up the story. The evening news was turned on in the living room, and I was carrying my dinner in on a plate from the kitchen. All of a sudden I heard the newscaster saying that the water at Camp Lejeune had been highly contaminated from 1968 to 1985, and that the chemicals it contained had been linked to childhood leukemia. When I heard that, I just dropped my plate right on the floor and began shaking. The next day . . . I started

Children are more vulnerable than adults to the harmful effects of most pollutants for a number of reasons. For instance, children drink more fluids, eat more food, and inhale more air per unit of body weight than do adults; in addition, crawling and a tendency to put their hands and other things in their mouths provide more opportunities for children to ingest chemical or heavy metal residues.

multiple chemical sensitivity Also known as "environmental illness," a condition whereby individuals experience adverse reactions when exposed to low levels of chemicals found in everyday substances.

Multiple Chemical Sensitivity Disorder. **Multiple chemical sensitivity** (MCS), also known as environmental illness, is a condition whereby individuals experience adverse reactions when exposed to low levels of chemicals found in everyday substances (vehicle exhaust, fresh paint, housecleaning products, perfume and other fragrances, synthetic building materials, and numerous other petrochemical-based products). Symptoms of MCS include headache, burning eyes, difficulty breathing, stomach distress or nausea, loss of mental

reading everything I could find and making contacts with everyone I knew. There are stages you go through when you lose a child to a catastrophic illness. First you go into shock; then you start wondering why it happened to your child. So, years ago, I checked my family history and her mother's and found there was nothing on either side. But that nagging question of why Janey got leukemia had stayed with me throughout her illness, her death, and for fourteen and a half years after it. And in that moment, that one moment, when I heard the newscast and dropped the plate of food right out of my hands, it all became clear. I suddenly knew why my little girl died.

I signed up with the Marines to serve my country, but I never signed anything that gave them the right to kill my child, to knowingly poison her. . . . I've said that publicly and the Marine Corps has never refuted anything I've said, including that the contamination went back to the 1950s. I'm sure they know that from the geological studies. . . .

The day she died . . . Janey was in a lot of pain, so they suggested that she take morphine. She didn't want to,

because she had tried it before and it made her so tired. But this time she just couldn't handle the pain. Janey went through hell for nearly two and a half years, and I went through hell with her. . . . I was there with her every step of the way. Her mother couldn't handle it. Every time Janey went to the hospital, I was the parent who went with her. Sometimes, she was screaming in my ear, "Daddy, don't let them hurt me." Like when she had the bone marrow transplant and the spinal taps. The last time she went into the hospital was the last day of July, just before her ninth birthday. She didn't come out until September 20, and that was in a casket.

About a week before she died, the head of hematology had come in talking about a new form of therapy. He said it would cause severe burns and ulcers, and they didn't recommend it, but Janey looked at them, blinking to control her tears, and said, "This is my life you are talking about, and I'm not giving up. Let's try it."

The ulcers were all over her mouth, her legs, inside her nose and her vagina. The day she died she was in such intense pain she could hardly speak,

but finally she managed to whisper, "I want to die peacefully." When Janey said that, I started sobbing. She hugged me and said, "Stop it, stop crying." I said, "I can't help it, I love you." "I know you do," she replied. "I love you too. But, Daddy, I hurt so bad." "Do you want some morphine?" I asked. She was already being given methadone. "Yes, Daddy," she said, "I'm ready."

When the nurse heard that Janey wanted morphine, she knew the time had come. Then, just as they started to give it to her, she said, "Wait. Stop. I want some for my daddy, too."

"This is a very powerful pain medicine. I can't give it to your daddy," the nurse said.

"But my daddy hurts, too," Janey answered. You see, I always took a little of whatever she took to show her I was with her. The morphine killed her. It's a respiratory depressant. . . .

The organization I served faithfully for 24 and a half years knew about this all along, and they never said anything, well, shame on them.

Source: From Hefling 2007; Sinks 2007; Schwartz-Nobel 2007.

concentration, and dizziness. The onset of MCS is often linked to acute exposure to a high level of chemicals or to chronic long-term exposure. Individuals with MCS often avoid public places and/or wear a protective breathing filter to avoid inhaling the many chemical substances in the environment. Some individuals with MCS build houses made from materials that do not contain the chemicals that are typically found in building materials.

In a national study on the prevalence of multiple chemical sensitivity, 11.2 percent of U.S. adults reported an unusual hypersensitivity to common chemical products such as perfume, fresh paint, and household cleaning products, and 2.5 percent said they had been diagnosed with MCS (Caress & Steinemann 2004). Two-thirds of those with hypersensitivity described their symptoms as either severe or moderately severe. More than a third (39.5 percent) of the sample reported having trouble shopping in public places due to chemical sensitivity.

Environmental Injustice

Although environmental pollution and degradation and depletion of natural resources affect us all, some groups are more affected than others. **Environmental injustice,** also referred to as **environmental racism,** refers to the tendency for socially and politically marginalized groups to bear the brunt of environmental ills.

Environmental Injustice in the United States. In the United States, polluting industries, industrial and waste facilities, and transportation arteries (that generate vehicle emissions pollution) are often located in minority communities (Bullard 2000). More than half (56 percent) of people living within 1.8 miles (3 km) of a commercial hazardous waste site are people of color (Bullard et al. 2007). Rates of poverty are also higher among households located near hazardous waste facilities. However, "racial disparities are more prevalent and extensive than socioeconomic disparities, suggesting that race has more to do with the current distribution of the nation's hazardous waste facilities than poverty" (Bullard et al. 2007, p. 60).

An area between New Orleans and Baton Rouge, Louisiana—known as Cancer Alley because of the high rates of cancer in this area—has about 140 chemical plants. The people who live and work in Cancer Alley are disproportionately poor and black. One resident of Norco, Louisiana, who lived next to a Shell Oil refinery and chemical plant, reported that nearly everyone in the community suffered from health problems caused by industry pollution (Bullard 2000).

In addition, in North Carolina, hog industries—and the associated environmental and health risks associated with hog waste—tend to be located in communities with large black populations, low voter registration, and low incomes (Edwards & Driscoll 2009).

Native Americans also experience the effects of environmental injustice. The U.S. federal government approved a plan by a private corporation to store tens of thousands of tons of radioactive nuclear waste on a Native American reservation in Utah (Vedantam 2005b). The reservation leaders agreed to the plan with the hope that the nuclear waste storage facility will provide jobs to Native Americans and bring in needed income.

Environmental Injustice Around the World. Environmental injustice affects marginalized populations around the world, including minority groups, indigenous peoples, peasants, and nomadic tribes (Renner 1996). These groups are often powerless to fight against government and corporate powers that sustain environmentally damaging industries. A member of the Igorot tribal people of the Philippines explained, "Indigenous people are no longer able to plant fruits or vegetables because the . . . mercury poisoning, produced from massive logging and mining operations, inhibits the growth of any plant life" (Sterritt 2005). Indigenous groups in Nigeria, such as the Urhobo, Isoko, Kalabare, and Ogoni, are facing environmental threats caused by oil production operations run by transnational corporations. Oil spills, natural gas fires, and leaks from toxic waste pits have polluted the soil, water, and air. "Many Ogoni suffer from respiratory diseases and cancer, and birth defects are frequent"

Laura BouShnak/AFP/Getty Images

After hatching out of their eggs on the beach, baby sea turtles are guided back to the ocean by the reflective glow of the water. But light from beachfront development can lead them to stray toward the artificial light, where they can be hit by cars or caught by predators.

environmental injustice Also known as environmental racism, the tendency for socially and politically marginalized groups to bear the brunt of environmental ills.

(Renner 1996, p. 57). Renner (1996) warned that "minority populations and indigenous peoples around the globe are facing massive degradation of their environments that threatens to irreversibly alter, indeed destroy, their ways of life and cultures" (p. 59).

Threats to Biodiversity

There are an estimated 8 to 14 million species of life on earth; only 1.8 million have been identified (IUCN 2008a). This enormous diversity of life, known as **biodiversity,** provides food, medicines, fibers, and fuel; purifies air and freshwater; pollinates crops and vegetation; and makes soils fertile.

One species of life on earth goes extinct every three hours (Leahy 2009). As shown in Table 13.1, 16,928 species worldwide are threatened with extinction. Most extinctions today result from habitat loss caused by deforestation and urban sprawl, overharvesting, global warming, and bioinvasion—environmental problems caused by human activity (Cincotta & Engelman 2000; Leahy 2009).

> One species of life on earth goes extinct every three hours.

TABLE 13.1 Threatened Species Worldwide, 2008

CATEGORY	NUMBER OF THREATENED* SPECIES IN 2008
Mammals	1,141
Birds	1,222
Amphibians	1,905
Reptiles	423
Fishes	1,275
Invertebrates (insects, mollusks, etc.)	2,496
Plants	8,457
Other (mushrooms, lichens, brown algae)	9
Total	16,928

*Threatened species include those classified as critically endangered, endangered, and vulnerable.
Source: IUCN 2008b.

Light Pollution

Light pollution refers to artificial lighting that is annoying, unnecessary, and/or harmful to life forms on earth. The United States, like much of the rest of the world, has become increasingly "lit up" with artificial light. Almost all people in developed societies use artificial light, reducing the natural period of darkness at night. Yet, "darkness is as essential to our biological welfare, to our internal clockwork, as light itself" (Klinkenborg 2008, p. 109). Some research suggests that exposure to artificial light at night contributes to sleep disorders, depression, and other mood disorders. Other studies have found a link between exposure to artificial light at night and breast cancer (Chepesiuk 2009).

Light pollution also has adverse effects on the migration, feeding, and reproductive patterns of many animal species (Chepesiuk 2009; Klinkenborg 2008). For example, frogs have been found to inhibit their mating calls and bats alter their feeding behavior when they are exposed to artificial light at night. Sea turtles, which have a natural predisposition to nest on dark beaches, find fewer of them to nest on. When sea turtle eggs hatch, the babies instinctually go toward the brighter, reflective sea horizon, but are confused by artificial lighting near the beach and may wander off toward the artificially lit beach development.

Social Causes of Environmental Problems

Various structural and cultural factors have contributed to environmental problems. These include population growth, industrialization and economic

biodiversity The diversity of living organisms on earth.

light pollution Artificial lighting that is annoying, unnecessary, and/or harmful to life forms on earth.

development, and cultural values and attitudes such as individualism, consumerism, and militarism.

Population Growth

The world's population is growing by about 80 million people a year (World Water Assessment Program 2009). Population growth places increased demands on natural resources and results in increased waste. As Hunter (2001) explained:

> Global population size is inherently connected to land, air, and water environments because each and every individual uses environmental resources and contributes to environmental pollution. While the scale of resource use and the level of wastes produced vary across individuals and across cultural contexts, the fact remains that land, water, and air are necessary for human survival. (p. 12)

However, population growth itself is not as critical as the ways in which populations produce, distribute, and consume goods and services. As shown in Figure 13.4, regions with the highest populations have the lowest environmental footprint.

Industrialization and Economic Development

Many of the environmental problems confronting the world are associated with industrialization and economic development. Industrialized countries, for example, consume more energy and natural resources and contribute more pollution to the environment than poor countries.

The relationship between level of economic development and environmental pollution is curvilinear rather than linear. For example, industrial emissions are minimal in regions with low levels of economic development and are high in the middle-development range as developing countries move through the early stages of industrialization. However, at more advanced stages of industrialization,

Figure 13.4 **Environmental Footprint and Population by Region, 2005**

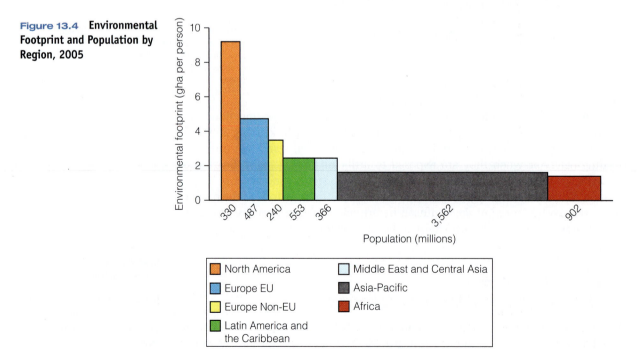

industrial emissions ease because heavy-polluting manufacturing industries decline, "cleaner" service industries increase, and because rising incomes are associated with a greater demand for environmental quality and cleaner technologies. However, a positive linear correlation has been demonstrated between per capita income and national carbon dioxide emissions (Hunter 2001).

In less developed countries, environmental problems are largely the result of poverty and the priority of economic survival over environmental concerns. Vajpeyi (1995) explained:

> Policymakers in the Third World are often in conflict with the ever-increasing demands to satisfy basic human needs—clean air, water, adequate food, shelter, education—and to safeguard the environmental quality. Given the scarce economic and technical resources at their disposal, most of these policymakers have ignored long-range environmental concerns and opted for short-range economic and political gains. (p. 24)

In considering the effects of economic development on the environment, note that the ways we measure economic development and the "health" of economies also influence environmental outcomes. Two primary measures of economic development and the health of economies are gross domestic product (GDP) and consumer spending. But these measures overlook the social and environmental costs of the production and consumption of goods and services. Until definitions and measurements of "economic development" and "economic health" reflect these costs, the pursuit of economic development will continue to contribute to environmental problems.

Cultural Values and Attitudes

Cultural values and attitudes that contribute to environmental problems include individualism, consumerism, and militarism.

Individualism. Individualism, which is a characteristic of U.S. culture, puts individual interests over collective welfare. Even though recycling is good for our collective environment, many individuals do not recycle because of the personal inconvenience involved in washing and sorting recyclable items. Similarly, individuals often indulge in countless behaviors that provide enjoyment and convenience for themselves at the expense of the environment: long showers, use of dishwashing machines, recreational boating, frequent meat eating, the use of air conditioning, and driving large, gas-guzzling SUVs, to name just a few.

Consumerism. Consumerism—the belief that personal happiness depends on the purchasing of material possessions—also encourages individuals to continually purchase new items and throw away old ones. The media bombard us daily with advertisements that tell us life will be better if we purchase a particular product. Consumerism contributes to pollution and environmental degradation by supporting polluting and resource-depleting industries and by contributing to waste.

Militarism. The cultural value of militarism also contributes to environmental degradation (see also Chapter 15). "It is generally agreed that the number one polluter in the United States is the American military. It is responsible each year for the generation of more than one-third of the nation's toxic waste . . . an amount greater than the five largest international chemical companies combined" (Blatt 2005, p. 25). Toxic substances from military vehicles, weapons materials, and munitions pollute the air, land, and groundwater in and around military bases and training areas. The Pentagon has asked Congress to loosen environmental laws for the military, and the EPA is forbidden to investigate or sue the military (Blatt 2005; Janofsky 2005).

Strategies for Action: Responding to Environmental Problems

Responses to environmental problems include environmental activism, environmental education, the use of green energy, modifications in consumer products and behavior, efforts to slow population growth, and government regulations and legislation. Sustainable economic development, international cooperation and assistance, and institutions of higher education also play important roles in alleviating environmental problems.

The U.S. environmental movement may be the largest single social movement in the United States.

Environmental Activism

With more than 6,500 national and 20,000 local environmental organizations with a combined membership of between 20 million and 30 million, the U.S. environmental movement may be the largest single social movement in the United States (Brulle 2009). A Gallup survey found that one in five (22 percent) of U.S. adults reports being an active participant in the environmental movement (Gallup Organization 2009) (see Table 13.2).

What Do You Think? Gallup Poll results indicate that women are more likely than men to worry about the environment, to be active in or sympathetic toward the environmental movement, and to give precedence to the environment over economic and energy concerns. Why do think this gender difference exists?

TABLE 13.2 Involvement in the Environmental Movement*

INVOLVEMENT	PERCENTAGE OF U.S. ADULTS
Active participant	22
Sympathetic but not active	47
Neutral	25
Unsympathetic	6
No opinion	1

*In a 2008 Gallup survey, a national sample of U.S. adults was asked: "Do you think of yourself as an active participant in the environmental movement, sympathetic towards the movement, but not active, neutral, or unsympathetic towards the movement?" Source: Gallup Organization 2009.

Environmental organizations exert pressure on government and private industry to initiate or intensify actions related to environmental protection. Environmentalist groups also design and implement their own projects and disseminate information to the public about environmental issues. In a national survey of college freshmen, nearly one-third (32 percent) said that "becoming involved in programs to clean up the environment" was personally "essential" or "very important" (Pryor et al. 2008).

In the United States, environmentalist groups date back to 1892, with the establishment of the Sierra Club, followed by the Audubon Society in 1905. Other environmental groups include the National Wildlife Federation, the World Wildlife Fund, the Environmental Defense, Rising Tide, Friends of the Earth, the Union of Concerned Scientists, Greenpeace, Earth First!, Environmental Action, the Natural Resources Defense Council, Worldwatch

Institute, World Resources Institute, the Rainforest Alliance, and Conservation International.

Online Activism. The Internet and e-mail provide important tools for environmental activism. For example, Action Network, an online environmental activism community sponsored by Environmental Defense, sends e-mail action alerts to 800,000 members informing them when Congress and other decision makers threaten the health of the environment. These members can then send e-mails and faxes to Congress, the president, and business leaders, urging them to support policies that protect the environment. Other environmental organizations have similar action alert features on their websites.

Religious Environmentalism. From a religious perspective, environmental degradation can be viewed as sacrilegious, sinful, and an offense against God (Gottlieb 2003a). "The world's dominant religions—as well as many people who identify with the 'spiritual' rather than with established faiths—have come to see that the environmental crisis involves much more than assaults on human health, leisure, or convenience. Rather, humanity's war on nature is at the same time a deep affront to one of the essentially divine aspects of existence" (Gottlieb 2003b, p. 489). This view has compelled religious groups to take an active role in environmental activism.

For example, the National Association of Evangelicals, an umbrella group of 51 church denominations, adopted a platform called For the Health of the Nation: An Evangelical Call to Civic Responsibility. This platform, which has been signed by nearly 100 evangelical leaders, calls on the government to "protect its citizens from the effects of environmental degradation" (Goodstein 2005). Larry Schweiger, president of the National Wildlife Federation, welcomes evangelicals as allies and explains that conservative lawmakers who might not pay attention to what environmental groups say may be more likely to pay attention to what the faith community is saying.

Radical Environmentalism. The **radical environmental movement** is a grassroots movement of individuals and groups that employ unconventional and often illegal means of protecting wildlife or the environment. Radical environmentalists believe in what is known as **deep ecology**: the view that maintaining the earth's natural systems should take precedence over human needs, that nature has a value independent of human existence, and that humans have no right to dominate the earth and its living inhabitants (Brulle 2009). The best-known radical environmental groups are the Earth Liberation Front (ELF) and the Animal Liberation Front (ALF). ALF and the ELF are international underground movements consisting of autonomous individuals and small groups who engage in "direct action" to (1) inflict economic damage on those profiting from the destruction and exploitation of the natural environment, (2) save animals from places of abuse (e.g., laboratories, factory farms, and fur farms), and (3) reveal information and educate the public on atrocities committed against the earth and all the species that populate it. Direct actions of ALF and ELF have included setting fire to a ski resort in Vail, Colorado, to express opposition to the resort's proposed expansion into the forest habitat of the Canada lynx, setting fire to gas-guzzling Hummers and other SUVs at car dealerships, vandalizing SUVs parked on streets and in driveways, setting fire to construction sites (in opposition to the development), and smashing windows at the homes of fur retailers.

radical environmental movement A grassroots movement of individuals and groups that employ unconventional and often illegal means of protecting wildlife or the environment.

deep ecology The view that maintaining the earth's natural systems should take precedence over human needs, that nature has a value independent of human existence, and that humans have no right to dominate the earth and its living inhabitants.

The opening of this chapter features the Kingsnorth Six Greenpeace activists who climbed a smokestack at a British power plant and were acquitted of criminal charges of property damage as jurors agreed that the defendants had a "lawful excuse" for their actions. But other radical environmentalists have been prosecuted as terrorists. **Ecoterrorism** is defined as any crime intended to protect wildlife or the environment that is violent, puts human life at risk, or results in damages of $10,000 or more (Denson 2000). Many environmentalists question whether "terrorist" is an appropriate label: "Where is the moral equivalence between burning an SUV in the dead of night (and doing as much as you can, given the nature of the business at hand, to see that no one gets hurt) and ramming a 767 into a skyscraper?" (Rasmussen 2007). Radical environmentalists argue that the real terrorists are corporations that plunder the earth.

What Do You Think? In Chapters 9 and 11, we discussed hate crimes, noting that hate crime laws impose harsher penalties on the perpetrator of a crime if the motive for that crime was hate or bias. Should motives be considered in imposing penalties on individuals who are convicted of acts of ecoterrorism? For example, should a person who sets fire to a business to protest that business's environmentally destructive activities receive the same penalty as a person who sets fire to a business for some other reason?

The Role of Corporations in the Environmental Movement. Corporations are major contributors to environmental problems and often fight against environmental efforts that threaten their profits. However, some corporations are joining the environmental movement for a variety of reasons, including pressure from consumers and environmental groups, the desire to improve their public image, genuine concern for the environment, and/or concern for maximizing current or future profits.

In 1994, out of concern for public and environmental health, Ray Anderson, founder and chairman of Interface carpet company, set a goal of being a sustainable company by 2020—"a company that will grow by cleaning up the world, not by polluting or degrading it" (McDaniel 2005, p. 33). Anderson envisioned recycling all the materials used, not releasing any toxins into the environment, and using solar energy to power all production. The company has made significant progress toward these goals, reducing use of fossil fuels by 45 percent, and reducing water and landfill use by as much as 80 percent.

Rather than hope that industry voluntarily engages in eco-friendly practices, corporate attorney Robert Hinkley suggested that corporate law be changed to mandate socially responsible behavior. Hinkley explained that corporations pursue profit at the expense of the public good, including the environment, because corporate executives are bound by corporate law to try to make a profit for shareholders (Cooper 2004). Hinkley suggested that corporate law should include a Code for Corporate Citizenship that would say the following: "The duty of directors henceforth shall be to make money for shareholders but not at the expense of the environment, human rights, public health and safety, dignity of employees, and the welfare of the communities in which the company operates" (quoted by Cooper 2004, p. 6).

Environmental Education

One goal of environmental organizations and activists is to educate the public about environmental issues and the seriousness of environmental problems.

ecoterrorism Any crime intended to protect wildlife or the environment that is violent, puts human life at risk, or results in damages of $10,000 or more.

Being informed about environmental issues is important because people who have higher levels of environmental knowledge tend to engage in higher levels of pro-environment behavior. For example, environmentally knowledgeable people are more likely to save energy in the home, recycle, conserve water, purchase environmentally safe products, avoid using chemicals in yard care, and donate funds to conservation (Coyle 2005).

A main source of information about environmental issues for most Americans is the media. However, because corporations and wealthy individuals with corporate ties own the media, unbiased information about environmental impacts of corporate activities may not readily be found in mainstream media channels. Indeed, the public must consider the source in interpreting information about environmental issues. Propaganda by corporations sometimes comes packaged as "environmental education." Hager and Burton (2000) explained: "Production of materials for schools is a growth area for public relations companies around the world. Corporate interests realize the value of getting their spin into the classrooms of future consumers and voters" (p. 107).

> Being informed about environmental issues is important because people who have higher levels of environmental knowledge tend to engage in higher levels of pro-environment behavior.

"Green" Energy

Increasing the use of **green energy**—energy that is renewable and nonpolluting—can help alleviate environmental problems associated with fossil fuels. Also known as clean energy, green energy sources include solar power, wind power, biofuel, and hydrogen. Although clean energy represents only about 1 percent of world energy production, this amount is expected to increase in coming years.

Solar Power. Solar power involves converting sunlight to electricity through the use of photovoltaic cells. Other forms of solar power include the use of solar thermal collectors, which capture the sun's warmth to heat building space and water, and "concentrating solar power plants," which use the sun's heat to make steam to turn electricity-producing turbines.

Wind Power. Wind turbines, which turn wind energy into electricity, are operating in 80 countries and produce more than 1.5 percent of world energy—up from 0.1 percent in 1997 (Sawin 2009). In 2008, the United States moved ahead of Germany to become the world's leading generator of wind energy (Makower et al. 2009). One disadvantage of wind power is that wind turbines have been known to result in bird mortality. However, this problem has been mitigated in recent years through the use of painted blades, slower rotational speeds, and careful placement of wind turbines.

Biofuel. Biofuels are fuels derived from agricultural crops. Two types of biofuels are ethanol and biodiesel.

Ethanol is an alcohol-based fuel that is produced by fermenting and distilling corn or sugar. Ethanol is blended with gasoline to create E85 (85 percent ethanol and 15 percent gasoline). Vehicles that run on E85, called flexible fuel vehicles, have been used by the government and in private fleets for years and have just recently become available to consumers. However, only a small fraction of passenger vehicles are flexible fuel vehicles, and many owners of these vehicles do

green energy Also known as clean energy, energy that is nonpolluting and/or renewable, such as solar power, wind power, biofuel, and hydrogen.

Wind energy is harnessed by turbines such as those pictured in this photo of a wind farm in Altamont Pass, California.

Morton Beebe/Encyclopedia/CORBIS

not know that their vehicle can operate with E85 and/or do not have access to a gas station that sells E85 (Price 2006).

Another problem associated with ethanol fuel is that increased demand for corn, which is used to make most ethanol in the United States, has driven up the price of corn, resulting in higher food prices (many processed food items contain corn, and animal feed is largely corn). And as corn prices rise, so too do those of rice and wheat because the crops compete for land. Rising food prices threaten the survival of the world's poorest 2 billion people who depend on grain to survive. The grain it takes to fill a 25-gallon tank with ethanol would feed one person for an entire year (Brown 2007).

Increased corn and/or sugar cane production to meet the demand for ethanol also has adverse environmental effects, including increased use and runoff of fertilizers, pesticides, and herbicides; depletion of water resources; and soil erosion. In addition, tropical forests are being clear-cut to make room for "energy crops," leaving less land for conservation and wildlife (Price 2006); biofuel refineries commonly run on coal and natural gas (which emit greenhouse gases); farm equipment and fertilizer production require fossil fuels; and the use of ethanol involves emissions of several pollutants. Finally, even if 100 percent of the U.S. corn crop were used to produce ethanol, it would only displace less than 15 percent of U.S. gasoline use (Food & Water Watch and Network for New Energy Choices 2007).

Biodiesel fuel is a cleaner-burning diesel fuel made from vegetable oils and/ or animal fats, including recycled cooking oil. Some individuals who make their own biodiesel fuel obtain used cooking oil from restaurants at no charge.

Hydrogen Power. Hydrogen, the most plentiful element on earth, is a clean burning fuel that can be used for electricity production, heating, cooling, and transportation. Many see a movement to a hydrogen economy as a long-term solution to the environmental and political problems associated with fossil fuels. Further research is needed, however, to develop nonpolluting and cost-effective ways to extract and transport hydrogen.

Carbon Capture and Storage. Coal-fired power plants emit more carbon dioxide—about a third of total U.S. emissions—than any other source. One proposal to reduce CO_2 emissions is carbon capture and storage (CCS)—a process of removing CO_2 from the smokestacks of coal-burning plants and storing it deep underground. The technology required for this process, which is still in the development stage, is expensive and requires large inputs of energy. The development of carbon capture and storage technology also promotes continued use of coal, and diverts or reduces investments in renewable energy such as solar, wind, and geothermal energy. Finally, scientists are concerned that stored carbon dioxide could leak out into the environment and cause sudden and drastic climate change (Miller & Spoolman 2009).

Modifications in Consumer Behavior

In the United States and other industrialized countries, many consumers are making "green" choices in their behavior and purchases that reflect concern for the environment. In some cases, these choices carry a price tag, such as paying more for organically grown food or for clothing made from organic cotton. Consumers are also motivated to make green purchases that save money. For example, after gas prices topped $4 a gallon in 2008, sales of gas-guzzling SUVs dropped, and sales of more fuel-efficient cars, such as hybrids, increased. Consumers often consider their utility bill when they choose energy-efficient appliances and electrical equipment. Although some eco-minded individuals choose "green" products and services, others choose to reduce their overall consumption and "buy nothing" rather than "buy green." For example, many consumers are choosing not to buy bottled water and to drink tap water instead (CBS News 2008). The switch from bottled to tap is partly fueled by the need to cut down on unnecessary spending in hard economic times, but environmental concerns are also a factor. The production and transportation of bottled water uses fossil fuels, and the disposal of plastic water bottles adds to our already overburdened landfills.

Although the average size of new housing in the United States has increased considerably, some homeowners are choosing to downsize their housing. For some, the driving force behind housing downsizing is economic, but others are moving into smaller dwellings out of concern for the environment.

Each person in the United States produces 22 tons of carbon dioxide each year. Table 13.3 presents tips for how consumers can reduce the amount of carbon dioxide each of us produces.

Green Building. The U.S. Green Building Council developed green building standards known as Leadership in Energy and Environmental Design (LEED). These standards consist of 69 criteria to be met by builders in six areas, including energy use and emissions, water use, materials and resource use, and sustainability of the building site. LEED buildings include the Pentagon Athletic Center, the Detroit Lions' football training facility, and the David L. Lawrence Convention Center in Pittsburgh.

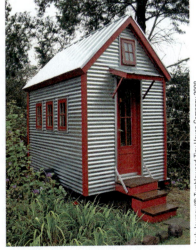

Tiny houses have less impact on the environment.

TABLE 13.3 Top Ten Things You Can Do to Fight Global Warming

National Geographic's *The Green Guide* (www.thegreenguide.com) lists the following tips for consumers to reduce the amount of carbon dioxide they produce each year. All of the following CO_2 reductions listed are on an annual basis:

1. Replace five incandescent light bulbs in your home with compact fluorescent bulbs (CFLs): Swapping those 75-watt incandescent bulbs with 19-watt CFLs can cut 275 pounds of CO_2.

2. Instead of short-haul flights of 500 miles or so, take the train and bypass 310 pounds of CO_2.

3. Sure it may be hot, but get a fan, set your thermostat to 75°F and blow away 363 pounds of CO_2.

4. Replace refrigerators more than 10 years old with today's more energy-efficient Energy Star models and save more than 500 pounds of CO_2.

5. Shave your 8-minute shower to 5 minutes for a savings of 513 pounds.

6. Caulk, weatherize, and insulate your home. If you rely on natural gas heating, you'll stop 639 pounds of CO_2 from entering the atmosphere (472 pounds for electric heating). And this summer, you'll save 226 pounds from air conditioner use.

7. Whenever possible, dry your clothes on a line outside or a rack indoors. If you air-dry half your loads, you'll dispense with 723 pounds of CO_2.

8. Trim down on the red meat. Because it takes more fossil fuels to produce red meat than fish, eggs, and poultry, switching to these foods will slim your CO_2 emissions by 950 pounds.

9. Leave the car at home and take public transportation to work. Taking the average U.S. commute of 12 miles by light-rail will leave you 1,366 pounds of CO_2 lighter than driving. The standard, diesel-powered city bus can save 804 pounds, and heavy rail subway users save 288.

10. Finally, support the creation of wind, solar, and other renewable energy facilities by choosing green power if offered by your utility. To find a green power program in your state, call your local utility or visit the U.S. Department of Energy's Green Power Markets page at http://apps3.eere.energy.gov/greenpower/marketing.

Source: National Geographic 2007.

One of the most environmentally friendly green buildings in the world is the Adam Joseph Lewis Center for Environmental Studies at Oberlin College in Oberlin, Ohio. The building is constructed with either recycled or sustainably produced nontoxic materials, bathroom and other liquid wastes are purified on-site and used again to flush toilets or to water outside vegetation, and power and heat are produced by solar and geothermal power (McDaniel 2005).

Slow Population Growth

As discussed in Chapter 12, slowing population growth is an important component of efforts to protect the environment. A recent study concluded that providing the 200 million women worldwide who want access to contraception but currently don't have it would reduce projected world population in 2050 by half a billion people, and would prevent the emission of at least 34 gigatons of carbon dioxide (a gigaton equals one billion tons) (Wire 2009).

Although Americans who are concerned about the environment may think about how their home energy use, travel, food choices, and other lifestyle behaviors affect the environment, they rarely consider the environmental impact of

TABLE 13.4 Carbon Reduction Figures for Various Lifestyle Behaviors,*

BEHAVIOR	AMOUNT OF CO_2 NOT RELEASED INTO THE ENVIRONMENT
Replace windows with energy-efficient windows	12 metric tons
Recycle newspapers, magazines, glass, plastic, aluminum, and steel cans	17 metric tons
Replace ten 75-watt incandescent bulbs with 25-watt energy-efficient lights	36 metric tons
Increase auto gas mileage from 20 mpg to 30 mpg	148 metric tons
Reduce number of children by one	9,441 metric tons

*Calculated over an 80-year period.
Source: Murtaugh and Schlax 2009.

their reproductive choices. Researchers at Oregon State University estimate that each child born in the United States, and the children that child has as an adult, adds 9,441 metric tons of carbon dioxide to the environment—the equivalent of burning 972,160 gallons of gas (Murtaugh & Schlax 2009). As shown in Table 13.4, reducing the number of children a woman has by one saves far more carbon dioxide from being released in the environment than do many other environmentally friendly lifestyle choices.

Government Policies, Programs, and Regulations

Worldwide, governments spend about $1 trillion (U.S. dollars) per year in subsidies that allow the prices of fuel, timber, metals, and minerals (and products using these materials) to be much lower than they otherwise would be, encouraging greater consumption (Renner 2004). Thus, governmental policies can contribute to environmental problems. Government policies and regulations can also play a role in protecting and restoring the environment, but enforcing these is critical. The Clean Water Act, passed in 1972, requires companies to disclose the toxins they dump into waterways and gives regulators the power to fine or jail polluters. A *New York Times* investigation found that, in the last five years, chemical and manufacturing plants, as well as small businesses such as gas stations and dry cleaners, violated water pollution laws more than half a million times. Violations range from failing to report emissions to dumping toxins that might cause cancer, birth defects, and other health problems. Most of these violations have gone unpunished; fewer than 3 percent of Clean Water Act violations from 2004 to 2007 resulted in fines or other punishments.

A few examples of more recent environmental government policies, programs, and regulations follow.

Cap and Trade Programs. Cap and trade programs are a free-market approach used to control pollution by providing economic incentives to power plants and other industries for achieving reductions in the emissions of pollutants. In a cap and trade system, a limit is set on the amount of carbon dioxide that can be released into the air. Polluters buy credits allowing them to emit a limited amount of carbon dioxide. They can sell leftover credits to other polluters, creating a monetary incentive to reduce emissions. Twenty-three U.S. states have joined regional agreements to lower carbon emissions through the cap and trade system (Pew Center on the States 2009).

Critics of the cap and trade approach argue that it fails to achieve the lowest possible emissions because it does not require all plants to use the best available technology to reduce emissions. By allowing some plants to have higher emissions, it also exposes populations living near these high-emissions plants to excessive air pollution.

Policies and Regulations on Energy. In 2004, more than 20 countries committed to specific targets for the renewable share of total energy use (UNEP 2007). A number of states have set goals of producing a minimum percentage of electricity from wind power, solar power, or other renewable sources (Prah 2007). In addition, more than 70 mayors and other local leaders from around the world signed the Urban Environmental Accords, pledging to obtain 10 percent of energy from renewable resources by 2012, and to reduce greenhouse gases by 25 percent by 2030 (Stoll 2005). Economic stimulus packages in the United States and elsewhere are targeting renewable energy, with the promise of investments in renewable energy as well as jobs. The American Recovery and Reinvestment Act of 2009 includes more than $70 billion in spending and tax credits for clean energy and transportation programs (Makower et al. 2009).

Taxes. Some environmentalists propose that governments use taxes to discourage environmentally damaging practices and products (Brown & Mitchell 1998). In the 1990s, a number of European governments increased taxes on environmentally harmful activities and products (such as gasoline, diesel, and motor vehicles) and decreased taxes on income and labor (Renner 2004). As a result of high gasoline taxes in Europe, gas there costs as much as $8 a gallon, which has increased consumer demand for small, fuel-efficient cars. Raising gasoline taxes in the United States is highly unpopular with voters and consumers. On the other hand, tax incentives and credits are used for renewable energy, plug-in hybrids, and energy efficiency.

Fuel Efficiency Standards. The Energy Independence and Security Act of 2007, which requires a 40 percent increase in fuel economy to 35 miles per gallon by 2020, lacked details instructing automakers how to comply. In 2009, President Obama signed an order requiring the Department of Transportation (DOT) to enforce the new fuel-economy standards.

What do you think the consequences will be of U.S. policies aimed at curbing global warming? Complete this chapter's Self and Society feature titled "Outcomes Expected from National Action to Reduce Global Warming," and compare your answers to those of other U.S. adults.

Policies on Chemical Safety. In 2003, the European Union drafted legislation known as Registration, Evaluation, and Authorization of Chemicals (REACH) that requires chemical companies to conduct safety and environmental tests to prove that the chemicals they are producing are safe. If they cannot prove that a chemical is safe, it will be banned from the market (Rifkin 2004). The European Union has become a world leader in environmental stewardship by

Directions: Please check all of the following answers that you believe are true.

If our nation takes steps to reduce global warming, it will . . .

(a) Provide a better life for our children and grandchildren _____

(b) Save many plant and animal species from extinction _____

(c) Improve people's health _____

(d) Create green jobs and a stronger economy _____

(e) Prevent the destruction of most life on the planet _____

(f) Help free us from dependence on foreign oil _____

(g) Protect God's creation _____

(h) Protect the poorest from environmental harm _____

(i) Save people around the world from poverty and starvation _____

(j) Improve our national security _____

(k) Lead to more government regulation _____

(l) Cause energy prices to rise _____

(m) Cost jobs and harm our economy _____

(n) Interfere with the free market _____

(o) Harm poor people more than it helps them _____

(p) Undermine American sovereignty _____

Comparison Data from a National Sample

Percentages of a nationally representative sample of U.S. adults who expected the various outcomes of national actions to curb global warming:

a. 67	b. 66	c. 61	d. 55
e. 51	f. 49	g. 48	h. 35
i. 35	j. 17	k. 44	l. 31
m. 17	n. 13	o. 11	p. 8

Source: Based on Maibach et al. 2009.

placing the "precautionary principle" at the center of EU regulatory policy. The precautionary principle requires industry to prove that their products are safe. In contrast, in the United States, chemicals are assumed to be safe unless proven otherwise, and the burden is put on the consumer, the public, or the government to prove that a chemical causes harm.

What Do You Think? In general, the European Union seems to be more concerned than the United States about health effects of chemicals—as evidenced by their stricter controls and bans on chemicals. If the United States had a national health insurance plan similar to that of European countries, in which the federal government paid for health care, do you think the U.S. government would enact tougher controls on hazardous chemicals and other environmental issues?

International Cooperation and Assistance

Global environmental concerns call for global solutions forged through international cooperation and assistance. For example, the 1987 Montreal Protocol on Substances That Deplete the Ozone Layer forged an agreement made by 70 nations to curb the production of CFCs (which contribute to ozone depletion and global warming).

In 1997, delegates from 160 nations met in Kyoto, Japan, and forged the **Kyoto Protocol**—the first international agreement to place legally binding limits on greenhouse gas emissions from developed countries. The United States, the

Kyoto Protocol The first international agreement to place legally binding limits on greenhouse gas emissions from developed countries.

world's largest producer of greenhouse gas emissions, rejected the Kyoto Protocol in 2001. As of June 2009, 184 countries and the European Union had ratified the Kyoto Protocol; the United States and Australia are the only industrialized nations who had refused to join.

Avoiding dangerous climate change will require rich countries to cut carbon emissions by at least 80 percent by the end of the 21st century, with cuts of 30 percent by 2020—significantly more than the cuts required under the Kyoto Protocol (United Nations Development Programme 2007). As this book goes to press, world leaders are planning to meet at the Climate Conference in Copenhagen, where, under President Obama's guidance, the United States is expected to participate in the negotiation of a new international climate treaty. Going beyond the Kyoto Protocol, the treaty forged at Copenhagen is expected to address not only the commitments to reduce greenhouse gases, but also financial mechanisms to support greenhouse gas reductions and provide assistance to help developing countries cope with climate change (Rosenthal 2009).

Some countries do not have the technical or economic resources to implement the requirements of environmental treaties. Wealthy industrialized countries can help less developed countries address environmental concerns through economic aid. Because industrialized countries have more economic and technological resources, they bear the primary responsibility for leading the nations of the world toward environmental cooperation.

Sustainable Economic Development

Achieving global cooperation on environmental issues is difficult, in part, because developed countries (primarily in the Northern Hemisphere) have different economic agendas from those of developing countries (primarily in the Southern Hemisphere). The northern agenda emphasizes preserving wealth and affluent lifestyles, whereas the southern agenda focuses on overcoming mass poverty and achieving a higher quality of life (Koenig 1995). Southern countries are concerned that northern industrialized countries—having already achieved economic wealth—will impose international environmental policies that restrict the economic growth of developing countries just as they are beginning to industrialize.

Development involves more than economic growth; it involves sustainability—the long-term environmental, social, and economic health of societies. **Sustainable development** involves meeting the needs of the present world without endangering the ability of future generations to meet their own needs. "The aim here is for those alive today to meet their own needs without making it impossible for future generations to meet theirs. . . . This in turn calls for an economic structure within which we consume only as much as the natural environment can produce, and make only as much waste as it can absorb" (McMichael et al. 2000, p. 1067).

Sustainable development requires the use of clean, renewable energy. Renewable energy projects in developing countries have demonstrated that providing affordable access to green energy helps to alleviate poverty by providing energy for creating business and jobs and by providing power for refrigerating medicine, sterilizing medical equipment, and supplying freshwater and sewer services needed to reduce infectious disease and improve health (Flavin & Aeck 2005).

sustainable development Societal development that meets the needs of current generations without threatening the future of subsequent generations.

The Role of Institutions of Higher Education

Colleges and universities can play an important role in efforts to protect the environment by encouraging use of bicycles on campus, using hybrid and electric vehicles, establishing recycling programs, using local and renewable building materials for new buildings, involving students in organic gardening to provide food for the campus, using clean energy, and incorporating environmental education into the curricula. An assessment of 200 public and private U.S. universities found that more than two of three schools showed improvement in their green practices and policies over the last year (Sustainable Endowments Institute 2008). Nevertheless, David Newport, director of the environmental center at the University of Colorado at Boulder, believes that institutions of higher education are not doing enough to promote sustainability. "We're supposed to be on the leading edge, and we're behind the curve. . . . There are, what 4,500 colleges in the United States, and how many of them are really doing something? Less than 100 or 200?" (quoted by Carlson 2006, p. A10).

Understanding Environmental Problems

Environmental problems are linked to corporate globalization, rapid and dramatic population growth, expanding world industrialization, patterns of excessive consumption, and reliance on fossil fuels for energy. Growing evidence of the irreversible effects of global warming and loss of biodiversity, and adverse health effects of toxic waste and other forms of pollution suggest that we cannot afford to ignore environmental problems. Researchers at Cornell University estimate that 40 percent of deaths worldwide are caused by water, air, and soil pollution (Pimentel et al. 2007).

Many Americans believe in a "technological fix" for the environment—that science and technology will solve environmental problems. Paradoxically, the same environmental problems that have been caused by technological progress may be solved by technological innovations designed to clean up pollution, preserve natural resources and habitats, and provide clean forms of energy. But leaders of government and industry must have the will to finance, develop, and use technologies that do not pollute or deplete the environment. When asked how companies can produce products without polluting the environment, Robert Hinkley suggested that, first, it must become a goal to do so:

> I don't have the technological answers for how it can be done, but neither did President John F. Kennedy when he announced a national goal to land a man on the moon by the end of the 1960s. The point is that, to eliminate pollution, we first have to make it our goal. Once we've done that, we will devote the resources necessary to make it happen. We will develop technologies that we never thought possible. But if we don't make it our goal, then we will never devote the resources, never develop the technology, and never solve the problem. (quoted by Cooper 2004, p. 11)

But the direction of technical innovation is largely in the hands of big corporations that place profits over environmental protection. Unless the global community challenges the power of transnational corporations to pursue profits at the expense of environmental and human health, corporate behavior will continue to take a heavy toll on the health of the planet and its inhabitants. Because oil has been implicated in political and military conflicts involving the Middle East (see Chapter 15), such conflicts are likely to continue as long as oil plays the lead role in providing the world's energy.

Global cooperation is also vital to resolving environmental concerns but is difficult to achieve because rich and poor countries have different economic development agendas: Developing poor countries struggle to survive and provide for the basic needs of their citizens; developed wealthy countries struggle to maintain their wealth and relatively high standard of living. Can both agendas be achieved without further pollution and destruction of the environment? Is sustainable economic development an attainable goal? With mounting concern about climate change, the health impacts of air pollution, rising oil prices, and the need to ensure energy access to all, governments worldwide have strengthened their commitment to sustainable, renewable energy policies and projects (UNEP 2007).

Our collective response to the precarious state of the environment constitutes a test that we cannot afford to fail. As environmentalist Bill McKibben (2008) notes,

> The next few years are a kind of final exam for the human species. Does that big brain really work or not? It gave us the power to build coal-fired power plants and SUVs and thereby destabilize the working of the earth. But does it give us the power to back away from those sources of power, to build a world that isn't bent on destruction? Can we think, and feel, our way out of this, or are we simply doomed to keep acting out the same set of desires for MORE that got us into this fix?

In 2004, the Nobel Peace Prize was given to Wangari Maathai for leading a grassroots environmental campaign to plant 30 million trees across Kenya. This was the first time ever that the Nobel Peace Prize was awarded to someone for accomplishments in restoring the environment. In her acceptance speech, Maathai explained, "A degraded environment leads to a scramble for scarce resources and may culminate in poverty and even conflict" (quoted by Little 2005, p. 2). With ongoing conflict around the globe, it is time for world leaders to recognize the importance of a healthy environment for world peace and to prioritize environmental protection in their political agendas.

The 2004 Nobel Peace Prize was awarded to Wangari Maathai for leading a grassroots environmental campaign called the Green Belt Movement, which is responsible for planting 30 million trees across Kenya. Maathai is the first person to be awarded the Nobel Peace Prize for environmental work.

CHAPTER REVIEW

■ **How do free trade agreements pose a threat to environmental protection?**
Free trade agreements such as NAFTA and the FTAA provide transnational corporations with privileges to pursue profits, expand markets, use natural resources, and exploit cheap labor in developing countries while weakening the ability of governments to protect natural resources or to implement environmental legislation.

■ **What are environmental refugees?**
Environmental refugees are individuals who have migrated because they can no longer secure a livelihood as a result of deforestation, desertification, soil erosion, and other environmental problems.

■ **What is greenwashing?**
Greenwashing refers to the ways in which environmentally and socially damaging companies portray their corporate image and products as being "environmentally friendly" or socially responsible.

■ **Where does most of the world's energy come from?**
Most of the world's energy comes from fossil fuels, which include oil, coal, and natural gas. This is significant because many of the serious environmental problems in the world today, including global warming and climate change, biodiversity loss, and pollution, stem from the use of fossil fuels.

■ **What are the major causes, and effects, of deforestation?**
The major causes of deforestation are the expansion of agricultural land, human settlements, wood harvesting, and road building. Deforestation displaces people and wild species from their habitats, contributes to global warming, and contributes to desertification, which results in the expansion of desert land that is unusable for agriculture. Soil erosion caused by deforestation can cause severe flooding.

■ **What are the effects of air pollution on human health?**
Air pollution, which is linked to heart disease, lung cancer, and respiratory ailments, such as emphysema, chronic bronchitis, and asthma kills about 3 million people a year.

■ **What are some examples of common household, personal, and commercial products that contribute to indoor pollution?**
Some common indoor air pollutants include carpeting, mattresses, drain cleaners, oven cleaners, spot removers, shoe polish, dry-cleaned clothes, paints, varnishes, furniture polish, potpourri, mothballs, fabric softener, caulking compounds, air fresheners, deodorizers, disinfectants, glue, correction fluid, printing ink, carbonless paper, and felt-tip markers.

■ **What is the primary cause of global warming?**
The prevailing view on what causes global warming is that greenhouse gases—primarily carbon dioxide, methane, and nitrous oxide—accumulate in the atmosphere and act like the glass in a greenhouse, holding heat from the sun close to the earth. The primary greenhouse gas is carbon dioxide, which is released into the atmosphere by burning fossil fuels.

■ **How does global warming contribute to further global warming?**
As global warming melts ice and snow, it exposes more land and ocean area, which absorbs more heat than ice and snow, further warming the planet. The melting of Siberia's frozen peat bog—a result of global warming— could release billions of tons of methane, a potent greenhouse gas, into the atmosphere and cause further global warming. This process, whereby the effects of global warming cause further global warming, is known as a positive feedback loop.

■ **What is the relationship between level of economic development and environmental pollution?**
There is a curvilinear relationship between level of economic development and environmental pollution. In regions with low levels of economic development, industrial emissions are minimal, but emissions rise in countries that are in the middle economic development range as they move through the early stages of industrialization. However, at more advanced stages of industrialization, industrial emissions ease because heavy-polluting manufacturing industries decline, "cleaner" service industries increase, and because rising incomes are associated with a greater demand for environmental quality and cleaner technologies.

■ **What are some of the concerns about nuclear energy?**
Nuclear waste contains radioactive plutonium, a substance linked to cancer and genetic defects. Nuclear waste in the environment remains potentially harmful to human and other life for thousands of years, and disposing of nuclear waste is problematic. Accidents at nuclear power plants, such as the 1986 accident at Chernobyl, and the potential for nuclear reactors to be targeted by terrorists add to the actual and potential dangers of nuclear power plants.

■ **How much garbage did each person in the United States generate in 1960? 1980? 2007?**
In 1960, each U.S. citizen generated 2.7 pounds of garbage on average every day. This figure increased to 3.7 pounds in 1980, and to 4.6 pounds in 2007.

■ **Why are women who may become pregnant, pregnant women, nursing mothers, and young children advised against eating certain types of fish?**
The U.S. Environmental Protection Agency advises women who may become pregnant, pregnant women, nursing mothers, and young children to avoid eating certain fish altogether (swordfish, shark, king mackerel, and tilefish) because of the high levels of mercury.

■ **Why are children more vulnerable than adults to the harmful effects of most pollutants?**
Children are more vulnerable than adults to the harmful effects of most pollutants for a number of reasons. For instance, children drink more fluids, eat more food, and inhale more air per unit of body weight than do adults; in addition, crawling and a tendency to put their hands and other things in their mouths provide more opportunities to ingest chemical or heavy metal residues.

■ **What does the term "environmental injustice" refer to?**
Environmental injustice, also referred to as environmental racism, refers to the tendency for socially and politically marginalized groups to bear the brunt of environmental ills. For example, in the United States, polluting industries, industrial and waste facilities, and transportation arteries (which generate vehicle emissions pollution) are often located in minority communities.

■ **How often does a species of life on earth become extinct?**
One species of life on earth goes extinct every three hours.

■ **What social and cultural factors contribute to environmental problems?**
Social and cultural factors that contribute to environmental problems include population growth, industrialization and economic development, and cultural values and attitudes such as individualism, consumerism, and militarism.

■ **What are some of the strategies for alleviating environmental problems?**
Strategies for alleviating environmental problems include efforts to lower fertility rates and slow population growth, environmental activism, environmental education, the use of "green" energy, modifications in consumer products and behavior, and government regulations and legislation. Sustainable economic development and international cooperation and assistance also play important roles in alleviating environmental problems.

■ **According to 2004 Nobel Peace Prize winner Wangari Maathai, why is environmental protection important for national and international security?**
In her acceptance speech for the 2004 Nobel Peace Prize, Wangari Maathai explained that "a degraded environment leads to a scramble for scarce resources and may culminate in poverty and even conflict."

TEST YOURSELF

1. In 2007, six Greenpeace activists were arrested after climbing and vandalizing a smokestack at a British coal power plant. What was their penalty?
   a. 800 hours of community service
   b. 6 months in prison
   c. 26 years in prison
   d. They were acquitted of charges
2. Red fire ants are an example of:
   a. A bioinvasion
   b. An extinct species
   c. A threatened species
   d. An alternative fuel
3. Which innovation has resulted in increased food prices?
   a. Solar power
   b. Bottled water
   c. Wind turbines
   d. Biofuels
4. If greenhouse gases were to be stabilized today, global air temperature and sea level would be expected to
   a. Remain at their current level
   b. Decrease immediately

   c. Begin to decrease within 20 years
   d. Continue to rise for hundreds of years
5. The United States has more operating nuclear reactors than any other country.
   a. True
   b. False
6. Most solid waste in the United States is recycled.
   a. True
   b. False
7. In spring 2007, which U.S. city became the first to ban plastic bags from supermarkets and chain pharmacies?
   a. Miami
   b. San Francisco
   c. Atlanta
   d. Portland
8. The average adult uses nine personal care products a day. How many chemicals, roughly, are in those nine personal care products?
   a. 12
   b. 20
   c. 120
   d. 1,200

9. Which of the following is the number one polluter in the United States?
   a. The military
   b. Dow Chemical
   c. Archer Daniels Midland
   d. ExxonMobil

10. Clean energy represents more than a quarter of world energy production.
   a. True
   b. False

Answers: 1: d; 2: a; 3: d; 4: d; 5: a; 6: b; 7: b; 8: c; 9: a; 10: b.

KEY TERMS

acid rain   480
biodiversity   491
bioinvasion   472
deep ecology   495
deforestation   478
desertification   478
ecosystems   475
ecoterrorism   496

environmental footprint   478
environmental injustice   490
environmental racism   490
environmental refugees   473
e-waste   485
global warming   480
green energy   497
greenhouse gases   480

greenwashing   475
Kyoto Protocol   503
light pollution   491
multiple chemical sensitivity   488
planned obsolescence   474
radical environmental movement   495
sustainable development   504

MEDIA RESOURCES

Understanding Social Problems,
Seventh Edition Companion Website
www.cengage.com/sociology/mooney
Visit your book companion website, where you will find flash cards, practice quizzes, Internet links, and more to help you study.

 Just what you need to know NOW! Spend time on what you need to master rather than on information you already have learned. Take a pretest for this chapter, and CengageNOW will generate a personalized study plan based on your results. The study plan will identify the topics you need to review and direct you to online resources to help you master those topics. You can then take a post-test to help you determine the concepts you have mastered and what you will need to work on. Try it out! Go to www.cengage.com/login to sign in with an access code or to purchase access to this product.

©Emrah Turudu/iStockphoto.com

14

Science and Technology

D'Zhana Simmons is no ordinary teenager. By the age of 14, she had had three different hearts. Today, however, D'Zhana can't wait to get back home to Clinton, South Carolina, to see her five brothers and sisters, and is looking forward to being outdoors again, including a boat trip off the coastline of Miami to celebrate her 15th birthday.

In 2007, D'Zhana was diagnosed with dilated cardiomyopathy, a life-threatening condition rendering the heart too weak to pump blood through the body (Madkour 2008). On July 2, 2008, D'Zhana received a heart transplant, but a clot developed in one of the ventricles and two days later the heart was removed (Paddock 2008; CBS News 2008). With no human heart beating inside of her, D'Zhana received two artificial pumps called ventricular assist devices (VADs), which acted as a "bridge" until a suitable heart was found and another transplant performed.

VADs are usually connected to a patient's own heart but, because children rarely have heart disease, pediatric artificial hearts are not profitable and therefore are commercially unavailable. Doctors at Holtz Children's Hospital in Miami had to craft artificial heart chambers for D'Zhana to which the VADs were attached (Madkour 2008).

The second transplant was a success and the prognosis is good although the "life expectancy of a transplanted heart . . . on the average . . . is about 15 years" (McFadden 2009, p. 2). In a television interview with D'Zhana, she said that she felt like a "fake person" because she didn't have a heart. In sports, the phrase "they've got heart" is used to mean they have the strength to carry on. Surely, D'Zhana "has heart" for she did something no other child has ever done—she lived 118 days without a heart—a true scientific and technological miracle.

Many of the technologies available today seem futuristic. But such technologies—from virtual reality, cloning, and teleportation to artificial hearts—are no longer just the stuff of popular science fiction movies. Virtual reality is now used to train workers in occupations as diverse as medicine, engineering, and professional football. The ability to genetically replicate embryos has sparked worldwide debate over the ethics of reproduction, California Institute of Technology scientists have transported a ray of light from one location to another, and D'Zhana Simmons lived for four months with an artificial heart. Just as the telephone, the automobile, the television, and countless other technological innovations have forever altered social life, so will technologies that are more recent.

Science and technology go hand in hand. **Science** is the process of discovering, explaining, and predicting natural or social phenomena. A scientific approach to understanding acquired immunodeficiency syndrome (AIDS), for example, might include investigating the molecular structure of the virus, the means by which it is transmitted, and public attitudes about AIDS. **Technology,** as a form of human cultural activity that applies the principles of science and mechanics to the solution of problems, is intended to accomplish a specific task—in this case, the development of an AIDS vaccine.

Societies differ in their level of technological sophistication and development. In agricultural societies, which emphasize the production of raw materials, the use of tools to accomplish tasks previously done by hand, or **mechanization,**

science The process of discovering, explaining, and predicting natural or social phenomena.

technology Activities that apply the principles of science and mechanics to the solutions of a specific problem.

mechanization Dominant in an agricultural society, the use of tools to accomplish tasks previously done by hand.

dominates. As societies move toward industrialization and become more concerned with the mass production of goods, automation prevails. **Automation** involves the use of self-operating machines, as in an automated factory in which autonomous robots assemble automobiles. Finally, as a society moves toward post-industrialization, it emphasizes service and information professions (Bell 1973). At this stage, technology shifts toward **cybernation,** whereby machines control machines—making production decisions, programming robots, and monitoring assembly performance.

What are the effects of science and technology on humans and their social world? How do science and technology help to remedy social problems, and how do they contribute to social problems? Is technology, as Postman (1992) suggested, both a friend and a foe to humankind? We address each of these questions in this chapter.

The Global Context:
The Technological Revolution

Less than 50 years ago, traveling across state lines was an arduous task, a long-distance phone call was a memorable event, and mail carriers brought belated news of friends and relatives from far away. Today, travelers journey between continents in a matter of hours, and for many, e-mail, faxes, instant messaging, texting, and electronic fund transfers have replaced previously conventional means of communication.

The world is a much smaller place than it used to be, and it will become even smaller as the technological revolution continues. In 2009, the Internet had 1.67 billion users in more than 200 countries with 220 million users in the United States alone (Internet World Statistics 2009). The percentage of Internet users is highest in North America (73.9 percent) followed by Oceania/Australia (60.1 percent), Europe (50.1 percent), Latin America and the Caribbean (30.3 percent), the Middle East (23.7 percent), Asia (18.1 percent), and Africa (6.7 percent). Worldwide, the average number of Internet users in a region is 24.7 percent (Internet World Statistics 2009).

Although penetration rates are higher in industrialized countries, there is some movement toward the Internet becoming a truly global medium as Africans, Middle Easterners, and Latin Americans increasingly "get online." For example, although Internet use in the United States grew 133 percent between 2000 and 2009, the number of Internet users in Africa increased by 1,360 percent during the same time period (Internet World Statistics 2009). Table 14.1 displays the averages of information and communication technology indicators (e.g., Internet users) for countries in worse (i.e., poorer countries) and better situations (i.e., richer countries) (Social Watch 2009). For example, the number of people with telephone lines is 22.5 per 1,000 population in poorer countries and 562.9 per 1,000 in richer countries. Although not shown in the table, an analysis of the data over time also suggests that the science and technology gap between poorer and richer nations is getting greater rather than shrinking (Social Watch 2009).

The movement toward globalization of technology is, of course, not limited to the use and expansion of the Internet. The world robot market and the U.S. share of it continues to expand, Microsoft's Internet platform and support products are sold all over the world, scientists collect skin and blood samples from remote islanders for genetic research, a global treaty regulating trade of genetically altered products has

automation Dominant in an industrial society, the replacement of human labor with machinery and equipment that are self-operating.

cybernation Dominant in postindustrial societies, the use of machines to control other machines.

TABLE 14.1 Averages by ICT (Information and Communication Technology) Indicator for Countries in Better and Worse Relative Situations*

	INTERNET USERS (PER 1,000 PEOPLE)	PERSONAL COMPUTERS (PER 1,000 PEOPLE)	TELEPHONE LINES (PER 1,000 PEOPLE)	EXPENDITURE ON ICT (% OF GDP)	EXPENDITURE ON R & D (% OF GDP)
Worse Situation (Poorer Countries)					
Average	28.7	20.6	22.5	3.9	0.0
Number of countries	68	77	67	17	46
Better Situation (Richer Countries)					
Average	668.8	678.4	562.9	8.7	7.2
Number of countries	41	23	35	11	6
Total Countries					
Average	258.8	167.9	217.2	6.0	1.1
Number of countries	201	186	203	74	107

*GDP = gross domestic product; R&D = research and development; differences in the total number of countries included in the study represent the availability or unavailability of data for varied years since 2000, for a particular information and communication technology indicator.

been signed by more than 100 nations, and Intel's central processing units (CPUs) power an estimated four-fifths of the world's personal computers (PCs).

To achieve such scientific and technological innovations, sometimes called research and development (R&D), countries need material and economic resources. *Research* entails the pursuit of knowledge; *development* refers to the production of materials, systems, processes, or devices directed to the solution of practical problems. According to the National Science Foundation (NFS), the United States spends over $340 billion a year on research and development, 40.0 percent of the world's R&D expenditures (NSF 2008; Galama & Hosek 2008). As in most other countries, U.S. funding sources are primarily from private industry (66 percent), followed by the federal government (28 percent) and nonprofit organizations such as research institutes at colleges and universities (6 percent) (NSF 2008).

The United States leads the world in science and technology, although there is some evidence that we are falling behind (Price 2008). A report by the Information Technology and Innovation Foundation (ITIF) concludes that the United States, when compared to 21 other countries and regions (ITIF 2009):

- Ranks 4th in the rate of science and technology scientists but 20th in the increase of science and technology scientists between 1999 and 2006.
- Ranks 5th in corporate R&D investments as a percentage of the gross domestic product but 17th in the growth of corporate R&D investments as a percentage of the gross domestic between 1999 and 2006.
- Ranks 2nd in a combined measure of the frequency and quality of scientific and technical publications, but 17th in the percentage increase of the frequency and quality of scientific and technical publications between 1999 and 2006.
- Ranks 2nd in the "utilization of digital technology in national government" (ITIF 2009, p. 17), but 22nd in utilization growth in e-government between 1999 and 2006.
- Ranks 7th in broadband quality and subscription rates per capita but 22nd in the percentage increase in subscription rates between 1999 and 2006.

The decline of U.S. supremacy in science and technology is likely to be the result of several interacting forces (Lemonick 2006a; ITIF 2009; Price 2008; World Bank 2009). First, the federal government has been scaling back its investment in research and development in response to fiscal deficits. Second, corporations, the largest contributors to research and development, have begun to focus on short-term products and higher profits as pressure from stockholders mounts. Third, developing countries, most notably China and India, are expanding their scientific and technological capabilities at a faster rate than the United States. For example, since 1991, China's R&D expenditures have doubled compared to those of the United States.

Fourth, there has been a drop in science and math education in U.S. schools, both in terms of quality and quantity. For example, although the United States still grants the highest proportion of science and engineering PhD degrees in the world, in recent years, the number has declined, whereas other countries' rates have increased. In response to such concerns, in a 2009 letter to President Obama, the chair of the National Science Board recommended that:

> ... the new Administration ... advance **STEM** (science, technology, engineering, and mathematics) education for all American students, to nurture innovation, and to ensure the long-term economic prosperity of the Nation. The urgency of this task is underscored by the need to ensure that the United States continues to excel in science and technology ... and [to] guarantee that all American students are provided the educational resources and tools needed to participate fully in the science and technology based economy of the 21st century. (Kalil 2009, p. 1)

Finally, Mooney and Kirshenbaum (2009) document "unscientific America"—the tremendous disconnect between the citizenry, media, politicians, religious leaders, education, and the entertainment industry (e.g., *CSI*, *Grey's Anatomy*) on the one hand, and science and scientists on the other. Post-World War II America, in part because of the Cold War, invested in R&D, leading to such scientific and technological advances as the space program, the development of the Internet, and the decoding of the genome. Yet, despite these significant contributions and primarily positive attitudes toward science and scientists, most Americans know very little about science (see this chapter's Self and Society) and there are divergent patterns of thinking in reference to some of today's most important issues (see Table 14.2). For example, 32 percent of the American public, compared to 87 percent of American scientists, believe that humans evolved from other living things due to a natural process (Pew 2009a).

What Do You Think? Scientific discoveries and technological developments require the support of a country's citizens and political leaders. For example, although abortion has been technically possible for years, millions of the world's citizens live in countries where abortion is either prohibited or permitted only when the life of the mother is in danger. Thus, the degree to which science and technology are considered good or bad, desirable or undesirable, is as social constructionists argue, a function of time and place. Can you name other scientific discoveries or technological developments that are technically possible, but likely to be rejected by large segments of the population?

STEM An acronym for science, technology, engineering and mathematics.

Answer each of the questions below. Count the number of questions that you answered correctly and then compare your score to a sample of 1,005 U.S. adults who were asked exactly the same questions during a phone survey in 2009.

1. Which over-the-counter drug do doctors recommend that people take to help prevent heart attacks?
   a. Antacids
   b. Cortisone
   c. Aspirin
   d. Allergy medications

2. According to most astronomers, which of the following is no longer considered a planet?
   a. Neptune
   b. Pluto
   c. Saturn
   d. Mercury

3. Which of the following may cause a tsunami?
   a. A very warm ocean current
   b. A large school of fish
   c. A melting glacier
   d. An earthquake under the ocean

4. The global positioning system, or GPS, relies on which of these to work?
   a. Satellites
   b. Stars
   c. Magnets
   d. Lasers

5. What gas do most scientists believe causes temperatures in the atmosphere to rise?
   a. Hydrogen
   b. Helium
   c. Carbon dioxide
   d. Radon

6. How are stem cells different from other cells?
   a. They can develop into many different types of cells
   b. They are found only in bone marrow
   c. They are found only in plants

7. What have scientists recently discovered on Mars?
   a. Platinum
   b. Plants
   c. Mold
   d. Water

8. The continents on which we live have been moving their location for millions of years and will continue to move in the future.
   a. True
   b. False

9. Antibiotics will kill viruses as well as bacteria.
   a. True
   b. False

10. Electrons are smaller than atoms.
    a. True
    b. False

11. Lasers work by focusing sound waves.
    a. True
    b. False

12. All radioactivity is man-made.
    a. True
    b. False

Answers: 1: c; 2: b; 3: d; 4: a; 5: e; 6: a; 7: d; 8: a; 9: b; 10: a; 11: b; 12: b.

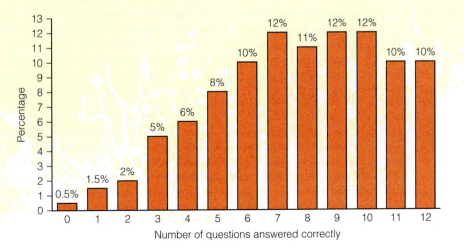

Number of questions answered correctly

TABLE 14.2 General Public's and Scientists' Views on Selected Science and Technology Issues, 2009

	PUBLIC (%)	SCIENTISTS (%)
Think that humans, other living things have evolved due to natural processes	32	87
Think that earth is getting warmer because of human activity	49	84
Favor use of animals in scientific research	52	93
Favor federal funding for embryonic stem cell research	58	93
Favor building more nuclear power plants	51	70
Say that all parents should be required to vaccinate their children	69	82

Source: Pew 2009a.

Postmodernism and the Technological Fix

Many Americans believe that social problems can be resolved through a **technological fix** (Weinberg 1966) rather than through social engineering. For example, a social engineer might approach the problem of water shortages by persuading people to change their lifestyle: use less water, take shorter showers, and wear clothes more than once before washing. A technologist would avoid the challenge of changing people's habits and motivations and instead concentrate on the development of new technologies that would increase the water supply.

Social problems can be tackled through both social engineering and a technological fix. In recent years, for example, social engineering efforts to reduce drunk driving have included imposing stiffer penalties for drunk driving and disseminating public service announcements, such as "Friends don't let friends drive drunk." An example of a technological fix for the same problem is the development of car airbags, which reduce injuries and deaths resulting from car accidents.

Not all individuals, however, agree that science and technology are good for society. **Postmodernism,** an emerging worldview holds that rational thinking and the scientific perspective have fallen short in providing the "truths" they were once presumed to hold. During the industrial era, science, rationality, and technological innovations were thought to hold the promises of a better, safer, and more humane world. Today, postmodernists question the validity of the scientific enterprise, often pointing to the unforeseen and unwanted consequences of resulting technologies. Automobiles, for example, began to be mass-produced in the 1930s in response to consumer demands. But the proliferation of automobiles has also led to increased air pollution and the deterioration of cities as suburbs developed, and, today, traffic fatalities are the number one cause of accident-related deaths.

technological fix The use of scientific principles and technology to solve social problems.

postmodernism A worldview that questions the validity of rational thinking and the scientific enterprise.

The world was made a smaller place in the late 1800s by the Pony Express. Today, the iPhone, combining a number of technological feats, makes the world even smaller.

Sociological Theories of Science and Technology

Each of the three major sociological frameworks helps us to better understand the nature of science and technology in society.

Structural-Functionalist Perspective

Structural functionalists view science and technology as emerging in response to societal needs—that "science was born indicates that society needed it" (Durkheim 1973). As societies become more complex and heterogeneous, finding a common and agreed-on knowledge base becomes more difficult. Science fulfills the need for an assumed objective measure of "truth" and provides a basis for making intelligent and rational decisions. In this regard, science and the resulting technologies are functional for society.

If society changes too rapidly as a result of science and technology, however, problems may emerge. When the material part of culture (i.e., its physical elements) changes at a faster rate than the nonmaterial part (i.e., its beliefs and values), a **cultural lag** may develop (Ogburn 1957). For example, the typewriter, the conveyor belt, and the computer expanded opportunities for women to work outside the home. With the potential for economic independence, women were able to remain single or to leave unsatisfactory relationships and/or establish careers. But although new technologies have created new opportunities for women, beliefs about women's roles, expectations of female behavior, and values concerning equality, marriage, and divorce have lagged behind.

Robert Merton (1973), a structural functionalist and founder of the subdiscipline sociology of science, also argued that scientific discoveries or

cultural lag A condition in which the material part of culture changes at a faster rate than the nonmaterial part.

technological innovations may be dysfunctional for society and may create instability in the social system. For example, the development of time-saving machines increases production, but it also displaces workers and contributes to higher rates of employee alienation. Defective technology can have disastrous effects on society. In 1994, a defective Pentium chip was discovered to exist in 2 million computers in aerospace, medical, scientific, and financial institutions, as well as in schools and government agencies. Replacing the defective chip was a massive undertaking but was necessary to avoid thousands of inaccurate computations and organizational catastrophe.

Conflict Perspective

Conflict theorists, in general, argue that science and technology benefit a select few. For some conflict theorists, technological advances occur primarily as a response to capitalist needs for increased efficiency and productivity and thus are motivated by profit. As McDermott (1993) noted, most decisions to increase technology are made by "the immediate practitioners of technology, their managerial cronies, and for the profits accruing to their corporations" (p. 93). In the United States, private industry spends more money on research and development than the federal government does. The Dalkon Shield (IUD) and silicone breast implants are examples of technological advances that promised millions of dollars in profits for their developers. However, the rush to market took precedence over thorough testing of the products' safety. Subsequent lawsuits filed by consumers who argued that both products had compromised the physical well-being of women resulted in large damage awards for the plaintiffs.

Science and technology also further the interests of dominant groups to the detriment of others. The need for scientific research on AIDS was evident in the early 1980s, but the required large-scale funding was not made available so long as the virus was thought to be specific to homosexuals and intravenous drug users. Only when the virus became a threat to mainstream Americans were millions of dollars allocated to AIDS research. Hence, conflict theorists argue that granting agencies act as gatekeepers to scientific discoveries and technological innovations. These agencies are influenced by powerful interest groups and the marketability of the product rather than by the needs of society.

When the dominant group feels threatened, it may use technology as a means of social control. For example, the use of the Internet is growing dramatically in China, the world's second largest Internet market. In 2005, the government of China announced that every website must be registered "and provide complete information on its organizers...or face being declared illegal" (Reuters 2005, p. 1). In 2006, Internet giant Google announced that it would censor search inquiries to conform to Chinese officials' requirements (BBC 2006), and today, it is estimated that over 30,000 "Internet police" monitor online traffic (Markoff 2008a). China is not alone, however. A study by OpenNet Initiative found that 25 of the 41 countries surveyed engaged in Internet censorship including Vietnam, Saudi Arabia, India, Singapore, and Thailand (Associated Press 2007a). Today, thousands of censors monitor videos, web pages, blogs, chat rooms, and voice and text transmissions (e.g., Skype), looking for politically incorrect statements.

Finally, conflict theorists as well as feminists argue that technology is an extension of the patriarchal nature of society that promotes the interests of men and ignores the needs and interests of women. As in other aspects of life, women play a subordinate role in reference to technology in terms of both its creation and its use. For example, washing machines, although time-saving devices, disrupted the communal telling of stories and the resulting friendships among women who gathered together to do their chores. Bush (1993) observed that, in a "society characterized by a sex-role division of labor, any tool or technique . . . will have dramatically different effects on men than on women" (p. 204).

Symbolic Interactionist Perspective

Knowledge is relative. It changes over time, over circumstances, and between societies. We no longer believe that the world is flat or that the earth is the center of the universe, but such beliefs once determined behavior because individuals responded to what they thought to be true. The scientific process is a social process in that "truths"—socially constructed truths—result from the interactions between scientists, researchers, and the lay public.

Kuhn (1973) argued that the process of scientific discovery begins with assumptions about a particular phenomenon (e.g., the world is flat). Because unanswered questions always remain about a topic (e.g., why don't the oceans drain?), science works to fill these gaps. When new information suggests that the initial assumptions were incorrect (e.g., the world is not flat), a new set of assumptions or framework emerges to replace the old one (e.g., the world is round). It then becomes the dominant belief or paradigm.

Symbolic interactionists emphasize the importance of this process and the effect that social forces have on it. Conrad (1997), for example, described the media's contribution in framing societal beliefs that alcoholism, homosexuality, and racial inequality are genetically determined. Social forces also affect technological innovations, and their success depends, in part, on the social meaning assigned to any particular product. As social constructionists argue, individuals socially construct reality as they interpret the social world around them, including the meaning assigned to various technologies. If claims makers can successfully define a product as impractical, cumbersome, inefficient, or immoral, the product is unlikely to gain public acceptance. Such is the case with RU-486, an oral contraceptive that is widely used in France, Great Britain, and China, but availability of which, although legal in the United States, is opposed by a majority of Americans (Gallup 2000; Gottlieb 2000). In 2009, over objections from the Vatican, RU-486 was approved for use in Italy (Rizzo 2009).

Not only are technological innovations subject to social meaning, but also who becomes involved in what aspects of science and technology is socially defined. Men, for example, far outnumber women in earning computer science degrees, as many as ten to one at some schools. Further, although women make up 46 percent of the general workforce, they make up just 28 percent of computer scientists and 29 percent of computer information systems managers (Stross 2008; U.S. Bureau of the Census 2009). Societal definitions of men as being rational, mathematical, and scientifically minded and as having greater mechanical aptitude than women are, in part, responsible for these differences. This chapter's Social Problems Research Up Close feature highlights one of the consequences of the masculinization of technology, as well as the ways in which computer hacker identities and communities are socially constructed.

Cyberstalking, pornography on the Internet, and identity theft are crimes that were unheard of before the computer revolution and the enormous growth of the Internet. One such "high-tech" crime, computer hacking, ranges from childish pranks to deadly viruses that shut down corporations. In this classic study, Jordan and Taylor (1998) enter the world of hackers, analyzing the nature of this illegal activity, hackers' motivations, and the social construction of the "hacking community."

Sample and Methods

Jordan and Taylor (1998) researched computer hackers and the hacking community through 80 semistructured interviews, 200 questionnaires, and an examination of existing data on the topic. As is often the case in crime, illicit drug use, and other similarly difficult research areas, a random sample of hackers was not possible. Snowball sampling is often the preferred method in these cases; that is, one respondent refers the researcher to another respondent, who then refers the researcher to another respondent, and so forth. Through their analysis, the investigators provide insight into this increasingly costly social problem and the symbolic interactionist notion of "social construction"—in this case, of an online community.

Findings and Conclusions

Computer hacking, or "unauthorized computer intrusion," is an increasingly serious problem, particularly in a society dominated by information technologies. Unlawful entry into computer networks or databases can be achieved by several means, including (1) guessing someone's password, (2) tricking a computer about the identity of another computer (called "IP spoofing"), or (3) "social engineering," a slang term referring to getting important access information by stealing documents, looking over someone's shoulder, going through their garbage, and so on.

Hacking carries with it certain norms and values, because, according to Jordan and Taylor (1998), the hacking community can be thought of as a culture within a culture. The two researchers identified six elements of this socially constructed community:

• *Technology*. The core of the hacking community is the technology that allows it to occur. As one professor who was interviewed stated, the young today have "lived with computers virtually from the cradle, and therefore have no trace of fear, not even a trace of reverence."

• *Secrecy*. The hacking community must, on the one hand, commit secret acts because their "hacks" are illegal. On the other hand, much of the motivation for hacking requires publicity to achieve the notoriety often sought. In addition, hacking is often a group activity that bonds members together. As one hacker stated, hacking "can give you a real kick some time. But it can give you a lot more satisfaction and recognition if you share your experiences with others."

• *Anonymity*. Whereas secrecy refers to the hacking act, anonymity refers to the importance of the hacker's identity remaining unknown. Thus, for example, hackers and hacking groups take on names such as Legion of Doom, the Inner Circle I, Mercury, and Kaos, Inc.

• *Membership fluidity*. Membership is fluid rather than static, often characterized by high turnover rates, in part, as a response to law enforcement pressures. Unlike more structured organizations, there are no formal rules or regulations.

• *Male dominance*. Hacking is defined as a male activity; consequently, there are few female hackers. Jordan and Taylor (1998) also note, after recounting an incident of sexual harassment, that "the collective identity hackers share and construct...is in part misogynist" (p. 768).

• *Motivation*. Contributing to the articulation of the hacking communities' boundaries are the agreed-upon definitions of acceptable hacking motivations, including (1) addiction to computers, (2) curiosity, (3) excitement, (4) power, (5) acceptance and recognition, and (6) community service through the identification of security risks.

Finally, Jordan and Taylor (1998, p. 770) note that hackers also maintain group boundaries by distinguishing between their community and other social groups, including "an antagonistic bond to the computer security industry (CSI)." Ironically, hackers admit a desire to be hired by the CSI, which would not only legitimize their activities but also give them a steady income.

Jordan and Taylor conclude that the general fear of computers and of those who understand them underlies the common, although inaccurate, portrayal of hackers as pathological, obsessed computer "geeks." When journalist Jon Littman asked hacker Kevin Mitnick if he was being demonized because of increased dependence on and fear of information technologies, Mitnick replied, "Yeah. . . . That's why they're instilling fear of the unknown. That's why they're scared of me. Not because of what I've done, but because I have the capability to wreak havoc" (Jordan & Taylor 1998, p. 776).

Technology and the Transformation of Society

A number of modern technologies are considerably more sophisticated than technological innovations of the past. Nevertheless, older technologies have influenced the nature of work as profoundly as the most mind-boggling modern inventions. Postman (1992) described how the clock—a relatively simple innovation that is taken for granted in today's world—profoundly influenced the workplace and with it the larger economic institution:

> The clock had its origin in the Benedictine monasteries of the twelfth and thirteenth centuries. The impetus behind the invention was to provide a more or less precise regularity to the routines of the monasteries, which required, among other things, seven periods of devotion during the course of the day. The bells of the monastery were to be rung to signal the canonical hours; the mechanical clock was the technology that could provide precision to these rituals of devotion. . . . What the monks did not foresee was that the clock is a means not merely of keeping track of the hours but also of synchronizing and controlling the actions of men. And thus, by the middle of the fourteenth century, the clock had moved outside the walls of the monastery, and brought a new and precise regularity to the life of the workman and the merchant. . . . In short, without the clock, capitalism would have been quite impossible. The paradox . . . is that the clock was invented by men who wanted to devote themselves more rigorously to God; it ended as the technology of greatest use to men who wished to devote themselves to the accumulation of money. (pp. 14–15)

Today, technology continues to have far-reaching effects not only on the economy but also on every aspect of social life. In the following section, we discuss societal transformations resulting from various modern technologies including workplace technology, computers, the Internet, and science and biotechnology.

. . . technology continues to have far-reaching effects not only on the economy but also on every aspect of social life.

Technology and the Workplace

All workplaces—from government offices to factories and from supermarkets to real estate agencies—have felt the impact of technology. Some technology lessens the need for supervisors and makes control by employers easier. For example, employees in the Department of Design and Construction in New York City must scan their hands each time they enter or leave the workplace. The use of identifying characteristics such as hands, fingers, and eyes is part of a technology called *biometrics*. Union leaders "called the use of biometrics degrading, intrusive and unnecessary and said that experimenting with the technology could set the stage for a wider use of biometrics to keep tabs on all elements of the workday" (Chan 2007, p. 1).

Technology can also make workers more accountable by gathering information about their performance. Further, through timesaving devices such as personal digital assistants (PDAs) and battery-powered store-shelf labels, technology can enhance workers' efficiency. New medical software marketed by Wal-Mart at a cost of $25,000 a year for one physician and $10,000 a year for each additional physician in a practice will not only save time and save money on costly record-keeping but is also likely to improve patient care (Lohr 2009).

However, technology can also contribute to worker error. In a recent study of a popular hospital computer system, researchers found several ways that the computerized drug-ordering program endangered the health of patients. For example, the software program warned a doctor of a patient's drug allergy only

Telepresencing, what used to be called tele-conferencing, has a decidedly high-tech feel nowadays, allowing participants, whether next door or around the world, to interact via audio and video links as though they were in the same room. Not just for business meetings anymore, telepresencing brings families together for holiday meals!

after the drug was ordered and, rather than showing the usual dose of a particular drug, the program showed the dosage available in the hospital pharmacy (DeNoon 2005).

Technology is also changing the location of work (see Chapter 7). Between 2006 and 2008, the number of employees who tele-commuted, including contract employees and the self-employed, increased from 28.7 million in 2006, to 33.7 million in 2008 (World at Work 2009). The increase in telecommuting is likely the result of several interacting social forces. First, a troubled economy and escalating gas prices make telecommuting a rational response to increase costs. Second, high-speed Internet and the increase in wireless access make telecommuting easier and more efficient. Finally, telecommuting is growing as an employee option as employers increasingly embrace the importance of balancing work and family obligations (World at Work 2009).

Telepresencing, a much more technologically sophisticated version of teleconferencing, allows life-sized participants in the virtual presence of one another to realistically communicate through broadcast quality sound and images (Houlahan 2006; Sharkey 2009). The telepresence industry includes such giants as Microsoft and Cisco Systems, both of which have invested millions of dollars into the R&D of this new technology.

Robotic technology has also revolutionized work. Ninety percent of robots work in factories, and more than half of these are used in heavy industry, such as automobile manufacturing (Tesler 2003). Robots are most commonly used for materials handling and spot welding although there has recently been a trend toward more fully integrating them into the manufacturing process (Pethokou-kis 2004). An employer's decision to use robotics depends on direct costs (e.g., initial investment) and indirect costs (e.g., unemployment compensation), the feasibility and availability of robots to perform the desired tasks, and the resulting increased rate of productivity. Use of robotics may also depend on whether labor unions resist the replacement of workers by machines.

What Do You Think? In Japan, "robots are more than mere gadgetry—they're practically family," says journalist Jonathan Skillings (2006, p. 1). And like many of today's family members, robots are increasingly unemployed (Hsu 2009). With a 40 percent reduction in manufacturing output because of the global recession, sales of industrial robots have declined significantly, as Japanese employers voluntarily sacrifice robots in favor of human workers. U.S. sales have declined as well but "robotic product manufacturers should do well . . . if they can expand beyond automakers and into consumer robotics, non-automated heavy industry, and service robotics . . ." (Robotic Trends 2009, p. 1). With soaring unemployment rates, particularly for unskilled workers who are the most likely to be replaced by automation, do you think state governments should impose limits on the number of robot-employees in ailing industries? Why or why not?

Telepresencing A sophisti-cated technology that allows life-sized participants in the virtual presence of one another to realistically communicate through broadcast quality sound and images.

Technology has also changed the nature of work. Federal Express not only created a FedEx intranet for its employees, they allow customers to enter their package-tracking database, saving the company millions of dollars a year. The used car industry has been revolutionized by the disabler—a remote device

Automation means that machines can now perform the labor originally provided by human workers, such as the robots that perform tasks on automobile assembly lines.

© Adam Lubroth/Riser/Getty Images

wired to a car's ignition system. When a customer fails to make a car payment on time, the device prevents the car from starting. Because the device has made it less likely that borrowers will default on their loans, there is some evidence that dealerships have become more willing to qualify low-income, low-credit customers (Welsh 2009).

The Computer Revolution

Early computers were much larger than the small machines we have today and were thought to have only esoteric uses among members of the scientific and military communities. In 1951, only about a half dozen computers existed (Ceruzzi 1993). The development of the silicon chip and sophisticated microelectronic technology allowed tens of thousands of components to be imprinted on a single chip smaller than a dime. The silicon chip led to the development of laptop computers, cellular phones, digital cameras, the iTouch, and portable DVDs. The silicon chip also made computers affordable. Although the first PC was developed only 20 years ago, today more than 74 percent of adult Americans use a computer, at least occasionally, compared to 54 percent of adult Americans in 1995 (U.S. Bureau of the Census 2009) (see Table 14.3).

Americans are more likely to use computers at home rather than at work (U.S. Bureau of the Census 2009). As with computer use in general, computer use in these two locations is associated with demographic variables. With the exception of age (computer use at home is highest for 15- to 24-year-olds whereas computer use at work is highest for 35- to 44-year-olds), computer use at home and at work follows the same pattern of overall computer use—the more educated you are and the higher your income, the higher the probability that you use a computer at home or at work (U.S. Bureau of the Census 2009). The most common computer activities include accessing the Internet, sending e-mail, using a search engine, getting the news or e-mail, followed by word processing,

TABLE 14.3 Computer and Internet Use by Select Characteristics, 1995 and 2008*

CHARACTERISTIC	ADULT COMPUTER USERS		ADULT INTERNET USERS	
	1995	2008	1995	2008
Total Adults	54	74	14	73
Age				
18 to 29 years old	70	90	21	90
30 to 49 years old	66	86	18	85
50 to 64 years old	46	70	9	70
65 years old and over	12	35	2	35
Sex				
Female	51	74	10	73
Male	58	74	18	73
Race/Ethnicity				
White, non-Hispanic	54	75	14	75
Black, non-Hispanic	50	65	11	59
English-speaking Hispanic	64	77	21	80
Educational Attainment				
Less than high school	17	46	2	44
High school graduate	46	65	8	63
Some college	72	83	20	84
College graduate or higher	82	92	29	91
Annual Household Income				
Less than $30,000	37	55	8	53
$30,000 to $49,999	61	78	15	76
$50,000 to $74,999	(NA)	84	23	85
$75,000 or more	(NA)	95	32	95

*Adults who use a computer or the Internet at home, work, school, or anywhere else at least on an occasional bases.
Source: U.S. Bureau of the Census 2009.

working with spreadsheets or databases, and accessing or updating calendars or schedules.

The U.S. ranks sixth in the world in terms of computer ownership, with 76.2 computers for every 100 people. Globally, Israel has the highest rate of computer ownership, 122 computers for every 100 people, and Russia has one of the lowest with just 12.2 computers for every 100 people (*The Economist* 2008). Further, computer usage varies by region of the world (Pew 2007a). Although increasing in 26 of the 35 countries in which comparable data is available (including in

Europe, where use has increased dramatically), computer use is nearly stagnant in many countries in Asia, Africa, and Latin America.

Not surprisingly, computer education has also mushroomed in the last two decades. In 1971, 2,388 U.S. college students earned a bachelor's degree in computer and information sciences; by 2006, that number had increased to 107,238 (U.S. Bureau of the Census 2009). Universities are moving toward requiring their students to have laptop computers, wireless corridors are increasingly common occurrences, and college and university spending on hardware and software is at an all-time high.

Computers are also big business, and the United States is one of the most successful producers of computer technology in the world, boasting several of the top companies—Dell, Hewlett-Packard, IBM, and Apple. Retail sales of computers exceed $20.3 billion annually, with Americans spending more than $7.6 billion on software alone (U.S. Bureau of the Census 2009). Spending on computers, however, has recently slowed down and, given the struggling economy, may not recover any time soon (Vance 2009a).

Computer software is, as noted, big business, and in some states, it is too big. In 2000, a federal judge found that Microsoft Corporation was in violation of antitrust laws, which prohibit unreasonable restraint of trade. At issue were Microsoft's Windows operating system and the vast array of Windows-based applications (e.g., spreadsheets, word processors, tax software)—applications that *only* work with Windows. The court held that the 70,000 programs written exclusively for Windows made "competing against Microsoft impractical" (Markoff 2000). In an agreement reached between the U.S. Department of Justice and Microsoft Corporation, Microsoft was to be divided into two companies—an operating systems company and an applications company. However, a federal appeals court overturned the lower court's order to split the company in half.

In 2007, Google claimed that Microsoft's latest operating system, Vista, was designed to make using the search engine more difficult and, therefore, was a violation of antitrust laws. In an about-face from the Clinton administration, officials appointed by President Bush dismissed the claim as unwarranted and recommended to state attorney generals that they do the same. Microsoft's problems are not over, however. Prosecutors in several states are pursuing Google's claim, regardless of the federal government's position (Labaton 2007). Further, Microsoft, which powers 90 percent of the world's computers, has been named in a class-action suit that contends that Microsoft provided misleading information to consumers about Windows-XP computers' ability to run Vista (Fiveash 2008). It is also a defendant in a patent violation case along with Apple, Inc., LG Electronics, and several other defendants. At issue is whether the accused violated a patent relating to touchpad technology (Magee 2009).

Information and Communication Technology and the Internet

Information and communication technology, or ICT, refers to any technology that carries information. Most information technologies were developed within a 100-year span: taking pictures and telegraphy (1830s), rotary power printing (1840s), the typewriter (1860s), transatlantic cable (1866), the telephone (1876), motion pictures (1894), wireless telegraphy (1895), magnetic tape recording (1899), radio (1906), and television (1923) (Beniger 1993). The concept of an "information society" dates back to the 1950s, when an economist identified a work sector he called "the production and distribution of knowledge." In 1958, 31 percent of the labor force was employed in this sector—today, more than 50 percent is. When this figure is combined with those in service occupations, more than 75 percent of the labor force is involved in the information society.

The **Internet** is an international information infrastructure—a network of networks—available through universities, research institutes, government agencies, libraries, and businesses. In 2008, 73 percent of all Americans used the Internet from some location compared to just 14 percent in 1995 (see Table 14.3). U.S. users are equally likely to be male or female but are more likely to be under 50 years of age, to have a college degree, and to have an annual household income of $75,000 or more (U.S. Bureau of the Census 2009). The most active Internet users are connected to broadband—that is, services that provide high-speed (DSL and cable) access rather than dial-up service—and are high-income young males (McGann 2005a). About 55 percent of all Internet users have used wireless connections at home, at the office, or at some other location to log on to the Internet (Pew 2009b).

> The growth of mobile Internet capabilities has led to "always present" connectivity as users working on laptops, Black-Berrys, and iPhones access the Internet from coffee shops, classrooms, and shopping malls.

The Internet, or the "web" as it is most commonly known, has evolved to what is now called **Web 2.0**—a platform for millions of users to express themselves online in the common areas of cyberspace (Grossman 2006; Pew 2007c). The development of Web 2.0 is a story about community and collaboration on a scale never seen before. It's about the cosmic compendium of knowledge that is Wikipedia and the million-channel people's network of YouTube and the online metropolis MySpace. It's about the many wresting power from the few and helping one another for nothing and how that will not only change the world, but also change the way the world changes. (Grossman 2006, p. 1)

Wireless access to the Internet is also altering Internet use. The growth of mobile Internet capabilities has led to "always present" connectivity as users working on laptops, BlackBerrys, and iPhones access the Internet from coffee shops, classrooms, and shopping malls (Horrigan 2009). Although varying significantly in motivation for use, attitudes toward the Internet, and demographic characteristics, 39 percent of Americans have wireless access to the Internet. Figure 14.1 displays daily Internet activities of American adults from 2000 through 2009 (Pew 2009c).

E-commerce. E-commerce is the buying and selling of goods and services over the Internet. Despite a slowdown in the economy, online retail sales, excluding travel, increased by 17 percent in 2008, to $204 billion, including $26.6 billion in clothing sales, $23.9 billion in computer sales, and $19.3 billion in car sales (Forrester Research 2008). Most online sales, about 35 percent, are initiated through online searches by two distinct groups of shoppers—bargain hunters looking for the best price and convenience shoppers who are more interested in saving time than money.

Internet An international information infrastructure available through universities, research institutes, government agencies, libraries, and businesses.

Web 2.0 A platform for millions of users to express themselves online in the common areas of cyberspace.

e-commerce The buying and selling of goods and services over the Internet.

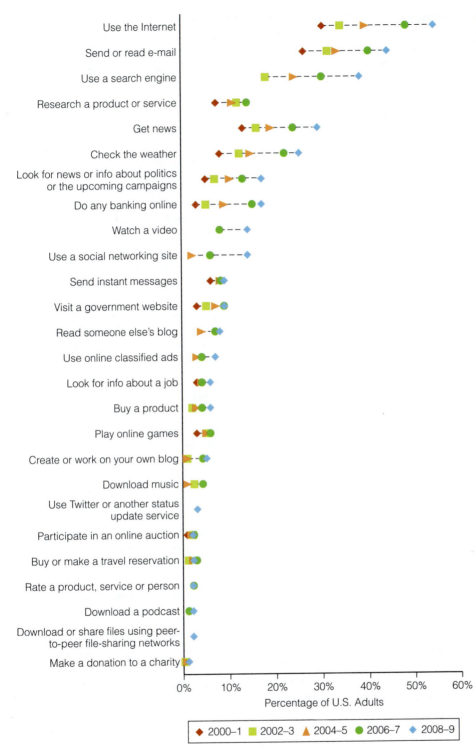

Figure 14.1 Daily Internet Activities, 2000–2009
Source: Pew 2009c.

Percentage of U.S. Adults

◆ 2000–1 ■ 2002–3 ▲ 2004–5 ● 2006–7 ◆ 2008–9

What Do You Think? The advent and proliferation of Wi-Fi (i.e., wireless access to the Internet) has facilitated a variety of Internet software and hardware innovations such as smartphones, the iTouch, netbooks, and thousands of down-loadable applications ("apps"). It has even led to institutional shifts; for example, e-commerce now includes m-commerce or the ability to make financial transactions (e.g., mobile banking) from a mobile device such as a cell phone. Newer versions of Wi-Fi will make it possible to move from "hotspot" to "hotspot" without losing connectivity, and future Wi-Fi technologies such as WiMAX will make entire cities wireless (Straubhaar et al. 2009). Do you think that the advantages of 24/7 access to the Internet outweigh the disadvantages that accompany "always present" technology?

Shopping online at retail websites, however, is only one component of e-commerce. In 2008, nearly half of online users reported visiting online classified sites such as Craig's list and, globally, 30 percent or more of Internet users do their banking on the Internet (Jones 2009; Budde 2009). Additionally, online consumers often use the Internet to research products and to chat with others about purchasing decisions. In a national survey, 56 percent of music buyers, 39 percent of cell phone purchasers, and 49 percent of prospective real estate investors, used the Internet to do product research. Interestingly, less than 12 percent of respondents in each of the three product categories reported that "online information had a major impact on their purchasing decision" (Horrigan 2008, p. 6).

Health and Digital Medicine. The Internet acts as the third most likely source of health information preceded only by health professionals, friends, or family members (Fox 2009) (also see Chapter 2). Most online searches result in useful information that affects health care decisions, including decisions about seeing a doctor, how to deal with a chronic disease, and diet and exercise information.

Besides providing medical information, there is considerable evidence that online medical records help improve medical care:

> A paper record is a passive, historical document. An electronic health record can be a vibrant tool that reminds and advises doctors. It can hold information on a patient's visits, treatments and conditions, going back years, even decades. It can be summoned with a mouse click, not hidden in a file drawer in a remote location and thus useless in medical emergencies. (Lohr 2008, p. 1)

Physicians who "get online" would not only have greater access to patient information, they would be linked to "online information on best practices, treatment recommendations, and harmful drug interactions" (Lohr 2008, p. 1). Doctors can also provide online services for patients. Beginning in 2010, the 700,000 members of Hawaii's state health plan will pay $10 per "virtual house call" via a webcam (Miller 2009).

Finally, there is some evidence that technology can help mediate the soaring cost of health care and save as much as $80 billion a year in the United States alone (Atkinson & Castro 2008). Much of the savings comes from "increases in efficiency, such as shorter hospital stays because of better coordination, better productivity for nurses, and more efficient drug utilization" (p. 27). Other important sources of cost effectiveness include electronic claims processing and reducing medical errors through more effective diagnostic and treatment interventions.

The Search for Knowledge and Information. The Internet, perhaps more than any other technology, is the foundation of the information society. Whether reading an online book, mapping directions, visiting the Louvre, or accessing Wikipedia, the Internet provides millions of surfers with instant answers to questions previously requiring a trip to the library. Even YouTube, at first a source of entertainment, has become a reference tool and is now the number-two search engine, edging out Yahoo (Helft 2009).

There is concern, however, that the very way in which the "Google generation" reads, thinks, and approaches problems has been altered by the new technology. Scholars at University College London, after examining computer logs of visitors to online research sites, found that visitors jumped from one site to the next, rarely returned to previous sites visited, and infrequently read more than a couple of pages of an article before jumping to another site. They concluded that:

> It is clear that users are not reading online in the traditional sense; indeed, there are signs that new forms of "reading" are emerging as users "power browse" horizontally through titles, contents pages, and abstracts going for quick wins. It almost seems that they go online to avoid reading in the traditional sense. (UCL 2008 p. 10)

Games and Entertainment. Over half of all Americans play video games, although less than a quarter play video games online; only 9 percent play massive multiplayer online games (MMOG) such as World of Warcraft, and less than 2 percent have visited a virtual world such as Second Life (Lenhart 2008).

Competition for the entertainment dollar is fierce on the Internet. YouTube, in competition with Hulu, has partnered with MGM, CBS, Sony, and Lions Gate to show full-length movies and television shows (Stelter & Helft 2009). MySpace recently launched MySpace Music, an online "jukebox," to try to loosen Apple's grip on the digital music business (Stone 2008). The entertainment industry, however, has also had its share of disputes with online entrepreneurs from illegal music downloads to the appearance of an uncut version of *X-Men Origins: Wolverine* on the Internet a month before it was released (Lynton 2009). These violations of intellectual property rights (i.e., piracy) remind us that the Internet remains unchartered territory.

Politics and e-Government. Technology is changing the world of politics. In 2008, approximately 60 percent of U.S. adult Internet users went online to get information about politics or the presidential election, or to send or receive political messages through e-mail, instant messaging, Twitter, and the like (Smith 2009).

When Iranian President-elect Mahmoud Ahmadinejad banned foreign media from the streets of Tehran, protestors contesting the election used Twitter, social networking sites, Flickr, and YouTube to stay in contact with one another and to transmit messages and images around the world. Senator Obama's campaign organizers

President Obama is called the first "digital president" because of his use of technology in his 2008 bid for the presidency as well as after taking office. He is shown here with his BlackBerry, which had to be retrofitted with additional security devices.

posted videos on YouTube, held virtual rallies, and curried favor with "friends" on Facebook and MySpace pages. Today, President Obama continues his use of technology. Although previous presidents released their weekly addresses to the nation to radio stations for the customary Saturday morning broadcast, President Obama posts video messages on YouTube and the White House website (Rutenberg & Nagourney 2009).

The United States is just one of 189 countries worldwide that hosts a government website. Annually, the United Nations ranks e-government readiness rankings—a composite measure used to assess a government's application of information and communication technologies. In 2008, Sweden, followed by the United States, Denmark, and Norway, received the highest e-government readiness ratings. Globally, regional differences indicate that Europe is the most e-government ready, followed by the Americas, Asia, Oceania, and Africa (United Nations 2008).

Social Networking and Blogging. Social network sites (e.g., Facebook, Twitter) and blogs comprise a sector of the Internet called **membership communities**. Membership communities have changed in recent years in three substantively significant ways. First, the *number of people* who visit membership communities has increased. For example, in 2008, two-thirds of the Internet population in the United States, Brazil, the United Kingdom, France, Germany, Italy, Spain, Switzerland, and Australia visited a membership community site at least once (Nielsen 2009a). Second, the *amount of time* members spend at a membership community site has grown dramatically. Between 2007 and 2008, time spent at membership community web pages grew three times faster than overall Internet growth, with 1 in 11 online minutes spent at social network sites or blogs (Nielson 2009a).

Finally, *who joins* membership communities is changing. Although it is true that adolescents and young adults remain more likely to use social networking sites than older adults, the percentage of older adults using social network sites has increased from 8 percent in 2005 to 35 percent in 2008 (Lenhart 2009). Similarly, Twitter, the micro-blogging site that asks "What are you doing?" had nearly 3 million visitors between the ages of 35 and 49 in 2008—42 percent of the site's total audience (Nielson 2009b).

Science and Biotechnology

Although recent computer innovations and the establishment of the Internet have led to significant cultural and structural changes, science and its resulting biotechnologies have produced not only dramatic changes but also hotly contested issues with public policy implications. In this section, we look at some of the issues raised by developments in genetics, food and biotechnology, and reproductive technologies.

membership communities Internet sites where participation requires membership and members regularly communicate with one another for personal and/or professional reasons.

genetic screening The use of genetic maps to detect predispositions to human traits or disease(s).

Genetics. Molecular biology has led to a greater understanding of the genetic material found in all cells—DNA (deoxyribonucleic acid)—and with it the ability for **genetic screening.**

> If you could uncoil a strip of DNA, it would reach 6 feet in length, a code written in words of four chemical letters: A, T, G, and C. Fold it back up, and it shrinks to trillionths of an inch, small enough to fit in any one of our 100 trillion cells, carrying the recipe for how to create human beings from scratch. (Gibbs 2003, p. 42)

Currently, researchers are trying to complete genetic maps that will link DNA to particular traits. There is some evidence that personality characteristics are inherited; other evidence links certain conditions previously thought of as psychological in nature (e.g., addiction, anorexia, and autism) as, at least in part, genetically induced (Harmon 2006). Already, specific strands of DNA have been identified as carrying physical traits such as eye color and height, as well as such diseases as sickle-cell disease, breast cancer, cystic fibrosis, prostate cancer, depression, and Alzheimer's (ORNL 2009).

The U.S. Human Genome Project (HGP), a 13-year effort to decode human DNA, is now complete. Conclusion of the project is transforming medicine:

> All diseases have a genetic component whether inherited or resulting from the body's response to environmental stresses like viruses or toxins. The successes of the HGP have . . . enabled researchers to pinpoint errors in genes—the smallest units of heredity—that cause or contribute to disease. The ultimate goal is to use this information to develop new ways to treat, cure, or even prevent the thousands of diseases that afflict humankind. (Human Genome Project 2007, p. 1)

The hope is that, if a defective or missing gene can be identified, possibly a healthy duplicate can be acquired and transplanted into the affected cell. This is known as **gene therapy.** Alternatively, viruses have their own genes that can be targeted for removal. Experiments are now under way to accomplish these biotechnological feats.

Food and Biotechnology. **Genetic engineering** is the ability to manipulate the genes of an organism in such a way that the natural outcome is altered. Genetically modified (GM) food, also known as genetically engineered food, and genetically modified organisms involve this process of DNA recombination—scientists transferring genes from one plant into the genetic code of another plant.

In the United States, in 2006, nearly 90 percent of soybeans, 83 percent of cotton, and 61 percent of corn crops were genetically modified varieties (Wegmann 2009). Further, an estimated 60 percent to 70 percent of processed foods in U.S. markets contain some form of GM ingredient. Yet a national survey of U.S. adults found that less than half (41 percent) were aware that foods containing GM ingredients are currently sold in stores, and although most Americans are likely to consume foods with GM ingredients every day, less than one-third of the survey sample (26 percent) said they had consumed food containing GM ingredients (Pew 2006).

Biotechnology companies and other supporters of GM foods commonly cite the alleviation of hunger and malnutrition as a main benefit, claiming that this technology can enable farmers to produce crops with higher yields. Critics of GM foods argue that the world already produces enough food for all people to have a healthy diet. According to the United Nations Development Programme, if all the food produced worldwide were distributed equally, every person would be able to consume 2,760 calories a day (UNDP 2003). Biotechnology, critics argue, will not alter the fundamental causes of hunger, which are poverty and lack of access to food and to land on which to grow food.

Biotechnology companies claim that GM foods approved by the Food and Drug Administration are safe for human consumption, and they even cite potential health benefits such as the use of genetic modification to remove harmful

gene therapy The transplantation of a healthy gene to replace a defective or missing gene.

genetic engineering The manipulation of an organism's genes in such a way that the natural outcome is altered.

Human health concerns [of genetically modified foods] include possible toxicity, carcinogenicity, food intolerance, antibiotic resistance buildup, decreased nutritional value, and food allergens...

allergens from foods or to improve nutritional benefits (Bailey 2005; Kaplan 2009). But critics claim that research on the effects of GM crops and foods on human health is inadequate, especially concerning long-term effects. Human health concerns include possible toxicity, carcinogenicity, food intolerance, antibiotic resistance buildup, decreased nutritional value, and food allergens in GM foods.

Biotechnology skeptics are also concerned about the environmental effects of GM crops. Biotechnology companies claim that crops that are genetically designed to repel insects negate the need for chemical (pesticide) control and thus reduce pesticide poisoning of land, water, animals, foods, and farm workers. However, critics are concerned that insect populations can build up resistance to GM plants with insect-repelling traits, which would necessitate increased rather than decreased use of pesticides.

GM seed contamination is another concern with regard to seed sterility technology. To maintain control over their products, biotechnology companies have developed "terminator" seeds, which cause the plant to produce sterile seeds. Because of public opposition to terminator seeds, in 1999, Monsanto agreed not to market its terminator technology. However, Monsanto later adopted a positive stance on genetic seed sterilization, suggesting that the commercialization of terminator technology may occur in the future (ETC Group 2003). Could the seed sterility trait in terminator crops inadvertently contaminate both traditional crops and wild plant life? The possible ramifications of widespread plant sterility could be devastating to life on earth.

Biotechnology critics also raise concerns about insufficient safeguards and regulatory mechanisms. In 2000, Taco Bell taco shells, made by Kraft Foods,

The first genetically engineered crop was introduced for commercial production in 1996. Today, more than 200 million acres are devoted to these crops, with the United States being the largest producer in the world.

© AP Photo/The Southern Illinoisan, Ceasar Maragni

were recalled after traces of Starlink corn were found in the taco shells. No one—from farmers to grain dealers to Kraft—could explain how the Starlink corn was mixed into corn meant for taco shells. The traces of unapproved corn were not found by the U.S. Department of Agriculture's Food Safety and Inspection Service nor by the Department of Health and Human Service's Food and Drug Administration. Rather, the traces were discovered by Genetically Engineered Food Alert—a coalition of biotechnology skeptics. In addition to taco shells, Starlink contamination caused a recall of more than 300 corn-based foods.

In 2000, worldwide concern about the safety of GM crops resulted in 130 nations signing the landmark Biosafety Protocol, which requires producers of a GM food to demonstrate that it is safe before it is widely used. The Biosafety Protocol also allows countries to ban the import of GM crops based on suspected health, ecological, or social risks. As of 2009, several countries have banned the importation of GM crops, and others had banned the commercial planting of GM crops.

Reproductive Technologies. The evolution of "reproductive science" has been furthered by scientific developments in biology, medicine, and agriculture. At the same time, however, its development has been hindered by the stigma associated with sexuality and reproduction, its link with unpopular social movements (e.g., contraception), and the feeling that such innovations challenge the natural order (Clarke 1990). Nevertheless, new reproductive technologies have been and continue to be developed.

In **in vitro fertilization (IVF),** an egg and a sperm are united in an artificial setting, such as a laboratory dish or test tube. Although the first successful attempt at IVF occurred in 1944, the first test-tube baby, Louise Brown, was not born until 1978. Perhaps the most famous test-tube babies are the Suleman octuplets born to an unemployed, single California mother who already had six children. The case of "Octomom" made national headlines and led to a debate over fertility clinics and the ethics of IVF. Presently, no laws restrict the number of embryos a woman may receive, although medical guidelines suggest that physicians take into consideration the age, environment, and mental and physical condition of the mother-to-be (Archibold 2009, p. 1).

Other concerns surround the disposal of the over 400,000 frozen embryos in U.S. fertility clinics. A survey of patients at nine fertility clinics revealed that, among patients who did not want any more children (i.e., had surplus embryos), 66 percent were willing to donate their embryos to research and 20 percent said they would likely keep the embryos indefinitely (Grady 2008). Further criticisms of IVF are often based on traditional definitions of the family and the legal complications created when a child can have as many as five potential parental ties—egg donor, sperm donor, surrogate mother, and the two people who raise the child (depending on the situation, IVF may not involve donors and/or a surrogate). Litigation over who the "real" parents are has already occurred. Over 50,000 children a year are born as a result of IVF technology.

Perhaps more than any other biotechnology, abortion epitomizes the potentially explosive consequences of new technologies. **Abortion** is the removal of an embryo or fetus from a woman's uterus before it can survive on its own. Globally, according to Singh et al. (2009):

- Of the estimated 208 million pregnancies in 2008, 40 percent are unintended; 33 million (16 percent) resulted in unintended births; and 41 million ended in abortions (20 percent).

in vitro fertilization (IVF) The union of an egg and a sperm in an artificial setting such as a laboratory dish.

abortion The intentional termination of a pregnancy.

- The lowest abortion rate in the world is in western Europe (12 abortions per 1,000 women between the ages of 15 and 44); the rate is 31 in Latin America, 29 in Africa and Asia, and 21 in North America.

- The rate of abortions are declining worldwide; they are declining faster in developed countries where abortions are generally safe and legal than in developing countries where more than half of abortions are unsafe and illegal.

- Five million women are hospitalized each year, and another 70,000 die from complications associated with unsafe abortions.

- Contraception use has increased in many parts of the world.

- Twenty-two countries or areas within countries have changed their abortion laws since 1997; 19 cases liberalized their laws and 3 made them more restrictive.

Abortion laws vary dramatically around the world. As Figure 14.2 indicates, the majority of women live in countries where abortion is allowed at least under some circumstances. However, 6 percent of women live in countries where abortion is banned under all circumstances. Nicaraguan law, for example, prohibits abortion even in the case of rape, incest, or deformity of the fetus. According to Amnesty International, the new Nicaraguan law has led to an increase in maternal deaths and hospital admissions associated with birth-related complications (Busari 2009). In Nicaragua, women and girls who seek an abortion and the doctors and nurses who provide abortion services receive prison sentences.

In the United States, since the U.S. Supreme Court's ruling in *Roe v. Wade* in 1973, abortion has been legal. However, recent Supreme Court decisions have limited the scope of the *Roe v. Wade* decision. For example, in *Planned Parenthood of Southeastern Pennsylvania v. Casey,* the court ruled that a state may restrict the conditions under which an abortion is granted, such as requiring a

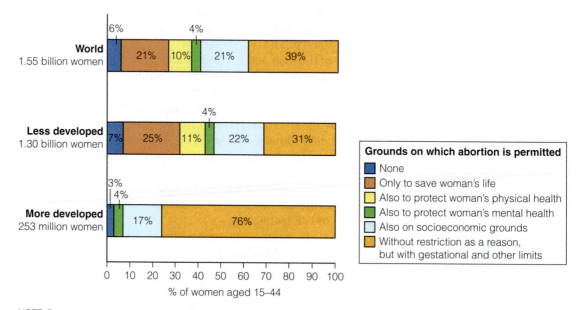

NOTE: Percentages may not add up to 100 because of rounding.

Figure 14.2 Abortion Laws by Restriction and Region, 2008
Source: Singh et al. 2009.

24-hour waiting period or parental consent for minors. In 2005, the Supreme Court agreed to hear a challenge to parental notification laws. Specifically, in *Ayotte v. Planned Parenthood,* the court considered "whether laws requiring parental notification before a minor can get an abortion must make an explicit exception when the minor's health is at stake" (Barbash 2005, p. 1). In 2006, the court's unanimous decision reaffirmed the need for a medical emergency exception for teenagers seeking an abortion (Greenhouse 2006).

Additional challenges *to Roe v. Wade* occur at the state level where some argue the real fight is being fought (Tumulty 2006). Missouri senators recently passed legislation that requires women seeking an abortion to be told the risks involved and the anatomical level of development of their fetus in a face-to-face conversation with their doctor at least 24 hours in advance of the procedure. If the proposed bill becomes law, it would also require that women have an opportunity to view an ultrasound of the fetus and to hear the fetal heartbeat (Blank 2009). A new Arizona law requires a 24-hour waiting period after visiting an abortion provider who must, in person, discuss the "procedures, risks, alternatives, and the fetuses' probable characteristics" (Davenport 2009, p. 1).

What Do You Think? In South Carolina, women seeking abortions are required to view an ultrasound image of their fetus before receiving the procedure. Presently, eleven states are considering bills that would require a woman to have ultrasound imaging before having an abortion. Sixteen states already have laws that require physicians offer a woman a pre-abortion ultrasound (Wiles 2009). A proposed Oklahoma law, although constitutionally challenged, requires that the woman look at the ultrasound monitor and that the doctor or technician describe the fetus (McKinley 2009). What do you think? Should women be required to view an ultrasound and/or listen to the fetal heartbeat before an abortion? What are the arguments for or against such practices?

Many of the state restrictions on abortion are a result of the pro-life movement's success in redefining the abortion issue as one concerning *fetal rights* rather than *women's rights* (Greenhouse 2007). For example, the Unborn Victims of Violence Act of 2004 protects all children in utero, regardless of stage of development. Whereas the original intent of the act was to protect the fetus from violent crimes, observers note that, if a fetus is defined as a human being with the same rights as women and men in law, then it may in effect overturn *Roe v. Wade.* Presently anti-abortion activists are raising money and pressuring state legislatures to get "personhood" measures on local and state ballots (Abcarian 2009).

Most recent debates concern intact dilation and extraction (D & E) abortions, which often take place in the second trimester of pregnancy. Opponents refer to such abortions as **partial birth abortions** because the limbs and the torso are typically delivered before the fetus has expired. However, former National Organization for Women president Kim Gandy states, "Try as you might, you won't find the term 'partial birth abortion' in any medical dictionary. That's because it doesn't exist in the medical world—it's a fabrication of the anti-choice machine" (U.S. Newswire 2003, p. 1). D & E abortions are performed because the fetus has a serious defect, the woman's health is jeopardized by the pregnancy, or both. In 2003, a federal ban on partial birth abortions was signed into law (White House 2003). Several constitutional challenges to the ban have occurred and, in 2004,

partial birth abortion Also called an intact dilation and extraction (D & E) abortion, the procedure may entail delivering the limbs and the torso of the fetus before it has expired.

a federal judge ruled that the ban was unconstitutional because it imposes an "undue burden on a woman's right to choose an abortion" (Willing 2005). However, in a major victory for the Bush administration, in 2007, the U.S. Supreme Court upheld the Partial-Birth Abortion Ban Act in a 5-4 decision (Greenhouse 2007). The significance of the case lies in the fact that it is the first time the U.S. Supreme Court has upheld a ban on any type of abortion procedure.

Feminists, including U.S. Supreme Court Justice Ruth Bader Ginsburg, strongly oppose the ban, arguing that it is just one step closer to making all abortions illegal. They are also quick to note that the ban was not supported by "the American Medical Association, the American College of Obstetricians and Gynecologists, the American Medical Women's Association, the American Nurses Association, or the American Public Health Association" (U.S. Newswire 2003, p. 1). Says Eleanor Smeal, president of the Feminist Majority Foundation, "[I]n upholding the Bush administration's abortion ban . . . the U.S. Supreme Court showed its true colors: that it does not care about the health, well-being, and safety of American woman. . . ." (Smeal 2007). Further, Justice Ginsburg, writing a strongly worded dissent, states, "this way of thinking reflects ancient notions of women's place in the family and under the Constitution—ideas that have long been discredited" (quoted in Greenhouse 2007, p. 1).

> Abortion is . . . a complex issue for societies, which must respond to the pressures of conflicting attitudes toward abortion and the reality of high rates of unintended and unwanted pregnancy.

Abortion is a complex issue for everyone, but especially for women, whose lives are most affected by pregnancy and childbearing. Women who have abortions are disproportionately poor, unmarried minority women who say that they intend to have children in the future. Abortion is also a complex issue for societies, which must respond to the pressures of conflicting attitudes toward abortion and the reality of high rates of unintended and unwanted pregnancy. The debate over abortion has also complicated passage of federal health care reform as conservatives, citing a 30-year ban on using taxpayer's money to pay for elective abortions, battle the Obama administration and abortion rights supporters (Kirkpatrick 2009).

In 2009, George R. Tiller, a physician who performed what some call "late-term abortions," was shot and killed while attending church in Wichita, Kansas. Scott Roeder has been charged with the killing.

Attitudes toward abortion tend to be polarized between two opposing groups of abortion activists—pro-choice and pro-life. In fact, in 2009, a survey of U.S. adults indicated that Americans are evenly divided between the two positions (Saad 2009). Advocates of the pro-choice movement hold that freedom of choice is a central human value, that procreation choices must be free of government interference, and that because the woman must bear the burden of moral choices, she should have the right to make such decisions. Alternatively, pro-lifers hold that the unborn fetus has a right to live and be protected, that abortion is immoral, and that alternative means of resolving an unwanted pregnancy should be found.

In July 1996, scientist Ian Wilmut of Scotland successfully cloned an adult sheep named Dolly. To date, cattle, goats, mice, pigs, cats, rabbits, and horses have also been cloned. This technological breakthrough has caused worldwide concern about the possibility of human cloning, leading the United Nations to adopt a declaration that calls for governments to ban all forms of cloning that are at odds with human dignity and the preservation of human life (Lynch 2005). One argument in favor of developing human cloning technology is its medical value; it may potentially allow everyone to have "their own reserve of therapeutic cells that would increase their chance of being cured of various diseases, such as cancer, degenerative disorders, and viral or inflammatory diseases" (Kahn 1997, p. 54). Human cloning could also provide an alternative reproductive route for couples who are infertile and for those in which one partner is at risk for transmitting a genetic disease.

Arguments against cloning are largely based on moral and ethical considerations. Critics of human cloning suggest that, whether used for medical therapeutic purposes or as a means of reproduction, human cloning is a threat to human dignity (Human Cloning Prohibition Act of 2007). For example, cloned humans would be deprived of their individuality, and as Kahn (1997, p. 119) pointed out, "creating human life for the sole purpose of preparing therapeutic material would clearly not be for the dignity of the life created." **Therapeutic cloning** uses stem cells from human embryos. **Stem cells** can produce any type of cell in the human body and thus can be "modeled into replacement parts for people suffering from spinal cord injuries or regenerative diseases, including Parkinson's and diabetes" (Eilperin & Weiss 2003, p. A6). In 2009, the U.S. Food and Drug Administration approved the first trials of embryonic stem cell therapy for paralyzed patients with spinal cord injuries (Park 2009, p. 2). This chapter's The Human Side recounts the use of stem cells to help a young boy recover from a life-threatening disease.

Today, the use of embryonic stem cells is quickly being replaced by induced pluripotent stem cells (iPS). Because of iPS, it is "now possible for researchers to churn out unlimited quantities of a patient's stem cells, which can then be turned into any of the cells that the body might need to repair or replace" (Park 2009, p. 3). Thus, iPS stem cells have "embryonic-like powers to morph into heart, cardiac, and other tissue types" (Svoboda 2009, p. 1).

In 2009, President Obama lifted the ban on federal funding of embryonic stem cell research and directed the National Institute of Health to develop research support guidelines. Abortion opponents criticized the executive order for publicly financing research that destroys embryos, something conservatives argue is morally wrong (see Figure 14.3). The research community in turn was critical of the administration's decision to limit federal financing to research on surplus embryos from fertility clinics (Harris 2009). The new guidelines allow for the expansion of stem cell lines previously restricted, but prohibit federal funding of stem cell lines created solely for the purpose of research or created through therapeutic cloning (Rovner & Gold 2009; Harris 2009).

Despite what appears to be a universal race to the future and the indisputable benefits of scientific discoveries such as the workings of DNA and the technology of IVF and stem cells, some people are concerned about the duality of science and technology. Science and the resulting technological innovations are often life assisting and life giving; they are also potentially destructive and life threatening.

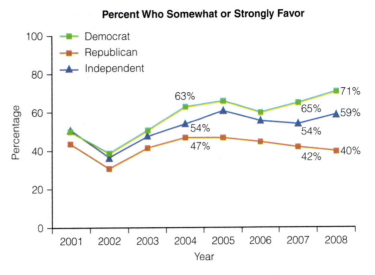

On the whole, how much do you favor or oppose medical research that uses stem cells from human embryos—Do you strongly favor, somewhat favor, somewhat oppose, or strongly oppose this?

Percent Who Somewhat or Strongly Favor

Figure 14.3 Partisan Differences in Attitudes Toward Embryonic Stem Cell Research, 2008
Source: VCU 2009.

therapeutic cloning Use of stem cells to produce body cells that can be used to grow needed organs or tissues; regenerative cloning.

stem cells Undifferentiated cells that can produce any type of cell in the human body.

For many people, the debate over the use of stem cells is a non-issue. This is particularly true for those with diseases or disabilities who might benefit from stem cell therapy and the people who love them. Here, new dad Andres Trevino recounts his son Andy's rare genetic condition and how his daughter Sofia saved Andy's life.

My wife is a carrier of a genetic condition that causes the immune system to fail. Ten years ago, we found out the hard way, after the birth of our son, Andy. He had his first life-threatening infection 48 hours after he was born in Mexico City and it took us 19 difficult months to find him a diagnosis.

The name of the disease is NEMO. We found out about NEMO at Children's Hospital Boston, around 2,280 miles away from our hometown and away from our families; Boston was our last resort. Out of the 20 thousand genes in the human body, researchers at Children's Hospital were able to identify the specific mutation (like finding a misspelled word in the *Encyclopedia Britannica*) that was causing the deadly problem. Until recently, the last time I checked, there were only about 72 cases registered in the medical journals of children with NEMO, and only half of the cases were alive. Most of them died of infections.

Andy had infections all over his body with different types of bacteria and viruses. He had no tools to fight them. He had infection in his central nervous system, stomach, skin, eyes, blood, and elsewhere, even in the bone on his right index finger.

At first, we didn't know the implications of knowing the roots of the disease, but now we feel fortunate. Knowing about NEMO gave us more options to keep him alive and try to find a cure.

Doctors recommended a bone marrow stem cell transplant to try to replace parts of his immune system and help him fight infections. New bone marrow seemed to be the only way he would be able to reach adulthood. The bone marrow's NEMO-free stem cells would help him fight infections. The success rate for the procedure greatly depended on the compatibility of the donor, so we searched at public registries worldwide. For two and a half years, we couldn't find a match.

"There's always the option of having another baby," someone told us. We could use the baby's umbilical cord stem cells for Andy. But we ran the risk of having another baby with NEMO, and on top of that, only one out of four siblings would be compatible with Andy.

We learned about using in vitro fertilization (IVF) to have babies, and pre-implantation genetic diagnosis (PGD) to check compatibility and rule out NEMO. It took us five IVF cycles and 36 embryos to find our daughter, Sofia, and we were blessed with a healthy baby.

Minutes after Sofia's birth, a doctor gathered umbilical cord blood in a plastic bag that was later sent via courier to a private umbilical cord blood storage location where stem cells were separated and frozen. A couple of weeks later, we learned that the number of stem cells obtained from it was not going to be enough for Andy's transplant. We were hoping to find a way to increase the numbers of cells without having to take more cells from Sofia. It was a difficult decision, but we knew that bone marrow stem cells are constantly replaced by the body, and we made sure that she would have no side effects. We waited six months until Sofia could donate an additional amount from her bone marrow, and she did fine, thanks to physicians at Children's Hospital.

After almost 1,000 days of hospitalization in Mexico City and Boston, a stem cell bone marrow transplant cured Andy.

My wife and I decided to donate our remaining embryos to the Stem Cell Research Program at Children's. We knew about the life-changing research done at Children's and we were hoping that with our cells, researchers could find ways to treat conditions like the one that affected Andy.

After signing at least 10 different consent forms, we transported the embryos from the fertility clinic to the Karp [Family Research] building at Children's. We knew some of the researchers there because of the time my son had been in the hospital.

Weeks later, I ran into Paul Lerou, MD, who gave me the news that they were able to obtain two viable stem cell lines from the donated embryos and he invited me to see them. He told me they had one female and one male line, both affected with NEMO.

I didn't know what to expect. Inside a shallow, circular glass container with a removable cover, I was able to see a group of silver-colored cells. It was amazing. Those cells are able to become any other cell of the human body.

I couldn't help remembering all the feelings we had while we were looking for compatible cells for Andy and how blessed we were with Sofia. Dr. Lerou told me that he didn't know of any other family with its own stem cell line.

I'm very happy to learn that the National Institutes of Health (NIH) recently released new guidelines for research using human stem cell lines that will allow Dr. Lerou and others to obtain public funding to continue much needed research to find cures.

My only hope is that he's able to obtain enough funding to be able to obtain stem cells for children with life-threatening diseases who are waiting for compatible cells.

We believe that embryonic stem cell research will revolutionize medicine and provide treatments to many life-threatening diseases. It won't take long until the Stem Cell Research program at Children's Hospital Boston is able to unleash the healing power of compatible stem cells for patients who desperately need them. Just as Sofia's stem cells allowed Andy to fight infections and live a normal life.

Source: Children's Hospital Boston Staff 2009. Available at http://childrenshospitalblog.org/stem-cell-research-a-fathers-story/.

The same scientific knowledge that led to the discovery of nuclear fission, for example, led to the development of both nuclear power plants and the potential for nuclear destruction. Thus, we now turn our attention to the problems associated with science and technology.

Societal Consequences of Science and Technology

Scientific discoveries and technological innovations have implications for all social actors and social groups. As such, they also have consequences for society as a whole. Figure 14.4, for example, displays the percentage of Americans, Canadians, and Europeans who believe the listed technologies will improve our way of life over the next 20 years (National Science Foundation 2007). Technology, however, also has negative consequences as we discuss in the following sections.

Social Relationships, Social Networking, and Social Interaction

Technology affects social relationships and the nature of social interaction. The development of telephones has led to fewer visits with friends and relatives; with the advent of DVRs and cable television, the number

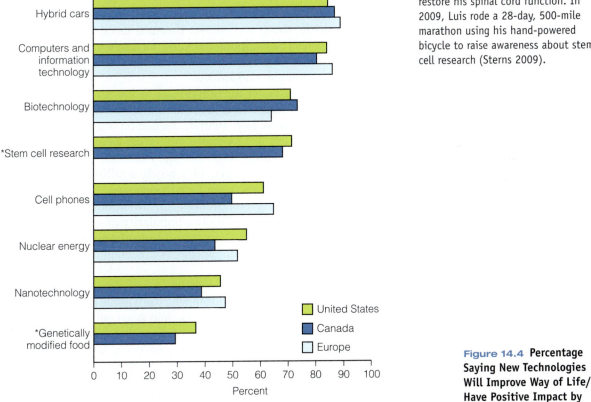

In 1994, Luis Gonzalez-Bunster, just one week out of high school, was seriously injured in a car accident. Paralyzed from the chest down, he remains in a wheelchair but is more hopeful than ever that stem cell therapy will restore his spinal cord function. In 2009, Luis rode a 28-day, 500-mile marathon using his hand-powered bicycle to raise awareness about stem cell research (Sterns 2009).

Courtesy of the Gonzalez-Bunster Family

Figure 14.4 Percentage Saying New Technologies Will Improve Way of Life/ Have Positive Impact by Location, 2005

*No data available for Europe.

Figure 14.5 What Teens Share in an Online Environment

Source: Lenhart, Amanda, and Mary Madden. 2007. *Teens, Privacy, and Online Social Networks*. Washington, DC: Pew Internet and American Life Project.

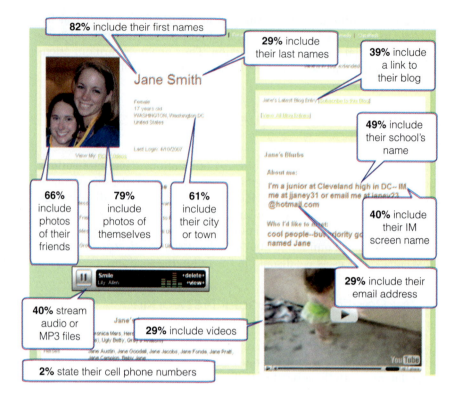

82% include their first names

29% include their last names

39% include a link to their blog

49% include their school's name

66% include photos of their friends

79% include photos of themselves

61% include their city or town

40% include their IM screen name

29% include their email address

40% stream audio or MP3 files

29% include videos

2% state their cell phone numbers

of places where social life occurs (e.g., movie theaters) has declined. Even the nature of dating has changed as computer networks facilitate instant messaging, cyber-dates, and "private" chat rooms. As technology increases, social relationships and human interaction are transformed.

Technology also makes it easier for individuals to live in a cocoon—to be self-sufficient in terms of finances (e.g., Quicken), entertainment (e.g., Hulu), work (e.g., telecommuting), news (e.g. Twitter), recreation (e.g., Wii), shopping (e.g., Amazon), communication (e.g., e-mail), family conferences (e.g. Skype), and many other aspects of social life. For example, Facebook, the fastest growing social networking site on the Internet, adds nearly a million users a day. It brings together 100 million users who spend, on the average, 2.5 hours a month responding to wall posts, updating their profiles, and "friending" other Facebook users (Hagel & Brown 2008; Stone 2009). The popularity of social networking sites has led to fears over the privacy and security of information posted (see Figure 14.5) and has led some to abandon membership communities (Heffernan 2009).

Although technology can bring people together, it can also isolate them from each other (Klotz 2004). For example, children who use a home computer "spend much less time on sports and outdoor activities than non-computer users" (Attewell et al. 2003, p. 277). In addition, a study of more than 1,500 U.S. Internet users between the ages of 18 and 64 found that, for every hour a respondent was on the Internet, there was a corresponding 23.5-minute reduction in face-to-face interaction with family members (Nie et al. 2004). Some technological innovations replace social roles—an answering machine may replace a secretary, a computer-operated vending machine may replace a waitperson, an automatic teller machine may replace a banker, and closed circuit television may replace a teacher. Parking attendants may even become obsolete as New York City opens its first robotic parking garage (Svensson 2007). These technologies may improve efficiency, but they also reduce necessary human contact.

Loss of Privacy and Security

Schools, employers, and the government are increasingly using technology to monitor individuals' performance and behavior. A 2005 study reported that 36 percent of companies use "keystroke monitoring so they can read what people type as well as track how much time they spend at the computer" and "55 percent retain and review e-mail messages" (MacMillan 2005, p. 1). Today, the legality of monitoring e-mails is under scrutiny. One in 20 companies has been sued for e-mail–related surveillance, and a federal appeals court has ruled that e-mails, like letters and phone conversations, are protected under the Fourth Amendment's prohibition against unreasonable search and seizure (Holding 2007). However, in 2009, the Wisconsin Supreme Court agreed to hear a case that pits the public's right to know against privacy rights of public school teachers, and "legal experts say the employees in question, and all public school employees in general, might not have a reasonable expectation of privacy" (Prabhu 2009, p. 1). Concern over the privacy of e-mails has led to a new software application called "Vanish," which destroys e-mail messages within a few hours of being sent (*The Economist* 2009a).

In addition to e-mail monitoring, high-tech machines monitor countless other behaviors, including counting a telephone operator's minutes online, videotaping a citizen walking down a city street, or tracking the whereabouts of a student or faculty member on campus. Employers and schools may subject individuals to drug-testing technology and, in 2008, identity theft was the number-one complaint filed with the Federal Trade Commission for the ninth year in a row (FTC 2009) (see Chapter 4). Through computers, individuals can obtain access to someone's phone bills, tax returns, medical reports, credit histories, bank account balances, and driving records. In 2009 alone:

- Personal data (names, social security numbers, etc.) for 30,000 Kaiser Permanente employees in Northern California was breached when a law enforcement agent found the data on a stolen computer.
- An Arkansas Department of Information Systems and Information Vaulting Service lost a computer storage tape containing copies of 807,000 criminal background checks.
- Hackers accessed the Federal Aviation Administration's computer system containing the names and social security numbers of 48,000 employees and retirees.
- A laptop stolen from an employee of the Oklahoma Department of Human Services contained the records of over 1,000,000 people who had received social services including names, social security numbers, and dates of birth.
- A civil employee of the New York City Police Department stole eight computer tapes that contained personal data on 80,000 current and retired police officers. (Privacy Rights Clearinghouse 2009)

Although just inconvenient for some, unauthorized disclosure of, for example, medical records, is potentially devastating for others. If a person's medical records indicate that he or she is human immunodeficiency virus (HIV)-positive that person

Customer service representatives answer queries at a call center in India. India's outsourcing industry employs more than 4 million workers—a number that is likely to grow as multinational corporations seek qualified personnel outside the United States at a lower cost.

could be in danger of losing his or her job or health benefits. If DNA testing of hair, blood, or skin samples reveals a condition that could make the person a liability in an insurer's or employer's opinion, the individual could be denied insurance benefits, medical care, or even employment. In response to such fears, the **Genetic Information Nondiscrimination Act of 2008** (GINA) was passed. GINA is a federal law that prohibits discrimination in health coverage or employment based on genetic information (Department of Health and Human Services 2009).

Technology has created threats not only to the privacy of individuals but also to the security of entire nations. Computers can be used (or misused) in terrorism and warfare to cripple the infrastructure of a society and to tamper with military information and communication operations (see Chapter 15). In 2009, cyber-attacks successfully stalled or slowed down 27 American and South Korean government and commercial websites, including websites of the Treasury Department, the New York Stock Exchange, the Secret Service, and the White House (Sang-Hun & Markoff 2009).

Unemployment, Immigration, and Outsourcing

Some technologies replace human workers—robots replace factory workers, word processors displace secretaries and typists, and computer-assisted diagnostics reduce the need for automobile mechanics. In 2009, Microsoft introduced "Laura," a personal assistant that resides on your desktop, senses moods and personality nuances, and follows directions (Vance 2009b). Unemployment rates can also increase when companies **outsource** (sometimes-called offshoring) jobs to lower wage countries:

> The globalization of work tends to start from the bottom up. The first jobs to be moved abroad are typically simple assembly tasks, followed by manufacturing, and later skilled work like computer programming. At the end of this progression is the work done by scientists and engineers in research and development laboratories. (Lohr 2006, p. 1)

For example, officials at Cisco Systems, the largest maker of information and communication equipment, have announced that 20 percent of its "top talent" will be in India over the next 5 years; Dow Chemical is building a new research center in Shanghai that will employ 600 technical workers; pharmaceutical company Eli Lilly is paying Indian scientists $500,000 to $1.5 million a year per scientist to develop new drugs for commercial use; and, since 1992, IBM has reduced its American labor force by 31,000, even as its Indian work force increased from 0 to 52,000 (Giridharadas 2007). As author Thomas Friedman explained in

Genetic Information Non-discrimination Act of 2008
A federal law that prohibits discrimination in health coverage or employment based on genetic information.

outsource A practice in which a business subcontracts with a third party, often in low-wage countries such as China and India, for services.

his book *The World Is Flat* (2005), ". . . it is now possible for more people than ever to collaborate and compete in real time with more other people on more different kinds of work from more different corners of the planet on a more equal footing than at any previous time in the history of the world . . ." (p. 8).

Outsourcing at a time when the U.S. employment rate is in double digit figures is, to some, reprehensible (Wadhwa 2009). Alan Blinder, former economic advisor to President Clinton and former vice chairman of the Federal Reserve, estimates that outsourcing "poses a risk to the employment of as many as 28 million to 42 million workers in the United States" (Giridharadas 2007, p. 3). When layoffs occur, older IT workers are more likely to remain unemployed, being less likely than younger workers to take a job outside the field.

Finally, there is some concern about the number of immigrant employees that are in the United States on H-1B visas. H-1B visas permit employers to temporarily hire foreign workers in certain specialty occupations including high-tech industries. The need for immigrant high-tech employees is a consequence of the lack of American counterparts in STEM occupations, that is, science, technology, engineering and mathematics (Price 2008). Currently, the number of H-1B visas issued a year by the federal government is limited to 65,000 (U.S. Citizenship and Immigration Services 2009). However, critics of the H-1B visa limitations are pressuring the federal government to increase the number of visas available annually.

> Several theorists hypothesize that, as technology displaces workers—most notably the unskilled and uneducated—certain classes of people will be irreparably disadvantaged—the poor, minorities, and women.

It should also be noted that the H-1B visa program is not without problems for the immigrant employee. Worker's visas are temporary—valid for a maximum of six years—at which time the visa holder must leave the United States unless permanent residency has been granted. Employees with H-1B visas are often paid less than American employees are, and cannot voluntarily quit their jobs for fear of deportation (Price 2008; U.S. Citizenship and Immigration Services 2009).

The Digital Divide

One of the most significant social problems associated with science and technology is the increased division between the classes. In a now oft-quoted statement Welter (1997) notes:

> It is a fundamental truth that people who ultimately gain access to, and who can manipulate, the prevalent technology are enfranchised and flourish. Those individuals (or cultures) that are denied access to the new technologies, or cannot master and pass them on to the largest number of their offspring, suffer and perish. (p. 2)

The fear that technology will produce a "virtual elite" is not uncommon. Several theorists hypothesize that, as technology displaces workers—most notably the unskilled and uneducated—certain classes of people will be irreparably disadvantaged—the poor, minorities, and women. There is even concern that biotechnologies will lead to a "genetic stratification," whereby genetic screening, gene therapy, and other types of genetic enhancements are available only to the rich.

The wealthier the family, for example, the more likely the family is to have broadband Internet access. Of American families with an annual income of

$75,000 or more, 85 percent of Americans report at least occasional computer use from some location compared to 55 percent of respondents with annual incomes of $30,000 of less. Furthermore, 84 percent of respondents with annual incomes of $75,000 or more have broadband Internet access (cable or DSL) compared to 32 percent of respondents with incomes of $30,000 a year or less (U.S. Bureau of the Census 2009).

Racial and ethnic disparities also exist in Internet access. Although 75 percent of whites occasionally use the Internet from some location, only 59 percent of blacks occasionally use the Internet from some location (U.S. Bureau of the Census 2009). Inner-city neighborhoods are disproportionately populated by racial and ethnic minorities and are simply less likely to be "wired," that is, to have the telecommunications hardware necessary for access to online services. In fact, cable and telephone companies are less likely to lay fiber optic cables in these areas—a practice called "information apartheid" or "electronic redlining."

In industrialized countries such as the United States and Japan, computer and Internet use is generally high, and there are few gender disparities. However, in developing countries, women play a subordinate role in information communication technologies, which affects their employability:

> The perception of women being passive consumers of ICT rather than producers extends to their work-related use as well, where one continues to see a feminization of lower level ICT jobs. . . . Women continue to be concentrated in tedious, repetitive tasks as when they were during the first wave of industrialization, in manufacturing sectors such as textiles, clothing, and electronics. The lower skilled ICT jobs that women typically find themselves in are word-processing and data entry. (Thas et al. 2007, p. 10)

Concern over accessibility to broadband connectivity has led to a debate over net neutrality. **Net neutrality** advocates hold that Internet users should be able to visit any website and access any content without Internet service providers (ISP) (e.g., cable or telephone companies) acting as gatekeepers by controlling, for example, the speed of downloads. Why would an ISP do that? Hypothetically, if Internet service provider company X signs an agreement with search engine Y, then it's in the best interest of Internet service provider X to slow down all other search engines' performances so that you will switch to search engine Y. Internet service providers argue that Internet users, be they individuals or corporations, who use more than their "fair share" of the Internet should pay more. Why should you pay the same monthly fee as your neighbor who nightly downloads full-length movie files? Others fear any government regulation of the Internet and/or prefer a strictly market model.

In response to the debate, the Internet Freedom Preservation Act of 2009 has been introduced into Congress. The introduction to the bill states that as "the Nation becomes more reliant upon such Internet technologies and services, unfettered access to the Internet to offer, access, and utilize content, services, and applications is vital" (Internet Freedom Preservation Act 2009, p. 1). The proposed legislation, among other things, requires that Internet service providers "not block, interfere with, discriminate against, impair, or degrade the ability of any person to use an Internet access service" (p. 1).

net neutrality A principle that holds that Internet users should be able to visit any website and access any content without Internet service provider interference.

Mental and Physical Health

Some new technologies have unknown risks. Biotechnology, for example, has promised and, to some extent, has delivered everything from life-saving drugs to hardier pest-free tomatoes. Limbs are being replaced by bionic devices controlled by the recipient's thoughts (Brown 2006), and micro-sized telescopes are being implanted in the eyes of people with damaged retinas (Eisenberg 2009). However, biotechnologies have also created **technology-induced diseases,** such as those experienced by Chellis Glendinning (1990). Glendinning, after using the pill and, later, the Dalkon Shield IUD, became seriously ill:

> Despite my efforts to get help, medical professionals did not seem to know the root of my condition lay in immune dysfunction caused by ingesting artificial hormones and worsened by chronic inflammation. In all, my life was disrupted by illness for twenty years, including six years spent in bed. . . . For most of the years of illness, I lived in isolation with my problem. Doctors and manufacturers of birth control technologies never acknowledged it or its sources. (p. 15)

Other technologies that pose a clear risk to a large number of people include nuclear power plants, the pesticide DDT, automobiles, X-rays, food coloring, breast implants and, according to the American Academy of Environmental Medicine, genetically modified foods, which have "more than a casual association" with "adverse health effects" (AAEM 2009, p. 1). Even cell phones, which are universally available, have health and safety risks. Research indicates that drivers using cell phones are four times more likely to cause an accident as those who do not use cell phones and as likely to cause a crash as a legally drunk driver (Richtel 2009).

Technological innovations are, for many, a cause of anguish, stress, and fear, particularly when the technological changes are far-reaching. About one-third of Americans believe that "science is going too far and hurting society," and 31 percent agree with the statement that "technology is making life too complicated for me" (Pew 2007c). Moreover, nearly 60 percent of workers report being "technophobes," that is, fearful of technology, and a recent editorial in a top-notch psychiatry journal had the author calling for Internet addiction to be added to the *Diagnostic and Statistical Manual of Mental Disorders* (Smith 2008).

What Do You Think? Some evidence suggests that technological gadgetry and the multitasking it engenders is taking a toll "on our ability to think clearly, work effectively, and function as healthy human beings" (Wallis & Steptoe 2006, p. 74). Psychiatrist Edward Hallowell, an expert in attention deficit disorder, has named this syndrome ADT—attention deficit trait. Dr. Hallowell explained that ADT takes "hold when we get so overloaded with incoming messages and competing tasks that we are unable to prioritize" (as quoted in Wallis & Steptoe 2006, p. 74). In the end we feel distracted, impulsive, guilty, and inadequate as we slip farther and farther behind in our work and in our lives. Do you think ADT exists, and if so, do you suffer from it? How much time do you think you lose a day from unwanted technologically based interruptions, and how much time does it takes you to get back on task?

technology-induced diseases Diseases that result from the use of technological devices, products, and/or chemicals.

Malicious Use of the Internet

Some Internet users access the Internet for malicious purposes including but not limited to cybercrime and prostitution (see Chapter 4), hacking (see this chapter's Social Problems Research Up Close), piracy, electronic aggression (e.g., cyber-bulling), and "questionable content" sites. **Internet piracy** entails illegally downloading or distributing copyrighted material (e.g., music, games, or software). Recent court cases indicate that trade organizations (e.g., the Recording Industry Association of American) are pursuing criminal cases against violators, and courts are imposing strict penalties. In 2009, a student at Boston University was assessed $675,000 in damages for sharing 30 songs (*The Economist* 2009b).

Malware is a general term that includes any spyware, crimeware, worms, viruses, and adware that is installed on owners' computers without their knowledge. Malware costs billions of dollars a year—$13 billion worldwide in 2006 (Thas et al. 2007). There is some evidence that, despite the best efforts of computer security experts, the proliferation of malicious software exceeds our knowledge or ability to stop it (Markoff 2008b).

Electronic aggression is defined as any kind of aggression that takes place with the use of technology (David-Ferdon & Hertz 2009). For example, **cyber-bullying** refers to the use of electronic devices (e.g., websites, e-mail, instant messaging, or text messaging) to send or post negative or hurtful messages or images about an individual or a group (Kharfen 2006). Estimates of the frequency of involvement in cyber-bullying either as victim, perpetrator, or both range dramatically, although electronic aggression researchers generally agree that texting is the most common means of cyber-bullying (David-Ferdon & Hertz 2009). Because cyber-bullying is capable of reaching wider audiences and thus doing more harm, many states and school districts have begun creating cyber-bullying disciplinary policies.

In 2008, a federal jury convicted Lori Drew of posing as a 13-year-old boy on a fraudulent MySpace account in order to get information from a 13-year-old friend of her daughter's, Megan Meier, who had allegedly spread lies about her. As "Josh Evans," Ms. Drew flirted with and then "broke up" with Megan with an email that read "The world would be a better place without you" (McCarthy & Michels 2009, p. 3). Despite the fact that Megan Meier hung herself after receiving the e-mail, the conviction was later dismissed.

Finally, Optenet (2008), a global Internet security vendor, analyzed the content of nearly three million randomly selected uniform resource locators (URLs) from around the world and classified them by content. Figure 14.6 shows the results. Note that the largest single category of Internet content is pornography (36.2 percent), which had over three times the volume of the next largest category, e-commerce/shopping (10.8 percent). All "questionable content" sites measured by the researchers increased between 2006 and 2007, including websites containing content related to the promotion of anorexia and bulimia (+470.0 percent), violence (+125.6 percent), racism (+70.3 percent), drug use (+62.3 percent), and child pornography (+18.0 percent). However, pornography websites in general, as a portion of the total number of websites, decreased 8.9 percent between 2006 and 2007.

The Challenge to Traditional Values and Beliefs

Technological innovations and scientific discoveries often challenge traditionally held values and beliefs, in part because they enable people to achieve goals

---

Internet piracy Illegally downloading or distributing copyrighted material (e.g., music, games, software).

malware A general term that includes any spyware, viruses, and adware that is installed on an owner's computer without their knowledge.

Cyber-bullying The use of electronic devices (e.g., websites, e-mail, instant messaging, text messaging) to send or post negative or hurtful messages or images about an individual or a group.

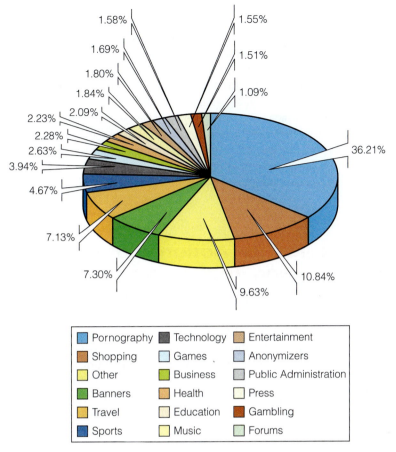

Figure 14.6 **Distribution of Internet Web Page Content, Percentages, 2007**
Source: Optenet 2008.

1.58% 1.55%
1.69% 1.51%
1.80% 1.09%
1.84% 36.21%
2.09%
2.23%
2.28%
2.63%
3.94%
4.67%
7.13% 10.84%
7.30% 9.63%

Pornography	Technology	Entertainment
Shopping	Games	Anonymizers
Other	Business	Public Administration
Banners	Health	Press
Travel	Education	Gambling
Sports	Music	Forums

that were previously unobtainable. Before recent advances in reproductive technology, for example, women could not conceive and give birth after menopause. Technology that allows postmenopausal women to give birth challenges societal beliefs about childbearing and the role of older women. The techniques of egg retrieval, in vitro fertilization, and gamete intrafallopian transfer make it possible for two different women to each make a biological contribution to the creation of a new life. Such technology requires society to re-examine its beliefs about what a family is and what a mother is. Should family be defined by custom, law, or the intentions of the parties involved?

Medical technologies that sustain life lead us to rethink the issue of when life should end. The increasing use of computers throughout society challenges the traditional value of privacy. New weapons systems make questionable the traditional idea of war as something that can be survived and even won. And cloning causes us to wonder about our traditional notions of family, parenthood, and individuality. Toffler (1970) coined the term **future shock** to describe the confusion resulting from rapid scientific and technological changes that unravel our traditional values and beliefs.

future shock The state of confusion resulting from rapid scientific and technological changes that unravel our traditional values and beliefs.

What Do You Think? Perhaps no technologies are as futuristic and as controversial as nanotechnologies. Nanotechnologies are a classification of different technologies, all of which have one thing in common—they are small, very, very small. A nanometer is one billionth of a meter. To put this in perspective, a piece of paper is 100,000 nanometers thick (NNI 2009). That nanotechnologies will change our lives and the social world we live in is indisputable. Would you wear a T-shirt that, through nanotechnology, converts the energy your movements generate to electricity that can then power your laptop? Would you be comfortable with paint that turns colors because you told it to? How about nano-sized medical robots that are injected into your blood stream to look for and fight off diseases? Do any of these innovations, all presently proposed by nanotechnology researchers, make you feel a little uneasy? Why or why not?

Strategies for Action: Controlling Science and Technology

As technology increases, so does the need for social responsibility. Nuclear power, genetic engineering, cloning, and computer surveillance all increase the need for social responsibility: "Technological change has the effect of enhancing the importance of public decision making in society, because technology is continually creating new possibilities for social action as well as new problems that have to be dealt with" (Mesthene 1993, p. 85). In the following sections, we address various aspects of the public debate, including science, ethics, the law, the role of corporate America, and government policy.

Science, Ethics, and the Law

Science and its resulting technologies alter the culture of society through the challenging of traditional values. Public debate and ethical controversies, however, have led to structural alterations in society as the legal system responds to calls for action. For example, several states now have what are called genetic exception laws. **Genetic exception laws** require that genetic information be handled separately from other medical information, leading to what is sometimes called patient shadow files (Legay 2001). The logic of such laws rests with the potentially devastating effects of genetic information being revealed to insurance companies, other family members, employers, and the like. Presently, 35 states prohibit genetic discrimination "in hiring, firing, and/or terms, conditions, or privileges of employment" (NCSL 2008a, p. 1).

Are such regulations necessary? In a society characterized by rapid technological and thus social change—a society in which custody of frozen embryos is part of the divorce agreement—many would say yes. Cloning, for example, is one of the most hotly debated technologies in recent years. Bioethicists and the public vehemently debate the various costs and benefits of this scientific technique. Despite such controversy, however, the chairman of the National Bioethics Advisory Commission warned nearly 15 years ago that human cloning will be "very difficult to stop" (McFarling 1998). At present, 15 states have laws pertaining to human cloning, with some prohibiting cloning for reproductive purposes, some prohibiting therapeutic cloning, and still others prohibiting both (NCSL 2008b).

genetic exception laws Laws that require that genetic information be handled separately from other medical information.

Should the choices that we make, as a society, be dependent on what we can do or what we should do? Whereas scientists and the agencies and corporations who fund them often determine what we *can* do, who should determine what we *should* do? Although such decisions are likely to have a strong legal component—that is, they must be consistent with the rule of law and the constitutional right of scientific inquiry—legality or the lack thereof often fails to answer the question, What should be done? *Roe v. Wade* (1973) did little to squash the public debate over abortion and, more specifically, the question of when life begins. Thus, it is likely that the issues surrounding the most controversial of technologies will continue into the 21st century with no easy answers.

Technology and Corporate America

As philosopher Jean-Francois Lyotard noted, knowledge is increasingly produced to be sold. The development of genetically altered crops, the commodification of women as egg donors, and the harvesting of regenerated organ tissues are all examples of potentially market-driven technologies. Like the corporate pursuit of computer technology, profit-motivated biotechnology creates several concerns.

First is the concern that only the rich will have access to life-saving technologies such as genetic screening and cloned organs. Such fears are justified. Companies with obscure names such as Progenitor International Research, Millennium Pharmaceuticals, Darwin Molecular, and Myriad Genetics have been patenting human life. Millennium Pharmaceuticals holds the patent on the melanoma gene and the obesity gene; Darwin Molecular controls the premature aging gene; deCODE controls the type 2 diabetes gene; Progenitor controls the gene for schizophrenia; and Myriad Genetics has nine patents on the breast and ovarian cancer genes (Mayer 2002; Lemonick 2006b; Bollier 2009).

These patents result in **gene monopolies,** which have led to astronomical patient costs for genetic screening and treatment. At present, more than 20 percent of human genes are privately owned (Bollier 2009). Because of gene monopolies today, the cost of a test for the breast cancer gene is $3,000. The biotechnology industry argues that such patents are the only way to recoup research costs that, in turn, lead to further innovations.

The commercialization of technology causes several other concerns, including issues of quality control and the tendency for discoveries to remain closely guarded secrets rather than collaborative efforts (Rabino 1998; Lemonick & Thompson 1999; Mayer 2002; Crichton 2007). In addition, industry involvement has made government control more difficult because researchers depend less and less on federal funding. More than 66 percent of research and development in the United States is supported by private industry using their own company funds (NSF 2008).

Although there is little doubt that profit acts as a catalyst for some scientific discoveries, other less commercially profitable but equally important projects may be ignored. As biologist Isaac Rabino states, "Imagine if early chemists had thrown their energies into developing profitable household products before the periodic table was discovered" (Rabino 1998, p. 112).

Runaway Science and Government Policy

Science and technology raise many public policy issues. Policy decisions, for example, address concerns about the safety of nuclear power plants, the privacy of electronic mail, the hazards of chemical warfare, and the ethics of cloning. In

gene monopoly Exclusive control over a particular gene as a result of government patents.

creating science and technology, have we created a monster that has begun to control us rather than the reverse? What controls, if any, should be placed on science and technology? And are such controls consistent with existing law? Consider the use of the file-sharing network BitTorrent to download music and movie files (the question of intellectual property rights and copyright infringement); a Utah law limiting children's access to material on the Internet (free speech issues); and Acxiom, the "cookie" collecting company that helps corporations customize advertising on websites by tracing clicks and keystrokes (Fourth Amendment privacy issues) (ACLU 2005; Kaplan 2000; Schaefer 2001; *The Economist* 2009b; Clifford 2009).

> Policy decisions...address concerns about the safety of nuclear power plants, the privacy of electronic mail, the hazards of chemical warfare, and the ethics of cloning.

The government, often through Congress, regulatory agencies, or departments, prohibits the use of some technologies (e.g., assisted-suicide devices) and requires others (e.g., seat belts). For example, in 2009, a bill was introduced in the U.S. House of Representatives that would (1) make human cloning illegal, (2) make any attempt to engage in human cloning illegal, (3) make knowingly shipping or receiving the product of a cloned egg cell illegal, and (4) set a criminal and civil penalty for violation of the statute (Human Cloning Prohibition Act of 2009).

Courtesy of OpenNet Initiative, http://opennet.net/

In China, the use of Green Dam, an Internet censorship software program, not only restricts access to offensive material, it also blocks information that is considered politically threatening. Pictured, a search for Tiananmen Square (see Chapter 1) results in a statement that the search is "inappropriate" (Faris et al. 2009).

Through financial support or the lack thereof, the government also promotes or discourages certain technologies. To help ameliorate the "digital divide," the 2009 economic stimulus package (see Chapter 7) provided $7.2 billion to expand broadband Internet services to underserved populations, primarily the 40 million poor and/or people living in rural areas who do not have broadband access (Cauley 2009). Lastly, the government may act symbolically by passing resolutions in support or condemnation of a particular technology or technology policy. In 2009, a resolution was introduced into Congress that expressed "grave concerns about the sweeping censorship, privacy, and cyber-security implications of China's *Green Dam* filtering software, and urges U.S. high-tech companies to promote the Internet as a tool for transparency, freedom of expression, and citizen empowerment around the world" (House Resolution 2009).

The federal government has also instituted several initiatives dealing with technology-related crime (see Chapter 4). Of late, the issue of online pornography has come to the forefront. A Federal Bureau of Investigation (FBI) report states that "computer telecommunications have become one of the most prevalent techniques used by pedophiles to share illegal photographic images of minors and to lure children into illicit sexual relationships" (FBI 2002, p. 1). For example, it is estimated that more than 100,000 websites offer child pornography, an industry that generates over $4.9 billion a year (Internet Pornography Statistics 2007). A second concern is the ease with which children can access and view online pornography. In response to such availability, the federal government enacted the Child Online Protection Act (COPA), which restricted minors' access to harmful material, "applying contemporary community standards," on the Internet (COPA 1998). The constitutionality of COPA was immediately contested and the bill was never enacted. After several lower court decisions that declared COPA as unconstitutional, in 2009, a government appeal to the U.S. Supreme Court was denied.

Finally, the government has several science and technology boards and initiatives, including the National Science and Technology Council, the Office of Science and Technology Policy, and the President's Council of Advisors on Science and Technology, and the National Nanotechnology Initiative (see Figure 14.7). These agencies advise the president on matters of science and technology, including research and development, implementation, national policy, and coordination of different initiatives.

What Do You Think? On July 21, 1969, the U.S. became the first country to put a person on the moon. Commander Neil Armstrong stepped onto the Sea of Tranquility region of the lunar surface and, as he put his left foot down declared, "That's one small step for man, one giant leap for mankind." (BBC 2009, p. 1). Yet, in a 2009 survey of Americans 18 years old or older, 42 percent responded that they did not think "the space program has brought enough benefits to justify costs" (Jones 2009, p. 1). Further, 40 percent of respondents thought that spending for the U.S. space program should be reduced or ended all together when asked, "Do you think spending on the U.S. space program should be increased, kept at the present level, reduced, or ended all together?" What do you think? Do you think the benefits of the space program outweigh the costs? Why or why not?

Understanding Science and Technology

What are we to understand about science and technology from this chapter? As structural functionalists argue, science and technology evolve as a social process and are a natural part of the evolution of society. As society's needs change, scientific discoveries and technological innovations emerge to meet these needs, thereby serving the functions of the whole. Consistent with conflict theory, however, science and technology also meet the needs of select groups and are characterized by political components. As Winner (1993) noted, the structure of science and technology conveys political messages, including "power is centralized," "there are barriers between social classes," "the world is hierarchically structured," and "the good things are distributed unequally" (p. 288).

The scientific discoveries and technological innovations that society embraces as truth itself are socially determined. Research indicates that science and the resulting technologies have both negative and positive consequences—a **technological dualism.** Technology saves lives, time, and money; it also leads to death, unemployment, alienation, and estrangement. Weighing the costs and benefits of technology poses ethical dilemmas, as does science itself. Ethics, however, "is not only concerned with individual choices and acts. It is also and, perhaps, above all concerned with the cultural shifts and trends of which acts are but the symptoms" (McCormick & Richard 1994, p. 16).

Thus, society makes a choice by the very direction it follows. These choices should be made on the basis of guiding principles that are both fair and just, such as those listed here (Goodman 1993; Winner 1993; Eibert 1998; Buchanan et al. 2000; Murphie & Potts 2003):

technological dualism The tendency for technology to have both positive and negative consequences.

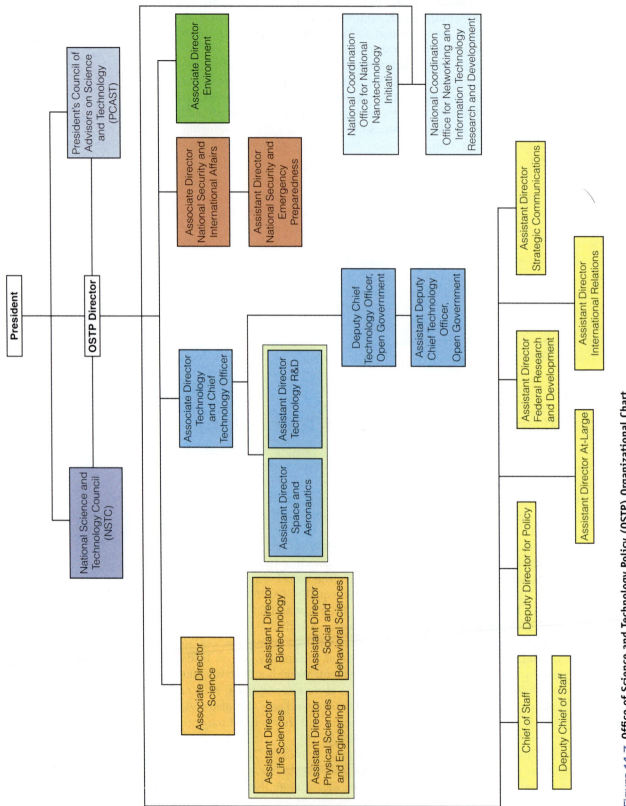

Figure 14.7 Office of Science and Technology Policy (OSTP) Organizational Chart
Source: Stine 2009.

1. Science and technology should be prudent. Adequate testing, safeguards, and impact studies are essential. Impact assessment should include an evaluation of the social, political, environmental, and economic factors.

2. No technology should be developed unless all groups, and particularly those who will be most affected by the technology, have at least some representation "at a very early stage in defining what that technology will be" (Winner 1993, p. 291). Traditionally, the structure of the scientific process and the development of technologies have been centralized (i.e., decisions have been made by a few scientists and engineers); decentralization of the process would increase representation.

3. Means should not exist without ends. Each new innovation should be directed to fulfilling a societal need rather than the more typical pattern in which a technology is developed first (e.g., high-definition television) and then a market is created (e.g., "You'll never watch a regular TV again!"). Indeed, from the space program to research on artificial intelligence, the vested interests of scientists and engineers, whose discoveries and innovations build careers, should be tempered by the demands of society.

What the 21st century will hold, as the technological transformation continues, may be beyond the imagination of most of society's members. Technology empowers; it increases efficiency and productivity, extends life, controls the environment, and expands individual capabilities. According to a National Intelligence Council report, "Life in 2015 will be revolutionized by the growing effort of multidisciplinary technology across all dimensions of life: social, economic, political, and personal" (NIC 2003, p. 1).

As we proceed into the first computational millennium, one of the great concerns of civilization will be the attempt to reorder society, culture, and government in a manner that exploits the digital bonanza yet prevents it from running roughshod over the checks and balances so delicately constructed in those simpler pre-computer years.

CHAPTER REVIEW

■ **What are the three types of technology?**
The three types of technology, escalating in sophistication, are mechanization, automation, and cybernation. Mechanization is the use of tools to accomplish tasks previously done by hand. Automation involves the use of self-operating machines, and cybernation is the use of machines to control machines.

■ **What are some of the reasons the United States may be "losing its edge" in scientific and technological innovations?**
The decline of U.S. supremacy in science and technology is likely to be the result of four interacting social forces. First, the federal government has been scaling back its investment in research and development. Second, corporations have begun to focus on short-term products and higher profits and third, there has been a drop in science and math education in U.S. schools

in terms of both quality and quantity. Finally, as documented in the book *Unscientific American,* there is a disconnect between American society and the principles of science.

■ **What are some Internet global trends?**
In 2009, the Internet had 1.67 billion users in more than 200 countries with 220 million users in the United States alone (Internet World Statistics 2009). The percentage of Internet users in a region is highest in North America followed by Oceania/Australia, Europe, Latin America and the Caribbean, the Middle East, Asia, and Africa. Worldwide, the average number of Internet users in a region is 24.7 percent (Internet World Statistics 2009).

■ **According to Kuhn, what is the scientific process?**
Kuhn describes the process of scientific discovery as occurring in three steps. First are assumptions about

a particular phenomenon. Next, because unanswered questions always remain about a topic, science works to start filling in the gaps. Then, when new information suggests that the initial assumptions were incorrect, a new set of assumptions or framework emerges to replace the old one. It then becomes the dominant belief or paradigm until it is questioned and the process repeats.

■ **What is meant by the computer revolution?**

The silicon chip made computers affordable. Today, 74 percent of Americans occasionally use a computer from some location, and 73 percent occasionally use the Internet from some location. The comparable statistics in 1995 for computer use and Internet use respectively are 54 percent and 14 percent.

■ **What is the Human Genome Project?**

The U.S. Human Genome Project is an effort to decode human DNA. The 13-year-old project is now complete, allowing scientists to "transform medicine" through early diagnosis and treatment as well as possibly preventing disease through gene therapy. Gene therapy entails identifying a defective or missing gene and then replacing it with a healthy duplicate that is transplanted to the affected area.

■ **What is the legal status of abortion in the United States?**

In the United States, since the U.S. Supreme Court's ruling in *Roe v. Wade* in 1973, abortion has been legal. However, recent Supreme Court and state court decisions have limited the scope of *Roe v. Wade*. For example, Missouri senators recently passed legislation that requires women seeking an abortion to be told, in a face-to-face conversation with their doctor at least 24 hours in advance of the procedure, the risks involved and the anatomical level of development of their fetus.

■ **How are some of the problems of the Industrial Revolution similar to the problems of the technological revolution?**

The most obvious example is in unemployment. Just as the Industrial Revolution replaced many jobs with technological innovations, so too has the technological revolution. Furthermore, research indicates that many of the jobs created by the Industrial Revolution, such as working on a factory assembly line, were characterized by high rates of alienation. Rising rates of alienation are also a consequence of increased estrangement as high-tech employees work in "white-collar factories."

■ **What is meant by outsourcing and why is it important?**

Outsourcing is the practice of a business subcontracting with a third party, often in low-wage countries such as China and India, for services. The problem with outsourcing is that it tends to lead to higher rates of unemployment in the export countries. It is estimated that the United States will lose between 28 million and 42 million jobs over the coming years due to outsourcing.

■ **What is the digital divide?**

The digital divide is the tendency for technology to be most accessible to the wealthiest and most educated. For example, some fear that there will be "genetic stratification," whereby the benefits of genetic screening, gene therapy, and other genetic enhancements will be available to only the richest segments of society.

■ **What is meant by the commercialization of technology?**

The commercialization of technology refers to profit-motivated technological innovations. Whether it be the isolation of a particular gene, genetically modified organisms, or the regeneration of organ tissues, where there is a possibility for profit, private enterprise will be there.

TEST YOURSELF

1. Which of the following technologies is associated with industrialization?
   a. Mechanization
   b. Cybernation
   c. Hibernation
   d. Automation
2. Analysis of the data over time suggests that the science and technology gap between poorer and richer nations is getting greater rather than shrinking.
   a. True
   b. False

3. The U.S. government, as part of the technological revolution, has recently increased spending on research and development.
   a. True
   b. False
4. Which theory argues that technology is often used as a means of social control?
   a. Structural functionalism
   b. Social disorganization
   c. Conflict theory
   d. Symbolic interactionism

5. The most common computer activity is accessing the Internet.
   a. True
   b. False
6. The ability to manipulate the genes of an organism to alter the natural outcome is called
   a. gene therapy
   b. gene splicing
   c. genetic engineering
   d. genetic screening
7. Genetically modified foods have been documented as harmless to humans by the Food and Drug Administration.
   a. True
   b. False
8. In 2007, the U.S. Supreme Court upheld the Partial Birth Abortion Ban in a 5 to 4 decision.
   a. True
   b. False

9. The practice of outsourcing entails
   a. creating high-tech jobs in the United States for immigrants
   b. hiring temporary workers to cover for employees who are absent
   c. allowing company workers to work from home
   d. subcontracting jobs often to workers in low-wage countries
10. The Genetic Information Nondiscrimination Act (GINA)
    a. makes human cloning illegal in the United States
    b. establishes a criminal penalty for human cloning in the United States
    c. is a federal law that prohibits discrimination in health coverage or employment based on genetic information.
    d. all of the above

Answers: 1: d; 2: a; 3: b; 4: c; 5: a; 6: c; 7: b; 8: a; 9: d; 10: c.

KEY TERMS

abortion 533
automation 512
cultural lag 517
cyber-bullying 546
cybernation 512
e-commerce 526
future shock 547
gene monopoly 549
gene therapy 531
genetic engineering 531
genetic exception laws 548
Genetic Information
    Nondiscrimination Act of 2008 542

genetic screening 530
in vitro fertilization (IVF) 533
Internet 526
Internet piracy 546
malware 546
mechanization 511
membership communities 530
net neutrality 544
outsource 542
partial birth abortion 535
postmodernism 516
science 511
STEM 514

stem cells 537
technological dualism 551
technological fix 516
technology 511
technology-induced diseases 545
telepresencing 522
therapeutic cloning 537
Web 2.0 526

MEDIA RESOURCES

Understanding Social Problems,
Seventh Edition Companion Website
www.cengage.com/sociology/mooney
Visit your book companion website, where you will find flash cards, practice quizzes, Internet links, and more to help you study.

Just what you need to know NOW! Spend time on what you need to master rather than on information you already have learned. Take a pre-test for this chapter, and CengageNOW will generate a personalized study plan based on your results. The study plan will identify the topics you need to review and direct you to online resources to help you master those topics. You can then take a post-test to help you determine the concepts you have mastered and what you will need to work on. Try it out! Go to www.cengage.com/login to sign in with an access code or to purchase access to this product.

Jim MacMillan/AP Photo

15

Conflict, War, and Terrorism

On a night patrol in Bagdad, Clifton Hicks, 19, and his fellow soldiers heard the sounds of an IED [improvised explosive device] blast, closely followed by gunfire. When arriving at the source of the explosion, Hicks saw several men in Humvees with machine guns like unruly hair, sticking out in every direction. They had been attacked, from two sides—from an open field and from an Iraqi civilian neighborhood. When the shooting stopped, some of the soldiers from the Humvee rushed into the field, looking for the insurgents but found nothing but empty shell casings. Others ran into the neighborhood looking for the house where the shots had come from. Kicking the door down, the soldiers were greeted by the extended family of a bride and groom celebrating their wedding day.

It was now clear what had happened—anyone who had been in Bagdad for any length of time would have known. The father of the bride had climbed onto the roof and fired his rifle in celebration of his daughter's wedding. At the same time, a patrol had passed by the father's house and, believing they were under fire from both sides of the road, the soldiers returned fire in both directions.

Three people in the wedding party had been hit—an adult man who was only slightly wounded and a young girl, about 10, who also had minor wounds. But, as Private Hicks recounts, "there was another girl who was six or seven, and she was dead. I was in the gunner's hatch of the Humvee. I didn't get out and go inside the house, but I looked through the doorway and that was the first time that I had ever seen a 6-year-old girl dead" (Glantz 2008, p. 31).

War is one of the great paradoxes of human history. It both protects and annihilates. It creates and defends nations but may also destroy them. **War,** the most violent form of conflict, refers to organized armed violence aimed at a social group in pursuit of an objective. Wars have existed throughout human history and continue in the contemporary world. Whether war is just or unjust, defensive or offensive, it involves the most horrendous atrocities known to humankind. This is especially true in the 21st century, when nearly all wars are fought in populated areas rather than on remote battlefields. As Private Hicks' eyewitness account attests, war is often conducted within the routine of daily life—and this can have deadly consequences. Thus, war is not only a social problem in and of itself; it contributes to a host of other social problems—death, disease, and disability, crime and immorality, psychological terror, loss of economic resources, and environmental devastation. In this chapter, we discuss each of these issues within the context of conflict, war, and terrorism, the most threatening of all social problems.

The Global Context: Conflict in a Changing World

As societies have evolved and changed throughout history, the nature of war has also changed. Before industrialization and the sophisticated technology that

war Organized armed violence aimed at a social group in pursuit of an objective.

resulted, war occurred primarily between neighboring groups on a relatively small scale. In the modern world, war can be waged between nations that are separated by thousands of miles as well as between neighboring nations. Increasingly, war is a phenomenon internal to states, involving fighting between the government and rebel groups or among rival contenders for state power. Indeed, Figure 15.1 documents that wars between states, that is, interstate wars, recently make up the smallest percentage of armed conflicts. In the following sections, we examine how war has changed our social world and how our changing social world has affected the nature of war in the industrial and postindustrial information age.

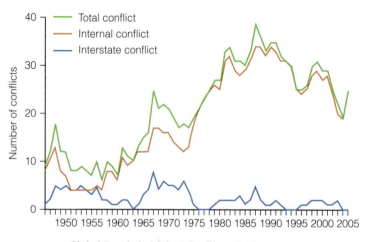

Figure 15.1 **Global Trends in Violent Conflict, 1946–2005**
Source: Hewitt et al. 2008, p. 12.

War and Social Change

The very act that now threatens modern civilization—war—is largely responsible for creating the advanced civilization in which we live. Before large political states existed, people lived in small groups and villages. War broke the barriers of autonomy between local groups and permitted small villages to be incorporated into larger political units known as chiefdoms. Centuries of warfare between chiefdoms culminated in the development of the state. The **state** is "an apparatus of power, a set of institutions—the central government, the armed forces, the regulatory and police agencies—whose most important functions involve the use of force, the control of territory, and the maintenance of internal order" (Porter 1994, pp. 5–6). The creation of the state in turn led to other profound social and cultural changes:

> And once the state emerged, the gates were flung open to enormous cultural advances, advances undreamed of during—and impossible under—a regimen of small autonomous villages. . . . Only in large political units, far removed in structure from the small autonomous communities from which they sprang, was it possible for great advances to be made in the arts and sciences, in economics and technology, and indeed in every field of culture central to the great industrial civilizations of the world. (Carneiro 1994, pp. 14–15)

Industrialization and technology could not have developed in the small social groups that existed before military action consolidated them into larger states. Thus war contributed indirectly to the industrialization and technological sophistication that characterize the modern world. Industrialization, in turn, has had two major influences on war. Cohen (1986) calculated the number of wars fought per decade in industrial and preindustrial nations and concluded that "as societies become more industrialized, their proneness to warfare decreases" (p. 265). Thus, for example, in 2008, there were 16 major armed conflicts, the majority of which were in less developed countries in Africa and Asia (SIPRI 2009a).

state The organization of the central government and government agencies such as the military, police, and regulatory agencies.

Although industrialization may decrease a society's propensity to war, it also increases the potential destruction of war. With industrialization, military technology became more sophisticated and more lethal. Rifles and cannons replaced the clubs, arrows, and swords used in more primitive warfare and, in turn, were replaced by tanks, bombers, and nuclear warheads. Today, the use of new technologies such as high-performance sensors, information processors, directed energy technologies, precision-guided munitions, and computer worms and viruses, have changed the very nature of conflict, war, and terrorism. For example, according to one defense analyst, "the U.S. military alone fields 7,000 unmanned drones in the air, like the Predators that fire missiles into Pakistan—and roughly another 12,000 on the ground, like the Packbots that hunt for roadside bombs in Iraq" (Singer 2009, p. 1).

In the postindustrial information age, computer technology has not only revolutionized the nature of warfare, it has made societies more vulnerable to external attacks. For example, in 2009, North Korea attacked computer systems in the United States and in South Korea, resulting in brief disruptions to government and major commercial networks (Sang-Hun & Markoff 2009).

> ... [T]he use of new technologies such as high-performance sensors, information processors, directed energy technologies, precision-guided munitions, and computer worms and viruses, have changed the very nature of conflict, war, and terrorism.

The Economics of Military Spending

The increasing sophistication of military technology has commanded a large share of resources, totaling $1.46 trillion worldwide in 2008, or about 2.4 percent of the total global domestic product (Perlo-Freeman et al. 2009). Global military spending has been increasing since 1998, with dramatic increases between 2002 and 2008, as a consequence of expenditures for U. S.-led operations after September 11. Analysts expect that U.S. funding for war since the 9/11 attacks will surpass $1 trillion in 2010 (Belasco 2009). Table 15.1 shows the government's estimates of the costs of the wars in Iraq and Afghanistan, as well as for Operation Noble Eagle, which deploys reservists for homeland defense.

The long-term costs of the Iraq war may be as high as 4 trillion dollars (Herszenhorn 2008). The Obama Administration's plan to send 30,000 additional troops to Afghanistan will cost $30 billion just for the first year of deployment (Associated Press 2009). In a recent survey about military spending, 24 percent of Americans responded "too little" was being spent, 41 percent that the amount was "about right," and 31 percent that "too much" was being spent (Saad 2009).

The **Cold War,** the state of political tension, economic competition, and military rivalry that existed between the United States and the former Soviet Union for nearly 50 years, provided justification for large expenditures for military preparedness. However, the end of the Cold War, along with the rising national debt, resulted in cutbacks in the U.S. military budget in the 1990s.

Today, military spending has nearly returned to the levels during the Cold War. In 2008, the United States accounted for 41.5 percent of the world's military spending, the largest single percentage of any nation—more than the combined total of the 32 next most powerful nations. The next highest military spenders

United States Air Force

A soldier prepares an unmanned Predator drone for a mission. In the 20th century, industrialization spurred technological innovations that transformed warfare more rapidly than at any other time in human history.

Cold War The state of military tension and political rivalry that existed between the United States and the former Soviet Union from the 1950s through the late 1980s.

TABLE 15.1 Annual Costs of Wars in Iraq and Afghanistan, and Enhanced Security, 2001–2010 (Estimated, in Billions)

OPERATION	2001 AND 2002	2003	2004	2005	2006	2007	2008	2001–2009	REQUESTED 2009	REQUESTED 2010
Iraq	0	53	75.9	85.5	101.6	133.6	140.9	684.0	93.5	130
Afghanistan	20.8	14.7	14.5	20	19	36.9	42.1	223.2	55.2	TBD
Domestic security	13	8	3.7	2.1	0.8	0.5	0.1	28.4	0	TBD
Unallocated		5.5					5.5	1	0	
TOTAL	33.8	81.1	94.1	107.6	121.4	171	183.3	941.3	149.1	130

NA = not applicable; TBD = to be determined. Totals may not add due to rounding.
Source: Belasco 2009.

are China, France, England, and Russia, which, along with the United States, account for 60 percent of the world's military expenditures (SIPRI 2009b). U.S. national defense outlays, estimated to be $692.8 billion in 2010, include expenditures for salaries of military personnel, research and development, and weapons, but does not include expenditures for overseas wars or veterans benefits (Office of Management and Budget 2009a).

What Do You Think? According to journalist Mark Thompson, "a volunteer army reflects the most central and sacred vow that citizens make to one another; soldiers protect and defend the country; in return the country promises to give them the tools they need to complete their mission . . ." (2007, p. 31). Some would argue, however, that the United States has not held up its side of the bargain, particularly when it comes to providing the quality and quantity of military personnel needed to win a war. For example, in an effort to boost numbers, recruitment standards have been lowered (meaning that more convicted felons and high school dropouts may be recruited), the maximum enlistment age has been raised from 35 to 42, and the Pentagon's so-called "stop-loss orders" have forced soldiers to remain in uniform and combat past the date of their contractual obligations for service. Those who serve get "abbreviated training," and time between deployments is now measured in months rather than years (Thompson 2007). Given these standards, should the military draft be reinstated? Should both men and women have to sign up for the draft? Alternatively, would you advocate universal military service upon high school graduation?

The U.S. government not only spends money on its own military and defense but also sells military equipment to other countries, either directly or by helping U.S. companies sell weapons abroad. Although the purchasing countries may use these weapons to defend themselves from hostile attack, foreign military sales may pose a threat to the United States by arming potential antagonists. For example, the United States, the world's leading arms-exporting nation, supplied weapons to Iraq to use against Iran during the war between 1980 and 1987. These same weapons were then used against Americans in the Gulf and Iraq wars (Silverstein 2007). Similarly, after the Soviet invasion of Afghanistan in 1979, the United States funded Afghan rebel groups. Years after the Soviets left Afghanistan, rebels continued to fight for control of the country. Using weapons supplied by the United States, the Taliban took over much of Afghanistan and sheltered al Qaeda and Osama bin Laden—also a former recipient of U.S. support—as they planned the attacks on September 11 (Rashid 2000; Bergen 2002).

The United States regularly transfers arms to countries in active conflict. For example, in 2006 and 2007, the United States sold about $9.8 billion in arms to allies for use in war zones in Pakistan, Iraq, Israel, Afghanistan, and Colombia (Berrigan 2009). A 2005 report titled *U.S. Weapons at War: Promoting Freedom or Fueling Conflict?* concluded that, far "from serving as a force for security and stability, U.S. weapons sales frequently serve to empower unstable, undemocratic regimes to the detriment of U.S. and global security" (Berrigan & Hartung 2005).

Sociological Theories of War

Sociological perspectives can help us understand various aspects of war. In this section, we describe how structural functionalism, conflict theory, and symbolic interactionism can be applied to the study of war.

Structural-Functionalist Perspective

Structural functionalism focuses on the functions that war serves and suggests that war would not exist unless it had positive outcomes for society. We have already noted that war has served to consolidate small autonomous social groups into larger political states. An estimated 600,000 autonomous political units existed in the world at about 1000 B.C. Today, that number has dwindled to fewer than 200.

Another major function of war is that it produces social cohesion and unity among societal members by giving them a "common cause" and a common enemy. For example, in 2005, *Newsweek* began a new feature about everyday American heroes called "Red, White, and Proud." Unless a war is extremely unpopular, military conflict also promotes economic and political cooperation. Internal domestic conflicts between political parties, minority groups, and special interest groups often dissolve as they unite to fight the common enemy. During World War II, U.S. citizens worked together as a nation to defeat Germany and Japan.

In the short term, war may also increase employment and stimulate the economy. The increased production needed to fight World War II helped pull the United States out of the Great Depression. The investments in the manufacturing sector during World War II also had a long-term impact on the U.S. economy. Hooks and Bloomquist (1992) studied the effect of the war on the U.S. economy between 1947 and 1972, and concluded that the U.S. government "directed, and in large measure, paid for a 65 percent expansion of the total investment in plant and equipment" (p. 304). War can also have the opposite effect, however. In a 2005 restructuring of the military, the Pentagon, seeking a "meaner, leaner fighting machine," recommended shutting down or reconfiguring nearly 180 military installations, "ranging from tiny Army reserve centers to sprawling Air Force bases that have been the economic anchors of their communities for generations" (Schmitt 2005), at a cost of thousands of civilian jobs.

Library of Congress Prints & Photographs Division, LC-USZC4-12183

As structural functionalists argue, a major function of war is that it produces unity among societal members. War provides a common cause and a common identity. Societal members feel a sense of cohesion, and they work together to defeat the enemy.

Wars also function to inspire scientific and technological developments that are useful to civilians. For example, innovations in battlefield surgery during World War II and the Korean War resulted in instruments and procedures that later became common practice in civilian hospital emergency wards (Zoroya 2006). Research on laser-based defense systems led to laser surgery, research in nuclear fission and fusion facilitated the development of nuclear power, and the Internet was created by the Pentagon for military purposes. In the U.S. airline industry, which owes much of its technology to the development of air power by the Department of Defense, the distinction between military and civilian technology is important because different government agencies regulate its export. The Department of Commerce regulates the export of parts produced for use on commercial airlines whereas the Department of State imposes stricter controls on parts produced for military aircraft to prevent sales to countries at odds with U.S. foreign policy objectives (Millman 2008). Other **dual-use technologies,** a term referring to defense-funded innovations that have commercial and civilian applications, include SLICE, a high-speed twin-hull water vessel originally made for the Office of Naval Research. SLICE has a variety of commercial applications, "including its use as a tour or sport fishing boat, oceanographic research vessel, oil spill response ship, and high-speed ferry" (State of Hawaii 2000).

War also serves to encourage social reform. After a major war, members of society have a sense of shared sacrifice and a desire to heal wounds and rebuild normal patterns of life. They put political pressure on the state to care for war victims, improve social and political conditions, and reward those who have sacrificed lives, family members, and property in battle. As Porter (1994) explained, "Since . . . the lower economic strata usually contribute more of their blood in battle than the wealthier classes, war often gives impetus to social welfare reforms" (p. 19).

Finally, the U.S. military has historically provided an alternative for the advancement of poor or disadvantaged groups who otherwise face discrimination or limited opportunities in the formal economy. The military's specialized training, tuition assistance programs for a college education, and preferential hiring practices improve the prospects of veterans to find a decent job or career after their service (Military 2007).

Conflict Perspective

Conflict theorists emphasize that the roots of war are often antagonisms that emerge whenever two or more ethnic groups (e.g., Bosnians and Serbs), countries (United States and Vietnam), or regions within countries (the U.S. North and South) struggle for control of resources or have different political, economic, or religious ideologies. In addition, conflict theory suggests that war benefits the corporate, military, and political elites. Corporate elites benefit because war often results in the victor taking control of the raw materials of the losing nations, thereby creating a bigger supply of raw materials for its own industries. Indeed, many corporations profit from defense spending. Under the Pentagon's bid-and-proposal program, for example, corporations can charge the cost of preparing proposals for new weapons as overhead on their Department of Defense contracts. Also, Pentagon contracts often guarantee a profit to the developing corporations. Even if the project's cost exceeds initial estimates, called a cost overrun, the corporation still receives the agreed-on profit. In the late 1950s,

dual-use technologies
Defense-funded technological innovations with commercial and civilian use.

President Dwight D. Eisenhower referred to this close association between the military and the defense industry as the **military-industrial complex.**

Contemporary examples of the military-industrial complex include the Bush administration's direct link to defense spending. Even as the U.S. government was deciding on the war in Iraq, "many former Republican officials and political associates of the Bush administration [were] associated with the Carlyle Group, an equity investment firm with billions of dollars in military and aerospace assets" (Knickerbocker 2002, p. 2). Further, news media have come to rely heavily on retired military professionals with ties to the defense establishment and military contractors for interpretation of the Iraq and Afghanistan wars. This close "intersection of network news and wartime commerce" blurs the line between security policy and private commercial interests (Barstow 2008, p. 1).

The military elite benefit because war and the preparations for it provide prestige and employment for military officials. For example, Military Professional Resources Inc. (MPRI), an organization staffed by former military, defense, law enforcement, and other professionals, serves "the national security needs of the U.S. government, selected foreign governments, international organizations, and the private sector" and lists capabilities such as war gaming, force development and management, and democracy transition assistance (Military Professional Resources Inc. 2009, p. 1). MPRI is one of four U.S. firms that are under a $300 million contract from the Department of Defense to produce public service announcements, news, and entertainment for the Iraqi media. The three-year project (2009 to 2011) supports Iraqi and U.S. political goals such as promoting reconciliation and nonsectarian nationalism, as well as pro-military, pro-police, and pro-U.S. attitudes (DeYoung & Pincus 2008).

According to some estimates, private contractors such as MPRI and many others contributed more than 180,000 civilians to the occupation of Iraq, about 20,000 more than the U.S. military and government employees deployed in country (Miller 2007). Private security companies—for-profit organizations contracted by the U.S. government to perform security functions that the military formerly provided—account for a significant portion of the U.S. forces. According to estimates from Congressional Research Service, 50 private security companies employ more than 30,000 personnel in Iraq. Twenty of these companies and at least 10,000 of their employees are under contract with the U.S. government, costing between $3 billion and $6 billion from 2003 to 2007 (CRS 2008).

The North Carolina-based firm Blackwater Worldwide—which changed its name to Xe, pronounced "z," in 2009—received national attention in 2004 when four of its employees in Iraq were killed by a Sunni mob in Fallujah, where their charred corpses were hung along public streets. In September 2007, Blackwater personnel guarding a U.S. diplomatic convoy opened fire at a traffic circle in Baghdad, killing 17 and wounding 24 Iraqi civilians. Company officials initially claimed that their contractors responded proportionately to a nearby attack. In response, and after many years of lodging complaints about alleged indiscriminate firings by private security contractors, the Iraqi government revoked Blackwater's license to operate in Iraq (Tavernise 2007). An FBI investigation of the incident found that most of the killings were in violation of the rules for use of deadly force (Johnston & Broder 2007). Five former Blackwater employees were charged with voluntary manslaughter in January 2009. They pleaded not guilty, and a tentative trial date is set for February 2010 (CNN 2009).

War also benefits the political elite by giving government officials more power. Porter (1994) observed that "throughout modern history, war has been the

military-industrial complex A term first used by Dwight D. Eisenhower to connote the close association between the military and defense industries.

level by which . . . governments have imposed increasingly larger tax burdens on increasingly broader segments of society, thus enabling ever-higher levels of spending to be sustained, even in peacetime" (p. 14). Political leaders who lead their country to a military victory also benefit from the prestige and hero status conferred on them.

Finally, feminists and many other analysts often note the overwhelming association between war and gender. By and large, active combat has historically been carried out by men. Nature-based arguments about gender—i.e., that men are innately aggressive or violent and women inherently peaceful—are not generally supported by social science research and do not adequately explain why men are more likely to kill than women. Feminists emphasize the social construction of aggressive masculine identities and their manipulation by elites as important reasons for the association between masculinity and militarized violence (Alexander & Hawkesworth 2008).

The recent entry of women into the U.S. armed forces is changing how women's roles in combat are perceived and even how the military conducts operations in war. The wars in Afghanistan and Iraq "are the first in which tens of thousands of American military women have lived, worked and fought with men for prolonged periods" (Myers 2009, p. 1). Although barred from most combat units, women have served nonetheless in most of the same capacities as male soldiers: in armed patrols, as gunners in vehicles, in bomb disposal units, and as officers leading male troops into combat (Alvarez 2009a). The military houses women in separate quarters, has separate showers and bathrooms for women, allows married couples to live together, and makes contraceptives available to soldiers (Myers 2009). Because women are already serving in combat roles (officially and unofficially) without the negative impact on discipline that opponents have feared, many analysts predict that the military's restrictions on women in combat will eventually be lifted (Alvarez 2009a).

Although some feminists view women's participation in the military as a matter of equal rights, others object because they see war as an extension of patriarchy and the subordination of women in male-dominated societies. Ironically, because protection of women is perceived as a feature of masculine identity, feminists also point out that war and other conflicts are often justified using "the language of feminism" (Viner 2002). For example, President Bush used respect for women's rights and protection of women subjugated under the Taliban as a partial justification for the attack on Afghanistan in 2001 (Viner 2002).

Symbolic Interactionist Perspective

The symbolic interactionist perspective focuses on how meanings and definitions influence attitudes and behaviors regarding conflict and war. The development of attitudes and behaviors that support war begins in childhood. American children learn to glorify and celebrate the Revolutionary War, which created our nation. Movies romanticize war, children play war games with toy weapons, and various video and computer games glorify heroes conquering villains.

Symbolic interactionism helps to explain how military recruits and civilians develop a mind-set for war by defining war and its consequences as acceptable and necessary.

Fans at the NASCAR Nextel Cup Food City 500 form the American flag in Bristol, Tennessee before an auto race. Although older Americans are the most patriotic age group, a recent survey of college graduates found that 83 percent defined themselves as patriotic.

The word *war* has achieved a positive connotation through its use in various phrases—the war on drugs, the war on poverty, and the war on crime. Positive labels and favorable definitions of military personnel facilitate military recruitment and public support of armed forces. In 2005, the Army National Guard launched a $38 million marketing campaign targeting young men and women with advertisements, showing "troops with weapons drawn, helicopters streaking and tanks rolling," all "in an attempt to remind people what the Guard has been about since Colonial Days: fighting wars and protecting the homeland." The new slogan? "The most important weapon in the war on terrorism. You" (Davenport 2005, p. 1).

Many government and military officials convince the masses that the way to ensure world peace is to be prepared for war. Patriotism is a popular sentiment in American society. For example, 62 percent of Americans say they display a U.S. flag at home, at the office, or on their car (Doherty 2007) (see Table 15.2).

Governments may use propaganda and appeals to patriotism to generate support for war efforts and to motivate individuals to join armed forces. Salladay (2003), for example, notes that those in favor of the war on Iraq have commandeered the language of patriotism, making it difficult but necessary for peace activists to use the same symbols or phrases. In their study of the U.S. peace movement, Woehrle et al. (2008) observed that the government and supporters of U.S. wars often framed the issue as "supporting our boys" or "supporting the troops." This made public discussion about whether a particular war is effective or justifiable appear to be a betrayal of soldiers. By analyzing public statements from leading peace movement groups that opposed the first Gulf War (1991) and the Iraq war, the researchers documented how opponents of these wars developed a counter-narrative that "peace is patriotic" and reframed the issue around how war itself endangered the troops, how government policies failed to provide for the welfare of troops, and how the well-being of civilians was negatively affected by war.

To legitimize war, the act of killing in war is not regarded as "murder." Deaths that result from war are referred to as casualties. Bombing military and civilian targets appears more acceptable when nuclear missiles are "peacekeepers" that are equipped with multiple "peace heads." Killing the enemy is more acceptable when derogatory and dehumanizing labels such as Gook, Jap, Chink, Kraut, and Haji convey the attitude that the enemy is less than human.

TABLE 15.2 Who Flies the Flag?

	DISPLAY FLAG AT HOME, OFFICE, OR ON CAR?*	
	YES	NO
Total	62	38
Men	65	34
Women	59	41
White	67	33
African American	41	59
18 to 29 years old	51	49
30 to 49 years old	63	37
50 to 64 years old	65	35
65 years old and older	71	28
Republican	73	26
Democrat	55	45
Independent	63	37
Northeast	69	31
Midwest	67	33
South	58	41
West	57	43
College graduate	61	39
Some college	65	35
High school graduate	64	36
Less than high school	54	45

*Numbers may not sum to 100 because of refusals or responses of "don't know."

Source: Doherty 2007.

> Governments may use propaganda and appeals to patriotism to generate support for war efforts and to motivate individuals to join armed forces.

Such labels are socially constructed as images, often through the media, and are presented to the public. Social constructionists, like symbolic interactionists in general, emphasize the social aspects of "knowing." Thus Li and Izard (2003) used content analysis to analyze newspaper and television coverage of the World Trade Center and Pentagon attacks on September 11. The researchers examined the first eight hours of coverage of the attacks presented on CNN, ABC, CBS, NBC, and Fox as well as in eight major U.S. newspapers (including the *Los Angeles Times,* the *New York Times,* and the *Washington Post*). Results of the analysis indicated that newspaper articles tended to have a "human interest" emphasis, whereas television coverage was more often "guiding and consoling." Other results suggested that both media relied most heavily on government sources, that newspapers and the networks were equally factual, and that networks were more homogeneous in their presentation than newspapers. One indication of the importance of the media lies in former President George W. Bush's creation of the Office of Global Communications—"a huge production company, issuing daily scripts on the Iraq war to U.S. spokesmen around the world, auditioning generals to give media briefings, and booking administration stars on foreign news shows" (Kemper 2003, p. 1).

Causes of War

The causes of war are numerous and complex. Most wars involve more than one cause. The immediate cause of a war may be a border dispute, for example, but religious tensions that have existed between the two combatant countries for decades may also contribute to the war. The following section reviews various causes of war.

Conflict Over Land and Other Natural Resources

Nations often go to war in an attempt to acquire or maintain control over natural resources, such as land, water, and oil. Michael Klare, author of *Resource Wars: The New Landscape of Global Conflict* (2001), predicted that wars will increasingly be fought over resources as supplies of the most needed resources diminish. Disputed borders have been common motives for war. Conflicts are most likely to arise when borders are physically easy to cross and are not clearly delineated by natural boundaries, such as major rivers, oceans, or mountain ranges.

Water is another valuable resource that has led to wars. Unlike other resources, water is universally required for survival. At various times, the empires of Egypt, Mesopotamia, India, and China all went to war over irrigation rights. In 1998, five years after Eritrea gained independence from Ethiopia, forces clashed over control of the port city Assab and with it, access to the Red Sea.

Not only do the oil-rich countries in the Middle East present a tempting target in themselves, but war in the region can also threaten other nations that are dependent on Middle Eastern oil. Thus, when Iraq seized Kuwait and threatened the supply of oil from the Persian Gulf, the United States and many other nations reacted militarily in the Gulf War. In a document prepared for the Center for Strategic and International Studies, Starr and Stoll (1989) warned that soon

water, not oil, will be the dominant resource issue of the Middle East. According to World Watch Institute, "despite modern technology and feats of engineering, a secure water future for much of the world remains elusive." The prognosis for Egypt, Jordan, Israel, the West Bank, the Gaza

Strip, Syria, and Iraq is especially alarming. If present consumption patterns continue, emerging water shortages, combined with a deterioration in water quality, will lead to more competition and conflict. (p. 1)

Despite such predictions, tensions in the Middle East have erupted into fighting repeatedly in recent years—but not over water. In July 2006, Israel and Lebanon fought a border war that killed a thousand people and displaced a million Lebanese. Civil war erupted in the Palestinian territories between rival parties when Hamas seized control of the Gaza Strip in response to Fatah's refusal to hand over the government after Hamas won legislative elections (BBC 2007a). Further, in the summer of 2007, hundreds were killed when Lebanese security forces attacked Palestinian militants based in Lebanon's refugee camps (BBC 2007b). In December 2008, the Israeli military launched air strikes against Gaza to prepare the way for its January 2009 ground invasion that lasted 22 days. Israeli authorities claimed the campaign was intended to stop Hamas's rocket attacks against civilians in southern Israel (BBC 2009a).

Conflict Over Values and Ideologies

Many countries initiate war not over resources but over beliefs. World War II was largely a war over differing political ideologies: democracy versus fascism. The Cold War involved the clash of opposing economic ideologies: capitalism versus communism.

Conflicts over values or ideologies are not easily resolved. They are less likely to end in compromise or negotiation because they are fueled by people's convictions. For example, when a representative sample of American Jews were asked, "Do you agree or disagree with the following statement? 'The goal of Arabs is not the return of occupied territories but rather the destruction of Israel,'" 81 percent agreed, 13 percent disagreed, and 6 percent were unsure (American Jewish Committee 2007).

If ideological differences can contribute to war, do ideological similarities discourage war? The answer seems to be yes; in general, countries with similar ideologies are less likely to engage in war with each other than countries with differing ideological values (Dixon 1994). Democratic nations are particularly disinclined to wage war against one another (Brown et al. 1996; Rasler & Thompson 2005).

Racial, Ethnic, and Religious Hostilities

Racial, ethnic, and religious groups vary in their cultural beliefs, values, and traditions. Thus, conflicts between racial, ethnic, and religious groups often stem from conflicting values and ideologies. Such hostilities are also fueled by competition over land and other scarce natural and economic resources. Gioseffi (1993) noted that "experts agree that the depleted world economy, wasted on war efforts, is in great measure the reason for renewed ethnic and religious strife. 'Haves' fight with 'have-nots' for the smaller piece of the pie that must go around" (p. xviii). Racial, ethnic, and religious hostilities sometimes are perpetuated by a wealthy minority to divert attention away from their exploitations and to maintain their own position of power. Such **constructivist explanations** of ethnic conflict—those that emphasize the role of leaders of ethnic groups in stirring up intercommunal hostility—differ sharply from **primordial explanations,** or those that emphasize the existence of "ancient hatreds" rooted in deep psychological or cultural differences between ethnic groups.

constructivist explanations
Those explanations that emphasize the role of leaders of ethnic groups in stirring up hatred toward others external to one's group.

primordial explanations
Those explanations that emphasize the existence of "ancient hatreds" rooted in deep psychological or cultural differences between ethnic groups, often involving a history of grievance and victimization, real or imagined, by the enemy group.

As described by Paul (1998), sociologist Daniel Chirot argued that the recent worldwide increase in ethnic hostilities is a consequence of "retribalization," that is, the tendency for groups, lost in a globalized culture, to seek solace in the "extended family of an ethnic group" (p. 56). Chirot identified five levels of ethnic conflict: (1) multiethnic societies without serious conflict (e.g., Switzerland), (2) multiethnic societies with controlled conflict (e.g., United States and Canada), (3) societies with ethnic conflict that has been resolved (e.g., South Africa), (4) societies with serious ethnic conflict leading to warfare (e.g., Sri Lanka), and (5) societies with genocidal ethnic conflict, including "ethnic cleansing" (e.g., Darfur).

Religious differences as a source of conflict have recently come to the forefront. An Islamic jihad, or holy war, has been blamed for the September 11 attacks on the World Trade Center and Pentagon as well as for bombings in Kashmir, Sudan, the Philippines, Indonesia, Kenya, Tanzania, Saudi Arabia, Spain, and Great Britain. Some claim that Islamic beliefs in and of themselves have led to recent conflicts (Feder 2003). Others contend that religious fanatics, not the religion itself, are responsible for violent confrontations. Wars over differing religious beliefs have led to some of the worst episodes of bloodshed in history, in part, because some religions lend themselves to martyrdom—the idea that dying for one's beliefs leads to eternal salvation. For example, Islamic leader Osama bin Laden claimed that unjust U.S. Middle East policies are responsible for "dividing the whole world into two sides—the side of believers and the side of infidels" (Williams 2003, p. 18).

> Wars over differing religious beliefs have led to some of the worst episodes of bloodshed in history, in part, because some religions lend themselves to martyrdom—the idea that dying for one's beliefs leads to eternal salvation.

What Do You Think? The charter of the United Nations obligates all member countries "to settle their international disputes by peaceful means" except in self-defense. In September 2002, the White House released a document describing the nation's new international security policy in light of the September 11 attacks. In situations of "imminent threat," "[W]e will not hesitate to act alone, if necessary, to exercise our right of self-defense by acting preemptively against such terrorists, to prevent them from doing harm against our people and our country" (National Security Council 2002). The U.S. government later asserted this preemptive right to self-defense in the justification for the war in Iraq. Do you agree with this policy? Should the United States attack countries that have not attacked it first? What if other countries adopt this policy?

Defense Against Hostile Attacks

The threat or fear of being attacked may cause the leaders of a country to declare war on the nation that poses the threat. This is an example of what experts in international relations refer to as the **security dilemma**: "actions to increase one's security may only decrease the security of others and lead them to respond in ways that decrease one's own security" (Levy 2001, p. 7). Such situations may lead to war inadvertently. The threat may come from a foreign country or from a group within the country. After Germany invaded Poland in 1939, Britain and France declared war on Germany out of fear that they would be Germany's next victims. Germany attacked Russia in World War I, in part out of fear that Russia had entered the arms race and would use its weapons against Germany. Japan bombed Pearl Harbor hoping to avoid a later confrontation with the U.S. Pacific fleet, which posed a threat

security dilemma A characteristic of the international state system that gives rise to unstable relations between states; as State A secures its borders and interests, its behavior may decrease the security of other states and cause them to engage in behavior that decreases A's security.

to the Japanese military. In 2001, a United States-led coalition bombed Afghanistan in response to the September 11 terrorist attacks. Moreover, in March 2003, the United States, Great Britain, and a loosely coupled "coalition of the willing" invaded Iraq in response to perceived threats of weapons of mass destruction and the reported failure of Saddam Hussein to cooperate with United Nations weapons inspectors. Yet, in 2005, a presidential commission concluded that the attack on Iraq was based on faulty intelligence and that, in fact, "America's spy agencies were 'dead wrong' in most of their judgments about Iraq's weapons of mass destruction" (Shrader 2005, p. 1). As a result, by 2007, many Americans, more than 60 percent, favored a partial or complete withdrawal from Iraq (CNN/Opinion Research Corporation Poll 2007). Many believe that the public's desire for withdrawal from Iraq was a key factor in the outcome of the 2008 U.S. presidential election.

Revolutions and Civil Wars

Revolutions and civil wars involve citizens warring against their own government and often result in significant political, economic, and social change. The difference between a revolution and a civil war is not always easy to determine. Scholars generally agree that revolutions involve sweeping changes that fundamentally alter the distribution of power in society (Skocpol 1994). The American Revolution resulted from colonists revolting against British control. Eventually, they succeeded and established a republic where none existed before. The Russian Revolution involved a revolt against a corrupt, autocratic, and out-of-touch ruler, Czar Nicholas II. Among other changes, the revolution led to wide-scale seizure of land by peasants who formerly were economically dependent on large landowners.

Civil wars may result in a different government or a new set of leaders but do not necessarily lead to such large-scale social change. Because the distinction between a revolution and a civil war depends upon the outcome of the struggle, it may take many years after the fighting before observers agree on how to classify it. Revolutions and civil wars are more likely to occur when a government is weak or divided, when it is not responsive to the concerns and demands of its citizens, and when strong leaders are willing to mount opposition to the government (Barkan & Snowden 2001; Renner 2000).

One of the world's longest running civil wars came to an end in May 2009. Since 1983, the government of Sri Lanka fought an insurgency led by the Liberation Tigers of Tamil Eelam (LTTE). Also known as the Tamil Tigers, the LTTE were separatist militants who sought to carve an independent state out of the northern and eastern portions of this island country. The war resulted in more than 68,000 deaths (Gardner 2007). The Sri Lankan Army defeated the last remnants of the LTTE and killed their leader in May 2009 (Buncombe 2009). Like many civil wars, the war in Sri Lanka was also a struggle between a majority community (in this case, Sinhalese Buddhists) and a relatively poor and disadvantaged minority community (Hindu Tamils). Despite the end of the war, a political settlement with the minority community has not been achieved. Civil wars have also erupted in newly independent republics created by the collapse of communism in eastern Europe, as well as in Rwanda, Sierra Leone, Chile, Uganda, Liberia, and Sudan.

Nationalism

Some countries engage in war in an effort to maintain or restore their national pride. For example, Scheff (1994) argued that "Hitler's rise to power was laid by

Arizona businessman Dan Frazier wears a version of the t-shirt banned in Arizona, Florida, Louisiana, Oklahoma, and Texas. In a case brought to a federal court by the ACLU in 2008, a judge overturned the ban in Arizona, finding that the shirt is a form of free speech protected by the Constitution.

the treatment Germany received at the end of World War I at the hands of the victors" (p. 121). Excluded from the League of Nations, punished by the Treaty of Versailles, and ostracized by the world community, Germany turned to nationalism as a reaction to material and symbolic exclusion.

In the late 1970s, Iranian militants seized the U.S. embassy in Tehran and held its occupants hostage for more than one year. President Carter's attempt to use military forces to free the hostages was not successful. That failure intensified doubts about America's ability to use military power effectively to achieve its goals. The hostages in Iran were eventually released after President Reagan took office, but doubts about the strength and effectiveness of the U.S. military still called into question America's status as a world power. Subsequently, U.S. military forces invaded the small island of Grenada because the government of Grenada was building an airfield large enough to accommodate major military armaments. U.S. officials feared that this airfield would be used by countries in hostile attacks on the United States. From one point of view, the large-scale "successful" attack on Grenada functioned to restore faith in the power and effectiveness of the U.S. military.

What Do You Think? To protest what he saw as an unjust war and to make money, businessman Dan Frazier printed and sold t-shirts that read "Bush Lied" on the front and "They Died" on the back over a list of the names of 4,058 soldiers who were killed in Iraq (Fischer 2007). In response to complaints from some families of soldiers, Frazier's home state of Arizona passed a law banning the shirts and authorizing family members to sue for damages. The American Civil Liberties Union petitioned a federal judge to suspend the law on the grounds that it violated First Amendment rights to free speech (Davenport 2007). In August 2008, a federal judge declared the ban an infringement of First Amendment rights (ACLU 2008). Did the federal judge make the right decision in this case? Is this a case of offensive and greedy profiteering from tragedy or an overzealous attack on the right to voice opposition to government policies?

Terrorism

Terrorism is the premeditated use, or threatened use, of violence against civilians by an individual or group to gain a political or social objective (Barkan & Snowden 2001; Brauer 2003). Terrorism may be used to publicize a cause, promote an ideology, achieve religious freedom, attain the release of a political prisoner, or rebel against a government. Terrorists use a variety of tactics, including assassinations, skyjackings, suicide bombings, armed attacks, kidnapping and hostage taking, threats, and various forms of bombings. Through such tactics, terrorists struggle to induce fear within a population, create pressure to change policies, or undermine the authority of a government they consider objectionable. Most analysts agree that, unlike war—where a clear winner is more likely—terrorism is unlikely to be completely defeated.

> There can be no final victory in the fight against terrorism, for terrorism (rather than full-scale war) is the contemporary manifestation of conflict, and conflict will not disappear from earth as far as one can look ahead and human nature has not undergone a basic change. But it will be in our power to make life for terrorists and potential terrorists much more difficult. (Laqueur 2006, p. 173)

Types of Terrorism

Terrorism can be either transnational or domestic. **Transnational terrorism** occurs when a terrorist act in one country involves victims, targets, institutions, governments, or citizens of another country. The 1988 bombing of Pan Am Flight 103 over Lockerbie, Scotland, exemplifies transnational terrorism. The incident took the lives of 270 people, including 35 Syracuse University undergraduates returning from an overseas studies program in London. After a 10-year investigation, Abdel Basset Ali al-Megrahi, a Libyan intelligence agent (CNN 2001), was sentenced to life imprisonment in Scotland for his role in preparing the bomb that brought down the plane. In 2003, the Libyan government agreed to pay $2.7 billion in compensation to the victims' families (Smith 2004). After a diagnosis of terminal cancer, the Scottish government released Megrahi "on compassionate grounds" to return to Libya where he received a hero's welcome organized by the Libyan government (Cowell & Sulzberger 2009).

The 2001 attacks on the World Trade Center, the Pentagon, and Flight 93—the most devastating in U.S. history—are also the deadliest examples of transnational terrorism. Al Qaeda, a global alliance of militant Sunni Islamic groups advocating jihad ("holy war") against the West, was also responsible for attacks on U.S. embassies in Kenya and Tanzania (1998) and the bombing of a naval ship, the USS *Cole,* moored in Aden Harbor, Yemen (2000). Al Qaeda has since been linked to deadly bombings in Bali, Indonesia (2002), Madrid (2004), and London (2005).

Many groups other than al Qaeda use terrorism to further their own social and political goals. In fact, the U.S. Department of State identifies 44 "foreign terrorist organizations . . . [that] threaten the security of U.S. nationals or the national security (national defense, foreign relations, or the economic interests) of the United States" (Office of the Coordinator for Counterterrorism, p. 283). One of the militant groups on this list includes Lashkar-e-Taiba, based in Pakistan. In November 2008, 10 members of Lashkar-e-Taiba laid siege to India's largest city,

STR/AFP/Getty Images

A policeman escorts an elderly man to his destination shortly after gunmen had opened fire in this crowded railway station in Mumbai, killing nearly 60 travelers. On November 26, 2008, 10 members of Lashkar-e-Taiba, a terrorist group based in Pakistan, went on a 3-day rampage through the streets of India's most populous city, Mumbai, killing and wounding over 460 people.

terrorism The premeditated use or threatened use of violence by an individual or group to gain a political objective.

transnational terrorism Terrorism that occurs when a terrorist act in one country involves victims, targets, institutions, governments, or citizens of another country.

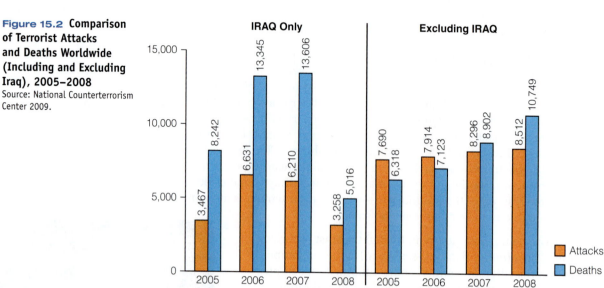

Figure 15.2 Comparison of Terrorist Attacks and Deaths Worldwide (Including and Excluding Iraq), 2005–2008
Source: National Counterterrorism Center 2009.

Mumbai, for 3 days. Armed with assault rifles, pistols, grenades, and at least one bomb, the gunmen stormed Mumbai, opening fire on civilians in public places, including two luxury hotels, a hospital, a railway station, a popular restaurant, a college, a cinema, and a residential compound. At least 166 people were killed during the attacks, and over 300 were wounded (Sengupta 2009).

Domestic terrorism, sometimes called insurgent terrorism (Barkan & Snowden 2001), is exemplified by the 1995 truck bombing of a nine-story federal office building in Oklahoma City, resulting in 168 deaths and the injury of more than 200 people. Gulf War veteran Timothy McVeigh and Terry Nichols were convicted of the crime. McVeigh is reported to have been a member of a paramilitary group that opposes the U.S. government. In 1997, McVeigh was sentenced to death for his actions, and he was executed in 2001 (Barnes 2004). The 2004 bombing of a Russian school by Chechen militants, which killed 324 people, nearly half of them children, is also an act of domestic terrorism, as Chechen rebels continue to fight for an independent state. In December 2006, the Basque separatist group ETA resumed its bombing campaign after a declaration of a permanent ceasefire that led to hopes of a peace process to settle the decades-old struggle (Espiau 2006).

Patterns of Global Terrorism

A report by the National Counterterrorism Center (2009) described patterns of terrorism around the world (see Figure 15.2). In 2008:

- There were approximately 11,770 domestic and international terrorist attacks around the world.
- 15,765 people lost their lives as a result of these attacks. Of this number, 33 were nonmilitary U.S. citizens.
- There was a 19 percent decrease in the number of attacks and a 30 percent decrease in the number of fatalities compared with 2007. However, if terrorist attacks in Iraq are excluded, the number of incidents and fatalities rose slightly from 2007 totals.
- An additional 34,000 individuals were wounded by terrorist attacks.
- About 55 percent of those killed or wounded lived in Iraq, Pakistan, or Afghanistan.

domestic terrorism Domestic terrorism, sometimes called insurgent terrorism, occurs when the terrorist act involves victims, targets, institutions, governments, or citizens from one country.

In 2009, a random sample of U.S. adults was asked, "How worried are you that you or someone in your family will become a victim of terrorism?" Of respondents, 36 percent said that they were "very worried" or "somewhat worried" (Morales 2009a), a sharp drop from 59 percent shortly after the September 11 attacks. At the same time, Americans continue to report that they are more concerned about international terrorism than any other international security matter, including the wars in Afghanistan and Iraq, Iran's and North Korea's nuclear capabilities, and the Israeli-Palestinian conflict (Morales 2009b).

> ... Americans continue to report that they are more concerned about international terrorism than any other international security matter including the wars in Afghanistan and Iraq, Iran's and North Korea's nuclear capabilities, and the Israeli-Palestinian conflict.

The Roots of Terrorism

In 2003, a panel of terrorist experts came together in Oslo, Norway, to address the causes of terrorism (Bjorgo 2003). Although not an exhaustive list, several causes emerged from the conference:

- A failed or weak state, which is unable to control terrorist operations
- Rapid modernization, when, for example, a country's sudden wealth leads to rapid social change
- Extreme ideologies—religious or secular
- A history of political violence, civil wars, and revolutions
- Repression by a foreign occupation (i.e., invaders to the inhabitants)
- Large-scale racial or ethnic discrimination
- The presence of a charismatic leader

Note that Iraq has several of the characteristics listed here, including rapid modernization (e.g., oil reserves), extreme ideologies (e.g., Islamic fundamentalism), a history of violence (e.g., invasion of Kuwait), large-scale ethnic discrimination (e.g., persecution of Kurdish minority), and a weak state that is unable to control terrorist operations (e.g., the newly elected Iraqi government).

The causes of terrorism listed here, however, are macro in nature. What of social-psychological variables? How do individuals choose to join terrorist organizations or use terrorist tactics? Borum (2003) suggested a four-stage micro-level process. The decision to commit a terrorist act begins with an individual's assessment that *something is not right* (e.g., government-imposed restrictions). Next, individuals *define the situation as unfair* in that the "not right" condition does not apply to everyone (e.g., government-imposed restrictions are imposed on some but not on others). Individuals then begin to *blame specific others for the injustice* (e.g., government leaders) and, finally, to *redefine those who are responsible for the injustice as bad or evil* (e.g., a fascist regime). This process of "ideological development," as portrayed in Figure 15.3, often leads to stereotyping and dehumanizing the enemy, which then facilitates violence. Although the process was developed as a heuristic device, Borum noted that "understanding the mind-set" of a terrorist can help in the fight against terrorism.

America's Response to Terrorism

A government can use both defensive and offensive strategies to fight terrorism. Defensive strategies include using metal detectors and X-ray machines at airports

Figure 15.3 **The Process of Ideological Development**
Source: Borum 2003.

It's Not Right | **It's Not Fair** | **It's Your Fault** | **You're Evil**

Social and Economic Deprivation → Inequality and Resentment → Blame/Attribution → Generalizing/Stereotyping / Dehumanizing/Demonizing the Enemy (Cause)

Context | **Comparison** | **Attribution** | **Reaction**

and strengthening security at potential targets, such as embassies and military command posts. The Department of Homeland Security (DHS) coordinates such defensive tactics for the U.S. government. DHS was created in 2002, from 22 domestic agencies (e.g., the U.S. Coast Guard, the Immigration and Naturalization Service, and the Secret Service) and has more than 180,000 employees. With a budget of $55.5 billion in 2009, the mission of DHS is as follows: "We will prevent and deter terrorist attacks and protect against and respond to threats and hazards to the Nation. We will secure our national borders while welcoming lawful immigrants, visitors, and trade" (U.S. Department of Homeland Security 2009).

Offensive strategies include retaliatory raids, such as the U.S. bombing of terrorist facilities in Afghanistan, group infiltration, and preemptive strikes. New legislation facilitates such offensive tactics. In October 2001, the USA PATRIOT Act (Uniting and Strengthening America by Providing Appropriate Tools Required to Intercept and Obstruct Terrorism) was signed into law. The act increases police powers both domestically and abroad. Advocates of the Patriot Act argue that, during war, some restrictions of civil liberties are necessary. Moreover, the legislation is not "a substantive shift in policy but a mere revitalization of already established precedents" (Smith 2003, p. 25). Critics hold that the act poses a danger to civil liberties. For example, the original act provides for the *indefinite* detention of immigrants if the immigrant group is defined as a "danger to national security" (Romero 2003). In 2007, Congress revised the act to address some of these concerns, particularly to limit the government's authority to conduct wiretaps. Unlike the earlier Patriot Act, which had "sunset" clauses mandating review by Congress before renewal, most of the provisions in the new act are now permanent features of the law.

Among the most controversial U.S. policies in the war on terrorism is the indefinite detention of "enemy combatants" at a military prison and interrogation camp in Guantanamo, Cuba. This is the primary detention center for the Taliban or their allies captured in Afghanistan, as well as suspected terrorists including al Qaeda members from other regions. Since 2002, as many as 775 detainees have been held at "Gitmo" (Dedman 2006). The Bush administration argued that, because these detainees were not members of a state's army, they were not covered by the **Geneva Conventions,** the principal international treaties governing the laws of war and in particular the treatment of prisoners of war and civilians during wartime. In 2006, the U.S. Supreme Court rejected this argument, ruling that the detainees were subject to minimal protections under the Conventions. Table 15.3 contains the results of a national survey of 1,464 Americans who were asked whether they thought the reports of mistreatment at Guantanamo Bay were "isolated incidents" or a "wider pattern" of abuse.

Geneva Conventions A set of international treaties that govern the behavior of states during wartime, including the treatment of prisoners of war.

TABLE 15.3 Reports of Prisoner Mistreatment at Guantanamo Bay

	ISOLATED (%)	WIDER PATTERN (%)	NEITHER/DON'T KNOW (%)
Total	54	34	12
Men	56	64	10
Women	52	34	14
18 to 29 years old	43	46	11
30 to 49 years old	55	36	9
50 to 64 years old	59	28	13
65 years old and older	56	25	19
White	57	31	12
Black	35	52	13
Hispanic	44	45	11
Republican	76	14	10
Democrat	43	45	12
Independent	45	44	11

Source: Adapted from Pew Research Center 2005.

On his second day in office, President Obama ordered the Pentagon to close Guantanamo Bay and all other detention facilities by January 2010. Concerns over security have led many to oppose bringing the detainees to the U.S. mainland for trial and possible incarceration. Nonetheless, in November 2009, the Obama administration announced that Khalid Shaikh Mohammed, the reputed mastermind of the September 11 attacks, would stand trial in a federal courtroom in Manahattan, a few blocks away from the site of the World Trade Center bombings (Savage 2009). In December 2009, the administration announced plans to convert an Illinois state prison to a federal detention center that would house dozens of Guantanamo terrorism suspects (Slevin 2009). Despite the closing of Guantanamo, experts expect that some "high-value" prisoners may be detained indefinitely without trial (BBC 2009b).

The treatment of detainees at Guantanamo and at secret detention centers (so-called "black sites") around the world has sparked public debate about interrogation techniques used on suspected terrorists. According to documents released by the Department of Justice in 2009, between 2002 and 2005, interrogators with the CIA and U.S. military intelligence used sleep deprivation, extended periods of standing, prolonged exposure to cold and noise, sexual degradation, and "waterboarding" (i.e., simulated drowning), among many other aggressive techniques (Mazzetti & Shane 2009a). The Bush administration and other supporters argued that such "enhanced interrogation techniques" were necessary to extract useful information that protected society from future terrorist attacks. Opponents, including many officials and security experts, claimed that these practices: (1) did not elicit useful information, (2) were counterproductive politically, (3) were banned by international treaties to which the United States is obligated, and (4) constituted torture. Congress and the White House have since

Demonstrators stage a mock "waterboarding" at a protest in front of the White House in March 2008. Such "enhanced interrogation techniques" were used repeatedly by the CIA against suspected terrorists since 9/11. Waterboarding sparked a public debate in the United States and around the world about what constitutes torture and whether the goal of getting important information from potentially dangerous detainees justifies the interrogation methods used.

banned these practices. In August 2009, the U.S. Attorney General Eric Holder appointed a federal prosecutor to investigate alleged abuses by the CIA and to consider whether a full criminal investigation is warranted (Mazzetti & Shane 2009b).

Public opinion on this topic is mixed. In response to a question about whether "torture to gain important information from suspected terrorists is justified," 48 percent of American adults said that they believed torture is "often" or "sometimes" justified, whereas 47 percent believed that torture is "rarely" or "never" justified (Pew Research Center for the People & the Press 2009). Republicans were twice as likely as Democrats to say that torture was "sometimes" justified (49 percent versus 24 percent), and Democrats were nearly three times more likely than Republicans to say that torture was "never" justified (38 percent versus 14 percent).

Combating terrorism is difficult, and recent trends will make it increasingly problematic (Zakaria 2000; Strobel et al. 2001). First, hackers who illegally gain access to classified information can easily acquire data stored on computers. For example, interlopers obtained the fueling and docking schedules of the USS *Cole*. Second, the Internet permits groups with similar interests, once separated by geography, to share plans, fund-raising efforts, recruitment strategies, and other coordinated efforts. Worldwide, thousands of terrorists keep in touch through Hotmail.com e-mail accounts, and virtually all terrorist groups maintain websites for recruitment, fund-raising, internal communication, and propagandizing (Weimann 2006). Third, globalization contributes to terrorism by providing international markets where the tools of terrorism—explosives, guns, electronic equipment, and the like—can be purchased. Finally, fighting terrorism under guerrilla warfare-like conditions is increasingly a concern. Unlike terrorist activity, which targets

The prisoner abuses at Guantanamo Bay (Cuba) and Abu Ghraib (Iraq) shocked the nation. When photos of the abuses at Abu Ghraib surfaced, they were transmitted around the world on the Internet and made into billboards in Arab countries. Army Pfc. Lynndie England, pictured, was convicted on six of seven counts involving prisoner mistreatment and was sentenced to three years in prison and given a dishonorable discharge.

An Iraqi insurgent takes aim on U.S. positions in Najaf during battle in January 2006. Although the U.S. Army has superior weapons and training, insurgent forces rely on deep knowledge of urban terrain and count on support from the local population.

Joao Silva/The New York Times/Redux Pictures

civilians and may be committed by lone individuals, **guerrilla warfare** is committed by organized groups opposing a domestic or foreign government and its military forces. The size of guerrilla forces is often hard to estimate and can vary widely. For example, in March 2007, an estimated 70,000 Iraqi insurgents were fighting U.S., allied, and Iraqi forces (O'Hanlon & Campbell 2007). In February 2009, officials in Afghanistan claimed that between 10,000 and 15,000 Taliban insurgents were fighting in the country (Afghanistan Conflict Monitor 2009).

The possibility of terrorists using weapons of mass destruction is the most frightening scenario of all and, as stated earlier, the motivation for the 2003 war with Iraq. **Weapons of mass destruction (WMD)** include chemical, biological, and nuclear weapons. Anthrax, for example, although usually associated with diseases in animals, is a highly deadly disease in humans and, although preventable by vaccine, has a "lethal lag time." In a hypothetical city of 100,000 people, delaying a vaccination program one day would result in 5,000 deaths; a delay of six days would result in 35,000 deaths. In 2001, trace amounts of anthrax were found in several letters sent to media and political figures, resulting in five deaths and the inspection and closure of several postal facilities (Baliunas 2004). Despite widespread speculation that al Qaeda or Saddam Hussein was responsible for the attacks, investigators soon began to suspect that the source was domestic. In 2008, shortly after the FBI informed Bruce Ivins, a microbiologist at a U.S. Army laboratory in Fort Detrick, Maryland, that they intended to charge him with the crime, the scientist committed suicide (Associated Press 2008).

Other examples of the use of WMD exist. On at least eight occasions, Japanese terrorists dispersed aerosols of anthrax and botulism in Tokyo (Inglesby et al. 1999), and in 2000, a religious cult, hoping to disrupt elections in an Oregon county, "contaminated local salad bars with salmonella, infecting hundreds" (Garrett 2001, p. 76). In 2004, the poison ricin was detected on a mail-opening machine in Senate majority leader Bill Frist's Washington, DC, office (Associated Press 2005). Furthermore, in 2006, the hit TV series *24* featured a fictional story about a Russian separatist group that stole chemical weapons from the U.S. military and planted them in a gas distribution facility in downtown

guerrilla warfare Warfare in which organized groups oppose domestic or foreign governments and their military forces; often involves small groups of individuals who use camouflage and underground tunnels to hide until they are ready to execute a surprise attack.

weapons of mass destruction (WMD) Chemical, biological, and nuclear weapons that have the capacity to kill large numbers of people indiscriminately.

TABLE 15.4 U.S. Military's Goals in Combating Weapons of Mass Destruction (WMD)
OVERALL COMBATING WMD DESIRED END STATES
The U.S. Armed Forces, in concert with other elements of U.S. national power deter WMD use.
The U.S. Armed Forces are prepared to defeat an adversary threatening to use WMD and prepared to deter follow-on use.
Existing worldwide WMD are secure, and the U.S. Armed Forces contribute, as appropriate, to secure, reduce, reverse, or eliminate them.
Current or potential adversaries are dissuaded from producing WMD.
Current or potential adversaries' WMD are detected and characterized, and elimination is sought.
Proliferation of WMD and related materials to current and/or potential adversaries is dissuaded, prevented, defeated, or reversed.
If WMD are used against the United States or its interests, the U.S. Armed Forces are capable of minimizing the effects in order to continue operations in a WMD environment and assist U.S. civil authorities, allies, and partners.
The U.S. Armed Forces assist in attributing the source of an attack, respond decisively, and/or deter future attacks.
Allies, partners, and U.S. civilian agencies are capable partners in combating WMD.

Source: Moroney et al. 2009.

Los Angeles, threatening to kill thousands unless special agent Jack Bauer (played by Kiefer Sutherland) could catch the terrorists in time. The show builds on real fears among experts, who estimate that Russia possesses about 40,000 metric tons of chemical weapons, the world's largest stockpile by far. The United States, Canada, and the European Union have pledged nearly $2 billion to support Russia's program to secure and destroy all chemical weapons by 2012 (Walker & Tucker 2006). Table 15.4 lists the U.S. government's military goals in combating WMD, including stopping their proliferation, securing existing stockpiles, and working with allies to prevent the use of WMD by potential adversaries (Moroney 2009).

Social Problems Associated with Conflict, War, and Terrorism

Social problems associated with conflict, war, and terrorism include death and disability; rape, forced prostitution, and displacement of women and children; social-psychological costs; diversion of economic resources; and destruction of the environment.

Death and Disability

Many American lives have been lost in wars, including 53,000 in World War I, 292,000 in World War II, 34,000 in Korea, and 47,000 in Vietnam (U.S. Census Bureau 2004). In Iraq, between March 2003 and July 2009, 30,200 U.S. troops were wounded and more than 4,200 were killed (Global Security 2009). Many civilians and enemy combatants also die or are injured in war. For example, based on a survey by the World Health Organization, an estimated 151,000 Iraqi civilians and insurgents died from war-related injuries between March 2003 and June 2006 (Wilson 2008). Despite thousands of Iraqi deaths, many Americans are unaware of this tremendous loss of life. In a program the Pentagon developed, American reporters were "embedded" into U.S. military units to provide "journalists with a detailed understanding of military culture and life on the frontlines" (Lindner 2009, p. 21). One of the byproducts of this program, however, was that 90 percent of the stories by embedded journalists were written from the perspective of the American soldier, focusing "on the horrors facing the troops, rather than upon the thousands of Iraqis who died" (p. 45).

The impact of war and terrorism extends far beyond those who are killed. Many of those who survive war incur disabling injuries or contract diseases. For example, one million people worldwide have been killed or disabled by land mines—a continuing problem in the aftermath of the 2003 war with Iraq (Renner 2005). In 1997, the Mine Ban Treaty, which requires that governments destroy stockpiles within 4 years and clear land mine fields within 10 years, became international law. To date, 156 countries have signed the agreement; 39 countries remain, including China, India, Israel, Russia, and the United States (ICBL 2009). War-related deaths and disabilities also deplete the labor force, create orphans and single-parent families, and burden taxpayers who must pay for the care of orphans and disabled war veterans (see Chapter 2 for a discussion of military health care).

The killing of unarmed civilians is also likely to undermine the credibility of armed forces and make their goals more difficult to defend. In Iraq, for example, the events in Haditha, a city in western Iraq, became international news, outraged Iraqis, and led to intense condemnation of the U.S. mission. In November 2005, after a roadside bomb killed one Marine and wounded two others, "Marines shot five Iraqis standing by a car and went house to house looking for insurgents, using grenades and machine guns to clear houses" (Watkins 2007, p. 1). Twenty-four Iraqis were killed, many of them women and children, "shot in the chest and head from close range" (McGirk 2006, p. 3). After *Time* magazine broke the story in March 2006 (McGirk 2006), the military investigated and reversed its claim that the civilians died as a result of the roadside bomb. Eight marines were accused of wrongdoing during the incident; charges against six of them were dismissed, one was cleared of all charges, and one awaits court martial proceedings (Reuters 2008; Puckett & Faraj 2009).

The remains of U.S. service members are flown to Dover Air Force Base in Delaware for transfer to loved ones. Between 1991 and 2009, the U.S. public was not permitted to see images of this procedure because of a government ban. In a reversal of that policy, Defense Secretary Gates announced in February 2009 that the Pentagon would consult with individual families about their wishes before allowing access to the media.

What Do You Think? When members of the U.S. military are killed, their remains are flown to Dover Air Force Base in Delaware to be returned to their loved ones. In 1991, during the First Gulf War, the U.S. government banned media coverage of coffins returning home. On February 26, 2009, Secretary Gates overturned the 18-year ban and announced that the Pentagon would consult families "on an individual basis" about their wishes, declaring "We ought not presume to make that decision in their place" (quoted in Tyson 2009, p. 2). Results from a CNN/Opinion Research Corp Poll indicate that 67 percent of Americans said they thought coverage should be allowed, whereas 31 percent were opposed. Do you think the ban should have been lifted? Is there a public interest in allowing media coverage even if families disagree?

A uniformed military escort accompanies the remains of all U.S. soldiers to ensure that they are returned safely and respectfully to the next of kin. In April 2004, Marine Corps Lieutenant Colonel Michael R. Strobl accompanied Private First Class Chance Phelps from Dover Air Force Base to Phelps's hometown of Dubois, Wyoming.

By 1400, most of the seats on the gym floor were filled and people were finding seats in the fixed bleachers high above the gym floor. There were a surprising number of people in military uniform. Many Marines had come up from Salt Lake City. Men from various VFW posts and the Marine Corps League occupied multiple rows of folding chairs. We all stood as Chance's family took their seats in the front.

It turned out that Chance's sister, a petty officer in the Navy, worked for a rear admiral—the chief of Naval intelligence—at the Pentagon. The admiral had brought many of the sailors on his staff with him to Dubois to pay respects to Chance and support his sister. After a few songs and some words from a Navy chaplain, the admiral took the microphone and told us how Chance had died.

Chance was an artillery cannoneer and his unit was acting as provisional military police outside of Baghdad. Chance had volunteered to man a .50 caliber machine gun in the turret of the leading vehicle in a convoy. The convoy came under intense fire but Chance stayed true to his post and returned fire with the big gun, covering the rest of the convoy, until he was fatally wounded.

Then the commander of the local VFW post read some of the letters Chance had written home. In letters to his mom he talked of the mosquitoes and the heat. In letters to his stepfather he told of the dangers of convoy operations and of receiving fire.

The service was a fitting tribute to this hero. When it was over, we stood as the casket was wheeled out with the family following. The casket was placed onto a horse-drawn carriage for the mile-long trip from the gym, down the main street, then up the steep hill to the cemetery. I stood alone and saluted as the carriage departed the high school. I found my car and joined Chance's convoy.

The town seemingly went from the gym to the street. All along the route, the people had lined the street and were waving small American flags. The flags that were otherwise posted were all at half-staff. For the last quarter mile up the hill, local boy scouts, spaced about 20 feet apart, all in uniform, held large flags. At the foot of the hill, I could look up and back and see the enormity of our procession. I wondered how many people would be at this funeral if it were in, say, Detroit or Los Angeles—probably not as many as were here in little Dubois, Wyoming.

The carriage stopped about 15 yards from the grave and the military pallbearers and the family waited until the men of the VFW and Marine Corps League were formed up and school busses had arrived carrying many of the people from the procession route. Once the entire crowd was in place, the pallbearers came to attention and began to remove the casket from the caisson. As I had done all week, I came to attention and executed a slow ceremonial salute as Chance was being transferred from one mode of transport to another.

From Dover to Philadelphia, Philadelphia to Minneapolis, Minneapolis to Billings, Billings to Riverton, and Riverton to Dubois, we had been together. Now, as I watched them carry him the final 15 yards, I was choking up. I felt that, as long as he was still moving, he was somehow still alive. Then they put him down above his grave. He had stopped moving.

. . . I left Dubois in the morning before sunrise for my long drive back to Billings. It had been my honor to take Chance Phelps to his final post. Now he is on the high ground overlooking his town.

I miss him.

Source: Strobl 2004.

Lastly, individuals who participate in experiments for military research may also suffer physical harm. U.S. representative Edward Markey of Massachusetts identified 31 experiments dating back to 1945, in which U.S. citizens were subjected to harm from participation in military experiments. Markey charged that many of the experiments used human subjects who were captive audiences or populations considered "expendable," such as elderly individuals, prisoners, and hospital patients. Eda Charlton of New York was injected with plutonium in 1945. She and 17 other patients did not learn of their poisoning until 30 years later. Her son, Fred Shultz, said of his deceased mother:

> I was over there fighting the Germans who were conducting these horrific medical experiments . . . at the same time my own country was conducting them on my own mother. (Miller 1993, p. 17)

Rape, Forced Prostitution, and Displacement of Women and Children

Half a century ago, the Geneva Convention prohibited rape and forced prostitution in war. Nevertheless, both continue to occur in modern conflicts.

Before and during World War II, the Japanese military forced 100,000 to 200,000 women and teenage girls into prostitution as military "comfort women." These women were forced to have sex with dozens of soldiers every day in "comfort stations." Many of the women died as a result of untreated sexually transmitted diseases, harsh punishment, or indiscriminate acts of torture.

Since 1998, Congolese government forces have fought Ugandan and Rwandan rebels. Women have paid a high price for this civil war, in which gang rape is "so violent, so systematic, so common . . . that thousands of women are suffering from vaginal fistula, leaving them unable to control bodily functions and enduring ostracism and the threat of debilitating health problems" (Wax 2003, p. 1). Though much less common than violence against women, aid workers also see increasing incidents of rape and sexual violence against men as "yet another way for armed groups to humiliate and demoralize Congolese communities into submission" (Gettleman 2009, p. 1). United Nations officials call the situation in Congo "the worst sexual violence in the world" (Gettleman 2008, p. 1).

Feminist analyses of wartime rape emphasize that the practice reflects not only a military strategy but also ethnic and gender dominance. For example, Refugees International, a humanitarian aid group, reports that rape is "a systematic weapon of ethnic cleansing" against Darfuris and is "linked to the destruction of their communities" (Boustany 2007, p. 9). Under Darfur's traditional law, prosecution of rapists is nearly impossible: Four male witnesses are required to accuse a rapist in court and single women risk severe corporal punishment for having sex outside of marriage.

A child soldier in Liberia points his gun at a cameraman while carting a teddy bear on his back. Although reliable figures are hard to obtain, the UN estimates that about 300,000 child soldiers are fighting in wars worldwide.

© Georgas Gobet/AFP/Getty Images

War and terrorism also force women and children to flee to other countries or other regions of their homeland. For example, since 1990, at least 17 million children have been forced to leave their homeland because of armed conflict (Save the Children 2005). Refugee women and female children are particularly vulnerable to sexual abuse and exploitation by locals, members of security forces, border guards, or other refugees. In refugee camps, women and children may also be subjected to sexual violation. A 2003 report by Save the Children examined the treatment of women and children in 40 conflict zones. The use of child soldiers was reported in 70 percent of the zones, and trafficking of women and girls was reported in 85 percent of the zones. Wars

The Effect of War on Young Women and Girls in Northern Uganda

In January 2006, Olara Otunnu, a former United Nation's advocate for child victims of war described Uganda as "the worst place in the world to be a child" (Large 2006). For 20 years, starting in 1986, northern Uganda has experienced a guerrilla war that has displaced nearly two million people and caused widespread poverty. For much of the war, the two main combatants were the government of Uganda and the Lord's Resistance Army (LRA), a rebel group led by Joseph Kony, a ruthless and unstable commander who fought to restore his ethnic group to power. Tens of thousands of Ugandan children were forcibly conscripted into the government's army or abducted by the LRA where they were sexually abused and forced into combat. Many thousands more were killed or wounded.

When the fighting subsided in 2006, more than 800,000 Ugandans were living in government-run camps (Large 2006). Dependent on international aid groups such as the United Nations World Food Programme, UNICEF, and Médecins Sans Frontières for food and medical care, other international groups intervened in the most war-torn areas to develop additional support projects. Among the most important of these projects were those designed to reintegrate child soldiers into the local communities.

Annan et al. (2008), in conducting the Survey of War Affected Youth (SWAY), wanted to provide information to the United Nations and other international agencies to help them develop and improve youth-related social services. Focusing, in part, on the impact of the war on young women and girls, SWAY provides reliable data about this population through a systematic random sample of households in the most affected areas.

Sample and Methods

Between October 2006 and August 2007, the SWAY team employed two methods of data collection. First, they administered a large-scale quantitative survey to young women and girls in two districts of Uganda. Relying on lists used by the World Food Programme for food distribution, 1,100 households were randomly selected. Household members were interviewed and asked to identify all youth living in the household in 1996. From this retrospective sample, Annan and colleagues identified 857 female youth between the ages of 14 and 35 in 2006 (who were between the ages of 4 and 25 in 1996, as the worst violence of the war began). The researchers then identified, located, and interviewed 619 respondents from this sample. Second, Annan et al. (2008) drew a nonrandom subsample of 30 girls from these respondents and conducted in-depth, qualitative interviews with them.

The data the researchers collected focused on the women's and girls' experiences during and after the war including their experiences with abduction, as "forced wives" assigned to LRA commanders, and with their families, communities, and nongovernmental organizations (NGO) upon returning home. In addition, the SWAY team collected data about the well-being of women and girls in the study, including their recent economic activity, education, physical and mental health, and community participation, among many other variables.

Findings and Conclusions

The data that SWAY researchers collected detailed the extent of violence and identified some important and surprising results. Because Annan and her colleagues had randomly sampled the population, they could use their data to estimate the total amount of violence that young women and girls

experienced. They concluded that about 66,000 (20 percent) of all women and girls between the ages of 14 and 30 in 2006 had been abducted by the LRA during the previous 10 years. The researchers report that, "The average age for females at first abduction is approximately 16 years of age, with the majority of females experiencing abduction between the ages of 10 to 18 years of age" (p. 46). Forty percent of abductees were held in captivity for less than 14 days, 25 percent between two weeks and one year, and 19 percent for over a year. Five percent never returned home and are presumed dead. Twenty-five percent were given to LRA members as "forced wives" and, of that number, half gave birth to at least one child as a result of their captivity.

Many of the abducted young women and girls witnessed horrendous acts of violence during their captivity, at rates well above what non-abductees experienced. They witnessed (1) the beating and torture of others (83 percent); (2) the violent death of a family member or friend (53 percent); (3) occupied houses being set on fire (42 percent); (4) multiple, simultaneous killings (38 percent); and (5) injuries from combat or a land mine (24 percent).

The abductees were also forced to commit a variety of brutal acts. Seventeen percent said they were forced to kill an opposing soldier; 21 percent were forced to kill an unknown civilian; and 5 percent were forced to kill a family member or friend. In addition, 21 percent were forced to beat or cut a civilian, and 25 percent were made to step on or otherwise abuse a corpse. In each of these categories, less than 1 percent of non-abducted girls and youth committed such acts.

The LRA commanders distributed the young women and girls as "forced wives," partly on the basis of seniority. Half of the

are particularly dangerous for the very young—"Nine out of ten countries with the highest under-5 mortality rates are experiencing, or emerging from, armed conflict." These include Sierra Leone, Angola, Afghanistan, Niger, Liberia, Somalia, Mali, Chad, and the Democratic Republic of the Congo (Save the

A researcher interviews a woman about her experiences during and after the civil war in Northern Uganda. The Survey of War Affected Youth gathered data to help aid agencies, community leaders, and the UN to improve their efforts to rebuild society after the war.

Khristopher Carlson

senior commanders took five or more wives, whereas low-ranking commanders took, on the average, two wives. When many young women and girls were abducted at the same time, they were distributed randomly among the rebels, as this account from a woman forcibly married after her abduction at 17 illustrates:

> After reaching the LRA camp, I was grouped with other females who were recently abducted. [The commander] gave the orders to his escorts to distribute the wives. Clothes were placed in bags and put into a pile. The men who wanted wives stood nearby and watched as girls were told to pick a bag of clothing. Whoever owned the clothing then became the man to that girl. The clothes I picked belonged to a 35-year-old fighter. (p. 40)

Other forced wives reported that commanders chose wives on the basis of their appearance and their educational level. As one abductee remembered: "The prettiest, more educated girls were the first to be chosen and commanders always chose before fighters. . . . [Commanders] preferred girls with education because they needed them to write down numbers when radio codes were coming in" (p. 43).

Most of the young women and girls (68 percent) who were held for more than two weeks escaped on their own. Of the remainder, the LRA captors released 27 percent and the Ugandan Army rescued 5 percent. Forced wives were more carefully guarded by their captors than other detainees—83 percent escaped, 10 percent were rescued by the Ugandan military, and only 7 percent were released by the LRA.

Surprisingly, only a small percentage of female abductees reported severe psychological distress. Forced wives exhibited a slightly higher rate of distress than other abductees. Problems were most pronounced when the captives returned home to their families, but generally lessened over time. In fact, 83 percent of abductees reported positive relationships with their families and communities. According to Annan et al. (2008), "resilience and acceptance rather than rejection or trauma" was the norm in most families. Despite the violence, they were forced to commit in captivity, former abductees were no more likely to commit violence, engage in fights, or report hostility after their release than non-abductees.

The SWAY team used this and other data they created to help aid agencies identify sub-groups within the population of abductees who were especially in need of assistance as they returned home. One of these groups was forced wives who gave birth during their abduction: "Girls who return from captivity with children are three times less likely to return to school than those who do not conceive children in captivity and 10 times less likely to return to school than girls who were never abducted" (p. 43). Also, because about one-third of female abductees returned home at the age of 18 or older, the researchers advised that age-based social service programs that targeted children under 18 were missing a significant percentage of the population in need.

Finally, NGO and government aid programs used categories like "abducted" and "forced wives" to determine categories of people requiring the most assistance. Based on the data, the research team concluded that these were "crude and poor predictors of vulnerability" among young women and girls affected by the war. In fact, regardless of whether they had been abducted or not, female youth who came from poor households, were unemployed, estranged from their families, suffered injuries or illness, and exhibited the highest levels of stress were ironically the least likely to receive assistance. By using rigorous methods of sampling and data collection, the *Survey of War Affected Youth* helped support groups revise their category-based approach in targeting the population to one based on "real needs" among those who suffered most from the devastation brought on by the war.

Source: Annan et al. 2008.

Children 2007, p. 13). Save the Children also estimated that, worldwide, 40 million children of primary school age—nearly one in three of all children in war zones—are prevented from attending school because of armed conflict (Save the Children 2009).

Social-Psychological Costs

Terrorism, war, and living under the threat of war disrupt social-psychological well-being and family functioning. For example, Myers-Brown et al. (2000) report that Yugoslavian children suffer from depression, anxiety, and fear as a response to conflicts in that region, emotional responses not unlike those Americans experienced after the events of 9/11 (NASP 2003). More recently, as a result of the war, there is evidence that many Iraqi children suffer from everything from "nightmares and bedwetting to withdrawal, muteness, panic attacks and violence towards other children, sometimes even to their own parents" (Howard 2007, p. 1). Further, a study of children in postwar Sierra Leone found that over 70 percent of boys and girls whose parents had been killed were at "serious risk" of suicide (Morgan & Behrendt 2009).

> The U.S. Army estimates that ... for every 100,000 soldiers deployed to Iraq, 20.2 committed suicide compared to 12 suicides per 100,000 soldiers who were not deployed to Iraq. This is the first time since the Vietnam War that the suicide rate for U.S. soldiers has surpassed that for civilians.

Guerrilla warfare is particularly costly in terms of its psychological toll on soldiers. In Iraq, soldiers were repeatedly traumatized as "guerrilla insurgents attack[ed] with impunity," and death was as likely to come from "hand grenades thrown by children, [as] earth-rattling bombs in suicide trucks, or snipers hidden in bombed-out buildings" (Waters 2005, p. 1). The U.S. Army estimates that, in 2008, for every 100,000 soldiers deployed to Iraq, 20.2 committed suicide compared to 12 suicides per 100,000 soldiers who were not deployed to Iraq (Alvarez 2009b; U.S. Army Medical Command 2007). This is the first time since the Vietnam War that the suicide rate for U.S. soldiers has surpassed that for civilians.

Military personnel who engage in combat and civilians who are victimized by war may experience a form of psychological distress known as **posttraumatic stress disorder (PTSD)**, a clinical term referring to a set of symptoms that can result from any traumatic experience, including crime victimization, rape, or war. Symptoms of PTSD include sleep disturbances, recurring nightmares, flashbacks, and poor concentration (NCPSD 2007). For example, Canadian Lieutenant General Romeo Dallaire, head of the United Nations peacekeeping mission in Rwanda, witnessed horrific acts of genocide. Four years after his return, he continued to have images of "being in a valley at sunset, waist deep in bodies, covered in blood" (quoted in Rosenberg 2000, p. 14). PTSD is also associated with other personal problems, such as alcoholism, family violence, divorce, and suicide.

Estimates of PTSD vary widely, although they are consistently higher among combat versus noncombat veterans. In a telephone study of 1,965 military personnel who had been deployed to Iraq and Afghanistan, 14 percent reported current symptoms of PTSD, 14 percent reported current symptoms of major depression, and 9 percent reported symptoms consistent with both conditions (Tanielian et al. 2008). If these estimates are correct, the researchers calculate that, as of April 2008, 303,000 Iraq and Afghanistan veterans were suffering from PTSD or major depression. Figure 15.4 describes the types of traumas to which Afghanistan and Iraq war veterans were exposed.

However, the rate of PTSD among soldiers is difficult to measure for several reasons. First, there is a lag, often of several years, between the time of exposure to trauma and the manifestation of symptoms. Second, soldiers are generally reluctant to report symptoms or to seek help (Wein 2009). When they do seek help, Army

posttraumatic stress disorder (PTSD) A set of symptoms that may result from any traumatic experience, including crime victimization, war, natural disasters, or abuses.

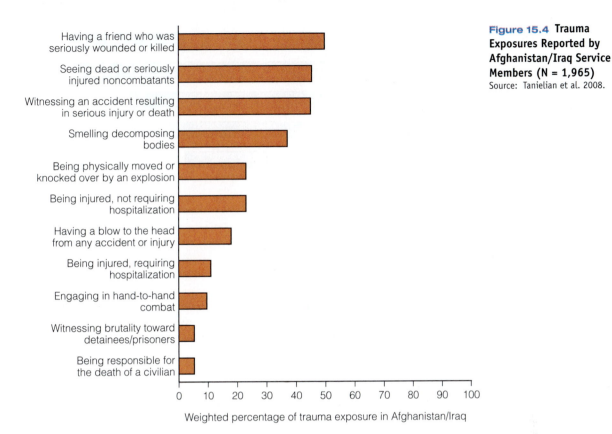

Figure 15.4 Trauma Exposures Reported by Afghanistan/Iraq Service Members (N = 1,965)
Source: Tanielian et al. 2008.

Having a friend who was seriously wounded or killed

Seeing dead or seriously injured noncombatants

Witnessing an accident resulting in serious injury or death

Smelling decomposing bodies

Being physically moved or knocked over by an explosion

Being injured, not requiring hospitalization

Having a blow to the head from any accident or injury

Being injured, requiring hospitalization

Engaging in hand-to-hand combat

Witnessing brutality toward detainees/prisoners

Being responsible for the death of a civilian

0 10 20 30 40 50 60 70 80 90 100

Weighted percentage of trauma exposure in Afghanistan/Iraq

officials say the Department of Veterans Affairs (VA) is slow to respond. With a backlog of 400,000 cases, an average claim takes 162 days to process (Dao 2009).

Diversion of Economic Resources

As discussed earlier, maintaining the military and engaging in warfare require enormous financial capital and human support. In 2008, worldwide military expenditures totaled more than $1.46 trillion (SIPRI 2009b). This amount exceeds the combined government research expenditures on developing new energy technologies, improving human health, raising agricultural productivity, and controlling pollution.

Money that is spent for military purposes could be allocated to social programs. The decision to spend $567 million, equal to the operating cost of the Smithsonian Institution (Center for Defense Information 2003), for one Trident II D-5 missile is a political choice. Similarly, allocating $2.3 billion for a "Virginia" attack submarine while our schools continue to deteriorate is also a political choice. Between 2001 and 2009, taxpayers will have paid $907 billion dollars for the cost of the Iraq and Afghanistan wars, the equivalent of providing (1) one year of health care for 267 million people, or (2) a year's salary for nearly 15 million elementary school teachers, or (3) one year of Head Start for 124 million children, or (4) 7 million affordable housing units, or (5) one year of scholarships for 140 million university students (National Priorities Project 2009). Estimated expenditures for the 2010 fiscal year include more money for national defense than for justice, transportation, veterans' benefits, and natural resources and the environment combined (Office of Management and Budget 2009b).

Chip East/Reuters/CORBIS

The sacrifices associated with military service often impose enormous psychological burdens on soldiers and their families.

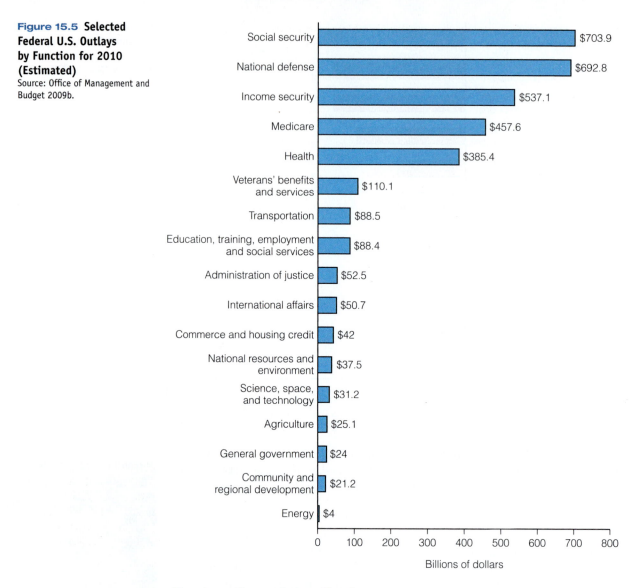

Figure 15.5 Selected Federal U.S. Outlays by Function for 2010 (Estimated)
Source: Office of Management and Budget 2009b.

Social security	$703.9
National defense	$692.8
Income security	$537.1
Medicare	$457.6
Health	$385.4
Veterans' benefits and services	$110.1
Transportation	$88.5
Education, training, employment and social services	$88.4
Administration of justice	$52.5
International affairs	$50.7
Commerce and housing credit	$42
National resources and environment	$37.5
Science, space, and technology	$31.2
Agriculture	$25.1
General government	$24
Community and regional development	$21.2
Energy	$4

Billions of dollars

Destruction of the Environment

The environmental damage that occurs during war devastates human populations long after war ends. As the casings for land mines erode, poisonous substances—often carcinogenic—leak into the ground (United Nations Association of the United States of America 2003). In 1991, during the Gulf War, Iraqi troops set 650 oil wells on fire, releasing oil, which covers the surface of the Kuwaiti desert and continues to seep into the ground, threatening underground water supplies. The smoke from the fires that hung over the Gulf region for eight months contained soot, sulfur dioxide, and nitrogen oxides—the major components of acid rain—and a variety of toxic and potentially carcinogenic chemicals and heavy metals. The U.S. Environmental Protection Agency estimates that, in March 1991, about 10 times as much air pollution was being emitted in Kuwait as by all U.S. industrial and power-generating plants combined (Renner 1993; Funke 1994; Environmental Media Services 2002).

Combatants often intentionally exploit natural resources to fuel their efforts. The local elephant population was heavily depleted during the civil war in

Thememoryhole.org/Getty Images News/Getty Images

The Human Side | Taking Chance

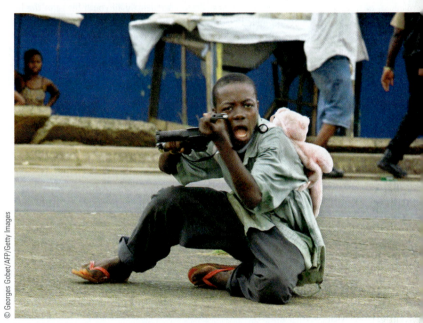
© Georges Gobet/AFP/Getty Images

southern Angola. The rebel group UNITA killed the animals to trade ivory for money to buy weapons. In addition, many were killed or fatally crippled by land mines planted by the guerrillas, though scientists report that, remarkably, the animals seemed to have learned to avoid mined areas (Marshall 2007). Between 1992 and 1997, UNITA also earned $3.7 billion to support its fighting from the sale of "conflict" or "blood" diamonds (GreenKarat 2007). Depending upon the location of diamonds, mining can be highly destructive to riverbed ecosystems or to areas surrounding open-pit mines.

The ultimate environmental catastrophe facing the planet is a massive exchange thermonuclear war. Aside from the immediate human casualties, poisoned air, poisoned crops, and radioactive rain, many scientists agree that the dust storms and concentrations of particles would block vital sunlight and lower temperatures in the Northern Hemisphere, creating a **nuclear winter**. In the event of large-scale nuclear war, most living things on earth would die. For example, a nuclear blast and the resulting blast wave create overpressure—the amount of pressure in excess of ordinary atmospheric levels as measured by pounds per inch (psi). As described in a report sponsored by the U.S. Air Force:

- At 20 psi of overpressure, even reinforced-concrete buildings are destroyed.
- 10 psi will collapse most factories and commercial buildings, as well as wood-frame and brick houses.
- 5 psi flattens most houses and lightly constructed commercial and industrial structures.
- 3 psi suffices to blow away the walls of steel-frame buildings.
- Even 1 psi will produce flying glass and debris sufficient to injure large numbers of people. (Ochmanek & Schwartz 2008, p. 6)

The fear of nuclear war has greatly contributed to the military and arms buildup, which, ironically, also causes environmental destruction even in times of peace. For example, in practicing military maneuvers, the armed forces demolish natural vegetation, disturb wildlife habitats, erode soil, silt up streams, and cause flooding. Bombs exploded during peacetime leak radiation into the atmosphere and groundwater. From 1945 to 1990, 1,908 bombs were tested—that is, exploded—at more than 35 sites around the world. Although underground testing has reduced radiation, some radioactive material still escapes into the atmosphere and is suspected of seeping into groundwater.

Finally, although arms control and disarmament treaties of the last decade have called for the disposal of huge stockpiles of weapons, no completely safe means of disposing of weapons and ammunition exist. Many activist groups have called for placing weapons in storage until safe disposal methods are found. Unfortunately, the longer the weapons are stored, the more they deteriorate, increasing the likelihood of dangerous leakage. In 2003, a federal judge gave permission, despite objections by environmentalists, to incinerate 2,000 tons of nerve agents and mustard gas left from the Cold War era. Although the army says that it is safe to dispose of the weapons, they have issued protective

Michel Lipschitz/AP Photo

Oil smoke from the 650 burning oil wells left in the wake of the Gulf War contains soot, sulfur dioxide, and nitrogen oxides, the major components of acid rain, along with a variety of toxic and potentially carcinogenic chemicals and heavy metals.

nuclear winter The predicted result of a thermonuclear war whereby thick clouds of radioactive dust and particles would block out vital sunlight, lower temperature in the Northern Hemisphere, and lead to the death of most living things on earth.

gear in case of an "accident" to the nearly 20,000 residents who live nearby (CNN 2003).

Strategies for Action: In Search of Global Peace

Various strategies and policies are aimed at creating and maintaining global peace. These include the redistribution of economic resources, the creation of a world government, peacekeeping activities of the United Nations, mediation and arbitration, and arms control.

Redistribution of Economic Resources

Inequality in economic resources contributes to conflict and war because the increasing disparity in wealth and resources between rich and poor nations fuels hostilities and resentment. Therefore, any measures that result in a more equal distribution of economic resources are likely to prevent conflict. John J. Shanahan (1995), retired U.S. Navy vice admiral and former director of the Center for Defense Information, suggested that wealthy nations can help reduce the social and economic roots of conflict by providing economic assistance to poorer countries. Nevertheless, U.S. military expenditures for national defense far outweigh U.S. economic assistance to foreign countries. For instance, the Obama administration's 2010 budget request for foreign affairs (including all nonmilitary foreign aid, the U.S. Department of State, the Peace Corps, and scholarly exchange programs) was $51.7 billion, or about 4 percent of the entire federal budget and 7.7 percent of the total request for the Department of Defense, the wars in Afghanistan and Iraq, and the global war on terrorism (Office of Management and Budget 2009a).

As discussed in Chapter 12, strategies that reduce population growth are likely to result in higher levels of economic well-being. Funke (1994) explained that "rapidly increasing populations in poorer countries will lead to environmental overload and resource depletion in the next century, which will most likely result in political upheaval and violence as well as mass starvation" (p. 326). Although achieving worldwide economic well-being is important for minimizing global conflict, it is important that economic development does not occur at the expense of the environment.

Finally, former United Nations Secretary General Kofi Annan, in an address to the United Nations, observed that it is not poverty per se that leads to conflict but rather the "inequality among domestic social groups" (Deen 2000). Referencing a research report completed by the Tokyo-based United Nations University, Annan argued that "inequality . . . based on ethnicity, religion, national identity, or economic class . . . tends to be reflected in unequal access to political power that too often forecloses paths to peaceful change" (Deen 2000).

The United Nations

Founded in 1945 after the devastation of World War II, the United Nations (UN) today includes 192 member states and is the principal organ of world governance. In its early years, the UN's main mission was the elimination of war from society.

In fact, the UN charter begins, "We the people of the United Nations—Determined to save succeeding generations from the scourge of war." During the past 65 years, the UN has developed major institutions and initiatives in support of international law, economic development, human rights, education, health, and other forms of social progress. The Security Council is the most powerful branch of the United Nations. Comprised of 15 member states, it has the power to impose economic sanctions against states that violate international law. It can also use force, when necessary, to restore international peace and security.

The United Nations has engaged in more than 60 peacekeeping operations since 1948 (see Table 15.5) (United Nations 2009).

United Nations peacekeepers—military personnel in their distinctive blue helmets or blue berets, civilian police and a range of other civilians—help implement peace agreements, monitor cease fires, create buffer zones, or support complex military and civilian functions essential to maintain peace and begin reconstruction and institution building in societies devastated by war. (United Nations 2003, p. 1)

Recently, the UN has been involved in overseeing multinational peacekeeping forces in Bosnia, East Timor, Sudan, Burundi, Haiti, Liberia, and Ethiopia (United Nations 2009).

In the last few years, the UN has come under heavy criticism. First, in recent missions, developing nations have supplied more than 75 percent of the troops while developed countries—United States, Japan, and Europe—have contributed 85 percent of the finances. As one UN official commented, "You can't have a situation where some nations contribute blood and others only money" (quoted by Vesely 2001, p. 8). Second, a review of UN peacekeeping operations noted several failed missions, including an intervention in Somalia in which 44 U.S. marines were killed (Lamont 2001). Third, as typified by the debate over the disarming of Iraq, the UN cannot take sides but must wait for a consensus of its members that, if not forthcoming, undermines the strength of the organization (Goure 2003). The consequences of delays can be staggering. For example, under the Genocide Convention of 1948, the UN is obligated to prevent instances of genocide, defined as "acts committed with the intent to destroy, in whole or in part, a national, racial, ethnical or religious group" (United Nations 1948). In 2000, the UN Security Council formally acknowledged its failure to prevent the 1994 genocide in Rwanda (BBC 2000). After the death of 10 Belgian soldiers in the days leading up to the genocide, and without a consensus for action among the members, the Security Council ignored warnings from the mission's commander about impending disaster and withdrew its 2,500 peacekeepers.

Finally, the concept of the UN is that its members represent individual nations, not a region or the world. And because nations tend to act in their own best economic and security interests, UN actions performed in the name of world peace may be motivated by nations acting in their own interests.

TABLE 15.5 United Nations Peacekeeping Operations: Summary Data, 2009	
Military personnel and civilian police serving in peacekeeping operations	92,876
Countries contributing military personnel and civilian police	118
International civilian personnel	5,742
Local civilian personnel	12,279
UN volunteers	2,326
Total number of fatalities in peacekeeping operations since 1948	2,599
Approved budgets for July 1, 2007, to June 30, 2008	$7.75 billion
Estimated total cost of operations from 1948 to June 30, 2008	$61 billion

Source: United Nations 2009.

A woman in Darfur scoops up grain that has spilled from bags dropped from a plane by the UN's World Food Programme. As in many of today's conflicts, the fighting in Darfur makes it very difficult for humanitarian aid agencies to run regular operations to feed, clothe, and shelter civilians.

Josphat Kasire/AP Photo

What Do You Think? The war in Darfur between the Sudanese Army and pro-government Arab militias on one side and Darfuri rebel groups on the other has already claimed at least 300,000 lives (BBC 2008). In March 2009, the International Criminal Court (ICC)—a permanent international body established to prosecute war crimes, mass atrocities, and genocide—issued a warrant for the arrest of Sudan's President Omar al-Bashir. The warrant charges the president with "intentionally directing attacks against an important part of the civilian population of Darfur, murdering, exterminating, raping, torturing and forcibly transferring large numbers of civilians and pillaging their property" (quoted in Charter et. al. 2009). This is the first time, since its founding in 2002, that the ICC has issued a warrant against a sitting head of state. Many observers in the West praised the move as a victory for international law, justice, and human rights. Sudan, on the other hand, does not recognize the court's authority, and other important actors (for example, China, Russia, the African Union, and the Arab League) reject the charges and object to this move as unlawful interference in Sudan's domestic affairs. Does the international community have the right to arrest a head of state and bring him to trial? What do you think?

As a result of such criticisms, outgoing UN Secretary General Kofi Annan called on the 191 members of the UN to approve the most far-reaching changes in the 60-year history of the organization (Lederer 2005). One of the most controversial recommendations concerns the composition of the Security Council, the most important decision-making body of the organization. Annan's recommendation that the 15 members of the Security Council—a body dominated by the United States, Great Britain, France, Russia, and China—be changed to include a more representative number of nations could, if approved, shift the global balance of power. Ban Ki-moon, former foreign minister of South Korea, was elected as the eighth Secretary General in October 2006 (MacAskill et al. 2006), after campaigning for the position on a platform that included support for the UN reforms. So far, the United Nations has not been able to reach agreement about whether and how to expand the membership of the Security Council.

Mediation and Arbitration

Most conflicts are resolved through nonviolent means (Worldwatch Institute 2003). Mediation and arbitration are just two of the nonviolent strategies used to resolve conflicts and to stop or prevent war. In mediation, a neutral third party intervenes and facilitates negotiation between representatives or leaders of conflicting groups. Good mediators do not impose solutions but rather help disputing parties generate options for resolving the conflict (Conflict Research Consortium 2003). Ideally, a mediated resolution to a conflict meets at least some of the concerns and interests of each party to the conflict. In other words, mediation attempts to find "win-win" solutions in which each side is satisfied with the solution.

Although mediation is used to resolve conflict between individuals, it is also a valuable tool for resolving international conflicts. For example, former U.S. Senator George Mitchell successfully mediated talks between parties to the conflict in Northern Ireland in 1998. The resulting political agreement continues to hold today. President Obama appointed Senator Mitchell as special envoy to the Middle East in order to pursue a negotiated political settlement. Also, in May 2008, the government of Qatar mediated talks between Lebanon's political parties that resulted in an agreement that averted civil war after an 18-month political crisis. Using mediation as a means of resolving international conflict is often difficult, given the complexity of the issues. However, research by Bercovitch (2003) shows that there are more mediators available and more instances of mediation in international affairs than ever before. For example, Bercovitch identified more than 250 separate mediation attempts during the Balkan wars of the early and mid-1990s.

Arbitration also involves a neutral third party who listens to evidence and arguments presented by conflicting groups. Unlike mediation, however, the neutral third party in arbitration arrives at a decision that the two conflicting parties agree in advance to accept. For instance, the Permanent Court of Arbitration—an intergovernmental organization based in The Hague since 1899—arbitrates disputes about territory, treaty compliance, human rights, commerce, and investment among any of its 107 member states who signed and ratified either of its two founding legal conventions. Recent cases include a dispute between France, Britain, and Northern Ireland about the Eurotunnel, a boundary dispute between Eritrea and Ethiopia, and a boundary dispute between the government of Sudan and the Sudan People's Liberation Army (Permanent Court of Arbitration 2009).

What Do You Think? Preventing Iran from acquiring nuclear weapons has been a central foreign policy goal for the past several U.S. presidential administrations. The second Bush administration refused to negotiate with Iran on any issue unless Iran first suspended its uranium enrichment program. Following on campaign promises, the Obama administration announced that it will pursue direct unconditional negotiations with Iran without any preconditions. Does setting preconditions make it more or less likely that the United States will change Iran's policies? What do you think?

arbitration Dispute settlement in which a neutral third party listens to evidence and arguments presented by conflicting groups and arrives at a decision that the parties have agreed in advance to accept.

Arms Control and Disarmament

In the 1960s, the United States and the Soviet Union led the world in a nuclear arms race, with each competing to build a larger and more destructive arsenal of nuclear weapons than its adversary. If either superpower were to initiate a full-scale war, the retaliatory powers of the other nation would result in the destruction of both nations as well as much of the planet. Thus, the principle of **mutually assured destruction (MAD)** that developed from nuclear weapons capabilities transformed war from a win-lose proposition to a lose-lose scenario. If both sides would lose in a war, the theory suggested, neither side would initiate war. At its peak year in 1966, the U.S. stockpile of nuclear weapons included more than 32,000 warheads and bombs (see this chapter's Self and Society feature).

As their arsenals continued to grow at an astronomical cost, both sides recognized the necessity for nuclear arms control including the reduction of defense spending, weapons production and deployment, and armed forces. Throughout the Cold War and even today, much of the behavior of the United States and the Soviet Union has been governed by major arms control initiatives. These initiatives include:

- The Limited Test Ban Treaty that prohibited testing of nuclear weapons in the atmosphere, underwater, and in outer space
- The Strategic Arms Limitation Treaties (SALT I and II) that limited the development of nuclear missiles and defensive antiballistic missiles
- The Strategic Arms Reduction Treaties (START I and II) that significantly reduced the number of nuclear missiles, warheads, and bombers
- The Strategic Offensive Reduction Treaty (SORT) that requires that the United States and the Russian Federation each reduce their number of strategic nuclear warheads to between 1,700 and 2,200 by 2012. (Atomic Archive 2009)

In 2009, the United States and the Russian Federation announced plans to negotiate further cuts in their nuclear stockpiles, reducing them to between 1,500 and 1,675 strategic nuclear warheads each (Baker and Cooper 2009).

With the end of the Cold War came the growing realization that, even as Russia and the United States greatly reduced their arsenals, other countries were poised to acquire nuclear weapons or expand their existing arsenals. Thus, the focus on arms control shifted toward the issue of **nuclear proliferation**, i.e., the spread of nuclear technology to nonnuclear states.

Nuclear Nonproliferation Treaty. The Nuclear Nonproliferation Treaty (NPT) was signed in 1970, and was the first treaty governing the spread of nuclear weapons technology from the original nuclear weapons states (i.e., the United States, the Soviet Union, the United Kingdom, France, and China) to nonnuclear countries. The NPT was renewed in 2000, and adopted by 187 countries. The NPT holds that countries without nuclear weapons will not try to get them; in exchange, the countries with nuclear weapons agree that they will not provide nuclear weapons to countries that do not have them. Signatory states without nuclear weapons also agree to allow the International Atomic Energy Agency to verify compliance with the treaty through on-site inspections (Atomic Archive 2009). Only India, Israel, and Pakistan have not signed the agreement, although each of these states is known to possess a nuclear arsenal. Further, many experts suspect that Iran and Syria—both signatories to the NPT—are developing nuclear weapons programs. Both countries claim that their nuclear reactors are for peaceful purposes (e.g., domestic power consumption), not to develop a nuclear weapons arsenal. However, a recent report by the International Atomic Energy Agency tentatively concluded that Iran has "sufficient information to be able to design and produce a workable" atomic bomb (Broad & Sanger 2009, p. 1).

mutually assured destruction (MAD) A Cold War doctrine referring to the capacity of two nuclear states to destroy each other, thus reducing the risk that either state will initiate war.

nuclear nonproliferation Efforts to prevent the spread of nuclear weapons, or the materials and technology necessary for the production of nuclear weapons.

Carefully read each of the following questions and select the correct answer. When you are finished, compare your answers to those provided and rank your performance using the scale at then end of this box.

1. How many nuclear weapons has the United States built since starting production in 1951?

    a. About 67,500
    b. About 100,000
    c. Nearly 25,000
    d. The number is classified

2. If you knew where to look, you could find 2,000 nuclear weapons in the state of Georgia.

    a. True
    b. False

3. How many nuclear weapons did the United States detonate in tests between 1945 and 1992 (the last year it detonated a nuclear weapon)?

    a. 13
    b. 130
    c. 1,030
    d. 10,300

4. What is the total volume (in cubic meters) of radioactive waste resulting from weapons activities?
    a. 650,000
    b. 104,000,000
    c. 426,000,000
    d. 1.2 billion

5. How many nuclear bombs has the United States lost (and not recovered) in accidents?

    a. 0
    b. 1
    c. 6
    d. 11

6. How many nuclear warheads and bombs does the United States currently possess in its stockpile?

    a. 3,416
    b. 7,982
    c. 10,600
    d. 32,750

7. With the end of the Cold War and the fall of the Soviet Union, most U.S. nuclear weapons are no longer targeted at specific sites around the world.

    a. True: Only about 100 weapons are targeted against other nuclear powers.
    b. False: The U.S. arsenal targets 2,500 sites for nuclear destruction around the world.

8. On U.S. soil, the government restricted its nuclear tests only to the Southwest desert because of the remote terrain and relatively low risks to human populations.

    a. True: There have been no nuclear tests on U.S. soil except those in Nevada and New Mexico.
    b. False: Nuclear weapons have been detonated in Alaska, Colorado, and Mississippi.

9. Which arm of the U.S. government supervises our nuclear weapons stockpile?

    a. Department of Defense
    b. Department of Energy
    c. National Security Council
    d. Department of State

10. Which was the last year that the United States detonated a nuclear bomb?

    a. 1945
    b. 1962
    c. 1992
    d. 1999

Answers: 1: a; 2: a; 3: c; 4: b; 5: d; 6: c; 7: b; 8: b; 9: b; 10: c.

Rank Your Performance: Number Correct

9 or 10:	Excellent
7 or 8:	Good
5 or 6:	Average
3 or 4:	Fair
2 or less:	Poor

*Since 1998, the time of this study, the United States has not detonated nuclear weapons for tests and has not added new nuclear weapons to its arsenal (Federation of American Scientists 2007).

Source: Adapted from U.S. Nuclear Weapons Cost Study Project 1998.

In January 2003, North Korea, under suspicion of secretly producing nuclear weapons, announced that it was withdrawing from the treaty, effective immediately (Arms Control Association 2003). In July 2005, after "more than a year of stalemate, North Korea agreed . . . to return to disarmament talks . . . and pledged to discuss eliminating its nuclear-weapons program" (Brinkley & Sanger 2005, p. 1). However, in May 2009, shortly after testing a long-range missile over Japan, North Korea detonated a nuclear device underground, heightening fears that it would launch a nuclear attack against South Korea, China, or Japan (NTI 2009).

> Even if military superpowers honor agreements to limit arms, the availability of black-market nuclear weapons and materials presents a threat to global security.

Even if military superpowers honor agreements to limit arms, the availability of black-market nuclear weapons and materials presents a threat to global security. One of the most successful nuclear weapons brokers was Pakistan's Abdul Qadeer Khan. Khan, the "father" of Pakistan's nuclear weapons program, sold the technology and equipment required to make nuclear weapons to rogue states such as Iran and North Korea. He is also suspected of selling nuclear weapons to al Qaeda (Powell & McGirk 2005).

In 2003, the United States accused Iran of operating a covert nuclear weapons program. Although Iranian officials responded that their nuclear program was solely for the purpose of generating electricity, concerns escalated with Iran's successful test of a 1,200-mile-range missile in 2009 (Dareini 2009). Hostile public statements about Israel by Iran's president Mahmoud Ahmadinejad deepen already grave concerns about Iran's possible acquisition of nuclear weapons. Analysts fear the development of nuclear weapons by Iran would spur a nuclear competition with Israel, the only other state in the Middle East to possess nuclear weapons, and provoke other states in the region to acquire them (Salem 2007). Nonetheless, analysts disagree about whether there is evidence that Iran, despite the capability, is designing a nuclear warhead (Broad et al. 2009).

Many observers consider South Asia the world's most dangerous nuclear rivalry. India and Pakistan are the only two nuclear powers that share a border and have repeatedly fired upon each other's armies while in possession of nuclear weapons (Stimson Center 2007). India first detonated nuclear weapons in 1974. Pakistan detonated six weapons in 1998, a few weeks after India's second round of nuclear tests. Although precise figures are hard to come by, experts estimate that India has about 50 nuclear bombs, and Pakistan has between 40 and 70 nuclear bombs (Center for Defense Information 2007). Pakistan, however, is aggressively expanding its nuclear weapons program (Shanker & Sanger 2009). Many fear that a conventional military confrontation between these two countries may some day escalate to an exchange of nuclear weapons. In addition, widespread social unrest in Pakistan coupled with clashes in 2009 between Taliban forces and Pakistan's army 60 miles from the capital, Islamabad, have heightened fears about the security of Pakistan's nuclear arsenal and its vulnerability to seizure by extremist forces (Kerr & Nikitin 2009).

As states that want to obtain nuclear weapons are quick to point out, nuclear states that advocate for nonproliferation possess well over 25,000 weapons, a huge reduction in the world's arsenal from Cold War days but still a massive potential threat to the earth. "Do as I say not as I do" is a weak bargaining position. Recognizing this situation and with concern about the possibility of new arms races, many high-level experts have begun to advocate a more comprehensive approach to banning nuclear weapons—with the United States leading the nuclear powers toward complete nuclear disarmament. In January 2007, three former U.S. Secretaries of State (George Schultz, William Perry, and Henry Kissinger) and former chairman of the Senate Armed Services Committee Sam Nunn published an editorial in the *Wall Street Journal* advocating that the United States "take the world to the next stage" of disarmament to "a world free of nuclear weapons" (Schultz et al. 2007). As a start, their proposal advocated the elimination of all short-range nuclear missiles, a complete halt in the production of weapons-grade uranium, and continued reduction of nuclear forces. President Obama visited Russia in July 2009, to sign an agreement that reduced the strategic nuclear arsenals of the United States and the Russian Federation by about 25 percent. This was reportedly "a first step in a broader effort intended to reduce the threat of such weapons drastically and to prevent their further spread to unstable regions" (Levy & Baker 2009).

The Problem of Small Arms

Although the devastation caused by even one nuclear war could affect millions, the availability of conventional weapons fuels many active wars around the world. Small arms and light weapons include handguns, submachine guns and automatic weapons, grenades, mortars, land mines, and light missiles. The Small Arms Survey estimated that, in 2006, 639 million firearms were in circulation, with about 59 percent of them being owned legally (Geneva Graduate Institute for International Studies 2006). Civilians purchased 80 percent of these weapons and were the victims of 90 percent of the casualties caused by them. Half a million people die each year as a result of small arms use, 200,000 in homicides and suicides and the rest during wars and other armed conflicts.

Unlike control of weapons of mass destruction such as chemical and biological weapons, controlling the flow of small arms is not easy because they have many legitimate uses by the military, by law enforcement officials, and for recreational or sporting activities (Schroeder 2007). Small arms are easy to afford, to use, to conceal, and to transport illegally. The small arms trade is also lucrative. Between 2000 and 2006, the top five exporters of military small arms were the United States ($228.5 million), Belgium ($27.1 million), France ($22.6 million), Germany ($16.2 million), and the United Kingdom ($13.6 million). The United States accounted for 54 percent of small arms exports during this time period (Small Arms Survey 2009).

The U.S. State Department's Office of Weapons Removal and Abatement (WRA) administers a program that has

Arnulfo Franco/AP Photo

A stockpile of assault rifles and light machine guns is ready to be inspected and prepared for destruction in 2006 through a U.S. government-funded program in Angola. Collected from ex-combatants and the Angolan army, these weapons are left over from Angola's 27-year-old civil war that formally ended in 2002.

supported the destruction of 1.3 million weapons and 50,000 tons of ammunition in 33 countries since its founding in 2001 (Office of Weapons Removal and Abatement 2009). The availability of these weapons fuels terrorist groups and undermines efforts to promote peace after wars have formally concluded. "If not expeditiously destroyed or secured, stocks of arms and ammunition left over after the cessation of hostilities frequently recirculate into neighboring regions, exacerbating conflict and crime" (U.S. Department of State 2007, p. 1). Operating on a budget of $16.1 million in 2007, the Bush administration increased the WRA budget to $44.3 million in 2008 (Office of Weapons Removal and Abatement 2009).

Understanding Conflict, War, and Terrorism

As we come to the close of this chapter, how might we have an informed understanding of conflict, war, and terrorism? Each of the three theoretical positions discussed in this chapter reflects the realities of global conflict. As structural functionalists argue, war offers societal benefits—social cohesion, economic prosperity, scientific and technological developments, and social change. Furthermore, as conflict theorists contend, wars often occur for economic reasons because corporate elites and political leaders benefit from the spoils of war—land and water resources and raw materials. The symbolic interactionist perspective emphasizes the role that meanings, labels, and definitions play in creating conflict and contributing to acts of war.

The September 11 attacks on the World Trade Center and the Pentagon and the aftermath—the battle against terrorism, the bombing of Afghanistan, and the war with Iraq—changed the world we live in. For some theorists, these events were inevitable. Political scientist Samuel P. Huntington argued that such conflict represents a **clash of civilizations.** In *The Clash of Civilizations and the Remaking of World Order* (1996), Huntington argued that in the new world order

> the most pervasive, important and dangerous conflicts will not be between social classes, rich and poor, or economically defined groups, but between people belonging to different cultural entities . . . the most dangerous cultural conflicts are those along the fault lines between civilizations . . . the line separating peoples of Western Christianity, on the one hand, from Muslim and Orthodox peoples on the other. (p. 28)

Although not without critics, the clash-of-civilizations hypothesis has some support. In an interview of almost 10,000 people from nine Muslim states representing half of all Muslims worldwide, only 22 percent had favorable opinions toward the United States (CNN 2003). Even more significantly, 67 percent saw the September 11 attacks as "morally justified," and the majority of respondents found the United States to be overly materialistic and secular and having a corrupting influence on other nations. Moreover, according to a poll of world public opinion taken every year since 2002, positive attitudes about the United States are consistently lowest in predominantly Muslim countries (Pew Research Center 2009). Although favorable views of the United States increased dramatically in a small number of Muslim countries after the election of President Obama, the 2009 poll found that only small minorities continued to see the United States favorably in most predominantly Muslim counties, for example, 14 percent in

clash of civilizations A hypothesis that the primary source of conflict in the 21st century has shifted away from social class and economic issues and toward conflict between religious and cultural groups, especially those between large-scale civilizations such as the peoples of Western Christianity and Muslim and Orthodox peoples.

Turkey, 15 percent in Palestinian territories, 16 percent in Pakistan, 25 percent in Jordan, and 27 percent in Egypt.

Conversely, according to a 2009 poll, 46 percent of Americans say they have unfavorable opinions of Muslim countries, a 5 percent increase from a similar poll taken in 2002 (Steinhauser 2009). In the same poll, only 24 percent of Americans reported favorable opinions of Muslim countries, and 30 percent said they had a neutral opinion. Further, 39 percent of Americans report being prejudiced against Muslims, 46 percent believe that they are too extreme in their religious beliefs, and 39 percent agree that Muslims in the United States should carry a special identification card (Saad 2006).

Ultimately, we are all members of one community—earth—and have a vested interest in staying alive and protecting the resources of our environment for our own and future generations. But, as we have seen, conflict between groups is a feature of social life that is not likely to disappear. What is at stake—human lives and the ability of our planet to sustain life—merits serious attention. World leaders have traditionally followed the advice of philosopher Karl von Clausewitz: "If you want peace, prepare for war." Thus, nations have sought to protect themselves by maintaining large military forces and massive weapons systems. These strategies are associated with serious costs, particularly in hard economic times. In diverting resources away from other social concerns, defense spending undermines a society's ability to improve the overall security and well-being of its citizens. Conversely, defense-spending cutbacks, although unlikely in the present climate, could potentially free up resources for other social agendas, including lowering taxes, reducing the national debt, addressing environmental concerns, eradicating hunger and poverty, improving health care, upgrading educational services, and improving housing and transportation. Therein lies the promise of a "peace dividend." The hope is that future dialogue on the problems of war and terrorism will redefine national and international security to encompass social, economic, and environmental well-being.

CHAPTER REVIEW

■ **What is the relationship between war and industrialization?**

War indirectly affects industrialization and techno-logical sophistication because military research and development advances civilian-used technologies. Industrialization, in turn, has had two major influences on war: The more industrialized a country is, the lower the rate of conflict, and if conflict occurs, the higher the rate of destruction.

■ **What are the latest trends in armed conflicts?**

Since World War II, wars between two or more states make up the smallest percentage of armed conflicts. In the contemporary era, the majority of armed conflicts have occurred between groups in a single state, who compete for the power to control the resources of the state or to break away and form their own state.

■ **In general, how do feminists view war?**

Feminists are quick to note that wars are part of the patriarchy of society. Although women and children may be used to justify a conflict (e.g., improving women's lives by removing the repressive Taliban in Afghanistan), the basic principles of male dominance and control are realized through war. Feminists also emphasize the social construction of aggressive masculine identities and their manipulation by elites as important reasons for the association between masculinity and militarized violence.

■ **What are some of the causes of war?**

The causes of war are numerous and complex. Most wars involve more than one cause. Some of the causes of war are conflict over land and natural resources; values or ideologies; racial, ethnic, and religious hostilities; defense against hostile attacks; revolution; and nationalism.

■ **What is terrorism, and what are the different types of terrorism?**
Terrorism is the premeditated use, or threatened use, of violence by an individual or group to gain a political or social objective. Terrorism can be either transnational or domestic. Transnational terrorism occurs when a terrorist act in one country involves victims, targets, institutions, governments, or citizens of another country. Domestic terrorism involves only one nation, such as the 1995 truck bombing of a nine-story federal office building in Oklahoma City.

■ **What are some of the macro-level "roots" of terrorism?**
Although not an exhaustive list, some of the macro-level "roots" of terrorism include (1) a failed or weak state, (2) rapid modernization, (3) extreme ideologies, (4) a history of violence, (5) repression by a foreign occupation, (6) large-scale racial or ethnic discrimination, and (7) the presence of a charismatic leader.

■ **How has the United States responded to the threat of terrorism?**
The United States has used both defensive and offensive strategies to fight terrorism. Defensive strategies include using metal detectors and X-ray machines at airports and strengthening security at potential targets, such as embassies and military command posts. The Department of Homeland Security coordinates such defensive tactics. Offensive strategies include retaliatory raids such as the U.S. bombing of terrorist

facilities in Afghanistan, group infiltration, and preemptive strikes.

■ **What is meant by "diversion of economic resources"?**
Worldwide, the billions of dollars used on defense could be channeled into social programs dealing with, for example, education, health, and poverty. Thus, defense monies are economic resources diverted from other needy projects.

■ **What are some of the criticisms of the United Nations?**
First, in recent missions, developing nations have supplied more than 75 percent of the troops. Second, several recent UN peacekeeping operations have failed. Third, the UN cannot take sides but must wait for a consensus of its members that, if not forthcoming, undermines the strength of the organization. Fourth, the concept of the UN is that its members represent individual nations, not a region or the world. Because nations tend to act in their own best economic and security interests, UN actions performed in the name of world peace may be motivated by nations acting in their own interests. Finally, the Security Council limits power to a small number of states.

■ **What problems do small arms pose?**
Even after a conflict ends, these weapons circulate in society, making crime worse or falling into the hands of terrorists. Trade in small arms is legal because they have many legitimate uses—for example, by the military, police, and hunters. Because they are small and simple to handle, these weapons are easily concealed and transported, making it difficult to control them.

TEST YOURSELF

1. War between states is still the most common form of warfare.
   a. True
   b. False
2. The rise of the modern state is most directly a result of
   a. industrialization and the creation of national markets
   b. innovations in communications technology
   c. the development of armies to control territory
   d. the development of police to control a population
3. Structural-functionalist explanations about war emphasize
   a. that war is a biological necessity
   b. that despite its destructive power, war persists because it fulfills social needs
   c. that war is an anachronism that will eventually disappear
   d. that war is necessary because it benefits political and military elites

4. On the whole, conflicts over values and ideologies are more difficult to resolve than those over material resources.
   a. True
   b. False
5. All wars are a result of unequal distribution of wealth.
   a. True
   b. False
6. Next to defense spending, transfers to foreign governments are the most expensive item in the U.S. government budget.
   a. True
   b. False
7. Which of the following factors is a likely cause of revolutions or civil wars?
   a. A weak or failed state
   b. An authoritarian government that ignores major demands from citizens
   c. The availability of strong opposition leaders
   d. All of the above

8. Primordial explanations of ethnic conflict suggest that
   a. ethnic leaders instigate hostilities to serve their own interests
   b. people become hostile when they blame their frustration with economic hardship on competing ethnic groups
   c. ancient hatreds compel ethnic groups to continue fighting
   d. None of the above
9. The consequences of war and the military for the environment
   a. are prevalent only during wartime
   b. persist in peacetime or for many years after a war is over
   c. do not significantly impact the environment
   d. have been mostly reduced by technological innovations

10. Advocates of nuclear nonproliferation seek to
    a. ban all nuclear weapons
    b. prevent construction of nuclear power plants
    c. prevent new states from acquiring nuclear weapons
    d. None of the above

Answers: 1: b; 2: c; 3: b; 4: a; 5: b; 6: b; 7: d; 8: c; 9: b; 10: c.

KEY TERMS

arbitration   591
clash of civilizations   596
Cold War   559
constructivist explanations   567
domestic terrorism   572
dual-use technologies   562
Geneva Conventions   574
guerrilla warfare   577

military-industrial complex   563
mutually assured destruction (MAD)   592
nuclear nonproliferation   592
nuclear winter   587
posttraumatic stress disorder (PTSD)   584
primordial explanations   567

security dilemma   568
state   558
terrorism   571
transnational terrorism   571
war   557
weapons of mass destruction (WMD)   577

MEDIA RESOURCES

Understanding Social Problems,
Seventh Edition Companion Website
www.cengage.com/sociology/mooney
Visit your book companion website, where you will find flash cards, practice quizzes, Internet links, and more to help you study.

 Just what you need to know NOW! Spend time on what you need to master rather than on information you already have learned. Take a pretest for this chapter, and CengageNOW will generate a personalized study plan based on your results. The study plan will identify the topics you need to review and direct you to online resources to help you master those topics. You can then take a posttest to help you determine the concepts you have mastered and what you will need to work on. Try it out! Go to **www.cengage.com/login** to sign in with an access code or to purchase access to this product.

Glossary

abortion The intentional termination of a pregnancy.

absolute poverty The lack of resources necessary for material well-being—most importantly food and water, but also housing, land, and health care.

acculturation The process of adopting the culture of a group different from the one in which a person was originally raised.

achieved status A status assigned on the basis of some characteristic or behavior over which the individual has some control.

acid rain The mixture of precipitation with air pollutants, such as sulfur dioxide and nitrogen oxide.

acquaintance rape Rape committed by someone known to the victim.

adaptive discrimination Discrimination that is based on the prejudice of others.

affirmative action A broad range of policies and practices to promote equal opportunity as well as diversity in the workplace and on campuses.

alienation A sense of powerlessness and meaninglessness in people's lives. The condition that results when workers perform repetitive, monotonous work tasks, and they become estranged from their work, the product they create, other people, and themselves.

alternative certification programs A program whereby college graduates with degrees in fields other than education can become certified if they have "life experience" in industry, the military, or other relevant jobs.

American Recovery and Reinvestment Act of 2009 Legislation that provided an economic stimulus package of $787 billion to create and save jobs and to reinvigorate the U.S. economy.

androgyny Having both feminine and masculine characteristics.

anomie A state of normlessness in which norms and values are weak or unclear.

antimiscegenation laws Laws banning interracial marriage until 1967, when the Supreme Court (in *Loving v. Virginia*) declared these laws unconstitutional.

arbitration Dispute settlement in which a neutral third party listens to evidence and arguments presented by conflicting groups and arrives at a decision or outcome that the parties have agreed in advance to accept.

ascribed status A status that society assigns to an individual on the basis of factors over which the individual has no control.

assimilation The process by which formerly distinct and separate groups merge and become integrated as one.

automation Dominant in an industrial society, the replacement of human labor with machinery and equipment that is self-operating.

aversive racism A subtle form of prejudice that involves feelings of discomfort, uneasiness, disgust, fear, and pro-white attitudes.

barrios In the United States, slums that are occupied primarily by Latinos.

behavior-based safety programs A strategy used by business management that attributes health and safety problems in the workplace to workers' behavior, rather than to work processes and conditions.

beliefs Definitions and explanations about what is assumed to be true.

bigamy The criminal offense in the United States of marrying one person while still legally married to another.

bilingual education In the United States, teaching children in both English and their non-English native language.

binge drinking As defined by the U.S. Department of Health and Human Services, drinking five or more drinks on the same occasion on at least 1 day in the past 30 days prior to the National Survey on Drug Use and Health.

biodiversity The diversity of living organisms on earth.

bioinvasion The intentional or accidental introduction of plant, animal, insect and other species in regions where they are not native.

biphobia Negative attitudes and emotions toward bisexuality and people who identify as bisexual.

bisexuality Emotional and sexual attraction to members of both sexes.

boy code A set of societal expectations that discourages males from expressing emotion, weakness, or vulnerability, or asking for help.

brownfields Abandoned or undeveloped sites that are located on contaminated land.

bullying Inherent in a relationship between individuals, groups, or individuals and groups, bullying entails an imbalance of power that exists over a long period of time in which the more powerful intimidate or belittle others.

capital punishment The state (the federal government or a state) takes the life of a person as punishment for a crime.

capitalism An economic system characterized by private ownership of the means of production and distribution of goods and services for profit in a competitive market.

character education Education that emphasizes the moral and interpersonal aspects of an individual.

charter schools Schools that originate in contracts, or charters, which articulate a plan of instruction that local or state authorities must approve.

chattel slavery An old form of slavery in which slaves are considered property that could be bought and sold.

chemical dependency A condition in which drug use is compulsive and users are unable to stop because of physical and/or psychological dependency.

child abuse The physical or mental injury, sexual abuse, negligent treatment, or maltreatment of a child younger than age 18 by a person who is responsible for the child's welfare.

child labor Involves children performing work that is hazardous, that interferes with a child's education, or that harms a child's health or physical, mental, social, or moral development.

civil union A legal status that entitles same-sex couples who apply for and receive a civil union certificate to nearly all of the benefits available to married couples.

clash of civilizations A hypothesis that the primary source of conflict in the 21st century has shifted away from social class and economic issues and toward conflict between religious and cultural groups, especially those between large-scale civilizations such as the peoples of Western Christianity and Muslim and Orthodox peoples.

classic rape Rape committed by a stranger, with the use of a weapon, resulting in serious bodily injury to the victim.

clearance rate The percentage of crimes in which an arrest and official charge have been made and the case has been turned over to the courts.

club drugs A general term for illicit, often synthetic, drugs commonly used at nightclubs or all-night dances called "raves."

Cold War The state of military tension and political rivalry that existed between the United States and the former Soviet Union from the 1950s through the late 1980s.

colonialism Occurs when a racial or ethnic group from one society takes over and dominates the racial or ethnic group(s) of another society.

coming out The process of LGBT individuals accepting and being open with others about their sexual orientation identity.

community development corporations (CDCs) Nonprofit groups formed by residents, small business owners, congregations, and other local stakeholders that work to create jobs and affordable housing and renovate parks and other community facilities.

comparable worth The belief that individuals in occupations, even in different occupations, should be paid equally if the job requires "comparable" levels of education, training, and responsibility.

comprehensive primary health care An approach to health care that focuses on the broader social determinants of health, such as poverty and economic inequality, gender inequality, environment, and community development.

comprehensive sexuality education Sex education that includes topics such as contraception, sexually transmitted diseases, HIV/AIDS, and disease-prevention methods as well as the benefits of abstinence.

compressed workweek A work arrangement that allows employees to condense their work into fewer days (e.g., four 10-hour days each week).

computer crime Any violation of the law in which a computer is the target or means of criminal activity.

constructivist explanations Those explanations that emphasize the role of leaders of ethnic groups in stirring up hatred toward others external to one's group.

contact hypothesis The idea that contact between groups is necessary for the reduction of prejudice.

conversion therapy See *reparative therapy.*

corporal punishment The intentional infliction of pain for a perceived misbehavior.

corporate violence The production of unsafe products and the failure of corporations to provide a safe working environment for their employees.

corporate welfare Laws and policies that benefit corporations.

corporatocracy A system of government that serves the interests of corporations and that involves ties between government and business.

couch-homeless Individuals who do not have a home of their own and who stay at the home of family or friends.

covenant marriage A type of marriage (offered in a few states) that requires premarital counseling and that permits divorce only under condition of fault or after a marital separation of more than two years.

crack A crystallized illegal drug product produced by boiling a mixture of baking soda, water, and cocaine.

crime An act, or the omission of an act, that is a violation of a federal, state, or local criminal law for which the state can apply sanctions.

crime rate The number of crimes committed per 100,000 population.

cultural imperialism The indoctrination into the dominant culture of a society.

cultural lag A condition in which the material part of culture changes at a faster rate than the nonmaterial part.

cultural sexism The ways in which the culture of society perpetuate the subordination of an individual or group based on the sex classification of that individual or group.

culture The meanings and ways of life that characterize a society including beliefs, values, norms, sanctions, and symbols.

culture of poverty The set of norms, values, beliefs, and self-concepts that contribute to the persistence of poverty among the underclass.

cybernation Dominant in postindustrial societies, the use of machines to control other machines.

cycle of abuse A pattern of abuse in which a violent or abusive episode is followed by a makeup period when the abuser expresses sorrow and asks for forgiveness and "one more chance," before another instance of abuse occurs.

date-rape drugs Drugs that are used to render victims incapable of resisting sexual assaults.

deconcentration The redistribution of the population from cities to suburbs and surrounding areas.

decriminalization The removal of criminal penalties for a behavior, as in the decriminalization of drug use.

deep ecology The view that maintaining the earth's natural systems should take precedence over human needs, that nature has a value independent of human existence, and that humans have no right to dominate the earth and its living inhabitants.

Defense of Marriage Act Federal legislation that states that marriage is a legal union between one man and one woman and denies federal recognition of same-sex marriage.

deforestation The conversion of forestland to nonforestland.

deinstitutionalization The removal of individuals with psychiatric disorders from mental hospitals and large residential institutions to outpatient community mental health centers.

demand reduction One of two strategies in the U.S. war on drugs (the other is supply reduction), demand reduction focuses on reducing the demand for drugs through treatment, prevention, and research.

demographic transition theory A theory that attributes population growth patterns to changes in birth rates and death rates associated with the process of industrialization.

dependent variable The variable that the researcher wants to explain; the variable of interest.

deregulation The reduction of government control over, for example, certain drugs.

desertification The degradation of semiarid land, which results in the expansion of desert land that is unusable for agriculture.

deterrence The use of harm or the threat of harm to prevent unwanted behaviors.

de-urbanization An urban survival strategy that involves completely leveling entire urban districts and returning them to nature.

devaluation hypothesis The hypothesis that women are paid less because the work they perform is socially defined as less valuable than the work men perform.

developed countries Countries that have relatively high gross national income per capita and have diverse economies made up of many different industries.

developing countries Countries that have relatively low gross national income per capita, with simpler economies that often rely on a few agricultural products.

discrimination Actions or practices that result in differential treatment of categories of individuals.

divorce mediation A process in which divorcing couples meet with a neutral third party (mediator) who assists the individuals in resolving issues such as property division, child custody, child support, and spousal support in a way that minimizes conflict and encourages cooperation.

domestic partners A status granted to unmarried couples, including gay and lesbian couples, by some states, counties, cities, and workplaces that conveys various rights and responsibilities.

domestic partnership A status that some states, counties, cities, and workplaces grant to unmarried couples, including gay and lesbian couples, which conveys various rights and responsibilities.

domestic terrorism Domestic terrorism, sometimes called insurgent terrorism, occurs when the terrorist act involves victims, targets, institutions, governments, or citizens from one country.

double or triple (multiple) jeopardy The disadvantages associated with being a member of two or more minority groups.

doubling time The time required for a population to double in size from a given base year if the current rate of growth continues.

drug Any substance other than food that alters the structure or functioning of a living organism when it enters the bloodstream.

drug abuse The violation of social standards of acceptable drug use, resulting in adverse physiological, psychological, and/or social consequences.

dual-use technologies Defense-funded technological innovations with commercial and civilian use.

earned income tax credit (EITC) A refundable tax credit based on a working family's income and number of children.

e-commerce The buying and selling of goods and services over the Internet.

e-learning Learning in which, by time or place, the learner is separated from the teacher.

economic institution The structure and means by which a society produces, distributes, and consumes goods and services.

ecosystems The complex and dynamic relationships between forms of life and the environments they inhabit.

ecoterrorism Any crime intended to protect wildlife or the environment that is violent, puts human life at risk, or results in damages of $10,000 or more.

elder abuse The physical or psychological abuse, financial exploitation, or medical abuse or neglect of the elderly.

emotion work Work that involves caring for, negotiating, and empathizing with people.

Employment Nondiscrimination Act (ENDA) A bill that would make it illegal to discriminate based on sexual orientation.

environmental footprint The demand that each person makes on the earth's natural resources, often measured in terms of global hectares (gha) per person.

environmental injustice Also known as **environmental racism,** the tendency for socially and politically marginalized groups to bear the brunt of environmental ills.

environmental racism See *environmental injustice.*

environmental refugees Individuals who have migrated because they can no longer secure a livelihood as a result of deforestation, desertification, soil erosion, and other environmental problems.

epidemiological transition A societal shift from low life expectancy and predominance

of parasitic and infectious diseases to high life expectancy and predominance of chronic and degenerative diseases.

Equal Employment Opportunity Commission (EEOC) A U.S. federal agency charged with ending employment discrimination in the United States that is responsible for enforcing laws against discrimination, including Title VII of the 1964 Civil Rights Act that prohibits employment discrimination on the basis of race, color, religion, sex, or national origin.

equal rights amendment (ERA) The proposed 28th amendment to the Constitution, which states that "equality of rights under the law shall not be denied or abridged by the United States, or by any state, on account of sex."

ethnicity A shared cultural heritage or nationality.

e-waste Discarded electrical appliances and electronic equipment.

experiments A research method that involves manipulating the independent variable to determine how it affects the dependent variable.

expressive roles Roles into which women are traditionally socialized (i.e., nurturing and emotionally supportive roles).

expulsion Occurs when a dominant group forces a subordinate group to leave the country or to live only in designated areas of the country.

extreme poverty Living on less than $1.25 a day.

Faith-Based and Neighborhood Partnerships A program in which faith-based and other neighborhood organizations receive federal funding for programs that serve the needy, such as homeless services and food aid programs.

familism The view that the family unit is more important than individual interests.

family A kinship system of all relatives living together or recognized as a social unit, including adopted people.

Family and Medical Leave Act (FMLA) A federal law that requires companies with 50 or more employees to provide eligible workers (who work at least 25 hours a week and have been working for at least 1 year) with up to 12 weeks of job-protected, unpaid leave so that they can care for an ill child, spouse, or parent; stay home to care for their newborn, newly adopted, or newly placed child; or take time off when they are seriously ill, and up to 26 weeks of unpaid leave to care for a seriously ill or injured family member who is in the Armed Forces, including the National Guard or Reserves.

feminism The belief that men and women should have equal rights and responsibilities.

feminist criminology An approach that focuses on how the subordinate position of women in society affects their criminal behavior and victimization.

feminization of poverty The disproportionate distribution of poverty among women.

fetal alcohol syndrome A syndrome characterized by serious physical and mental handicaps as a result of maternal drinking during pregnancy.

field research Research that involves observing and studying social behavior in settings in which it occurs naturally.

flextime A work arrangement that allows employees to begin and end the workday at different times so long as 40 hours per week are maintained.

forced labor Also known as slavery, any work that is performed under the threat of punishment and is undertaken involuntarily.

four assurances The four assurances refer to a state's commitment to improving teacher quality, raising academic standards, intervening in failing schools, and developing assessment databases in return for federal stimulus dollars.

free trade agreements Pacts between two countries or among a group of countries that make it easier to trade goods across national boundaries by reducing or eliminating restrictions on exports and tariffs (or taxes) on imported goods and protecting intellectual property rights.

future shock The state of confusion resulting from rapid scientific and technological changes that unravel our traditional values and beliefs.

gateway drug A drug (for example, marijuana) that is believed to lead to the use of other drugs (for example, cocaine).

gay-straight alliances (GSAs) School-sponsored clubs for gay teens and their straight peers.

gender The social definitions and expectations associated with being female or male.

gender tourism The recent tendency for definitions of masculinity and femininity to become less clear, resulting in individual exploration of the gender continuum.

gene monopolies Exclusive control over a particular gene as a result of government patents.

gene therapy The transplantation of a healthy gene to replace a defective or missing gene.

genetic engineering The manipulation of an organism's genes in such a way that the natural outcome is altered.

genetic exception laws Laws that require that genetic information be handled separately from other medical information.

Genetic Information Nondiscrimination Act of 2008 A federal law that prohibits discrimination in health coverage or employment based on genetic information.

genetic screening The use of genetic maps to detect predispositions to human traits or disease(s).

Geneva Conventions A set of international treaties that govern the behavior of states during wartime, including the treatment of prisoners of war.

genocide The deliberate, systematic annihilation of an entire nation, people, or ethnic group.

gentrification A type of neighborhood revitalization in which middle- and upper-income individuals buy and rehabilitate older homes in an economically depressed neighborhood.

ghettos In the United States, slums that are occupied primarily by African Americans.

glass ceiling An invisible barrier that prevents women and other minorities from moving into top corporate positions.

glass escalator effect The tendency for men seeking or working in traditionally female occupations to benefit from their minority status.

GLBT See *LGBT*.

global economy An interconnected network of economic activity that transcends national borders.

global warming The increasing average global air temperature, caused mainly by the accumulation of various gases (greenhouse gases) that collect in the atmosphere.

globalization The growing economic, political, and social interconnectedness among societies throughout the world.

green energy Also known as clean energy, energy that is nonpolluting and/or renewable, such as solar power, wind power, biofuel, and hydrogen.

greenhouse gases Gases (primarily carbon dioxide, methane, and nitrous oxide) that accumulate in the atmosphere and act like the glass in a greenhouse, holding heat from the sun close to the earth.

green schools Schools that are "in harmony with the natural environment."

greenwashing The way in which environmentally and socially damaging companies portray their corporate image and products as being "environmentally friendly" or socially responsible.

guerrilla warfare Warfare in which organized groups oppose domestic or foreign governments and their military forces; often involves small groups of individuals who use camouflage and underground tunnels to hide until they are ready to execute a surprise attack.

harm reduction A recent public health position that advocates reducing the harmful consequences of drug use for the user as well as for society as a whole.

hate crime An unlawful act of violence motivated by prejudice or bias.

Head Start Begun in 1965 to help preschool children from the most disadvantaged homes, Head Start provides an integrated program of health care, parental involvement, education, and social services for qualifying children.

health According to the World Health Organization, "a state of complete physical, mental, and social well-being."

heterosexism The institutional and societal reinforcement of heterosexuality as the privileged and powerful norm, based on the belief that heterosexuality is superior to homosexuality.

heterosexuality The predominance of emotional and sexual attraction to individuals of the other sex.

homeschooling The education of children at home instead of in a public or private school.

homophobia Negative attitudes and emotions toward homosexuality and those who engage in same-sex sexual behavior.

homosexuality The predominance of emotional and sexual attraction to individuals of the same sex.

honor killings Murders, often public, as a result of a female dishonoring, or perceived to have dishonored, her family or community.

human capital The skills, knowledge, and capabilities of individuals.

human capital hypothesis The hypothesis that female-male pay differences are a function of differences in women's and men's levels of education, skills, training, and work experience.

human poverty index (HPI) A measure of poverty based on measures of deprivation of a long, healthy life; deprivation of knowledge; and deprivation in decent living standards.

hypothesis A prediction or educated guess about how one variable is related to another variable.

identity theft The use of someone else's identification (for example, social security number, birth date) to obtain credit or other economic rewards.

in vitro fertilization (IVF) The union of an egg and a sperm in an artificial setting such as a laboratory dish.

incapacitation A criminal justice philosophy that argues that recidivism can be reduced by placing offenders in prison so that they are unable to commit further crimes against the general public.

incumbent upgrading Aid programs that help residents of depressed neighborhoods buy or improve their homes and stay in the community.

independent variable The variable that is expected to explain change in the dependent variable.

index offenses Crimes identified by the FBI as the most serious including personal or violent crimes (homicide, assault, rape, and robbery) and property crimes (larceny, motor vehicle theft, burglary, and arson).

individual discrimination The unfair or unequal treatment of individuals because of their group membership.

individualism The tendency to focus on one's individual self-interests and personal happiness rather than on the interests of one's family and community.

industrialization The replacement of hand tools, human labor, and animal labor with machines run by steam, gasoline, and electric power.

infant mortality rate The number of deaths of live-born infants under 1 year of age per 1,000 live births (in any given year).

infrastructure The underlying foundation that enables a city to function, including such things as water and sewer lines, phone lines, electricity cables, sidewalks, streets, bridges, curbs, lighting, and storm drainage systems.

institution An established and enduring pattern of social relationships.

institutional discrimination Discrimination in which the normal operations and procedures of social institutions result in unequal treatment of minorities.

instrumental roles Roles into which men are traditionally socialized (i.e., task-oriented roles).

integration hypothesis A theory that the only way to achieve quality education for all racial and ethnic groups is to desegregate the schools.

intergenerational poverty Poverty that is transmitted from one generation to the next.

internalized homophobia A sense of personal failure and self-hatred among some lesbians and gay men due to social rejection and stigmatization.

Internet An international information infrastructure available through universities, research institutes, government agencies, libraries, and businesses.

Internet piracy Entails illegally downloading or distributing copyrighted material (e.g., music, games, software).

intimate partner violence (IPV) Actual or threatened violent crimes committed against individuals by their current or former spouses, cohabiting partners, boyfriends, or girlfriends.

Jim Crow laws Between 1890 and 1910, a series of U.S. laws enacted to separate blacks from whites by prohibiting blacks from using "white" buses, hotels, restaurants, and drinking fountains.

job burnout Prolonged job stress that can cause or contribute to high blood pressure, ulcers, headaches, anxiety, depression, and other health problems.

job exportation The relocation of jobs to other countries where products can be produced more cheaply.

Kyoto Protocol The first international agreement to place legally binding limits on greenhouse gas emissions from developed countries.

labor unions Worker advocacy organizations originally developed to protect workers and represent them at negotiations between management and labor.

latent functions Consequences that are unintended and often hidden.

least developed countries The poorest countries of the world.

legalization Making prohibited behaviors legal; for example, legalizing drug use or prostitution.

LGBT or GLBT Terms often used to refer collectively to lesbians, gays, bisexuals, and transgendered individuals.

life expectancy The average number of years that individuals born in a given year can expect to live.

light pollution Artificial lighting that is annoying, unnecessary, and/or harmful to life forms on earth.

living apart together (LAT) relationships An emerging family form in which couples—married or unmarried—live apart in separate residences.

living wage laws Laws that require state or municipal contractors, recipients of public subsidies or tax breaks, or, in some cases, all businesses to pay employees wages that are significantly above the federal minimum, enabling families to live above the poverty line.

long-term unemployment rate The share of the unemployed who have been out of work for 27 weeks or more.

malware A general term that includes any spyware, viruses, and adware that is installed on an owner's computer without their knowledge.

managed care Any medical insurance plan that controls costs through monitoring and controlling the decisions of health care providers.

manifest functions Consequences that are intended and commonly recognized.

marital decline perspective A pessimistic view of the current state of marriage that includes the beliefs that (1) personal happiness has become more important than marital commitment and family obligations, and (2) the decline in lifelong marriage and the increase in single-parent families have contributed to a variety of social problems.

marital resiliency perspective A view of the current state of marriage that includes the beliefs that (1) poverty, unemployment, poorly funded schools, discrimination, and the lack of basic services (such as health insurance and child care) represent more serious threats to the well-being of children and adults than does the decline in married two-parent families, and (2) divorce provides a second chance for happiness for adults and an escape from dysfunctional and aversive home environments for many children.

master status The status that is considered the most significant in a person's social identity.

maternal mortality rate A measure of deaths that result from complications associated with pregnancy, childbirth, and unsafe abortion.

McDonaldization The process by which principles of the fast food industry (efficiency, calculability, predictability, and control through technology) are being applied to more sectors of society, particularly the workplace.

means-tested programs Assistance programs that have eligibility requirements based on income.

mechanization Dominant in an agricultural society, the use of tools to accomplish tasks previously done by hand.

Medicaid A public health insurance program, jointly funded by the federal and state governments, that provides health insurance coverage for the poor who meet eligibility requirements.

medicalization Defining or labeling behaviors and conditions as medical problems.

Medicare A federally funded public insurance program that provides health insurance benefits to the elderly, disabled workers and their dependents, and people with advanced kidney disease.

megacities Urban areas with 10 million residents or more.

membership communities Internet sites where participation requires membership and members regularly communicate with one another for personal and/or professional reasons.

mental health The successful performance of mental function, resulting in productive activities, fulfilling relationships with other people, and the ability to adapt to change and to cope with adversity.

mental illness All mental disorders, which are health conditions that are characterized by alterations in thinking, mood, and/or behavior associated with distress and/or impaired functioning and that meet specific criteria (such as level of intensity and duration) specified in *The Diagnostic and Statistical Manual of Mental Disorders*.

meta-analysis Meta-analysis combines the results of several studies addressing a research question; that is, it is the analysis of analyses.

metropolitan area Also known as a metropolis, a densely populated core area together with adjacent communities.

microcredit programs The provision of loans to people who are generally excluded from traditional credit services because of their low socioeconomic status.

micropolitan area A small city (between 10,000 and 50,000 people) located beyond congested metropolitan areas.

military-industrial complex A term used by Dwight D. Eisenhower to connote the close association between the military and defense industries.

Millennium Development Goals Eight goals that comprise an international agenda for reducing poverty and improving lives.

minority group A category of people who have unequal access to positions of power, prestige, and wealth in a society and who tend to be targets of prejudice and discrimination.

modern racism A subtle form of racism that involves the belief that serious discrimination in America no longer exists, that any continuing racial inequality is the fault of minority group members, and that the demands for affirmative action for minorities are unfair and unjustified.

monogamy Marriage between two partners; the only legal form of marriage in the United States.

morbidity Illnesses, symptoms, and the impairments they produce.

mortality Death.

motherhood penalty The tendency for women with children, particularly young children, to be disadvantaged in hiring, wages, and the like compared to women without children.

multicultural education Education that includes all racial and ethnic groups in the school curriculum, thereby promoting awareness and appreciation for cultural diversity.

multiple chemical sensitivity Also known as "environmental illness," a condition whereby individuals experience adverse reactions when exposed to low levels of chemicals found in everyday substances.

mutually assured destruction (MAD) A Cold War doctrine referring to the capacity of two nuclear states to destroy each other, thus reducing the risk that either state will initiate war.

National Coming Out Day Celebrated on October 11, this day is recognized in many countries as a day to raise awareness of the LGBT population and foster discussion of gay rights issues.

naturalized citizens Immigrants who apply for and meet the requirements for U.S. citizenship.

needle exchange programs Programs designed to reduce transmission of HIV by providing intravenous drug users with new, sterile syringes in exchange for used, contaminated syringes.

neglect A form of abuse involving the failure to provide adequate attention, supervision, nutrition, hygiene, health care, and a safe and clean living environment for a minor child or a dependent elderly individual.

net neutrality A principle that holds that Internet users should be able to visit any web site and access any content without Internet service provider interference.

write what MADD means
Mothers against drunk driving

new urbanism A movement in urban planning that approaches the idea of sustainable urban communities with the goal of raising the quality of life for all those in the community by creating compact communities with a sustainable infrastructure.

no-fault divorce A divorce that is granted based on the claim that there are irreconcilable differences within a marriage (as opposed to one spouse being legally at fault for the marital breakup).

norms Socially defined rules of behavior including folkways, mores, and laws.

nuclear nonproliferation Efforts to prevent the spread of nuclear weapons, or the materials and technology necessary for the production of nuclear weapons.

nuclear winter The predicted result of a thermonuclear war whereby thick clouds of radioactive dust and particles would block out vital sunlight, lower temperature in the Northern Hemisphere, and lead to the death of most living things on earth.

objective element of a social problem Awareness of social conditions through one's own life experiences and through reports in the media.

occupational sex segregation The concentration of women in certain occupations and men in other occupations.

organized crime Criminal activity conducted by members of a hierarchically arranged structure devoted primarily to making money through illegal means.

outsource See *outsourcing.*

outsourcing A practice in which a business subcontracts with a third party, often in low-wage countries such as China and India, for services.

overt discrimination Discrimination that occurs because of an individual's own prejudicial attitudes.

parental alienation syndrome (PAS) An emotional and psychological disturbance in which children engage in exaggerated and unjustified denigration and criticism of a parent.

parity In health care, a concept requiring equality between mental health care insurance coverage and other health care coverage.

parole Parole entails release from prison, for a specific time period and subject to certain conditions, before the inmate's sentence is finished.

partial birth abortions Also called an intact dilation and extraction (D & E) abortion, the procedure may entail delivering the limbs and the torso of the fetus before it has expired.

patriarchy A male-dominated family system that is reflected in the tradition of wives taking their husband's last name and children taking their father's name.

pink-collar jobs Jobs that offer few benefits, often have low prestige, and are disproportionately held by women.

planned obsolescence The manufacturing of products that are intended to become inoperative or outdated in a fairly short period of time.

pluralism A state in which racial and ethnic groups maintain their distinctness but respect each other and have equal access to social resources.

polyandry The concurrent marriage of one woman to two or more men.

polygamy A form of marriage in which one person may have two or more spouses.

polygyny A form of marriage in which one husband has more than one wife.

population density The number of people per unit of land area.

population momentum Continued population growth as a result of past high fertility rates that have resulted in a large number of young women who are currently entering their childbearing years.

postindustrialization The shift from an industrial economy dominated by manufacturing jobs to an economy dominated by service-oriented, information-intensive occupations.

postmodernism A worldview that questions the validity of rational thinking and the scientific enterprise.

post-traumatic stress disorder (PTSD) A set of symptoms that may result from any traumatic experience, including crime victimization, war, natural disasters, or abuses.

prejudice Negative attitudes and feelings toward or about an entire category of people.

primary deviance Deviant behavior committed before a person is caught and labeled an offender.

primary group Usually small numbers of individuals characterized by intimate and informal interaction.

primordial explanations Those explanations that emphasize the existence of "ancient hatreds" rooted in deep psychological or cultural differences between ethnic groups, often involving a history of grievance and victimization, real or imagined, by the enemy group.

probation The conditional release of an offender who, for a specific time period and subject to certain conditions, remains under court supervision in the community.

pronatalism A cultural value that promotes having children.

public housing Federally subsidized housing that is owned and operated by local public housing authorities (PHAs).

race A category of people who are believed to share distinct physical characteristics that are deemed socially significant.

racial profiling The law enforcement practice of targeting suspects on the basis of race.

racism The belief that race accounts for differences in human character and ability and that a particular race is superior to others.

Racism 2.0 A form of racism that allows for and celebrates the achievements of individuals of color who are viewed as having "transcended" their minority status.

radical environmental movement A grassroots movement of individuals and groups that employ unconventional and often illegal means of protecting wildlife or the environment.

recession A significant decline in economic activity spread across the economy and lasting for at least 6 months.

recidivism A return to criminal behavior by a former inmate, most often measured by re-arrest, re-conviction, or re-incarceration.

refined divorce rate The number of divorces per 1,000 married women.

regionalism A form of collaboration among central cities and suburbs that encourages local governments to share common responsibility for common problems.

registered partnerships Federally recognized relationships that convey most but not all the rights of marriage.

rehabilitation A criminal justice philosophy that argues that recidivism can be reduced by changing the criminal through such programs as substance abuse counseling, job training, education, and so on.

relative poverty The lack of material and economic resources compared with some other population.

reparative therapy or conversion therapy Various therapies that are aimed at changing homosexuals' sexual orientation.

replacement-level fertility The level of fertility at which a population exactly replaces itself from one generation to the next; currently, the number is 2.1 births per woman (slightly more than 2 because not all female children will live long enough to reach their reproductive years).

restorative justice A philosophy primarily concerned with reconciling conflict between the victim, the offender, and the community.

revolving door The practice of employees cycling between roles in an industry, and roles in government that influence that industry.

role The set of rights, obligations, and expectations associated with a status.

sample A portion of the population, selected to be representative so that the information from the sample can be generalized to a larger population.

sanctions Social consequences for conforming to or violating norms.

school vouchers Tax credits that are transferred to the public or private school that parents select for their child.

science The process of discovering, explaining, and predicting natural or social phenomena.

second shift The household work and child care that employed parents (usually women) do when they return home from their jobs.

secondary deviance Deviance that results from being caught and labeled as an offender.

secondary group Involving small or large numbers of individuals, groups that are task oriented and are characterized by impersonal and formal interaction.

Section 8 housing A housing assistance program in which federal rent subsidies are provided either to tenants (in the form of certificates and vouchers) or to private landlords.

security dilemma A characteristic of the international state system that gives rise to unstable relations between states; as State A secures its borders and interests, its behavior may decrease the security of other states and cause them to engage in behavior that decreases A's security.

segregation The physical separation of two groups in residence, workplace, and social functions.

selective primary health care An approach to health care that focuses on using specific interventions to target specific health problems.

self-fulfilling prophecy A concept referring to the tendency for people to act in a manner consistent with the expectations of others.

serial monogamy A succession of marriages in which a person has more than one spouse over a lifetime but is legally married to only one person at a time.

sex A person's biological classification as male or female.

sexism The belief that innate psychological, behavioral, and/or intellectual differences exist between women and men and that these differences connote the superiority of one group and the inferiority of the other.

sexual harassment In reference to workplace harassment, when an employer requires sexual favors in exchange for a promotion, salary increase, or any other employee benefit and/or the existence of a hostile environment that unreasonably interferes with job performance.

sexual orientation The classification of individuals as heterosexual, bisexual, or homosexual, based on their emotional and sexual attractions, relationships, and self-identity.

shaken baby syndrome A form of child abuse whereby the caretaker shakes a baby to the point of causing the child to experience brain or retinal hemorrhage.

single-payer health care A health care system in which a single tax-financed public insurance program replaces private insurance companies.

slums Concentrated areas of poverty and poor housing in urban areas.

smart growth A strategy for managing urban sprawl that serves the economic, environmental, and social needs of communities.

social group Two or more people who have a common identity, interact, and form a social relationship.

social movement An organized group of individuals with a common purpose to either promote or resist social change through collective action.

social problem A social condition that a segment of society views as harmful to members of society and in need of remedy.

socialism An economic system characterized by state ownership of the means of production and distribution of goods and services.

sociological imagination The ability to see the connections between our personal lives and the social world in which we live.

sodomy Oral and anal sexual acts.

state The organization of the central government and government agencies such as the military, police, and regulatory agencies.

State Children's Health Insurance Program (SCHIP) A public health insurance program, jointly funded by the federal and state governments, that provides health insurance coverage for children whose families meet income eligibility standards.

status A position that a person occupies within a social group.

STEM An acronym for science, technology, engineering, and mathematics.

stem cells Undifferentiated cells that can produce any type of cell in the human body.

stereotype threat The tendency of minorities and women to perform poorly on high-stakes tests because of the anxiety created by the fear that a negative performance will validate societal stereotypes about one's member group.

stereotypes Exaggerations or generalizations about the characteristics and behavior of a particular group.

stigma A discrediting label that affects an individual's self-concept and disqualifies that person from full social acceptance.

structural sexism The ways in which the organization of society, and specifically its institutions, subordinate individuals and groups based on their sex classification.

structure The way society is organized including institutions, social groups, statuses, and roles.

structured choice Choices that are limited by the structure of society.

subjective element of a social problem The belief that a particular social condition is harmful to society, or to a segment of society, and that it should and can be changed.

subprime mortgages High-interest or adjustable-rate mortgages that require little money down and are issued to borrowers with poor credit ratings or limited credit history.

suburbanization The development of urban areas outside central cities, and the movement of populations into these areas.

suburbs Urban areas surrounding central cities.

Supplemental Nutrition Assistance Program (SNAP) The largest U.S. food assistance program.

supply reduction One of two strategies in the U.S. war on drugs (the other is demand reduction), supply reduction concentrates on reducing the supply of drugs available on the streets through international efforts, interdiction, and domestic law enforcement.

survey research A research method that involves eliciting information from respondents through questions.

sustainable development Societal development that meets the needs of current generations without threatening the future of subsequent generations.

sweatshops Work environments that are characterized by less-than-minimum-wage pay, excessively long hours of work (often without overtime pay), unsafe or inhumane working conditions, abusive treatment of workers by employers, and/or the lack of worker organizations aimed to negotiate better working conditions.

symbol Something that represents something else.

technological dualism The tendency for technology to have both positive and negative consequences.

technological fix The use of scientific principles and technology to solve social problems.

technology Activities that apply the principles of science and mechanics to the solutions of a specific problem.

technology-induced diseases Diseases that result from the use of technological devices, products, and/or chemicals.

telepresencing A sophisticated technology which allows life-sized participants in the virtual presence of one another to realistically communicate through broadcast quality sound and images.

telework A work arrangement involving the use of information technology that allows employees to work part- or full-time at home or at a satellite office.

Temporary Assistance for Needy Families (TANF) A federal cash welfare program that involves work requirements and a 5-year lifetime limit.

terrorism The premeditated use or threatened use of violence by an individual or group to gain a political objective.

therapeutic cloning Use of stem cells from human embryos to produce body cells that can be used to grow needed organs or tissues.

therapeutic communities Organizations in which approximately 35 to 500 individuals reside for up to 15 months to abstain from drugs, develop marketable skills, and receive counseling.

Title I Part of the Elementary and Secondary Education Act of 1965, Title I provides assistance to school districts with a high proportion of poor children.

total fertility rates The average lifetime number of births per woman in a population.

transgender individual A person whose sense of gender identity (masculine or feminine) is inconsistent with his or her birth sex (male or female).

transnational corporations Also known as multinational corporations, corporations that have their home base in one country and branches, or affiliates, in other countries.

transnational crime Criminal activity that occurs across one or more national borders.

transnational terrorism Terrorism that occurs when a terrorist act in one country involves victims, targets, institutions, governments, or citizens of another country.

under-5 mortality rate The rate of deaths of children under age 5.

underclass A persistently poor and socially disadvantaged group.

underemployment Unemployed workers as well as (1) those working part-time but who wish to work full-time, (2) those who want to work but have been discouraged from searching by their lack of success, and (3) others who are neither working nor seeking work but who want and are available to work and have looked for employment in the last year. Also refers to the employment of workers with high skills and/or educational attainment working in low-skill or low-wage jobs.

unemployment To be currently without employment, actively seeking employment, and available for employment, according to U.S. measures of unemployment.

unemployment insurance A federal-state program that temporarily provides laid-off workers with a portion of their paychecks.

union density The percentage of workers who belong to unions.

universal health care Health care to all citizens.

urban area A spatial concentration of people whose lives are centered around nonagricultural activities.

urban sprawl The ever-increasing outward growth of urban areas that results in the loss of green open spaces, the displacement and endangerment of wildlife, traffic congestion and noise, and pollution.

urbanization The transformation of a society from a rural to an urban one.

values Social agreements about what is considered good and bad, right and wrong, desirable and undesirable.

variable Any measurable event, characteristic, or property that varies or is subject to change.

victimless crimes Illegal activities that have no complaining participant(s) and are often thought of as crimes against morality, such as prostitution.

war Organized armed violence aimed at a social group in pursuit of an objective.

wealth The total assets of an individual or household minus liabilities.

wealthfare Laws and policies that benefit the rich.

weapons of mass destruction (WMD) Chemical, biological, and nuclear weapons that have the capacity to kill large numbers of people indiscriminately.

Web 2.0 A platform for millions of users to express themselves online in the common areas of cyberspace.

white-collar crime Includes both *occupational crime,* in which individuals commit crimes in the course of their employment, and *corporate crime,* in which corporations violate the law in the interest of maximizing profit.

workers' compensation Also known as workers' comp, an insurance program that provides medical and living expenses for people with work-related injuries or illnesses.

Workforce Investment Act Legislation passed in 1998 that provides a wide array of programs and services designed to assist individuals to prepare for and find employment.

working poor Individuals who spend at least twenty-seven weeks per year in the labor force (working or looking for work) but whose income falls below the official poverty level.

References

Chapter 1

Anwar, Yasmin. 2009. "Fighting Global Poverty Is Fastest-Growing Minor." *UC Berkeley News*, March 10. Available at http://www.berkeley.edu/news

Associated Press. 2006 (September 7). "Florida Appeals Court Upholds Ban of Veil in Driver's License Photo." Religious News. Pew Forum on Religion and Public Life. Available at http://pewforum.org/

Batsell, Jake. 2002. USA: "Students Campaign for Coffee in Good Conscience." *The Seattle Times*, March 17.

BBC (British Broadcasting Company). 1989. "On This Day: Massacre at Tiananmen Square." Available at http://news.bbc.co.uk

Beeby, Dean. 2005. "Census Includes Same-Sex Marriage Question." *Globe and Mail*, April 18. Available at http://globeandmail.com

Berkowitz, Elana. 2005. "They Say Tomato, Students Say Justice." *Campus Progress*, March 18.

Blumer, Herbert. 1971. "Social Problems as Collective Behavior." *Social Problems* 8(3):298–306.

Burns, Laura. 2009. "Students Oppose Contracts with Russell Athletic." *The Phoenix*, March 25. Available at http://loyolaphoenix.com

Caldas, Stephen, and Carl L. Bankston III. 1999. "Black and White TV: Race, Television Viewing, and Academic Achievement." *Sociological Spectrum* 19:39–61.

Canedy, Dana. 2002. "Lifting Veil for Photo ID Goes Too Far, Driver Says." *New York Times*, June 27. Available at http://www.nytimes.com

Centers for Disease Control and Prevention. 2008 (August 1). *Trends in HIV- and STD-Related Risk Behaviors Among High School Students—United States, 1991–2007.* Morbidity and Mortality Weekly Report 57(30):817–822.

Dolan, Maura. 2009. "California Supreme Court Looks Unlikely to Kill Proposition 8." *Los Angeles Times*, March 6. Available at http://www.latimes.com

The Eleanor Roosevelt Papers. Teaching Eleanor Roosevelt: American Youth Congress. Available at http://www.nps.gov

Engel, Robin Shepard, and Robert E. Worden. 2003. "Police Officers' Attitudes, Behavior, and Supervisory Influences: An Analysis of Problem Solving." *Criminology* 41:131–166.

Ferrara, Leigh, Ann Friedman, April Rabkin, Amaya Rivera, Cameron Scott, Marisa Taylor, and Marcus Wohlsen. 2006. "Extra Credit: Campus Activism 2006." *Mother Jones News*, September/October. Available at http://motherjones.com

Fleming, Zachary. 2003. "The Thrill of It All." In *In Their Own Words*, ed. Paul Cromwell, 99–107. Los Angeles, CA: Roxbury.

Garrison, Jessica. 2009. "Gay Couples Hold Vigils Urging Justices to End Prop. 8." *Los Angeles Times*, March 5. Available at http://articles.latimes.com

Gee, Gilbert, Barbara Curbow, Margaret Ensminger, Joan Griffin, et al. 2006. "Are You Positive? The Relationship of Minority Composition to Workplace Drug and Alcohol Testing." *Journal of Drug Issues* 35(4):755–779.

GSAN (Gay-Straight Alliance Network). 2009. Available at http://gsanetwork.org

Hass, Christopher. 2009. "A Call to Serve: President Obama Signs the Edward M. Kennedy Serve America Act." Available at http://my.barackobama.com

Hewlett, Sylvia Ann. 1992. *When the Bough Breaks: The Cost of Neglecting Our Children.* New York: Harper Perennial.

Jacobs, Bruce A. 2003. "Researching Crack Dealers." In *In Their Own Words*, ed. Paul Cromwell, 1–11. Los Angeles, CA: Roxbury.

Jacobs, Dennis. 2008 (March 13). "Economy Widely Viewed as Most Important Problem." Available at http://www.gallup.com

James, Susan Donaldson. 2007. "Students Use Civil Rights Tactics to Combat Global Warming." *ABC News*, January 19. Available at http://abcnews.go.com

Jekielek, Susan M. 1998. "Parental Conflict, Marital Disruption, and Children's Emotional Well-Being." *Social Forces* 76:905–935.

KillerCoke.org. 2005. "Coke Campaign at Grinell College, Iowa." Available at http://killercoke.org

Kmec, Julie A. 2003. "Minority Job Concentration and Wages." *Social Problems* 50:38–59.

Liebowitz, Debra K. 2009 (May 5). "Student Recycling Activist Raises the Bar in Miami Beach." Available at http://www.miamiherald.com

Merton, Robert K. 1968. *Social Theory and Social Structure.* New York: Free Press.

Mills, C. Wright. 1959. *The Sociological Imagination.* London: Oxford University Press.

Newman, Jessica Clark, Don Des Jarlais, Charles F. Turner, Jay Gribble, Philip Cooley, and Denice Paone. 2002. "The Differential Effects of Face-to-Face and Computer Interview Modes." *American Journal of Public Health* 92(2):294–297.

Obama, Barack. 2009. "Inaugural Address." Available at http://www.whitehouse.gov

Palacios, Wilson R., and Melissa E. Fenwick. 2003. "'E'" Is for Ecstasy." In *In Their Own Words*, ed. Paul Cromwell, 277–283. Los Angeles, CA: Roxbury.

Pryor, John, Sylvia Hurtado, Linda DeAngelo, Jessica Sharkness, Laura C. Romero, William S. Korn, and Serge Tran. 2008. *The American Freshman: National Norms for Fall 2008.* Los Angeles, CA: Higher Education Research Institute.

Reiman, Jeffrey. 2007. *The Rich Get Richer and the Poor Get Prison.* 8th ed. Boston, MA: Allyn and Bacon.

Rifkind, Hugo. 2009. "Student Activism Is Back. *Times Online*, February 16. Available at http://women.timesonline.co.uk

Saad, Lydia. 2008 (August 13). "U.S. Satisfaction Steady at a Dismal 17%." Available at http://www.gallup.com

Saad, Lydia. 2007 (January 5). "Local TV Is No. 1 Source of News for Americans." Gallup Poll. The Gallup Organization. Available at http://www.galluppoll.com

Schlosser, Jim. 2000. "Activist Recalls 'Catalyst for Civil Rights.'" *Greensboro News and Record*, February 2. Available at http://www.sitins.com

*The authors and Wadsworth acknowledge that some of the Internet sources may have become unstable; that is, they are no longer hot links to the intended reference. In that case, the reader may want to access the article through the search engine or archives of the homepage cited (e.g., fbi.gov, cbsnews.com).

The Scranton Report. 1970. *The Report of the President's Commission on Campus Unrest.* Washington, DC: U.S. Government Printing Office.

Shaw, Desair. 2009. "Survey Says: Many Newspapers Won't Be Missed." *USA Today,* March 13. Available at http://content .usatoday.com

Simi, Pete, and Robert Futrell. 2009. "Negotiating White Power." *Social Problems* 56(1):98–110.

Stark, Lisa, Brian Hartman, and Kate Barrett. 2009. "Major Expansion Announced in Peanut Recall." ABC News, January 28. Available at http://abcnews.go.com

Sykes, Marvin. 1960. "Negro College Students Sit at Woolworth Lunch Counter." *Greensboro Record,* February 2. Available at http://www.sitins.com

Thomas, W. I. 1931/1966. "The Relation of Research to the Social Process." In *W. I. Thomas on Social Organization and Social Personality,* ed. Morris Janowitz, 289–305. Chicago: University of Chicago Press.

Troyer, Ronald J., and Gerald E. Markle. 1984. "Coffee Drinking: An Emerging Social Problem." *Social Problems* 31:403–416.

Ukers, William H. 1935. *All About Tea,* vol. 1. New York: Tea and Coffee Trade Journal Co.

United Nations Development Programme. 2008. *2008 Human Development Report.* Available at http://hdr.undp.org

Weir, Sara, and Constance Faulkner. 2004. *Voices of a New Generation: A Feminist Anthology.* Boston, MA: Pearson Education Inc.

Wilson, John. 1983. *Social Theory.* Englewood Cliffs, NJ: Prentice Hall.

Chapter 2

Alan Guttmacher Institute. 2004. *New Evidence from Africa, Asia, and Latin America Underscores Impact of Violence against Women.* News release, December 10. Available at http://www.guttmacher.org

Allen, P. L. 2000. *The Wages of Sin: Sex and Disease, Past and Present.* Chicago: University of Chicago Press.

Allen, Terry J. 2007 (March 7). "Merck's Murky Dealings: HPV Vaccine Lobby Backfires." *CorpWatch.* Available at www. corpwatch.org

Alvarez, Lizette, and Erik Eckholm. 2009. "Purple Heart Is Ruled Out for Traumatic Stress." *New York Times,* January 27, p. A1.

American College Health Association. 2008. "American College Health Association National College Health Assessment Fall 2007 Reference Group Data Report." Baltimore: American College Health Association.

American Psychiatric Association. 2000. *Diagnostic and Statistical Manual of Mental Disorders.* 4th ed., text revision.

Washington, DC: American Psychiatric Association.

Associated Press. 2007. "Mom: Social workers threaten to take away 254-pound son." *The Daily Reflector,* March 23, p. B2.

AVERT. 2004 (November 30). *HIV and AIDS Discrimination and Stigma.* Available at http://www.avert.org/aidsstigma.htm

Barker, Kristin. 2002. "Self-Help Literature and the Making of an Illness Identity: The Case of Fibromyalgia Syndrome (FMS)." *Social Problems* 49(3):279–300.

Beckel, Michael. 2009. "Will 1.2 Million a Day Convince Congress to Buy Big Pharma's Rx for Change?" Capitol Eye Blog, June 25. Center for Responsive Politics. Available at http://www.opensecrets.org

Capriccioso, Rob. 2006. "Counseling Crisis." *Inside Higher Ed,* March 13. Available at http://insidehighered.com/news

Centers for Disease Control and Prevention. 2009. *HIV/AIDS Surveillance Report, 2007.* vol. 19. Available at http://www.cdc.gov

Centers for Disease Control and Prevention. 2007. (November 9). "Syringe Exchange Programs—United States, 2005." *Morbidity and Mortality Weekly Report* 56(44): 1164–1167.

Centers for Medicare and Medicaid Services. 2009 (May 17). *National Health Expenditures: Overview. U.S. Department of Health and Human Services.* Available at http://www.cms.hhs.gov

Chandra, Anita, and Cynthia S. Minkovitz. 2006. "Stigma Starts Early: Gender Differences in Teen Willingness to Use Mental Health Services." *Journal of Adolescent Health* 38(6):754.e1–754.e8.

Children's Defense Fund. 2000. *The State of America's Children Yearbook 2000.* Available at http://www.childrensdefense.org

CNN Heroes. 2009 (May 25). "Once-Suicidal Man Becomes Veterans' Lifeline." Available at http://www.cnn.com

Cockerham, William C. 2007. *Medical Sociology.* 10th ed. Upper Saddle River, NJ: Prentice Hall.

Combes, Marie-Laure. 2007. "French Government Wants Food Warning." Associated Press, March 2. Available at http://www.washingtonpost.com

Cotton, Ann, Kenneth R. Stanton, Zoltan J. Acs, and Mary Lovegrove. 2007. *2006 UB Obesity Report Card.* Baltimore, MD: University of Baltimore. Available at http:// www.ubalt.edu

Davis, Karen, Cathy Schoen, Stephen C. Schoenbaum, Michelle M. Doty, Alyssa L. Holmgren, Jennifer L. Kriss, and Katherine K. Shea. 2007 (May). *Mirror, Mirror on the Wall: An International Update on the Comparative Performance of American Health Care.* Commonwealth Fund. Available at http://commonwealthfund.org

DeNavas-Walt, Carmen, Bernadette D. Proctor, and Jessica Smith. 2009 (September). *Income, Poverty, and Health Insurance in the United States: 2008.* Current Population Reports P60-236. U.S. Census Bureau. Available at http://www.census.gov

Diamond, Catherine, and Susan Buskin. 2000. "Continued Risky Behavior in HIV-Infected Youth." *American Journal of Public Health* 90(1):115–118.

Dingfelder, Sadie F. 2009a (June). "Stigma: Alive and Well." *Monitor on Psychology* 40(6):56.

Dingfelder, Sadie F. 2009b (June). "The Military's War on Stigma." *Monitor on Psychology* 40(6):52.

Egwu, Igbo N. 2002. "Health Promotion Challenges in Nigeria: Globalization, Tobacco Marketing, and Policies." In *Health Communication in Africa: Contexts, Constraints, and Lessons,* eds. A. O. Alali and B. A. Jinadu, 40–55. Lanham, MD: University Press of America.

Epstein, Samuel S. 2006. *What's in Your Milk?* Victoria, BC, Canada: Trafford Publishing.

Families USA. 2007 (January 9). "No Bargain: Medicare Drug Plans Deliver Higher Prices." Available at http://www.familiesusa.org

Feachum, Richard G. A. 2000. "Poverty and Inequality: A Proper Focus for the New Century." *International Journal of Public Health* 78:1–2.

Feldman, Debra S., Dennis H. Novack, and Edward Gracely. 1998. "Effects of Managed Care on Physician Patient Relationships, Quality of Care, and the Ethical Practice of Medicine: A Physician Survey." *Archives of Internal Medicine* 158:1626–1632.

Fischer, Edward, and Amerigo Farina. 1995. "Attitudes Toward Seeking Professional Psychological Help: A Shortened Form and Considerations for Research." *Journal of College Student Development* 36(4):368–373.

Gallagher, Robert P. 2006. *National Survey of Counseling Center Directors.* The International Association of Counseling Services. Available at http://www.iacsinc.org

Goldberg, Lisa. 2005. "Educating Seniors on Safe Sex." *Baltimore Sun,* March 1. Available at http://www.baltimoresun.com

Goldstein, Michael S. 1999. "The Origins of the Health Movement." In *Health, Illness, and Healing: Society, Social Context, and Self,* eds. Kathy Charmaz and Debora A. Paterniti, 31–41. Los Angeles, CA: Roxbury.

Grand Island Independent. 2001 (April 20). "Access to Free Health Care Means Crime Does Pay for Some Sick Inmates." Available at http://theIndependent.com

Grossman, Amy. 2009. "A Birth Pill." *New York Times,* May 9. Available at http://www. nytimes.com

Gruttadaro, Darcy. 2005. "Federal Leaders Call on Schools to Help." *NAMI Advocate,* Winter, pp. 7–9.

Hadley, Jack. 2007. "Insurance Coverage, Medical Care Use, and Short-Term Health Changes Following an Unintentional Injury or the Onset of a Chronic Condition." *JAMA* 297(10):1073–1084.

Halperin, D. T., M. J. Steiner, M. M. Cassell, E. C. Green, N. Hearst, D. Kirby, H. D. Gayle, and W. Cates. 2004. "The Time Has Come for Common Ground on Preventing Sexual Transmission of HIV." *The Lancet* 364:1913–1915.

Hawkes, Corinna. 2006. "Uneven Dietary Development: Linking the Policies and Processes of Globalization with the Nutrition Transition, Obesity, and Diet-Related Chronic Diseases." *Globalization and Health* 2:4.

Harris, Paul. 2009. "Whistleblower Tells of America's Hidden Nightmare for Its Sick Poor." The Observer, July 26. Available at http://www.guardian.co.uk

Harrison, Joel A. 2008. "How Much Is the Sick U.S. Health Care System Costing You?" *Dollars and Sense*, May/June. Available at http://dollarsandsense.com

Harrison, Mary M. 2002. "Silence That Promotes Stigma." *Teaching Tolerance* 21:52–53.

Himmelstein, D. U., D. Thorne, E. Warren, and S. Woolhandler. 2009. "Medical Bankruptcy in the United States, 2007: Results of a National Study." *The American Journal of Medicine* 122(8):741–46.

Hippert, Christine. 2002. "Multinational Corporations, the Politics of the World Economy, and Their Effects on Women's Health in the Developing World: A Review." *Health Care for Women International* 23:861–869.

Hollander, D. 2005. "Unprotected Anal Sex is Not Uncommon Among Men with HIV Infection." *Perspectives on Sexual and Reproductive Health* 37(1). Available at http://www.guttmacher.org

Hollander, D. 2003. "Having Two Parents Helps." *Perspectives on Sexual and Reproductive Health* 35(2):60.

Honberg, Ron. 2005. "Decriminalizing Mental Illness." *NAMI Advocate*, Winter, pp. 4–5, 7.

Hudson, C. G. 2005. "Socioeconomic Status and Mental Illness: Tests of the Social Causation and Selection Hypothesis." *American Journal of Orthopsychiatry* 75(1):3–18.

Hull, Anne, and Dana Priest. 2007. "'It Is Just Not Walter Reed': Soldiers Share Troubling Stories of Military Health Care across U.S." *Washington Post*, March 5. Available at http://washingtonpost.com

Human Rights Watch. 2005 (March 3). *U.S. Gag on Needle Exchange Harms U.N. AIDS Efforts*. Available at http://www.hrw.org

Human Rights Watch. 2003. *Ill-Equipped: U.S. Prisons and Offenders with Mental Illness*. Available at http://www.hrw.org

Institute of Medicine. 2004. *Insuring America's Health: Principles and Recommendations*. Washington, DC: National Academies Press.

James, Cara, Megan Thomas, Marsha Lillie-Blanton, and Rachel Garfield. 2007. *Key Facts: Race, Ethnicity, and Medical Care*. Kaiser Family Foundation. Available at http://www.kff.org

Jenkins, Chris L. 2008. "Law Equalizes Coverage for Mental, Physical Care." *Washington Post*, October 10, p. B01.

Johnson, Cary Alan. 2007. *Off the Map: How HIV/AIDS Programming Is Failing Same-Sex Practicing People in Africa*. New York: International Gay and Lesbian Human Rights Commission.

Johnson, Tracy L., and Elizabeth Fee. 1997. "Women's Health Research: An Introduction." In *Women's Health Research: A Medical and Policy Primer*, eds. Florence P. Haseltine and Beverly Greenberg Jacobson, 3–26. Washington, DC: Health Press International.

Kaiser Commission on Medicaid and the Uninsured. 2004 (November). *The Uninsured: A Primer Key Findings About Americans Without Health Insurance*. Washington, DC: Kaiser Family Foundation.

Kaiser Health Poll Report. 2004 (July-August). *Views of Managed Care and Customer Service*. Available at http://www.kff.org

Kaiser Family Foundation. 2009a. *The Global HIV/AIDS Epidemic Fact Sheet 2009*. Available at http://www.kff.org.

Kaiser Family Foundation. 2009b (June). *Health Tracking Poll*. Available at http:www.kff.org.

Kaiser Family Foundation. 2008. *Employer Health Benefits 2008 Annual Survey*. Available at http://www.kff.org

Kaiser Family Foundation. 2007. *How Changes in Medical Technology Affect Health Care Costs*. Available at http://www.kff.org

Kaiser Family Foundation. 2004a (June). *The HIV/AIDS Epidemic in the U.S.* HIV/AIDS Policy Fact Sheet. Available at http://www.kff.org

Kaiser Family Foundation. 2004b (August). *Survey of Americans on HIV/AIDS. Part 3. Experiences and Opinions by Race/Ethnicity and Age*. Available at http://www.kff.org

Kolata, Gina. 2007. "A Surprising Secret to a Long Life: Stay in School." *New York Times*, January 3, A1, A16.

Komiya, Noboru, Glenn E. Good, and Nancy B. Sherrod. 2000. "Emotional Openness as a Predictor of College Students' Attitudes Toward Seeking Psychological Help." *Journal of Counseling Psychology* 47(1):138–143.

Lalasz, Robert. 2004 (November). *World AIDS Day 2004: The Vulnerability of Women and Girls*. Population Reference Bureau. Available at http://www.prb.org

Lantz, Paula M., James S. House, James M. Lepkowski, David R. Williams, Richard P. Mero, and Jieming Chen. 1998. "Socioeconomic Factors, Health Behaviors, and Mortality: Results from a Nationally Representative Prospective Study of U.S. Adults." *Journal of the American Medical Association* 279:1703–1708.

Lee, Kelley. 2003. "Introduction." In *Health Impacts of Globalization*, ed. Kelley Lee, 1–10. New York: Palgrave MacMillan.

Lerer, Leonard B., Alan D. Lopez, Tord Kjellstrom, and Derek Yach. 1998. "Health for All: Analyzing Health Status and Determinants." *World Health Statistics Quarterly* 51:7–20.

Link, Bruce G., and Jo Phelan. 2001. "Social Conditions as Fundamental Causes of Disease." In *Readings in Medical Sociology*, 2nd ed., eds. William C. Cockerham, Michael Glasser, and Linda S. Heuser, 3–17. Upper Saddle River, NJ: Prentice Hall.

Mahar, Maggie. 2006. *Money-Driven Medicine*. New York: Harper-Collins.

Martin, J. K., B Pescosolido, and S. A. Tuch. 2000. "Of Fear and Loathing: The Role of 'Disturbing Behavior' Labels and Causal Attributions in Shaping Public Attitudes Toward Persons with Mental Illness." *Journal of Health and Social Behavior* 41(2):208–223.

Mastny, L., and R. P. Cincotta. 2005. "Examining the Connection Between Population and Security." In *State of the World 2005: Redefining Global Security*, eds. M. Renner, H. French, and E. Assadourian, 22–39. New York: W. W. Norton.

Mayer, Lindsay Renick. 2009. "Insurers Fight Public Health Plan." Capitol Eye Blog, June 18. Center for Responsive Politics. Available at http://www.opensecrets.org

McGinn, Anne P. 2003. "Combating Malaria." In *State of the World 2003*, ed. L. Starke, 62–83. New York: W. W. Norton.

Mirowsky, John, and Catherine E. Ross. 2003. *Social Causes of Psychological Distress*. 2nd ed. New York: Walter de Gruyter.

MMWR (Morbidity and Mortality Weekly Report). 2009 (January 30). "QuikStats Percentage of Adults Aged ≥ 18 Years Who Had Ever Been Tested for Human Immunodeficiency Virus (HIV), by Age Group and Sex—National Health Interview Survey, United States, 2007." Available at http://www.cdc.gov/mmwr

Moyers, Bill, and Michael Winship. 2009 (July 27). "Bill Moyers: Dangerous Alliance of Health Industry and Right-Wingers Will Stop at Nothing to Derail Progressive Reforms." AlterNet. Available at http://www.alternet.org

Murphy, Elaine M. 2003. "Being Born Female Is Dangerous for Your Health." *American Psychologist* 58(3):205–210.

Murray, C., and A. Lopez, eds. 1996. *The Global Burden of Disease.* Boston: Harvard University Press.

Nader, Ralph. 2009 (July 25). "Health Care Hypocrisy." Common Dreams. Available at http://www.commondreams.org

Nadin, Rachel, Himmelstein, D., and Woolhandler, S. 2009 (February 19). "Massachusetts' Plan: A Failed Model for Health Care Reform." Physicians for a National Health Program. Available at http://www.pnhp.org

National Center for Chronic Disease Prevention and Health Promotion. 2004. *Physical Activity and Good Nutrition: Essential Elements to Prevent Chronic Diseases and Obesity.* Available at http://www.cdc.gov/nccdphp

National Center for Health Statistics. 2008. *Health, United States, 2008 with Chartbook on Trends in the Health of Americans.* Hyattsville, MD: U.S. Government Printing Office.

The National Coalition on Health Care. 2009. "Health Insurance Costs." Available at http://www.nchc.org

National Institute of Mental Health. 2008. *The Numbers Count: Mental Disorders in America.* Available at http://www.nimh.nih.gov

Niemeier, H. M., H. A. Raynor, El E. Lloyd-Richardson, M. L. Rogers, and R. R. Wing. 2006. "Fast Food Consumption and Breakfast Skipping: Predictors of Weight-Gain from Adolescence to Adulthood in a Nationally Representative Sample." *Journal of Adolescent Health* 39(6):842–849.

Ninan, Ann. 2003 (March). *"Without My Consent": Women and HIV-Related Stigma in India.* Population Reference Bureau. Available at http://prb.org

Guardian. 2007. "Obese Boy Stays with Mother." *Guardian Unlimited*, February 27. Available at http://www.guardian.co.uk

Olshansky, S. J., D. J. Pasaro, R. C. Hershow, J. Layden, B. A. Carnes, J. Brody, L. Hayflick, R. N. Butler, D. B. Allison, and D. S. Ludlow. 2005. "A Potential Decline in Life Expectancy in the United States in the 21st Century." *New England Journal of Medicine* 352(11):1138–1145.

Orr, Andrea. 2009 (July 14). "Insured . . . and Broke." Economic Policy Institute. Available at http://www.epi.org

Oxfam GB. 2004 (March 19). *The Cost of Childbirth.* Press release. Available at http://www.oxfam.org.uk/

Parsons, Talcott. 1951. *The Social System.* New York: Free Press.

Peeno, Linda. 2000. "Taking on the System." *Hope* (22):18–21.

Pereira, M. A., A. I. Kartashov, C. B. Ebbeling, L. Van Horn, M. I. Slattery, D. R. Jacobs, and D. S. Ludwig. 2005. "Fast-Food Habits, Weight Gain, and Insulin Resistance." *The Lancet* 365(9453):36–42.

Pew Research Center. 2009 (March 19). "Support for Health Care Overhaul, but It's Not 1993." Available at http://people-press.org

Pollan, Michael. 2006. *The Omnivore's Dilemma.* New York: Penguin Press.

Potter, Wendell. 2009 (June 24). "The Health Care Industry vs. Health Reform." Center for Media and Democracy. Available at http://www.prwatch.org

Quadagno, Jill. 2004. "Why the United States Has No National Health Insurance: Stakeholder Mobilization Against the Welfare State 1945–1996." *Journal of Health and Social Behavior* 45:25–44.

Robbins, John. 2006. *Healthy at 100.* New York: Random House.

Sachs, Jeffrey D. 2005. "A Practical Plan to End Poverty." *Washington Post*, January 17, p. A17.

Sack, Kevin. 2009. "Defying Slump, 13 States Insure More Children." *New York Times*, July 18. Available at http://www.nytimes.com

Sager, Alan, and Deborah Socolar. 2004 (October 28). *2003 U.S. Prescription Drug Prices 81 Percent Higher Than in Other Wealthy Nations.* Data Brief No. 7. Boston University School of Public Health.

Sampson, Zinie Chen. 2007. "Bill Would Bar Punishing Students Solely for Attempting Suicide." Associated Press, February 23. Available at http://www.timesdispatch.com

Sanders, David, and Mickey Chopra. 2003. "Globalization and the Challenge of Health for All: A View from Sub-Saharan Africa." In *Health Impacts of Globalization*, ed. Kelley Lee, 105–19. New York: Palgrave Macmillan.

Sanders, Jim. 2005. "Bill Would OK Condoms in Prisons." *The Sacramento Bee*, February 26. Available at http://www.sacbee.com

Save the Children. 2002. *State of the World's Mothers, 2002.* Available at http://www.savethechildren.org

Scal, Peter, and Robert Town. 2007. "Losing Insurance and Using the Emergency Department: Critical Effect of Transition to Adulthood for Youth with Chronic Conditions." *Journal of Adolescent Health* 40 (2 Suppl. 1):S4.

Schlosser, Eric. 2002. *Fast Food Nation.* New York: HarperCollins.

Schoenborn, Charlotte A. 2004 (December 15). "Marital Status and Health: United States, 1999–2002." Advance data from *Vital and Health Statistics*, no. 351. Hyattsville, MD: National Center for Health Statistics.

Sered, Susan Starr, and Rushika Fernandopulle. 2005. *Uninsured in America: Life and Death in the Land of Opportunity.* Berkeley and Los Angeles, CA: University of California Press.

Sidel, Victor W., and Barry S. Levy. 2002. "The Health and Social Consequences of Diversion of Economic Resources to War and Preparation for War." In *War or Health: A Reader*, eds. Ilkka Taipale, P. Helena Makela, Kati Juva, and Vappu Taipale, 208–21. New York: Palgrave MacMillan.

Singer, Peter, and Jim Mason. 2006. *The Way We Eat: Why Our Food Choices Matter.* Emmaus, PA: Rodale.

Single Payer Action. 2009 (May 26). "Single Payer Trial for the Baucus 13." Available at http://www.singlepayeraction.org

SmokeFree Educational Services. 2003 (June 5). *5 Million Deaths a Year Worldwide from Smoking.* Corp-Watch. Available at http://www.corpwatch.org

Specter, Mike, and John D. Stoll. 2007. "GM, Ford to See Escalating Health-Care Costs in 2007." *MarketWatch*, February 26. Available at http://www.marketwatch.com

Springen, Karen. 2006. "Health Hazards: How Mounting Medical Costs Are Plunging More Families into Debilitating Debt and Why Insurance Doesn't Always Keep Them Out of Bankruptcy." *Newsweek*, August 25. Available at www.msnbc.com/newsweek

"Stark Introduces Constitutional Amendment to Establish Right to Health Care." 2005. Press release, March 3. Available at http://www.house.gov

Stein, R., and C. Connolly. 2004. "Medicare Changes Policy on Obesity, Some Treatments May Be Covered." *Washington Post*, July 16. Available at http://www.washingtonpostonline.com

Sturm, Roland. 2005 (April). "Childhood Obesity: What We Can Learn from Existing Data on Social Trends, Part 2." *Preventing Chronic Disease* 2(2). Available at http://www.cdc.gov

Substance Abuse and Mental Health Services Administration. 2008. *Results from the 2007 National Survey on Drug Use and Mental Health: National Findings.* Rockland, MD: Department of Health and Human Services.

Suris, J. C., Ralph Thomas, Jean-Jacques Cheseaux, Pierre-Andre Michaud, and Swiss Mother-Child HIV Cohort Study. 2007. "'I Don't Want To Be Disappointed': HIV(1) Adolescents' Reasons for Not Disclosing Their Disease: A Qualitative Study." *Journal of Adolescent Health* 40 (2 Suppl. 1):S41–S42.

Szasz, Thomas. 1961/1970. *The Myth of Mental Illness: Foundations of a Theory of Personal Conduct.* New York: Harper & Row.

Thomson, George, and Nick Wilson. 2005. "Policy Lessons from Comparing Mortality from Two Global Forces: International Terrorism and Tobacco." *Globalization and Health* 1:18.

Trust for America's Health. 2008. *F as in Fat: How Obesity Policies Are Failing in America*. Available at http://healthyamericans.org

UNICEF. 2008. *The State of the World's Children 2009: Maternal and Newborn Health*. Available at http://www.unicef.org

UNICEF. 2006. *The State of the World's Children 2007*. Available at http://www.unicef.org

U.S. Department of Health and Human Services. 2001. *Mental Health: Culture, Race, and Ethnicity—A Supplement to Mental Health: A Report of the Surgeon General*. Rockville, MD: U.S. Government Printing Office.

U.S. Department of Health and Human Services. 1999. *Mental Health: A Report of the Surgeon General—Executive Summary*. Rockville, MD: U.S. Government Printing Office.

Vaccarino, V., S. S. Rathore, N. K. Wenger, P. D. Frederick, J. L. Abramson, H. V. Barron, A. Manhapra, S. Mallik, and H. M. Krumholz. 2005 (August 18). "Sex and Racial Differences in the Management of Acute Myocardial Infarction, 1994 Through 2002." *New England Journal of Medicine* 353(7):671–682.

Weeks, Jennifer. 2007 (January 12). "Factory Farms." *The CQ Researcher* 17(2).

Weitz, Rose. 2010. *The Sociology of Health, Illness, and Health Care: A Critical Approach*. 5th ed. Belmont, CA: Wadsworth/Cengage.

White, Frank. 2003. "Can International Public Health Law Help to Prevent War?" *Bulletin of the World Health Organization* 81(3):228.

Williams, David R. 2003. "The Health of Men: Structured Inequalities and Opportunities." *American Journal of Public Health* 93(5):724–731.

World Health Organization. 2006. *World Health Statistics 2006*. Available at http://www.who.int

World Health Organization. 2005. *The World Health Report 2005: Make Every Mother and Child Count*. Available at http://www.who.int

World Health Organization. 2004. *The World Health Report 2004: Changing History*. Available at http://www.who.int

World Health Organization. 2002. *The World Health Report 2002*. Available at http://www.who.int

World Health Organization. 2000. *The World Health Report 2000*. Available at http://www.who.int

World Health Organization. 1946. *Constitution of the World Health Organization*. New York: World Health Organization Interim Commission.

Zeldin, Cindy, and Mark Rukavina. 2007. *Borrowing to Stay Healthy: How Credit Card Debt Is Related to Medical Expenses*. New York: Demos.

Chapter 3

A Better Way Foundation. 2009. Available at http://www.abwf-ct.org/

Abadinsky, Howard. 2008. *Drugs: An Introduction*. Belmont, CA: Wadsworth.

Alcoholics Anonymous. 2007. *AA at a Glance*. Available at http://www.alcoholics-anonymous.org

Aloise-Young, Patricia, Michael D. Slater, and Courtney C. Cruickshank. 2006. "Mediators and Moderators of Magazine Advertisement Effects on Adolescent Cigarette Smoking." *Journal of Health Communication* 11:281–300.

American Cancer Society. 2009. "Tobacco and Cancer." Available at http://www.cancer.org

American College Health Association. 2009. National College Health Assessment II. "Reference Group Executive Summary, Fall 2008." Baltimore, Maryland: American College Health Association.

American Heart Association (AHA). 2008a. "Support Illinois SB 1455 Prohibit the Sale of Candy/Alcohol Flavored Tobacco Products." Available at http://www.americanheart.org

American Heart Association (AHA). 2008b. "Targeting Minorities." Available at http://www.americanheart.org

Archibold, Randal C. 2009a. "Mexican Drug Cartel Violence Spills Over, Alarming U.S." *The New York Times*, March 23. Available at http://www.nytimes.com

Archibold, Randal C. 2009b. "U.S. Plans Border 'Surge' Against Any Drug Wars." *The New York Times*. January 8. Available at http://www.nytimes.com

Armstrong, Elizabeth, and Christina McCarroll. 2004. "Girls Lead in Teen Alcohol Use." *Seattle Times*, August 14. Available at http://seattletimes.nwsource.com

Becker, H. S. 1966. *Outsiders: Studies in the Sociology of Deviance*. New York: Free Press.

Behrendt, S, H.-U. Wittchen, M. Höfler, R. Lieb, and K. Beesdo. 2009. "Transitions from First Substance Use to Substance Use Disorders in Adolescence: Is Early Onset Associated with a Rapid Escalation?" *Drug and Alcohol Dependence* 99:68–78.

Brownell, Ginanne. 2009. "'Special K' Goes Global." *Global Post*, May 22. Available at http://huffingtonpost.com

Bureau of Justice Statistics. 2008. *Drugs and Crime Facts*, ed. Tina L. Dorsey. NCJ165148. Washington, DC: U.S. Department of Justice.

Burke, Jason. 2006. "Europeans Turn to Cocaine and Alcohol as Cannabis Loses Favour." *Guardian Unlimited*, October 15. Available at http://observer.guardian.co.uk

Carroll, Joseph. 2007. *Little Change in Public's View of the U.S. Drug Problem*. Gallup Poll, October 19. Available at http://www.gallup.com

CASA (National Center on Addiction and Substance Abuse). 2009. "The Impact of Substance Abuse on Federal, State, and Local Budgets." New York: Columbia University.

Caulkins, Jonathan, Peter Reuter, Martin Y. Iguchi, and James Chiesa. 2005. "How Goes the 'War on Drugs'? An Assessment of U.S. Drug Problems and Policy." Rand Drug Policy Research Center. Available at http://www.rand.org/pubs

CDC (Centers for Disease Control). 2006 "Fetal Alcohol Spectrum Disorders." Available at http://www.cdc.gov

CDC. 2008 (August 6). "Alcohol-Related Disease Impact (ARDI) Software." Available at http://www.cdc.gov

CDC (Centers for Disease Control). 2009 (January 29). "Tobacco Use: Targeting the Nation's Leading Killer." *Chronic Disease Prevention and Health Promotion*. Available at http://www.cdc.gov/tobacco

CNN. 2009. "Penalties for Drug-Related Crime in Asia." Available at http://edition.cnn.com

Copes, Heigh, Andy Hochstetler, and J. Patrick Williams. 2008. "'We Weren't Like No Regular Dope Fiends': Negotiating Hustler and Crackhead Identities." *Social Problems* 55(2):254–270.

Corso, Regina A. 2009. "Three in Ten Americans Drink Alcohol Weekly." *The Harris Poll 48*, May 14. Available at http://www.harrisinteractive.com

Degenhardt, Louisa, Wai-Tat Chiu, Nancy Sampson, Ronald C. Kessler, James C. Anthony, Matthias Angermeyer, Ronny Bruffaerts, Giovanni de Girolamo, Oye Gureje, Yueqin Huang, Aimee Karam, Stanislav Kostyuchenko, Jean Pierre Lepine, Maria Elena Medina Mora, Yehuda Neumark, J. Hans Ormel, Alejandra Pinto-Meza, José Posada-Villa, Dan J. Stein, Tadashi Takeshima, and J. Elisabeth Wells. 2008. "Toward a Global View of Alcohol, Tobacco, Cannabis, and Cocaine Use: Findings from the WHO World Mental Health Surveys." *PLoS Medicine* 5(1):1053–1077.

Dekel, Rachel, Rami Benbenishty, and Yair Amram. 2004. "Therapeutic Communities for Drug Addicts: Prediction of Long-Term Outcomes." *Addictive Behaviors* 29(9):1833–1837.

Dinan, Stephen, and Ben Conery. 2009. "DEA Pot Raids Go On; Obama Opposes." *The Washington Times*, February 5. Available at http://www.washingtontimes.com

Dinno, Alexis, and Stanton Glantz. 2009. "Tobacco Control Policies are Egalitarian: A Vulnerabilities Perspective on Clean Indoor Air Laws, Cigarette Prices, and Tobacco Use Disparities." *Social Science & Medicine* 68:1439–1447.

Drug Policy Alliance. 2007. *State by State.* Available at http://www.drugpolicy.org/statebystate/

Drug Policy Alliance. 2003. *Drug Policy around the World: The Netherlands.* Available at http://www.drugpolicy.org/global

Duke, Steven, and Albert C. Gross. 1994. *America's Longest War: Rethinking Our Tragic Crusade Against Drugs.* New York: G. P. Putnam.

Elias, Marilyn. 2009. "Secondhand Smoke May Double Likelihood of Depression." *USA Today*, March 4. Available at http://www.usatoday.com/news.health

EMCDDA (European Monitoring Centre for Drugs and Drug Addiction). 2008. *The State of the Drugs Problem in Europe.* Luxembourg: Office for Official Publications of the European Communities.

ESPAD (European School Survey Project on Alcohol and Other Drugs). 2009. *Stockholm: The Swedish Council for Information on Alcohol and Other Drugs.* Available at www.espad.org

Fact Sheet. 2009 (June 22). "Fact Sheet: The Family Smoking Prevention and Tobacco Control Act of 2009." Available at http://www.whitehouse.gov/the_press_office

Feagin, Joe R., and C. B. Feagin. 1994. *Social Problems.* Englewood Cliffs, NJ: Prentice Hall.

Filkins, Dexter. 2009. "Poppies a Target in Fight Against Taliban." *The New York Times*, April 29. Available at http://www.nytimes.com

Firshein, Janet. 2003. "The Role of Biology and Genetics." *Moyers on Addiction.* Public Broadcasting System (PBS). Available at http://www.pbs.org

Flanzer, Jerry P. 2005. "Alcohol and Other Drugs Are Key Causal Agents of Violence." *In Current Controversies on Family Violence*, 2nd ed., Donileen R. Loseke, Richard J. Gelles, and Mary M. Cavanaugh, eds. (pp. 163–174). City of publication?

Freeman, Dan, Merrie Brucks, and Melanie Wallendorf. 2005. "Young Children's Understanding of Cigarette Smoking." *Addiction* 100(10):1537–1545.

Friedman-Rudovsky, Jean. 2009. "Red Bull's New Cola: A Kick from Cocaine?" *Time/CNN*, May 25. Available at http://www.time.com

Gilbert, R., and C. S.Widom, K. Browne, D. Fergusson, E. Webb, and S. Janson. 2009. "Burden and Consequence of Child Maltreatment in High-income Countries." *The Lancet* 73(9657):68–81.

Gusfield, Joseph. 1963. *Symbolic Crusade: Status Politics and the American Temperance Movement.* Urbana, IL: University of Illinois Press.

Heinrich, Henry. 2009. "Obama Drug Policy to Do More to Ease Health Risks." Reuters, March 16 Available at: http://www.reuters.com

HHS (U.S. Department of Health & Human Services). 2009. "Alcohol: Side Effects." Available at http://ncadistore.samhsa.gov

Hingson, Ralph W., Timothy Heeren, and Michael R. Winter. 2006. "Age at Drinking Onset and Alcohol Dependence." *Archives of Pediatrics & Adolescent Medicine* 160:739–746.

Hsu, Spencer S. 2009 (April 16). "Obama Targets Drug Cartels." Available at http://www.washingtonpost.com

Human Rights Watch. 2007. "Reforming the Rockefeller Drug Laws." Available at http://www.hrw.org/campaigns/drugs

Jarvik, M. 1990. "The Drug Dilemma: Manipulating the Demand." *Science* 250:387–392.

Jervis, Rick. 2009. "YouTube Riddled with Drug Cartel Videos, Messages." *USA Today*, April 9. Available at http://www.usatoday.com

Johnson, David T. 2009 (March 10). "The Merida Initiative." Bureau of International Narcotics and Law Enforcement Affairs. Available at http://www.state.gov

Jones, Jeffrey. 2009. "Most Americans Concerned About Mexico's Drug Violence." Gallup Poll, April 3. Available at http://www.gallup.com

Keegan, Rebecca Winters, and Meaghan Haire. 2008. "Beer Pong's Big Splash." *Time/CNN*, August 7. Available at http://www.time.com/time

Kelly, J. F., J. W. Finney, and R. Moos. 2005. "Substance Use Disorder Patients Who Are Mandated to Treatment: Characteristics, Treatment Process, and 1 and 5 Year Outcomes." *Journal of Substance Abuse Treatment* 28(3):213–223.

Kelly, John, Robert Stout, William Zywiak, and Robert Schneider. 2006. "A 3-Year Study of Addiction Mutual-Help Group Participation Following Intensive Outpatient Treatment." *Alcoholism: Clinical and Experimental Research* 30:1381–1392.

Kraft, Scott. 2009. "Pursuing Smugglers, Border Agents Become Trackers." *Los Angeles Times*, May 12. Available at http://www.latimes.com/news

Lee, Yoon, and M. Abdel-Ghany. 2004. "American Youth Consumption of Licit and Illicit Substances." *International Journal of Consumer Studies* 28(5):454–465.

Lombardi, Frank. 2009. "City Council Bills Would Ban Candy-Flavored Tobacco Smoking Outside of Hospitals." *NY Daily News*, May 22. Available at http://www.nydailynews.com

MacCoun, Robert J., and Peter Reuter. 2001. "Does Europe Do It Better? Lessons from Holland, Britain and Switzerland." In *Solutions to Social Problems*, eds. D. Stanley Eitzen and Craig S. Leedham, pp. 260–264. Boston: Allyn and Bacon.

Marin Institute. 2006. "Lawsuits Against the Industry." Available at http://www.marininstitute.org

Mauer, Marc. 2009. "The Changing Racial Dynamics of the War on Drugs." The Sentencing Project, April 2009. Washington, DC. Available at http://www.sentencingproject.org

Mears, Bill. 2007. "Justices: Judges Can Slash Crack Sentences." CNN. Available at http://www.cnn.com

Mickleburgh, Rod, and Gloria Galloway. 2007. "Storm Brews over Drug Strategy." *The Globe and Mail*, January 15. Available at http://www.theglobeandmail.com

Mieszkowski, Katherine. 2009. "Everybody Must Get Stoned." *Salon*, March 3. Available at http://www.salon.com

Mock, Vanessa. 2008. "Dutch Mayors Back Plan to Cultivate Cannabis." Available at http://www.nzherald.co.nz

MTF (Monitoring the Future). 2009. *National Results on Adolescent Drug Use.* The University of Michigan Institute for Social Research. Available at http://monitoringthefuture.org

Morgan, Patricia A. 1978. "The Legislation of Drug Law: Economic Crisis and Social Control." *Journal of Drug Issues* 8:53–62.

Myers, Matthew L. 2009a (June 22). "President Obama Delivers Historic Victory for America's Kids and Health over Tobacco." Available at http://www.tobaccofreekids.org

Myers, Matthew L. 2009b. "U.S. Court of Appeals Affirms 2006 Lower Court Ruling That Tobacco Companies Committed Fraud for Five Decades and Lied About the Dangers of Smoking." Press Office Release, May 22. Available at http://www.tobaccofreekids.org

National Institute on Alcohol Abuse and Alcoholism. 2000 (April). *Mechanisms of Addiction.* Alcohol Alert. National Institute on Alcohol Abuse and Alcoholism no. 47. Available at http://www.niaaa.nih.gov

NHTSA (National Highway Traffic Safety Administration). 2008a. "Alcohol-Impaired

Driving." *Traffic Safety Facts 2007 Data.* DOT HS 810 985. Available at http://www.nhtsa.gov

NHTSA. 2008b. "New Data Show Drinking Age Laws Saved 4,441 Lives Over 5 Years." *In the News*, November 6. Available at http://www.nhtsa.gov

NIDA (National Institute on Drug Abuse). 2009. "Commonly Abused Drugs." Available at http://www.drugabuse.gov/DrugPages/DrugsofAbuse.html

NIDA. 2006. *Principles of Drug Addiction Treatment: A Research Based Guide.* Available at http://drugabuse.gov/PODAT

NIDA. 2005. *Heroin Abuse and Addiction.* Research Report Series, NIH Publication 05-4165. Available at http://www.nida.nih/gov

NSDUH (National Survey on Drug Use and Health). 2008 (September). *Results from the 2007 National Survey on Drug Use and Health: National Findings.* Office of Applied Studies, NSDUH Series H-34, DHHS Publication No. SMA 08-4343). Rockville, MD. Available at http:// oas.samhsa.gov

NSDUH (National Survey on Drug Use and Health). 2004 (February 13). *Alcohol Dependence or Abuse Among Parents with Children Living at Home.* Office of Applied Studies, Substance Abuse and Mental Health Services Administration. Available at http://www.oas.samhsa.gov

Obama, Barack. 2009 (April 17). "Remarks by the President at the Summit of the Americas Opening Ceremony." Available at http://www.whitehouse.gov

ONDCP (Office of National Drug Control Policy). 2009. *National Youth Anti-Drug Media Campaign.* Available at http://www.theantidrug.com/about.asp

ONDCP. 2008. *National Drug Control Strategy.* Available at http://www.whitehousedrugpolicy.gov

ONDCP. 2006. *Methamphetamine.* Available at http://www.whitehousedrugpolicy.gov/drugfact/methamphetamine.

Osborn, Claire. 2009. "Loss of Son Lead Family to Warn of Alcohol Poisoning at Cap 10K." *Austin American Statesman*, March 28. Available at http://www.statesman.com

Peters, Jeremy W. 2009. "Albany Takes Step to Repeal '70s-Era Drug Laws." *The New York Times*, March 5. Available at http://www.nytimes.com

Pew Research Center. 2002. *Illegal Drugs.* Pew Research Center for the People and Press Survey. Available at http://www.pollingreport.com

Phoenix House. 2008. "Phoenix House Facts." *Phoenix House Annual Report.* Available at http://www.phoenixhouse.org

Primack, Brian A., James E. Bost, Stephanie R. Land and Michael J. Fine. 2007. "Volume of Tobacoo Advertising in African American Markets: Systematic Review and Meta-Analysis." *Public Health Reports* 122(5):607–615.

Rorabaugh, W. J. 1979. *The Alcoholic Republic: An American Tradition.* New York: Oxford University Press.

SAMHSA (Substance Abuse and Mental Health Services Administration). 2009. "Children Living with Substance-Dependent or Substance-Abusing Parents: 2002–2007." *The NSDUH Report*, April 16. Available at http://oas.samhsa.gov

SAMHSA. 2007. "Parental Substance Abuse Raises Children's Risk." *Practice What You Preach*, February 20. Available at http://www.family.samhsa.gov

Sargant, James, Thomas Wills, Mike Stoolmiller, Jennifer Gibson, and Frederick Gibbons. 2006. "Alcohol Use in Motion Pictures and Its Relation with Early Onset Teen Drinking." *Journal of Studies on Alcohol* 77(1):54–66.

Sheldon, Tony. 2000. "Cannabis Use Among Dutch Youth." *British Medical Journal* 321:655.

Siegel, Larry. 2006. *Criminology.* Belmont, CA: Wadsworth.

Snyder, Leslie B., Frances Fleming Miller, Michael Slater, Helen Sun, and Yuliya Strizhakova. 2006. "Effects of Alcohol Advertising Exposure on Drinking Among Youth." *Archives of Pediatrics & Adolescent Medicine* 160(1):18–24.

Stateman, Alison. 2009. "Can Marijuana Help Rescue California's Economy?" *Time*, March 13. Available at http://www.time.com

Szalavitz, Maia. 2009. "Drugs in Portugal: Did Decriminalization Work?" *Time*, April 26. Available at: http://www.time.com

Taifa, Nkechi. 2006 (May). "The 'Crack/Powder Disparity': Can the International Race Convention Provide a Basis for Relief?" American Constitution Society for Law and Policy White Paper. Available at http://acslaw.org

Tarter, Ralph E., Michael Vanyukov, Levent Kirisci, Maureen Reynolds, and Duncan B. Clark. 2006. "Predictors of Marijuana Use in Adolescents Before and After Licit Drug Use: Examination of the Gateway Hypothesis." *American Journal of Psychiatry* 163:2134–2140.

Terry-McElrath, Yvonne M., Melanie A. Wakefield, Sherry Emery, Henry Saffer, Glen Szczypka, Patrick M. O'Malley, Lloyd D. Johnston, Frank J. Chaloupka, and Brian R. Flay. 2007. "State Anti-Tobacco Advertising and Smoking Outcomes by Gender and Race/Ethnicity." *Ethnicity and Health* 12(4):339–362.

The Economist. 2008 (August 7). "Needle Match." Available at http://www.economist.com

The Economist. 2009 (March 7). "A Toker's Guide." Available at http://www.economist.com

Thio, Alex. 2007. *Deviant Behavior.* Boston: Allyn and Bacon.

Timeline. 2001. "Timeline of Tobacco Litigation." Fox News, March 8. Available at http://www.foxnews.com

Tobacco Free Kids. 2009 (February 18). "Deadly in Pink- Big Tobacco Steps Up Its Targeting of Women and Girls." Available at http://www.tobaccofreekids.org

Tobacco Free Kids. 2005 (September 19). "Big Tobacco Still Targeting Kids." Available at http://www.tobaccofreekids.org

U.S. Department of Justice. 2007 (December 19). "Circumstances." Bureau of Justice Statistics Intimate Partner Violence in the U.S. Available at http://www.ojp.usdoj.gov/bjs/intimate/circumstances.htm

Van Dyck, C., and R. Byck. 1982. "Cocaine." *Scientific American* 246:128–141.

Wakefield, Melanie A., Sarah Durkin, Matthew J. Spittal, Mohammad Siahpush, Michelle Scollo, Julie A. Simpson, Simon Chapman, Victoria White, and David Hill. 2008. "Impact of Tobacco Control Policies and Mass Media Campaigns on Monthly Adult Smoking Prevalence." *American Journal of Public Health* 98(8):1443–1450.

WDR (World Drug Report). 2008. *Executive Summary.* Office of Drug Control and Crime Prevention. United Nations. Available at www.unodc.org/documents/wdr/WDR_2008/WDR_2008_eng_web.pdf

Wechsler, William, and Toben F. Nelson. 2008. "What We Have Learned from the Harvard School of Public Health College Alcohol Study: Focusing Attention on College Student Alcohol Consumption and the Environmental Conditions That Promote It." *Journal of Alcohol Studies* (July):1–9.

Weitzman, Elissa, Henry Wechsler, and Toben F. Nelson. 2003 (January 21). *Environment, Not Education, a Stronger Predictor of Binge Drinking Behavior Among College Freshman.* Press Release, Harvard School of Public Health.

Werner, Erica. 2009. "Do Smokers Cost Society Money?" *USA Today*, April 8. Available at http://www.usatoday.com/news/health

West, Steven L. and Keri K. O'Neal. 2004. "Project D.A.R.E. outcome effectiveness revisited." *American Journal of Public Health* 94(6):1027–1029.

Wheeler, Mark. 2008. "UCLA Issues New Report on Prop. 36." *UCLA Newsroom*, October 14. Available at http://newsroom.ucla.edu

Williams, Jenny, Frank J. Chaloupka, and Henry Wechsler. 2005. "Are There Differential Effects of Price and Policy on College Students' Drinking Intensity?"

Contemporary Economic Policy 23(1): 78–90.

Willing, Richard. 2002. "Study Shows Alcohol Is Main Problem for Addicts." *USA Today*, October 3, p. B4.

Willing, Richard. 2004. "Lawsuits Target Alcohol Industry." *USA Today*, May 13. Available at http://www.usatoday.com

Wilson, Joy Johnson. 1999. "Summary of the Attorneys General Master Tobacco Settlement Agreement." National Conference of State Legislators—AFI Health Committee. March. Available at http://academic.udayton.edu/health

Wilson, Nick. 2009. "Alcohol Poisoning Killed Cal Poly Freshman Carson Starkey." San Luis Obispo Tribune, January 21. Available at www.sanluisobispo.com

Witters, Weldon, Peter Venturelli, and Glen Hanson. 1992. *Drugs and Society*, 3rd ed. Boston: Jones & Bartlett.

WHO (World Health Organization). 2008. *WHO Report on the Global Tobacco Epidemic, 2008: The MPOWER Package*. Geneva: World Health Organization, 2008.

Wysong, Earl, Richard Aniskiewicz, and David Wright. 1994. "Truth and Dare: Tracking Drug Education to Graduation and as Symbolic Politics." *Social Problems* 41:448–468.

Zailckas, Koren. 2005. *Smashed: Story of a Drunken Girlhood*. New York: Viking.

Zickler, Patrick. 2003. "Study Demonstrates That Marijuana Smokers Experience Significant Withdrawal." *NIDA Notes* 17: 7, 10.

Chapter 4

Amnesty International. 2009. *Death Sentences and Executions*. Available at http://web.amnesty.org

Associated Press. 2009a (April 25). "New York City Crime Rate Still Falling." Available at http://www.nydailynews.com

Associated Press. 2007. "Austria Uncovers Vast Pornography Ring." *International Herald Tribune*, February 7. Available at http://www.iht.com

Associated Press. 2005. "Fla. Gov. Jeb Bush Signs Lunsford Act." Fox News, May 2. Available at http://www.foxnews.com

Barkan, Steven. 2006. *Criminology*. Upper Saddle River, NJ: Prentice Hall.

Baze v. Rees, 553 U.S. ___2008. U.S. Supreme Court. Available at http://www.supremecourtus.gov

Barboza, David. 2009. "Death Sentences Given in Chinese Milk Scandal." *The New York Times*, February 2. Available at http://www.nytimes.com

Becker, Howard S. 1963. *Outsiders: Studies in the Sociology of Deviance*. New York: Free Press.

Beiser, Vince. 2009. "Study in Contrepreneurship." *Miller-McCune Magazine*, April 18. Available at http://www.miller-mccune.com/culture_society/study-in-contrepreneurship-1098

Bell, Kerryn E. 2009. "Gender and Gangs: A Quantitative Comparison." *Crime and Delinquency* 55(3):363–387.

BJS (Bureau of Justice Statistics). 2006a. *Four Measures of Serious Violent Crime.* Available at http://www.ojp.gov

BJS. 2006b. *Homicide Trends in the U.S.: Trends by Gender*. Available at http://ojp.usdoj.gov/bjs/homicide/gender.htm

BJS. 2008a. *Crime and Victims Statistics*. U.S. Department of Justice. Available at http://www.ojp.usdoj.gov/bjs/cvict.htm

BJS. 2008b. "Justice Expenditure and Employment Extracts." U.S. Department of Justice. Available at http://www.ojp.gov/bjs

Brennan Center for Justice. 2009 (April 22). "Challenging Global AIDS Funding Restrictions." Available at http://www.brennancenter.org

Brunswick, Mark. 2009. "Shutting the Door to Sexual Predators." *Star Tribune*, May 6. Available at http://www.startribune.com

Carlson, Darren K. 2005. "Americans Deal with Crime by Steering Clear." Gallup Poll, November 22. Available at http://www.gallup.com

Carlson, Jr., Joseph R. 2009. "Prison Nurseries: A Pathway to Crime-Free Futures." *Corrections Compendium* 34(1):17–24.

CDCP (Centers for Disease Control and Prevention). 2008. "Youth Risk Behavior Surveillance." *Morbidity and Mortality Weekly Report* June 6, pp. 1–131.

Chen, Elsa Y. 2008. "The Liberation Hypothesis and Racial and Ethnic Disparities in the Application of California's Three Strikes Law." *Journal of Ethnicity in Criminal Justice* 6 (2):83–102.

Chesney-Lind, Meda, and Randall G. Shelden. 2004. *Girls, Delinquency, and Juvenile Justice*. Belmont, CA: Wadsworth.

CNN. 2007. "After Two Years, Sex Offender to Stand on Trial for Girl's Death." *CNN*, February 12. Available at http://www.cnn.com

Coates, Sam. 2005. "Rader Gets 175 Years for BTK Slayings." *Washington Post*, August 19, p. A3. Available at http://www.washingtonpost.com

Conklin, John E. 2007. *Criminology*, 9th ed. Boston: Allyn and Bacon.

COPS (Community Oriented Policing Services). 2009. *Who We Are*. Available at http://www.cops.usdoj.gov

Correctional Association. 2009. *Education from the Inside Out: The Multiple Benefits of College Programs in Prison*. The Correctional Association of New York. Available at http://www.correctionalassociation.org

D'Alessio, David, and Lisa Stolzenberg. 2002. "A Multilevel Analysis of the Relationship Between Labor Surplus and Pretrial Incarceration." *Social Problems* 49:178–193.

District of Columbia v. Heller, 554 U.S. ____, 2008. Available at http://supreme.justia.com

Durose, Matthew R., Erica L. Smith, and Patrick A. Langan. 2007. "Contacts Between Police and the Public, 2005." *Bureau of Justice Statistics Special Report*, April. U.S. Department of Justice. NCJ 215243.

The Economist. 2008 (October 30). "The Oldest Conundrum." Available at http://www.economist.com

Erikson, Kai T. 1966. *Wayward Puritans*. New York: Wiley.

Europol. 2009. "Frequently Asked Questions." Available at http://www.europol.europa.eu

Fact Sheet. 2009. "Sen. Webb's National Criminal Justice Commission Act of 2009." Available at http://webb.senate.gov

FBI (Federal Bureau of Investigation). 2009a. *National Incident-Based Reporting System*. April. Available at http://www.fbi.gov

FBI. 2009b. *Crime in the United States, 2008. Preliminary Annual Uniform Crime Report*. Washington, DC: U.S. Government Printing Office.

FBI. 2008a. *Crime in the United States, 2007. Uniform Crime Report*. Washington, DC: U.S. Government Printing Office.

FBI. 2008b. "Innocence Lost Sting II." *Headline Archives*, October 27. Available at http://www.fbi.gov

FBI. 2008c. *Innocent Images National Initiative*. Available at http://www.fbi.gov

FDA (U.S. Food and Drug Administration). 2008. "Melamine Pet Food Recall of 2007." Available at http://www.fda.gov/Animal Veterinary

Felson, Marcus. 2002. *Crime and Everyday Life*, 3rd ed. Thousand Oaks, CA: Sage.

Fight Crime. 2009. *Fight Crime: Invest in Kids*. Available at http://www.fightcrime.org

Florida Criminal Code. 2008. Available at http://www.leg.state.fl.us/statutes

Ford, Jason. 2005. "Substance Use, Social Bond, and Delinquency." *Sociological Inquiry* 75 (1):109–128

FTC (Federal Trade Commission). 2009 (February). "Consumer Sentinel Network Data Book." Available at http://www.ftc.gov

Gallup Poll. 2007. *Gallup's Pulse of Democracy: Crime*. Available at http://www.galluppoll.com

Gallup Poll. 2009a. *Crime*. Available http://www.gallup.com

Gallup Poll. 2009b. *Confidence in Institutions*. Available http://www.gallup.com

Gallup Poll. 2009c. *Before Recent Shooting, Gun Control Support Was Fading*. Available http://www.gallup.com

Gettleman, Jeffrey.2009. "The West Turns to Kenya as Piracy Criminal Court." *The New York Times*, April 24. Available at http://www.nytimes.com

Greenblatt, Alan. 2008. "Second Chance Programs Quietly Gain Acceptance." *Congressional Quarterly Weekly*, September 15. Available at www.cq.com

Greenburg, Zack O'Malley. 2009. "America's Most Dangerous Cities." *Forbes*, April 23. Available at http://www.forbes.com

Hartney, Christopher, and Linh Vuong. 2009. *Created Equal: Racial and Ethnic Disparities in the U.S. Criminal Justice System*. Oakland, CA: National Council on Crime and Delinquency.

Harvard School of Public Health. 2002. "American Females at Highest Risk of Murder." Press release, April 17. Harvard School of Public Health.

Hayes, Tom, and Larry Neumeister. 2009. "Madoff Sentenced to 150 Years in Prison." *The Huffington Post*, June 29. Available at http://www.huffingtonpost.com

Heimer, Karen, Stacy Wittrock and Halime Unal. 2005. "Economic Marginalization and the Gender Gap in Crime." In *Gender and Crime: Patterns of Victimization and Offending*, Karen Heimer and Candace Kruttschnitt, eds. (pp. 115–136). New York University Press.

Henriques, Diana B. 2009. "Madoff Will Plead Guilty; Faces Life for Vast Swindle." *The New York Times*, March 11. Available at http://www.nytimes.com

The Herald Sun. 2008 (January 17). "Jessica's Victim Impact Statement." Available at http://www.news.com.au/heraldsun

Hirschi, Travis. 1969. *Causes of Delinquency*. Berkeley, CA: University of California Press.

ICCC (Internet Crime Complaint Center). 2009. *2008 Internet Crime Report*. Washington, DC: National White Collar Crime Center and the Federal Bureau of Investigation. Available at http://www.ic3.gov/media/annualreport/2008

ICPC (International Centre for the Prevention of Crime). 2009. *About ICPC*. Available at http://www.crime-prevention-intl.org/

IMB (International Maritime Bureau). 2009. "IMB Reports Unprecedented Rise in Maritime Hijackings." *ICC Commercial Crime Services*, January 16. Available at http://www.icc-ccs.org

The Innocence Project. 2009. "News and Information." Available at http://www.innocenceproject.org

Interpol. 2009. *About Us*. Available at http://www.interpol.int/

Jacobs, David, Zhenchao Qian, Jason Carmichael, and Stephanie Kent. 2007. "Who Survives on Death Row? An Individual and Contextual Analysis." *American Sociological Review* 72:610–632.

Jessica Lunsford Foundation. 2006. *About Jessica Marie Lunsford*. Available at http://www.jmlfoundation.com

Kaplan, Esther. 2005. "Just Say Não." *The Nation*, May 30. Available at http://www.thenation.com

Kerley, Kent, Michael Benson, Matthew Lee, and Francis Cullen. 2004. "Race, Criminal Justice Contact, and Adult Position in the Social Stratification System." *Social Problems* 51(4):549–568.

Klapper, Bradley. 2005. "UN Told Governments Must Combat Internet Child Pornography." Associated Press, April 14. Available at http://informationweek.com

Krisberg, Barry, Christopher Hartney, Angela Wolf, and Fabiana Silva. 2009 (February 12). "Youth Violence Myths and Realities. A Tale of Three Cities." National Council on Crime and Delinquency. Anne E. Casey Foundation.

Kruse, Michael. 2007. "Jessie's Story: An Ordinary Kid." *St. Petersburg Times*, February 10. Available at http://www.sptimes.com

Kubrin, Charis E. 2005. "Gangsters, Thugs, and Hustlas: Identity and the Code of the Street in Rap Music." *Social Problems* 52(3):360–378.

Kubrin, Charis, and Ronald Weitzer. 2003. "Retaliatory Homicide: Concentrated Disadvantage and Neighborhood Culture." *Social Problems* 50:157–180.

Lacey, Marc. 2009. "In Mexico, Curbing Violence Before It Is Learned." *The New York Times*, January 11. Available at http://www.nytimes.com

Latimer, Jeff, Craig Dowden, and Danielle Muise. 2005. "The Effectiveness of Restorative Justice Practices: A Meta-Analysis." *The Prison Journal* 85:127–144.

Lawrence, Alison. 2009 (March). "Crime Costs." *National Conference of State Legislatures Magazine*. Available at http://www.ncsl.org

Levy, Steven. 2006. "An Identity Heist the Size of Texas." *Newsweek*, June 12, p. 18.

Lichtblau, Eric, David Johnston, and Ron Nixon. 2008. "F.B.I. Struggles to Handle Financial Fraud Cases." *The New York Times*, October 19. Available at http://www.nytimes.com

Liptak, Adam. 2009. "Justices Agree to Take up Sentencing for Young Offenders." *The New York Times*, May 5. Available at http://www.nytimes.com

Lott, John R., Jr. 2003. "Guns Are an Effective Means of Self-Defense." In *Gun Control*, Helen Cothran, ed. (pp. 86–93). Farmington Hills, MI: Greenhaven Press.

MAD DADS. 2009. *Taking It to the Streets*. Available at http://www.maddads.com

Marks, Alexandria. 2006. "Prosecutions Drop for US White Collar Crime." *The Christian Science Monitor*, August 31. Available at http://www.csmonitor.com

Maxwell, Lesli A. 2006. "The Long Arm of the Law." *Education Weekly*, October 18, pp. 25–28.

McCormick, Lisa Wade. 2007 (December). "Veterinarians Solve Pet Food Death Puzzle." Available at http://www.consumeraffairs.com

Merton, Robert. 1957. *Social Theory and Social Structure*. Glencoe, IL: Free Press.

Moore, Solomon. 2009. "Prison Spending Outpaces All but Medicaid." *The New York Times*, March 3. Available at http://www.nytimes.com

Moses, Marilyn, and Cindy J. Smith. 2007. "Factories Behind Fences: Do Prison 'Real Work' Programs Work?" *National Institute of Justice Journal* 257:32–35.

MSNBC. 2007. "Jury Selection in Child Killer Case." MSNBC, February 12. Available at http://www.msnbc.msn.com

National Gang Intelligence Center. 2009. *National Gang Threat Assessment 2009*. Product No. 2009-M0335-001. Washington, DC.

National Night Out. 2009. *What Is National Night Out?* Available at http://www.nationaltownwatch.org/nno/about.html

National Research Council. 1994. *Violence in Urban America: Mobilizing a Response*. Washington, DC: National Academy Press.

NCJRS (National Criminal Justice Reference Service). 2009. *Drug Courts: Facts and Figures. In the Spotlight*. Available at http://www.ncjrs.gov/spotlight/drug_courts

NCMEC (National Center on Missing and Exploited Children). 2007. *Online Enticement Laws Vary Between States*. Available at http://www.missingkids.com

NCPC (National Crime Prevention Council). 2008. "Crime Prevention Matters." Available at http://www.ncpc.org

NCPC. 2005. "Preventing Crime Saves Money." Available at http://www.ncpc.org

Nelson, Katie. 2009 (June 5). "Johann Justin-Jinich: Wesleyan Student Killed in Shooting." Available at http://www.huffingtonpost.com

Nored, Lisa S., and Philip E. Carlan. 2008. "Success of Drug Court Programs: Examination of the Perceptions of Drug Court Personnel." *Criminal Justice Review* 33(3):329–342.

NWCCC (National White Collar Crime Center). 2006. *2005 NW3C National Survey*. Available at http://www.nw3c/pressroom/current_releases.cfm

Ovaska, Sarah, and Lynn Bonner. 2009. "Ruling Lifts a Barrier to Execution: A Judge Strikes Down Inmates' Objections to Death-Penalty Procedures." The News and Observer, May 15. Available at http://www.newsobserver.com

PBB (Puppies Behind Bars). 2009. *About Us*. Available at http://www.puppiesbehindbars.com/

PEP (Prison Entrepreneurship Program). 2009. *About PEP*. Available at http://www.prisonentrepreneurship.org/

Pertossi, Mayra. 2000 (September 27). *Analysis: Argentine Crime Rate Soars*. Available at http://news.excite.com

Pew. 2009. "One in 31: The Long Reach of American Correction." Washington, DC: The Pew Charitable Trust. Available at http://www.pewtrusts.org

Pew. 2008. "One in 100: Behind Bars in America 2008." Washington, DC: The Pew Charitable Trust. Available at http://www.pewtrusts.org

Pew. 2007. "Public Safety, Public Spending." Washington, DC: The Pew Charitable Trust. Available at http://www.pewtrusts.org

PUPS (Prison University Project). 2009. "Associates of Art Degree Program." Available at http://www.prisonuniversityproject.org

Reiman, Jeffrey, and Paul Leighton. 2010. *The Rich Get Richer and the Poor Get Prison*. Boston: Allyn and Bacon.

Richtel, Matt. 2003. "Mayhem, and Far from the Nicest Kind." *New York Times*, February 10. Available at http://www.nytimes.com

Romano, Lois. 2005. "More Complete Portrait of BTK Suspect Is Emerging." *Washington Post*, March 5. Available at http://www.washingtonpost.com

Rosin, Hanna. 2008. "American Murder Mystery." *The Atlantic*, July/August. Available at http://www.theatlantic.com

Rosoff, Stephen, Henry Pontell, and Robert Tillman. 2002. "White Collar Crime." In *Social Problems: Readings with Four Questions*, ed. Joel M. Charon, pp. 339–350. Belmont, CA: Wadsworth.

Rothfeld, Michael. 2009 (January 14). "Panel. Take Another Look at Sex Offender Restrictions." Available at http://articles.latimes.com

Rubin, Paul H. 2002. "The Death Penalty and Deterrence." *Forum*, Winter, pp. 10–12.

Salow, Julie. 2009. "AIG's Six Year Saga of Alleged Fraud." *Huffington Post*, April 2. Available at http://www.huffingtonpost.com

Sampson, Robert J., Jeffrey D. Morenoff and Stephen W. Raudenbush. 2005. "Social Anatomy of Racial and Ethnic Disparities in Violence." *American Journal of Public Health* 95(2):224–232.

Schecter, Anna, Brian Ross, and Justin Rood. 2009. "The Executive Who Brought Down AIG." ABC News, March 30. Available at http://www.abcnews.go.com

Schelzig, Erik. 2007 (September 19). "Court Ruling Halts Tennessee Executions." Available at http://www.wral.com

Schmit, Julie. 2009. "Broken Links in Food-Safety Chain Hid Peanut Plants' Risks." USA Today, April 26. Available at http://www.usatoday.com

Schweinhart, Lawrence J. 2007. "Crime Prevention by the High/Scope Perry Preschool Program." *Victims and Offenders* 2:141–160.

Shapland, Joanna, and Matthew Hall. 2007. "What Do We Know About the Effects of Crime on Victims?" University of Sheffield: Great Britain. *International Review of Victimology* 14:175–217.

Shelley, Louise. 2007. "Terrorism, Transnational Crime and Corruption Center." American University. Available at http://www.american.edu/traccc/

Sherman, Lawrence. 2003. "Reasons for Emotion." *Criminology* 42:1–37.

Siegel, Larry. 2006. *Criminology*. 9th ed. Belmont, CA: Wadsworth.

Siegel, Larry. 2009. *Criminology*. 10th ed. Belmont, CA: Cengage Learning Wadsworth.

Smalley, Suzanne. 2009. "Bringing Up Baby in the Big House." *Newsweek*, May 14. Available at http://www.newsweek.com/id/197275

Stafford, Rob and Alex Johnson. 2009. "Year After Killings, Many Prayers, Few Answers." NBC News, February 2. Available at http://www.msnbc.com

Steinhauer, Jennifer. 2009. "To Cut Costs, States Relax Prison Policies." *The New York Times*, March 25. Available at http://www.nytimes.com

Stone, Brad. 2008. "Global Trail of an Online Crime Ring." *The New York Times*, April 12. Available at http://www.nytimes.com

Surgeon General. 2002. "Cost-Effectiveness." In *Youth Violence: A Report of the Surgeon General*. Available at http://www.mentalhealth.org

Sutherland, Edwin H. 1939. *Criminology*. Philadelphia: Lippincott.

Thio, Alex. 2007. *Deviant Behavior*, 9th ed. Boston: Allyn and Bacon.

Thio, Alex. 2004. *Deviant Behavior*, 7th ed. Boston: Allyn and Bacon.

Turner, Wendy. 2007. "Experiences of Offenders in Prison Canine Programs." *Federal Probation* 71(1):38–43.

Urbina, Ian. 2009. "Citing Cost, States Consider End to Death Penalty." *The New York Times*, February 25. Available at http://www.nytimes.com

USA Today. 2009 (February 20). "Court Strikes Down California Video Game Law." Available at http://www.usatoday.com

U.S. Census Bureau. 2009. *Statistical Abstract of the United States*, 129th edition. Washington, DC: U.S. Government Printing Office.

U.S. Census Bureau. 2007. *Statistical Abstract of the United States*, 127th edition. Washington, DC: U.S. Government Printing Office.

U.S. Congress. 2009. "Pending Legislation." Available at http://www.congress.org

U.S. Department of Justice. 2009 (April). "100 Day Progress Report." Available at http://www.usdoj.gov/ag/progress-report.htm#fraud

U.S. Department of Justice. 2008. "Serial Murder: Multi-Disciplinary Perspectives for Investigators." Washington, DC: Behavioral Analysis Unit, National Center for the Analysis of Violent Crime.

U.S. Department of Justice. 2003. "Global Crime Issues." In *International Center Global Crimes Issues*, National Institute of Justice. Washington, DC: U.S. Government Printing Office.

U.S. Department State. 2008 (June 4). "Trafficking in Persons Report." Available at http://www.state.gov/g/tip/rls/tiprpt/2008/105376.htm

Van de Kamp, John. 2009. "California Can't Afford the Death Penalty." *Los Angeles Times*, June 10. Available at http://www.latimes.com/news/opinion/commentary

VORP (Victim-Offender Reconciliation Program). 2009. *About Victim-Offender Mediation and Reconciliation*. Available at http://www.vorp.com

Vu, Pauline. 2007. "Executions Halted as Doctors Balk." *Stateline*, March 21. Available at http://www.stateline.org

Wagley, John R. 2006. *Transnational Organized Crime: Principal Threats and U.S. Responses*. Congressional Research Service: The Library of Congress.

Weed and Seed. 2009. *Weed and Seed*. Available at http://www.ojp.usdoj.gov/ccdo/ws/welcome.html

Williams, Linda. 1984. "The Classic Rape: When Do Victims Report?" *Social Problems* 31:459–467.

Williams, Pete. 2007. "Court Overturns D.C. Handgun Ban." MSNBC, March 9. Available at http://www.msnbc.msn.com/id/17538139

Winslow, Robert W., and Sheldon X. Zhang. 2008. *Criminology: A Global Perspective*. Upper Saddle River, NJ: Prentice Hall.

Women's Prison Association (WPA). 2009. "Mothers, Infants, and Imprisonment: A National Look at Prison Nurseries and Community-Based Alternatives." New York: Institute on Women and Criminal Justice.

Wright, Darlene, and Kevin Fitzpatrick. 2006. "Violence and Minority Youth: The Effects of Risk and Asset Factors among African American Children and Adolescents." *Adolescence* 41(162):251–263.

Chapter 5

Administration for Children and Families. 2003. *Prevention Pays: The Costs of Not Preventing Child Abuse and Neglect.* U.S. Dept. of Health and Human Services. Available at http://www.acf.hhs.gov

Ahrons, C. 2004. *We're Still Family: What Grown Children Have to Say About Their Parents' Divorce.* New York: HarperCollins.

Amato, Paul. 2004. "Tension Between Institutional and Individual Views of Marriage." *Journal of Marriage and Family* 66:959–965.

Amato, Paul. 2003. "The Consequences of Divorce for Adults and Children." In *Family in Transition*, 12th ed., Arlene S. Skolnick and Jerome H. Skolnick, eds. (pp. 190–213). Boston: Allyn and Bacon.

Amato, Paul. 1999. "The Postdivorce Society: How Divorce Is Shaping the Family and Other Forms of Social Organization." In *The Postdivorce Family: Children, Parenting, and Society*, R. A. Thompson and P. R. Amato, eds. (pp. 161–190). Thousand Oaks, CA: Sage.

Amato, P. R., and J. Cheadle. 2005 "The Long Reach of Divorce: Divorce and Child Well-Being Across Three Generations." *Journal of Marriage and the Family* 67:191–206.

Amato, P. R., A. Booth, D. R. Johnson, and S. J. Rogers. 2007. *Alone Together: How Marriage in America Is Changing.* Cambridge MA: Harvard University Press.

American College Health Association. 2008. American College Health Association National College Health Assessment Fall 2007 reference group data report. Baltimore: American College Health Association.

Anderson, Kristin L. 1997. "Gender, Status, and Domestic Violence: An Integration of Feminist and Family Violence Approaches." *Journal of Marriage and the Family* 59: 655–669.

Applewhite, Ashton. 2003. "Covenant Marriage Would Not Benefit the Family." In *The Family: Opposing Viewpoints*, Auriana Ojeda, ed. (pp. 189–195). Farmington Hill, MI: Greenhaven Press.

Babcock, J. C., S. A. Miller, and C. Siard. 2003. "Toward a Typology of Abusive Women: Differences Between Partner-Only and Generally Violent Women in the Use of

Violence." *Psychology of Women Quarterly* 27:153–161.

Bernstein, Nina. 2007. "Polygamy, Practiced in Secrecy, Follows Africans to New York." *New York Times*, March 23, p. A1.

Block, Nadine. 2003. "Disciplinary Spanking Should Be Banned." In *Child Abuse: Opposing Viewpoints*, L. I. Gerdes, ed. (pp. 182–190). Farmington Hills, MI: Greenhaven Press.

Bonach, Kathryn. 2009. "Empirical Support for the Application of the Forgiveness Intervention Model to Postdivorce Coparenting." *Journal of Divorce & Remarriage* 50(1):38–54.

Boonstra, Heather D. 2009 (Winter). "Advocates Call for a New Approach After the Era of 'Abstinence-Only' Sex Education." *Guttmacher Policy Review* 12(1):6–11.

Bureau of Labor Statistics. 2008. *Employment Characteristics of Families in 2007.* Available at http://www.bls.gov

Busby, D. M., T. B. Holman, E. Walker 2008. "Pathways to Relationship Aggression between Adult Partners." *Family Relations* 57:72–83.

Carrington, Victoria. 2002. *New Times: New Families.* Dordrecht, the Netherlands: Kluwer Academic.

Catalano, Shannan. 2006. *Intimate Partner Violence in the United States.* Bureau of Justice Statistics. Available at http://www.ojp.usdoj.gov

Centers for Disease Control and Prevention. 2008. "Surveillance Summaries: Youth Risk Behavior Surveillance—United States, 2007." *Morbidity and Mortality Weekly Report* 57(SS-4).

Cherlin, Andrew J. 2009. *The Marriage-Go-Round: The State of Marriage and Family in America Today.* New York: Alfred A. Knopf.

Coontz, Stephanie. 2005a. *Marriage, a History.* New York: Penguin Books.

Coontz, Stephanie. 2005b. "For Better, for Worse." *Washington Post*, May 1. Available at http://www.washingtonpost.com

Coontz, Stephanie. 2004. "The World Historical Transformation of Marriage." *Journal of Marriage and Family* 66(4): 974–979.

Coontz, Stephanie. 2000. "Marriage: Then and Now." *Phi Kappa Phi Journal* 80:10–15.

Coontz, Stephanie. 1997. *The Way We Really Are.* New York: Perseus.

Coontz, Stephanie. 1992. *The Way We Never Were: American Families and the Nostalgia Trap.* New York: Basic.

Daniel, Elycia. 2005. "Sexual Abuse of Males." In *Sexual Assault: The Victims, the Perpetrators, and the Criminal Justice System*, Frances P. Reddington and Betsy

Wright Kreisel, eds. (pp. 133–140). Durham, NC: Carolina Academic Press.

Davis, Lisa Selin. 2009. "Everything but the Ring." *Time* (May 25):57–58.

de Anda, Diane. 2006. "Baby Think It Over: Evaluation of an Infant Simulation Intervention for Adolescent Pregnancy Prevention." *Health and Social Work* 31(1):26–35.

Decuzzi, A., D. Knox, and M. Zusman. 2004. "The Effect of Parental Divorce on Relationships with Parents and Romantic Partners of College Students." Roundtable Discussion, Southern Sociological Society, Atlanta, April 17.

Demo, David H., Mark A. Fine, and Lawrence H. Ganong. 2000. "Divorce as a Family Stressor." In *Families and Change: Coping with Stressful Events and Transitions*, 2nd ed., P. C. McKenry and S. J. Price, eds. (pp. 279–302). Thousand Oaks, CA: Sage.

Dennison, R. P., and S. Koerner 2008. "A Look at Hopes and Worries about Marriage: The Views of Adolescents Following a Parental Divorce." *Journal of Divorce & Remarriage* 48:91–107.

DiLillo, D., G. C. Tremblay, and L. Peterson. 2000. "Linking Childhood Sexual Abuse and Abusive Parenting: The Mediating Role of Maternal Anger." *Child Abuse and Neglect* 24:767–769.

Doyle, Joseph. 2007. "Child Protection and Child Outcomes: Measuring the Effects of Foster Care." *American Economic Review* 97(5):1583–1610.

Edin, Kathryn. 2000. "What Do Low-Income Single Mothers Say About Marriage?" *Social Problems* 47(1):112–133.

Emery, Robert E. 1999. "Postdivorce Family Life for Children: An Overview of Research and Some Implications for Policy." In *The Postdivorce Family: Children, Parenting, and Society*, R. A. Thompson and P. R. Amato, eds. (pp. 3–27). Thousand Oaks, CA: Sage.

Emery, Robert E., David Sbarra, and Tara Grover. 2005. "Divorce Mediation: Research and Reflections." *Family Court Review* 43(1):22–37.

Family Court Reform Council of America. 2000. *Parental Alienation Syndrome.* Rancho Santa Margarita, CA: Family Court Reform Council of America.

Family Violence Prevention Fund. 2005 (February 24). *Testimony of the Family Violence Prevention Fund on Welfare Reform and Marriage Promotion Initiatives: Submitted to the House Ways and Means Committee.* Available at http://endabuse.org

Federal Bureau of Investigation. 2008. *Crime in the United States*, 2007. Available at http://www.fbi.gov

Fincham, F. D., J. Hall, and S. R. H. Beach. 2006. "Forgiveness in Marriage: Current Status and Future Directions." *Family Relations* 55:415–427.

Fogle, Jean M. 2003. "Domestic Violence Hurts Dogs, Too." *Dog Fancy*, April, p. 12.

Gadalla, Tahany M. 2009. "Impact of Marital Dissolution on Men's and Women's Income: A Longitudinal Study." *Journal of Divorce & Remarriage* 50(1):55–65.

Gallup Organization. 2009. "Marriage." Available at http://www.gallup.com

Gelles, Richard J. 2000. "Violence, Abuse, and Neglect in Families." In *Families and Change: Coping with Stressful Events and Transitions*, 2nd ed., P. C. McKenry and S. J. Price, eds. (pp. 183–207). Thousand Oaks, CA: Sage.

Gilbert, Neil. 2003. "Working Families: Hearth to Market." In *All Our Families*, 2nd ed., M. A. Mason, A. Skolnick, and S. D. Sugarman, eds. (pp. 220–243). New York: Oxford University Press.

Global Initiative to End All Corporal Punishment of Children. 2009. *Global Progress Towards Prohibiting All Corporal Punishment*. Available at http://www.endcorporalpunishment.org

Gore, Al, and Tipper Gore. 2002. *Joined at the Heart*. New York: Henry Holt.

Grogan-Kaylor, Andrew, and Melanie Otis. 2007. "The Predictors of Parental Use of Corporal Punishment." *Family Relations* 56:80–91.

Grych, John H. 2005. "Interparental Conflict as a Risk Factor for Child Maladjustment: Implications for the Development of Prevention Programs." *Family Court Review* 43(1):97–108.

Guttmacher Institute. 2009a (May). *State Policies in Brief: Sex and STD/HIV Education*. Available at http://www.guttmacher.org

Guttmacher Institute. 2009b (May). *State Policies in Brief: Minors' Access to Contraceptive Services*. Available at http://www.guttmacher.org

Guttmacher Institute. 2009c (May). *State Policies in Brief: Refusing to Provide Health Services*. Available at http://www.guttmacher.org

Guttmacher Institute. 2006. *Facts on Sex Education in the United States*. Available at http://www.guttmacher.org

Hackstaff, Karla B. 2003. "Divorce Culture: A Quest for Relational Equality in Marriage." In *Family in Transition*, 12th ed., Arlene S. Skolnick and Jerome H. Skolnick, eds. (pp. 178–190). Boston: Allyn and Bacon.

Hamilton, Brady E., Joyce A. Martin, and Stephanie Ventura. 2009. "Births: Preliminary Data for 2007." *National Vital Statistics Reports* 57(12). Available at http://www.cdc.gov/nchs

Hawkins, Alan J., Jason S. Carroll, William J. Doherty, and Brian Willoughby. 2004. "A Comprehensive Framework for Marriage Education." *Family Relations* 53(5): 547–558.

Hewlett, Sylvia Ann, and Cornel West. 1998. *The War Against Parents: What We Can Do for Beleaguered Moms and Dads*. Boston: Houghton Mifflin.

Heyman, R. E., and A. M. S. Slep. 2002. "Do Child Abuse and Interpersonal Violence Lead to Adult Family Violence?" *Journal of Marriage and Family* 64:864–870.

Hochschild, Arlie Russell. 1997. *The Time Bind: When Work Becomes Home and Home Becomes Work*. New York: Henry Holt.

Hochschild, Arlie Russell. 1989. *The Second Shift: Working Parents and the Revolution at Home*. New York: Viking.

Huffstutter, P. J. 2007. "States Reject Abstinence Classes, Funds." *The News & Observer*, April 9. Available at http://www.newsobserver.com

Jackson, Shelly, Lynette Feder, David R. Forde, Robert C. Davis, Christopher D. Maxwell, and Bruce G. Taylor. 2003 (June). *Batterer Intervention Programs: Where Do We Go from Here?* U.S. Department of Justice. Available at http://www.usdoj.gov

Jalovaara, M. 2003. "The Joint Effects of Marriage Partners' Socioeconomic Positions on the Risk of Divorce." *Demography* 40:67–81.

Jasinski, J. L., L. M. Williams, and J. Siegel. 2000. "Childhood Physical and Sexual Abuse as Risk Factors for Heavy Drinking Among African-American Women: A Prospective Study." *Child Abuse and Neglect* 24:1061–1071.

Jekielek, Susan M. 1998. "Parental Conflict, Marital Disruption, and Children's Emotional Well-Being." *Social Forces* 76:905–935.

Johnson, Michael P. 2001. "Patriarchal Terrorism and Common Couple Violence: Two Forms of Violence against Women." In *Men and Masculinity: A Text Reader*, T. F. Cohen, ed. (pp. 248–260). Belmont, CA: Wadsworth.

Johnson, Michael P., and Kathleen Ferraro. 2003. "Research on Domestic Violence in the 1990s: Making Distinctions." In *Family in Transition*, 12th ed., A. S. Skolnick and J. H. Skolnick, eds. (pp. 493–514). Boston: Allyn and Bacon.

Kalmijn, Matthijs, and Christiaan W. S. Monden. 2006. "Are the Negative Effects of Divorce on Well-Being Dependent on Marital Quality?" *Journal of Marriage and the Family* 68:1197–1213.

Kaufman, Joan, and Edward Zigler. 1992. "The Prevention of Child Maltreatment: Programming, Research, and Policy." In *Prevention of Child Maltreatment:*

Developmental and Ecological Perspectives, Diane J. Willis, E. Wayne Holden, and Mindy Rosenberg, eds. (pp. 269–295). New York: John Wiley.

Kimmel, Michael S. 2004. *The Gendered Society*. 2nd ed. New York: Oxford University Press.

Kitzmann, K. M., N. K. Gaylord, A. R. Holt, and E. D. Kenny. 2003. "Child Witnesses to Domestic Violence: A Meta-Analytic Review." *Journal of Clinical and Consulting Psychology* 71:339–352.

Knox, David (with Kermit Leggett). 1998. *The Divorced Dad's Survival Book: How to Stay Connected with Your Kids*. New York: Insight Books.

Koch, Wendy. 2009. "Fees Cut Down Private Adoptions." *USA Today*, April 27, 1A.

LaFraniere, Sharon. 2005. "Entrenched Epidemic: Wife-Beatings in Africa." *New York Times*, August 11, pp. A1 and A8.

Lara, Adair, 2005. "One for the Price of Two: Some Couples Find Their Marriages Thrive When They Share Separate Quarters." *San Francisco Chronicle*, June 29. Available at http://www.sfgate.com

Laungani, P. 2005. "Changing Patterns of Family Life in India." In *Families in Global Perspective*, J. L. Roopnarine and U. P. Gielen, eds. (pp. 85–103). Boston: Pearson, Allyn and Bacon.

Levin, Irene. 2004. "Living Apart Together: A New Family Form." *Current Sociology* 52(2):223–240.

Levine, Judith A., Clifton R. Emery, and Harold Pollack. 2007. "The Well-Being of Children Born to Teen Mothers." *Journal of Marriage and Family* 69:105–122.

Lewin, Tamar. 2000. "Fears for Children's Well-Being Complicates a Debate over Marriage." *New York Times*, November 4. Available at http://www.nytimes.com

Lindsey, Linda L. 2005. *Gender Roles: A Sociological Perspective*. 4th ed. Upper Saddle River, NJ: Pearson Prentice Hall.

Lloyd, Sally A. 2000. "Intimate Violence: Paradoxes of Romance, Conflict, and Control." *National Forum* 80(4):19–22.

Lloyd, Sally A., and Beth C. Emery. 2000. *The Dark Side of Courtship: Physical and Sexual Aggression*. Thousand Oaks, CA: Sage.

Mason, Mary Ann. 2003. "The Modern American Step-Family: Problems and Possibilities." In *All Our Families*, 2nd ed., Mary Ann Mason, Arlene Skolnick, and Stephen D. Sugarman, eds. (96–116). New York: Oxford University Press.

Mason, Mary Ann, Arlene Skolnick, and Stephen D. Sugarman. 2003. "Introduction." In *All Our Families*, 2nd ed., Mary Ann Mason, Arlene Skolnick, and Stephen D. Sugarman, eds. (pp. 1–13). New York: Oxford University Press.

Mauldon, Jane. 2003. "Families Started by Teenagers." In *All Our Families*, 2nd ed., Mary Ann Mason, Arlene Skolnick, and Stephen D. Sugarman, eds. (pp. 40–65). New York: Oxford University Press.

McKenry, P. C., J. M. Serovich, T. L. Mason, and K. E. Mosack. 2004 (November). *Perpetration of Gay and Lesbian Violence: A Disempowerment Perspective.* Paper presented at the Annual Conference of the National Council on Family Relations, Orlando, Florida.

Mental Health America. 2003. *Effective Discipline Techniques for Parents: Alternatives to Spanking.* Strengthening Families Fact Sheet. Available at http://www.nmha.org

Mindel, Charles H., Robert W. Habenstein, and Roosevelt Wright, Jr. 1998. *Ethnic Families in America: Patterns and Variations.* Upper Saddle River, NJ: Prentice Hall.

Mollborn, Stephanie. 2007. "Making the Best of a Bad Situation: Material Resources and Teenage Parenthood." *Journal of Marriage and Family* 69:92–104.

National Center for Injury Prevention and Control. 2007. *Child Maltreatment: Fact Sheet.* Available at http://www.cdc.gov/ncipc

National Center for Injury Prevention and Control. 2006. *Intimate Partner Violence: Overview.* Available at http://www.cdc.gov/ncipc

National Marriage Project. 2009. *The State of Our Unions: 2008.* Rutgers, the State University of New Jersey: The National Marriage Project.

Nelson, B. S., and K. S. Wampler. 2000. "Systemic Effects of Trauma in Clinic Couples: An Exploratory Study of Secondary Trauma Resulting from Childhood Abuse." *Journal of Marriage and Family Counseling* 26:171–184.

Nock, Steven L. 1995. "Commitment and Dependency in Marriage." *Journal of Marriage and the Family* 57:503–514.

Parker, Marcie R., Edward Bergmark, Mark Attridge, and Jude Miller-Burke. 2000. "Domestic Violence and its Effect on Children." *National Council on Family Relations Report* 45(4):F6–F7.

Pasley, Kay, and Carmelle Minton. 2001. "Generative Fathering after Divorce and Remarriage: Beyond the 'Disappearing Dad.'" In *Men and masculinity: A Text Reader*, T. F. Cohen, ed. (pp. 239–248). Belmont CA: Wadsworth.

Popenoe, David. 2008. *The State of Our Unions 2007: The Social Health of Marriage in America.* Rutgers, The State University of New Jersey: The National Marriage Project.

Rand, Michael R. 2008. *Criminal Victimization, 2007.* National Crime Victimization Survey. Bureau of Justice Statistics Bulletin.

Ricci, L., A. Giantris, P. Merriam, S. Hodge, and T. Doyle. 2003. "Abusive Head Trauma in Maine Infants: Medical, Child Protective, and Law Enforcement Analysis." *Child Abuse and Neglect* 27:271–283.

Rubin, D. M., C. W. Christian, L. T. Bilaniuk, K. A. Zaxyczny, and D. R. Durbin. 2003. "Occult Head Injury in High-Risk Abused Children." *Pediatrics* 111:1382–1386.

Russell, D. E. 1990. *Rape in Marriage.* Bloomington: Indiana University Press.

Saad, Lydia. 2006 (May 30). "Americans Have Complex Relationship with Marriage." *Gallup Poll Briefing.* Gallup Organization. Available at http://www.gallup.com

Schacht, Thomas E. 2000. "Protection Strategies to Protect Professionals and Families Involved in High-Conflict Divorce." *UALR Law Review* 22(3):565–592.

Scott, K. L., and D. A. Wolfe. 2000. "Change among Batterers: Examining Men's Success Stories." *Journal of Interpersonal Violence* 15:827–842.

Shepard, Melanie F., and James A. Campbell. 1992. "The Abusive Behavior Inventory: A Measure of Psychological and Physical Abuse." *Journal of Interpersonal Violence* 7(3):291–305.

Sigle-Rushton, W., and S. McLanahan. 2002. "The Living Arrangements of New Unmarried Mothers." *Demography* 39: 415–433.

Simmons, T., and M. O'Connell. 2003. *Married-Couple and Unmarried Partner Households: 2000.* Census 2000 Special Reports. Available at http://www.census.gov

Simonelli, C. J., T. Mullis, A. N. Elliott, and T. W. Pierce. 2002. "Abuse by Siblings and Subsequent Experiences of Violence within the Dating Relationship." *Journal of Interpersonal Violence* 17:103–21.

Singh, Susheela, and Jacqueline E. Darroch. 2000. "Adolescent Pregnancy and Childbearing: Levels and Trends in Developed Countries." *Family Planning Perspectives* 32(1):14–23.

Smith, J. 2003. "Shaken Baby Syndrome." *Orthopaedic Nursing* 22:196–205.

Steimle, Brynn M., and Stephen F. Duncan. 2004. "Formative Evaluation of a Family Life Education Web Site." *Family Relations* 53(4):367–376.

Stone, R. D. 2004. *No Secrets, No Lies: How Black Families Can Heal from Sexual Abuse.* New York: Broadway Books.

Straus, Murray. 2000. "Corporal Punishment and Primary Prevention of Physical Abuse." *Child Abuse and Neglect* 24:1109–1114.

Swan, S. C., L. J. Gambone, J. E. Caldwell, T. P. Sullivan, and D. L Snow. 2008. "A Review of Research on Women's Use of Violence with Male Intimate Partners." *Violence and Victims* 23:301–315.

Swiss, Liam, and Celine Le Bourdais. 2009. "Father-Child Contact After Separation: The Influence of Living Arrangements." *Journal of Family Issues* 30(5):623–652.

Teaster, Pamela B., Tyler A. Dugar, Marta S. Mendiondo, Erin L. Abner, Kara A. Cecil, and Joanne M. Otto. 2006 (February). *The 2004 Survey of State Adult Protective Services: Abuse of Adults 60 Years of Age and Older.* National Center on Elder Abuse. Washington, DC.

Terry-Humen, Elizabeth, Jennifer Manlove, and Kristin A. Moore. 2005. *Playing Catch-Up: How Children Born to Teen Mothers Fare.* Washington, DC: National Campaign to Prevent Teen Pregnancy.

Thornton, A., and L. Young-DeMarco. 2001. "Four Decades of Trends in Attitudes toward Family Issues in the United States: The 1960s Through the 1990s." *Journal of Marriage and the Family* 63:1009–1037.

Trenholm, Christopher, Barbara Devaney, Ken Fortson, Lisa Quay, Justin Wheeler, and Melissa Clark. 2007 (April). *Impacts of Four Title V, Section 510 Abstinence Education Programs.* Princeton, NJ: Mathematica Policy Research, Inc.

Trinder, L. 2008. "Maternal Gate Closing and Gate Opening in Postdivorce Families." *Journal of Family Issues* 29:1298–1298.

Ulman, A. 2003. "Violence by Children Against Mothers in Relation to Violence between Parents and Corporal Punishment by Parents." *Journal of Comparative Family Studies* 34:41–56.

Umberson, D., K. L. Anderson, K. Williams, and M. D. Chen. 2003. "Relationship Dynamics, Emotion State, and Domestic Violence: A Stress and Masculine Perspective." *Journal of Marriage and the Family* 65:233–247.

United Nations Development Programme. 2000. *Human Development Report 2000.* New York: Oxford University Press.

U.S. Census Bureau. 2009. *America's Families and Living Arrangements: 2008.* Available at http://www.census.gov

U.S. Department of Health and Human Services. Administration on Children, Youth, and Families. 2009. *Child Maltreatment 2007.* Washington, DC: U.S. Government Printing Office.

U.S. Department of Health and Human Services. 2000 (June 17). *HHS Fatherhood Initiative.* Available at http://www.hhs.gov/news/press/2000pres/20000617.html

Ventura, S. J., and Christine A. Bachrach. 2000 (October 18). "Nonmarital Childbearing in the United States, 1940–1999." *National Vital Statistics Report* 48(16).

Ventura, Stephanie J., Sally C. Curtin, and T. J. Mathews. 2000 (April 24). "Variations in Teenage Birth Rates, 1991–1998: National and State Trends." *National Vital Statistics Report* 48(6).

Walker, Alexis J. 2001. "Refracted Knowledge: Viewing Families Through the Prism of Social Science." In *Understanding Families into the New Millennium: A Decade in Review*, Robert M. Milardo, ed. (pp. 52–65). Minneapolis, MN: National Council on Family Relations.

Wallerstein, Judith S. 2003. "Children of Divorce: A Society in Search of Policy." In *All Our Families*, 2nd ed., Mary Ann Mason, Arlene Skolnick, and Stephen D. Sugarman, eds. (pp. 66–95). New York: Oxford University Press.

Waxman, Henry A. 2004 (December). *The Context of Federally Funded Abstinence-Only Education Programs.* U.S. House of Representatives Committee on Government Reform, Minority Staff, Special Investigation Division. Available at http://www.democrats.reform.house.gov

Whiffen, V. E., J. M. Thompson, and J. A. Aube. 2000. "Mediators of the Link between Childhood Sexual Abuse and Adult Depressive Symptoms." *Journal of Interpersonal Violence* 15:1100–1120.

Whitehurst, Dorothy H., Stephen O'Keefe, and Robert A. Wilson. 2008. "Divorced and Separated Parents in Conflict: Results from a True Experiment Effect of a Court Mandated Parenting Education Program." *Journal of Divorce & Remarriage* 48(3/4):127–144.

Williams, K., and A. Dunne-Bryant. 2006. "Divorce and Adult Psychological Well-Being: Clarifying the Role of Gender and Child Age." *Journal of Marriage and the Family* 68:1178–1196.

Zeitzen, Miriam K. 2008. *Polygamy: A Cross-Cultural Analysis.* Oxford: Berg.

Chapter 6

Administration for Children and Families. 2002. *Early Head Start Benefits Children and Families.* U.S. Department of Health and Human Services. Available at http://www.acf.hhs.gov

Albelda, Randy, M. V. Lee Badgett, Alyssa Schneebaum, and Gary J. Gates. 2009. "Poverty in the Lesbian, Gay, and Bisexual Community." The Williams Institute. Available at http://www.law.ucla.edu/williamsinstitute

Albelda, Randy, and Chris Tilly. 1997. *Glass Ceilings and Bottomless Pits: Women's Work, Women's Poverty.* Boston: South End Press.

Alex-Assensoh, Yvette. 1995. "Myths About Race and the Underclass." *Urban Affairs Review* 31:3–19.

Anderson, Sarah, John Cavanagh, Chris Hartman, Scott Klinger, and Stacey Chan. 2004. *Executive Excess: 11th Annual CEO Compensation Survey.* Boston: Institute for Policy Studies and United for a Fair Economy.

Anderson, Sarah, John Cavanagh, Chuck Collins, Sam Pizzigati, and Mike Lapham. 2008. *Executive Excess: 15th Annual CEO Compensation Survey.* Boston: Institute for Policy Studies and United for a Fair Economy.

Bickel, G., M. Nord, C. Price, W. Hamilton, and J. Cook. 2000. *United States Department of Agriculture Guide to Measuring Household Food Security.* Alexandria, VA: U.S. Department of Agriculture, Food and Nutrition Service.

Boston, Rob. 2005 (February 15). "Faith-Based 'Flim-Flam' Initiative Didn't Have a Prayer Says Former White House Aide." Americans United for a Separation of Church and State. Available at http://blog.au.org

Bowhay, Samuel. 2009. "Video Exploitation of Homeless People." In *Hate, Violence, and Death on Main Street USA: A Report on Hate Crimes and Violence Against People Experiencing Homelessness*, p. 34. National Coalition for the Homeless. Available at http://www.nationalhomeless.org

Briggs, Vernon M., Jr. 1998. "American-Style Capitalism and Income Disparity: The Challenge of Social Anarchy." *Journal of Economic Issues* 32(2):473–481.

Cauthen, Nancy K., and Sarah Fass. 2007 (April). *Measuring Income and Poverty in the United States.* National Center for Children in Poverty. Available at http://nccp.org/

Center on Budget and Policy Priorities. 2005 (May 31). *New Study Finds Poor Medicaid Beneficiaries Face Growing Out-of-Pocket Medical Costs.* Available at http://www.cbpp.org

Chen, Shaohua, and Martin Ravallion. 2008. *The Developing World Is Poorer Than We Thought, but No Less Successful in the Fight Against Poverty.* Policy Research Working Paper 4703. Washington, DC: World Bank.

Children's Defense Fund. 2003. *Children in the United States.* Available at http://www.childrensdefense.org

Corak, Miles. 2006. "Do Poor Children Become Poor Adults? Lessons from a Cross-Country Comparison of Generational Earnings Mobility. In *Dynamics of Inequality and Poverty: Research on Economic Inequality*, eds. J. Creedy and G. Kalb 13:143-188. Elsevier.

Cunningham, Mary. 2009 (February). *Poverty and Ending Homelessness—Next Steps.* Washington, DC: The Urban Institute.

Davies, James B., Susanna Sandstrom, Anthony Shorrocks, and Edward N. Wolff. 2006 (December 5). *The World Distribution of Household Wealth.* United Nations University-World Institute for Development Economics Research. Available at http://www.wider.unu.edu

Davis, Kingsley, and Wilbert Moore. 1945. "Some Principles of Stratification." *American Sociological Review* 10:242–249.

DeNavas-Walt, Carmen, Bernadette D. Proctor, and Jessica C. Smith. 2009. *Income, Poverty, and Health Insurance in the United States: 2008.* U.S. Census Bureau, Current Population Reports P60-236. Washington, DC: U.S. Government Printing Office.

Deng, Francis M. 1998. "The Cow and the Thing Called 'What': Dinka Cultural Perspectives on Wealth and Poverty." *Journal of International Affairs* 52(1): 101–115.

Dordick, Gwendolyn. 1997. *Something Left to Lose: Personal Relations and Survival Among New York's Homeless.* Philadelphia: Temple University Press.

Dowd, Maureen. 2005. "United States of Shame." *New York Times*, September 3. Available at http://www.nytimes.com

Dvorak, Petula. 2009 (February 5). "Increase Seen in Attacks on Homeless." *Washington Post*, p. DZ01.

Edin, Kathryn, and Laura Lein. 1977. *Making Ends Meet.* New York: Russell Sage Foundation.

Edsall, Thomas B. 2009. "Barack Obama: King of Corporate Welfare." *Huffington Post*, April 25. Available at http://www.huffingtonpost.com

Epstein, William M. 2004. "Cleavage in American Attitudes Toward Social Welfare." *Journal of Sociology and Social Welfare* 31(4):177–201.

FAO (Food and Agricultural Organization). 2009 (June 19). "1.02 Billion People Hungry." Available at http://www.fao.org

Forster, Michael, and Marco Mira d'Ercole. 2005 (March 10). "Income Distribution and Poverty in OECD Countries in the Second Half of the 1990s." Organization for Economic Cooperation and Development. Available at http://www.oecd.org

Goesling, Brian. 2001. "Changing Income Inequalities Within and Between Nations: New Evidence." *American Sociological Review* 66:745–761.

Gonzalez, David. 2005. "From Margins of Society to Center of Tragedy." *New York Times*, September 2. Available at http://www.nytimes.com

Grunwald, Michael. 2006 (August 27). "The Housing Crisis Goes Suburban." *Washington Post*, August 27. Available at http://www.washingtonpost.com

Hill, Lewis E. 1998. "The Institutional Economics of Poverty: An Inquiry into the Causes and Effects of Poverty." *Journal of Economic Issues* 32(2):279–286.

Hoback, Alan, and Scott Anderson. 2007. "Proposed Method for Estimating Local Population of Precariously Housed." National Coalition for the Homeless. Available at http://www.nationalhomeless.org

hooks, bell. 2000. *Where We Stand: Class Matters*. New York: Routledge.

International Labour Organization. 2008a. *Global Wage Report 2008/2009*. Geneva: International Labour Office.

International Labour Organization. 2008b. *World of Work Report 2008: Income Inequalities in the Age of Financial Globalization*. Geneva: International Labour Office.

Jargowsky, Paul A. 1997. *Poverty and Place: Ghettos, Barrios, and the American City*. New York: Russell Sage Foundation.

Jones, Jeffrey M. 2007 (February 9). "Public: Family of 4 Needs to Earn Average of $52,000 to Get By." Gallup News Service. Available at http://www.galluppoll.com

Kennedy, Bruce P., Ichiro Kawachi, Roberta Glass, and Deborah Prothrow-Stith. 1998. "Income Distribution, Socioeconomic Status, and Self-Rated Health in the U.S.: Multilevel Analysis." *British Medical Journal* 317(7163):917–921.

Kraut, Karen, Scott Klinger, and Chuck Collins. 2000. *Choosing the High Road: Businesses That Pay a Living Wage and Prosper*. Boston: United for a Fair Economy.

Leftin, J., and K. Wolkwitz. 2009 (June). "Trends in Supplemental Nutrition Assistance Program Participation Rates: 2000-2007." Food and Nutrition Service, USDA. Available at http://www.fns.usda.gov

Leventhal, Tama, and Jeanne Brooks-Gunn. 2003. "Moving to Opportunity: An Experimental Study of Neighborhood Effects on Mental Health." *American Journal of Public Health* 93(9):1576–1585.

Levin, Brian. 2009. "Recognizing Anti-Homeless Violence as Hate Crime." In *Hate, Violence, and Death on Main Street USA: A Report on Hate Crimes and Violence against People Experiencing Homelessness*, pp. 35–38. National Coalition for the Homeless. Available at http://www.nationalhomeless.org

Lewan, Todd. 2007 (April 8). "Unprovoked Beatings of Homeless Soaring." Associated Press. Available at http://www.breitbart.com

Lewis, Oscar. 1998. "The Culture of Poverty: Resolving Common Social Problems." *Society* 35(2):7–10.

Lewis, Oscar. 1966. "The Culture of Poverty." *Scientific American* 2(5):19–25.

Lichtblau, Eric. 2009. "Attacks of Homeless Bring Push on Hate Crimes Laws." *New York Times*, August 7. Available at http://www.nytimes.com

Llobrera, Joseph, and Bob Zahradnik. 2004. *A HAND UP: How State Earned Income Tax Credits Helped Working Families Escape Poverty in 2004*. Center on Budget and Policy Priorities. Available at http://www.cbpp.org

Luker, Kristin. 1996. *Dubious Conceptions: The Politics of Teenage Pregnancy*. Cambridge, MA: Harvard University Press.

Malatu, Mesfin Samuel, and Carmi Schooler. 2002. "Causal Connections Between Socioeconomic Status and Health: Reciprocal Effects and Mediating Mechanisms." *Journal of Health and Social Behavior* 43:22–41.

Mann, Judy. 2000 (May 15). "Demonstrators at the Barricades Aren't Very Subtle, but They Sometimes Win." *Washington Spectator* 26(10):1–3.

Massey, D. S. 1991. "American Apartheid: Segregation and the Making of the American Underclass." *American Journal of Sociology* 96:329–357.

Mayer, Susan E. 1997. *What Money Can't Buy: Family Income and Children's Life Chances*. Cambridge, MA: Harvard University Press.

McIntyre, Robert. 2005a (February 2). *Corporate Tax Avoidance in the States Even Worse Than Federal*. Citizens for Tax Justice. Available at http://www.ctj.org

McIntyre, Robert. 2005b (April 12). *Tax Cheats and Their Enablers*. Citizens for Tax Justice. Available at http://www.ctj.org

Mishel, Lawrence, Jared Bernstein, and Heidi Shierholz. 2009. *The State of Working America 2008–2009*. New York: Cornell University Press.

Narayan, Deepa. 2000. *Voices of the Poor: Can Anyone Hear Us?* New York: Oxford University Press.

National Association of Child Care Resource and Referral Agencies. 2008. *2008 Price of Child Care*. Available at http://www.naccrra.org

National Coalition for the Homeless. 2009. *Hate, Violence, and Death on Main Street USA: A Report on Hate Crimes and Violence against People Experiencing Homelessness*. National Coalition for the Homeless. Available at http://www.nationalhomeless.org

National Coalition for the Homeless. 2008 (June). *A Dream Denied: The Criminalization of Homelessness in U.S. Cities*. Available at http://www.national homeless.org

Nord, Mark, Margaret Andrews, and Steven Carlson. 2008. *Household Food Security in the United States, 2007*. Economic Research Report No. 66. USDA Economic Research Service. Available at http://www.ers.usda.gov

OECD. 2009 (March 3). "Development Aid at its Highest Level Ever in 2008." Available at http://www.oecd.org

Office of Family Assistance. 2009. *Temporary Assistance for Needy Families Program (TANF): Eighth Annual Report to Congress*. Available at http://www.acf.hhs.gov

Oxfam. 2006 (November). *Our Generation's Choice*. Oxfam Briefing Paper. Available at http://www.oxfam.org

Oxfam. 2005. *Paying the Price: Why Rich Countries Must Invest Now in a War on Poverty*. Available at http://www.oxfam.org

Pew Research Center. 2009 (May 21). *Trends in Political Values and Core Attitudes: 1987–2009*. Washington, DC: Pew Research Center for People & the Press.

Popenoe, David. 2008. *The State of Our Unions 2007: The Social Health of Marriage in America*. Rutgers, The State University of New Jersey: The National Marriage Project.

Pugh, Tony. 2007 (February 22). "U.S. Economy Leaving Record Numbers in Severe Poverty." *McClatchy Newspapers*. Available at http://www.mcclatchydc.com

Ramos, Alcida Rita, Rafael Guerreiro Osorio, and Jose Pimenta. 2009. "Indigenising Development." *Poverty in Focus*, 17 (May), pp. 3–5. International Policy Centre for Inclusive Growth.

Roberts, John. 2005 (January 17). *Thai Government Puts Tourism Ahead of the Poor in Tsunami Relief Effort*. World Socialist Web Site. Available at http://www.wsws.org

Robinson, Phyllis. 2009 (June 9). "Urgent Action: Protest Massacre of Indigenous People in Peru!" Small Farmers. Big Change. Available at http://smallfarmersbigchange.coop

Roseland, Mark, and Lena Soots. 2007. "Strengthening Local Economies." In *2007 State of the World*, Linda Starke, ed., (152–169). New York: W. W. Norton & Co.

Rothstein, Richard. 2004. *Class and Schools*. Washington, DC: Economic Policy Institute.

Satterthwaite, David, and Gordon McGranahan. 2007. "Providing Clean Water and Sanitation." In *2007 State of the World: Our Urban Future*, L. Starke, ed. (pp. 26–45). New York: W. W. Norton & Company.

Schifferes, Steve. 2004. "Can Globalization Be Tamed?" *BBC News Online*, February 24. Available at http://www.bbc.co.uk

Seccombe, Karen. 2001. "Families in Poverty in the 1990s: Trends, Causes, Consequences, and Lessons Learned." In *Understanding Families into the New Millennium: A Decade in Review*, Robert M. Milardo, ed. (pp. 313–332). Minneapolis, MN: National Council on Family Relations.

Sherman, Arloc. 2009 (April 17). "Income Gaps Hit Record Levels in 2006, New

Data Show." Center on Budget and Policy Priorities. Available at http://cbpp.org

Shierholz, Heidi. 2009 (July 7). "Nearly Six Unemployed Worker Per Available Job." Economic Policy Institute. Available at http://www.epi.org

Sobolewski, Juliana M., and Paul R. Amato. 2005. "Economic Hardship in the Family of Origin and Children's Psychological Well-Being in Adulthood." *Journal of Marriage and Family* 67(1):141–156.

Stocking, Barbara. 2005 (January 5). *The Tsunami and the Bigger Picture*. Oxfam. Available at http://www.oxfam.org

Susskind, Yifat. 2005 (May). *Ending Poverty, Promoting Development: MADRE Criticizes the United Nations Millennium Development Goals*. Available at http://www.madre.org

Turner, Margery Austin, Susan J. Popkin, G. Thomas Kingsley, and Deborah Kaye. 2005 (April). *Distressed Public Housing: What It Costs to Do Nothing*. The Urban Institute. Available at http://www.urban.org

UNDP (United Nations Development Programme). 2009. *Human Development Report 2009*. New York: Palgrave Macmillan.

UNDP. 2006. *Human Development Report 2006*. New York: Palgrave Macmillan.

UNDP. 2000. *Human Development Report 2000*. New York: Oxford University Press.

UNDP. 1997. *Human Development Report 1997*. New York: Oxford University Press.

UNICEF. 2006 (September). *Progress for Children: A Report Card on Water and Sanitation*. Number 5. Available at http://www.unicef.org

United Nations. 2009. *The Millennium Development Goals Report 2009*. New York: United Nations.

United Nations. 2005. *Report on the World Social Situation 2005*. New York: United Nations.

United Nations Population Fund. 2002. *State of World Population 2002: People, Poverty, and Possibilities*. New York: United Nations.

U.S. Census Bureau. 2009a. *Poverty Thresholds for 2008 by Size of Family and Number of Related Children Under 18 Years*. Available at http://www.census.gov

U.S. Census Bureau. 2009b. *Current Population Survey, Annual Social and Economic Supplement*. Table POV29. Available at http://www.census.gov

U.S. Census Bureau. 2008. *Historical Poverty Tables: People*. Available at http://www.census.gov

U.S. Census Bureau. 2006. Annual Social and Economic Supplement. *Current Population Survey*. Available at http://www.census.gov

U.S. Conference of Mayors–Sodexho, Inc. 2008. *Hunger and Homelessness Survey: A Status Report on Hunger and Homelessness in America's Cities*. Washington, DC: U.S. Conference of Mayors.

Van Kempen, Eva T. 1997. "Poverty Pockets and Life Chances: On the Role of Place in Shaping Social Inequality." *American Behavioral Scientist* 41(3):430–450.

Wilson, William J. 1996. *When Work Disappears: The World of the New Urban Poor*. New York: Knopf.

Wilson, William J. 1987. *The Truly Disadvantaged: The Inner City, the Underclass, and Public Policy*. Chicago: University of Chicago Press.

World Bank. 2009. *Global Monitoring Report 2009*. Washington DC: The World Bank.

World Bank. 2005. *Global Monitoring Report 2005*. Available at http://www.worldbank.org

World Bank. 2001. *World Development Report: Attacking Poverty, 2000/2001*. Herndon, VA: World Bank and Oxford University Press.

World Health Organization. 2002. *The World Health Report 2002*. Available at http://www.who.int/pub/en

World Population News Service. 2003. "Reducing Poverty Is Key to Global Stability." *Popline*, May–June, p. 4.

Zedlewski, Sheila R. 2003. *Work and Barriers to Work Among Welfare Recipients in 2002*. Urban Institute. Available at http://www.urban.org

Zedlewski, Sheila R., and Kelly Rader. 2005 (March 31). *Feeding America's Low-Income Children*. New Federalism: National Survey of America's Families, No. B-65. Urban Institute. Available at http://www.urban.org

Chapter 7

AFL-CIO. 2009. "Number of OSHA Inspectors by State Compared with ILO Benchmark Number of Labor Inspectors." Available at http://www.aflcio.org

AFL-CIO. 2005. *Death on the Job: The Toll of Neglect*, 14th ed. Available at http://www.aflcio.org

Aron-Dine, Aviva, and Isaac Shapiro. 2007 (March 29). "Share of National Income Going to Wages and Salaries at Record Low in 2006." Center for Budget and Policy Priorities. Available at http://www.cbpp.org

Austin, Colin. 2002. "The Struggle for Health in Times of Plenty." In *The Human Cost of Food: Farmworkers' Lives, Labor, and Advocacy*, C. D. Thompson Jr. and M. F. Wiggins, eds. (pp. 198–217). Austin: University of Texas Press.

Baily, Martin Neil, and Douglas J. Elliott. 2009 (June 15). "The U.S. Financial and Economic Crisis: Where Does It Stand and Where Do We Go from Here?" The Brookings Institution. Available at http://www.brookings.edu

Bales, Kevin. 1999. *Disposable People: New Slavery in the Global Economy*. Berkeley: University of California Press.

Barstow, David, and Lowell Bergman. 2003. "Deaths on the Job, Slaps on the Wrist." *New York Times Online*, January 10. Available at http://www.nytimes.com

Bassi, Laurie J., and Jens Ludwig. 2000. "School-to-Work Programs in the United States: A Multi-Firm Case Study of Training, Benefits, and Costs." *Industrial and Labor Relations Review* 53(2):219–239.

Benjamin, Medea. 1998. *What's Fair About Fair Labor Association (FLA)?* Sweatshop Watch. Available at http://www.sweatshopwatch.org

Bond, James T., Cindy Thompson, Ellen Galinsky, and David Prottas. 2002. *Highlights of the National Study of the Changing Workforce. Families and Work Institute*. Available at http://www.familiesandwork.org

Bonior, David. 2006. "Undermining Democracy: Worker Repression in the United States." *Multinational Monitor* 27(4). Available at http://www.essential.org/monitor

Brand, Jennie E., and Sarah A. Burgard. 2008. "Job Displacement and Social Participation over the Lifecourse; Findings for a Cohort of Joiners." *Social Forces* 87(1):211–242.

Brenner, Mark. 2007 (June). "Give Your Union Dues a Checkup." *Labornotes*, No. 339. Available at http://labornotes.org

Brown, Heidi. 2009 (May 14). "Hi Mom! I'm . . . Back!" *Forbes*. Available at http://www.Forbes.com

Buncombe, Andrew. 2007 (March 14). *Independent: Halliburton: From Bush's Favorite to a National Disgrace*. CorpWatch. Available at http://www.corpwatch.org

Bureau of Labor Statistics. 2009a (July 2). "The Employment Situation Summary" June 2009." Available at http://www.bls.gov

Bureau of Labor Statistics. 2009b (July 2). Table A-12. Alternative Measures of Labor Utilization. Available at http://www.bls.gov

Bureau of Labor Statistics. 2009d (March). Nonfatal Occupational Injuries and Illnesses Requiring Days away from Work, 2007. Available at http://www.bls.gov

Bureau of Labor Statistics. 2009e. *Employment Characteristics of Families in 2008*. Available at http://www.bls.gov

Bureau of Labor Statistics. 2009f. *Union Members in 2008*. Available at http://www.bls.gov

Bureau of Labor Statistics. 2009c. *National Census of Fatal Occupational Injuries in 2008.* Available at http://www.bls.gov

Cantor, David, Jane Waldfogel, Jeffrey Kerwin, Mareena McKinley Wright, Kerry Levin, John Rauch, Tracey Hagerty, and Martha Stapelton Kudela. 2001. *Balancing the Needs of Families and Employers: The Family and Medical Leave Surveys, 2000 Update.* U.S. Department of Labor. Available at http://www.dol.gov

Caston, Richard J. 1998. *Life in a Business-Oriented Society: A Sociological Perspective.* Boston: Allyn and Bacon.

Center for Responsive Politics. 2009 (February 4). "TARP Recipients Paid Out $114 Million for Politicking Last Year." Capital Eye Blog. Available at http://www.opensecrets.org

Cernasky, Rachel. 2003 (December). "Slavery: Alive and Thriving in the World Today—The Satya Interview with Kevin Bales." In *Law & Ethics in the Business Environment.* 6th ed., Terry Halbert and Elaine Ingulli, eds. (pp. 172–173). Mason, Ohio: South-Western Cengage Learning.

Cockburn, Andrew. 2003. "21st Century Slaves." *National Geographic*, September, pp. 2–11, 18–24.

Collins, Kristin. 2008 (September 11). "Ag-Mart Workers Testify." *The News & Observer:* 1A; 6A. North Carolina: Raleigh.

Cooper, Arnie. 2004 (September). "Twenty-Eight Words That Could Change the World." *The Sun*, pp. 4–9.

CorpWatch. 2002 (April 22). "Unions Forge Global Network." Available at http://www.corpwatch.org

Davis, Amy E., and Arne L. Kalleberg. 2006. "Family-Friendly Organizations? Work and Family Programs in the 1990s." *Work and Occupations* 33(2):191–223.

Ebeling, Richard M. 2009 (February 12). "Capitalism the Solution, Not Cause of the Current Economic Crisis." American Institute for Economic Research. Available at http://www.aier.org

Edwards, Kathryn. 2009 (June 10). "Commencing Unemployment." Economic Policy Institute. Available at http://epi.org

Faux, Jeff. 2008 (February 29). "Overhauling NAFTA." Economic Policy Institute. Available at http://www.epi.org

FLA Watch. 2007a. *FLA Watch: Monitoring the Fair Labor Association.* Available at http://www.flawatch.org

FLA Watch. 2007b. *About FLA Watch.* Available at http://www.flawatch.org

Fram, Alan. 2007 (June 3). Poll: *A Fifth Vacation with Laptop.* Associated Press. Yahoo News. Available at http://news.yahoo.com

Frederick, James, and Nancy Lessin. 2000. "Blame the Worker: The Rise of Behavior-Based Safety Programs." *Multinational Monitor* 21(11). Available at http://www.essential.org/monitor

Galinsky, E., J. T. Bond, and E. J. Hill. 2004. *When Work Works: A Status Report on Workplace Flexibility.* New York: Families and Work Institute.

Galinsky, Ellen, James T. Bond, Stacy S. Kim, Lois Backon, Erin Brownfield, and Kelly Sakai. 2005. *Overwork in America: When the Way We Work Becomes Too Much.* New York: Families and Work Institute..

Galinsky, Ellen, James T. Bond, Kelly Sakai, Stacy S. Kim, and Nicole Giuntoli. 2008. *2008 National Study of Employers.* New York: Families and Work Institute.

Galinsky, Ellen, Kerstin Aumann, James T. Bond. 2009. *Times Are Changing: Gender and Generation at Work and Home.* New York: Families and Work Institute.

Gallup Organization. 2009. *Work and Workplace.* Available at http://www.galluppoll.com

"The Garment Industry." 2001. Sweatshop Watch. Available at http://www.change.org

George, Kathy. 2003 (December 1). *Myanmar: Unocal Faces Landmark Trial over Slavery.* CorpWatch. Available at http://www.corpwatch.org

Gordon, David M. 1996. *Fat and Mean: The Corporate Squeeze of Working Americans and the Myth of Managerial "Downsizing."* New York: Free Press.

Grabell, Michael, and Jennifer La Fleur. 2009. "Stimulus Spending Fails to Follow Unemployment, Poverty." *Huffington Post*, August 6. Available at http://www.huffingtonpost.com

Greenhouse, Steven. 2008. *The Big Squeeze.* New York: Alfred A. Knopf.

Greenhouse, Steven. 2000. "Anti-Sweatshop Movement Is Achieving Gains Overseas." *New York Times*, January 26. Available at http://www.nytimes.com

Hagenbaugh, Barbara. 2009. "The Future Holds More Job Losses, Survey Says." *USA Today*, April 27, p. 1B.

Hall, Charles A. S., and John W. Day, Jr. 2009. "Revising the Limits to Growth After Peak Oil." *American Scientist* (May–June): 230–237.

Harris Poll. 2009 (June 30). "Americans Are Purchasing More Generic Brands and Brown Bagging It to Save Money." Available at http://www.harrisinteractive.com

Heymann, Jody. Alison Earle, and Jeffrey Hayes. 2007. *The Work, Family and Equity Index.* Montreal, QC: The Project on Global Working Families and The Institute for Health and Social Policy.

Huffstutter, P. J. 2009. "Struggling Cities Cancel Fourth of July Fireworks." *Los Angeles Times*, June 29. Available at http://www.latimes.com

Human Rights Watch. 2009 (January). *The Employee Free Choice Act: A Human Rights Imperative.* Available at http://www.hrw.org

Human Rights Watch. 2007 (May). *Discounting Rights: Wal-Mart's Violation of US Workers' Right to Freedom of Association.* Volume 19, No. 2 (G). Available at http://hrw.org

Human Rights Watch. 2003. *Child Farmworkers.* Available at http://www.hrw.org/campaigns

Human Rights Watch. 2000. *Unfair Advantage: Workers' Freedom of Association in the United States Under International Human Rights Standards.* Available at http://www.hrw.org

Hunter, Donna, and Gail Deutsch. 2009 (March 19). "Down but Not Out: From Hedge Funds to Pizza Delivery." ABC News. Available at http://www.abcnews.go.com

International Confederation of Free Trade Unions. 2009. *2009 Annual Survey of Violations of Trade Union Rights.* Available at http://www.icftu.org

ILO (International Labour Office). 2009. *Global Employment Trends 2009.* Geneva: International Labour Office.

ILO. 2006. "The End of Child Labour: Within Reach." International Labour Conference, 95th Session, Report I (B). Geneva, Switzerland.

International Programme on the Elimination of Child Labour. 2006. "Education as an Intervention Strategy to Eliminate and Prevent Child Labour." International Labour Organization. Available at http://www.ilo.org

Jensen, Derrick. 2002. "The Disenchanted Kingdom: George Ritzer on the Disappearance of Authentic Culture." *The Sun*, June, pp. 38–53.

Johnson, Richard W., and Corina Mommaerts. 2009 (March). "Unemployment Rate Hits All-Time High for Adults Age 65 and Older." Urban Institute, Retirement Policy Program. Available at http://www.urban.org

Kenworthy, Lane. 1995. *In Search of National Economic Success.* Thousand Oaks, CA: Sage.

"Labor's 'Female Friendly' Agenda." 1998. *Labor Relations Bulletin* no. 690, p. 2.

Lee, S., D. McCann, and J. C. Messenger. 2007. *Working Time Around the World: Trends in Working Hours, Laws, and Policies in a Global Comparative Perspective.* New York: Routledge and International Labour Organization.

Lenski, Gerard, and J. Lenski. 1987. *Human Societies: An Introduction to Macrosociology.* 5th ed. New York: McGraw-Hill.

Leonard, Bill. 1996 (July). "From School to Work: Partnerships Smooth the Transition." *HR Magazine* (Society for Human Resource Management). Available at http://www.shrm.org

Leopold, Les. 2009. "How to Create a Million Jobs in One Month." *The Huffington Post,* July 14. Available at http://www.huffingtonpost.com

MacEnulty, Pat. 2005 (September). "An Offer They Can't Refuse: John Perkins on His Former Life as an Economic Hit Man." *The Sun* 357:4–13.

Martinson, Karin, and Pamela Holcomb. 2007. *Innovative Employment Approaches and Programs for Low-Income Families.* Washington, DC: The Urban Institute.

Mehta, Chirag, and Nik Theodore. 2005 (December). *Undermining the Right to Organize: Employer Behavior During Union Representation Campaigns.* American Rights at Work. Available at http://www.americanrightsatwork.org

Mendenhall, Ruby, Ariel Kalil, Laurel J. Spindel, and Cassandra M. D. Hart. 2008. "Job Loss at Mid-Life: Managers and Executives Face the 'New-Risk Economy." *Social Forces* 87(1):185–209.

Miers, Suzanne. 2003. *Slavery in the Twentieth Century: The Evolution of a Global Problem.* Walnut Creek, CA: AltaMira Press.

Mishel, Lawrence, Jared Bernstein, and Heidi Shierholz. 2009. *The State of Working America 2008–2009.* New York: Cornell University Press.

Moloney, Anastasia. 2005. "Terror as Anti-Union Strategy: The Violent Suppression of Labor Rights in Colombia." *Multinational Monitor* 26(3–4).

Multinational Monitor. 2003. "Workers at Risk." *Multinational Monitor* 24(6). Available at http://www.essential.org/monitor

Multinational Monitor. 2000. "Big Business for Reform." *Multinational Monitor* 21(11). Available at http://www.essential.org

National Labor Committee. 2007. "Senate Minority Leader Senator Harry Reid, Congressman Bernie Sanders, AFL-CIO and Others Endorse Anti-Sweatshop Bill." Available at http://www.nlcnet.org

National Labor Committee. 2001 (January 16). *Nightmare at J. C. Penney Contractor.* Available at http://www.nlcnet.org

National Partnership for Women and Families. 2008. "Family and Medical Leave Act." Available at http://www.nationalpartnership.org

"New OSHA Policy Relieves Employees." 1998. *Labor Relations Bulletin* no. 687, p. 8.

O'Rourke, Dara. 2003. "Outsourcing Regulation: Analyzing Nongovernmental Systems of Labor Standards and Monitoring." *Policy Studies Journal* 31(1):1–29.

Parenti, Michael. 2007 (February 16). *Mystery: How Wealth Creates Poverty in the World.* Common Dreams NewsCenter. Available at http://www.commondreams.org

Perkins, John. 2004. *Confessions of an Economic Hit Man.* San Francisco: Berrett-Koehler Publishers, Inc.

Pew Research Center. 2009. *Trends in Political Values and Core Attitudes: 1987–2009.* Washington DC: The Pew Research Center for People & the Press.

Pugh, Tony. 2009 (June 25). "Recession's Toll: Most Recent College Grads Working Low-skill Jobs." McClatchy News. Available at http://www.mcclatchydc.com

Ritzer, George. 1995. *The McDonaldization of Society: An Investigation into the Changing Character of Contemporary Social Life.* Thousand Oaks, CA: Pine Forge Press.

Roach, Brian. 2007. *Corporate Power in a Global Economy.* Medford, MA: Global Development and Environment Institute, Tufts University.

Runyan, Carol W., Michael Schulman, Janet Dal Santo, Michael Bowling, Robert Agans, and Ta Myduc. 2007. "Work Related Hazards and Workplace Safety of U.S. Adolescents Employed in the Retail and Service Sectors." *Pediatrics* 119(3):526–534.

Schaeffer, Robert K. 2003. *Understanding Globalization: The Social Consequences of Political, Economic, and Environmental Change.* 2nd ed. Lanham, MD: Rowman & Littlefield.

Scott, Robert E., and David Ratner. 2005 (July 20). *NAFTA'S Cautionary Tale.* Economic Policy Institute Briefing Paper 214. Available at http://www.epi.org

"Sex Trade Enslaves Millions of Women, Youth." 2003. *Popline* 25:6.

Shierholz, Heidi. 2009 (July 7). "Nearly Six Unemployed Worker per Available Job." Economic Policy Institute. Available at http://www.epi.org

Shipler, David K. 2005. *The Working Poor.* New York: Vintage Books.

Skinner, E. Benjamin. 2008. *A Crime So Monstrous: Face-to-Face with Modern-Day Slavery.* New York: Free Press.

Soto, Mauricio. 2009 (March 9). "How Is the Financial Crisis Affecting Retirement Savings?" Urban Institute, Retirement Policy Program. Available at http://www.urban.org

Strully, Kate .W. 2009. "Job Loss and Health in the U.S. Labor Market." *Demography* 46(2):221–247.

Students and Scholars Against Corporate Misbehavior. 2005 (August 12). *Looking for Mickey Mouse's Conscience: A Survey of the Working Conditions of Disney Factories in China.* Available at http://www.nlcnet.org

SweatFree Communities. n.d. "Adopted Policies." Available at http://www.sweatfree.org.

Tate, Deborah. 2007 (February 14). *U.S. Lawmakers Seek to Crack Down on Foreign Sweatshops.* The National Labor Committee. Available at http://www.nlcnet.org

Thompson, Charles D., Jr. 2002. "Introduction." In *The Human Cost of Food: Farmworkers' Lives, Labor, and Advocacy,* C. D. Thompson, Jr., and M. F. Wiggins, eds. (pp. 2–19). Austin: University of Texas Press.

Turner, Anna, and John Irons. 2009 (July). "Mass Layoffs at Highest Level Since at Least 1995." Economic Policy Institute. Available at http://epi.org

Uchitelle, Louis. 2006. *The Disposable American: Layoffs and Their Consequences.* New York: Knopf.

UNICEF. 2009. *Children and Conflict in a Changing World.* New York: UNICEF.

United Nations. 2005. *The Millennium Development Goals Report.* New York: United Nations.

Vandivere, Sharon, Kathryn Tout, Martha Zaslow, Julia Calkins, and Jeffrey Cappizzano. 2003. *Unsupervised Time: Family and Child Factors Associated with Self-Care. Assessing the New Federalism.* Occasional Paper No. 71. Urban Institute. Available at http://www.urbaninstitute.org

Watkins, Marilyn. 2002. *Building Winnable Strategies for Paid Family Leave in the United States.* Economic Opportunity Institute. Available at http://www.econop.org

Weiss, Tara. 2009 (May 27). "Some New Grads Are Glad There Are No Jobs." *Forbes.* Available at http://www.Forbes.com

Chapter 8

AASA (American Association of School Administrators). 2009. *Bullying at School and Online.* Education.com. Available at http://www.education.com

ACLU (American Civil Liberties Legal Union). 2009. *Missing the Mark.* New York: American Civil Liberties Union. Available at http://www.aclu.org

AFT (American Federation of Teachers). 2009. *Building Minds, Minding Buildings: A Union's Roadmap to Green and Sustainable Schools.* Available at http://www.aft.org

AFT. 2008a. *Survey and Analysis of Teacher Salary Trends.* Available at http://www.aft.org

AFT. 2008b. *American Academic: The State of the Higher Education Workforce 1997–2007.* Available at http://www.aft.org

AFT. 2006. *Building Minds, Minding Buildings: Turning Crumbling Schools into Environments for Learning.* Available at http://www.aft.org

Allen, Brooks. 2009. *Williams v. California: A Progress Update.* ACLU Foundation of Southern California. Available at http://www.decentschools.org

Associated Press. 2009. "Budget Crisis Forces Deep Cuts at Calif. Schools." *Education Weekly*, June 22. Available at http://www.edweekly.org

Barton, Paul E. 2005. *One Third of a Nation: Rising Drop-Out Rates and Declining Opportunities.* Princeton, NJ: Educational Testing Service.

Barton, Paul E. 2004. "Why Does the Gap Persist?" *Educational Leadership* 62(3): 9–13.

Bausell, Carole Viongrad, and Elizabeth Klemick. 2007. "Tracking U.S. Trends." *Education Week* 26(30):42–44.

Benenson Strategy Group. Benenson, Joel. 2009. "Hi-Tech Cheating: Cell Phones and Cheating in Schools: A National Poll." Benenson Strategy Group Commissioned by Common Sense Media. Available at http://www.commonsensemedia.org

Berger, Joseph. 2007. "Fighting over When Public Pay Private Tuition for Disabled." *New York Times*, March 21. Available at http://www.nytimes.com

BLS (Bureau of Labor Statistics). 2009. "Education Pays in Higher Earnings and Lower Unemployment Rates." Available at http://www.bls.gov

Blankinship, Donna Gordan. 2009. "New CEO: Gates Foundation Learns from Experiments." *Forbes*, May 28. Available at http://www.forbes.com

Bowens, Dan. 2009. "North Carolina High Court Upholds Mandatory Year-Around Schools." Available at http://www.wral.com

Boyd, Donald, and Pamela Grossman, Hamilton Lankford, Susanna Loeb, and James Wyckoff. 2009. "Who Leaves? Teacher Attrition and Student Achievement." Working Paper No. 23. National Center for the Analysis of Longitudinal Data in Education Research.

Bridgeland, John M., John DiIulio, Jr., and Karen Burke Morison. 2006. *The Silent Epidemic.* A Report by Civic Enterprises in association with Peter D. Hart Research Association. The Bill & Melinda Gates Foundation.

CDC (Centers for Disease Control and Prevention). 2008. Understanding School Violence Fact Sheet 2008." National Center for Injury Prevention and Control. Available at http://www.cdc.gov/injury

CDF. 2008. *Cradle to Prison Pipeline Campaign.* Available at http://www.childrensdefense.org

Chaney, B., and L. Lewis. 2007. *Public School Principals Report on Their School Facilities: Fall 2005.* U.S. Department of Education. Washington, DC: National Center for Educational Statistics.

Coleman, James S., J. E. Campbell, L. Hobson, J. McPartland, A. Mood, F. Weinfield, and R. York. 1966. *Equality of Educational Opportunity.* Washington, DC: U.S. Government Printing Office.

Corbett, Christianne, Catherine Hill and Andresse St. Rose. 2008. *Where the Girls Are.* Washington, D.C.: American Association of University Women.

CNCS (Corporation for National Community Service). 2008. "The Impact of Service-Learning: A Review of Current Research." Issue Brief (January). Available at http://www.learnandserve.gov

DeBell, Matthew, and Chris Chapman. 2006. *Computer and Internet Use by Students in 2003.* Washington DC: U.S. Department of Education, NCES 2006-065.

DeNeui, Daniel, and Tiffany Dodge. 2006. "Asynchronous Learning Networks and Student Outcomes." *Journal of Instructional Psychology* 33(4):256–259.

Dillon, Sam. 2009. "Education Secretary Says Aid Hinges on New Data." *The New York Times*, April 2. Available at http://www.nytimes.com

Dobbs, Michael. 2005. "Youngest Students Most Likely to Be Expelled." *Washington Post*, May 16. Available at http://www.washingtonpost.com

Duncan, Arne. 2009 (March 20). "Speech to the National Science Teachers Association Conference." Available at http://www.ed.gov

EFA Global Monitoring Report. 2007. *Education for All: The Quality Imperative.* UNESCO Publishing. Available at http://unesdoc.unesco.org

ELT (Expanded Learning Time). 2009. "Guiding Principles." Available at http://www.mass2020.org

ERC (Educational Research Center). 2009. *Quality Counts.* Editorial Products in Educational Research Center. Available at http://www.edweek.org/rc

Office of Management and Budget. 2009. "Federal Budget for Fiscal Year 2010." Available at http://www.whitehouse.gov

Fletcher, Robert S. 1943. *History of Oberlin College to the Civil War.* Oberlin, OH: Oberlin College Press.

Flexner, Eleanor. 1972. *Century of Struggle: The Women's Rights Movement in the United States.* New York: Atheneum.

Forster, Greg. 2009. *A Win-Win Solution: The Empirical Evidence on How Vouchers Affect Public Schools.* Friedman Foundation for Educational Choice. Available at http://www.friedmanfoundation.org

Gallup Poll. 2007. *Education.* Available at http://www.gallup.com

Goldenberg, Claude. 2008. "Teaching English Language Learners." *American Educator* (Summer): 8–11, 14–19, 22–23, 42–44.

Gonzales, Roberto. 2009 (April). *Young Lives on Hold: The College Dream of Undocumented Students.* College Board Advocacy. Available at http://www.professionals.collegeboard.com

Goodnough, Abby. 2009. "States Turning to Last Resorts in Budget Crisis." *The New York Times.* June 22. Available at http://www.nytimes.com

Greenhouse, Linda. 2007. "Supreme Court Votes to Limit the Use of Race in Integration Plans." *New York Times*, June 29. Available at http://www.nytimes.com

Hall, Daria, and Shana Kennedy. 2006. *Primary Progress, Secondary Challenge: A State-by-State Look at Student Achievement Patterns.* Washington, DC: The Education Trust.

Hanushek, Eric A., Steven G. Rivkin, and John J. Kain. 2005. "Teachers, Schools and Academic Achievement." *Econometrics* 73(2):417–458.

Head Start Act. 2008. *Head Start Act.* Available at http://www.acf.hhs.gov

Human Development Report. 2008. "Adult Literacy Rate Percent Aged 15 and Older." Available at http://hdrstats.undp.org

Hurst, Marianne. 2005. "When It Comes to Bullying, There Are No Boundaries." *Education Week*, February 8. Available at http://www.edweek.org

Independent Sector. 2002. *Engaging Youth in Lifelong Service.* Newsroom. Available at http://www.independentsector.org

ICAS (Institute for College Access and Success). 2009. *Proposed Cal Grant Cuts Would Hit Community College Students Hardest.* Issue Brief. Available at http://www.ticas.org

Jacobs, Joanne. 1999. "Gov. Davis to Make Volunteering Mandatory for Students." *San Jose Mercury News*, April 23.

Johnson, Nicholas, Jeremy Koulish, and Phil Ollif. 2009 (February 10). "Most States Are Cutting Education." Center on Budget and Policy Priorities. Available at http://www.cbpp.org

Kahlenberg, Richard D. 2006. "A New Way on School Integration." *Issue Brief.* The Century Foundation. Available at http://www.equaleducation.org

Kanter, Rosabeth Moss. 1972. "The Organization Child: Experience

Management in a Nursery School." *Sociology of Education* 45:186–211.

Klein, Alyson, and Stephen Sawchuk. 2008. "Congress OKs Renewal of Higher Education Act." *Education Weekly*, August 1. Available at http://www.edweek.org

Klein, Alyson. "2009 Stimulus Aid to Schools a Management Challenge." *Education Week Online*. February 17. Available at http://www.edweek.org

Kozol, Jonathan. 2005. *The Shame of the Nation*. New York: Crown.

Kozol, Jonathan. 1991. *Savage Inequalities: Children in America's Schools*. New York: Crown.

Lewin, Tamar. 2009. "Community Colleges Challenge Hierarchy with 4-Year Degrees." *The New York Times*, May 3. Available at http://www.nytimes.com

Lickona, Thomas, and Matthew Davidson. 2005. *A Report to the Nation: Smart and Good High Schools*. Available at http://www.cortland.edu

Lubienski, Sarah Theule, and Christopher Lubienski. 2006. "School Sector and Academic Achievement: A Multi-Level Analysis of NAEP Mathematics Data." *American Educational Research Journal* 43(4):651–698.

Manzo, Kathleen K. 2005. "College-Based High Schools Fill Growing Need." *Education Week*, May 25. Available at http://www.edweek.org

Mau, Rosalind Y. 1992. "The Validity and Devolution of a Concept: Student Alienation." *Adolescence* 27(107): 739–740.

McDonnell, Sanford. 2009 (October 3). "America's Crisis of Character—And What to Do About It." Available at http://www.edweek.org

McKinsey and Company 2009. *The Economic Impact of the Achievement Gap in America Schools*. Social Sector Office: McKinsey & Company. Available at http://www.mckinsey.com

McNeal, Michelle. 2009. "46 States Agree to Common Academic Standards Effort." *Education Weekly*, June 1. Available at http://www.edweek.org

Mead, Sara. 2006. *The Truth About Boys and Girls. Education Sector*, June. Available at http://www.educationsector.org

Merton, Robert K. 1968. *Social Theory and Social Structure*. New York: Free Press.

MetLife. 2008. *MetLife Survey of the American Teacher: Past, Present and Future*. Available at http://www.metlife.com

Morse, Jodie. 2002. "Learning While Black." *Time*, May 27, pp. 50–52.

MPR (Mathematica Policy Research, Inc.). 2007. "Early Head Start Research and Evaluation." Available at http://mathematica-mpr.com

Muller, Chandra, and Katherine Schiller. 2000. "Leveling the Playing Field?" *Sociology of Education* 73:196–218.

NBPTS (National Board for Professional Teaching Standards). 2007. *Impact of National Board Certification*. Available at http://www.nbpts.org

NCD (National Council on Disability). 2000. *Executive Summary*. Available at http://www.ncd.gov

NCDPI (Department of Public Instruction). 2007. *No Child Left Behind: Overview*. Available at http://www.ncpublicschools.org

NCES (National Center for Educational Statistics). 2009a. *Digest of Education Statistics, 2008*. U.S. Department of Education. Available at http://nces.ed.gov

NCES. 2009b. *The Condition of Education, 2008*. U.S. Department of Education. Available at http://nces.ed.gov

NCES. 2009c. *Indicators of School Crime and Safety 2008*. U.S. Department of Education. Available at http://nces.ed.gov

NCES. 2007. *Effectiveness of Reading and Mathematics Software Products*. Report to Congress. Washington, DC: U.S. Department of Education. NCES 2007-4005.

NCES. 2003. *Young Children's Access to Computers in the Home and at School in 1999 and 2000: Executive Summary*. Available at http://nces.ed.gov

NIEER (The National Institute for Early Education Research). 2008. *The State of Preschool 2008*. Rutgers Graduate School of Education. Available at http://www.nieer.org

NCPPHE (The National Center for Public Policy and Higher Education). 2009. *Measuring Up 2008: The National Report Card on Higher Education*. Available at http://highereducation.org

NSBA (National School Board Association). 2007. *Cleveland Voucher Program*. Available at http://www.nsba.org

OECD (Organization for Economic Cooperation and Development). 2009. *Education at a Glance 2008*. Available at http://www.oecd.org

Office of Head Start. 2008. *Head Start Program Fact Sheet*. Available at http://www.acf.hhs.gov

Olson, Lynn. 2007. "Paying Attention Early On." *Education Week* 26(17):29–31.

Olson, Lynn, and David Hoff. 2006. "Framing the Debate." *Education Week* 26(15):22.

Orfield, Gary, and Chungmei Lee. 2006. *Racial Transformation and the Changing Nature of Segregation*. Cambridge, MA: Harvard University: The Civil Rights Project.

Peskin, Melissa Fleschler, Susan R. Tortolero, and Christine M. Markham. 2006. "Bullying and Victimization among Black and Hispanic Adolescents." *Adolescence* 41(163):467–484.

Picciano, Anthony G., and Jeff Seaman. 2009. *K-12 Online Learning: A 2008 Follow-up of the Survey of U.S. School District Administrators*. Available at http://www.sloanconsortium.org

Poliakoff, Anne Rogers. 2006. "Closing the Gap: An Overview." *ASCD InfoBrief* 44 (January):1–10. The Association for Supervision and Curricular Development.

Ramierz-Valles, Jesus, and Amanda Brown. 2003. "Latinos' Community Involvement in HIV/AIDS: Organizational and Individual Perspectives on Volunteering." *AIDS Education and Prevention* 15(Suppl. 1):90–104.

Rand. 2009. *Are Charter Schools Making a Difference? A Study of Student Outcomes in Eight States*. Available at http://www.rand.org

Ravitch, Diane. 2009. "Is Arne Duncan Really Margaret Spellings in Drag?" *Education Weekly*, February 24. Available at http://blogs.edweek.org/edweek

Reardon, Sean F., Allison Atteberry, Nicole Arshan, and Michal Kurlaend. 2009. *Effects of the California High School Exit Exam on Student Persistent, Achievement and Graduation*. Stanford: Institute for Research on Education Policy & Practice. Available at http://www.stanford.edu

Riehl, Carolyn. 2004. "Bridges to the Future: Contributions of Qualitative Research to the Sociology of Education." In *Schools and Society*, Jeanne Ballantine and Joan Spade, eds. (pp. 56–72). Belmont, CA: Thomson Wadsworth.

Robbins, Charles Robbins, and Eliza Byard. 2009. "Bullied to Death." *The Advocate*, April 23. Available at http://www.advocate.com

Robelen, Erik W. 2009 (April 6). "Obama Echoes Bush on Education Ideas." Available at http://www.edweek.org

Rodriguez, Richard. 1990. "Searching for Roots in a Changing World." In *Social Problems Today*, James M. Henslin, ed. (pp. 202–213). Englewood Cliffs, NJ: Prentice-Hall.

Rosenthal, Robert, and Lenore Jacobson. 1968. *Pygmalion in the Classroom: Teacher Expectations and Pupils' Intellectual Development*. New York: Holt, Rinehart & Winston.

Sadker, David, and Karen Zittleman. 2009. *Still Failing at Fairness: How Gender Bias Cheats Boys and Girls in Schools*. New York: Simon and Schuster.

Sadovnik, Alan. 2004. "Theories in the Sociology of Education." In *Schools and*

Society, Jeanne Ballantine and Joan Spade, eds. (pp. 7–26). Belmont, CA: Thomson Wadsworth.

Save the Children. 2009. "An Uneducated Girl Is a Girl in Darkness." Available at http://www.savethechildren.org

Schemo, Diana. 2006. "Federal Rules Back Single-Sex Public Education." *New York Times*, October 25. Available at http://www.nytimes.com

Schott Report. 2008. *Given Half a Chance: The Schott 50 State Report on Public Education and Black Males*. Schott Foundation for Public Education. Available at http://www.blackboysreport.org

Shin, Annys. 2007. "Removing Schools' Soda Is Sticky Point." *The Washington Post*, March 22, p. D3.

Spelling, Margaret. 2008 (September 15). "U.S. Secretary of Education Margaret Spellings Unveils Indicators to Track Nation's Education Progress at Aspen Institute Summit in Washington, D.C." Press Release. Available at http://www.ed.gov

Spielhagen, Frances R. 2007. *Debating Single-Sex Education: Separate and Equal?* Lanham, MD: Rowman and Littlefield Education.

Stotsky, Sandra. 2009. "The Academic Quality of Teachers: A Civil Rights Issue." *Education Weekly*, June 26. Available at ttp://www.edweek.org

Strauss, Valerie. 2009. "Colleges Consider 3-Year Degrees to Save Undergrads Time, Money." *The Washington Post*, May 23. Available at http://www.washingtonpost.com

Tanner, C. K. 2008. "Explaining relationships Among Student Outcomes and the School's Physical Environment." *Journal of Advanced Academics* (19): 444–471.

Toppo, Greg. 2006. "Student 'Superstar' on Brink of Dropping Out." *USA Today*, December 4. Available at http://usatoday.com

Toppo, Greg. 2004. "Low-Income College Students Are Increasingly Left Behind." *USA Today*, January 14. Available at http://usatoday.com

Troy, Forrest (Frosty). 2001. "You Want Heroes?" *Oklahoma Observer*, April 23.

Tyler, John H., and Magnus Lofstrom. 2009. "Finishing High School: Alternative Pathways and Dropout Recovery." *The Future of Children* 19(1):78–102.

Tyre, Peg. 2008. *The Trouble with Boys: A Surprising Report Card on Our Sons, Their Problems at School, and What Parents and Educators Must Do*. New York: Crown Publishing Group.

UNESCO (United Nations Educational Scientific and Cultural Organization). 2009. "Why the Literacy Decade?" Available at http://www.unesco.org/

U.S. Census Bureau. 2009. *Statistical Abstracts of the United States*. Washington, DC: U.S. Government Printing Office.

Vaznis, James. 2009. "Boston Students Struggle with English-Only Rule." The Boston Globe, April 7. Available at http://www.boston.com

Viadero, Debra. 2006. "Rags to Riches in U.S. Largely a Myth, Scholars Write." *Education Week*, October 25. Available at http://www.edweek.org

Viadero, Debra. 2005. "Study Sees Positive Effects of Teacher Certification." *Education Week*, April 27. Available at http://www.edweek.org

Waggoner, Martha. 2009. "Judge: 'Academic genocide' in Halifax Schools." *News & Observer*, March 19. Available at http://www.newsobserver.com/

Walsh, Mark. 2009. "Full 6th Circuit Weighs NEA Suit Against NCLB." *Education Weekly*, December 10. Available at http://www.edweek.org

Walsh, Mark, and Erik W. Robelen. 2009. "Supreme Court Backs Reimbursement for Private Tuition." *Education Weekly*, June 22. Available at http://www.edweek.org

Wei, C. C., L. Berkner, S. He, S. Lew, M. Cominole, and P. Siegel. 2009. *National Postsecondary Student Aid Study: Student Financial Aid Estimates for 2007–08: First Look* (NCES 2009-166). Washington DC: National Center for Education Statistics, Institute of Education Sciences, U.S. Department of Education.

White House. 2009. Available at http://www.whitehouse.gov

Winters, Rebecca. 2001. "From Home to Harvard." *Time*, September 11. Available at http://www.time.com

Wong, Jennifer S. 2009. *No Bullies Allowed: Understanding Peer Victimization, the Impacts on Delinquency, and the Effectiveness of Prevention Programs*. Rand Corporation. Available at http://www.rand.org/pubs/rgs_dissertations

Xu, Zeyu, Jane Hannaway, and Colin Taylor. 2007. *Making a Difference? The Effects of Teach for America in High School*. Working Paper No. 17. National Center for the Analysis of Longitudinal Data in Education Research.

Zehr, Mary Ann. 2009a. "Mock Graduation Takes Place on Behalf of DREAM Act." *Education Week*, June 24. Available at http://www.edweek.org

Zehr, Mary Ann. 2009b. "Community-Service Opportunities Expanded." *Education Week*, April 29. Available at http://www.edweek.org

Zigler, Edward, Sally Styfco, and Elizabeth Gilman. 2004. "The National Head Start Program for Disadvantaged Preschoolers." In *Schools and Society*, Jeanne Ballantine and Joan Spade, eds. (pp. 341–346). Belmont, CA: Thomson Wadsworth.

Chapter 9

Allport, G. W. 1954. *The Nature of Prejudice*. Cambridge, MA: Addison-Wesley.

Amato, P. R., A. Booth, D. R. Johnson, and S. J. Rogers. 2007. *Alone Together: How Marriage in America Is Changing*. Cambridge MA: Harvard University Press.

American Council on Education and American Association of University Professors. 2000. *Does Diversity Make a Difference? Three Research Studies on Diversity in College Classrooms*. Washington, DC: American Council on Education and American Association of University Professors.

Balko, Radley. 2009 (July 6). "The El Paso Miracle." Reasononline. Available at http://www.reason.com

Bauer, Mary, and Sarah Reynolds. 2009. *Under Siege: Life for Low-Income Latinos in the South*. Montgomery, AL: The Southern Poverty Law Center.

Bazar, Emily. 2007. "Local Laws Target Immigrant Ills." *USA Today*, July 11. Available at http://www.usatoday.com

Beeman, Mark, Geeta Chowdhry, and Karmen Todd. 2000. "Educating Students About Affirmative Action: An Analysis of University Sociology Texts." *Teaching Sociology* 28(2):98–115.

Beirich, Heidi. 2007 (Summer). "Getting Immigration Facts Straight." *Intelligence Report*. Available at http://www.splcenter.org

Beirich, Heidi, Michelle Bramblett, Angela Freeman, Anthony Griggs, Janet Smith, and Laurie Wood. 2007 (Summer). "The Year in Hate." *Intelligence Report*. Available at http://www.splcenter.org

Brace, C. Loring. 2005. *"Race" Is a Four-Letter Word*. New York: Oxford University Press.

Brooks, Roy. 2004. *Atonement and Forgiveness: A New Model for Black Reparations*. Berkeley: University of California Press.

Brown University Steering Committee on Slavery and Justice. 2007. *Slavery and Justice*. Providence, RI: Brown University.

Capps, Randolph, Michael E. Fix, Jason Ost, Jane Reardon-Anderson, and Jeffrey S. Passel. 2005 (February 8). *The Health and Well-Being of Young Children of Immigrants*. Urban Institute. Available at http://www.urban.org

CBCNews.ca. 2009 (May 25). "Girl Watched Skinhead Videos and Talked of How to Kill, Hearing Told." Available at http://www.cbc.ca/canada

CNN. 2009 (June 18). "Senate Approves Resolution Apology for Slavery." Available at http://www.cnn.com

Cohen, Mark Nathan. 1998. "Culture, Not Race, Explains Human Diversity." *Chronicle of Higher Education* 44(32):B4–B5.

Conley, Dalton. 2002. "The Importance of Being White." *Newsday*, October 13. Available at http://www.newsday.com

Conley, Dalton. 1999. *Being Black, Living in the Red: Race, Wealth, and Social Policy in America*. Berkeley: University of California Press.

Dees, Morris. 2000 (December 28). Personal correspondence. Morris Dees, co-founder of the Southern Poverty Law Center, 400 Washington Avenue, Montgomery, AL 36104.

EEOC (Equal Employment Opportunity Commission). 2009. "Charge Statistics FY 1997–FY 2008." Available at http://www.eeoc.gov

EEOC. 2007b. "EEOC and Walgreens Resolve Lawsuit." Press release. Available at http://www.eeoc.gov

EEOC. 2004. *Retail Distribution Centers: How New Business Processes Impact Minority Labor Markets*. U.S. Equal Employment Opportunity Commission. Available at http://www.eeoc.gov

Etzioni, Amitai. 1997. "New Issues: Re-Thinking Race." *Public Perspective*, June–July, pp. 39–40. Available at http://www.ropercenter.uconn.edu

FBI (Federal Bureau of Investigation). 2008a. *Hate Crime Statistics 2007*. Available at http://www.fbi.gov

FBI. 2008b (Fall). "Intelligence Briefs." FBI report confirms extremist activity in U.S. military. Intelligence Reports (131):5–6.

Fry, Richard. 2009. "Sharp Growth in Suburban Minority Enrollment Yields Modest Gains in School Diversity." Pew Hispanic Center. Available at http://www.pewhispaniccenter.org

Gaertner, Samuel L., and John F. Dovidio. 2000. *Reducing Intergroup Bias: The Common Ingroup Identity Model*. Philadelphia: Taylor & Francis.

Gallup Organization. 2009a. *Race Relations*. Available at http://www.gallup.com

Gallup Organization. 2009b. *Immigration*. Available at http://www.gallup.com

Glaser, Jack, Jay Dixit, and Donald P. Green. 2002. "Studying Hate Crime with the Internet: What Makes Racists Advocate Racial Violence?" *Journal of Social Issues* 58(1):177–193.

Glaubke, Christina Roman, and Katharine Heintz-Krowles. 2004. *Fall Colors: Prime Time Diversity Report 2003–04*. Oakland, CA: Children Now & the Media Program.

Goldstein, Joseph. 1999. "Sunbeams." *The Sun* 277, January, p. 48.

Goodnough, Abby. 2001. "New York City Is Short-Changed in School Aid, State Judge Rules." *New York Times*, January 11. Available at http://www.nytimes.com

Greenberg, Daniel S. 2003. "Supreme Court Sets Showdown on Affirmative Action." *The Lancet* 361(March 1):762. Available at http://www.thelancet.com

Greenhouse, Steven. 2005. "Wal-Mart to Pay U.S. $11 Million in Lawsuit on Illegal Workers." *New York Times*, March 19. Available at http://www.nytimes.com

Greenhouse, Steven. 2003. "Suit Claims Discrimination Against Hispanics on Job." *New York Times*, February 9. Available at http://www.nytimes.com

Gurin, Patricia. 1999. "New Research on the Benefits of Diversity in College and Beyond: An Empirical Analysis." *Diversity Digest*, Spring, pp. 5–15.

Healey, Joseph F. 1997. *Race, Ethnicity, and Gender in the United States: Inequality, Group Conflict, and Power*. Thousand Oaks, CA: Pine Forge Press.

"Hear and Now." 2000 (Fall). *Teaching Tolerance*, p. 5.

Hodgkinson, Harold L. 1995. "What Should We Call People? Race, Class, and the Census for 2000." *Phi Delta Kappa*, October, pp. 173–179.

Holthouse, David. 2009. "The Year in Hate.' *Intelligence Report*, Spring (133):48–50.

Holthouse, David. 2006 (July 7). "A Few Bad Men." *Intelligence Report*. Available at http://www.splcenter.org

Holzer, Harry. 2005 (May 14). *New Jobs in Recession and Recovery: Who Are Getting Them and Who Are Not?* Urban Institute. Available at http://www.urban.org

Holzer, Harry, and David Neumark. 2000. "Assessing Affirmative Action." *Journal of Economic Literature* 38(3):483–568.

hooks, bell. 2000. *Where We Stand: Class Matters*. New York: Routledge.

Humphreys, Debra. 2000 (Fall). "National Survey Finds Diversity Requirements Common Around the Country." *Diversity Digest*. Available at http://www.diversityweb.org

Humphreys, Debra. 1999. "Diversity and the College Curriculum: How Colleges and Universities Are Preparing Students for a Changing World." *Diversity-Web*. Available at http://www.inform.umd.edu

Jay, Gregory. 2005 (March 17). "Introduction to Whiteness Studies." Available at http://www.uwm.edu

Jensen, Derrick. 2001. "Saving the Indigenous Soul: An Interview with Martin Prechtel." *The Sun* 304(April):4–15.

Kaplan, David E., and Lucian Kim. 2000. "Nazism's New Global Threat." *U.S. News Online*, September 25. Available at http://www.usnews.com

Keller, Larry. 2009. "White Heat." *Intelligence Report*, Spring (133):36–37.

Keita, S. O. Y., and Rick A. Kittles. 1997. "The Persistence of Racial Thinking and the Myth of Racial Divergence." *American Anthropologist* 99(3):534–544.

King, Joyce E. 2000 (Fall). "A Moral Choice." *Teaching Tolerance* 18:14–15.

Kovac, Amy. 2009. "Transcript of Rev. Lowery's Inaugural Benediction." *Washington Post*, January 20. Available at http://voices.washingtonpost.com

Kozol, Jonathan. 1991. *Savage Inequalities: Children in America's Schools*. New York: Crown.

Kremer, James D., Kathleen A. Moccio, and Joseph W. Hammell. 2009. *Severing a Lifeline: The Neglect of Citizen Children in America's Immigration Enforcement Policy*. Minneapolis: Dorsey & Whitney LLP.

Lawrence, Sandra M. 1997. "Beyond Race Awareness: White Racial Identity and Multicultural Teaching." *Journal of Teacher Education* 48(2):108–117.

Leadership Council on Civil Rights Education Fund. 2009. *Confronting the New Faces of Hate: Hate Crimes in America*. Available at http://www.civilrights.org

Lee, Christopher. 2006. "EEOC Is Hobbled, Groups Contend." *Washington Post*, June 14, p. A21.

Levin, Jack, and Jack McDevitt. 1995. "Landmark Study Reveals Hate Crimes Vary Significantly by Offender Motivation." *Klanwatch Intelligence Report*, August, pp. 7–9.

Lollock, Lisa. 2001 (March). *The Foreign-Born Population in the United States: March 2000*. Current Population Reports P20-534. Washington DC: U.S. Bureau of the Census.

Mahony, Edmund H., and Josh Kovner. 2009 (June 30). "U.S. Supreme Court Rules in Favor of New Haven Firefighters." *The Hartford Courant*. Available at http://www.courant.com

Maril, Robert Lee. 2004. *Patrolling Chaos: The U.S. Border Patrol in Deep South Texas*. Lubbock, TX: Texas Tech University Press.

Massey, Douglas, and Nancy Denton. 1993. *American Apartheid: Segregation and the Making of an American Under-Class*. Cambridge, MA: Harvard University Press.

Massey, Douglas S., and Garvey Lundy. 2001. "Use of Black English and Racial Discrimination in Urban Housing Markets: New Methods and Findings." *Urban Affairs Review* 36(4):452–469.

McIntosh, Peggy. 1990 (Winter). "White Privilege: Unpacking the Invisible Knapsack." *Independent School* 49(2):31–35.

Mock, Brentin. 2007 (Spring). "Sharing the Hate." *Intelligence Report* 125:15–16.

Morrison, Pat. 2002. "September 11: A Year Later—American Muslims Are Determined

Not to Let Hostility Win." *National Catholic Reporter*, 38(38):9–10.

Moser, Bob. 2004. "The Battle of 'Georgiafornia.'" *Intelligence Report* 116:40–50.

Mukhopadhyay, Carol C., Rosemary Henze, and Yolanda T. Moses. 2007. *How Real Is Race?* Lanham, MD: Rowman & Littlefield Education.

National Immigration Law Center. 2009. "Basic Facts About In-State Tuition for Undocumented Immigrant Students." Available at http://www.nilc.org

Navarro, Mireya. 2003. "For New York's Black Latinos, a Growing Racial Awareness." *New York Times*, April 28. Available at http://www.nytimes.com

Nemes, Irene. 2002. "Regulating Hate Speech in Cyberspace: Issues of Desirability and Efficacy." *Information and Communication Technology Law* 11(3):193–220.

New York Times. 2005. "National Briefing." *New York Times*, August 11, p. A19.

NPR, Kaiser Family Foundation, and Kennedy School of Government. 2005. *Immigration: Summary of Findings. NPR/Kaiser/Kennedy School Poll.* Available at http://www.npr.org/news/

Olson, James S. 2003. *Equality Deferred: Race, Ethnicity, and Immigration in America Since 1945.* Belmont, CA: Wadsworth/Thomson.

Orfield, Gary. 2001 (July). *Schools More Separate: Consequences of a Decade of Resegregation.* Cambridge, MA: Harvard University, Civil Rights Project.

Ossorio, Pilar, and Troy Duster. 2005. "Race and Genetics." *American Psychologist* 60(1):115–128.

Pager, Devah. 2003. "The Mark of a Criminal Record." *American Journal of Sociology* 108(5):937–975.

Passel, Jeffrey S., and D'Vera Cohn. 2009. "A Portrait of Unauthorized Immigrants in the United States. Pew Hispanic Research Center." Available at http://www.pewhispanic.org

Passel, Jeffrey S., and Paul Taylor. 2009. "Who's Hispanic?" Pew Hispanic Research Center. Available at http://www.pewhispanic.org

Paul, Pamela. 2003. "Attitudes Toward Affirmative Action." *American Demographics* 25(4):18–19.

Perryman, M. Ray. 2008 (April). "An Essential Resource: An Analysis of the Economic Impact of Undocumented Workers on Business Activity in the U.S. with Estimated Effects by State and by Industry." Waco, Texas: The Perryman Group.

Pew Research Center. 2009. *Trends in Political Values and Core Attitudes: 1987–2009.* Washington, DC: Pew Research Center for People & the Press.

Pew Research Center. 2005 (September 8). *Huge Racial Divide over Katrina and Its Consequences.* Available at http://www.people-press.org

Picca, Leslie, and Joe R. Feagin. 2007. *Two-Faced Racism.* New York: Routledge.

Plous, S. 2003. "Ten Myths About Affirmative Action." In *Understanding Prejudice and Discrimination*, S. Plous, ed. (pp. 206–212). New York: McGraw-Hill.

Polling Report, Inc. 2009. "Race and Ethnicity." Available at http://www.pollingreport.com

Posner, Eric A., and Adrian Vermeule. 2003. "Reparations for Slavery and Other Historical Injustices." *Columbia Law Review* 103(689):689–748.

Potok, Mark. 2005 (September 20). "In Katrina's Wake, White Supremacists Spew Hatred." Southern Poverty Law Center. Available at http://www.splcenter.org/news/new.jsp

Pryor, John H., Sylvia Hurtado, Linda DeAngelo, Jessica Sharkness, Laura C. Romero, William S. Korn, and Serge Tran. 2008. *The American Freshman: National Norms Fall 2008.* Los Angeles: Higher Education Research Institute, UCLA.

Rosenfeld, Michael J., and Byung-Soo Kim. 2005 (August). "The Independence of Young Adults and the Rise of Interracial and Same-Sex Unions." *American Sociological Review* 70:541–562.

Rothenberg, Paula S. 2002. *White Privilege.* New York: Worth.

Schaefer, Richard T. 1998. *Racial and Ethnic Groups.* 7th ed. New York: HarperCollins.

Schiller, Bradley R. 2004. *The Economics of Poverty and Discrimination.* 9th ed. Upper Saddle River, NJ: Pearson Education.

Schmidt, Peter. 2004 (January 30). "New Pressure Put on Colleges to End Legacies in Admissions." *Chronicle of Higher Education* 50(21):A1.

Schuman, Howard, and Maria Krysan. 1999. "A Historical Note on Whites' Beliefs About Racial Inequality." *American Sociological Review* 64:847–855.

Shipler, David K. 1998. "Subtle vs. Overt Racism." *Washington Spectator* 24(6):1–3.

SPLC (Southern Poverty Law Center). 2009a. "SPLC Files Complaint After Worker Sexually Assaulted, Brutalized by Manager." *SPLC Report* 39 (2):5.

SPLC. 2009b. "Immigrants Scapegoated After Swine Flue Outbreak." *SPLC Report* 39 (2):3.

SPLC. 2007. *Close to Slavery: Guestworker Programs in the United States.* Montgomery, AL: Southern Poverty Law Center.

SPLC. 2007 (Spring). "Hate Websites." *Intelligence Report* 125:59–65.

SPLC. 2005a (June). "Center Wins $1.35 Million Judgment Against Violent Border Vigilantes." *SPLC Report* 35(2):3.

SPLC. 2005b (June). "New Lawsuits Seek Reform of Abusive Labor Practice." *SPLC Report* 35(2):1.

SPLC. 2004 (Winter). "Neo-Nazi Label Woos Teens with Hate-Music Sampler." *Intelligence Report* 116:5.

SPLC. 2002 (January). "White Power Bands." *Hate in the News.* Southern Poverty Law Center. Available at http://www.tolerance.org

SPLC. 2000 (Spring). "Hate on Campus." *Intelligence Report* 98:6–15.

Tanneeru, Manav. 2007 (May 11). "Asian-Americans' Diverse Voices Share Similar Stories." CNN.com. Available at http://www.cnn.com

Thompson, Ginger. 2009. "After Losing Freedom, Some Immigrants Face Loss of Custody of Their Children." *New York Times*, April 23. Available at http://nytimes.com

Tolbert, Caroline J., and John A. Grummel. 2003. "Revisiting the Racial Threat Hypothesis: White Voter Support for California's Proposition 209." *State Politics and Policy Quarterly* 3(2):183–202, 215–216.

Turner, Margery Austin, Stephen L. Ross, George Galster, and John Yinger. 2002. *Discrimination in Metropolitan Housing Markets.* Washington, DC: Urban Institute.

Turner, Margery Austin, and Felicity Skidmore. 1999. *Mortgage Lending Discrimination: A Review of Existing Evidence.* Washington, DC: Urban Institute.

Turner, Margery Austin, and Karina Fortuny. 2009. *Residential Segregation and Low-Income Working Families.* Washington, DC: Urban Institute.

U.S. Census Bureau. 2009a. "2010 Census Materials." Available at http://www.census.gov

U.S. Census Bureau. 2009b. *2007 American Community Survey.* Available at http://www.census.gov

U.S. Census Bureau. 2009c. *Current Population Survey.* Available at http://www.census.gov

U. S. Census Bureau. 2008a (August 14). "An Older and More Diverse Nation by Mid-Century." Press Release. Available at http://www.census.gov

U.S. Census Bureau. 2008b (August 14). *2008 National Population Projections, Tables and Charts.* Available at http://www.census.gov

U.S. Census Bureau. 2007. *Current Population Survey, Annual Social and Economic Supplement 2006.* Available at http://www.census.gov

U.S. Citizenship and Immigration Services. 2009. *A Guide to Naturalization.* Available at http://www.uscis.gov

U.S. Customs and Border Protection. 2009. *U.S. Customs and Border Protection*

Performance and Accountability Report, FY 2008. Available at http://www.cbp.gov

U.S. Department of Labor. 2002. *Facts on Executive Order 11246 Affirmative Action.* Available at http://www.dol.gov

Van Ausdale, Debra, and Joe R. Feagin. 2001. *The First R: How Children Learn Race and Racism.* Lanham, MD: Rowman & Littlefield.

Washington Post. 2009. "Washington Post-ABC News Poll." Available at http://www.washingtonpost.com

Williams, Eddie N., and Milton D. Morris. 1993. "Racism and Our Future." In *Race in America: The Struggle for Equality,* Herbert Hill and James E. Jones Jr., eds. (pp. 417–424). Madison: University of Wisconsin Press.

Williams, Richard, Reynold Nesiba, and Eileen Diaz McConnell. 2005. "The Changing Face of Inequality in Home Mortgage Lending." *Social Problems* 52(2):181–208.

Willoughby, Brian. 2003. "Hate on Campus." *Tolerance in the News,* June 13. Available at http://www.tolerance.org

Wilson, William J. 1987. *The Truly Disadvantaged: The Inner City, the Underclass and Public Policy.* Chicago: University of Chicago Press.

Winter, Greg. 2003a. "Schools Resegregate, Study Finds." *New York Times,* January 21. Available at http://www.nytimes.com

Winter, Greg. 2003b. "U. of Michigan Alters Admissions Use of Race." *New York Times,* August 29. Available at http://www.nytimes.com

Wise, Tim. 2009. *Between Barack and a Hard Place: Racism and White Denial in the Age of Obama.* San Francisco: City Light Books.

Zack, Naomi. 1998. *Thinking About Race.* Belmont, CA: Wadsworth.

Zinn, Howard. 1993. "Columbus and the Doctrine of Discovery." In *Systemic Crisis: Problems in Society, Politics, and World Order,* William D. Perdue, ed. (pp. 351–357). Fort Worth, TX: Harcourt Brace Jovanovich.

Chapter 10

AAUW. 2007a. *Gender Pay Gap: State by State.* Available at http://www.aauw.org

AAUW. 2007b (April 23). *Pay Gap Exists as Early as One Year out of College, Research Says.* Available at http://www.aauw.org

AAUW (American Association of University Women). 2006 (January 24). *Drawing the Line: Sexual Harassment on Campus.* Available at http://www.aauw.org

Abernathy, Michael. 2003. *Male Bashing on TV.* Tolerance in the News. Available at http://www.tolerance.org

ACLU (American Civil Liberties Union). 2009. "Affirmative Action: Get the Facts." Available at http://www.aclu.org

Adams, Jimi. 2007. "Stained Glass Makes the Ceiling Visible." *Gender and Society* 21(1):80–105.

Anderson, Margaret L. 1997. *Thinking About Women.* 4th ed. New York: Macmillan.

Begley, Sharon. 2000. "The Stereotype Trap." *Newsweek,* November 6, pp. 66–68.

Bennett, Rosemary. 2009 (April 24). "Compulsory Audits on Equal Pay Will Force Firms to Give Women More." Available at http://business.timesonline.co.uk

Bertrand, Marianne, Claudia Goldin, and Lawrence F. Katz. 2009 (January). "Dynamics of the Gender Gap for Young Professionals in the Financial and Corporate Sectors." Working Paper. Available at http://www.economics.harvard.edu

Bianchi, Suzanne M., Melissa A. Milkie, Liana C. Sayer, and John Robinson. 2000. "Is Anyone Doing the Housework? Trends in the Gender Division of Household Labor." *Social Forces* 79:191–228.

Billitteri, Thomas J. 2008. "Gender Pay Gap." *CQ Researcher* 8(11):241–264. Available at http://www.cqresearcher.com

Bittman, Michael, and Judy Wajcman. 2000. "The Rush Hour: The Character of Leisure Time and Gender Equity." *Social Forces* 79:165–189.

Bly, Robert. 1990. *Iron John: A Book About Men.* Boston: Addison-Wesley.

Bourin, Lenny, and Bill Blakemore. 2008. "More Men Take Traditionally Female Jobs." ABC World News, September 1. Available at http://abcnews.go.com

Boylan, Jennifer Finney. 2009. "Is My Marriage Gay?" *New York Times,* May 11. Available at http://www.nytimes.com

Brady, David, and Denise Kall. 2008. "Nearly Universal, but Somewhat Distinct: The Feminization of Poverty in Affluent Western Democracies, 1969–2000." *Social Science Research* 37:976–1007.

Burnett, Victoria. 2008. "Defense Minister's New Baby Confirms Symbolism of Parity in Spain." *New York Times,* May 20. Available at http://www.nytimes.com

CAWP. 2009. *Women in Elected Offices, 2009.* Available at http://www.cawp.rutgers.edu

Ceci, Stephen J., Wendy M. Williams, and Susan M. Barnett. 2009. "Women's Underrepresentation in Science: Socio-cultural and Biological Considerations." *Psychological Bulletin* 135 (2):218–261.

Clothier, Peter. 2009. "Men's Warrior Weekend." *The Huffington Post,* March 9. Available at http://www.huffingtonpost.com

Cohen, Theodore. 2001. *Men and Masculinity.* Belmont, CA: Wadsworth.

Coltrane, Scott, and Michelle Adams. 2008. *Gender and Families.* Lanham, MD: Rowman and Littlefield.

Common Dreams. 2005 (March 24). *Media Advisory: Women's Opinions Also Missing on Television.* Available at http://www.commondreams.org

Connolly, Ceci. 2005. "Access to Abortion Pared at State Level." *Washington Post,* August 29, p. A1.

Cook, Carolyn. 2009. "ERA Would End Women's Second-Class Citizenship: Only Three More States Are Needed to Declare Gender Bias Unconstitutional." *The Philadelphia Inquirer,* April 12. Available at http:www.philly.com

Correll, Shelly J. Stephen Benard, and In Paik. 2007. "Getting a Job: Is There a Motherhood Penalty?" *American Journal of Sociology* 112 (5):1297–1338.

Current Legislation. 2009. National Committee on Pay Equity. Available at http://www.pay-equity.org

Darrow, Whitney. 1970. *I'm Glad I'm a Boy? I'm Glad I'm a Girl.* Windmill Books. New York: Simon and Schuster.

Davis, James A., Tom W. Smith, and Peter V. Marsden. 2002. *General Social Surveys, 1972–2002: 2nd ICPSR Version.* Chicago: National Opinion Research Center.

Dee, Thomas S. 2006 (Fall). "How a Teacher's Gender Affects Boys and Girls." *Education Next.* Available at http://www.educationnext.org

EEOC (Equal Employment Opportunity Commission). 2009a. *Sex-Based Charges: 1997–2008.* Available at http://www.eeoc.gov

EEOC. 2009b. "Sexual Harassment." Available at http://www.eeoc.gov

EEOC. 2009c (February 26). "Big Vanilla Athletic Club to Pay $161,000 to Settle EEOC Lawsuit for Sexual Harassment." http://eeoc.gov

Egelko, Bob. 2009. "Wal-Mart Wins Chance to Block Class-Action Suit." *San Francisco Chronicle,* February 14. Available at http://www.sfgate.com

ERA (Equal Rights Amendment). 2009. *The Equal Rights Amendment.* Available at http://www.equalrightsamendment.org

Faiola, Anthony. 2007. "Japanese Working Women Still Serve the Tea." *The Washington Post,* March 2, p. A09.

Faludi, Susan. 2008. "Think the Gender War Is Over? Think Again." *The New York Times,* June 15. Available at http://www.nytimes.com

Faludi, Susan. 1991. *Backlash: The Undeclared War Against American Women.* New York: Crown.

Filkins, Dexter. 2009. "Afghan Women Protest New Law on Home Life." *The New York*

Times, April 16. Available at http://www.nytimes.com

Foerstel, Karen. 2008. "Women's Rights: Are Violence and Discrimination Against Women declining?" *Global Researcher* 2(5):115–147.

Forbes. 2009 (June 3). "The 2009 Celebrity 100." Available at http://www.forbes.com

Gallagher, Margaret. 2006. *Who Makes the News? Global Media Monitoring Project 2005.* Available at http://www.whomakesthenews.org

Gallagher, Sally K. 2004. "The Marginalization of Evangelical Feminism." *Sociology of Religion* 65:215–237.

Goffman, Erving. 1963. *Stigma.* Englewood Cliffs, NJ: Prentice Hall.

Goldman, Tom. 2009 (April 8). "A First-Timer at Augusta National Golf Club." *A Reporter's Notebook.* Available at http://www.npr.org/

Goodstein, Laurie. 2009."U.S. Nuns Facing Vatican Scrutiny." *The New York Times,* July 2. Available at http://www.nytimes.com

Grogan, Sarah. 2008. *Body Image: Understanding Body Dissatisfaction in Men, Women and Children.* New York: Routledge.

Gupta, Sanjay. 2003. "Why Men Die Young." *Time,* May 12, p. 84.

Guy, Mary Ellen, and Meredith A. Newman. 2004. "Women's Jobs, Men's Jobs: Sex Segregation and Emotional Labor." *Public Administration Review* 64:289–299.

Haines, Errin. 2006 (November 1). *Father Convicted in Genital Mutilation.* Available at http://www.breitbart.com/news

Hamilton, Mykol C., David Anderson , Michelle Broaddus, and Kate Young. 2006. "Gender Stereotyping and Under-representation of Female Characters in 200 Popular Children's Picture Books: A Twenty-first Century Update." *Sex Roles* 55:757–765.

Hausmann, Ricardo, Laura Tyson, and Saadia Zahidi. 2009. *The Global Gender Gap Report 2008.* Geneva: World Economic Forum.

Heyzer, Noeleen. 2003. "Enlisting African Women to Fight AIDS." *Washington Post,* July 8. Available at http://www.globalpolicy.org

Hochschild, Arlie. 1989. *The Second Shift.* London: Penguin.

HRC (Human Rights Campaign). 2005. *Transgender Americans: A Handbook for Understanding.* Available at http://www.hrc.org

IASC (Inter -Agency Standing Committee). 2009. "IASC Policy Statement Gender Equality in Humanitarian Action." Available at http://www.humanitarianinfo.org

IDEA (International Institute for Democracy and Electoral Assistance). 2009. "Quotas for Women." Available at http://www.idea.int

ILO (International Labour Organization). 2009 (March). *Global Employment Trends for Women.* Available at http://www.ilo.org

ILO News. 2009. "ILO Warns Economic Crisis Could Generate up to 22 Million More Unemployed Women in 2009, Jeopardize Equality Gains at Work and at Home." Available at http://www.ilocarib.org.tt

IPC (International Poverty Centre). 2008. *Poverty in Focus: Gender Equality.* No. 13. Available at http://www.undp-povertycentre.org

ITUC (International Trade Union Commission). 2009. *Gender (In)Equality in the Labour Market: An Overview of Global Trends and Development.* Available http://www.ituc-csi.org

IWPR (Institute for Women's Policy Research). 2009a (April). *Fact Sheet: The Gender Wage Gap, 2008.* Available at http://www.iwpr.org

IWPR. 2009b (April). *Fact Sheet: The Gender Wage Gap by Occupation.* Available at http://www.iwpr.org

Jackson, Robert, and Meredith A. Newman. 2004. "Sexual Harassment in the Federal Workplace Revisited: Influences on Sexual Harassment." *Public Administration Review* 64(6):705–717.

Jose, Paul, and Isobel Brown. 2009. "When Does the Gender Difference in Rumination Begin? Gender and Age Differences in the Use of Rumination by Adolescents." *Journal of Youth and Adolescence* 37:180–192.

Judd, Terri. 2008. "Barbaric 'Honour Killings' Become the Weapon to Subjugate Women in Iraq." *The Independent,* April 28. Available at http://www.independent.co.uk

Kaufman, David. 2009. "Introducing America's First Black, Female Rabbi." *Time,* June 6. Available at http://www.time.com

Kimmel, Michael. 2008. *Guyland.* New York: Harper Collins.

Kravets, David. 2007 (February 6). *Court Says Wal-Mart Must Face Bias Trial.* Available at http://www.breitbart.com

Lacey, Marc. 2008. "A Lifestyle Distinct: The Muxe of Mexico." *The New York Times*, Dec. 7. Available at http://www.nytimes.com

LCCREF (Leadership Conference on Civil Rights Education Fund). 2009. "Confronting the New Faces of Hate: Hate Crimes in America." Available at http://www.civilrights.org

Lepkowska, Dorothy. 2008. "Playing Fair?" *The Guardian,* December 16. Available at http://www.guardian.co.uk/

Leslie, David W. 2007 (March). "The Reshaping of America's Academic Workforce." *Research Dialogue* 87. New York: TIAA-CREF Institute. Available at http://www.tiaa-crefinstitute.org

Levin, Diane E., and Jean Kilbourne. 2008. *So Sexy So Soon: The New Sexualized Childhood and What Parents Can Do to Protect Their Kids.* New York: Random House.

Lopez-Claros, Augusto, and Saadia Zahidi. 2005. *Women's Empowerment: Measuring the Global Gender Gap.* World Economic Forum.

Lutheran World Federation. 2009 (January 29). "Lutheran Woman Bishop Jeruma-Grinberga Succeeds Jagucki in Great Britain." Available at http://www.lutheranworld.org

Mahapatra, Dhananjay. 2008. "Customary payments, gifts not dowry: SC." *The Times of India.* February. Available at http://timesofindia.indiatimes.com

Mannino, Celia Anna, and Francine M. Deutsch. 2007. "Changing the Division of Household Labor: A Negotiated Process between Partners." *Sex Roles* 56:309–324.

McBrier, Debra Branch. 2003. "Gender and Career Dynamics Within a Segmented Professional Labor Market: The Case of Law Academia." *Social Forces* 81:1201–1266.

Media Matters. 2009."Limbaugh Asserts 'Chicks . . . Have Chickified the News'; Again Refers to Female Reporter as Infobabe." Available at http://mediamatters.org

Mehmood, Isha. 2009 (January 29). "Lilly Ledbetter Fair Pay Act becomes Law." Available at http://www.civilrights.org

Messner, Michael A., and Jeffrey Montez de Oca. 2005. "The Male Consumer as Loser: Beer and Liquor Ads in Mega Sports Media Events." *Signs* 30:1879–1909.

MKP (ManKind Project). 2009a. "Our Values." Available at http://www.mkp.org

MKP. 2009b. "Frequently Asked Questions About the New Warrior Training Adventure." Available http://www.mkp.org

Mnet (Media Awareness Network). 2009. "Beauty and Body Image in the Media." Available http://www.media-awareness.ca

Moen, Phyllis, and Yan Yu. 2000. "Effective Work/Life Strategies: Working Couples, Working Conditions, Gender, and Life Quality." *Social Problems* 47:291–326.

Moreno, Pedro C. 2009 (July 3). "Report from Rio: Men and Boys in Gender Equality." Available at http://www.huffingtonpost.com

Morin, Rich, and Paul Taylor. 2008 (September 15). *Revisiting the Mommy Wars: Politics, Gender and Parenthood.* Available at http://pewsocialtrends.org

Nanda, Serena. 2000. *Gender Diversity: Cross-Cultural Variations.* Long Grove, IL: Wavelane Press.

NCFM. 2009. "Philosophy." Available at http://www.ncfm.org

NCES (National Center for Educational Statistics). 2009a. *The Condition of Education, 2008.* U.S. Department of Education. Available at http://nces.ed.gov

NCES. 2009b. "Fast Facts: What Is Title IX?" Available at http://nces.ed.gov

NCWGE (National Coalition for Women and Girls in Education). 2008. *Title IX at 35: Beyond the Headlines*. Available at http://www.ncwge.org

NOMAS (National Organization for Men Against Sexism). 2009. "Statement of Principles." Available at http://www.nomas.org

Nosek, B. A., F. L. Smyth, N. Sriram, N. M. Lindner, T. Devos, A. Ayala, Y. Bar-Anan, R. Bergh, H. Cai, K. Gonsalkorale, S. Kesebir, N. Maliszewski, F. Neto, E. Olli, J. Park, K. Schnabel, K. Shiomura, B. Tulbure, R. W. Wiers, N. Somogyi, N. Akrami, B. Ekehammar, M. Vianello, M. R. Banaji, and A. G. Greenwald. 2009. "National Differences in Gender-Science Stereotypes Predict National Sex Differences in Science and Math Achievement." *Proceedings of the National Academy of Sciences* 106: 10593–10597.

NOW (National Organization for Women). 2007 (April 24). "Women Lose Millions Due to Wage Gap, NOW Calls for Passage of Pay Equity Legislation." Available at http://now.org

NWLC. 2009. "Action Alerts." Available at: http://womensrights.change.org

Olkon, Sara. 2009. "Power Move by Male Students Ruffles U. of C." *Chicago Tribune*, May 27. Available at http://www.chicagotribune.com

Padavic, Irene, and Barbara Reskin. 2002. *Men and Women at Work.* 2nd ed. Thousand Oaks, CA: Pine Forge Press.

Parker, Kathleen. 2000. "It's Time for Women to Get Angry but Not at Men." *Greensboro News Record,* March 14, p. F4.

Parks, Janet B., and Mary Ann Roberton. 2000. "Development and Validation of an Instrument to Measure Attitudes Toward Sexist/Nonsexist Language." *Sex Roles* 42(5/6):415–438.

Parks, Janet B. and Mary Ann Roberton. 2001. "Inventory of Attitudes Toward Sexist/Nonsexist Language—General (IASNL-G): A Correction in Scoring Procedures." Erratum. *Sex Roles* 44: (3/4): 253.

Partlow, Joshua, and Lonnae O'Neal Parker. 2006. "West Point Mourns a Font of Energy, Laid to Rest by War." *The Washington Post*, September 27, A01.

Pew Research Center. 2007 (July 12). "From 1997 to 2007 Fewer Mothers Prefer Full-time Work." Available at http://www.pewresearch.org

Polachek, Soloman W. 2006. "How the Life-Cycle Human Capital Model Explains Why the Gender Gap Narrowed." In *The Declining Significance of Race,* Francine D. Blau, Mary C. Brinton, and David B. Grusky, eds. (pp. 102–124). New York: Russell Sage.

Pollack, William. 2000a. *Real Boys' Voices.* New York: Random House.

Pollack, William. 2000b. "The Columbine Syndrome." *National Forum* 80:39–42.

Portraits of Sacrifice. 2006. "U.S. Causalities in Iraq." *San Francisco Chronicle.* Available at http://www.sfgate.com

Pozner, Jennifer L. 2009 (July 8). "Hot and Bothering: Media Treatment of Sarah Palin." National Public Radio. Available at http://www.npr.org

Preston, Julia. 2009. "New Policy Permits Asylum for Battered Women." *The New York Times,* July 16. Available at http://www.nytimes.com

Quist-Areton, Ofeibea. 2003. "Fighting Prejudice and Sexual Harassment of Girls in Schools." *All Africa,* June 12. Available at http://www.globalpolicy.org

Ramesh, Randeep. 2007. "Fetuses Aborted and Dumped Secretly as India Shuns Baby Girls." *The Guardian,* July 28. Available at http://www.guardian.co.uk

Rao, Kavitha. 2006. "Missing Daughters on an Indian Mother's Mind." *Women's e-News,* March 16. Available at http://www.womensenews.org

Renzetti, Claire, and Daniel Curran. 2003. *Women, Men and Society.* Boston: Allyn and Bacon.

Reskin, Barbara, and Debra McBrier. 2000. "Why Not Ascription? Organizations' Employment of Male and Female Managers." *American Sociological Review* 65:210–233.

Rose, Stephen J., and Heidi Hartman. 2008 (February). "Still a Man's Labor Market: The Long-Term Earnings Gap." IWPR# C366. New York: Institute for Women's Policy Research.

Rubin, Alissa J., and Sam Dagher. 2009. "Changes in Iraq Election Law Weaken Quota for Women." *New York Times,* January 13. Available at http://www.nytimes.com

Rutenberg, Jim. 2009. "New Protections for Transgender Federal Workers." *New York Times,* June 24. Available at http://www.nytimes.com

Sadker, David, and Karen Zittleman. 2009. *Still Failing at Fairness: How Gender Bias Cheats Boys and Girls in Schools.* New York: Simon and Schuster.

Saltarelli, Steve. 2009. "Men in Power." *Chicago Maroon,* March 2. Available at http://www.chicagomaroon.com

Sanchez, Diana T., and Jennifer Crocker. 2005. "How Investment in Gender Ideals Affects Well-Being: The Role of External Contingencies of Self-Worth." *Psychology of Women Quarterly* 29:63–77.

Sherrill, Andrew. 2009 (April 28). "Women's Pay." Testimony before the Joint Economic Committee. U.S. Congress. Washington, DC: United States Government Accountability Office.

Simpson, Ruth. 2005. "Men in Non-Traditional Occupations: Career Entry, Career Orientation, and Experience of Role Strain." *Gender Work and Organization* 12(4):363–380.

Smith, Melanie. 2006. "Is Church Too Feminine for Men?" *The Decatur Daily*, July1. Available at http://www.decaturdaily.com

Snyder, Karrie Ann, and Adam Isaiah Green. 2008. "Revisiting the Glass Escalator: The Case of Gender Segregation in a Female Dominated Occupation." *Social Problems* 55(2):271–299.

Social Watch. 2009. "Gender: 20th Century Debts, 21st Century Shame." *Social Watch Report 2008.* Available at http://www.socialwatch.org

Sonnert, Gerhard, Mary Frank Fox, and Kristen Adkins. 2007. "Undergraduate Women in Science and Engineering: Effects of Faculty, Fields, and Institutions Over Time." *Social Science Quarterly* 88 (5):1333–1356.

Tenenbaum, Harriet R. 2009. "You'd Be Good at That: Gender Patterns in Parent-Child Talk about Courses." *Social Development* 18(2):447–463.

Thornburgh, Nathan. 2006. "A Death in the Class of 9/11." *Time,* September 28. Available at http://time.com

Uggen, C., and A. Blackstone. 2004. "Sexual Harassment as a Gendered Expression of Power." *American Sociological Review* 69:64–92.

UNESCO (United Nations Educational, Scientific and Cultural Organization). 2009. "Why the Literacy Decade?" Available at http://www.unesco.org

UNICEF. 2009. *The State of the World's Children: 2009.* United Nations: United Nations Children's Fund. Available at http://www.unicef.org

UNICEF. 2007. *The State of the World's Children: 2007.* United Nations: United Nations Children's Fund. Available at http://www.unicef.org

UNIFEM (United Nations Development Fund for Women). 2007. "Harmful Traditional Practices." Available at http://www.unifem.org

United Nations. 2008. *The Millennium Development Goals Report 2008.* Available at http://www.un.org

U.S. Census Bureau. 2009. *Statistical Abstract of the United States: 2008.* 128th ed. Washington, DC: U.S. Government Printing Office.

Vandello, Joseph A., Jennifer K. Bosson, Dov Cohen, Rochelle M. Burnaford, and Jonathan R. Weaver. 2008. "Precarious

Manhood." *Journal of Personality and Social Psychology* 95 (6):1325–1339.

Venkatesan, J. 2009. "Supreme Court Anguish over Dowry Deaths." *The Hindu,* June 2. Available at http://www.hindu.com

Vincent, Norah. 2006. *Self-Made Man.* New York: Viking Penguin.

Vogel, Chris. 2008. "A New Retreat for the ManKind Project Houston." *Houston Press,* June 26. Available at http://www.houstonpress.com

White, Alan, and Karl Witty. 2009. "Men's Under Use of Health Services—Finding Alternative Approaches." *Journal of Men's Health* 6(2):95–97.

WHO (World Health Organization). 2009. "Ten Facts About Women's Health." Available at http://www.who.int

WHO. 2006. *Gender and Women's Mental Health.* Available at http://www.who.int/mental_health

WHO. 2008. *Eliminating Female Genital Mutilation: An Interagency Statement.* Available at http://whqlibdoc.who.int

Williams, Christine L. 2007. "The Glass Escalator: Hidden Advantages for Men in the 'Female' Occupations." In *Men's Lives,* 7th ed., Michael S. Kimmel and Michael Messner, eds. (pp. 242–255). Boston: Allyn and Bacon.

Williams, Joan. 2000. *Unbending Gender: Why Family and Work Conflict and What to Do About It.* Oxford: Oxford University Press.

Williams, Kristi, and Debra Umberson. 2004. "Marital Status, Marital Transitions, and Health: A Gendered Life Course Perspective." *Journal of Health and Social Behavior* 45:81–98.

Wood, Wendy and Alice H. Eagly. 2002. "A Cross-Cultural Analysis of the Behavior of Women and Men: Implications for the Origins of Sex Differences." *Psychological Bulletin* 128(5):699–727.

Yoder, P. Stanley, N. Abderrahim, and A. Zhuzhuni. 2004. *Female Genital Cutting in the Demographic and Health Surveys: A Critical and Comparative Analysis.* DHS Comparative Reports 7. Calverton, MD: ORC Macro.

York, E. Anne. 2008. "Gender Differences in the College and Career Aspiration of High School Valedictorians." *Journal of Advanced Academics* 19(4):578–580.

Yumiko, Ehara. 2000. "Feminism's Growing Pains." *Japan Quarterly* 47:41–48.

Zakrzewski, Paul. 2005. "Daddy, What Did You Do in the Men's Movement?" *Boston Globe,* June 19. Available at http://www.bostonglobe.com

Chapter 11

ACLU. 1999 (January 14). *The Rights of Lesbian, Gay, Bisexual, and Transgendered People.* Available at http://www.aclu.org

The Advocate. 2002 (August 20). *Advocate Sex Poll,* pp. 28–43.

Allport, G. W. 1954. *The Nature of Prejudice.* Cambridge, MA: Addison-Wesley.

Amato, Paul R. 2004. "Tension Between Institutional and Individual Views of Marriage." *Journal of Marriage and Family* 66:959–965.

American Anthropological Association. n.d. *Statement on Marriage and Family from the American Anthropological Association.* Available at http://www.aaanet.org

American College Health Association. 2008. "American College Health Association National College Health Assessment Fall 2007 Reference Group Data Report." Baltimore: American College Health Association.

Amnesty International. 2005. *Stonewalled: Police Abuse and Misconduct Against Lesbian, Gay, Bisexual, and Transgender People in the United States.* Available at http://www.amnesty.org

Badgett, Lee. 2009. "Domestic Partnership Benefits and Obligations Act." Williams Institute. Available at http://www.law.ucla.edu/williamsinstitute

Bayer, Ronald. 1987. *Homosexuality and American Psychiatry: The Politics of Diagnosis.* 2nd ed. Princeton, NJ: Princeton University Press.

Besen, Wayne. 2000. "Introduction." In *Feeling Free: Personal Stories—How Love and Self-Acceptance Saved Us from "Ex-Gay" Ministries,* Human Rights Campaign, ed. (p. 7). Washington, DC: Human Rights Campaign Foundation.

Bobbe, Judith. 2002. "Treatment with Lesbian Alcoholics: Healing Shame and Internalized Homophobia for Ongoing Sobriety." *Health and Social Work* 27(3):218–223.

Bontempo, Daniel E., and Anthony R. D'Augelli. 2002. "Effects of At-School Victimization and Sexual Orientation on Lesbian, Gay, or Bisexual Youths' Health Risk Behavior." *Journal of Adolescent Health* 30:364–374.

Bradford, Judith, Kirsten Barrett, and Julie A. Honnold. 2002. *The 2000 Census and Same-Sex Households: A User's Guide.* New York: National Gay and Lesbian Task Force Policy Institute, Survey and Evaluation Research Laboratory, and Fenway Institute. Available at http://www.thetaskforce.org

Buhl, Larry. 2005 (June 16). "Youth's Blog Stirs Uproar Over 'Ex-Gay' Camp." *Planet Out.* Available at http://www.planetout.com

Bumiller, Elizabeth. 2009. "In Military, New Debate over Policy Toward Gays." *New York Times,* May 1. Available at http://www.nytimes.com

Button, James W., Barbara A. Rienzo, and Kenneth D. Wald. 1997. *Private Lives, Public Conflicts: Battles over Gay Rights in American Communities.* Washington, DC: CQ Press.

Cahill, Sean. 2007. "The Coming GLBT Senior Boom." *The Gay & Lesbian Review,* January–February, pp. 19–21.

Cahill, Sean, Ellen Mitra, and Sarah Tobias. 2002. *Family Policy: Issues Affecting Gay, Lesbian, Bisexual, and Transgender Families.* National Gay and Lesbian Task Force Policy Institute. Available at http://www.thetaskforce.org

Cahill, Sean, and Kenneth T. Jones. 2003. *Child Sexual Abuse and Homosexuality: The Long History of the "Gays as Pedophiles" Fallacy.* National Gay and Lesbian Task Force. Available at http://www.thetaskforce.org

Carney, Michael P. 2007 (September 5). "The Employment Non-Discrimination Act: Testimony by Officer Michael P. Carney." Springfield Massachusetts. Available at http://edlabor.house.gov

Chamberlain, Jared, Monica K. Miller, and Brian H. Bornstein. 2008. "The Rights and Responsibilities of Gay and Lesbian Parents: Legal Developments, Psychological Research, and Policy Implications." *Social Issues and Policy Review* 2(1):103–126.

Chase, Bob. 2000. *NEA President Bob Chase's Historic Speech from 2000 GLSEN Conference.* Available at http://www.glsen.org

Cianciotto, Jason, and Sean Cahill. 2007. "Anatomy of a Pseudo-Science." *The Gay & Lesbian Review,* July–August, pp. 22–24.

Cianciotto, Jason, and Sean Cahill. 2003. *Educational Policy: Issues Affecting Lesbian, Gay, Bisexual, and Transgender Youth.* Washington, DC: National Gay and Lesbian Task Force Policy Institute.

Courage Campaign. n.d. "President Obama: Don't Fire Dan Choi." Available at http://www.couragecampaign.org

Curtis, Christopher. 2005 (August 31). "Group Links Katrina to Annual Gay Party." *PlanetOut.* Available at http://www.planetout.com

Curtis, C. 2004. "Poll: 1 in 20 High School Students Is Gay." *PlanetOut.* Available at http://www.planetout.com

Curtis, Christopher. 2003 (October 7). "Poll: U.S. Public is 50–50 on Gay Marriage." *PlanetOut.* Available at http://www.planetout.com

Dozetos, Barbara. 2001 (March 7). "School Shooter Taunted as 'Gay.'" *PlanetOut.* Available at http://www.planetout.com

Durkheim, Emile. 1993. "The Normal and the Pathological." In *Social Deviance,* Henry N. Pontell, ed. (pp. 33–63). Englewood Cliffs, NJ: Prentice-Hall. (Originally published in *The Rules of Sociological Method,* 1938.)

Esterberg, K. 1997. *Lesbian and Bisexual Identities: Constructing Communities,*

Constructing Selves. Philadelphia: Temple University Press.

FBI (Federal Bureau of Investigation). 2008. *Hate Crime Statistics 2007*. Available at http://www.fbi.gov

Fidas, Deena. 2009. "At the Water Cooler." *Equality* (Spring): 23, 29.

Firestein, B. A. 1996. "Bisexuality as Paradigm Shift: Transforming Our Disciplines." In *Bisexuality: The Psychology and Politics of an Invisible Minority*, B. A. Firestein, ed. (pp. 263–291). Thousand Oaks, CA: Sage.

Fone, Byrne. 2000. *Homophobia: A History*. New York: Henry Holt.

Frank, Barney. 1997. "Foreword." In *Private Lives, Public Conflicts: Battles over Gay Rights in American Communities*, by J. W. Button, B. A. Rienzo, and K. D. Wald, eds. (pp. i–xi). Washington, DC: CQ Press.

Frank, David John, and Elizabeth H. McEneaney. 1999. "The Individualization of Society and the Liberalization of State Policies on Same-Sex Relations, 1984–1995." *Social Forces* 77(3):911–944.

Franklin, Karen. 2000. "Antigay Behaviors among Young Adults." *Journal of Interpersonal Violence* 15(4):339–362.

Gallup Organization. 2009. "Pulse on Democracy: Homosexual Relations." Available at http://www.gallup.com

Garnets, L., G. M. Herek, and B. Levy. 1990. "Violence and Victimization of Lesbians and Gay Men: Mental Health Consequences." *Journal of Interpersonal Violence* 5:366–383.

Gates, Gary. 2009 (September). "Same-Sex couples in the 2008 American Community Survey." Los Angeles: The Williams Institute on Sexual Orientation Law and Public Policy.

Gates, Gary J., M. V. Lee Badgett, Jennifer Ehrle Macomber, and Kate Chambers. 2007 (March). *Adoption and Foster Care by Gay and Lesbian Parents in the United States*. Washington, DC: The Urban Institute.

Gilman, Stephen E., Susan D. Cochran, Vickie M. Mays, Michael Hughes, David Ostrow, and Ronald C. Kessler. 2001. "Risk of Psychiatric Disorders Among Individuals Reporting Same-Sex Sexual Partners in the National Comorbidity Survey." *American Journal of Public Health* 91(6):933–939.

GLSEN. 2007 (July 2). *Number of Gay-Straight Alliance Registrations Passes 3,500*. Available at http://www.glsen.org

GLSEN (Gay, Lesbian, and Straight Education Network). 2000. *Homophobia 101: Teaching Respect for All*. Gay, Lesbian, and Straight Education Network. Available at http://www.glsen.org

Green, John C. 2004. *The American Religious Landscape and Political Attitudes: A Baseline for 2004*. Pew Forum on Religion and Public Life. Available at http://pewforum.org

Harris Interactive and GLSEN. 2005. *From Teasing to Torment: School Climate in America, A Survey of Students and Teachers*. New York: GLSEN.

Held, Myka. 2005 (March 16). *Mix It Up: T-Shirts and Activism*. Available at http://www.tolerance.org/teens

Human Rights Campaign. 2009a. *The State of the Workplace for Lesbian, Gay, Bisexual, and Transgendered Americans, 2007–2008*. Washington, DC: Human Rights Campaign. Available at http://www.hrc.org

Human Rights Campaign. 2009b (June 4). *State Prohibitions on Marriage for Same-Sex Couples*. Available at http://www.hrc.org

Human Rights Campaign. 2009c. *State Hate Crime Laws*. Available at http://www.hrc.org

Human Rights Campaign. 2009d. *Statewide Laws or Policies Affecting Schools & Educational Institutions*. Available at http://www.hrc.org

Human Rights Campaign. 2009e. *An Introduction to Welcoming Schools*. Washington, DC: Human Rights Campaign Foundation.

Human Rights Campaign. 2005 (March 22). "Recruiting Age Bump Highlights Consequences of 'Don't Ask, Don't Tell.'" HRC press release. Available at http://www.hrc.org

Human Rights Campaign. 2000a. *Feeling Free: Personal Stories—How Love and Self-Acceptance Saved Us from "Ex-Gay" Ministries*. Washington, DC: Human Rights Campaign Foundation.

Human Rights Campaign. 2000b. "Custody and Visitation." FamilyNet. Available at http://familynet.hrc.org

Human Rights Watch. 2001 (March). *World Report 2001*. Available at http://www.hrw.org

Intelligence Briefs. 2006. "Phelps' Latest Tactic Shocks Americans." *SPLC Report*, p. 3.

International Gay and Lesbian Human Rights Commission. 1999. *Antidiscrimination Legislation*. Available at http://www.iglhrc.org

Israel, Tania, and Jonathan J. Mohr. 2004. "Attitudes Toward Bisexual Women and Men: Current Research, Future Directions." In *Current Research on Bisexuality*, Ronald C. Fox, ed. (pp. 117–134). New York: Harrington Park Press.

Kilman, Carrie. 2007 (Spring). "This Is Why We Need a GSA." *Teaching Tolerance*, pp. 30–37.

Kinsey, A. C., W. B. Pomeroy, and C. E. Martin. 1948. *Sexual Behavior in the Human Male*. Philadelphia: W. B. Saunders.

Kinsey, A. C., W. B. Pomeroy, C. E. Martin, and P. H. Gebhard. 1953. *Sexual Behavior in the Human Female*. Philadelphia: W. B. Saunders.

Kirkpatrick, R. C. 2000. "The Evolution of Human Sexual Behavior." *Current Anthropology* 41(3):385–414.

Kosciw, Joseph G., and Elizabeth M. Diaz. 2005. *The 2005 National School Climate Survey*. New York: GLSEN.

Landis, Dan. 1999 (February 17). "Mississippi Supreme Court Made a Tragic Mistake in Denying Custody to Gay Father, Experts Say." *American Civil Liberties Union News*. Available at http://www.aclu.org

LaSala, M. 2005. "Monogamy of the Heart: A Qualitative Study of Extradyadic Sex Among Gay Male Couples." *Journal of Gay and Lesbian Social Services* 17:1–24.

LaSala, M. 2004. "Extradyadic Sex and Gay Male Couples: Comparing Monogamous and Nonmonogamous Relationships." *Families in Society: The Journal of Contemporary Human Services* 85:405–415.

LAWbriefs. 2005a (April). "Recent Developments in Sexual Orientation and Gender Identity Law." *LAWbriefs* 7(1).

LAWbriefs. 2005b (July). "Recent Developments in Sexual Orientation and Gender Identity Law." *LAWbriefs* 7(2).

LAWbriefs. 2003 (Spring). "Recent Developments in Sexual Orientation and Gender Identity Law." *LAWbriefs* 6(1).

LAWbriefs. 2000 (Fall). "Recent Developments in Sexual Orientation and Gender Identity Law." *LAWbriefs* 3(3).

Lever, Janet. 1994. "The 1994 Advocate Survey of Sexuality and Relationships: The Men." *The Advocate*, August 23, pp. 16–24.

Loftus, Jeni. 2001. "America's Liberalization in Attitudes Toward Homosexuality, 1973 to 1998." *American Sociological Review* 66:762–782.

Louderback, L. A., and B. E. Whitley. 1997. "Perceived Erotic Value of Homosexuality and Sex-Role Attitudes as Mediators of Sex Differences in Heterosexual College Students' Attitudes Toward Lesbians and Gay Men." *Journal of Sex Research* 34:175–182.

Lyall, Sarah. 2005. "New Course by Royal Navy: A Campaign to Recruit Gays." *New York Times*, February 22. Available at http://www.nytimes.com

McGregor, Nanette R. n.d. "A Letter to My Gay Son." Available at GayFamilySupport.com (retrieved May 2, 2009).

McKinney, Jack. 2004 (February 8). "The Christian Case for Gay Marriage." Pullen Memorial Baptist Church. Available at http://www.pullen.org

Michael, Robert T., John H. Gagnon, Edward O. Laumann, and Gina Kolata. 1994. *Sex in America: A Definitive Survey*. Boston: Little, Brown.

Miller, Patti, McCrae A. Parker, Eileen Espejo, and Sarah Grossman-Swenson. 2002. *Fall*

Colors: Prime Time Diversity Report 2001–02. Oakland CA: Children Now and The Media Program.

Mohipp, C., and M. M. Morry. 2004. "Relationship of Symbolic Beliefs and Prior Contact to Heterosexuals' Attitudes Toward Gay Men and Lesbian Women." *Canadian Journal of Behavioral Science* 36(1):36–44.

National Coalition of Anti-Violence Programs. 2008. *Anti-Lesbian, Gay, Bisexual and Transgender Violence in 2007.* New York: National Coalition of Anti-Violence Programs.

National Gay and Lesbian Task Force. 2009a (July 1). "State Nondiscrimination Laws in the U.S." Available at http://www.thetaskforce.org

National Gay and Lesbian Task Force. 2009b (July). "Relationship Recognition for Same-Sex Couples in the U.S." Available at http://www.thetaskforce.org

Neidorf, Shawn, and Rich Morin. 2007 (May 23). "Four-in-Ten Americans Have Close Friends or Relatives Who Are Gay." Pew Research Center for People & the Press. Available at http://pewresearch.org

Newport, Frank. 2002 (September). "In-Depth Analysis: Homosexuality." Gallup Organization. Available at http://www.gallup.com

Ottosson, Daniel. 2008. *A World Survey of Laws Prohibiting Same-Sex Activity between Consenting Adults.* International Lesbian and Gay Association. Available at http://www.ilga.org

Page, Susan. 2003. "Gay Rights Tough to Sharpen into Political 'Wedge Issue.'" *USA Today,* July 28, p. 10A.

Patel, S. 1989. *Homophobia: Personality, Emotional, and Behavioral Correlates.* Master's thesis, East Carolina University, Greenville, NC.

Patel, S., T. E. Long, S. L. McCammon, and K. L. Wuensch. 1995. "Personality and Emotional Correlates of Self-Reported Antigay Behaviors." *Journal of Interpersonal Violence* 10:354–366.

Pathela, Preeti, Anjum Hajat, Julia Schillinger, Susan Blank, Randall Sell, and Farzad Mostashari. 2006. "Discordance Between Sexual Behavior and Self-Reported Sexual Identity: A Population-Based Survey of New York City Men." *Annals of Internal Medicine* 145(6):416–425.

Patterson, Charlotte J. 2001. "Family Relationships of Lesbians and Gay Men." In *Understanding Families into the New Millennium: A Decade in Review,* Robert M. Milardo, ed. (pp. 271–288). Minneapolis, MN: National Council on Family Relations.

Paul, J. P. 1996. "Bisexuality: Exploring/Exploding the Boundaries." In *The Lives of Lesbians, Gays, and Bisexuals: Children to Adults,* R. Savin-Williams and K. M.

Cohen, eds. (pp. 436–461). Fort Worth, TX: Harcourt Brace.

Pelham, Brett, and Steve Crabtree. 2009 (March 10). "Religiosity and Perceived Intolerance of Gays and Lesbians." Gallup Organization. Available at http://www.gallup.com

Pew Research Center for People & the Press. 2009 (March 25). "Americans Favor Carbon Cap, Gays in the Military and Renewing U.S.-Cuba Ties." Pew Research Center Publications. http://pewresearch.org

Pew Research Center. 2007 (March 22). *Trends in Political Values and Core Attitudes: 1987–2007.* Washington, DC: Pew Research Center for People and the Press.

Platt, Leah. 2001. "Not Your Father's High School Club." *American Prospect* 12(1):A37–A39.

Plugge-Foust, C., and George Strickland. 2000. "Homophobia, Irrationality, and Christian Ideology: Does a Relationship Exist?" *Journal of Sex Education and Therapy* 25:240–244.

Potok, Mark. 2005 (Spring). "Vilification and Violence." *Intelligence Report* 117:1.

Price, Jammie, and Michael G. Dalecki. 1998. "The Social Basis of Homophobia: An Empirical Illustration." *Sociological Spectrum* 18:143–159.

Pryor, John H., Sylvia Hurtado, Linda DeAngelo, Jessica Sharkness, Laura C. Romero, William S. Korn, and Serge Tran. 2008. *The American Freshman: National Norms Fall 2008.* Los Angeles: Higher Education Research Institute, UCLA.

Puccinelli, Mike. 2005 (April 19). "Students Support, Decry Gays with T-Shirts." CBS 2 Chicago. Available at http://cbs2chicago.com

Rankin, Susan R. 2003. *Campus Climate for Gay, Lesbian, Bisexual, and Transgendered People: A National Perspective.* New York: National Gay and Lesbian Task Force Policy Institute. Available at http://www.thetaskforce.org

Roderick, T. 1994. *Homonegativity: An Analysis of the SBS-R.* Master's thesis, East Carolina University, Greenville, NC.

Roderick, T., S. L. McCammon, T. E. Long, and L. J. Allred. 1998. "Behavioral Aspects of Homonegativity." *Journal of Homosexuality* 36:79–88.

Rosin, Hanna, and Richard Morin. 1999 (January 11). "In One Area, Americans Still Draw a Line on Acceptability." *Washington Post National Weekly Edition* 16(11):8.

Ryan, Caitlin, David Huebner, Rafael M. Diaz, and Jorge Sanchez. 2009. "Family Rejection as a Predictor of Negative Health Outcomes in White and Latino Lesbian, Gay, and Bisexual Young Adults." *Pediatrics* 123(1): 346-352.

Saad, Lydia. 2007 (May 29). "Tolerance for Gay Rights at High Water Mark." Gallup News Service. Available at http://www.gallup.com

Saad, Lydia. 2005 (May 20). "Gay Rights Attitudes a Mixed Bag." Gallup Organization. Available at http://www.gallup.com

Safe Zone Resources. 2003. *National Consortium of Directors of LGBT Resources in Higher Education.* Available at http://www.lgbtcampus.org

Sanday, Peggy R. 1995. "Pulling Train." In *Race, Class, and Gender in the United States,* 3rd ed., P. S. Rothenberg, ed. (pp. 396–402). New York: St. Martin's Press.

Savin-Williams, R. C. 2006. "Who's Gay? Does It Matter?" *Current Directions in Psychological Science* 15:40–44.

Schiappa, E., P. B. Gregg, and D. E. Hewes. 2005. "The Parasocial Contact Hypothesis." *Communication Monographs* 72(1):92–115.

Shernoff, Michael. 2006. "Negotiated Nonmonogamy and Male Couples." *Family Process* 45(4):407–418.

Shilts, Randy. 1982. *The Mayor of Castro Street: The Life and Times of Harvey Milk.* New York: St. Martin's Press.

SIECUS (Sexuality Information and Education Council of the United States). 2008. *A Portrait of Sexuality Education and Abstinence-Only-Until-Marriage Programs in the States.* Available at http://www.siecus.org

SIECUS. 2000. *Fact Sheets: Sexual Orientation and Identity.* New York: SIECUS.

Simmons, Tavia, and Martin O'Connell. 2003 (February). *Married-Couple and Unmarried-Partner Households: 2000.* Washington, DC: U.S. Census Bureau.

Singh, Daniel P., and Heather D. Wathington. 2003. "Valuing Equity: Recognizing the Rights of the LGBT Community." *Diversity Digest* 7(1–2):8–9.

Smith, D. M., and G. J. Gates. 2001. *Gay and Lesbian Families in the United States: Same-Sex Unmarried Partner Households—Preliminary Analysis of 2000 U.S. Census Data.* Washington, DC: Human Rights Campaign. Available at http://www.hrc.org

Sullivan, A. 1997. "The Conservative Case." In *Same-Sex Marriage: Pro and Con,* A. Sullivan, ed. (pp. 146–154). New York: Vintage Books.

Sylivant, S. 1992. *The Cognitive, Affective, and Behavioral Components of Adolescent Homonegativity.* Master's thesis, East Carolina University, Greenville, N.C.

Thompson, Cooper. 1995. "A New Vision of Masculinity." In *Race, Class, and Gender in the United States,* 3rd ed., P. S. Rothenberg, ed. (pp. 475–481). New York: St. Martin's Press.

Tobias, Sarah, and Sean Cahill. 2003. *School Lunches, the Wright Brothers, and Gay Families*. National Gay and Lesbian Task Force. Available at http://www.thetaskforce.org

Ontario Consultants on Religious Tolerance. 1999. "The United Methodist Church and Homosexuality." Available at www.religioustolerance.org/hom_umc.htm

White, Mel. 2005 (Spring). "A Thorn in Their Side." *Intelligence Report* 117:27–30.

Wilcox, Clyde, and Robin Wolpert. 2000. "Gay Rights in the Public Sphere: Public Opinion on Gay and Lesbian Equality." In *The Politics of Gay Rights*, Craig A. Rimmerman, Kenneth D. Wald, and Clyde Wilcox, eds. (pp. 409–432). Chicago: University of Chicago Press.

Yang, Alan. 1999. *From Wrongs to Rights 1973 to 1999: Public Opinion on Gay and Lesbian Americans Moves Toward Equality*. New York: Policy Institute of the National Gay and Lesbian Task Force.

Chapter 12

Belden, Russonello & Stewart Research and Communications. 2004. *American Community Survey*. Washington, DC.

Belden Russonello & Stewart Research and Communications. 2003 (April). *Americans' Attitudes Toward Walking and Creating Better Walking Communities*. Washington, DC.

Bishaw, Alemayehu, and Jessica Semega. 2008. (August). *Income, Earnings, and Poverty Data from the 2007 American Community Survey*. Washington, D.C.: U.S. Government Printing Office.

Bongaarts, John, and Susan Cotts Watkins. 1996. "Social Interactions and Contemporary Fertility Transitions." *Population and Development Review* 22(4):639–682.

Brown, Lester. 2006. *Plan B 2.0: Rescuing a Planet Under Stress and a Civilization in Trouble*. New York: W. W. Norton.

"China Is Softening Messages." 2007. *Popline*, September–October, pp. 1, 3.

Cincotta, Richard, Robert Engelman, and Daniele Anastasion. 2003. *The Security Demographic: Population and Civil Conflict After the Cold War*. Washington, DC: Population Action International.

Clark, David. 1998. "Interdependent Urbanization in an Urban World: An Historical Overview." *Geographical Journal* 164(1):85–96.

Cohen, Susan A. 2006. "The Global Contraceptive Shortfall: U.S. Contributors and U.S. Hindrances." *Guttmacher Policy Review* 9(2):15–18.

Cowherd, Phil. 2001. "What Is the Business Case for Investing in Inner-City Neighborhoods?" *Public Management* 83(1):12–14.

Davis, Mike. 2006. *Planet of Slums*. New York: Verso.

Douthwaite, M., and P. Ward. 2005. "Increasing Contraceptive Use in Rural Pakistan: An Evaluation of the Lady Health Worker Programme." *Health Policy and Planning* 20(2):117–123.

Durning, Alan. 1996. *The City and the Car*. Northwest Environment Watch. Seattle, WA: Sasquatch Books.

Egan, Timothy. 2005. "Vibrant Cities Find One Thing Missing: Children." *New York Times*, March 24. Available at http://www.nytimes.com

Ewing, Reid, Tom Schmid, Richard Killingsworth, Amy Zlot, and Stephen Raudenbush. 2003. "Relationship Between Urban Sprawl and Physical Activity, Obesity, and Morbidity." *American Journal of Health Promotion* 18(1):47–57.

Farrell, Paul. 2009 (January 26). "Peak oil? Global Warming? No, It's 'Boomsday'!" MarketWatch. Available at http://www.marketwatch.com

Frost, Ashley E. and F. Nii-Amoo Dodoo. 2009. "Men Are Missing from African Family Planning." *Contexts: Understanding People in Their Social Worlds*. Vol. 8(1):44–49.

Hollander, Dore. 2007. "Women Who Are Fecund but Do Not Wish to Have Children Outnumber the Involuntarily Childless." *Perspectives on Sexual and Reproductive Health* 39(2):120.

Johnson, William C. 1997. *Urban Planning and Politics*. Chicago: American Planning Association and Planners Press.

Kimuna, Sitawa R., and Donald J. Adamchak. 2001. "Gender Relations: Husband-Wife Fertility and Family Planning Decisions in Kenya." *Journal of Biosocial Science* 33:13–23.

Lalasz, Robert. 2005 (July). *Baby Bonus Credited with Boosting Australia's Fertility Rate. Population Reference Bureau*. Available at http://www.prb.org

Leahy, Elizabeth. 2003. "As Contraceptive Use Rises, Abortions Decline." *Popline*, November-December, pp. 3, 8.

Leonard, Matt. 2009. "The Kindest Cut." *Earth Island Journal*. Available at http://www.earthislandjournal.org

Leonard, Tom. 2009. "U.S. Cities May Have To Be Bulldozed in Order to Survive." Telegraph, June 12. Available at http://www.telegraph.co./uk

Lindstrom, Matthew J., and Hugh Bartling. 2003. "Introduction." In *Suburban Sprawl: Culture, Theory, and Politics*, M. J. Lindstrom and H. Bartling, eds. (pp. xi–xxvii). Lanham, MD: Rowman & Littlefield.

Livernash, Robert, and Eric Rodenburg. 1998. "Population Change, Resources, and the Environment." *Population Bulletin* 53(1):1–36.

Lohn, Martiga. 2007. "Divers Locate Remains at Bridge Collapse Site." *The Boston Globe*, August 13. Available at http://www.boston.com

Mackinnon, Mark. 2009. "China's One-Child Policy Gets a Second Thought." The Globe and Mail, August 21. Available at http://www.theglobeandmail.com

McFalls, Joseph A. 2007 (March). "Population: A Lively Introduction, 5th ed." *Population Bulletin* 62(1).

McLaughlin, Eliot C. 2007 (August 3). "Experts: Leadership, Money Keys to Building Bridges." CNN. Available at http://www.CNN.com

Millennium Ecosystem Assessment. 2005. *Ecosystems and Human Well-Being: Synthesis*. Washington, DC: Island Press.

Newman, Peter, and Jeff Kenworthy. 2007. "Greening Urban Transportation." In L. Starke, ed. *2007 State of the World: Our Urban Future* (pp. 66–85). New York: W.W. Norton.

Office of Management and Budget. 2008 (November 20). *OMB Bulletin No. 09-01*. Available at http://www.whitehouse.gov

Parks, Kristin. 2005. "Choosing Childlessness: Weber's Typology of Action and Motives of the Voluntarily Childless." *Sociological Inquiry* 75(3):372–402.

Pelley, Janet. 1999. "Building Smart-Growth Communities." *Environmental Science and Technology News* 33(1):28A–32A.

Population Reference Bureau. 2009. *World Population Data Sheet*. Washington, DC: Population Reference Bureau. Available at http:www.prb.org

Population Reference Bureau. 2007. *World Population Data Sheet*. Washington, DC: Population Reference Bureau. Available at http://www.prb.org

Population Reference Bureau. 2004. "Transitions in World Population." *Population Bulletin* 59(1).

Schrank, David, and Tim Lomax. 2007. *The 2007 Urban Mobility Report*. Texas Transportation Institute and Texas A&M University System. Available at http://mobility.tamu.edu

Schueller, Jane. 2005 (August). "Boys and Changing Gender Roles." YouthNet. *YouthLens* 16. Available at http://www.fhi.org

Sheehan, M. O. 2003. "Uniting Divided Cities." In *State of the World 2003*, L. Starke, ed. (pp. 130–151). New York: W. W. Norton.

Sheehan, M. O. 2001. "Making Better Transportation Choices." In *State of the World 2001*, L. Starke, ed. (pp. 103–122). New York: W. W. Norton.

Shevis, Jim. 1999. "More Affluent Than Their Inner-City Neighbors, Suburbanites Still Have Growth Problems." *Washington Spectator* 25(19):1–3.

Smart Growth Network. 2009. *About Smart Growth.* Available at http://www.smartgrowth.org

Smith, Gar. 2009. "Planet Girth." *Earth Island Journal* 24(2):15.

Sullivan, Daniel Monroe. 2007. "Reassessing Gentrification: Measuring Residents' Opinions Using Survey Data." *Urban Affairs Review* 42(4):583–592.

SUNY College of Environmental Science and Forestry. 2009. "Worst Environmental Problem? Overpopulation, Experts Say." *ScienceDaily,* April 20. Available at http://www.sciencedaily.com

UNEP (United Nations Environment Programme). 2005. *GEO Yearbook: An Overview of Our Changing Environment 2004/5.* Available at http://www.unep.org

United Nations. 2009. *World Population Prospects: The 2008 Revision.* New York: United Nations.

United Nations. 2008. *Millennium Development Goals 2008.* New York: United Nations.

United Nations. 2007. *World Population Prospects: The 2006 Revision.* New York: United Nations.

United Nations Population Division. 2009. "What Would It Take to Accelerate Fertility Decline in the Least Developed Countries?" United Nations Population Division Policy Brief No. 2009/1. New York: United Nations.

United Nations Population Fund. 2007. *State of World Population 2007: Unleashing the Potential of Urban Growth.* New York: United Nations.

U.S. Conference of Mayors. 2001. *Traffic Congestion and Rail Investment.* Washington, DC: Global Strategy Group.

U.S. Department of Housing and Urban Development. 2004. "Revitalizing Neighborhoods by Redeveloping Brownfields: The HUD Brownfields Economic Development Initiative." *Research Works* 1(3):2.

Weeks, John R. 2008. *Population: An Introduction to Concepts and Issues,* 10th ed. Belmont, CA: Wadsworth, Cengage Learning.

Weiland, Katherine. 2005. *Breeding Insecurity: Global Security Implications of Rapid Population Growth.* Washington, DC: Population Institute.

Wiseman, Paul. 2005. "Towns Hope Cash-for-Babies Incentives Boost Populations." *USA Today,* July 28. Available at http://www.usatoday.com

Women's Studies Project. 2003. *Women's Voices, Women's Lives: The Impact of Family Planning.* Family Health International. Available at http://www.fhi.org

World Health Organization. 2003. *World Health Report 2003: Shaping the Future.* Geneva: World Health Organization.

World Watch Institute. 2004 (January 7). "State of the World 2004: Consumption by the Numbers." Press Release. Available at http://www.worldwatch.org

Chapter 13

American Lung Association. 2009. *State of the Air: 2009.* Available at http://lungaction.org

American Lung Association. 2005. *Lung Disease Data in Culturally Diverse Communities: 2005.* Available at http://www.lungusa.org

Associated Press. 2007 (May 13). "Report: Students Exposed to Pesticides Used Near Schools." CNN.com. Available at http://www.cnn.com

Barbosa, Luiz C. 2009. "Theories in Environmental Sociology." In *Twenty Lessons in Environmental Sociology,* Kenneth A. Gould and Tammy L. Lewis, eds. (pp. 25–44). New York: Oxford University Press.

Beinecke, Frances. 2009 (Spring). "Debunking the Myth of Clean Coal." Onearth. Available at http://www.onearth.org

Blatt, Harvey. 2005. *America's Environmental Report Card: Are We Making the Grade?* Cambridge, MA: MIT Press.

Brody, Julia Gree, Kirsten B. Moysich, Olivier Humblet, Kathleen R. Attfield, Gregory P. Beehler, and Ruthann A. Rudel. 2007. "Environmental Pollutants and Breast Cancer." *Cancer* 109(S12):2667–2711.

Brown, Lester R. 2007. "Distillery Demand for Grain to Fuel Cars Vastly Understated: World May Be Facing Highest Grain Prices in History." *Earth Policy News,* January 4. Available at http://www.earthpolicy.org

Brown, Lester R., and Jennifer Mitchell. 1998. "Building a New Economy." In *State of the World 1998,* Lester R. Brown, Christopher Flavin, and Hilary French, eds. (pp. 168–187). New York: W. W. Norton.

Brulle, Robert J. 2009. "U.S. Environmental Movements." In *Twenty Lessons in Environmental Sociology,* Kenneth A. Gould and Tammy L. Lewis, eds. (pp. 211–227). New York: Oxford University Press.

Bruno, Kenny, and Joshua Karliner. 2002. *Earthsummit.biz: The Corporate Takeover of Sustainable Development.* CorpWatch and Food First Books. Available at http://www.corpwatch.org

Bullard, Robert D. 2000. *Dumping in Dixie: Race, Class, and Environmental Quality,* 3rd ed. Boulder, CO: Westview Press.

Bullard, Robert D., Paul Mohai, Robin Saha, and Beverly Wright. 2007 (March). *Toxic Wastes and Race at Twenty 1987–2007.* Cleveland, Ohio: United Church of Christ.

Caress, Stanley M., and Anne C. Steinemann. 2004. "A National Population Study of the Prevalence of Multiple Chemical Sensitivity." *Archives of Environmental Health* 59(6):300–305.

Carlson, Scott. 2006. "In Search of the Sustainable Campus." *The Chronicle of Higher Education* LIII(9):A10–A12, A14.

CBS News. 2008 (June 17). "Tapped Out Consumers Spurn Bottled Water." Available at http://www.cbsnews.com

Chafe, Zoe. 2006. "Weather-Related Disasters Affect Millions." In *Vital Signs,* L. Starke ed. (pp. 44–45). New York: W. W. Norton & Co.

Chafe, Zoe. 2005. "Bioinvasions." In *State of the World 2005,* L. Starke, ed. (pp. 60–61). New York: W. W. Norton.

Cheeseman, Gina-Marie. 2007. "Plastic Shopping Bags Being Banned." *The Online Journal,* June 27. Available at http://www.onlinejournal.com

Chepesiuk, Ron. 2009. "Missing the Dark: Health Effects of Light Pollution." *Environmental Health Perspectives* 117(1):20, A20–A27.

Cincotta, Richard P., and Robert Engelman. 2000. *Human Population and the Future of Biological Diversity.* Washington, DC: Population Action International.

Clarke, Tony. 2002. "Twilight of the Corporation." In *Social Problems, Annual Editions 02/03,* 30th ed., Kurt Finster-Busch, ed. (pp. 41–45). Guilford, CT: McGraw-Hill/Dushkin.

Cooper, Arnie. 2004. "Twenty-Eight Words That Could Change the World: Robert Hinkley's Plan to Tame Corporate Power." *The Sun* 345(September):4–11.

Coyle, Kevin. 2005. *Environmental Literacy in America.* Washington, DC: The National Environmental Education and Training Foundation.

Denson, Bryan. 2000. "Shadowy Saboteurs." *IRE Journal* 23(May-June):12–14.

Duhigg, Charles. 2009 (September 12). "Clean Water Laws Are Neglected, at a Cost in Suffering." *New York Times.* Available at http://www.nytimes.com

Edwards, Bob, and Adam Driscoll. 2009. "From Farms to Factories: The Environmental Consequences of Swine Industrialization in North Carolina." In *Twenty Lessons in Environmental Sociology,* Kenneth A. Gould and Tammy L. Lewis, eds. (pp. 153–175). New York: Oxford University Press.

Electronics TakeBack Coalition. 2009 (February 21). "Facts and Figures on E Waste and Recycling." Available at http://www.electronicstakeback.com

Energy Information Administration. 2008. *World Energy Overview: 1996–2006.* Washington DC: Energy Information Administration.

Environmental Defense Fund. 2001. "A Prescription for Reducing the Damage Caused by Dams." *Environmental Defense* 32(2).

Environmental Working Group. 2005 (July 14). *Body Burden: The Pollution in Newborns.* Available at http://www.ewg.org

EPA (U.S. Environmental Protection Agency). 2009. *NPL Site Totals by Status and Milestone.* Available at http://www.epa.gov/superfund/sites

EPA. 2008. *Municipal Solid Waste Generation, Recycling, and Disposal in the United States: Facts and Figures for 2007.* Available at http://www.epa.gov

EPA. 2004. *What You Need to Know About Mercury in Fish and Shellfish.* Available at http://www.epa.gov

Fisher, Brandy E. 1999. "Focus: Most Unwanted." *Environmental Health Perspectives* 107(1). Available at http://ehpnet1.niehs.nih.gov/docs/1999/107-1/focus-abs.html

Fisher, Brandy E. 1998. "Scents and Sensitivity." *Environmental Health Perspectives* 106(12). Available at http://ehpnet1.niehs.nih.gov/docs/1998/106-12/focus-abs.html

Flavin, Christopher, and Robert Engelman. 2009. "The Perfect Storm." In *State of the World 2009: Into a Warming World*, L. Starke ed. (pp. 5–12). New York: Oxford University Press.

Flavin, Chris, and Molly Hull Aeck. 2005 (September 15). "Cleaner, Greener, and Richer." Tom.Paine.com. Available at http://www.tompaine.com/articles/2005/09/15/cleaner_greener_and_richer.php

Food and Drug Administration. 2008. *Pesticide Residue Monitoring Program Results and Discussion FY 2006.* Available at http://www.fda.gov

Food & Water Watch and Network for New Energy Choices. 2007. *The Rush to Ethanol: Not All Biofuels Are Created Equal.* Available at http://www.newenergychoices.org

French, Hilary. 2000. *Vanishing Borders: Protecting the Planet in the Age of Globalization.* New York: W. W. Norton.

Gallup Organization. 2009. *Environment.* Available at http://www.gallup.com/poll

Gardner, Gary. 2005. "Forest Loss Continues." In *Vital Signs 2005*, Linda Starke, ed. (pp. 92–93). New York: W. W. Norton.

Gearhart, Jeff. 2009 (September 16). "Toxic Chemicals in Everyday Products: Find Out What's in the Stuff You Use." Alternet. Available at http://www.alternet.org

Global Humanitarian Forum. 2009. *Human Impact Report: Climate Change—The Anatomy of a Silent Crisis.* Geneva: Global Humanitarian Forum.

Global Invasive Species Database. 2006. "Felis Catus." Available at http://www.issg.org

Goodstein, Laurie. 2005. "Evangelical Leaders Swing Influence Behind Effort to Combat Global Warming." *New York Times*, March 10. Available at http://www.nytimes.com

Gottlieb, Roger S. 2003a. "Saving the World: Religion and Politics in the Environmental Movement." In *Liberating Faith*, Roger S. Gottlieb, ed. (pp. 491–512). Lanham, MD: Rowman & Littlefield.

Gottlieb, Roger S. 2003b. "This Sacred Earth: Religion and Environmentalism." In *Liberating Faith*, Roger S. Gottlieb, ed. (pp. 489–490). Lanham, MD: Rowman & Littlefield.

Hager, Nicky, and Bob Burton. 2000. *Secrets and Lies: The Anatomy of an Anti-Environmental PR Campaign.* Monroe, ME: Common Courage Press.

Hefling, Kimberly. 2007. "Hearing Planned Today in Lejeune Water Case." *Marine Corps Times*, June 12. Available at http://www.marinecorpstimes.com

Hilgenkamp, Kathryn. 2005. *Environmental Health: Ecological Perspectives.* Sudbury, MA: Jones and Bartlett Publishers.

Hunter, Lori M. 2001. *The Environmental Implications of Population Dynamics.* Santa Monica, CA: Rand Corporation.

Intergovernmental Panel on Climate Change. 2007a. *Climate Change 2007: The Physical Science Basis.* United Nations Environmental Programme and the World Meteorological Organization. Available at http://www.ipcc.ch/

Intergovernmental Panel on Climate Change. 2007b. *Climate Change 2007: Impacts, Adaptation and Vulnerability.* United Nations Environmental Programme and the World Meteorological Organization. Available at http://www.ipcc.ch/

International Atomic Energy Agency. 2008. *Energy, Electricity and Nuclear Power for the Period up to 2030.* Available at http://www.iaea.org

IUCN (International Union for Conservation of Nature). 2008a. *State of the World's Species.* Available at http://www.iucn.org

IUCN. 2008b. *2008 IUCN Red List of Threatened Species.* Available at http://www.iucnredlist.org

Janofsky, Michael. 2005. "Pentagon Is Asking Congress to Loosen Environmental Laws." *New York Times*, May 11. Available at http://www.nytimes.com

Jensen, Derrick. 2001. "A Weakened World Cannot Forgive Us: An Interview with Kathleen Dean Moore." *The Sun* 303 (March):4–13.

Kaplan, Sheila, and Jim Morris. 2000. "Kids at Risk." *U.S. News and World Report*, June 19, pp. 47–53.

Klinkenborg, Verlyn. 2008. "Our Vanishing Night." *National Geographic* (November):102–123.

Knoell, Carly. 2007 (August 9). "Malaria: Climbing in Elevation as Temperature Rises." *Population Connection.* Available at www.populationconnection.org

Koenig, Dieter. 1995. "Sustainable Development: Linking Global Environmental Change to Technology Cooperation." In *Environmental Policies in the Third World: A Comparative Analysis*, O. P. Dwivedi, and Dhirendra K. Vajpeyi, eds. (pp. 1–21). Westport, CT: Greenwood Press.

Lamm, Richard. 2006. "The Culture of Growth and the Culture of Limits." *Conservation Biology* 20(2):269–271.

Lavelle, Mariane. 2009 (April 20). "The 'Clean Coal' Blitz." Center for Public Integrity. Available at http://publicintegrity.org

Leahy, Stephen. 2009 (May 21). "Alien Species Eroding Ecosystems and Livelihoods." Interpress Service News Agency. Available at http://www.ipsnews.net

Little, Amanda Griscom. 2005. "Maathai on the Prize: An Interview with Nobel Peace Prize Winner Wangari Maathai." *Grist Magazine*, February 15. Available at http://www.grist.org

Makower, Joel, Ron Pernick, and Clint Wilder. 2009. *Clean Energy Trends 2009.* Clean Edge. Available at http://www.cleanedge.com

Maibach, E., C. Roser-Renouf, and A. Leiserowitz. 2009. *Global Warming's Six Americas 2009: An Audience Segmentation Analysis.* Yale Project on Climate Change and the George Mason University Center for Climate Change Communication. Available at http://www.americanprogress.org

Malcolm, Jay R., Canran Liu, Ronald P. Neilson, Lara Hansen, and Lee Hannah. 2006. "Global Warming and Extinctions of Endemic Species from Biodiversity Hotspots." *Conservation Biology* 20(2): 538–548.

Margesson, Rhoda. 2005. "Environmental Refugees." In *State of the World 2005*, Linda Starke, ed. (pp. 40–41). New York: W. W. Norton.

McCarthy, Michael. 2008. "Cleared: Jury Decides That Threat of Global Warming Justifies Breaking the Law." *The Independent*, September 11. Available at http://www.independent.co.uk

McDaniel, Carl N. 2005. *Wisdom for a Livable Planet.* San Antonio, TX: Trinity University Press.

McGinn, Anne Platt. 2000. "Endocrine Disrupters Raise Concern." In *Vital Signs 2000,* Lester R. Brown, Michael Renner, and Brian Halweil, eds. (pp. 130–131). New York: W. W. Norton.

McKibben, Bill. 2008 (October 24). "Meltdown: A Global Warming Travelogue." CNN.com. Available at http://www.cnn.com

McMichael, Anthony J., Kirk R. Smith, and Carlos F. Corvalan. 2000. "The Sustainability Transition: A New Challenge." *Bulletin of the World Health Organization* 78(9):1067.

Mead, Leila. 1998. "Radioactive Wastelands." *The Green Guide* 53(April 14):1–3.

Millennium Ecosystem Assessment. 2005. *Ecosystems and Human Well-Being: Synthesis.* Washington, DC: Island Press.

Miller, G. Tyler, Jr., and Scott E. Spoolman. 2009. *Living in the Environment.* 16th ed. Belmont, CA: Brooks/Cole, Cengage Learning.

Miranda, Marie Lynn, Dohyeong Kim, Alicia Overstreet Galeano, Christopher J. Paul, Andrew P. Hull, and S. Philip Morgan. 2007. "The Relationship Between Early Childhood Blood Lead Levels and Performance on End-of-Grade Tests." *Environmental Health Perspectives* 115(8):1242–1247.

Mooney, Chris. 2005. "Some Like It Hot." *Mother Jones,* May-June. Available at http://www.MotherJones.com

Murtaugh, Paul, and Michael Schlax. 2009. "Reproduction and the Carbon Legacies of Individuals." *Global Environmental Change* 19:14–20.

NASA. 2009. *Ozone Hole Watch.* Available at http://ozonewatch.gsfc.nasa.gov/index.html

National Assessment Synthesis Team. 2000. *Climate Change Impacts on the United States: The Potential Consequences of Climate Variability and Change.* Washington, DC: U.S. Global Change Research Program.

National Geographic. 2007 (June 5). Top ten tips to fight global warming. *Green Guide.* Available at http://www.thegreenguide.com

National Snow and Ice Data Center. 2007 (October 1). *Arctic Sea Ice Shatters All Previous Record Lows.* Available at http://www.nsidc.org

"Nukes Rebuked." 2000. *Washington Spectator* 26(13):4.

Pearce, Fred. 2005. "Climate Warming as Siberia Melts." *New Scientist,* August 11. Available at http://www.NewScientist.com

Pew Center on the States. 2009. *State of the States 2009.* Available at http://www.stateline.org

Pimentel, D., S. Cooperstein, H. Randell, D. Filiberto, S. Sorrentino, B. Kaye, C. Nicklin, J. Yagi, J. Brian, J. O'Hern, A. Habas and C. Weinstein. 2007. "Ecology of Increasing Diseases: Population Growth and Environmental Degradation." *Human Ecology 35(6)*:653–668.

Prah, Pamela M. 2007 (July 30). "States Forge Ahead on Immigration, Global Warming." *Stateline.* Available at http://www.stateline.org

Price, Tom. 2006. "The New Environmentalism." *CQ Researcher* 16(42):987–1007.

Pryor, John H., S. Hurtado, L. DeAngelo, J. Sharkness, L. C. Romero, W. S. Korn, and S. Tran. 2008. *The American Freshman: National Norms Fall 2008.* Los Angeles: Higher Education Research Institute, UCLA.

Public Citizen. 2005 (February). *NAFTA Chapter 11 Investor-State Cases: Lessons for the Central America Free Trade Agreement. Public Citizens Global Trade Watch Publication E9014.* Available at http://www.citizen.org

Rasmussen, Matt. 2007 (January 25). "When Does Green Rage Become Ecoterrorism?" Alternet. Available at http://www.alternet.org

Reese, April. 2001 (February). *Africa's Struggle with Desertification.* Population Reference Bureau. Available at http://www.prb.org

Renner, Michael. 2004. "Moving Toward a Less Consumptive Economy." In *State of the World 2004,* Linda Starke, ed. (pp. 96–119). New York: W. W. Norton.

Renner, Michael. 1996. *Fighting for Survival: Environmental Decline, Social Conflict, and the New Age of Insecurity.* New York: W. W. Norton.

Rifkin, Jeremy. 2004. *The European Dream: How Europe's Vision of the Future Is Quietly Eclipsing the American Dream.* New York: Tarcher/Penguin.

Rogers, Sherry A. 2002. *Detoxify or Die.* Sarasota, FL: Sand Key.

Rosenthal, Elisabeth. 2009. "Obama's Backing Raises Hopes for Climate Pact." *New York Times,* March 1, p. A1.

Sampat, Payal. 2003. "Scrapping Mining Dependence." In *State of the World 2003,* Linda Starke, ed. (pp. 110–129). New York: W. W. Norton.

Sawin, Janet L. 2009 (May 7). "Wind Power in 2008 Exceeds 10-Year Average Growth Rate." WorldWatch Institute. Available at http://www.worldwatch.org

Sawin, Janet. 2004. "Making Better Energy Choices." In *State of the World 2004,* Linda Starke, ed. (pp. 24–43). New York: W. W. Norton.

Schapiro, Mark. 2007. *Exposed: The Toxic Chemistry of Everyday Products and What's at Stake for American Power.* White River Junction, Vermont: Chelsea Green Publishing.

Schapiro, Mark. 2004. "New Power for 'Old Europe.'" *The Nation,* December 27, pp. 11–16.

Schwartz-Nobel, Loretta. 2007. *Poisoned Nation.* New York: St. Martin's Press.

Siddall, Mark, Thomas F. Stocker, and Peter U. Clark. 2009. "Constraints on Future Sea-Level Rise from Past Sea-Level Reconstructions." *Nature Geoscience* 2:571–575.

Silicon Valley Toxics Coalition. 2001. *Just Say No to E-Waste: Background Document on Hazards and Waste from Computers.* Available at http://www.svtc.org

Sinks, Thomas. 2007 (June 12). *Statement by Thomas Sinks, PhD, Deputy Director, Agency for Toxic Substances and Disease Registry on ATSDR's Activities at U.S. Marine Corps Base Camp Lejeune Before Committee on Energy and Commerce Subcommittee on Oversight and Investigations United States House of Representatives.* Available at http://www.hhs.gov

Sterritt, Angela. 2005. "Indigenous Youth Challenge Corporate Mining." *Cultural Survival, Weekly Indigenous News,* July 15. Available at http://www.culturalsurvival.org

Stoll, Michael. 2005 (September-October). "A Green Agenda for Cities." *E Magazine* 16(5). Available at http://www.emagazine.com

Sustainable Endowments Institute. 2008. *College Sustainability Report Card 2008.* Available at http://www.endowmentinstitute.org

TerraChoice Group Inc. 2009. *The Seven Sins of Greenwashing: Environmental Claims in Consumer Markets.* Available at http://sinsofgreenwashing.org

United Nations Development Programme. 2007. *Human Development Report 2007/2008: Fighting Climate Change: Human Solidarity in a Divided World.* New York: Palgrave Macmillan.

UNEP (United Nations Environment Programme). 2008. *Vital Forest Graphics.* UNEP/GRID-Arendal. Available at http://www.fao.org

UNEP. 2007. *GEO Yearbook 2007: An Overview of Our Changing Environment.* Available at http://www.unep.org

U.S. Department of Health and Human Services. 2004. *11th Report on Carcinogens.* Washington, DC: Public Health Service.

U.S. Geological Survey. 2007 (September 7). "Future Retreat of Arctic Ice Will Lower Polar Bear Populations and Limit Their Distribution." *USGS Newsroom.* Available at http://www.usgs.gov

Vajpeyi, Dhirendra K. 1995. "External Factors Influencing Environmental Policymaking: Role of Multilateral Development Aid

Agencies." In *Environmental Policies in the Third World: A Comparative Analysis,* O. P. Dwivedi and Dhirendra K. Vajpeyi, eds. (pp. 24–45). Westport, CT: Greenwood Press.

Vedantam, Shankar. 2005a. "Nuclear Plants Not Keeping Track of Waste." *Washington Post,* April 19. Available at http://www.washingtonpost.com

Vedantam, Shankar. 2005b. "Storage Plan Approved for Nuclear Waste." *Washington Post,* September 10. Available at http://www.washingtonpost.com

Vidal, John. 2008. "Not Guilty: The Greenpeace Activists Who Used Climate Change as a Defence." *The Guardian,* September 11. Available at http://www.guardian.co.uk

Westerling, A. L., H. G. Hidalgo, D. R. Cayan, and T. W. Swetnam. 2006. "Warming and Earlier Spring Increase Western U.S. Forest Wildfire Activity." *Science* 313 (5789): 940–943.

Willett, Katharine M., Nathan P. Gillett, Philip D. Jones and Peter W. Thorne. 2007. "Attribution of Observed Surface Humidity Changes to Human Influence." *Nature* 449(October 11):710–712.

Wire, Thomas. 2009 (August.) "Fewer Emitters, Lower Emissions, Less Cost." Optimum Population Trust. Available at http://www.optimumpopulation.org

Woodward, Colin. 2007. "Curbing Climate Change." *CQ Global Researcher* 1(2):27–50. Available at http://www.globalresearcher.com

World Health Organization. 2005. *Indoor Air Pollution and Health.* Fact Sheet. Available at http://www.who.org

World Resources Institute. 2000. *World Resources 2000–2001: People and Ecosystems—The Fraying Web of Life.* Washington, DC: World Resources Institute.

World Water Assessment Program. 2009. *World Water Development Report 3: Water in a Changing World.* Available at http://www.unesco.org

WWF (World Wildlife Federation). 2008. *Living Planet Report.* World Wildlife Fund, Zoological Society of London, and Global Footprint Network. Available at http://www.panda.org

Zandonella, Catherine. 2007. "The Dirty Dozen Chemicals in Cosmetics." *The Greene Guide* No. #122 (October/November). Available at http://www.thegreenguide.com

Chapter 14

AAEM (American Academy of Environmental Medicine). 2009 (May 8). "Genetically Modified Foods." Available at http://www.aaemonline.org

Abcarian, Robin. 2009. "A New Push to Define 'Person,' and to Outlaw Abortion in the Process." *Los Angeles Times,* September 28. Available at http://www.latimes.com

ACLU (American Civil Liberties Union). 2005. *Utah Businesses, Free Speech Groups, and Individuals Challenge Restrictions on Internet Speech. Privacy and Technology.* Available at http://www.aclu.org/Privacy

Archibold, Randal C. 2009. "Octuplets, 6 Siblings, and Many Questions." *New York Times,* February 3. Available at http://www.nytimes.com

Associated Press. 2007a (June 24). *Google Fights Global Internet Censorship.* Available at http://www.breitbart.com

Atkinson, Robert D., and Daniel D. Castro. 2008 (October). "Digital Quality of Life: Understanding the Personal and Social Benefits of the Information Technology Revolution." The Information and Technology Foundation. Available at http://www.itif.org

Attewell, Paul, Belkis Suazo-Garcia, and Juan Battle. 2003. "Computers and Young Children: Social Benefit or Social Problem?" *Social Forces* 82:277–296.

Bailey, Ronald. 2005. "Scientific Arguments Against Biotechnology Are Fallacious." In *Genetically Engineered Foods,* Nancy Harris, ed. (pp. 80-93). Farmington Hills, MI: Greenhaven Press.

Barbash, Fred. 2005. "High Court Takes up Abortion Consent Case." *Washington Post,* May 25. Available at http://www.washingtonppost.com

BBC. 2009. "On This Day: July 21, 1969." Available at http://news.bbc.co.uk/onthisday

BBC. 2006. "Google Censors Itself for China." BBC News, January 25. Available at http://newsvote.bbc.co.uk

Bell, Daniel. 1973. *The Coming of Post-Industrial Society: A Venture in Social Forecasting.* New York: Basic Books.

Beniger, James R. 1993. "The Control Revolution." In *Technology and the Future,* Albert H. Teich, ed. (pp. 40– 65). New York: St. Martin's Press.

Blank, Chris. 2009. "Mo. Senate Approves Abortion Legislation." Available at http://www.wkrg.com

Bollier, David. 2009 (July 22). "Deadly Medical Monopolies." On the Commons. Available at http://onthecommons.org

Brown, David. 2006. "For 1st Woman with Bionic Arm, a New Life Is Within Reach." *Washington Post,* September 14, p. A01.

Budde, Paul. 2009. "2008 Global Broadband—M-Commerce, E-Commerce & E-Payments." Available at http://www.researchandmarkets.com

Buchanan, Allen, Dan Brock, Norman Daniels, and Daniel Wikler. 2000. *From Chance to Choice: Genetics and Justice.* New York: Cambridge University Press.

Busari, Stephanie. 2009 (July 28). "Nicaragua Abortion Ban 'Cruel and Inhuman Disgrace.'" CNN Health. Available at http://www.cnn.com/2009/HEALTH

Bush, Corlann G. 1993. "Women and the Assessment of Technology." In *Technology and the Future,* Albert H. Teich, ed. (pp. 192–214). New York: St. Martin's Press.

Cauley, Leslie. 2009. "What's Broadband? Billions in Stimulus Funds are at Stake." *USA Today,* April 7. Available at http://www.usatoday.com/money

CBS News. 2008. "Artificial Heart Never Skipped a Beat." November 19. Available at http://www.cbsnews.com

Ceruzzi, Paul. 1993. "An Unforeseen Revolution." In *Technology and the Future,* Albert H. Teich, ed. (pp. 160–174). New York: St. Martin's Press.

Chan, Sewell. 2007. "New Scanners for Tracking City Workers." *New York Times,* January 23. Available at http://www.nytimes.com

Children's Hospital Boston staff. 2009. Available at http://childrenshospitalblog.org/stem-cell-research-a-fathers-story

Clarke, Adele E. 1990. "Controversy and the Development of Reproductive Sciences." *Social Problems* 37(1):18–37.

Clifford, Stephanie. 2009. "Ads Follow Web Users, and Get More Personal." *The New York Times,* July 31. Available at http://www.nytimes.com

Conrad, Peter. 1997. "Public Eyes and Private Genes: Historical Frames, New Constructions, and Social Problems." *Social Problems* 44:139–154.

COPA (Child Online Protection Act). 1998. "Title XIV—Child Online Protection." Available at http://epic.org/free_speech/censorship/copa.html

Crichton, Michael. 2007. "Patenting Life." *New York Times,* February 13. Available at http://www.nytimes.com

Davenport, Paul. 2009. "Arizona Governor Approves Abortion Constraints." Associated Press, July 13. Available at http://www.abcnews.go.com

David-Ferdon, Corrine, and Marci Feldman Hertz. 2009. *Electronic Media and Youth Violence: A CDC Issue Brief for Researchers.* Atlanta, GA: Centers for Disease Control.

DeNoon, Daniel. 2005 (March 8). *Study: Computer Design Flaws May Create Dangerous Hospital Errors.* Available at http://my.webmd.com

Department of Health and Human Services. 2009 (April 6). "The Genetic Information Nondiscrimination Act of 2008: Information for Researchers and Health Care Professionals." Available at http://www.genome.gov

Durkheim, Emile. 1973/1925. *Moral Education*. New York: Free Press.

Eibert, Mark D. 1998. "Clone Wars." *Reason* 30(2):52–54.

Eilperin, Juliet, and Rick Weiss. 2003. "House Votes to Prohibit All Human Cloning." *Washington Post,* February 28. Available at http://www.washingtonpost.com

Eisenberg, Anne. 2009. "Better Vision, with a Telescope Inside the Eye." *The New York Times,* July 19. Available at http://www.nytimes.com

ETC Group. 2003. *Contamination by Genetically Modified Maize in Mexico Much Worse Than Feared*. Available at http://www.etcgroup.org

Faris, Robert, Hal Roberts, and StephanieWang. 2009. "China's Green Dam: The Implications of Government Control Encroaching on the Home PC." *OpenNet Initiative Bulletin*. Available at http://opennet.net

FBI (Federal Bureau of Investigation). 2002. *Online Child Pornography: Innocent Images National Initiative*. Available at http://www.fbi.gov

Fiveash, Kelly. 2008 (February 25). "Judge Greenlights Lawsuit Against Microsoft, Sticky Issue of Deceptive 'Vista Capable' Labels." Available at http://www.channelregister.com

Forrester Research. 2008. "The State of Retailing Online 2008: Marketing Report." Available at http://www.forrester.com

Fox, Susannah. 2009 (October 26). "The Social Life of Health Information." Pew Internet and the American Life Project. Available at http://www.pewinternet.org

Friedman, Thomas L. 2005. *The World Is Flat*. New York: Farrar, Strauss, and Giroux.

FTC (Federal Trade Commission) 2009. *The Consumer Sentinel Data Book*. Washington, DC. Available at http://www.ftc.gov

Galama, Titus, and James Hosek. 2008. *U.S. Competitiveness in Science and Technology*. Rand Corporation. Available at http://www.rand.org

Gallup. 2000. *Abortion Issues*. Available at http://www.gallup.com/poll/indicators

Gibbs, Nancy. 2003. "The Secret of Life." *Time,* February 17, pp. 42–45.

Giridharadas, Anand. 2007. "India's Edge Goes Beyond Outsourcing." *New York Times,* April 4. Available at http://www.nytimes.com

Glendinning, Chellis. 1990. *When Technology Wounds: The Human Consequences of Progress*. New York: William Morrow.

Goodman, Paul. 1993. "Can Technology Be Humane?" In *Technology and the Future,* Albert H. Teich, ed. (pp. 239–255). New York: St. Martin's Press.

Gottlieb, Scott. 2000. "Abortion Pill Is Approved for Sale in United States." *British Medical Journal* 321:851.

Grady, Denise. 2008. "Parents Torn over Fate of Frozen Embryos." *New York Times,* December 4. Available at http://www.nytimes.com

Greenhouse, Linda. 2007. "Justices Back Ban on Method of Abortion." *New York Times,* April 19. Available at http://www.nytimes.com

Greenhouse, Linda. 2006. "Justices Reaffirm Emergency Access to Abortion." *New York Times,* January 19. Available at http://www.nytimes.com

Grossman, Lev. 2006. "Time's Person of the Year: You." *Time,* December 13. Available at http://www.time.com

Hagel, John, and John Seely Brown. 2008. "Life on the Edge: Learning from Facebook." *BusinessWeek,* April 2. Available at http://www.businessweek.com

Harmon, Amy. 2006. "That Wild Streak? Maybe It Runs in the Family." *New York Times,* June 15. Available at http://www.nytimes.com

Harris, Gardiner. 2009. "Some Stem Cell Research Limits Lifted." *The New York Times,* April 18. Available at http://www.nytimes.com

Heffernan, Virginia. 2009. "Facebook Exodus." *The New York Times*, August 30. Available at http://www.nytimes.com

Helft, Miguel. 2009. "At First, Funny Videos. Now, a Reference Tool." *The New York Times,* January 18. Available at http://www.nytimes.com

Holding, Reynolds. 2007. "E-Mail Privacy Gets a Win in Court." *Time,* June 21. Available at http://www.time.com

Horrigan, John. 2009 (May 7). " Wireless Connectivity Has Drawn Many Users More Deeply into Digital Life." Washington, DC: Pew Internet and American Life Project.

Horrigan, John. (May 18). 2008. "The Internet and Consumer Choice." Washington, DC: Pew Internet and American Life Project.

Houlahan, Brent. 2006. (September 18) "TelePresence Defined by Brent Houlahan with HSL's Thoughts and Analysis." Human Productivity Lab. Available at http://www.humanproductivitylab.com/archive

House Resolution. 2009. "H. Res 590." Available at http://thomas.loc.gov

Hsu, Jimmy. 2009 (July 13). "Robot Unemployment Rate Soars in Japan." Available at http://www.popsci.com

Human Cloning Prohibition Act of 2009. U.S. House of Representatives, Washington, DC. Available at http://www.govtrack.us

Human Genome Project. 2007. *Medicine and the New Genetics*. Available at http://www.ornl.gov

ITIF (Information Technology and Innovation Foundation). 2009. "Benchmarking EU & U.S. Innovation and Competitiveness." European-American Business Council (February). Available at http://www.itif.org

Internet Freedom Preservation Act. 2009. Available at http://www.congress.org

Internet Pornography Statistics. 2007. Available at http://www.Internet-filter-review.toptenreviews.com

Internet World Statistics. 2009. *Internet Usage Statistics: The Big Picture*. Available at http://www.Internetworldstats.com

Jones, Jeffrey M. 2009 (July 17). "Majority of Americans Say Space Program Costs Justified." Gallup Poll. Available at http://www.gallup.com

Jordan, Tim, and Paul Taylor. 1998. "A Sociology of Hackers." *Sociological Review* 46(4):757–778.

Kahn, A. 1997. "Clone Mammals . . . Clone Man?" *Nature* 386:119.

Kalil, Thomas. 2009. "Letter to President-Elect Obama." Available at http://www.nsf.gov/nsb/publications/2009/01_10_stem_rec_obama.pdf

Kaplan, Karen. 2009. "Corn Fortified with Vitamins Devised by Scientists." *Los Angeles Times,* April 29. Available at http://www.latimes.com

Kaplan, Carl S. 2000. (December 22). "The Year I Technology Law." Available at http://www.nytimes.com

Kharfen, Michael. 2006. "1 of 3 and 1 in 6 Pre-Teens Are Victims of Cyber-Bulling." Available at http://www.fightcrime.org

Kirkpatrick, David D. 2009. "Abortion Fight Complicates Debate on Health Care." *The New York Times,* September 29. Available at http://www.nytimes.com

Klotz, Joseph. 2004. *The Politics of Internet Communication*. Lanham, MD: Rowman and Littlefield.

Kuhn, Thomas. 1973. *The Structure of Scientific Revolutions*. Chicago: University of Chicago Press.

Labaton, Stephen. 2007. "Microsoft Finds Legal Defender in Justice Department." *New York Times,* June 10. Available at http://www.time.com

Legay, F. 2001 (January). "Genethics: Should Genetic Information Be Treated Separately?" *Virtual Mentor,* p. E5. Available at http://virtualmentor.ama-assn.org/site/archives.html

Lemonick, Michael. 2006a. "Are We Losing our Edge?" *Time,* February 13, pp. 22–33.

Lemonick, Michael. 2006b. "The Iceland Experiment." *Time,* February 20, pp. 49–52.

Lemonick, Michael, and Dick Thompson. 1999. "Racing to Map Our DNA." *Time*

Daily 153:1–6. Available at http://www.time.com

Lenhart, Amanda. 2009 (January 29). "Adults and Social Network Sites." Available at http://pewinternet.org/Reports

Lenhart, Amanda. 2008 (December 1). "Adults and Video Games." Available at http://www.pewinternet.org

Lohr, Steve. 2009. "Wal-Mart Plans to Market Digital Health Records System." *The New York Times*, March 11. Available at http://www.nytimes.com

Lohr, Steve. 2008. "Health Care That Puts a Computer on the Team." *The New York Times*, December 27. Available at http://www.nytimes.com

Lohr, Steve. 2006. "Outsourcing Is Climbing Skills Ladder." *New York Times*, February 16. Available at http://www.nytimes.com

Lynch, Colum. 2005. "U.N. Backs Human Cloning Ban." *Washington Post*, March 8. Available at http://www.washingtonpost.com

Lynton, Michael. 2009 (May 26). "Guardrails for the Internet: Preserving Creativity Online." Available at http://www.huffingtonpost.com

MacMillan, Robert. 2005. "My Cubicle, My Cell." *Washington Post*, May 19. Available at http://www.washingtonpost.com

Madkour, Rasha. 2008. "Girl Lives 118 Days Without a Heart." November 20. Available at http://www.boston.com

Magee, Mike. 2009 (July 17). "Apple, Microsoft, Others Sued over Touchpad Products." Available at http://www.tgdaily.com

Markoff, John. 2008a. "Surveillance of Skype Messages Found in China." *The New York Times*, October 2. Available at http://www.nytimes.com

Markoff, John. 2008b. "Thieves Winning Online War, Maybe Even in Your Computer." *The New York Times*, December 5. Available at http://www.nytimes.com

Markoff, John. 2000. "Report Questions a Number in Microsoft Trial." *New York Times*, August 28. Available at http://www.nytimes.com

Mayer, Sue. 2002. "Are Gene Patents in the Public Interest?" *BIO-IT World*, November 12. Available at http://www.bio-itworld.com

McCarthy, Tom, and Scott Michels. 2009. "Lori Drew MySpace Suicide Hoax Conviction Thrown Out." ABC News, July 2. Available at http://abcnews.go.com

McCormick, S. J., and A. Richard. 1994. "Blastomere Separation." *Hastings Center Report*, March-April, pp. 14–16.

McDermott, John. 1993. "Technology: The Opiate of the Intellectuals." In *Technology and the Future*, Albert H. Teich, ed.

(pp. 89–107). New York: St. Martin's Press.

McFadden, Maureen. 2009. "Girl Lives Months Without a Heart." August 14. Available at http://www.wndu.com

McFarling, Usha L. 1998. "Bioethicists Warn Human Cloning Will Be Difficult to Stop." *Raleigh News and Observer*, November 18, p. A5.

McGann, Rob. 2005. (January 6). "Most Active Web Users Are Young, Affluent." Jupiter Research. Available at http://www.clickz.com

McKinley, Jr., James C. 2009. "Abortion Law Backers Vow Oklahoma Appeal." *The New York Times*, August 20. Available at http://www.nytimes.com

Merton, Robert K. 1973. "The Normative Structure of Science." In *The Sociology of Science*, Robert K. Merton, ed. Chicago: University of Chicago Press.

Mesthene, Emmanuel G. 1993. "The Role of Technology in Society." In *Technology and the Future*, Albert H. Teich, ed. (pp. 73–88). New York: St. Martin's Press.

Miller, Claire Cain. 2009. "Doctors Will Make Web Calls in Hawaii." *The New York Times*, January 6. Available at http://www.nytimes.com

Mooney, Chris, and Sheril Kirshenbaum. 2009. *Unscientific America: How Scientific Literacy Threatens Our Future*. Philadelphia PA: Basic Books.

Murphie, Andrew, and John Potts. 2003. *Culture and Technology*. New York: Palgrave Macmillan.

NSF (National Science Foundation). 2008. "Science and Engineering Indicators, 2008." Available at http://www.nsf.gov

National Science Foundation. 2007. "Science and Engineering Indicators, 2006." Available at http://www.nsf.gov

NCSL (National Conference of State Legislatures). 2008a. "Human Cloning Laws." Available at http://ncsl.org

NCSL. 2008b. "Genetic Employment Laws." Available at http://www.ncsl.org

NIC (National Intelligence Council). 2003. "The Global Technology Revolution, Preface and Summary." Rand Corporation. Available at http://www.rand.org

Nie, Norman, Alberto Simpser, Irena Stepanikova, and Lu Zheng. 2004 (December). *Ten Years After the Birth of the Internet: How Do Americans Use the Internet in Their Daily Lives?* Stanford Institute for the Quantitative Study of Society.

Nielsen. 2009a (March). "Global Faces and Networked Places: A Nielsen Report on Social Networking's New Global Footprint." Available at http://blog.nielsen.com

Nielsen. 2009b (February). "Twitter Posts Meteoric 1,382% YoY Growth." Available at http://www.marketingcharts.com

NNI (National Nanotechnology Initiative). 2009. Available at http://www.nano.gov

Ogburn, William F. 1957. "Cultural Lag as Theory." *Sociology and Social Research* 41:167–174.

Optenet. 2008. "2008 International Internet Trends Study." Available at http://www.optenet.com

ORNL (Office of Biological and Environmental Research). 2009. "Medicine and the New Genetics." Available at http://www.ornl.gov

Paddock, Catharine. 2008. "American Teenager Survives 4 Months Without Heart." *Medical News Today*, November 20. Available at http://www.medicalnewstoday.com

Park, Alice. 2009. "Stem-Cell Research: The Quest Resumes." *Time*, January 29. Available at http://www.time.com

Pethokoukis, James M. 2004. "Meet Your New Co-Worker." *U.S. News and World Report*, March 7. http://www.usnews.com

Pew. 2009a (July 9). "Public Praises Science; Scientists Fault Public, Media Scientific Achievements Less Prominent Than a Decade Ago." Available at http://people-press.org/reports/pdf

Pew. 2009b (July 29). "56% of All Americans Have Accessed the Internet by Wireless Means." Available at http://www.pewinternet.org

Pew. 2009c. "Trend Data: Daily Internet Activities, 2000-2009." Available at http://www.pewinternet.org/Static-Pages/Trend-Data

Pew. 2007a. *Demographics of Internet Users*. Available at http://www.pewinternet.org

Pew. 2007b. "34% of Internet Users Have Logged on with a Wireless Internet Connection." Available at http://www.pewinternet.org

Pew. 2007c. *Trends in Political Values and Core Attitudes: 1987–2007*. Washington, DC: The Pew Research Center.

Pew. 2006. *Awareness of Genetically Modified Food Has Declined over the Last Five Years*. Available at http://pewagbiotech.org/research

Postman, Neil. 1992. *Technopoly: The Surrender of Culture to Technology*. New York: Alfred A. Knopf.

Prabhu, Maya. 2009. "Is Personal eMail Subject to Open-Records Law? *eSchool News*, September 14. http://www.eschoolnews.com

Price, Tom 2008. "Science in America: Are We Falling Behind in Science and Technology?" *CQ Researcher* 18(2): 24-48.

Privacy Rights Clearinghouse. 2009. "A Chronology of Data Breaches." Available at http://www.privacyrights.org

Rabino, Isaac. 1998. The Biotech Future." *American Scientist* 86(2):110–112.

Reuters. 2005. *Report: China's New Bid to Gag Web.* Available at http://www.cnn.com/technology

Richtel, Matt. 2009. "Drivers and Legislators Dismiss Cellphone Risks." *The New York Times,* July 19. Available at http://www.nytimes.com

Rizzo, Alessandra, 2009. "RU-486 Abortion Drug to Be Allowed in Italy Despite Vatican's Protests." *The Huffington Post,* July 31. Available at http://www.huffingtonpost.com

Robotic Trends. 2009 (March 17). "Orders Drop for U.S. Robot Makers; World Outlook Remains Strong." Available at http://www.roboticstrends.com

Roe v. Wade. 1973. 410 U.S. 113.

Rovner, Julie, and Jenny Gold. 2009 (July 25). "Obama Lifts Limit on Funding Stem Cell Research." NPR. Available at http://www.npr.org

Rutenberg, Jim, and Adam Nagourney. 2009. "Melding Obama's Web to a YouTube Presidency." *The New York Times,* January 26. Available at http://www.nytimes.com

Saad, Lydia. 2009 (August 4). "U.S. Abortion Attitudes Closely Divided." Available at http://www.gallup.com

Sang-Hun, Choe, and John Markoff. 2009. "Cyberattacks Jam Government and Commercial Web Sites in U.S. and South Korea." *The New York Times,* July 9. Available at http://www.nytimes.com

Schaefer, Naomi. 2001. "The Coming Internet Privacy Scrum." *American Enterprise* 12:50–51.

Sharkey, Joe. 2009. "A Meeting in New York? Can't We Videoconference?" *The New York Times,* May 12. Available at http://www.nytimes.com

Singh, Susheela, Deirdre Wulf, Rubina Hussain, Akinrinola Bankole, and Gilda Sedgh. 2009. *Abortion Worldwide: A Decade of Uneven Progress.* New York, NY: Guttmacher Institute Available at http://www.guttmacher.org

Skillings, Jonathan. 2006. "In Japan, Robots Are People, Too." C/Net News.Com, October 11. Available at http://news.com.com

Smeal, Eleanor. 2007. "Supreme Court Ruling in Abortion Ban Case Disastrous for Women's Health and Safety." Press Release, April 18. Available at http://www.feminist.org

Smith, Aaron. 2009 (April 15). "The Internet's Role in Campaign 2008." Available at http://www.pewinternet.org/Reports

Smith, David. 2008. "Addiction to Internet 'Is An Illness.'" *The Observer,* March 23. Available at http://www.guardian.co.uk

Social Watch. 2009. "Information, Science and Technology: The Gap Is Widening Faster." Available at http://www.socialwatch.org

Stelter, Brian, and Miguel Helft. 2009. "Deal Brings TV Shows and Movies to YouTube." *The New York Times,* April 17. Available at http://www.nytimes.com

Sterns, Olivia. 2009. "Paralyzed from the Chest Down, Man Bikes 500 Miles for Stem Cell Research." Huffington Post, July 29. Available at http://www.huffingtonpost.com

Stine, Deborah. 2009 (June 3). "President's Office of Science and Technology Policy (POST)." Congressional Research Service. Available at http://www.crs.gov

Stone, Brad. 2009. "Is Facebook Growing up Too Fast?" *The New York Times,* March 29. Available at http://www.nytimes.com

Stone, Brad. 2008. "A Jukebox on MySpace That Takes Aim at Apple." *The New York Times,* September 16. Available at http://www.nytimes.com

Straubhaar, Joseph, Robert LaRose, and Lucinda Davenport. 2009. *Media Now: Understanding Media, Culture and Technology.* Belmont, CA: Wadsworth Publishing Company.

Stross, Randall. 2008 (November 15). "What Has Driven Women out of Computer Science?" Available at http://www.nytimes.com

Svensson, Peter. 2007 (January 30). *Robot Parking Garage to Open in New York.* Available at http://www.breitbart.com

Svoboda, Elizabeth. 2009. "The Essential Guide to Stem Cells." *Popular Science,* June 1. Available at http://www.popsci.com

Tesler, Pearl. 2003. "Universal Robots: The History and Workings of Robots." Tech Museum of Innovation. Available at http://www.thetech.org/robotics

Thas, Angela, Chat Garcia Ramilo, and Cheekay Garcia Cinco. 2007. *Gender and ICT.* United Nations Development Programme. Bangkok, Thailand: Pacific Development Information Programme. Available at http://www.apdip.net

The Economist. 2009a (August 4). "This Message Will Self-Destruct." Available at http://www.economist.com

The Economist. 2009b (September 3). "Keeping Pirates at Bay." Available at http://www.economist.com

The Economist. 2008 (December 18). "Pocket World in Figures." Available at http://www.economist.com

Toffler, Alvin. 1970. *Future Shock.* New York: Random House.

Tumulty, Karen. 2006. "Where the Real Action Is. . . ." *Time,* January 30, pp. 50–53.

UCL (University College London). 2008 (July 11). "Information Behaviour of the Researcher of the Future." Available at http://www.bl.uk/news/pdf/googlegen.pdf

UNDP (United Nations Development Programme) 2003. *Human Development Report 2003.* New York: Oxford University Press.

United Nations. 2008. "UN E-Government Survey 2008." Available at http://unpan1.un.org/intradoc/groups/public

U.S. Bureau of the Census. 2009. *Statistical Abstract of the United States, 2008,* 126th ed. Washington, DC: U.S. Government Printing Office.

U.S. Citizenship and Immigration Services. 2009. "H-1B Specialty Occupations and Fashion Models." Available at http://www.uscis.gov

U.S. Newswire. 2003 (October 2). *Feminists Condemn House Passage of Deceptive Abortion Ban, Urge Activists to March on Washington.* Available at http://www.usnewswire.com

Vance, Ashlee. 2009a. "Microsoft Profit Falls for First Time in 23 Years." *The New York Times,* April 24. Available at http://www.nytimes.com

Vance, Ashlee. 2009b. "Microsoft Mapping Course to a Jetsons-Style Future." *The New York Times.* March 2. Available at http://www.nytimes.com

VCU (Virginia Commonwealth University). 2009. "VCU Life Sciences Survey 2008." Available at http://www.vcu.edu

Wadhwa, Vivek. 2009. "Outsourcing Benefits U.S. Workers, Too." *BusinessWeek,* August 2. Available at http://www.businessweek.com

Wallis, Claudia, and Sonja Steptoe. 2006. "E-Mail and Cell Phones Help Us Multi-Task. . . ." *Time,* January 16, pp.73–76.

Wegmann, M. J. 2009. "Is Your Food Genetically (GM) Modified? How to Tell." *Huffington Post,* May 30. Available at http://www.huffingtonpost.com

Weinberg, Alvin. 1966. "Can Technology Replace Social Engineering?" *University of Chicago Magazine* 59(October):6–10.

Welsh, Jonathan. 2009. "Late on a Car Loan? Meet the Disabler." *The Wall Street Journal,* March 25. Available at http://www.online.wsj.com

Welter, Cole H. 1997. "Technological Segregation: A Peek Through the Looking Glass at the Rich and Poor in an Information Age." *Arts Education Policy Review* 99(2):1–6.

White House. 2003 (November 5). *President Bush Signs Partial Birth Abortion Ban Act of 2003.* Available at http://www.whitehouse.gov

Wiles, Taylor. 2009. "Eleven States Enter New Abortion Debate." *Mother Jones,* February 9. Available at http://www.motherjones.com

Willing, Richard. 2005. "Federal Judge Blocks 'Partial Birth Abortion' Ban." *USA Today,* June 1. Available at http://www.usatoday.com

Winner, Langdon. 1993. "Artifact/Ideas as Political Culture." In *Technology and the Future,* Albert H. Teich, ed. (pp. 283–294). New York: St. Martin's Press.

World at Work. 2009. "Telework Trendlines 2009." Available at www.worldatwork.org

World Bank. 2009. *Global Economic Prospects: Technology Diffusion in the Developing World.* Washington DC: The World Bank.

Chapter 15

ACLU (American Civil Liberties Union). 2008. ACLU-TN Defends Peace Activist's Free-Speech Rights, December 12. Available at http://www.aclu.org

Afghanistan Conflict Monitor. 2009. *Up to 15,000 Taliban in Afghanistan: Minister,* February 27. Human Security Report Project, Simon Fraser University. Available at http://www.afghanconflictmonitor.org/

Alexander, Karen, and Mary E. Hawkesworth, eds. 2008. *War and Terror: Feminist Perspectives.* Chicago: University of Chicago Press.

Alvarez, Lizette. 2009a. "G.I. Jane Breaks the Combat Barrier." *The New York Times,* August 15. Available at http://www.nytimes.com

Alvarez, Lizette. 2009b. "Suicides of Soldiers Reach High of Nearly 3 Decades." *New York Times,* January 29. Available at http://www.nytimes.com

American Jewish Committee. 2007. *2006 Annual Survey of American Jewish Opinion.* Available at http://www.ajc.org

Annan, Jeannie, Christopher Blattman, Khristopher Carlson, and Dyan Mazurana. 2008. "Findings from the Survey of War-Affected Youth (SWAY) Phase II." Available at https://wikis.uit.tufts.edu/confluence/display

Arms Control Association. 2003 (May). *The Nuclear Proliferation Treaty at a Glance.* Available at http://www.armscontrol.org/factsheets

Associated Press. 2009 (December 1). "Obama Details Afghan War Plan, Troop Increases." MSNBC. Available at http://www.msnbc.msn.com

Associated Press. 2008 (August 6). "U.S. Officials: Scientist Was Anthrax Killer." MSNBC. Available at http://www.msnbc.msn.com

Associated Press. 2005 (March 15). *Pentagon Tests Negative for Anthrax.* MSNBC. Available at http://www.msnbc.msn.com

Atomic Archive. 2009. *Arms Control Treaties.* Available at http://www.atomicarchive.com

Baker, Peter, and Helene Cooper. 2009. "U.S. and Russia to Consider Reductions of Nuclear Arsenals in Talks for New Treaty." *New York Times*, March 31. Available at http://www.nytimes.com

Baliunas, Sallie. 2004. "Anthrax Is a Serious Threat." In *Biological Warfare,* William Dudley, ed. (pp. 53–58). Farmington Hills, MA: Greenhaven Press.

Barkan, Steven, and Lynne Snowden. 2001. *Collective Violence.* Boston: Allyn and Bacon.

Barnes, Steve. 2004. "No Cameras in Bombing Trial." *New York Times,* January 29, p. 24.

Barstow, David. 2008. "One Man's Military-Industrial-Media Complex." *New York Times,* November 29. Available at http://www.nytimes.com

BBC. 2009a. "Gaza Crisis: Key Maps & Timelines." BBC News, January 18. Available at http://news.bbc.co.uk

BBC. 2009b. "Guantanamo Detainee Denies Guilt." BBC News, June 9. Available at http://news.bbc.co.uk

BBC. 2008. "Darfur Deaths 'Could Be 300,000.'" BBC News, April 23. Available at http://www.bbc.co.uk

BBC. 2007a. "Hamas Takes Full Control of Gaza." BBC News, June 15. Available at http://news.bbc.co.uk

BBC. 2007b. "Fresh Clashes Engulf Lebanon Camp." BBC News, June 1. Available at http://news.bbc.co.uk

BBC. 2007c. "Q & A: Sudan's Darfur Conflict." BBC News, May 29. Available at http://news.bbc.co.uk

BBC. 2000. "UN Admits Rwanda Genocide Failure." BBC News, April 15. Available at http://news.bbc.co.uk

Belasco, Amy. 2009. *The Cost of Iraq, Afghanistan, and Other Global War on Terror Operations Since 9/11.* Congressional Research Service. Available at http://www.opencrs.com

Bercovitch, Jacob, ed. 2003. *Studies in International Mediation: Advances in Foreign Policy Analysis.* New York: Palgrave Macmillan.

Bergen, Peter. 2002. *Holy War, Inc.: Inside the Secret World of Osama bin Laden.* New York: Free Press.

Berrigan, Frida. 2009. "We Arm the World." *In These Times,* January 2. Available at http://www.inthesetimes.com

Berrigan, Frida, and William Hartung. 2005. *U.S. Weapons at War: Promoting Freedom or Fueling Conflict?* World Policy Institute Report. Available at http://www.worldpolicy.org/projects

Bjorgo, Tore. 2003. *Root Causes of Terrorism.* Paper presented at the International

Expert Meeting, June 9–11. Oslo, Norway: Norwegian Institute of International Affairs.

Borum, Randy. 2003. "Understanding the Terrorist Mind-Set." *FBI Law Enforcement Bulletin*, July, pp. 7–10.

Boustany, Nora. 2007. "Janjaweed Using Rape as 'Integral' Weapon in Darfur, Aid Group Says." *Washington Post,* July 3. Available at http://www.washingtonpost.com

Brauer, Jurgen. 2003. "On the Economics of Terrorism." *Phi Kappa Phi Forum*, Spring, pp. 38–41.

Brinkley, Joel, and David E. Sanger. 2005. "North Koreans Agree to Resume Nuclear Talks." *New York Times*, July 10. Available at http://www.nytimes.com

Broad, William J., Mark Mazzetti, and David E. Sanger. 2009. "A Nuclear Debate: Is Iran Designing Warheads?" *The New York Times*, September 28. Available at http://www.nytimes.com

Broad, William J., and David E. Sanger. 2009 (October 3). "Report Says Iran Has Data to Make a Nuclear Bomb." Available at http://www.nytimes.com

Brown, Michael E., Sean M. Lynn-Jones, and Steven E. Miller, eds. 1996. *Debating the Democratic Peace*. Cambridge, MA: MIT Press.

Buncombe, Andrew. 2009. "End of Sri Lanka's Civil War Brings Back Tourists." *The Independent,* August 16. Available at http://www.independent.co.uk

Carneiro, Robert L. 1994. "War and Peace: Alternating Realities in Human History." In *Studying War: Anthropological Perspectives*, S. P. Reyna and R. E. Downs, eds. (pp. 3–27). Langhorne, PA: Gordon & Breach.

Center for Defense Information. 2007 (April 30). *The World's Nuclear Arsenals.* Available at http://www.cdi.org

Center for Defense Information. 2003. *Military Almanac.* Available at http://www.cdi.org

Charter, David, Catherine Philp, and Rob Crilly. 2009. "Darfur War Crimes Court Orders Arrest of President Omar Al-Bashir." *The Times,* March 5. Available at http://www.timesonline.co.uk

CNN. 2009 (January 6). *Blackwater Defendants Plead Not Guilty.* Available at http://www.cnn.com

CNN. 2003 (February 26). *Poll: Muslims Call U.S. Ruthless, Arrogant.* Available at http://www.cnn.com

CNN. 2001 (January 31). *Libyan Bomber Sentenced to Life.* Available at http://www.europe.cnn.com

CNN/Opinion Research Corporation Poll. 2007 (June 22–24). *Iraq.* Available at http://www.pollingreport.com/iraq3.htm

Cohen, Ronald. 1986. "War and Peace Proneness in Pre- and Post-industrial

States." In *Peace and War: Cross-Cultural Perspectives*, M. L. Foster and R. A. Rubinstein, eds. (pp. 253–267). New Brunswick, NJ: Transaction Books.

Conflict Research Consortium. 2003. *Mediation.* Available at http://www.colorado.edu/conflict/peace

CRS (Congressional Research Service). 2008 (August 25). *Private Security Contractors in Iraq: Background, Legal Status, and Other Issues.* Available at http://www.opencrs.com

Cowell, Alan, and A. G. Sulzberger. 2009. "Lockerbie Convict Returns to Jubilant Welcome." *The New York Times*, August 20. Available at http://www.nytimes.com

Dao, James. 2009. "Veterans Affairs Faces Surge of Disability Claims." *New York Times*, July 12. Available at http://www.nytimes.com

Dareini, Ali Akbar. 2009. "Iran Missile Test: Ahmadinejad Says It's Within Israel's Range." *Huffington Post*, May 20. Available at http://www.huffingtonpost.com

Davenport, Christian. 2005. "Guard's New Pitch: Fighting Words." *Washington Post*, April 28. Available at http://www.washingtonpost.com

Davenport, Paul. 2007. "States Crack Down on T-Shirts with Fallen Soldiers' Names." *Associated Press*, May 18. Available at http://hamptonroads.com/node/268071

Dedman, Bill. 2006 (October 24). *In Limbo: Cases Are Few Against Gitmo Detainees.* Available at http://www.msnbc.msn.com

Deen, Thalif. 2000 (September 9). *Inequality Primary Cause of Wars, Says Annan.* Available at http://www.hartford-hwp.com/archives

DeYoung, Karen, and Walter Pincus. 2009 (October 3). *Iraq: U.S. to Fund Pro-American Publicity in Iraqi Media.* Available at http://www.corpwatch.org

Dixon, William J. 1994. "Democracy and the Peaceful Settlement of International Conflict." *American Political Science Review* 88(1):14–32.

Doherty, Carroll. 2007 (June 27). "Who Flies the Flag? Not Always Who You Might Think." Pew Research Center for the People and the Press. Available at http://www.pewresearch.org

Environmental Media Services. 2002 (October 7). *Environmental Impacts of War.* Available at http://www.ems.org

Espiau, Gorka. 2006. *The Basque Conflict: New Ideas and Prospects for Peace. Special Report*, No. 161, United States Institute of Peace. Available at http://www.usip.org

Feder, Don. 2003. "Islamic Beliefs Led to the Attack on America." In *The Terrorist Attack on America*, Mary E. Williams, ed. (pp. 20–23). Farmington Hills, MA: Greenhaven Press.

Federation of American Scientists. 2007. *Estimate of the U.S. Nuclear Weapons Stockpile, 2007 and 2012.* Available at http://www.fas.org

Fischer, Howard. 2007. "ACLU Seeks to End Ban on Sales of Shirt Listing Iraq War Dead." *The Arizona Daily Star*, June 29. Available at http://www.azstarnet.com

Funke, Odelia. 1994. "National Security and the Environment." In *Environmental Policy in the 1990s: Toward a New Agenda*, 2nd ed., Norman J. Vig and Michael E. Kraft, eds. (pp. 323–345). Washington, DC: Congressional Quarterly.

Gardner, Simon. 2007 (February 22). *Sri Lanka Says Sinks Rebel Boats on Truce Anniversary.* Available at http://www.reuters.com

Garrett, Laurie. 2001. "The Nightmare of Bioterrorism." *Foreign Affairs* 80:76.

Gettleman, Jeffrey. 2009. "Symbol of Unhealed Congo—Male Rape Victims." The New York Times, August 4. Available at http://www.nytimes.com

Gettleman, Jeffrey. 2008. "Rape Victims' Words Help Jolt Congo into Change." *New York Times*, October 18. Available at http://www.nytimes.com

Geneva Graduate Institute for International Studies. 2006. *Small Arms Survey, 2006.* New York: Oxford University Press.

Gioseffi, Daniela. 1993. "Introduction." In *On Prejudice: A Global Perspective*, Daniela Gioseffi, ed. (pp. xi–l). New York: Anchor Books, Doubleday.

Goure, Don. 2003. *First Casualties? NATO, the U.N.* MSNBC News, March 20. Available at http://www.msnbc.com/news

GreenKarat. 2007. *Mining for Gems.* Available at http://www.greenkarat.com

Herszenhorn, David. 2008. "Estimates of Iraq War Cost Were Not Close to Ballpark." *New York Times*, March 19. Available at http://www.nytimes.com

Hewitt, J. Joseph, Jonathan Wilkenfeld, and Ted Robert Gurr. 2008. *Peace and Conflict 2008: Executive Summary.* College Park, MD: Center for International Development and Conflict Management.

Hooks, Gregory, and Leonard E. Bloomquist. 1992. "The Legacy of World War II for Regional Growth and Decline: The Effects of Wartime Investments on U.S. Manufacturing, 1947–72." *Social Forces* 71(2):303–337.

Howard, Michael. 2007. "Children of War: The Generation Traumatised by Violence in Iraq." *The Guardian,* February 6. Available at http://www.guardian.co.uk

Huntington, Samuel. 1996. *The Clash of Civilizations and the Remaking of World Order.* New York: Simon and Schuster.

Inglesby, Thomas, Donald Henderson, John Bartlett, Michael Archer, et al. 1999. "Anthrax as a Biological Weapon: Medical and Public Health Management." *Journal of the American Medical Association* 281:1735–1745.

ICBL (International Campaign to Ban Landmines). 2009. *Making the Treaties Universal.* Available at http://www.icbl.org

Glantz, Aaron (eds). 2008. *Winter Soldier, Iraq and Afghanistan: Eyewitness Accounts of the Occupations.* Chicago, IL: Haymarket Books.

Johnston, David, and John Broder. 2007. "F.B.I. Says Guards Killed 14 Iraqis Without Cause." *The New York Times*, November 14. Available at http://www.nytimes.com

Kemper, Bob. 2003. "Agency Wages Media Battle." *Chicago Tribune*, April 7. Available at http://www.chicagotribune.com

Kerr, Paul K., and Mary Beth Nikitin. 2009. *Pakistan's Nuclear Weapons: Proliferation and Security Issues.* Congressional Research Service, June 12. Available at http://www.opencrs.com

Klare, Michael. 2001. *Resource Wars: The New Landscape of Global Conflict.* New York: Metropolitan Books

Knickerbocker, Brad. 2002. "Return of the Military-Industrial Complex?" *Christian Science Monitor*, February 13. Available at http://www.csmonitor.com

Lamont, Beth. 2001. "The New Mandate for UN Peacekeeping." *The Humanist* 61:39–41.

United Nations Association of the United States of America. 2003. *The Problem: Impact of Landmines.* Available at http://www.landmines.org

Laqueur, Walter. 2006. "The Terrorism to Come." In *Annual Editions 05–06*, Kurt Finsterbusch, ed. (pp. 169–176). Dubuque, IA: McGraw-Hill/Dushkin.

Large, Tim. 2006. *Crisis Profile: What's Going on in Northern Uganda?* Thomas Reuters Foundation. Available at http://www.alertnet.org

Lederer, Edith. 2005 (May 20). "Annan Lays out Sweeping Changes to U.N." Associated Press. Available at http://www.apnews.com

Levy, Jack S. 2001. "Theories of Interstate and Intrastate War: A Levels of Analysis Approach." In *Turbulent Peace: The Challenges of Managing International Conflict*, Chester A. Crocker, Fen Osler Hampson, and Pamela Aall, eds. (pp. 3–27). Washington, DC: U.S. Institute of Peace.

Levy, Clifford J., and Peter Baker. 2009. "U.S.-Russian Nuclear Agreement Is First Step in Broad Effort." *The Washington Post,* July 6. Available at http://www.washingtonpost.com

Li, Xigen, and Ralph Izard. 2003. "Media in a Crisis Situation Involving National Interest: A Content Analysis of Major U.S.

Newspapers' and TV Networks' Coverage of the 9/11 Tragedy." *Newspaper Research Journal* 24:1–16.

Lindner, Andrew M. 2009. "Among the Troops: Seeing the Iraq War Through Three Journalistic Vantage Points." Social Problems 56 (1): 21–48.

MacAskill, Ewen, Ed Pilkington, and Jon Watts. 2006. "Despair at UN over Selection of 'Faceless' Ban Ki-moon as General Secretary." *The Guardian*, October 7. Available at http://www.guardian.co.uk

Marshall, Leon. 2007 (July 16). *Elephants "Learn" to Avoid Land Mines in War-Torn Angola.* National Geographic News. Available at http://www.nationalgeographic.com

Mazzetti, Mark, and Scott Shane. 2009a. "Interrogation Memos Detail Harsh Tactics by the C.I.A." *New York Times*, April 16. Available at http://www.nytimes.com

Mazzetti, Mark, and Scott Shane. 2009b. "C.I.A. Abuse Cases Detailed in Report on Detainees." *The New York Times*, August 24. Available at http://www.nytimes.com

McGirk, Tim. 2006. "Collateral Damage or Civilian Massacre in Haditha?" *Time*, March 19. Available at http://www.time.com

Military. 2007. *Tuition Assistance (TA) Program Overview.* Available at http://education.military.com

Military Professional Resources Inc. 2007. *About MPRI.* Available at http://www.mpri.com

Miller, Susan. 1993. "A Human Horror Story." *Newsweek*, December 27, p. 17.

Miller, T. Christian. 2007. "Private Contractors Outnumber U.S. Troops in Iraq." *Los Angeles Times*, July 4. Available at http://www.latimes.com

Millman, Jason. 2008. "Industry Applauds New Dual-Use Rule." *Hartford Business Journal,* September 29. Available at http://www.hartfordbusiness.com

Morales, Lymari. 2009a (July 2). *Americans' Worry About Terrorism Nears 5-Year Low.* Available at http://www.gallup.com

Morales, Lymari. 2009b (April 6). *Americans See Newer Threats on Par with Ongoing Conflicts.* Available at http://www.gallup.com

Morgan, Jenny, and Alic Behrendt. 2009. *Silent Suffering: The Psychological Impact of War, HIV, and Other High-Risk Situations on Girls and Boys in West and Central Africa.* Woking: Plan. Available at http://plan-international.org

Moroney, Jennifer D. P., Joe Hogler, Benjamin Bahney, Kim Cragin, David R. Howell, Charlotte Lynch, and Rebecca Zimmerman. 2009. "Building Partner Capacity to Combat Weapons of Mass Destruction." Rand National Defense Research Institute. Available at http://www.rand.org

Myers, Steven Lee. 2009. "Women at Arms: Living and Fighting Alongside Men, and Fitting In." *The New York Times*, August 16. Available at http://www.nytimes.com

Myers-Brown, Karen, Kathleen Walker, and Judith A. Myers-Walls. 2000. "Children's Reactions to International Conflict: A Cross-Cultural Analysis." Paper presented at the National Council of Family Relations, Minneapolis, November 20.

NASP (National Association of School Psychologists). 2003. *Children and Fear of War and Terrorism.* Available at http://www.nasponline.org

NCPSD (National Center for Posttraumatic Stress Disorder). 2007. *What Is Post Traumatic Stress Disorder?* Available at http://www.ncptsd.va.gov

National Counterterrorism Center. 2009. *2008 Report on Terrorism.* Available at http://www.nctc.gov

National Priorities Project. 2009. *Federal Budget Trade-Offs.* Available at http://www.nationalpriorities.org/tradeoffs

National Security Council. 2002. *The National Security Strategy of the United States of America.* Available at http://www.whitehouse.gov/nsc

NTI (Nuclear Threat Initiative). 2009. *North Korea Profile.* Available at http://www.nti.org

Office of Management and Budget. 2009a. Table S-7: Funding Levels for Appropriated ("Discretionary") Programs by Agency. Available at http://www.whitehouse.gov

Office of Management and Budget. 2009b. Table 26-1: Policy Function and Category Program—Summary, Budget Authority by Function, Category, and Program. Available at http://www.whitehouse.gov

Office of the Coordinator for Counterterrorism. 2008. "Chapter 6: Terrorist Organizations," in *Country Reports on Terrorism 2008,* April. Washington, DC: U.S. Department of State. Available at http://www.state.gov/s/ct

Office of Weapons Removal and Abatement. 2009. *To Walk the Earth in Safety: The United States' Commitment to Humanitarian Mine Action and Conventional Weapons Destruction.* Available at http://www.state.gov

O'Hanlon, Michael E., and Jason H. Campbell. 2007 (July 16). *Iraq Index: Tracking Variables of Reconstruction and Security in Post-Saddam Iraq.* Available at http://www.brookings.edu

Ochmanek, David and Lowell H. Schwartz. 2008. "The Challenge of Nuclear- Armed Regional Adversaries." Rand Project Air Force. Available at http://www.rand.org

Paul, Annie Murphy. 1998. "Psychology's Own Peace Corps." *Psychology Today* 31:56–60.

Perlo-Freeman, Sam, Catalina Perdomo, Elisabeth Skons, and Petter Stalenheim. 2009. *Chapter 5: Military Expenditure.* Available at http://www.sipri.org/yearbook/2009/05

Permanent Court of Arbitration. 2009. *About Us* and *Cases.* Available at http://www.pca-cpa.org

Pew Research Center. 2005 (July 13). *Guantanamo Prisoner Mistreatment Seen as an Isolated Incident.* Available at http://www.pewtrust.org

Pew Research Center. 2009 (July 23). *Confidence in Obama Lifts U.S. Image Around the World: Most Muslim Publics Not So Easily Moved.* Pew Global Attitudes Project. Available at http://www.pewglobal.org

Pew Research Center for the People & the Press. 2009 (April 24). *Public Remains Divided over Use of Torture.* Available at http://www.people-press.org

Porter, Bruce D. 1994. *War and the Rise of the State: The Military Foundations of Modern Politics.* New York: Free Press.

Powell, Bill, and Tim McGirk. 2005. "The Man Who Sold the Bomb." *Time*, February 14, pp. 22–31.

Puckett, Neal, and Haytham Faraj. 2009 (June 25). *Navy-Marine Corps Court Hears Appeal in Wuterich v. U.S.* Available at http://www.puckettfaraj.com

Rashid, Ahmed. 2000. *Taliban: Militant Islam, Oil, and Fundamentalism in Central Asia.* New Haven, CT: Yale University Press.

Rasler, Karen, and William R. Thompson. 2005. *Puzzles of the Democratic Peace: Theory, Geopolitics, and the Transformation of World Politics.* New York: Palgrave Macmillan.

Renner, Michael. 2005. "Disarming Postwar Societies." In *State of the World: 2005*, Linda Starke, ed. (pp. 122–123). New York: W. W. Norton.

Renner, Michael. 2000. "Number of Wars on Upswing." In *Vital Signs: The Environmental Trends That Are Shaping Our Future*, Linda Starke, ed. (pp. 110–111). New York: W. W. Norton.

Renner, Michael. 1993. "Environmental Dimensions of Disarmament and Conversion." In *Real Security: Converting the Defense Economy and Building Peace*, Karl Cassady and Gregory A. Bischak, eds. (pp. 88–132). Albany: State University of New York Press.

Reuters. 2008. "Case Dropped Against Officer Accused in Iraq Killings." *New York Times*, June 18. Available at http://www.nytimes.com

Romero, Anthony. 2003. "Civil Liberties Should Not Be Restricted During Wartime." In *The Terrorist Attack on America*, Mary Williams, ed. (pp. 27–34). Farmington Hills, MA: Greenhaven Press.

Rosenberg, Tina. 2000. "The Unbearable Memories of a U.N. Peacekeeper." *New York Times*, October 8, pp. 4, 14.

Saad, Lydia. 2006 (August 10). *Anti-Muslim Sentiments Fairly Commonplace*. Available at http://poll.gallup.com

Saad, Lydia. 2009 (March 26). *Americans More Upbeat About U.S. Defense Readiness*. Available at http://www.gallup.com

Salem, Paul. 2007. "Dealing with Iran's Rapid Rise in Regional Influence." *The Japan Times*, February 22. Available at http://www.carnegieendowment.org

Salladay, Robert. 2003 (April 7). *Anti-War Patriots Find They Need to Reclaim Words, Symbols, Even U.S. Flag from Conservatives*. Available at http://www.commondreams.org

Sang-Hun, Choe, and John Markoff. 2009. "Cyberattacks Jam Government and Commercial Web Sites in U.S. and South Korea." *The New York Times*, July 8. Available at http://www.nytimes.com

Savage, Charlie. 2009. "Accused 9/11 Mastermind to Face Civilian Trial in N.Y." *New York Times*, November 13. Available at http://www.nytimes.com

Save the Children. 2009. *Last in Line, Last in School 2009*. Available at http://www.savethechildren.org

Save the Children. 2007. *State of the World's Mothers: Saving the Lives of Children Under Five*. Available at http://www.savethechildren.org

Save the Children. 2005. *One World, One Wish: The Campaign to Help Children and Women Affected by War*. Available at http://www.savethechildren.org

Scheff, Thomas. 1994. *Bloody Revenge*. Boulder, CO: Westview Press.

Schmitt, Eric. 2005. "Pentagon Seeks to Shut Down Bases Across Nation." *New York Times*, May 14. Available at http://www.nytimes.com

Schroeder, Matt. 2007. "The Illicit Arms Trade." Washington, DC: Federation of American Scientists. Available at http://www.fas.org

Schultz, George P., William J. Perry, Henry A. Kissinger, and Sam Nunn. 2007. "A World Free of Nuclear Weapons." *Wall Street Journal*, January 4 Available at http://www.wsj.com

Sengupta, Somini. 2009. "Dossier Gives Details of Mumbai Attacks." *New York Times*, January 6. Available at http://www.nytimes.com

Shanahan, John J. 1995. "Director's Letter." *Defense Monitor* 24(6):8.

Shanker, Thom, and David E. Sanger. 2009. "Pakistan Is Rapidly Adding Nuclear Arms, U.S. Says." *New York Times*, May 17. Available at http://www.nytimes.com

Shrader, Katherine. 2005. "WMD Commission Releases Scathing Report." *Washington Post*, March 31. Available at http://www.washingtonpost.com

Silverstein, Ken. 2007. "Six Questions for Joost Hiltermann on Blowback from the Iraq-Iran War." *Harper's Magazine*, July 5. Available at http://www.harpers.org

Singer, Peter W. 2009. "Attack of the Military Drones." *The National,* June 29. Available at http://www.brookings.edu

SIPRI (Stockholm International Peace Research Institute). 2009a. *SIPRI Yearbook 2009: Armaments, Disarmament, and International Security*. Oxford, UK: Oxford University Press.

SIPRI. 2009b. *Recent Trends in Military Expenditure*. Available at http://www.sipri.org

Skocpol, Theda. 1994. *Social Revolutions in the Modern World*. Cambridge, UK: Cambridge University Press.

Slevin, Peter. 2009. "U.S. to Announce Transfer of Detainees to Ill. Prison." *Washington Post,* December 15. Available at http://www.washingtonpost.com

Small Arms Survey. 2009. *Shadows of War*. Available at http://www.smallarmssurvey.org

Smith, Craig. 2004. "Libya to Pay More to French in '89 Bombing." *New York Times*, January 9, p. 6.

Smith, Lamar. 2003. "Restricting Civil Liberties During Wartime Is Justifiable." In *The Terrorist Attack on America*, Mary Williams, ed. (pp. 23–26). Farmington Hills, MA: Greenhaven Press.

Starr, J. R., and D. C. Stoll. 1989. *U.S. Foreign Policy on Water Resources in the Middle East*. Washington, DC: Center for Strategic and International Studies.

State of Hawaii. 2000. *Dual-Use Technologies*. Available at http://www.state.hi.us

Steinhauser, Paul. 2009. *Poll: Few Americans Have Favorable View of Muslim World*. CNN, June 2. Available at http://www.cnn.com

Stimson Center. 2007. *Reducing Nuclear Dangers in South Asia*. Available at http://www.stimson.org

Strobel, Warren, David Kaplan, Richard Newman, Kevin Whitelaw, and Thomas Grose. 2001. "A War in the Shadows." *U.S. News and World Report* 130:22.

Strobl, Michael R. 2004. "Taking Chance." Available at http://www.blackfive.net

Tanielian, Terri, Lisa H. Jaycox, Terry L. Schell, Grant N. Marshall, Audrey Burnam, Christine Eibner, Benjamin R. Karney, Lisa S. Meredith, Jeanne S. Ringe, Mary E. Vaiana, and the Invisible Wounds Study Team. 2008. "Invisible Wounds of War: Summary and Recommendations for Addressing Psychological and Cognitive Injuries." Available at http://www.rand.org

Tavernise, Sabrina. 2007. "U.S. Contractor Banned by Iraq over Shootings." *New York Times*, September 18. Available at http://www.nytimes.com

Thompson, Mark. 2007. "Broken Down" *Time*, April 16, pp. 28–33.

Tyson, Ann Scott. 2009. "Media Ban Lifted for Bodies' Return." *Washington Post*, February 27. Available at http://www.washingtonpost.com

United Nations. 2009. *United Nations Peacekeeping Operations*. Available at http://www.un.org/Depts/dpko/dpko/bnote.htm

United Nations. 2003. *Some Questions and Answers*. Available at http://www.unicef.org

United Nations. 1948. *Convention on the Prevention and Punishment of the Crime of Genocide*. Available at http://www.un.org

U.S. Army Medical Command. 2007 (May). *Mental Health Advisory Team IV Findings*. Available at http://www.armymedicine.army.mil

U.S. Census Bureau. 2004. *Statistical Abstract of the United States: 2004–2005*, 124th ed. Washington, DC: U.S. Government Printing Office.

U.S. Department of Homeland Security. 2009. *Budget-in Brief: Fiscal Year 2010*. Available at http://www.dhs.gov

U.S. Department of State. 2007. *Small Arms/ Light Weapons Destruction*. Available at http://www.state.gov

U.S. Nuclear Weapons Cost Study Project. 1998. *50 Facts About U.S. Nuclear Weapons*. Available at http://www.brookings.edu

Vesely, Milan. 2001. "UN Peacekeepers: Warriors or Victims?" *African Business* 261:8–10.

Viner, Katharine. 2002. "Feminism as Imperialism." *The Guardian,* September 21. Available at http://www.guardian.co.uk

Walker, Paul F., and Jonathan B. Tucker. 2006. "The Real Chemical Threat." *The Los Angeles Times*, April 1. Available at http://cns.miis.edu

Waters, Rob. 2005. "The Psychic Costs of War." *Psychotherapy Networker*, March– April, pp. 1–3.

Watkins, Thomas. 2007. *Haditha Hearings Enter Fourth Day*. Time, May 11. Available at http://www.time.com

Wax, Emily. 2003. "War Horror: Rape Ruining Women's Health." *Miami Herald*, November 3. Available at http://www.miami.com

Weimann, Gabriel. 2006. "Terror on the Internet: The New Arena, the New Challenges." Washington, DC: U.S. Institute of Peace Press.

Wein, Lawrence. 2009. "Counting the Walking Wounded." *New York Times*, January 25. Available at http://www.nytimes.com

Williams, Mary E., ed. 2003. *The Terrorist Attack on America*. Farmington Hills, MA: Greenhaven Press.

Wilson, Brenda. 2008. "WHO Estimates Iraqi Death Toll at 151,000," January 9. Available at http://www.npr.org

Woehrle, Lynne M., Patrick G. Coy, and Gregory M. Maney. 2008. *Contesting Patriotism: Culture, Power, and Strategy in the Peace Movement*. Lanham, Md.: Rowman & Littlefield Publishers, Inc.

Worldwatch Institute. 2003. *Vital Signs: The Trends That Are Shaping Our Future*. New York: W. W. Norton.

Zakaria, Fareed. 2000. "The New Twilight Struggle." *Newsweek,* October 12. Available at http://www.msnbc.com/news

Zoroya, Gregg. 2006. "Lifesaving Knowledge, Innovation Emerge In War Clinic." *USA Today,* March 27. Available at http://www.usatoday.com

Credits

Name Index

Note: Page numbers in *italics* refer to information in tables, figures, illustrations and captions.

Subject Index

Note: Page numbers in **boldface** type denote definitions. Page numbers in *italics* denote tables, figures, illustrations and captions.